Sustainable Engineering Forum 2016

Core Programming Area at the 2016 AIChE Annual Meeting

San Francisco, California, USA
13 - 18 November 2016

ISBN: 978-1-5108-3449-1

Printed from e-media with permission by:

Curran Associates, Inc.
57 Morehouse Lane
Red Hook, NY 12571

Some format issues inherent in the e-media version may also appear in this print version.

Printed by Curran Associates, Inc. (2017)

For permission requests, please contact AIChE
at the address below.

AIChE
120 Wall Street, FL 23
New York, NY 10005-4020

Phone: (800) 242-4363
Fax: (203) 775-5177

www.aiche.org

Additional copies of this publication are available from:

Curran Associates, Inc.
57 Morehouse Lane
Red Hook, NY 12571 USA
Phone: 845-758-0400
Fax: 845-758-2633
Email: curran@proceedings.com
Web: www.proceedings.com

TABLE OF CONTENTS

450979 Chemistry and Engineering of Energy, Environment, and Health

Abstract:

Wicked challenges, such as providing for safe available water, ensuring clean-sustainable energy, and providing for health solutions will require STEM, arts, leadership, and chemistry based solutions! My goal is to provide for progressive education and research via fundamental chemistry and engineering sciences to prepare future generations to solve grand challenges and explore multidisciplinary unknowns with holistic consciousness. To provide for SMART experiential learning, I have developed two teaching tools that can be applied in "choose your own experience" fashion (Jensen 2016, Jensen 2016A). Applied environmental research (Jensen and Müeller 2015, Jensen and Gujarathi 2015), energy research (Jensen *et al*. 2013, Jensen and Lewinski), and Android development (Jensen 2015) contributed to my increasing interest in nanotechnology and nanoinformatics. Synthesis and the evaluation of nanoscale features are state of the art areas of scientific inquiry for the fields of renewable chemicals-fuels, catalysis, medicine and composites to name a few. Nanoinformatics integrates information technology and big data science for multidisciplinary theoretical study and can be used in coordination with laboratory results to provide for layers of validation (*e.g.*, methods, applications, complex computational processes).

References

Jensen CD, et al. (2013) Characterization of Biomass Particles & Geometry During Past Pyrolysis for Systems Models." Proceedings of Conference on Earth & Energy Research. Colorado School of Mines, Golden CO, USA.

Jensen, CD (2015). ChE App Help. Proceedings of the AIChE Annual Meeting, Salt Lake City, UT.

Jensen CD, Gujarathi N. (2015) "Characterization of a Macrophyte Microcosm as a Surface Water Treatment System for Antibiotics." *Environmental Progress and Sustainable Energy*. DOI 10.1002/ep.12158.

In progress, Jensen and Lewinski, Mesoporous Sulphur Doped Titania Photocatalytic Conversion of Oils to Biodiesel.

Jensen CD, Müeller C. (2015) "Benchmark Characteristics of pH based Optoelectronic Enzymatic Biosensors with Non-modified E. coli BL21 pGELAF+." *Applied Biochemistry and Biotechnology*, DOI 10.1007/s12010-015-1623-1.

In press, Jensen, CD. (2016) "Graduate Student Action Research to Support Development of Engineering for Sustainable Development Degree Programs. Part I: A Collaborative Community Action Research Vignette." *Journal of Cleaner Production.*

In submission Jensen, CD (2016A). The Loped: a Multidisciplinary Teaching and Research Tool.

Research Interests: Energy, environment, sustainability, sensors-systems, nano-synthesis and informatics, applications to health, citizen science, and Android apps.

Teaching Interests: Professional development, outreach, innovation, experiential learning, SMART assessment.

460601 A Stabilized, Intrinsically Safe, 10% Efficient, Solar-Driven Water-Splitting Cell Incorporating Earth-Abundant Electrocatalysts with Steady-State pH Gradients and Product Separation Enabled By a Bipolar Membrane

Abstract:

Research Interests:

Materials Innovation and Electrochemistry for Water, Energy and Electronics

- Synthesis, integration, and assembly of 1D/2D nanomaterials with controlled multi-functions, multi-dimensions and multi-compositions for direct production of fuels from sunlight, water and air (N2 and CO2)
- 2D nanomaterial-based electronic, optoelectronic, photonic devices for photodetectors, field-effect transistors, photoelectrochemical biosensors, and electrochromic display
- Materials synthesis and characterization for water quality monitoring, electrochemical water treatment, chemical detection/sensing and seawater desalination driven by renewable energy

Teaching Interests:

- Material Science: Structure and Analysis of Inorganic Solids, Electronic and Optical Properties of Materials, Nanomaterials: Properties and Applications, Semiconductor Physics and Devices
- Electrochemistry and solar energy conversion: Introduction to Analytical Electrochemistry, Electrochemistry for Solar Energy Conversion, Advanced Catalysis for Energy Conversion/Storage, Electrochemical Impedance Spectroscopy, Semiconductor Physics for Photovoltaics and Photoelectrochemistry
- MEMS Materials and Fabrication, Electron Microscopy, Design of Experiments

469261 The Formation of Choline-Lignin Complex By Choline Based Ionic Liquid

Abstract:

Research Interests:

I would dedicate my efforts to lignin valorization to complement the existing strengths of the department in biotechnology and sustainable energy. Three unique characteristics (abundance, energy density and aromatic structure) allow lignin to have great potentials for industrial application.

Teaching Interests:

Teaching Biofuels and Bioenergy is an incredible challenge, because students may have to deal with systematic problems that not only require interdisciplinary knowledge, but also require a long-term effort. My course will address questions such as: Why bioenergy? How will biofuels be produced? What biofuels' challenges need to be resolved? I will also teach students chemistry of biomass and lignin and how to collect and analyze data that will be used in their future careers.

The Formation of Choline-Lignin Complex by Choline Based Ionic Liquid

Jijiao Zeng[1], Michael Kent[2], Jian Sun[1], Feng Xu[1], Tanmoy Dutta[1], Kai Deng[1], Seema Singh[1], Blake Simmons[3] and Kenneth L. Sale[1*]

(1) Deconstruction Division, Joint BioEnergy Institute / Sandia National Laboratories, Emeryville, CA, (2) Sandia National Laboratories, Albuquerque, NM, (3) Lawrence Berkeley National Laboratories, Berkeley, CA

Abstract

Recently ionic liquids have been extensively studied for biomass pretreatment, as they are capable of selectively solubilizing and isolating cell wall components. The deconstruction division within the Joint BioEnergy Institute (JBEI) is developing cheaper, biocompatible and adaptable ionic liquid technologies for use in current biorefineries. Our work has demonstrated that choline-based ionic liquids are highly effective in removing lignin from herbaceous crops such as switch grass. However, the dissociation mechanism of lignin from different biomass by choline-based ionic liquids is still unclear. In this study, modification of the lignin structure by choline glutamate ([Ch][Glu]) and choline α-ketoglutarate ([Ch][aKg]) was investigated. Interestingly, the results showed that choline-based ionic liquids results in the formation a new phenyl-choline ether (4-O-C) bond. Although the yield of isolated lignin from switch grass was significantly higher than for softwood and hardwood, the newly identified signals were present in all isolated lignins. NMR and MALDI-TOF-MS were used to analyze the mechanisms of generating phenyl-choline ether. It is hypothesized that cholination of lignin is a fingerprint for modification of lignin by choline-based ionic liquids.

472159 Can We Use Short Rotation Hybrid Poplar Coppice for Fuels and Chemicals Production Via Biochemical and Thermochemical Conversion?

Abstract:

The economic success of a potential biorefinery is directly related to the use of low cost biomass. Due to the high cost, it is unlikely that a wood-based biorefinery would use only whitewood chips from mature trees in conversion. Using a more heterogeneous raw material is less costly, but efficacy of the conversion using whole short ration coppice (including leaves, bark, branches and white wood) has not been investigated. This study investigated the influence of using whole hybrid poplar coppice on the overall sugar yield via biochemical conversion and the bio-oil production via thermochemical conversion. Results show that converting the short rotation hybrid poplar coppice may be challenging, with up to 50% lower sugar yield and close to 30% lower bio-oil yield than mature poplar wood chips. However, the removal of leaves significantly increased the overall sugar recovery and bio-oil production by 20% and 15%, respectively. These results show that leaf removal from short rotation hybrid poplar coppice prior to harvest may enhance the biorefinery production. With techno-economic analysis and life cycle assessment, we further evaluate the feasibility of converting short rotation hybrid poplar coppice to sugar and bio-oil via biochemical and thermochemical conversions.

Research Interests: Renewable fuels, bioprocess engineering, biorefinery, techno-economic analysis, life cycle assessment, mathematic and mechanistic modeling, fermentation engineering, biochemical engineering

Teaching Interests: Biochemical engineering, bioprocess and biosystem enegineering, natural chemistry, analytical chemistry

475575 Environmental Sustainability through Process Integration, Control and Optimization

Abstract:

Environmental Sustainability through Process Integration, Control and Optimization

Monzure-Khoda Kazi

Post-Doctoral Researcher

Department of Chemical Engineering | Qatar University

Email: kazi.khoda@gmail.com; kazi0001@qu.edu.qa |Tel: (+974) 6640 8300

KEY CONTRIBUTION AREAS

-

§ Process Integration	§ Process Optimization	§ Process Control
§ Process Design	§ Process Intensification	§ Process Simulation
§ Financial Modelling	§ Ad hoc Reporting	§ Market Analysis

VISION AND PERSPECTIVE

The vision of my research is aimed at the improvement of environmental sustainability focusing on energy, water and food security. My research will promote sustainable engineered systems that support human well-being and that lead to economic growth, environmental protection, and social progress for the current generation without compromising the potential of future generations to have an ecosystem that meets their needs. Innovations in industrial ecology, green engineering and large scale engineering research are the prime focuses of my long time research horizons. The ultimate target of my research activities is to conserve natural resources (mass and energy), recycle and reuse materials, prevent pollution, enhance yield, improve quality, advance inherent safety, and increase profitability.

Given the rise in human population and the inevitable consequences of climate change, the challenges of achieving secure and sustainable supplies of energy, water and food are Herculean. Within the next 20 years we will need to produce 50% more food and we will also need 50% more energy, and 30% more fresh water. There is an urgent need to develop long-term strategic programmes of research with integrated and sustainable approaches that will be able to offer potential solutions for managing such challenges. To achieve the desire goals a systematic framework and associated concepts and tools are necessary to guide designer methodically with the global insight of the process, identify root causes of the problems or key areas of opportunities, benchmark the performance of the process, and develop a set of design recommendations that can attain the true potential of the process.

My research background of process integration, process control and specially process optimization assist me for developing new concepts and tools to systematically enhance and reconcile various process objectives, such as cost effectiveness, yield enhancement, energy efficiency, and pollution prevention. I am working last few years to develop such methodical sustainable approaches for various processes starting from small scale production (producing valuable pharmaceutical drags using liquid chromatography) to large scale facility (managing industrial flares during abnormal situations). As my specialization is to work with different kinds of optimization algorithms and control concepts, I would also like to face the challenges from process uncertainty and from the black box problems which can be mitigated using sustainable process integration.

My knowledge of research is not limited to the industrial processes, I have also implemented my acquire optimization knowledge to develop some adaptable financial modelling/algorithms which can be used in the most challenging and unpredictable global financial markets. The unique point of my research activity is that I can use my long experience of capital market to develop long term view for energy and commodity prices; this knowledge can be integrated with my developed framework to make the system/design more robust to face different kind of challenges including unpredictable price fluctuations (price fluctuations of oil, natural gas etc.), unpredictable incidents (industrial flaring) or unknown disturbances (sudden changes in process parameters).

Future Plans

Following are three principal research areas and corresponding projects that I plan to advance in my research group during the first several years:

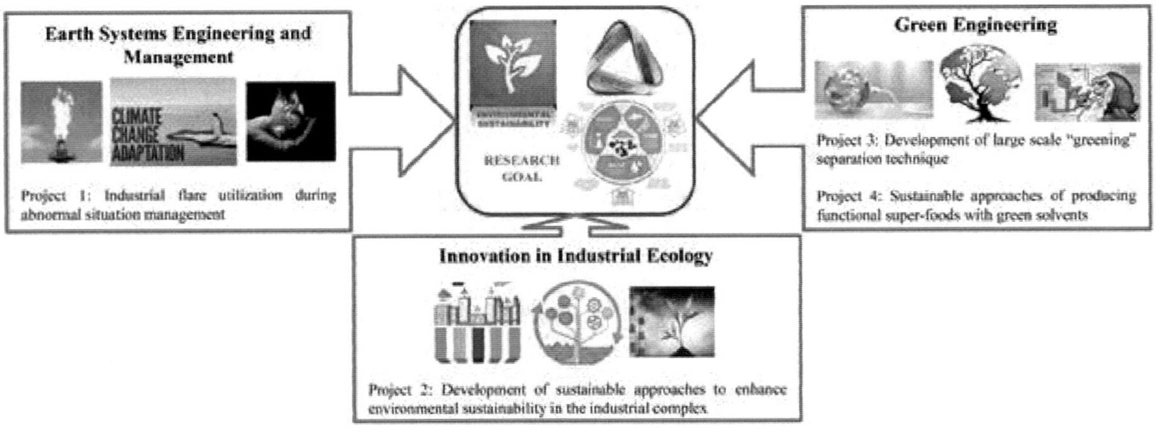

Figure 1. Schematic of proposed research projects and the research goal. Each of these rectangles (representing

research projects) will interact with one another in an iterative manner.

Research Interests:

§ Industrial flare utilization strategies through process integration and energy optimization for abnormal situation management by economic and environmental friendly alternatives

§ Development of energy alternative tools and sustainable process configurations to manage excess resources (i.e., waste energy streams, waste water) in the industrial complex for environmental sustainability

§ Development of fundamentals/technologies/alternatives for large scale "greening" separation technique (i.e., chromatography) to move towards environmental sustainability

§ Sustainable approaches of producing functional super-foods with green solvents for global food security through process integration and intensification

Teaching Interests:

I am comfortable teaching the Chemical Engineering core courses offered within the department at both the undergraduate and graduate level. I would be well suited to teach fundamental course such as: chemical process control, chemical reaction engineering, particle technology, separation technology, chemical & engineering thermodynamics, numerical methods, unit operations, mass transfer technology, process integration etc. I appreciate the need for faculty flexibility and embrace the opportunity to cover a broad spectrum of courses. Based on my research, I also have a strong interest in process integration, control and optimization. I look forward to incorporating fundamental engineering concepts with cutting-edge technologies.

SCHOLARLY ACHIEVEMENTS

SUMMARY OF RECORDS

Peer Reviewed Journal articles: **5** published (**3** first- and 2 co-authored) and **1** (first-authored) in review

Peer Reviewed Conference Proceedings: **3**

Conference Presentations: **14**

Selected Peer Reviewed Publications (*Citations: 16; h-index: 2; i-10 index: 1; total impact factor: 13.90, as in April 2016*)

· **Monzure-Khoda Kazi**, Fadwa Eljack, Nesreen A. Elsayed and Mahmoud M. El-Halwagi. Integration of energy and wastewater treatment alternatives with process facilities to manage industrial flares during normal and abnormal operations – A multi-objective extendible optimization framework. *Journal of Industrial and Engineering Chemistry Research,* ACS publication, 2016, 55, (7), 2020-2034. (doi:10.1021/acs.iecr.5b03938)

· **Monzure-Khoda Kazi**, Fahd Mohammed, Ahmed Mhd Nabil AlNouss and Fadwa Eljack. Multi-objective optimization methodology to size cogeneration systems for managing flares from uncertain sources during abnormal process operation. *Computers and Chemical Engineering,* Elsevier, 2015 vol 76, p.76-86 (doi:10.1016/j.compchemeng.2015.02.012)

· **Monzure-Khoda Kazi**, Bijan Medi, and M. Amanullah. Optimization of an Improved Single-Column Chromatographic Process for the Separation of Enantiomers. *Journal of Chromatography A,* Elsevier, 2012. 1231: p. 22-30. (doi:10.1016/j.chroma.2012.01.057).

GRANT WRITING EXPERIENCE

1. "Water recovery from power plant flue gas: a hybrid technology as a novel source of water for Qatar", PI: Asst. Prof. Mohammad Aman Ullah, 2015 (NPRP no.: 8-2191-2-855)

2. "Experimental evaluation of the optimal operation of a single column continuous chromatography process under overloaded condition: profile-profile interaction", Source: Qatar University (individual grant), Budget: US$ 35,715, PI: Asst. Prof. Mohammad Aman Ullah, 2013

3. "Optimization of a single column continuous chromatographic speration process for the separation of optical isomers", Source: 13th cycle of the QNRF's undergraduate research experience program (UREP), Budget: US$ 58,850, Lead PI: Asst. Prof. Mohammad Aman Ullah, 2013

4. "Design of a simulated moving bed (SMB) process for continuous binary separation of chiral compounds", Source: Qatar University (student grant), Budget: US$ 2,715, PI: Asst. Prof. Mohammad Aman Ullah, 2012

5. "Super food production using supercritical carbon dioxide technology: extraction, micronization, encapsulation and optimization", PI: Assoc. Prof. Timothy Tan, 2012

TEACHING AND MENTORING EXPERIENCE

Lecturer (Green University, Bangladesh)

1. CSE 2102 - Design and Innovation with Computer Simulation
2. TEX 2201 - Fundamental of Textile Testing
3. TEX 2209 – Environmental & Pollution Control

Graduate Teaching Assistant (Nanyang Technological University, Singapore)

1. CH3101 – Chemical & Bioengineering Process Control and Dynamics (AY 2011-2012 Sem 2)
2. CH3001 – Process Control Lab (AY 2011-2012 Sem 2)
3. MATLAB workshop on Numerical Methods (AY 2011-2012 Sem 2)
4. CH1801 - Chemical/Bioengineering Lab (AY 2011-2012 Sem 1)
5. CH1701 - Chemical/Bioengineering Lab (AY 2010-2011 Sem 1)
6. CH3102 – Chemical Reaction Engineering (AY 2009-2010 Sem 2)

Student Advising

1. Fahd Mohammed, MSc Student, Qatar University (2015)
2. Fah Keen Chong, PhD Student, University of Nottingham Malaysia (2015)
3. Viknesh Andiappan, PhD Student, University of Nottingham Malaysia (2015)
4. Calvin Wiratama, Undergraduate Student, Nanyang Technological University Singapore (2012)
5. Xueyuan Tian, Undergraduate Student, Nanyang Technological University Singapore (2010)

REFERENCES

Dr. Mahmoud El-Halwagi

Professor, Artie McFerrin Department of Chemical Engineering, Texas A&M University, College Station, TX 77843-3251

Tel: (979) 845-3484

Email: el-halwagi@tamu.edu

(He is my current research collaborator and mentored me in a workshop)

Dr. Fadwa T. Eljack

Assistant Professor, Department of Chemical Engineering, Qatar University, P.O. Box: 2713 – Doha – Qatar

Tel: +974 4403 4141 Fax: +974 4403 4131 Email: Fadwa.Eljack@qu.edu.qa

(She is my current supervisor)

Dr. Mohammad Aman Ullah

Assistant Professor, Department of Chemical Engineering, Qatar University, P.O. Box: 2713 – Doha – Qatar

Tel: +974 4403 4162 Fax: +974 4403 4131

Email: Aman@qu.edu.qa

(He was my PhD supervisor in Singapore)

Dr. M. M. Faruque Hasan

Assistant Professor, Artie McFerrin Department of Chemical Engineering, Texas A&M University, College Station, TX 77843-3251 Tel: (979) 862-1449

Email: hasan@tamu.edu

(He was my senior colleague in Singapore)

-

476048 Life-Cycle Environmental and Techno-Economic Analysis of Novel Biomass Thermochemical and Electrochemical Energy Systems

Abstract:

Research Interests:

Our focus has shifted from fossil energy to renewable energy sources due to environmental pollution, global warming, and sustainable development issues caused by the consumption of fossil fuels. As a promising renewable alternative, biomass has been playing a critical role in enhancing the nationsâ€™ energy security and mitigating carbon emissions. My research has focused on thermochemical conversion of biomass into fuels, electric power, and chemicals through experimental investigation, process modeling using Aspen Plus, and life-cycle environmental and techno-economic analysis.

I have been involved in several projects working as a post-doc researcher at Iowa State University. We have proposed a novel strategy for reducing carbon emissions from coal-fired power plants based on the utilization of bio-oil and biochar sequestration. This approach provides a sustainable and economical pathway to meeting life-cycle carbon reduction targets in the power sector. We have developed new methods for optimizing revenues from various product portfolios via fast pyrolysis and upgrading in an integrated biorefinery including drop-in transportation biofuels, biobased products, and hydrocarbon chemicals production. The findings from this analysis shed light on product selection and risk mitigation for investors from the bioproduct industry. In more recent work, we explore hybrid thermochemical and electrochemical methods for electricity, fuels, and chemicals generation and focus on three novel system designs: direct biomass fuel cell, biomass electrolyzer, and CO_2electrolyzer. The potential economic advantages and challenges are identified from this analysis.

During my Ph.D. program, I explored transportation biofuel production via thermochemical conversion of agriculture residues (rice husk and corn stover). My work spanned from helping setting up the experimental platform (reactor with a volume of 100mL, 300mL and 10L respectively) to conducting experiments on the catalytic upgrading of fast pyrolysis bio-oil in supercritical ethanol/methanol system. The optimal reaction conditions and catalysts for this bio-oil upgrading method were proposed. I further enhanced my experimental experience by converting model compounds of algae biomass to biofuels through hydrothermal liquefaction. In addition, I focused on employing modeling tools such as GREET and SimaPro to evaluate the environmental performance of different bio-oil upgrading pathways.

My future research interests are applying a combination of experimental and modeling tools to better understand novel biomass conversion technologies and help advance the development of these technologies. Specific research interests include experimental investigation of biofuels, chemicals and power generation from thermochemical processes, process modeling, and evaluation of bioenergy systems from environmental and techno-economic perspectives.

Teaching Interests:

I would like to teach both undergraduate and graduate level courses such as introduction to mechanical engineering, energy and power system, biorenewable technologies, chemical process modeling, life-cycle environmental and techno-economic analysis which are closely related to my research. By integrating my research in my teaching, I can enhance the engagement of students in certain types of research. I hope my students can get some hands-on experience and a better understanding of core engineering concepts by employing state-of-the-art engineering tools to solve real engineering problems after finishing the courses.

476131 Exploring New Boundaries for Sustainable Water and Industrial Wastewater Treatment Technology

Abstract:

Clean and safe water is a basic human need, but it is not available to 1.4 -1.8 billion people around the world. It is essential to direct current research trends toward sustainable water and wastewater treatment technologies that can solve the existing industrial and environmental issues, especially when it comes to solutions that can be successfully commercialized on the global scale. Membrane application technologies are the most effective and sustainable methods of addressing environmental problems in treating water and wastewater to meet or exceed stringent environmental standards. Nevertheless, membrane fouling is one of the primary operational concerns that is currently hindering a widespread application of membrane technology. My major research plans are to optimize synthesis of biomimetic and bioinspired membranes designed with antifouling, self-cleaning, and selective permeation that will pave the way for the production of clean water; to address all of the existing deficiencies of the Forward Osmosis systems; and to control membrane morphology by zeolites incorporation.

Research Interests: Membrane Science and Technology in Water and Wastewater Treatment; Biomimetic and Bioinspired Membranes for Next Generation of Biomaterials and Nanomaterials Engineering; Process Modelling, Simulation, and Optimization of complex systems.

Teaching Interests:

- Received a Formal recognition of completion of the Teaching Development Professional Program (UTDP), accredited by the Staff and Educational Development Association of the United Kingdom (SEDA UK).
- Participated in the curriculum development processes by designing a new effective course for graduate students

  at Ryerson University.
- Strong record of Teaching & Educational Engineering publications

475052 Renewable Transportation Biofuel Production from Wet Biowaste

Abstract:

Research Interests:

- **Summary**

Increasing wet organic waste production has become an unavoidable issue in the US. Although often considered as environmental pollutants, these wet organic feedstocks contain large amounts of unharnessed energy and nutrients. My research aims to recover the energy and nutrients from the wet organic waste with appropriate waste remediation systems. Hydrothermal treatments, including catalytic hydrothermal gasification (CHG) and hydrothermal liquefaction (HTL), is a new energy efficient thermochemical process that can generate high-energy content biogas/biocrude and simultaneously reduce total organic carbon in wastewater. In contrast, anaerobic digestion (AD) is a widely utilized biochemical process that similarly produces renewable biogas. Application of different catalysts and adsorbents in each scenario can increase energy yields and nutrient recovery. Running both systems in parallel on identical wet organic wastes can compare energy returns and pollutant reduction through techno-economic analysis. In addition, effluent waters from each process containing high concentrations of nitrogen and phosphorous, amongst other dilute nutrients, can be further recycled as a desirable substrate for production of crops (e.g. algal biomass or crops grown in hydroponic systems). Secondary algal treatment of wastewaters from CHG/HTL and AD will clean water for environmental discharge meeting EPA standards and also generate renewable resources for further application. Simultaneously, using isotopes with representative model compounds in CHG/HTL process can further elucidate the reaction kinetics, network, and mechanism in this biomass conversion system. Successful completion of these objectives will provide clearer understanding of how advanced thermochemical and biochemical waste remediation technologies can alleviate the ***food-water-energy nexus***.

- **Rationale:**

Wet biowastes have been an increasing interest to researchers for the potentials to produce bioenergy, reuse nutrients and improve the environment. Each year, the United States produces 79 million dry tons of wet biowastes including food-processing waste and livestock manure (DOE, 2011; FAO, 2016; USDA, 2010). These wet biowaste exceeds the â€˜Impactful tonnageâ€™ (>50million ton per year) as a bioenergy feedstock according to DOE, and containing nutrients (N&P) that can substitute about 1.5 million tons of reactive nitrogen as fertilizers. Carbon, nitrogen and phosphorous, along with other minerals contained in wet biowastes are often discharged to environment as pollutants because they cannot be efficiently recovered and utilized. Based on data provided by USDA (USDA, 2011), it is estimated that as much as 45% of nitrogen and 70% of phosphorous could be recycled from such agricultural waste streams. One of the main reasons this energy and nutrients are not effectively harvested is the large water content (20-98%), which greatly reduces the number of bioenergy conversion technologies that can have a net positive energy balance. On the other hand, energy content approximately over 10 quads, equal to all current US renewable energy production, could be recovered for beneficial reuse with appropriate technologies or systems. Moreover, the growth of the global population has placed an ever-increasing stress on food, energy, and water (UN, 2015). Advancing knowledge/technologies that foster more efficient, safe, and secure use of resource within the food-energy-water nexus remains a significant challenge (NSF, 2016).

- **Research Topics**

My research enthusiasm was influenced by my past and present research which encompasses thermochemical conversion processes, biofuel production, biowaste treatment, biomass pretreatment, value-added chemical extraction, techno-economic analysis of the nutrient and bioenergy recovery processes, and reaction network under supercritical fluid. Combining both laboratory and computational methods to link local processes to up-scaled system in our dynamic environment can help address the following questions: (1) How to energetically harness available energy and nutrients from wet organic waste and what are the most relevant factors affecting the biomass conversion process? (2) How to control and fine tune the biomass conversion processes more precisely to obtain more valuable products? (3) How do we combine new technologies with multiple scale (e.g. from laboratory to pilot to demonstration scale) of investigation to demonstrate an affordable system, minimize greenhouse gas emission, maximize the recovered energy, and identified research gaps to strengthen a robust food-energy-water nexus? To answer these questions, my research closely couples innovative thermochemical, biological, environmental, and computational approaches to provide an integrated perspective to establish a solid energy-water-food nexus.

1. ***How to energetically harness available energy and nutrients from wet organic waste?***

Catalytic hydrothermal gasification (CHG) is an emerging new technology alternative for wet biowaste conversions with much faster reaction times (minutes), high conversion efficiencies (>90%), and a cost trajectory that is projected to be competitive with current fossil fuels when the feedstock has a low or negative cost. However, there are no full-scale applications of this technology because commercial-scale equipment has not been developed, and the data available from the current literature highlights several potential practical concerns (e.g. catalysts fouling issues) that need to be addressed before large investments in the technology are justified. Anaerobic digestion (AD) is traditionally a dominant conversion technology used for wet wastes. In order to achieve a higher organic loading rate and energy recovery, a two-stage AD process is proposed in my research program and the results will be compared to those from a conventional AD system. By optimizing the conditions of sludge loading rate, operating temperatures (30-55°C), the dosage of adsorbents, and the types of adsorbents, the yields of hydrogen and methane (hythane) and the energy recovery of the proposed two-stage AD system can be improved. Conventional AD process produces mainly methane, while two-stage AD can produce more hydrogen because it can better control the reaction environment (e.g., pH) and pathways by dividing the fermentation process into hydrolysis, acidogensis, and methanogensis separately (Massanet-Nicolau et al., 2013; Schievano et al., 2014). According to recent studies, CHG and AD can both substantially recover the carbon nutrients in swine manure, but not nitrogen and phosphate (Breinl & Zhang, 2015; Shin & Schideman, 2015; Tommaso et al., 2014). On the other hand, aquatic biomass such as macrophytes and algae has been proven to have removal efficiencies of greater than 80 and 95% for nitrogen and phosphorous, respectively (Möller & Müller,

2012; Polomski et al., 2009; Woertz et al., 2009). In addition, crops (e.g. lettuce and tomatoes) that can grow in the hydroponic system also have the potential to further assimilate the residual nutrients in the CHG and AD effluents. Therefore, a hydroponic farming system is proposed in my research program to simultaneously uptakes the nutrients and remediates the wastewaters coming from AD and CHG. In this case, research efforts on hydroponic cropping development and algal biotechnology for concentrated agricultural wastewater treatment can be advanced to satisfy growing need in the food-water-energy nexus, while the potential of these concentrated agricultural effluents as nutrient sources for aquatic biomass grown in hydroponic systems can be realized. In short, my research program aims to investigate the efficacy of using AD and CHG to harness the chemical energy stored in wet biowastes (e.g. animal manure and citrus processing waste containing value-added chemicals such as flavonoids). The ultimate goal of my research program is to demonstrate the feasibility of a novel concept for integrated bioenergy and biomass production, and wastewater reuse via parallel treatment of CHG and a two-stage AD of wet organic wastes.

2. How to control and fine tune the biomass conversion processes more precisely to obtain more valuable products?
No full-scale applications of the hydrothermal treatment technology have been realized so far, because the data available from the current literature highlights several potential practical concerns that need to be addressed before large investments in the technology are justified.

(i) Catalysts development for effective CHG. Previous literature (Elliott, 2008; Stucki et al., 2009) has extensively focused on using noble or transitional metal catalysts such as Pt, Ru, and Ni for catalytic hydrothermal gasification. To make CHG a more affordable technology for the public, other economic catalysts such as activated carbons, zeolites, calcium carbonates derived from waste shells, and iron recovered from slag will be explored and their catalysts performance (e.g. yield of hydrogen) will be evaluated.

(ii) Catalyst activity under CHG. Fouling of catalyst is another issue under hydrothermal processes and will be investigated in my research program. The previous study (Ong, 2013) using CHG to catalyze waste newspaper shows that unless the catalysts can be regenerated and recycled, CHG can produce more net-energy and present a more efficient energy consumption ratio (ECR, defined as energy input divided by energy output) than anaerobic digestions of waste newspaper. Thereby, in my research program, the catalysts will be characterized before and after the reactions to understand the deactivation mechanisms (e.g. catalysts fouling or catalysts poison) through SEM, TEM, TGA/DSC, XRD, and XRF analyses. Different regeneration methods for catalysts will be explored and the morphology of the regenerated catalysts will be studied. Meanwhile, the life of the catalysts will be understood. Ultimately, methodologies to minimize the catalysts deactivation will be established through my research program.

(iii) *Reaction Kinetics and Network for bioactive chemicals under CHG.* Previous studies have demonstrated that CHG can effectively degrade bioactive chemicals, such as environment hormones (e.g. estrogens) and antibiotics (e.g. florfenicol), entrained in swine manure (Shin & Schideman, 2015). However, the reaction kinetics and reaction network of these bioactive chemicals remained unknown. Understanding the reaction kinetics and network for bioactive chemicals may help design the up-scaled CHG reactor in a more practical way and make CHG a more versatile tool for treating different types of bioactive chemicals (e.g. virus and pathogens). In order to achieve this goal, CHG of model bioactive compounds with heavy water (i.e., deuterium oxide) will be conducted at 150-400 °C for 0-2 hr. The CHG products obtained at different reaction time will be analyzed under GC/MS, NMR (1D and 2D), and FTIR essay to further elucidate the reaction pathways of wet biowaste. Sensitivity analysis will also be conducted to explore significant factors governing CHG processes.

3. How do we combine new technologies with multiple scale (e.g. from lab to pilot scale) of investigation to demonstrate an affordable system, minimize greenhouse gas emission, and maximize the recovered energy to strengthen the current food-energy-water nexus?

System analysis of the proposed parallel CHG and AD systems can help understand and identify research gaps to establish a roust food-energy-water nexus. For example, the necessity of reusing residual nutrients in the aqueous products (PHWW) obtained from hydrothermal liquefaction (HTL) processes has been identified through a multicomponent mass balance. Conducting the mass balance, energy balance, and economic analyses of the integrated system using the STELLA® and SuperPro software can further elucidate and identify the revenue streams, quantification of cost functions, capital allocations, upstream suppliers, and downstream customers. Simultaneously, life cycle analysis (LCA) through quantitative and computational modeling will be utilized to optimize this system and significantly advance the understanding of the proposed food-energy-water nexus.

- **Indication and Expected Outcomes:**

Successful results of this research program will lead to bridging gaps between biowaste management and energy/nutrient recovery. Modification of two-stage AD processes including fermentation concentration and operating temperatures will help optimize treatment efficiency and yield greater energy content in the form of methane and hydrogen, which is underdeveloped. More recently investigated thermochemical technologies like HTL and CHG show great promise to efficiently convert wet organic feedstocks into high-energy content products such as biocrude oil and biogas, and the combination of different catalysts and reaction parameters can improve overall energy recovery and at the same time lowering energy inputs. Comparison of improvement in two-stage AD and CHG systems will demonstrate the ability to remediate different types of wet biowaste. Once carbonaceous contents are recovered as biogas, the remaining nitrogen and phosphorous will be assimilated by aquatic biomass for more renewable feedstock and/or food production. The reuse of nutrient can amplify the biomass thus the bioenergy production. An overall techno-economic assessment will elucidate the sustainability of the proposed parallel CHG and AD systems.

Teaching Interests:

- **Summary**

My teaching philosophy was molded by my education as an engineer. For me, education is significantly more than imparting knowledge found in textbooks. As an engineering student involved in a multidisciplinary research group, I had the opportunity to gain a great deal of know-hows by participating in different research activities. This experience also motivated and reinforced my teaching philosophy. I truly believe that by involving students with hands-on projects and real-world solid examples, learning new knowledge can be more effective while creative thinking and innovative solutions will be strongly encouraged. As Benjamin Franklin has said, "Tell me and I forget. Teach me and I remember. Involve me and I learn."

- **Teaching Philosophy**

I am deeply committed to effective teaching and mentorship. I have been very fortunate to have had many positive and influential mentors during my education. This experience incentivizes and eagerly drives me to be a supportive and helpful mentor to my students. My objective when teaching and mentoring is to create a learning environment that simultaneously develops critical thinking skills and fosters an open mind to all possibilities by emphasizing the rich conceptual issues that defines sustainable energy science and engineering and how this discipline impacts our lives. By expanding not only the studentsâ€™ scientific understanding of engineering problems but most especially how they apply it to the real world, we can critically review more available sustainable engineering practices and/or stewardship. Advantages and limitations of the established methods would be identified during this critical thinking process, and valid alternative state-of-the-art methods can then be developed. Ultimately, I hope my students can think â€œoutside the boxâ€ and develop their confidence in proposing solution to existing and new problems.

- **Teaching experience**

As I have served as a teaching assistant (TA) in the class of *Transportation Phenomenon*(ABE 341), I have designed weekly worksheets for students, conducted office hours, worked with students on their lab projects, and graded homework and worksheets. Moreover, I realized that effective communication is a key element to help students understand fundamental theories (e.g. fluid dynamics) and accomplish lab projects successfully. A smooth communication can inspire studentsâ€™ potential and establish their confidence. For example, I always encouraged students to come to office hours and work on those difficult problems together. Textbook and homework are not something to beat them but tools to help them learn. Working as a team to solve problem sets usually can help them develop critical thinking skills and address problems more efficiently. For instance, one of my students learnt that units may help her check if the answer makes sense in a fluid dynamics problem. Additionally, I have served as a guest TA for several class projects (ABE 100, ABE 225, CEE 398 and TSM 431, please refer to my CV). As a mentor, I usually ask students some questions and give some solid examples to initiate their interest. Literature review and experimental design are generally required in class projects. Through lab projects, students can gain knowledge on bioenvironmental engineering design, particularly about biomass conversion and sustainable energy engineering, while I always benefit from the mentoring and discussion. Teaching and mentoring are not only a give-and-take process but also an interactive and mutual beneficial discussion with the mentored undergraduates at the same time.

- **Mentoring experience**

By participating a multidisciplinary research programs in University of Illinois, joining the algae club in my home department, and involving with SWE (Society of Women Engineers) on campus, I was very fortunate to have ample opportunities to mentor or work with students from elementary school, high school, to different levels of undergraduate students (please refer to my CV). As providing technical training and helping them develop research topics, I always benefit from the mutual beneficial discussion. Students often inspire me from a different point of view and pay more attention to details than I do. This mutually beneficial discussion has been a cornerstone of my research career. For example, when I worked with my first undergraduate, sparked by her previous experience on centrifugation, we began to develop a more sustainable pretreatment technique, combining centrifugation and ultrasonic processes, for improved bioenergy conversion efficiency from algal biomass. Another instance is that I have further advanced my knowledge on gas-chromatography mass-spectrometry (GC-MS) by instructing several undergraduate and graduate students on using this equipment for analyzing biocrude oil and volatile aqueous productsâ €”mentoring itself is an intensive learning experience.

- **Teaching methods**

When I served as a TA four years ago, I found it was drastically different between knowing the knowledge and transferring to others. For instance, demonstrating a practice example is likely more helpful to the novice than throwing a challenging problem at him/her, which may spoil a studentâ€™s interests and confidence. During my work with undergraduate students from different departments and culture background in University of Illinois, I realized that working on projects can significantly motivate students to learn a specific topic and inspire studentsâ€™ potential as well as establish their confidence. Therefore, I will combine lecture and lab projects in my classes. For fundamental classes (e.g. transport phenomenon and heat and mass transfer), I plan to apply in-class discussion to one of the weekly lectures with a worksheet. I will also separate students into different groups and assign different groups of students to design the worksheet every week. This type of student-centered learning method can further help students think about why the lecturer/textbook asks questions like that way and why those questions matter. For advanced classes (e.g. classes provided for senior undergraduate students or graduate students), I will combine lectures, lab projects, field trips (e.g. visit a zero energy residence building, a bio-ethanol plant, or a pilot scale continuous reactor to generate biofuels/biogas), term-paper writing, and practice proposal writing together. When I took classes and served as a TA in University of Illinois, I found out lab projects usually can make students effectively learn hands-on skills and apply what they learnt in classes to real-world problems while field trips can greatly initiate studentsâ€™ interest on a specific topic. In addition, term paper and practice proposal writing can help students synthesize their thoughts on specific topics and present their ideas in a professional way. For instance, when I wrote my practice proposal for *Green Chemistry* (CHEM 460) and *Polymer Chemistry*(CHEM 480), I read a lot of papers about biorefienries and reviewed different biomass conversion techniques. In the end, I gained deeper insights about renewable energy and value-added chemical production processes, which ultimately contribute to part of my PhD thesis. Furthermore, I will conduct an early feedback, which will happen in the one third or half of the semester, in my classes so that I will have time to fine tune my teaching style or adjust course content during the semester. Overall, by using versatile teaching methods and techniques, I hope to give students a great incentive to learn different levels of courses and understand how they may change and impact the world.

References:

Breinl, J., Zhang, Y. 2015. Hydrothermal Gasification of HTL wastewater.

DOE. 2011. US billion-ton update: Biomass supply for a bioenergy and bioproducts industry. Oak Ridge National Laboratory (ORNL).

Elliott, D.C. 2008. Catalytic hydrothermal gasification of biomass. *Biofuels, Bioproducts and Biorefining*, **2**(3), 254-265.

FAO. 2016. Database of Food and Agriculture Organization of the United Nations

Massanet-Nicolau, J., Dinsdale, R., Guwy, A., Shipley, G. 2013. Use of real time gas production data for more accurate comparison of continuous single-stage and two-stage fermentation. *Bioresource technology*, **129**, 561-567.

Möller, K., Müller, T. 2012. Effects of anaerobic digestion on digestate nutrient availability and crop growth: A review. *Engineering in Life Sciences*, **12**(3), 242-257.

NSF. 2016. Innovations at the Nexus of Food, Energy and Water Systems (INFEWS).

Ong, M. 2013. Evaluation of anaerobic membrane bioreactors and hydrothermal catalytic gasification for enhanced conversion of organic wastes to renewable fuels, University of Illinois at Urbana-Champaign.

Polomski, R.F., Taylor, M.D., Bielenberg, D.G., Bridges, W.C., Klaine, S.J., Whitwell, T. 2009. Nitrogen and phosphorus remediation by three floating aquatic macrophytes in greenhouse-based laboratory-scale subsurface constructed wetlands. *Water, air, and soil pollution*, **197**(1-4), 223-232.

Schievano, A., Tenca, A., Lonati, S., Manzini, E., Adani, F. 2014. Can two-stage instead of one-stage anaerobic digestion really increase energy recovery from biomass? *Applied Energy*, **124**, 335-342.

Shin, Y.H., Schideman, L. 2015. Characterizing the fate and transport of Chemicals of Emerging Concerns (CECs) from integrated bioenergy and manure management system. *2015 ASABE Annual International Meeting*. American Society of Agricultural and Biological Engineers. pp. 1.

Stucki, S., Vogel, F., Ludwig, C., Haiduc, A.G., Brandenberger, M. 2009. Catalytic gasification of algae in supercritical water for biofuel production and carbon capture. *Energy & Environmental Science*, **2**(5), 535-541.

Tommaso, G., Chen, W.-T., Li, P., Schideman, L., Zhang, Y. 2014. Chemical Characterization and Anaerobic Biodegradability of Hydrothermal Liquefaction Aqueous Products from Mixed-culture Wastewater Algae. *Bioresource Technology*.

UN. 2015. *World Population Prospects: The 2015 Revision, Volume I: Comprehensive Tables.* United Nations Publications.

USDA. 2010. Manure and Byproduct Utilization.

USDA. 2011. U.S. consumption of nitrogen, phosphate, and potash, (1960-2011).

Woertz, I., Feffer, A., Lundquist, T., Nelson, Y. 2009. Algae grown on dairy and municipal wastewater for simultaneous nutrient removal and lipid production for biofuel feedstock. *Journal of Environmental Engineering*.

Zhou, Y., Schideman, L., Yu, G., Zhang, Y. 2013. A synergistic combination of algal wastewater treatment and hydrothermal biofuel production maximized by nutrient and carbon recycling. *Energy & Environmental Science*, **6**(12), 3765-3779.

476331 Energy-Dense Liquids from Renewable Energy

Abstract:

Those in the global chemical enterprise now agree that the future of the chemical industry depends on becoming more sustainable. Such a goal includes developing energy sources that are not based on fossil fuels, and that do not release large quantities of climate-changing gases. This goal implies developing technologies that effectively collect stranded renewable energy - solar and wind - and enable long-term energy storage and long-distance energy delivery from remote areas. Solar and wind energy are continuing to get cheaper, but they are far from population centers and storage/transmission is difficult. Hence, small-scale distributed manufacturing processes that directly convert the renewable energy to transportable chemicals are important.

Research Interests:

At the University of Minnesota, I have been working on designing small scale Haber-Bosch process powered by stranded wind-generated electricity. The electricity generated by windmills is used to split water to make hydrogen, and to separate nitrogen from air by pressure swing adsorption. The hydrogen and nitrogen are then reacted to make ammonia. This ammonia is made without the use of natural gas or coal, and hence without emission of climate-changing carbon dioxide. Our analysis of this small process shows that its performance is governed by three characteristic times, those for chemical reaction, for ammonia removal by condensation, and for unreacted gas recycle. To date, our proposed process is limited by the time for ammonia removal, that is, by how long it takes the condenser to liquefy the ammonia product. While this acceleration of ammonia removal should be straightforward, we believe it will require considerable capital investment. As a result, we have been trying to simplify and intensify the Haber-Bosch process. We have not tried to improve the catalytic steps in the reactor at all, because we believe that the existing reaction, optimized by a century of intensive effort, has produced a very rapid potential reaction as good as can be hoped. We have focused instead on reducing the operating pressure and rapidly separating the ammonia produced. In my research, I have showed that the conversion could be increased from 20% to over 90% by coupling reaction and absorption. I have also showed how the operating pressure could be cut from 130 bar to 20 bar, while increasing the specific rate of reaction by cutting the product concentration in the reactor.

My future research has its origin in efforts to develop novel small-scale distributed production processes from stranded renewable energy, for production of carbon-neutral liquid fuels. In particular, I am interested in developing: devices that enable low-pressure ammonia synthesis via reactive-separation in a single vessel; processes for enhanced synthesis of renewable methanol for sustainable chemical industry; and novel supported adsorbents for facile mass and heat transfer for chemical synthesis.

Teaching Interests:

My philosophy of teaching was shaped and developed from my experience as a teaching assistant and guest lecturer in various undergraduate and graduate level courses. Whether I am teaching or learning, I am strongly committed and strive to be student-centered, engaging with students for active learning, and to support diversity as well as cooperation among everyone.

Prior Teaching Experience

My passion for academia and learning culminated into a longing to teach - a way to help others by imparting my knowledge or method of understanding. As a result, my aspirations to teach have pressed me to seek the most unique and challenging positions related to teaching. I have taught as a guest lecturer and teaching assistant in multiple undergraduate and graduate courses in the following subjects: Chemical Engineering Reactions, Mass Transfer, Senior Chemical Engineering Design, and Advanced Membrane Separations.

Teaching in academia requires continual development of one's teaching ability. This aptitude to constantly improve has been my motivation to stay informed not only with the latest practices in chemical engineering, but also the art of teaching. In regards to chemical engineering practices, I am continually interested in exploring and pursuing different strategies for teaching chemical engineering principles. Additionally, my enthusiasm propelled me to take part in the Senior Chemical Engineering Product Design course at the University of Minnesota. In this activity, we are piloting new undergraduate product design projects using chemical engineering principles familiar to students, but have unique design objectives that are innovative for students. The outcomes of our effort was presented at the 2016 American Society of Educational Engineering in New Orleans. In addition, a manuscript based on this work is currently under preparation that will explain the lessons we as faculty members learned, practices we recommend, and objectives that we accomplished.

My fervent desire for teaching extends beyond the classroom as I incorporate teaching principles in my daily life when working with family members, friends, and to people I have never met. [Add More]

Future Teaching Plans

I am prepared to teach the following courses:

- Reaction Kinetics and Reaction Engineering

- Chemical Engineering Design

- Mass Transfer

- Separation Processes

- Junior and Senior Chemical Engineering Lab

As a well-trained chemical engineer, I am more than capable and willing to teach other courses based on the department's needs. I am also personally interested in developing an elective course called "Novel Separation Processes for Sustainability," that I plan to offer to graduate students and upper level undergraduate students.

479779 Upgrading Lignocellulosic Biomass to Biofuels and Value-Added Materials for a Sustainable Future

Abstract:

Research Interests:

With the progressive depletion of fossil fuel resources and the necessity to combat climate change through the reduction of greenhouse gas emissions, there is an emergent need to identify alternative environmentally sustainable energy sources. A potential green energy source is derived from a variety of biomass and industrial waste streams, which can be converted to fungible fuels

Lignin, after cellulose, is the second most abundant biopolymer on Earth, comprising 30% of the organic carbon in the biosphere. Although lignin exhibits a high-energy content, it is typically destined for combustion in biorefineries due to its heterogenic and recalcitrant nature. However, cellulose and hemicellulose can be converted to renewable fuels and value-add products.

Currently, biorefineries under value the potential of using lignin for addressing the worldâ€™s need for fuel and materials. I am interested in genetically modifying bacteria and fungi that are able to utilize lignin and industrial waste streams as their sole carbohydrate source to increased lipid titers for biofuels, generate useful platform chemicals and value-added products for materials, and lignin degradation. I am also interested in developing techniques for enhancing the sustainability of biofuel production, efficiency of carbon conversion, and economic viability.

Teaching Interests:

Coupled with my research interests in sustainability, I am interested in teaching green engineering, bioprocess engineering, engineering economics, and life cycle analysis. I am also interested in applying an entrepreneurial aspect to process and product design.

456032 Systematic Framework for Carbon Dioxide Capture and Utilization Processes to Reduce the Global Carbon Dioxide Emissions

Abstract:

In the year 2013, 9.5 billion metric tons of carbon dioxide gas was emitted into the air, and each year this amount is increasing [1]. Carbon dioxide emissions are of particular concern as they represent 80% of greenhouse gas emissions and therefore are a large contributor to global warming. Among the two approaches that are currently being investigated, carbon capture and storage (CCS) and carbon capture and utilization (CCU) [1] to address this issue, the later approach is more promising as it reuses captured carbon dioxide, as a fuel, reactant, solvent, and others, to produce valuable products. There is not only a need for technologies for capture and utilization, via conversion, but also there are numerous questions that need to be resolved. For example, which higher value chemicals can be produced, what are their current demands and costs of production, and, how much of the captured carbon dioxide would be utilized? Also, how much carbon dioxide would be indirectly generated due to the capture and utilization efforts? Can the regulated carbon dioxide emission reduction targets be achieved only through the CCU and/or CCS efforts? Therefore, there is a need for a systematic computer-aided framework through which the issue of global carbon dioxide emissions can be investigated in terms of different available capture-utilization technologies, solution methods, and benefit scenarios, with the objective to determine more sustainable solutions within an appropriate application boundary. The framework would need to provide, amongst other options: useful data from in-house databases on carbon dioxide emission sources; mathematical models from a library of process-property models; numerical solvers from library of implemented solvers; and, work-flows and data-flows for different benefit scenarios to be investigated.

It is useful to start by developing a prototype framework and then augmenting its application range by increasing the contents of its databases, libraries and work-flows and data-flows. The objective is to present such a prototype framework with its implemented database containing collected information-data on various carbon dioxide emission sources and available capture-utilization technologies; the model and solution libraries [2]; and the generic 3-stage approach for determining more sustainable solutions [3] through superstructure (processing networks) based optimization – adopted for global carbon dioxide emission problems. More specifically, the prototype framework and the three-stage approach adopted for systematic and sustainable design of carbon capture and utilization processes incorporates (i) process synthesis stage, the determination of a processing path from a network of alternatives; (ii) process design, the design and analysis of a process or generated processing path in terms of process "hot-spots" or deficiencies to set design targets for improvement; and (iii) innovative and more sustainable design, determination of solutions matching the design targets. In this way, the starting point is an analysis of the current carbon dioxide emission status and the end point is an analysis of the more sustainable solutions in terms of one or more carbon dioxide benefit scenarios.

The developed framework contains a database constructed based on a developed knowledge representation structure, which has collected data on carbon dioxide emission sources, capture technologies, carbon dioxide-based reaction paths, and known utilization technologies leading to various products. The database currently has 5 carbon dioxide emission sources, 3 capture technologies, 10 conversion routes leading to 8 products, and over 50 reaction paths for many more products. The framework also integrates a software tool (the Super-O interface [2]) containing a library of process models and links to numerical solvers that enables the generation of processing networks (superstructure) and the determination of the optimal processing route. Finally, the framework has access to, a collection of tools for analysis, such as economic analysis, sustainability and life cycle assessment, and links to detailed process simulation (process simulators).

Following the work-flow and data-flow implemented in the framework, data for the carbon capture and utilization alternatives are retrieved from the database and linked to form a network using the interface tool, Super-O. For each alternative, models for economic and environmental impacts are available; the promising utilization products considered are methanol, dimethyl carbonate, and succinic acid. The optimal utilization process (or processes) is determined via a superstructure-based method (using solvers in GAMS). Then, the carbon dioxide emission values for the utilization routes are compared with the current industrial production routes for the listed products in order to identify those that have zero or negative emission values (objective function). Subsequently, the identified utilization processes are designed rigorously and further improved with unique technology and integration options. Using this approach, the utilization processes for methanol, dimethyl carbonate and succinic acid give the following interesting result: by converting at least 70% (methanol), 40 % (dimethyl carbonate) and 20% (succinic acid) of the current industrial processes to the corresponding utilization processes, the emissions from these processes can be neutralized, representing over 100 million metric tons of emissions that are neutralized. Note that the issue of product demand increase has not been considered. Also, on a global scale, this reduction of carbon dioxide emissions represents a very small but very valuable reduction. A more complete sustainability analysis is needed to enlarge the boundary of the analysis; for example, the boundary can change in terms of reduction of non-renewable resources and the use of renewable energy sources, as well as more efficient and sustainable designs of current production routes. While a combination of methods is needed to globally address carbon dioxide emissions and other aspects of sustainability, this systematic approach to designing carbon capture and utilization processes shows the capacity for such processes to reduce emissions and improve sustainability while producing valuable products. The current prototype framework with its implemented methods and tools is a small but important step. Collaboration and integration of data, methods and tools is necessary to provide a more sustainable solution to the global carbon dioxide emission problem.

References

[1] Yuan, Z., Eden, M.R., Gani, R., 2016. Toward the Development and Deployment of Large-Scale Carbon Dioxide Capture and Conversion Processes. Ind. Eng. Chem. Res. 55, 3383–3419. doi:10.1021/acs.iecr.5b03277

[2] Bertran, M.-O., Frauzem, R., Zhang, L., Gani, R., 2016. A generic methodology for superstructure optimization of different processing networks, in: Kravanja, Z. (Ed.), Proceedings of the 26th European Symposium on Computer Aided Process Engineering – ESCAPE 26. Elsevier B.V., Portorož, Slovenia.

[3] Babi, D.K., Holtbruegge, J., Lutze, P., Gorak, A., Woodley, J.M., Gani, R., 2015. Sustainable process synthesis–intensification. Comput. Chem. Eng. 81, 218–244. doi:10.1016/j.compchemeng.2015.04.030

462685 Design of Energy Saving CO2 Separation Process Using Circulating Fluidized Bed

Abstract:

By fossil fuel combustion, large amounts of greenhouse gases (GHGs) are generated and emitted into the atmosphere. The reduction of carbon dioxide (CO2) emission has become a major target in efforts to suppress global warming. Carbon capture and storage (CCS) has attracted attention in the past decades to reduce greenhouse gas (GHG) emissions. CO2 separation from gas mixtures requires huge amounts of energy. In fact, CO2 separation from gas mixtures is one of the most energy intensive process in CCS. Therefore it is necessary to reduce the energy consumption of CO2 separation for propagation of CCS.

The most widely used method among CO2 separation is a chemical absorption using amine based solution such as monoethanolamine (MEA). In the conventional CO2 separation by absorption using amine based solution, the thermal energy is supplied from the bottom of regenerator which utilizes reaction heat for regeneration of the solutions and heat of vaporization for stripping. To reduce the supply of the thermal energy, authors have developed CO2 gas separation processes based on self-heat recuperation. Applying the self-heat recuperation to thermal processes, not only the latent heat but also the sensible heat of the process stream can be circulated into the processes without any heat addition, leading to minimizing exergy destruction of heat transfer and reduction of the process energy consumption. However, the absorption using solution physically and chemically requires additional energy for separation of liquid/gas, vaporization for stripping.

In this paper, an innovative energy saving CO2 separation process using chemical absorption was proposed for energy saving point of view. In this process, metal oxide particles were selected as absorbents not for requiring separation energy of liquid/gas. These particles were fluidized by feed gas mixture for increasing segregation effect in the bed and heat transfer rate from chemical reaction to a heat pump system. Heat of absorption was recuperated and supplied to the heat of thermal decomposition through the heat pump by following exergy recuperation. In addition, some of the heat was circulating with absorbents. The energy saving performance of the proposed process was evaluated by simulations.

464620 Systems Design and Economic Analysis of Direct Air Capture of CO2 Â through Temperature Vacuum Swing Adsorption on Metal Organic Frameworks

Abstract:

The rising levels of CO_2 concentration in the atmosphere poses a serious threat to global climate stability. To mitigate this threat, efforts needs to be made for reduction of CO_2 levels. One of the strategy is to capture CO_2 from stationary, point, sources. These point sources accounts for approximately only one-third of the global emissions. Another strategy is to capture CO_2 directly from ambient air which, if successfully implemented, could result in capture of CO_2 from disperse emission sources. In contrast to CO_2 capture from stationary sources, Direct Air Capture (DAC) plants are location independent and can be scaled more easily. Thus, the DAC plant can be set up near a sequestration or utilization site, eliminating the need for CO_2transportation infrastructure.

Our work proposes comparison of two metal organic frameworks, namely, MIL-101(Cr)-PEI and mmen-Mg_2(dobpdc) [1] for DAC through temperature vacuum swing adsorption (TVSA). Isotherm parameter estimation for CO_2 adsorption was carried out for both the adsorbents using published and original data and a kinetic study was performed through empirical correlations [2] to estimate the mass transfer rates for the process . The adsorbents are grown as films inside 900 cpsi (cells per square inch) monolith. To achieve a temperature swing, steam is used as a stripping agent during the desorption step. In this thermal desorption operation, it is believed that the amines groups present in the metal organic frameworks are prone to oxidative degeneration at higher temperatures[3]. To overcome this challenge, we propose a five step process: adsorption, evacuation, pressurization, desorption and cooling. In the proposed process, evacuation and cooling steps are implemented through vacuum swing to remove the oxygen in the channels when the temperature is high. In order to model this system, partial differential algebraic equations have been implemented in gPROMS, which is a commercial dynamic process modeling and optimization software. The linear driving force (LDF) model is used to approximate the rate of CO_2adsorption [2].

This talk will discuss the energy requirements and overall economics of this DAC process. Different components which contribute to consumption of energy are identified and the TVSA model is simulated to estimate the net energy requirement for MIL-101-PEI-50 and mmen-Mg_2(dobpdc). Operating costs during the five steps (adsorption, evacuation, pressurization, desorption and cooling) and capital costs for metal organic frameworks and monoliths have been estimated for both the adsorbents. It has been shown that lifetime of the adsorbent is critical in estimating the net cost and a sensitivity analysis of the adsorbentâ€™s lifetime has been performed to analyze the DAC economics for both the adsorbents.Â We have identified sensitive parameters such as gas flow rate, cycle time, film thickness etc. which effects the overall energy requirements and net economics of the DAC process.

1. McDonald, T.M., et al., *Capture of carbon dioxide from air and flue gas in the alkylamine-appended metal-organic framework mmen-Mg2(dobpdc).* J Am Chem Soc, 2012. **134**(16): p. 7056-65.

2. Y. Y. Li, S.P.P., B. D. Crittenden, *Zeolite monoliths for air separation Part 2: Oxygen Enrichment, Pressure Drop and Pressurization.* Trans IChemE, 1998. **76**: p. 931-941.

3. Heydari-Gorji, A. and A. Sayari, *Thermal, Oxidative, and CO2-Induced Degradation of Supported Polyethylenimine Adsorbents.* Industrial & Engineering Chemistry Research, 2012. **51**(19): p. 6887-6894.

466081 New Performance Indicators for Adsorbents Used in CO_2Â Capture Swing Adsorption Processes

Abstract:

As adsorption process has increasingly received attention in CO_2 capture area, a number of studies have suggested a variety of adsorbents such a metal-organic frameworks (MOFs), zeolitic imidazolate frameworks (ZIFs) and many modified versions of conventional adsorbents. However, because of their limited scope of the studies, they have evaluated their adsorbents by using conventional performance indicators: working capacity and selectivity. The performance indicators rarely consider the situation of the adsorbents' practical application in the adsorption process. In order to evaluate the adsorbents with the consideration of their eventual application, e.g., pressure swing adsorption (PSA) process, dynamic simulations with a rigorous model of the adsorption process are needed, but the simulations cannot be carried out conveniently by the developers of new adsorbents. Furthermore, the high computation cost caused by the nonlinearity and cyclic operation of the PSA system hinders the evaluation of adsorbentsâ€™ suitability for their process-level application. For those reasons, this work derives simple formulas for the process-level performance of an idealized PSA system. The efficiency and the CO_2 purity of produced gas in the idealized PSA system are defined as performance indicators for the evaluation of a target adsorbent. For further simplicity, the performance indicators are derived in the explicit form of an analytic solution. The derived explicit equations are suggested as formulas of the new performance indicators for the evaluation. With the formulas, the performance of a target adsorbent in the PSA process can be obtained with simple calculation only from lab-scale measurement data such as adsorption isotherm data. Because the suggested formulas are based on the idealized PSA process where the theoretically maximum performance of each adsorbent is achieved, the suggested way is fairer than comparing the simulations of different adsorbents with some prior design and operating conditions which may be preferred by particular adsorbents. In order to show the usefulness of the formulas, a case study involving several real adsorbents is presented. The results from the formulas about the new indicators are then compared both to the simulation results of a rigorous dynamic model and to the conventional performance indicators.

Acknowledgement: The authors would like to acknowledge the financial support from the grant (G01160051) funded by Saudi Aramco-KAIST CO2 management center.

466627 Optimal Retrofit of a Post Combustion CO_2 Capture Process Using Reduced Superstructure and Rate-Based Models

Abstract:

We propose a superstructure-based methodology for optimal retrofit of a CO_2 capture pilot plant using rigorous rate-based model for the reactive distillation. The optimal retrofit for existing plants is particular challenging because rigorous process models used for describing the currently operating process often result very complex optimization problems.[1] Herein, we preselect process alternatives implemented in the superstructure based on thermodynamic analysis. Thermodynamic analysis identifies unit operations that have high energy saving potential and process alternatives improving these unit processes are considered in the superstructure. Consequently, the problem size of the superstructure can be substantially reduced and optimization can be performed using commercially available process simulators and optimization solvers. The optimum process configuration and operating conditions are obtained stochastically using a genetic algorithm. Near global optimum solution is obtained by solving the problem multiple times with different crossover fraction. Results indicate that the optimum retrofit process includes three stages solvent recirculation, lean vapor recompression, and mechanical vapor recompression. Thermal energy and total energy consumption in the optimal retrofit process are reduced about 59% and 27%, respectively.

471576 Dynamic Modeling with Uncertainty Quantification of Solid Sorbent Based CO_2 Capture Processes

Abstract:

Post-combustion CO_2 capture is a potential approach to mitigate buildup of CO_2 concentration in the environment. Significant research is currently being carried out all over the world to reduce the penalty for CO_2 capture. Due to the large amount of CO_2 that needs to be captured, significant scale-up is needed, starting from the lab scale investigation of a potential capture technology to its commercial implementation for large-scale power plants. Since scale-up of novel chemical technologies generally extends over 20-30 years, it is imperative that the time required be significantly reduced for the accelerated commercialization of the potential technologies. Rigorous process models of these novel capture processes can be very useful for this purpose. However, process models suffer from uncertainty. The parameters for submodels are typically calibrated deterministically for each submodel. Further, the models are imperfect and the experimental data are not only imperfect, but are often inadequate. The resulting parametric and model-form uncertainties can lead to high discrepancy in the model outputs. Therefore, process models should be developed with quantified uncertainty, which can also be instrumental in identifying where additional experimental data should be collected and/or where different or modified model form should be considered to reduce the discrepancy.

With these motivations, a 1-D dynamic, non-isothermal and non-adiabatic fixed bed adsorption reactor model is developed for novel solid sorbent-based CO_2 capture processes. Rigorous models for mass transfer, heat transfer and chemical kinetics, if applicable, of both physisorption- as well as chemisorption-dominated processes are developed. A Bayesian calibration approach is used for quantifying parametric uncertainty. Model-form discrepancy is described by a set of orthogonal functions. Posterior distribution of the parameters as well as the model-form discrepancies are propagated through the process model to obtain the bounded confidence on the modelâ€™s predictions of variables of interest such as bed pressure drop, breakthrough curve, and bed temperature profile. Our study shows that the output discrepancy strongly depends on the operating conditions and helps to identify approaches that can be used to reduce the scale-up uncertainty of the novel CO_2 capture processes.

468716 CO_2 Capture and Conversion to Chemicals Via Syngas: Rigorous Modeling, Intensification, and Superstructure-Based Process Synthesis

Abstract:

Recent global warming and climate change are attributed to increased CO_2 emission from fossil fuels. Significant efforts have been made in the past to develop CO_2 capture and sequestration (CCS) technologies. However, large-scale CCS is still not deployed in reality for reasons including high cost, technological barriers and uncertainty in geological storage. A promising alternative to CCS is to convert CO_2 to value-added chemicals via syngas (CO and H_2), which is an intermediate for many hydrocarbon-based fuels and chemicals. As much as 2 Gt/yr of fuels and 200 Mt/yr of chemicals can be produced from CO_2 utilization [1].

In this work, we explore alternative technologies and routes for the thermochemical conversion of CO_2 to syngas and then syngas to various oxygenates. We postulate a flowsheet superstructure which includes layers with various alternatives for separation, conversion and upgrading. We allow both pure and dilute raw materials, such as CO_2 from flue gas, methane from stranded sources, oxygen from air and water, and hydrogen from renewables and other sources. A novel feature of our superstructure is that it also includes various intensified alternatives. One such example is the inclusion of a tri-reforming membrane reactor which, instead of using pure CO_2, uses the mixed flue gas as feed, produces syngas, and separates nitrogen from the product at the same time. Therefore, it does not need a separate CO_2 capture section at the upstream. For each alternative, we develop a rigorous model detailing the transport and kinetics to accurately predict the process performance. We use a pseudo-homogenous model for plug-flow reactors (PFR) [2] to describe various alternatives for CO_2 reforming (e.g., steam methane reforming, dry-reforming, tri-reforming, partial oxidation, and variants of combined reforming). Using the rigorous simulation platform, we also develop efficient

surrogate models [3] for conversion as it changes with reactor type, design and operating conditions. The replacement of the ODE models with their algebraic surrogates allows us to pose the overall synthesis problem as a mixed-integer nonlinear optimization (MINLP) problem, which we solve to optimality using ANTIGONE [4]. We consider different objectives such as maximizing the overall utilization of CO_2, minimizing the total annualized cost, and maximizing profit. We also take into account the auxiliary carbon dioxide emissions associated with various processing tasks in the network. In this presentation, we will discuss our overall synthesis framework and the optimization results.

References:

[1] Quadrelli, E. A.; Centi, G.; Duplan, J.-L.; Perathoner, S. Carbon Dioxide Recycling: Emerging Large-Scale Technologies with Industrial Potential. *ChemSusChem*. 2011, 4:1194â€"1215.

[2] Aboosadi, Z. A.; Jahanmiri, A. H.; Rahimpour, M. R. Optimization of tri-reformer reactor to produce synthesis gas for methanol production using differential evolution (DE) method. *Applied Energy*, 88(8):2691â€"2701, 2011.

[3] Bajaj, I.; Hasan, M. M. F. Effective Sampling, Modeling and Optimization of Constrained Black-Box Problems. Accepted for publication in proceedings of ESCAPE26, Slovenia, 2016.

[4] Misener, R.; Floudas, C. A. ANTIGONE: algorithms for continuous/integer global optimization of nonlinear equations. *Journal of Global Optimization*. 59 (2-3), 503-526, 2014.

472251 Viability of Low Pressure Ammonia Synthesis Via Reactive Separation

Abstract:

Viability of Low Pressure Ammonia Synthesis via Reactive Separation

Mahdi Malmali, Alon McCormick, and Edward L. CusslerDepartment of Chemical Engineering and Materials Science, University of Minnesota, Minneapolis, MN 55455

Ammonia is a crucial chemical primarily used as an agricultural fertilizer; it is the second most manufactured chemical in the world with more than 160 million tons produced per year[1]. Ammonia is produced via the Haber-Bosch process, where hydrogen (from thermochemical steam-reforming of methane) and nitrogen are combined and react over a promoted iron-based catalyst at temperatures ranging from 380° to 520° C and pressures from 150 to 300 bar[2]. Reaction at such an elevated pressures only achieves partial conversions (typically less than 30%), which necessitates the recycle of unreacted gas using costly phase separation processes. After more than a century of research and optimization, this process is still highly capital- and energy-intensive[3], hence only economically viable in large-scale centralized plants.

We seek to develop technologies for small-scale synthesis of ammonia with zero-carbon footprint[4,5]. This process is powered by electricity produced from stranded intermittent wind or solar energy and can make hydrogen from electrolysis of water and nitrogen from pressure swing adsorption of air. A major drawback for widespread application of such a process is the high cost of the synthesized ammonia. The pressure in this process can be reduced to improve the economics of the process. This is achieved by simultaneous removal of the ammonia from the reaction system and circumventing the reaction equilibrium limitation.

We will initially present the ammonia reaction rate kinetics at low pressures. Then we discuss our findings of the reaction-separation process, which consists of a catalytic reactor and absorber unit; we report the absorption of ammonia on $CaCl_2$, which is utilized to selectively remove the ammonia from the system at high temperatures. Our experimental results demonstrate the high capacity of the absorption material at close-to-reaction temperature, which suppresses the equilibrium and increases production rates in our reaction-separation system; conversions as high as 80% at significantly reduced pressure (< 20 bar) are achieved. The effect of different operating conditions, including reaction temperature, reaction pressure, absorption temperature, gas transport, etc. on the kinetics of ammonia synthesis will be discussed in detail. For proof-of-concept, we compare the results obtained from high pressure reaction and low pressure reaction-separation process.

(1) http://minerals.usgs.gov/minerals/pubs/commodity/nitrogen/mcs-2011-nitro.pdf. (2) *Catalytic Ammonia Synthesis: Fundamentals and Practice*; Springer Science & Business Media, 2013. (3) Vojvodic, A.; Medford, A. J.; Studt, F.; Abild-Pedersen, F.; Khan, T. S.; Bligaard, T.; Nørskov, J. K. Exploring the Limits: A Low-Pressure, Low-Temperature Haberâ€"Bosch Process. *Chem. Phys. Lett.* **2014**, *598*, 108.

Â (4) Reese, M.; Marquart, C.; Malmali, M.; Wagner, K.; Buchanan, E.; McCormick, A.; Cussler, E. L. Performance of a Small-Scale Haber Process. *Ind. Eng. Chem. Res.* **2016**, *55*(13), 3742.

Â (5) Himstedt, H. H.; Huberty, M. S.; McCormick, A. V.; Schmidt, L. D.; Cussler, E. L. Ammonia Synthesis Enhanced by Magnesium Chloride Absorption. *AIChE J.* **2015**, *61* (4), 1364.

472974 Production of Acetic Acid from Natural Gas with Zero Carbon Emissions

Abstract:

In this work, the design of a process plant flow sheet that produces acetic acid from natural gas with zero carbon emissions is developed. Two such cases are developed - the production of acetic acid alongside power generation and hydrogen respectively.

The UNISIM software (Honeywell trademark) is used to create the flow sheet for the various modules of each plant based on established technologies of industrial acetic acid and hydrogen production processes. The various modules, when coupled for each process plant, create the net effect of zero-carbon emission.

Energy self-sufficiency, feasibility and operability limits of both cases are established via a thermodynamic analysis, while profitability is ascertained via economic and revenue studies based on current prices of the various commodities. The results are compared to stand-alone power, hydrogen and acetic production plants.

469218 Ionic Liquid Incorporation for Self Healing Purposes: New Measuring Methods for Bitumen Binder, Applications to Asphalt Concrete and Understanding of Chemical Interactions

Abstract:

Asphalt concrete is the most commonly used material in pavement industry. It is prepared by mixing bitumen, which is the bottom product of crude oil distillation, and aggregates at temperatures around 180 â°C. Asphalt concrete, once paved in the roads, experiences traffic loading along with changes in the ambient temperature throughout its service life. Thereby, failures in asphalt pavement such as fatigue cracking, rutting and thermal cracking, occur due to these continuous loading and climatic conditions. The mentioned failures in the roads give rise to building of new roads or maintaining the damaged roads. Building new roads or maintaining the already built ones requires preparation of asphalt mixtures, which consume large amounts of hydrocarbons, bitumen, together with excess energy due to high mixing temperatures. In addition to the large resource and energy consumptions, heating the mixture to high temperatures results in emission of greenhouse gases and bitumen fumes into the air. Therefore, extending the service life of asphalt pavements is very important both from an economic and environmental point of view. Such an improvement would also mean more efficient usage of the fossil resources.

Bitumen, which is the binder of asphalt concretes, is known to have intrinsic self- healing properties. It has the ability to restore its mechanical strength by closing the cracks at high temperatures or during long rest periods. However, during its service life, the asphalt pavement may not always have enough resting periods or the high temperatures which are needed to heal itself. Therefore, improving the self- healing capabilities of bitumen/asphalt is a necessity to minimize the mentioned environmental and economic problems.

In this study, the effects of ionic liquids, which are also known as green solvents, on the self- healing properties of bitumen is studied. 50/70 penetration grade bitumen from Tupras Ä°zmit Refinery is used in the studies. As part of the ionic liquid characterization, Thermogravimetric Analysis (TGA), Differential Scanning Calorimetry and Fourier Transform Infrared Spectroscopy are carried out. The bitumen is characterized by means of TGA, FTIR, DSC, Elemental Analysis and Saturate- Aromatic- Resin- Asphalthene (SARA) fractionation. Two new test methods are developed in order to measure the self- healing capability of bitumen. One of these methods is developed to see the effects of rest periods when intermittent loading is applied on the sample at high temperatures. The other method is developed to investigate the effects of long rest periods on the self- healing capability of bitumen at low temperatures. Measurements are carried out both on unmodified and ionic liquid modified bitumen to investigate the effects of ionic liquids with different side chain lengths on the self- healing properties of the bitumen. In addition to the measurements by the newly-developed techniques, stripping tests are carried out to find out about the effects of different ionic liquids on bitumen- aggregate interactions. Zeta potential measurements are done to quantify the electrostatic interactions between bitumen and aggregates. The asphalt performance of the modified bitumen is measured by Universal Testing Machine with Fatigue test, in which a loading- rest period cycle is applied to the sample until the sample is cracked, and Stiffness tests, which measures the hardness of the asphalt sample.

The results indicated that different ionic liquids have different effects on the bitumen/asphalt properties. With the addition of ionic liquids into bitumen, a 25% improvement in stripping properties, 30% improvement in high temperature self-healing performance, 200% improvement in low temperature performance and 120% improvement in asphalt fatigue life is obtained, meaning that different modification recipes can be used in road pavements depending on the climatic conditions and traffic density.

470079 Environmentally Responsible Fabrication of Efficient Perovskite Solar Cells from Recycled Car Batteries

Abstract:

Environmentally Responsible Fabrication of Efficient Perovskite Solar Cells

from Recycled Car Batteries

Po-Yen Chen[1,2]*, **Paula T. Hammond**[1], **Angela M. Belcher**[3]

[1]*Department of Chemical Engineering, Massachusetts Institute of Technology, Cambridge, MA, 02139*

[2]*School of Engineering, Brown University, Providence, RI 02912*

[3]*Department of Materials Science and Engineering, Massachusetts Institute of Technology, Cambridge, MA, 02139*

Email: pychen@mit.edu

Organolead halide perovskite solar cells (PSCs) show great promise as a new large-scale and cost-competitive photovoltaic technology. Power conversion efficiencies over 15% to over 20% have been achieved within 18 to 36 months of development, and thus perovskite materials have attracted great attention in photovoltaic research. However, the manufacture of PSCs raises environmental concerns regarding the over-production of raw lead ore, which has harmful health and ecological effects. In this study, we have demonstrated an environmentally responsible synthetic pathway to reuse car batteries for the fabrication of efficient PSCs (**Figure 1**). Both the anodes and cathodes of car batteries serve as material sources for the synthesis of lead iodide perovskite materials. In contrast to the traditional lead extraction process, our synthesis pathway from recycled battery materials occurs at a lower temperature (600 degree C) and does not include the hazardous emission of lead vapor/dust and CO_2 to the environment. The lead iodide perovskite materials synthesized from car batteries and high-purity reagents demonstrate identical material characteristics. The photovoltaic performance of the PSCs synthesized by each route is the same, which demonstrates that the device quality does not suffer from the materials sourced from spent car batteries. Also, electrochemical impedance spectra (EIS) measurements reveal that each device type displays the same resistances of electron recombination, indicating that the electron-trans- port properties of the lead iodide perovskite are identical. Finally, a simple economic analysis reveals that a single lead- acid car battery can supply enough lead material for the fabrication of 709-m^2 PSCs, which can provide enough electricity to power 30.2 US residential units in Las Vegas, Nevada. With these techniques, the time required to find a lead replacement for PSCs can be further increased. The environmentally responsible fabrication is expected to be broadly applicable not only to the PSC technology but also other applications.

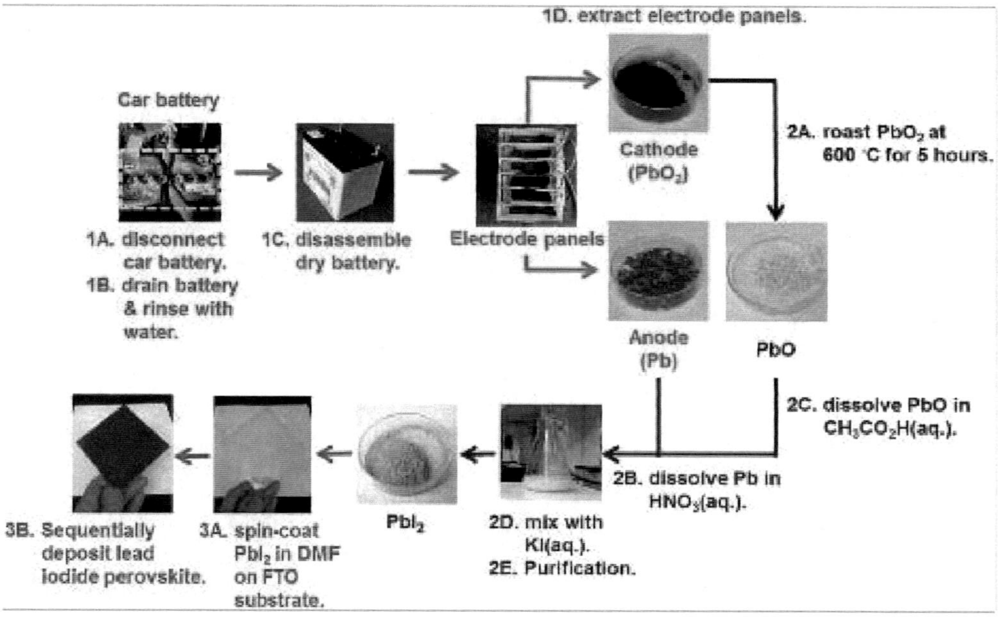

Figure 1. Synthetic process of lead iodide perovskite materials from a lead-acid car battery. The process includes three steps: (1) harvesting material from the anodes and cathodes of a car battery (red); (2) synthesizing PbI_2 from the collected materials (blue); (3) depositing lead iodide perovskite nanocrystals (green).

471578 Recycling & Garbage - a Separations & Materials Challenge

Abstract:

The field of recycling and garbage is largely overlooked by the Chemical Engineering community and has been the realm of mechanical, civil and even electrical engineers. But there is much to learn and apply in terms of Chemical Engineering from this field. Some surprising facts is that 99.9% of the separations at recycling facilities are done by machines - using the densities and shapes of materials to perform the separation. There are also major opportunities to increase the profit in garbage and recycling through better separation techniques and innovative material applications.

San Francisco has the goal of not sending anything to the landfill by 2020 - only 4 years away. Right now, they are currently around a 70% diversion rate. As the deadline approaches, innovations in this field cannot be just limited to the realm of mechanical engineering. For instance, in the past 10 years, composting technology has increased the amount of materials which are able to be composted. First only yard trimmings were accepted, then food scraps, now San Francisco is able to accept food stained paper products. The ultimate goal of composting is â€œanything which was a plant.â€

The garbage stream is also seeing innovations. Earlier this year, San Francisco adopted an organic press, which squeezes all the liquids from the garbage with a hydraulic press. Liquid from the process is sent to a biogas plant to convert into methane & then electricity.

On the recycling side, more materials and separation advances are required to keep the recycling industry afloat.

Currently, the driving force for places where recycling makes sense is areas with not a large amount of space - therefore landfill tipping rates (the cost to dump at a landfill) increase. The Pacific Northwest is home to the highest tipping fee in the nation ($142/ton) and land rich Texas has the lowest at $5/ton. Recycling programs therefore follow the incentives of the state they reside. Diversion from the landfill makes economic sense when instead of paying

$40/ton (Bay Area price), separating and giving away material for free. The $40 gap is saved but no profit is made.

The real profit from recycling comes from one material - aluminum. It accounts for 2% of the weight of the recyclate but 40% of the revenue. And paper and plastic are shipped overseas to be manufactured. Advances in material separations and new applications of recycled material is required to make recycling a truly sustainable business model.

Glass is the major loser of recycling. It is such a heavy material so the cost of transportation and disposal is very high. Additionally, remanufacturing options are limited - making new glass bottles is much cheaper than using recycled glass. New, business-sustainable applications of glass are required to make recycling of glass not a losing proposition. Companies are putting recycled glass into concrete and pavement - which is good - but these are not high margin applications. So, glass needs to be and is heavily subsidized by the State of California - which charges a fee to distributors (and then end customers) for glass products.

These opportunities in the garbage and recycling industries are ripe for the chemical engineering community. The focus must turn towards novel applications of materials and advanced chemical separation techniques â€“ both well within the chemical engineering realm.

463231 Techno-Economic and Life Cycle Analysis of Chemical Production from Biomass Feedstocks

Abstract:

Techno-economic and Life Cycle Analysis of Chemical Production from Biomass Feedstocks

Abhay Athaley[1], Praneeth Annam[1], Basudeb Saha[2], Marianthi Ierapetritou[1]

[1] Department of Chemical and Biochemical Engineering, Rutgers - The State University of New Jersey

[2]·Catalysis Center for Energy Innovation & Department of Chemical & Biomolecular Engineering, University of Delaware

This work focuses on the techno-economic analysis and life cycle analysis on the production of chemicals from lignocellulose feedstocks including three types of hydrolysis of lignocellulose to produce sugar mixtures to convert to furan derivatives and the production of p-Xylene. The development of efficient and economically sustainable routes for biobased chemical production has raised a lot of interest especially for the platform chemicals such as hydromethylfurfural (HMF), furfural, levulinic acid, etc.[1,2] The aforementioned chemicals can be produced from glucose or xylose solutions. Different type of hydrolysis is used to produce sugar mixtures such as enzymatic hydrolysis, thermo-chemical hydrolysis etc.[3,4,5] It is found that it is more economically and environmentally attractive to manufacture HMF and furfural from C5 and C6 mixtures that are produced from hydrolysis of 2nd generation (lignocellulosic) biomass.[6,7]

The acceptance of bio-products in the market place depends on their competitiveness compared with petroleum-based products not only in terms of economics but also in terms of process sustainability[8]. In our previous work we evaluated the economics and environmental impacts of bio-products such as p-Xylene[9-11]; however this study compares different hydrolysis processes and the integrated flowsheet to produce p-Xylene. This is significant since different hydrolysis processes produce different mixtures of C5 and C6. Present day, water and energy systems are interconnected with each other- water-carbon-energy nexus. Hence life cycle analysis is also carried out in this study to check the competitiveness of the above process in terms of sustainability.

This work uses techno-economic analysis and life cycle assessment to design and evaluate the production alternatives. The detailed process flowsheet is developed and simulated using Aspen Plus.[12]Three different hydrolysis steps are used to produce sugar mixtures: 1) Dilute Acid Hydrolysis 2) Concentrated Acid Hydrolysis and 3) Hydrolysis of Biomass using Molten Salt Hydrates (add references to each one). At the next stage heat integration is carried out and economic analysis is done. Also sensitivity analysis is carried out to compare the cost of p-Xylene with fluctuating cost of raw materials and to compare the cost of p-Xylene when HMF is considered as by-product to produce different chemicals. It is seen that utility cost is directly related to the solvent requirement and thus related to price of p-Xylene and also the cost of p-Xylene increases when the amount of HMF considered as by-product increases. Finally, life cycle analysis is carried out to check for process sustainability with respect to carbon production and water consumption. High requirement of water is needed for the process and thus sustainability related to water consumption is also presented.

References:

1. Bozell JJ, Petersen GR. Technology development for the production of biobased products from biorefinery carbohydrates-the US Department of Energy's "Top 10" revisited. *Green Chemistry.* 2010;12(4):539-554.

2. Dutta S, De S, Saha B, Alam MI. Advances in conversion of hemicellulosic biomass to furfural and upgrading to biofuels. *Catalysis Science & Technology.* 2012;2(10):2025-2036.

3. Harris JF, Laboratory FP. *Two-stage, dilute sulfuric acid hydrolysis of wood: an investigation of fundamentals*: U.S. Dept. of Agriculture, Forest Service, Forest Products Laboratory; 1985.

4.Weydahl K.R., Process for the production of Alcohol, 2012, Patent application no. 13/318302.

5. Deng W., Kennedy J.R., Tsilomelekis G., Zheng W., Nikolakis V., Cellulose Hydrolysis in Acidified LiBr Molten Salt Hydrate Media, *Industrial & Engineering Chemistry Research*;2015;54:5226-5236.

6. Chheda JN, Roman-Leshkov Y, Dumesic JA. Production of 5-hydroxymethylfurfural and furfural by dehydration of biomass-derived mono- and poly-saccharides. *Green Chemistry.* 2007;9(4):342-350.

7. Zhang J, Lin L, Liu S. Efficient Production of Furan Derivatives from a Sugar Mixture by Catalytic Process. *Energy & Fuels.* 2012/07/19 2012;26(7):4560-4567.

8.Hawkins ACG, David A.; Buelter, Thomas; Wade, James; Meinhold, Peter; Peters, Matthew W.; Gruber, Patrick R.; Evanko, William A.; Aristidou, Aristos A., Assignee, Inventors; Gevo Inc. USA, assignee. METHODS FOR THE ECONOMICAL PRODUCTION OF BIOFUEL PRECURSOR THAT IS ALSO A BIOFUEL FROM BIOMASS 2009.

9. Lin Z, Ierapetritou M, Nikolakis V. Aromatics from Lignocellulosic Biomass: Economic Analysis of the Production of p-Xylene from 5-Hydroxymethylfurfural. *AIChE Journal.* 2013;59(6):2079-2087.

10. Lin Z, Nikolakis V, Ierapetritou M. Alternative Approaches for p-Xylene Production from Starch: Techno-Economic Analysis. *Industrial & Engineering Chemistry Research.* 2014/07/02 2014;53(26):10688-10699.

11. Lin Z, Nikolakis V, Ierapetritou M. Life Cycle Assessment of Biobased p-Xylene Production. *Industrial & Engineering Chemistry Research.* 2015/03/04 2015;54(8):2366-2378.

12. Torres AI, Daoutidis P, Tsapatsis M., Continuous Production of 5- hydroxymethylfurfural from fructose: a design case study. Energy & Environmental Science, 2010;3(10):1560-1572.

467521 Carbon Footprint Analysis of Biomethane from Anaerobic Digestion of a Food Waste / Manure Mixture in Colorado

Abstract:

This study determined the life cycle greenhouse gas (GHG) emissions of bio-methane produced from anaerobic digestion of a mixture of food waste and cow manure in Colorado. Included in the system boundary for this analysis are long distance semi-truck transport of collected food waste from the Denver metropolitan area, avoid landfilling (and composting in a separate scenario) emissions for the food waste, short haul transport of manure waste from local dairy farms, anaerobic digestion process and biogas purification, distribution of the bio-methane into the natural gas pipeline grid, and combustion of bio-methane. The carbon accounting included both fossil and biogenic carbon. The effects of co-products solid compost sold to retail gardening market and nutrient rich liquid digestate distributed to local farm fields were accounted for in the analysis.

After including emission credits for avoiding landfill emissions of CO_2 and CH_4, bio-methane emits over its life cycle -3.5 kg CO_2 equivalents / kg bio-methane, thus reducing global warming effect with each unit of production of bio-methane. This favorable emission intensity is compared to that of fossil methane of between 3.3 to 5.1 kg CO_2 equivalents / kg fossil methane. Effects of life cycle system parameters in the model on GHG results are investigated in scenario analyses and also in time dependent analysis of landfill emissions.

469717 Multistage Torrefaction of Biomass and in Situ Catalytic Upgrading to Hydrocarbon Biofuels and Specialty Biochemicals

Abstract:

Thermochemical conversion of lignocellulosic biomass via pyrolysis and catalytic upgrading of bio-oil is touted as an environmental panacea, capable of producing fuel, energy, and value-added bio-based chemicals with an improved environmental profile compared to conventional fossil routes. However, bio-oil contains a host of compounds with myriad chemical functionalities, making selective catalytic upgrading to fuel and/or chemicals challenging and a major technical bottleneck. This work explores an innovative design strategy, in which the primary constituents of lignocellulosic biomass (i.e. hemicellulose, cellulose, lignin) are selectively decomposed via a series of torrefaction and pyrolysis reactors. Staged thermal decomposition produces several fractionated bio-oil streams comprised primarily of light oxygenates, furan derivatives, and phenolic species respectively. The purposed multistage design permits tailored catalytic upgrading of the fractionated bio-oil streams to hydrocarbon fuels and/or specialty bio-based chemicals, and thus holds promise for improved environmental and economic performance over the current state of the art (i.e. fast pyrolysis and hydroprocessing fuel platforms).

This work performs a prospective life cycle assessment to determine the environmental profile of concurrent hydrocarbon biofuel and specialty biochemical production via a multistage torrefaction and *in situ* catalytic upgrading platform. Several multistage design strategies consisting of different catalytic strategies and targeted platform chemicals (i.e. 2-methylfuran, tetrahydrofuran, and cyclopentanone) are considered. Detailed material and energy balances for multistage systems are developed using a combination of experimental data and ASPEN process simulations. Several critical sustainability metrics including life cycle greenhouse gas (GHG) emissions and Energy Return on Investment (EROI) are used to compare the environmental performance of multistage systems, and benchmark against a base-case fast pyrolysis and hydroprocessing platform. Preliminary results reveal that multistage systems have the capacity of achieving over 75% GHG reductions relative to petroleum diesel, and produce high quality hydrocarbon transportation fuels and specialty bio-chemicals. The influence of different LCA allocation schemes and coproduct scenarios for the produced biochar on the environmental profile of hydrocarbon biofuels and biochemicals will be discussed.

469925 Optimizing Land Use Change and Life Cycle Greenhouse Gas Emissions of Biofuels

Abstract:

Biofuels are a potential source of renewable liquid fuel, in contrast to the nonrenewable fossil fuels we rely on today. Another driver for the interest in biofuels is the possibility that life cycle greenhouse gas (GHG) emissions for biofuels might be smaller than fossil fuels, providing an environmental benefit in addition to the benefits of a secure and renewable fuel supply. However, there still exist GHG emissions sources throughout the biofuel life cycle, and many of them do not exist at a reasonable level for fossil fuels. For example, there are cultivation GHG emissions associated with cultivating the biomass feedstocks, land use change (LUC) GHG emissions incurred by transforming land into cropland to grow the biomass feedstocks, etc. GHG emissions from LUC due to biofuels are perhaps the most difficult to characterize and quantify; they are not released from a smokestack or a tailpipe. A host of researchers propose different characterization and quantification methods to calculate GHG emissions from LUC due to biofuels, but there is still significant disagreement on just how many GHG emissions from LUC biofuels are responsible for.[1,2] In other words, no one has been able to show which mix of biofuel production pathways and biomass sourcing strategies are best to minimize life cycle GHG emissions including LUC emissions. It is imperative to reduce uncertainty on this issue as current biofuels legislation and policies are based on these estimates. Furthermore, previous studies explore fixed biofuel production scenarios, e.g. 14 Bgal/year of bioethanol from corn, and are not able to identify better pathways with diminished GHG emissions from LUC. Thus, not only are current policies and biofuel production and biomass sourcing strategies based on uncertain LUC GHG emissions estimates, but they are also not founded on optimized stratagems.

In this work, we develop a first-ever LUC life cycle optimization (LCO) model that incorporates cutting-edge characterization and calculation methods for bioethanol LUC GHG emissions within an optimization framework. This modeling framework identifies an optimal bioethanol production and biomass sourcing strategy that minimizes LUC GHG emissions by selecting any combination and size of 14 bioethanol production pathways and 5 biomass feedstocks. We choose bioethanol production pathways that are either already commercialized or are commercialization-ready, using data from technoeconomic analyses/reports from the U.S. DOE or the peer-reviewed scientific community.[3-6] LUC and direct processing GHG emissions are considered as well as excess electricity production credits from processing. We calculate LUC GHG emissions from exhaustive data bases for soil organic carbon (SOC), carbonaceous biomass cover, foregone CO_2 sequestration by forested land, and the net primary productivity (NPP) of each land type in each region around the world.[7-10] The CGE is integrated into the LCO modeling framework with the land carbon data to determine LUC GHG emissions around the world due to economic forces arising from increased bioethanol production in the US. Thus, for the first time, cutting-edge LUC characterization and calculation methods are integrated into an LCO modeling framework. Furthermore, this study presents the first time an integrated CGE-LCO model has been developed and utilized.

We consider a case study of US bioethanol production levels ranging from 10 – 100 Bgal/year. We find that minimized LUC GHG emissions from our results are lower than previous estimates, indicating that these previous estimates are at least possible. Corn stover is always preferred, as LUC due to its utilization is negligible. Switchgrass is then selected when the sustainably harvestable corn stover resource base is exhausted. Two-stage dilute acid pretreatment followed by distillation or dilute acid pretreatment followed by pervaporation are preferred technologies for corn stover conversion. Dilute acid pretreatment followed by distillation is always the selected bioethanol production pathway for switchgrass. Forested land is largely preserved around the world whenever possible, and pastureland is by far preferred for conversion to cropland. However, some of the recent LUC GHG emissions estimates for biofuels are close to our minimized value, indicating that some might be optimistic. Nonetheless, if the optimal bioethanol production and biomass sourcing strategies produced by our work are followed, then life cycle GHG emissions including LUC GHG emissions of bioethanol can be managed.

References

1. Qin Z, Dunn JB, Kwon H, Mueller S, Wander MM. Influence of spatially dependent, modeled soil carbon emission factors on life cycle greenhouse gas emissions of corn and cellulosic ethanol. *GCB Bioenergy.* 2016.

2. Plevin RJ, Beckman J, Golub AA, Witcover J, O'Hare M. Carbon accounting and economic model uncertainty of emissions from biofuels-induced land use change. *Environmental science & technology.* 2015;49(5):2656-2664.

3. Pfromm PH, Amanor-Boadu V, Nelson R, Vadlani P, Madl R. Bio-butanol vs. bio-ethanol: a technical and economic assessment for corn and switchgrass fermented by yeast or Clostridium acetobutylicum. *Biomass and Bioenergy.* 2010;34(4):515-524.

4. Zhu Y, Gerber MA, Jones SB, Stevens DJ. *Analysis of the Effects of Compositional and Configurational Assumptions on Product Costs for the Thermochemical Conversion of Lignocellulosic Biomass to Mixed Alcohols – FY 2007 Progress Report,*. Richland, Washington: PNNL;2009.

5. Zhu Y, Jones SB. *Techno-economic Analysis for the Thermochemical Conversion of Lignocellulosic Biomass to Ethanol via Acetic Acid Synthesis.* Richland, Washington: PNNL;2009.

6. Dal-Mas M, Giarola S, Zamboni A, Bezzo F. Strategic design and investment capacity planning of the ethanol supply chain under price uncertainty. *Biomass and Bioenergy.* 2011;35(5):2059-2071.

7. Change IPoC. *2006 IPCC guidelines for national greenhouse gas inventories*: Intergovernmental Panel on Climate Change; 2006.

8. Monfreda C, Ramankutty N, Hertel TW. Global agricultural land use data for climate change analysis. *Economic analysis of land use in global climate change policy.* 2009;14:33.

9. West TO, Brandt CC, Baskaran LM, et al. Cropland carbon fluxes in the United States: increasing geospatial resolution of inventory-based carbon accounting. *Ecological Applications.* 2010;20(4):1074-1086.

10. Gibbs H, Yui S, Plevin R. *New estimates of soil and biomass carbon stocks for global economic models*: Center for Global Trade Analysis, Department of Agricultural Economics, Purdue University;2014.

470623 Developing a Systemic Methodological Framework for the Sustainability Assessment of Biobased Fuels and Chemicals

Abstract:

In the last two decades, there has been some effort in academia and industrial practice to provide tools that can deal with the challenges of designing sustainable chemical processes. Taking into account the diversity of the sustainability objectives, data availability and, therefore, modelling detail, it is inevitable that the structure of the developed tools varies significantly, ranging from simple indicators to holistic frameworks for an overall process assessment. These sustainability frameworks have also been applied in the field of biorefineries, in particular for biofuels, bioethanol, methanol and syngas production, incorporating process integration and life cycle assessment (LCA), exergy and inherent safety analysis, and life cycle energy and environmental analysis. However, perhaps the most striking difference between biobased and conventional chemical production in terms of sustainability assessment is the availability of process data, even for single biobased production lines, not to mention for integrated biorefinery concepts.

From this perspective, the aim of this study is (i) to present the methodological steps for the development of a model-based database of LCA related sustainability impacts in biomass based production, (ii) to demonstrate the use of this database in single- and integrated multi-production lines and (iii) to provide a first generation of surrogate models that correlate available input to desirable output process parameters and assessment metrics with computationally inexpensive algebraic forms.

In particular, this study presents detailed cradle-to-gate inventory analysis in biomass based production using protocols for process flowsheeting and filling of data gaps (e.g., for cradle-to-gate production steps and waste treatment of process effluents), considers various environmental impact allocation methods for the resulting joint and co-production biorefinery systems, applies a wide list of sustainability impact assessment metrics and tests multiple surrogate modelling approaches (i.e., from simple linear regression to partial least squares, Kriging and neural network approaches) on the basis of model generated data and sampling plans.

The overall methodological framework comprises an accessible, modular and expandable database of biobased production lines, including mass and energy balances, other process related information and various allocation approaches for the calculation of life cycle inventories in single- and multi-production systems. The resulting data structures are integrated with a reporting mechanism which offers information about the environmental impacts expressed by relevant indicators proposed by life cycle impact assessment methods such as the RECIPE and Cumulative Energy Demand methods and provides an environment for benchmarking alternative choices.

Molreover, metrics of "green chemistry" assessing the efficiency of processes (yield, mass intensity, E-factor etc.) as well as other process-related variables are extracted from the data structures aiming at the development of a first generation of surrogate models to provide guidance for a fast screening of innovative biobased processes and products.

The data structures, assessment results and surrogate modelling performance is demonstrated for a series of platform chemicals (e.g., syngas, sugars and lignin) and biofuels (e.g., biodiesel, biogas, and alcohols), starting from diverse biomass sources (e.g., wood chips, wheat straw, vegetable oil) and biomass availability scenarios, in single and integrated multi-production lines.

471048 Assessing Sustainability By Life Cycle Assessment Versus Techno-Ecological Synergy: A Case Study for Biofuel Production

Abstract:

Limitations on resource availability and the increasing consumption patterns have necessitated the quest for sustainable development. Many methods and tools have been developed to assess and design for sustainability. Attempts have been made to improve process efficiency to reduce resource consumption and waste generation. Paradoxically, even though the environmental impacts associated with per unit product have been reduced significantly, our ecosystem has been degraded further. This is mainly due to the following two reasons, 1) problems may shift outside the selected boundary, and 2) the role of ecosystem goods and services for supporting human activities is largely ignored despite their essential role in sustaining human activities. Therefore, for a methodology to claim its contribution to guiding sustainable decision making, both of these aspects need to be accounted for.

The most widely used sustainability assessment method is Life Cycle Assessment (LCA). It is used to estimate and assess the environmental impacts of resources used and emissions created during the main stages of a product's life cycle. Even though this method can potentially expand the system boundary to include the upstream and downstream effects, it largely ignores the role of ecological goods and services. LCA encourages decisions that do â€œless badâ€. Thus, the results from LCA are comparative in nature. The alternative that has the lowest environmental impacts is considered sustainable. This can potentially lead to misleading results and perverse decisions [1].

Techno-Ecological Synergy (TES) framework aims to assess and encourage synergies between systems in technological and ecological spheres [1]. The methodology considers systems at multiple spatial scales. Technological systems, ranging from individual processes to supply chains and life cycles can be evaluated along with their relevant supporting ecological systems. The key concept for TES is that resource uses and emissions are permitted but must stay in line with the capacity of relevant ecosystems. The necessary but not sufficient condition to be satisfied for reaching the goal of sustainability is to maintain the demand for ecosystem services within the supply at the largest ecological scale. By directly considering the role of ecosystem services, different insights and novel ecological solutions can be obtained.

The potential benefits achievable with TES methodology are illustrated by an application to a biofuel life cycle, including the production of selected direct inputs into farming, the farming phase and the ethanol manufacturing. Three ecosystem services have been considered to imply the general applicability of TES, namely carbon sequestration, air regulation and water provisioning. From this multi-scale and multi-objective analysis, we can identify processes that demand more from nature than can be supplied. This insight can be used to understand trade-offs and synergies between various ecosystem services, and identify opportunities for technological improvements and ecological restoration. Thus, by using TES approach, we not only have the option to reduce resource inputs and emissions as with the LCA approach, but also the option of increasing the supply of ecosystem services that can be achieved by proper land use. This solution might be potentially more cost-effective and environmentally friendlier than solutions developed by traditional methods such as LCA.

References:

[1] Bakshi, Bhavik, Guy Ziv, and Michael Lepech. "Techno-ecological synergy: A framework for sustainable engineering." Environmental Science & Technology 49.3 (2015): 1752-1760.

467847 Indirect Liquefaction of Biomass to Transportation Fuels Via Mixed Oxygenated Intermediates

Abstract:

This paper presents a comparative techno-economic analysis of five conversion pathways from biomass to gasoline-, jet-, and diesel-range hydrocarbons via indirect liquefaction with specific focus on pathways utilizing oxygenated intermediates (derived either via thermochemical or biochemical conversion steps). The four emerging pathways of interest are compared with one conventional pathway (Fischer-Tropsch) for the production of the hydrocarbon blendstocks. The processing steps of the four emerging pathways include: biomass-to-syngas via indirect gasification, gas cleanup, conversion of syngas to alcohols/oxygenates, followed by conversion of alcohols/oxygenates to hydrocarbon blendstocks via dehydration, oligomerization, and hydrogenation. We show that the emerging pathways via oxygenated intermediates have the potential to be cost competitive with the conventional Fischer-Tropsch process. The evaluated pathways and the benchmark process generally exhibit similar fuel yields and carbon conversion efficiencies. The resulting minimum fuel selling prices are comparable to the benchmark at approximately $3.60 per gallon-gasoline equivalent, with potential for two new pathways to be more economically competitive. Additionally, the coproduct values can play an important role in the economics of the processes with oxygenated intermediates derived via syngas fermentation. Major cost drivers for the integrated processes are tied to achievable fuel yields and conversion efficiency of the intermediate steps, i.e., the production of oxygenates/alcohols from syngas and the conversion of oxygenates/alcohols to hydrocarbon fuels.

465137 Incorporating Unsteady-State Fermentation Models in Aspen Plus: Fed-Batch and *in Situ* Gas Stripping Simulation and Analyses

Abstract:

Process simulation can predict the most informative experiments to conduct for process design, control and optimization. Traditional refinery processes have been extensively simulated and analyzed using the thermodynamic models and separation unit operations in process simulation tools, such as Aspen Plus. However, these essential tools for optimizing refinery processes have not been fully extended to biorefineries as a result of the inherent unsteady-state nature of fermentation processes. One of the approaches used to reduce substrate and product inhibition to fermentation microorganisms and increase fermentation productivity is fed-batch fermentation and *in situ* product recovery techniques, such as gas stripping. In the fed-batch fermentation, a bioreactor is charged with a relatively low concentration of substrate and as the substrate is consumed, fresh substrate is supplied at a rate that keeps the substrate concentration in the reactor below the inhibitory levels to the microorganisms. Gas stripping as a chemical separation method allows for the selective removal of the volatile components from the aqueous fermentation solution using a gas bubbled through the fermentation broth.

In our previous study, the batch acetone-butanol-ethanol (ABE) fermentation was used as a model system to simulate an unsteady-state batch ABE fermentation using a Fortran user kinetics subroutine linked to the batch reactor (RBatch) in Aspen Plus as a starting point for a biorefinery. This study extends the Aspen Plus batch model to fed-batch fermentation and *in situ* gas stripping with the aim of comparing the trends in the simulation to their corresponding laboratory fed-batch and *in situ* gas stripping experiments. An existing ABE fermentation model was reformulated to describe cell growth and inhibition using Monod kinetics, making it sensitive to changes in the fermentation environment as a result of fresh substrate addition (fed-batch) and product accumulation. This study provides a platform that can be used to guide integrated fed-batch fermentation and *in situ* gas stripping experiments and assess the accuracy of literature data. Additionally, a universally accepted process simulator, Aspen Plus, can be used in the simulation of fermentation processes linked with separation unit operations to provide strategies for the design and optimization of biorefineries.

463187 Techno-Economic Analysis of Direct Coal-Biomass to Liquids (CBTL) Plants with Shale Gas Utilization and CO_2 Capture and Storage (CCS)

Abstract:

Direct coal liquefaction (DCL) technology is a promising technology for producing transportation fuels from non-petroleum sources. However, a large amount of H_2 is required in the DCL system because of the low H/C ratio in coal. A recent study shows that about 0.48 tonne of CO_2 is released per barrel of transportation fuels produced in the Shenhua DCL plant, where about 80% of CO_2 is produced in the gasification-based H_2 production unit. Addition of biomass and applying carbon capture and storage (CCS) technologies are two possible solutions to reduce the carbon footprint, but would lead to higher capital and operating costs. Furthermore, H_2 can also been produced from shale gas instead of coal with less capital investment and CO_2 emission, as shale gas is currently abundant in most of the continental US. Even though co-processing of coal and biomass and inclusion of CCS have been widely studied for indirect liquefaction and power generation system, limited studies have been conducted for direct liquefaction processes. With this motivation, we will present a techno-economic study of direct coal-biomass to liquid (CBTL) plants with shale gas utilization and CCS.

In this study, the direct CBTL plants is designed based on the catalytic two-stage liquefaction process investigated by HTI, where inline hydrotreating is considered.[1] Four different plant configurations, named SMR_CCS, SMR_VT, CG_CCS, CG_VT, are considered in this work based on different extent of CCS and H_2 sources. In the SMR_CCS and SMR_VT processes, hydrogen is produced from shale gas steam methane reforming (SMR) and partial oxidation of liquefaction residue. In the CG_CCS and CG_VT processes, hydrogen is produced from co-gasification (CG) of coal/biomass/residue. High extent of CCS is considered in the SMR_CCS and CG_CCS processes, while all CO_2 is vented (VT) in the SMR_VT and CG_VT processes. Both chemical and physical absorption processes are considered for carbon capture depending on the sources and partial pressure of CO_2 in the CO_2-containing stream. The process model and economic model are developed mainly in Aspen Plus® and Aspen Process Economic Analyzer® (APEA®), respectively, and are validated by comparing with the data obtained from the open literature.

In this presentation, we will focus on the following aspects: (1) development of plant-wide process model and rigorous economic model in Aspen Plus®, Aspen Exchanger Design and Rating®, APEA® and Excel, (2) sensitivity studies to analyze the impact of key design parameters (i.e. biomass/coal ratio, H_2 sources, extent of CCS) and investment parameters (i.e. price of coal, biomass, shale gas, project life, plant contingency and plant capacity) on key economic measures, such as net present value (NPV), internal rate of return (IRR), break-even oil price (BEOP) and equivalent oil price (EOP), (3) analysis of the potential environmental credits due to use of biomass and CCS, (4) comparison between indirect and direct CBTL plants with or without CCS.

Reference

(1) Comolli, A.G., Lee, L.K., Pradhan, V.R., Stalzer, T.H., Harris, E.C., Mountainland, D.M., Karolkiewicz, W.F. and Pablacio, R.M., Direct Liquefaction Proof-of-Concept Facility, Technical Progress Report POC Run 01, Contract No. AC22-92PC92148, Hydrocarbon Research Inc., 1995.

469616 Techno-Economic (TEA) and Life Cycle Analysis (LCA) of the Pyrolysis-Bioenergy-Biochar Pathway for Carbon-Negative Energy

Abstract:

Techno-economic (TEA) and Life Cycle Analysis (LCA) of the Pyrolysis-Bioenergy-Biochar Pathway for Carbon-Negative Energy

Wenqin Li.[b], Qi Dang [ab], Wright M. M.[ab], Brown, R. C.[ab], Laird, D.

[a]Bioeconomy Institute

[b]Department of Mechanical Engineering

[c]Department of Agronomy

Iowa State University

The objective of this project is to evaluate the economic and environment impacts of the Pyrolysis-Bioenergy-Biochar Pathway for carbon-negative energy. Biochar, a carbon rich coproduct of pyrolysis, has the potential to improve soil quality by sequestering carbon thus helping to achieve a carbon negative energy goal. In addition to biochar, fast pyrolysis of biomass can also produce biofuels, bio-power and chemicals. Techno-economic Analysis (TEA) and Life Cycle Analysis (LCA) results will vary with the selection of pyrolysis products, feedstock materials, and facility locations.

Economic analysis was conducted on two pyrolysis product utilization scenarios: biochar & biofuel and biochar & bio-power. In the biochar & biofuel scenario, we estimate that the minimum fuel-selling price (MFSP) varies between $2.91 and $3.03 per gallon if biochar can be sold for $60 to $20 per tonne. On the other hand, if the biofuel can be sold at market prices of $2.00 to $5.00 per gallon, the biochar can be sold for $364 to -$637 per metric tonne. The biochar & bio-power scenario yields minimum electricity-selling prices of 9.85 to 10.07 ¢/kWh for biochar prices of $60 to $20 per tonne. Biochar can be sold for -$336 to $400 per tonne at electricity prices of 12 to 8 ¢/kWh. These prices do not include environmental credits or incentives. More pyrolysis product combinations scenarios will be assessed and compared.

Biochar properties and yields vary from different types of biomass. Regression models of relationship between biochar & biofuel yield and biomass carbon, ash & oxygen contents have been built based on series of fast pyrolysis experimental yields data. The regression model was validated for different types of feedstock and used to predict the yield for various feedstocks. Process models for fast pyrolysis of various feedstocks will be built in Aspen Plus™, with the products prediction data. Mass and energy balance and subsequent techno-economics analysis will be assessed and compared to elect the best economic performance feedstock.

The pyrolysis facility locations might affect various factors, such as feedstock cost, feedstock type, capital cost, and operating cost. Biochar type and the demand for biochar will also vary with the plant locations. Three specific land areas of US are investigated depending on the selection of pyrolysis facility locations: Upper Mississippi River Basin (UMRB), California, and U.S. Southeast. The capital cost, operating cost and minimum fuel selling price was estimated to indicate the economic benefits of employing integrated Pyrolysis-Bioenergy-Biochar pathway in various regions. Life cycle analysis will also be conducted using GREET to evaluate the GHG emissions of the integrated Pyrolysis-Bioenergy-Biochar production facility in different regions.

467412 Techno-Economic and Environmental Life Cycle Assessments of Hydrocarbon Biofuel from Loblolly Pine

Abstract:

Thermochemical conversion of wood to biofuel via fast pyrolysis is regarded as a promising conversion pathway for producing biofuels. The process involves rapid thermal degradation of wood in the absence of air at a temperature of approximately 500-550°C with a short residence time of less than 2 seconds. Hydrocarbon biofuel is then produced through catalytic upgrading of the pyrolysis bio-oil product. Drying and size reduction of wood are major contributors to energy consumption of this process. Torrefaction, often referred to as mild pyrolysis, tends to enhance bio-oil properties by reducing water content, minimizing acidity, and reducing size reduction energy. The development of a two-stage fast pyrolysis process that involves a torrefaction pretreatment step prior to pyrolysis was investigated as an approach to minimize the energy consumption associated with size reduction. The impacts of different torrefaction temperatures have on the cost of production of hydrocarbon biofuel as well as the environmental impacts was investigated by carrying out techno-economic analysis (TEA) and life cycle assessment (LCA), respectively, with two-stage processing compared to the one-stage pyrolysis-based processing and to fossil fuels. Effects of the use of renewable co-products such as char and non-condensable gases from pyrolysis to replace fossil energy inputs (natural gas for process heat) to the process are explored in scenario analyses.

Aspen Plus process simulation package was used to model the two stage torrefaction-fast pyrolysis process with catalytic upgrading of pyrolysis bio-oil to hydrocarbon fuel blends. The effect of torrefaction severity on product composition, yield of pyrolysis bio-oil and subsequently yield of hydrocarbon fuel were obtained from the literature.[1-6] The effect of torrefaction on the energy requirement for size reduction was also included in our model based on literature data.[7] Using these data sources, mass and energy balances were obtained, and used in sizing the equipment, with equipment prices estimated from a number of sources such as the Aspen Economic Process Analyzer, previous

works and equipment vendors. A Discounted Cash Flow Rate of Return spreadsheet was used to obtain the gate cost of production, and data obtained from the simulation also served as inputs for the LCA carried out using the LCA software SimaPro 8.0 with greenhouse gas emissions (GHG) as the impact category.

From our model simulations, having a torrefaction step as a pretreatment step prior to fast pyrolysis reduces the cost of bio-oil production from loblolly pine, by reducing energy cost associated with size reduction. Results from the model also show a reduction in GHG emissions associated with the bio-oil production and catalytic upgrading as a result of the torrefaction pretreatment step. The effect of substituting renewable energy (char) for process heat demand rather than natural gas yielded a cost of production decrease from $4.26/gallon hydrocarbon fuel to $4.15/gallon for the one-step pyrolysis-based process. Full results on the TEA and LCA for all scenarios will be presented.

References

1. Park, J., Meng, J., Lim, K.H., Rojas, O.J. & Park, S. Transformation of lignocellulosic biomass during torrefaction. *Journal of Analytical and Applied Pyrolysis* **100**, 199-206 (2013).

2. Westerhof, R.J. et al. Stepwise fast pyrolysis of pine wood. *Energy & fuels* **26**, 7263-7273 (2012).

3. Jones, S. et al. (National Renewable Energy Laboratory (NREL), Golden, CO., 2013).

4. Vispute, T. Pyrolysis oils: characterization, stability analysis, and catalytic upgrading to fuels and chemicals. (2011).

5. Huber, G.W., Chheda, J.N., Barrett, C.J. & Dumesic, J.A. Production of liquid alkanes by aqueous-phase processing of biomass-derived carbohydrates. *Science* **308**, 1446-1450 (2005).

6. Furimsky, E. Catalytic hydrodeoxygenation. *Applied Catalysis A: General* **199**, 147-190 (2000).

7. Phanphanich, M. & Mani, S. Impact of torrefaction on the grindability and fuel characteristics of forest biomass. *Bioresource technology* **102**, 1246-1253 (2011).

458183 Novel Process of Biofuel Production from Acid Oil and Hydrous Bioethanol Using Ion-Exchange Resin; I. Experimental Optimization of Operating Conditions

Abstract:

Introduction

Liquid biofuels offer attractive advantages over other forms of biomass energy resources in terms of high energy density and ease of storage and transport. Bioethanol is useful as a gasoline additive (up to 10%) and biodiesel, or fatty acid methyl ester (FAME), is also used as a fuel alternative (100%) to petroleum diesel. However, there has been no significant increase in the production of these fuels in the past few years. The reason is that there is no profitability for producers owing to high production cost and no benefits for consumers because of unstable product quality and lower fuel economy and power. Moreover, emergence of next-generation automobiles will lower the demand for gasoline. The water content of ethanol used as the gasoline additive should be less than 0.5 wt% and the dewatering process requires a large energy input. Thus, the necessity of anhydrous ethanol as gasoline additive fades and new technology which can use hydrous ethanol is desired. On the other hand, the diesel fuel continues to be used for heavy duty on-road commercial transportation, non-road construction and various types of ocean vessels. If new technology to economically produce high quality biodiesel can be developed, the necessity of biodiesel utilization will increase.

In this research (Part I), biodiesel is produced in the form of fatty acid ethyl ester (FAEE) using hydrous ethanol by esterification of free fatty acid (FFA). The FAEE has higher cetane number and fuel economy and power than the traditional fatty acid methyl ester (FAME) and FAEEâ€™s low temperature fluidity is also improved[1)]. However, the conventional technologies for esterification of FFA never use hydrous ethanol because the water in the reaction solution must be completely removed to obtain a high conversion. We have reported to produce FAME at more than 95% conversion by esterification of acid oil using a cation-exchange resin catalyst without dewatering and adding excess alcohol[2)]. This means that the catalytic activity of the resin is not inhibited by water. Thus, using the cation-exchange resin catalyst the batch and continuous esterification experiments with hydrous ethanol are performed. The maximum water content in hydrous ethanol to obtain high conversion is clarified.

Materials and methods

The cation-exchange resin, Diaion PK208LH (Mitsubishi Chemical Co., Ltd., Tokyo, Japan) with a high catalytic activity is used[2)]. The resin was supplied in an activated H-form swelled with water. For pretreatment, the resin was packed with the column and ethanol is supplied to remove water in the resin before the esterification. In the batchwise esterification, the preheated mixed-solution at the molar ratio of oleic acid and ethanol, 1:3, and 33 wt% of the resin were added to the glass bottle. The bottle was well shaken in a thermal bath at 50 °C. At specific time intervals, samples of the solution were collected and the concentrations of the reactants and products were analyzed. The water content in ethanol is set at 0, 5, and 10 wt%.

In the continuous esterification, the column packed with the resin was kept constant at 50 °C. For the pretreatment, ethanol with a certain water content was supplied similarly to the batchwise system. Then, the mixed solution of oleic acid and the ethanol at the same molar ratio as that in the batchwise system was supplied to the bottom of the column by a pump. Under each water content in ethanol, the residence time through the resin bed was regulated by changing the feed flow rate. The effluent solutions from the top of the column were collected and the concentrations of the reactants and the products in the solution were also determined.

Results and Discussion

In the batchwise system, using anhydrous ethanol for both pretreatment and esterification, the complete conversion was reached after 50 h. Using ethanol with 5wt% water content, the esterification rate became slower and more than 90% conversion was obtained at 120 h. After the pretreatment using ethanol with 5wt% water content, the water content in the resin was higher than that in the resin pretreated with anhydrous ethanol, so that the lipid soluble reactant, FFA, was hard to transfer into the resin to lower the esterification rate.

In the continuous system, the conversion was determined after steady state was reached under each condition. Using anhydrous ethanol, the complete conversion was obtained at the residence time through the bed of 10 h. Using ethanol with 5wt% water content, more than 95% conversion was attained at the residence time of 35 h and the complete conversion was also obtained at 60 h. Using ethanol with 10wt% water content, more than 90% conversion was obtained at 60 h. In the continuous system, the esterification rate became slower with increasing water content in ethanol, but its effect was smaller than that in the batchwise system. This was because the water formed by esterification was accumulated in the batch system, whereas the water was removed by continuous flow in the continuous system. Therefore, the continuous esterification using the column reactor packed with the resin is very effective for utilization the hydrous ethanol.

Lower esterification rate leads to an increase in energy consumption, thus contributes negatively from a life cycle perspective by inducing more environmental impacts associated with energy. On the other hand, increased water content means lower duty for distillation of ethanol. In another part of this research (Part II), to allow a system-wide optimization, modeling of the system with a scope that contains all the relevant product life cycles is performed. Using the data acquired from the experiments with continuous configuration, the abovementioned trade-off is highlighted using a case study with a Japanese example.

Reference

1)N.Shibasaki-Kitakawa et al., Kagaku Kogaku Ronbunshu, 42 (2016) 30-36

2)N.Shibasaki-Kitakawa et al., Fuel, 139 (2015), 11-17.

<u>**Economic and Environmental Assessment of Methane to Ethylene Via Oxidative Coupling**</u>

Abstract ID# 469784

Kenneth O'Neill and Benjamin J. Davis, The Cooper Union (New York, NY)

One possible separation train for the purification of ethylene from reactor effluent is proposed, with interest in meeting economic and environmental (via LCA) performance targets. The system consists of two flash tanks and four distillation columns which serve to purify polymer-grade ethylene and reactor recycle. The reaction for producing ethylene is the oxidative coupling of methane on nanowire catalysts, and low conversions mean a large percentage of the methane enters the separation train and must be separated. The four columns following the two initial flash tanks are a carbon monoxide removal column for the separation of carbon monoxide from methane for reactor recycle, a demethanizer column for separation of the remaining methane from the heavier hydrocarbons, a C_2-splitter for the separation of ethylene from ethane, and a carbon dioxide removal tower which operates at vacuum to break the azeotrope which forms between ethylene and carbon dioxide to obtain polymer grade ethylene. The purified ethylene distillate from the carbon dioxide removal tower is 99.481 % pure and the separation system has purchase and bare module costs of \$776,220 and \$3,300,000, respectively (based on 2015 CEPCI values).

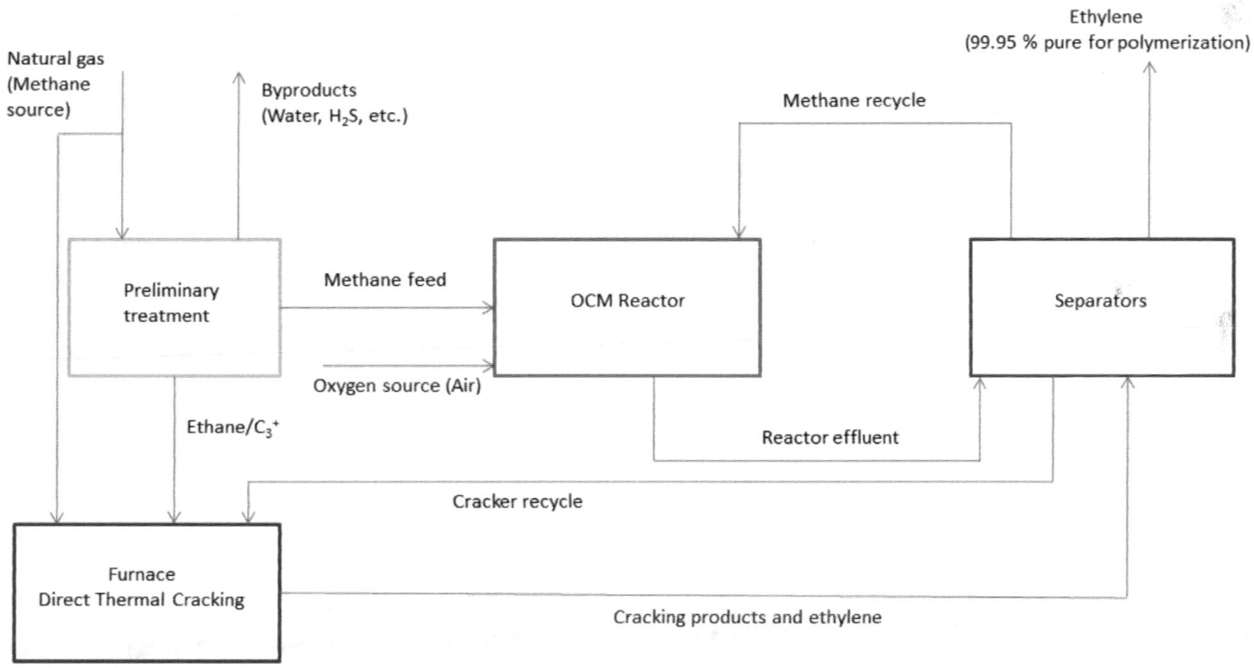

Figure 1: BFD for Siluria's OCM process integrated into an existing ethane cracker

Approximately 90,000 tons of ethylene were produced in United States (US) gas plants in 2014[i], making it one of the world's largest volume commodity chemicals. Ethylene is valuable as a precursor to a variety of products, including low and high density polyethylene, vinyl chloride monomer (VCM), ethylene glycol, and styrene; more than 30 million tons of these commodities were sold in 2012[ii]. Because of its high volume, finding cheaper feedstocks is one of the few ways to have an impact on the ethylene production market. Traditionally ethylene is made from cracking ethane or naptha, which are less abundant species in natural gas and crude oil, respectively. As of February 16, 2016, the sales price of natural gas was $1.95 per MMBTU,[iii] while the total sales of ethylene based polymers decreased 4 % to $69.4 billion in 2014[iv].

Due to the importance of ethylene in the chemical industry, means of revolutionizing and optimizing its production are constantly sought after. Siluria Technologies has created and patented nanowire catalysts[v] and processes[vi] related to the oxidative coupling of methane (OCM) to make ethylene; this is a more advantageous route because methane is cheaper and more abundant than naptha and ethane. A simplified flowsheet from their website is shown below as Figure 2.

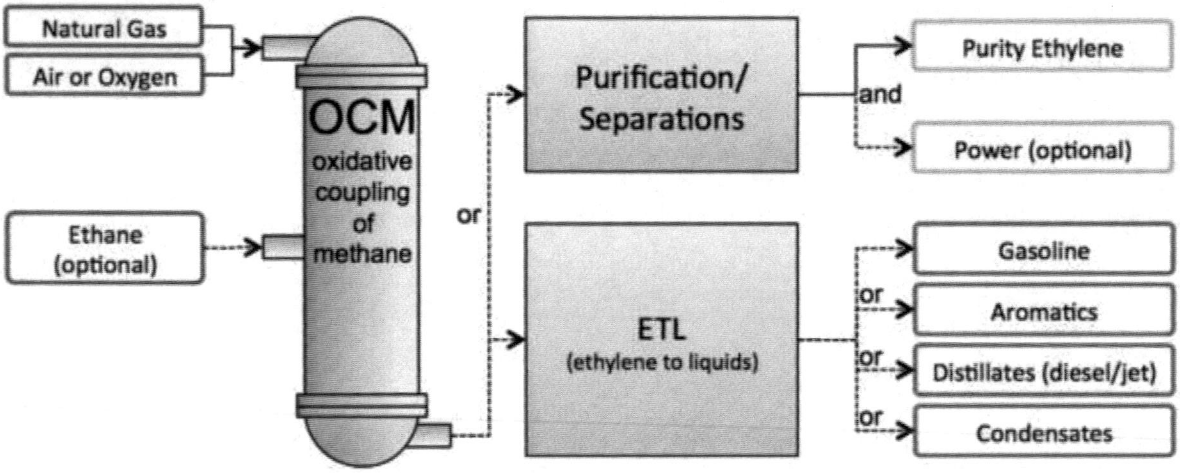

Figure 2: Simplified flowsheet of Siluria's OCM process[vii]

In the Siluria process, the feed to the OCM reactor consists of a methane source (natural gas) and an oxygen source (air or O_2 in the Siluria process). The reactor effluent can then be routed to one of two steps: a separation train for the recovery of high-purity ethylene, or to an ethylene to liquids process to produce gasoline/aromatics/distillates/condensates. The present investigation attempts to propose a more efficient means of separating the reactor effluent for the acquisition of high-purity ethylene.

Traditionally, methane conversions in OCM reactors with ordinary catalysts have been found to vary between 1 % and a theoretical upper bound around 40 % according to Bhasin *et al.* and Lunsford *et al.*, with the remaining 60 % of methane either unreacted, burned, or coked. Selectivities for ethane, ethylene, and acetylene (C_2) tend to inversely with methane conversion, from highs of 90 % to some reports of as low as 1 % have been reported[viii,ix]. At 799 °C, one catalyst patented by Siluria exhibits methane conversión around 21 % with a 70 % selectivity for C_2; on a similar catalyst over an SiO_2 support, the conversion is about 28 % with a 60 % selectivity for at 800 °C. In this study the properties of the former catalyst were used.

With the goal of having an industrially viable OCM system, Siluria has one of two options: construct a stand-alone facility focused on OCM-based ethylene production or integrate the process into a preexisting ethylene cracking facility. With integration into an existing ethylene cracker in mind, the preliminary design investigated herein was one in which the OCM reactor produces about 40 metric tons/day of product (40,000 kg/day).This corresponds to overall plant production of 740 metric tons/day, with 5.4 % of the cracking facility's ethylene from OCM and the remainder from the cracker.
(700 metric tons/day from the cracker).

The primary goal of the research thus far has been to separate the OCM reactor effluent. The difficulty in the separation of high-purity ethylene comes from the complicated OCM reactor effluent mixture. This mixture was assumed to consist of the following species:

Species and Boiling Points	
Component	**Normal Boiling Point (K)**
N_2	77.35
CO	81.15
O_2	90.15
CH_4	111.5
C_2H_4	169.15
C_2H_6	184.15
C_2H_2	189.15
CO_2	194.15
C_3H_6	225.15

Table 1: Components in OCM reactor effluent in order of increasing boiling points

The following assumptions were used in simplifiying the number of components in the reactor effluent:

- Air served as the oxygen source
- Hydrogen was considered light enough to be separated relatively easily
- Water's distinct chemical differences from the hydrocarbon mixture allowed for the assumption that it too could be easily removed
- Propylene was used as a proxy for hydrocarbons of chain length 3 or more, e.g. propanes, butanes, etc., due to those species' relatively higher boiling points allowing for relatively easy separations

In addition to the similarities in properties between the hydrocarbon components, these compounds are also soluble in one another, making separations via distillation more difficult than a simple boiling point difference, making determining appropriate operating conditions for the required separation units difficult. These four simplifying assumptions, along with the results of a material balance around the reactor given below, served as the basis for the simulation.

Component	Mass Flow Rate (MT/day)	Mass %	Molar Flow Rate (kmol/day)	Mole % (OCM Effluent)
CH_4	205.4	42	12800	57
O_2	60.1	0.12	18.8	.083
N_2	226	46	8070	36
CO	3.97	8.2E-01	142	0.63
CO_2	12.3	2.5	279	1.2
C_2H_6	28.1	5.8	938	4.2
C_2H_4	10.079	2.1	360	1.6
C_2H_2	0.041	8.4E-03	1.58	0.0070
C_3H_6	0.082	1.7E-02	1.95	0.0086
Total	487	100	22600	100

Table 2: Mass and molar flow rates and percents for the feed to the separation process simulated via ASPEN Plus V8.0

Figure 3 shows the current separation process as simulated in ASPEN Plus V8.0. The Peng-Robinson equation (P-R EOS) of state was used for the simulation because it works well for light gases[x].

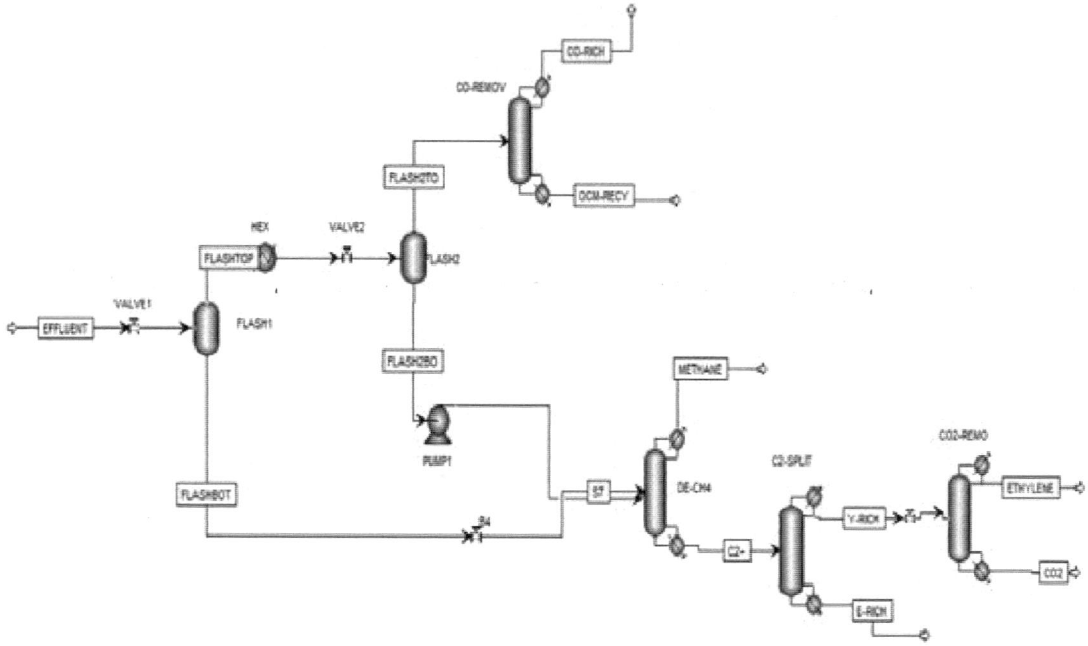

Figure 3: Separation process flowsheet from ASPEN Plus V8.0

The rationale for the flowsheet above was to aim for a good initial separation of light and heavy components, where the ideal case has all of the methane and carbon monoxide leaving as vapor with oxygen and nitrogen, and the remaining components leaving as liquid bottoms from flasg 1. The vapor overhead from the initial flash is cooled, throttled, and flashed again at atmospheric pressure to condense the remaining ethylene; the vapor product is sent to a cryogenic distillation column for the separation of carbon monoxide from the methane and oxygen to be sent as reactor recycle (above the higher flammability limit of methane in air).

The specifications on the individual units are as follows in Table 3 and Table 4.

	Carbon Monoxide Removal	Demethanizer	C2-Splitter	Carbon Dioxide Removal
# of Theoretical Stages	30	10	50	30
Stage Efficiency	0.79	0.77	0.79	0.9
Actual Stages	38	13	63	32
Feed Stage (from top)	17	8, 8	15	24
Reflux Ratio	9	3	3.5	11
Distillate:Feed Ratio	0.14	0.505	0.4	0.544
Condenser P (atm)	2	9.5	9.5	0.08
Condenser T (K)	84	148	220	135
Reboiler P (atm)	2.18	9.555	9.8	0.11
Reboiler T (K)	113	227	241	143

Table 3: Distillation column specifications

Unit	Flash 1	Flash 2	Valve 1	Valve 2	Pump
P or ΔP (atm)	10	2.1	10	1	9.75
T (K)	165	120	-	-	-

Table 4: Flash tank, valve, and pump specifications

From the heavy components (20 % of the feed), CO_2 would preferably be removed as early as possible due to its abundance in the mixture, but due to its high solubility in hydrocarbons and its azeotrope with ethylene, vacuum distillation was necessary to separate it from the ethylene product. The bottoms of the two flash tanks are fed into the demethanizer column to separate the remaining methane (the bottoms of the second flash tank would be pumped back up to 9.75 atm). The distillate from the demethanizer is 99 % pure methane, with the remainder oxygen and nitrogen (above the higher flammability limit) and can be recycled back to the reactor. The bottoms of the tower containing all C_2^+ species would then be sent to a C_2-tower for the separation of ethylene from ethane. Finally, the ethylene and carbon dioxide rich distillate from the C_2-tower would be sent to a vacuum distillation column to break the azeotrope and get a 99.95 % pure ethylene product to be used for polymerization to various grades of polyethylene.

The ethylene separated in the CO_2-removal tower was 99.481 % pure, close to the required purity limit, and would need to be sent through an adsorber prior to polymerization. The CO-Removal tower is processing 90 % of the process flow at cryogenic temperatures and will most likely be the most expensive unit in the system with the consideration of utilties. In addition, the vacuum distillation column for breaking the azeotrope between ethylene and CO_2 will require expensive auxiliary equipment to generate vacuum in the column. The CO-free methane recovered from the bottoms of the CO-Removal tower and the remaining methane recovered as overhead from the demethanizer can be recycled to the OCM Reactor to decrease feed costs. Ethane, acetylene, and other heavier products can be used as feedstock for a traditional cracking furnace to further increase ethylene yields. The current focus of the study is to optimize the designs of the distillation columns in terms of utilities required for the heat exchangers and the power requirements for the vacuum pump.

The columns were sized with 14 inches between trays, 3 feet at the top of the column for vapor disengagement, and 10 feet at the bottom for liquid holdup and structural support. The reflux drums were sized in the same way as the flash tanks previously and the pumps were priced based on the operating pressures and necessary powers.

Distillation Columns				
Specification	**CO-Removal**	**Demethanizer**	**C2-Splitter**	**CO2-Removal**
Theoretical Trays	30	10	50	25
Tray Efficiency	0.8	0.8	0.8	0.8
Actual Trays	38	13	63	32
R (mole ratio)	9	3.5	3.5	11
Diameter (ft)	6	2.5	3	4.3
Height (ft)	70	32.5	107.5	61
Condenser Duty (MMBTU/h)	-14.3	-1.7	-1.4	-1.5
Reboiler Duty (MMBTU/h)	10.2	2.6	1.4	1.5
Cond. Utility (K)	83.7	147.9	220.4	135.1
Reboil. Utility (K)	112.7	226.9	240.9	142.6
P condenser (atm)	2	9.5	9.5	0.08
P reboiler (atm)	2.18	9.55	9.8	0.11
Purchase Cost	$154,000	$24,100	$243,000	$83,000
Bare Module Cost	$334,000	$58,800	$701,000	$177,000
Reflux Drums				
L/D	5	5	5	5
Volume (ft^3)	40.33	13.87	9.64	4.52
D (ft)	5.71	1.96	1.36	0.64
L (ft)	28.53	9.81	6.82	3.20
Thickness (in)	0.07	0.08	0.05	0.00
Purchase Cost	$23,100	$4,910	$3,390	$2,650
Bare Module Cost	69500	15300	10200	7960
Pumps				
	Demethanizer Pump		**Vacuum Pump**	
Purchase Cost	$6,970		$6,900	
Bare Module Cost	$27,800		$20,700	

Table 5: Sizing and costing of the distillation columns, their corresponding reflux drums, and the two pumps required in the flowsheet

Finally, after a possible optimal design is made, a pinch analysis for heat integration will be performed to determine the number heat exchangers and amount of utilities required, a full cost-analysis, and LCA will be done. When this is complete, a comparison as a stand-alone plant versus a supplementary ethylene production facility to a traditional ethane cracking plant. Alternative means of separating the CO_2—including membrane separation and the deposition of solid CO_2—will be investigated.

References

[i] U.S. Energy Information Administration. "U.S. Gas Plant Production of Ethane-Ethylene." https://www.eia.gov/dnav/pet/hist/LeafHandler.ashx?n=pet&s=metfpus1&f=a. Feb. 2, 2016. Web.

[ii] Esposito, Frank. "American Chemistry Council reports 2012 resin statistics." http://www.plasticsnews.com/article/20130220/NEWS/130229994/american-chemistry-council-reports-2012-resin-statistics. Feb. 2, 2016. Web.

[iii] U.S. Energy Information Administration. "Natural Gas." http://www.eia.gov/naturalgas/. Feb. 14, 2016. Web.

[iv] UN Comtrade Database. "Polymers of ethylene, in primary forms." http://comtrade.un.org/.Feb. 8, 2016. Web.

[v] United States Patent Application Publication, Scher*et al*. Publication No.: US 2012/0041246 A1, Feb. 16, 2012.

[vi] United States Patent Application Publication, Iyer*et al*. Publication No.: US 2014/0012053 A1, Jan. 9, 2014.

[vii] http://siluria.com/Technology/Overview

[viii] Bhasin, M. M., and Keller, G. E. "Synthesis of Ethylene via Oxidative Coupling of Methane I. Determination of Active Catalysts."*Journal of Catalysis* 73.1 (1982): 9-19. Print.

[ix] Clearfield, Abraham, Hinson, Paul G., and Lunsford, Jack H. "The Oxidative Coupling of Methane on Chlorinated Lithium-doped Magnesium Oxide."*Journal of the Chemical Society*. 20.1 (1991): 1430-1432. Print.

[x] Henley, Ernest J., Junior D. Seader, and D. Keith Roper. *Separation Process Principles*. 3rd ed. Hoboken, NJ: Wiley, 2011. Print.

472330 Highly Efficient Formate Production By Hydrogenation of Captured CO_2

Abstract:

Climate change resulting from the emission of CO2 has become a widespread concern in the recent years. Carbon Capture and Storage (CCS) is a promising technology that has the potential to address the 40% of emission emanating from large-point sources such as power plants. The utilization of CO2 as a raw material in the synthesis of chemicals and liquid energy carriers offers a direct way to mitigate the increasing CO2 build-up but catalytic reduction of CO2 to fuels or value added chemicals is still one of the greatest challenges. Recently, increasing efforts have been devoted to the transformation of CO2 into value added substances such as formic acid or formates, which are highly attractive as commodity chemicals used as preservative and antibacterial agents, as well as clean sources for H2 generation. Herein, an effective low-temperature process of hydrogenating liquid ammonia or amine captured CO2 over supported palladium nano-catalysts, which can leverage the existing industrial infrastructure to produce value-added formate by utilizing the CO2 derivatives in the commercial CCS process, has been developed. For the first time, we found that the yield of formates is highly dependent on the ethyl carbonate ion intermediates in a solution of ammonium carbonate or amine-CO2 adduct. We further designed the rechargeable â€œhydrogen batteryâ€ based on the redox equilibrium between liquid ammonia captured CO2 and the corresponding ammonium formates, where hydrogen was stored in chemical bonds. The stored hydrogen can be catalytically liberated under mild conditions as the feed to fuel cells for regenerating electricity.

Allothermal gasification of lignite and biomass with integrated syngas cleaning and CO_2 removal

Peter Treiber, Michael Neubert, Jonas Leimert, Juergen Karl
Friedrich-Alexander-University, Erlangen-Nuremberg, Germany

Introduction

For the production of synthetic natural gas (SNG) from lignite or biomass in general the following process steps are required [1].

- Gasification of the solid feedstock (allothermal or autothermal)
- Syngas cleaning from particles, tars, and off-gases
- Methanation
- Optional product gas conditioning for grid injection

For small- and mid-scale, decentralized plants the main challenge is to keep the plant complexity to a minimum while still reaching the requirements for gas grid injection. The Heatpipe Reformer technology is a suitable way for decentralized SNG production [2], [3]. A hydrogen rich syngas with a nearly ideal composition for methanation can be produced in the Heatpipe Reformer using allothermal steam gasification. A syngas cleaning unit with integrated removal of CO_2 would reduce the plant complexity and provide syngas for methanation without elaborate gas cleaning steps.

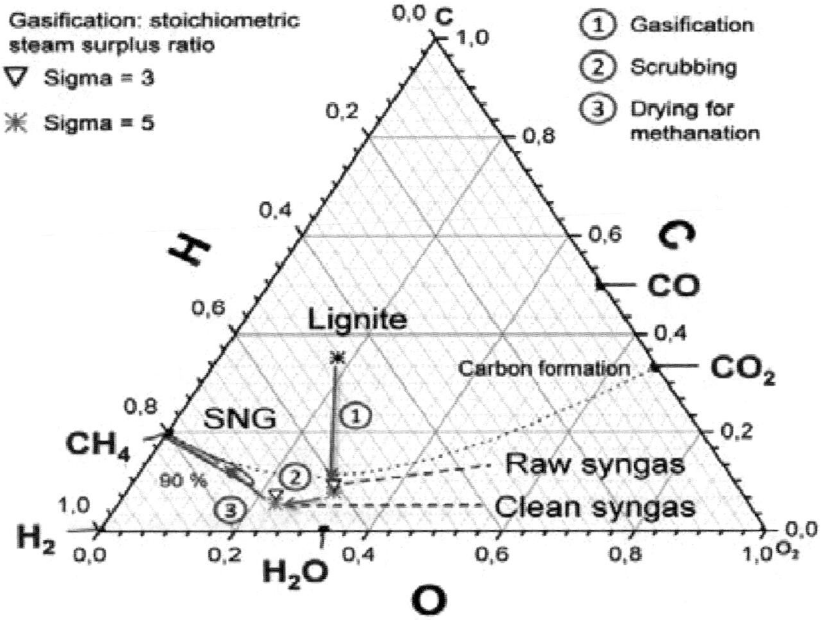

Figure 1: Coal-to-SNG process chain in the ternary diagram

With the EU research project CO2freeSNG2.0 (RFCR-CT-2013-00008), funded by Research Fund for Coal and Steel (RFCS), the research focus of the Chair of Energy Process Engineering at the Friedrich-Alexander-University Erlangen-Nürnberg for the subjects of gasification, syngas cleaning, and methanation is intensified [4]. In addition to the formerly proposed hot gas cleaning [5], [6] a hot carbonate scrubbing unit based on the Benfield process [7], [8] is integrated in the process chain for the combined bulk removal of CO_2, sulfur, and tar species. The proposed process chain and its structure in the ternary diagram are depicted in Figure 1. The tests stands and the results of the parameter variation tests are presented in the following chapters.

5 kW Bench scale process chain

For more flexibility and a reduced test time frame, a process chain for the production of SNG was designed and manufactured in the bench scale for 5 kW fuel input. The schematic setup of the single test rigs and the possible combinations are shown in Figure 2.

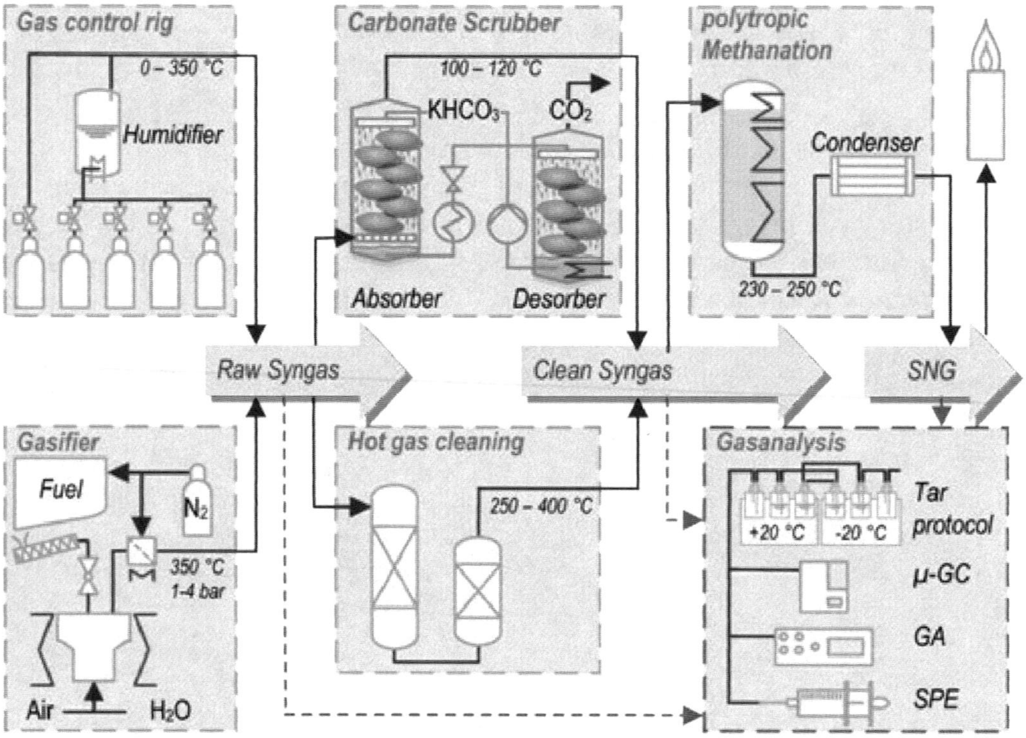

Figure 2: Schematic layout of the EVT 5 kW Coal-to-SNG process chain

Besides real syngas from coal or biomass also bottled gas mixtures can be used in the test rigs. The whole process chain is designed for an operating pressure of 5.2 bar_a. Furthermore, gas sampling ports have been integrated between each of the process steps to analyze gas composition, tar loading and sulfur species in the gas streams.

The basis for the bench scale SNG production is the 5 kW bench scale gasifier. Lignite or biomass can be gasified in a pressurized, steam blown fluidized bed. The regulation of the fluidization behavior as well as operating pressure is integrated in the fully automated test rig which allows for long term tests of over 100 h. The raw syngas from the gasifier passes a hot gas filter where particles and alkali are removed from the gas stream. Subsequently the syngas is passed to the next test rig (scrubber or methanation unit with hot gas cleaning) reducing the pressure between each test stand by 200 mbar to enable an independent pressure regulation for each unit. At the end, the syngas or product gas is disposed over a lab scale flare. The setup and function schematics of the bench scale gasifier are shown in Figure 3.

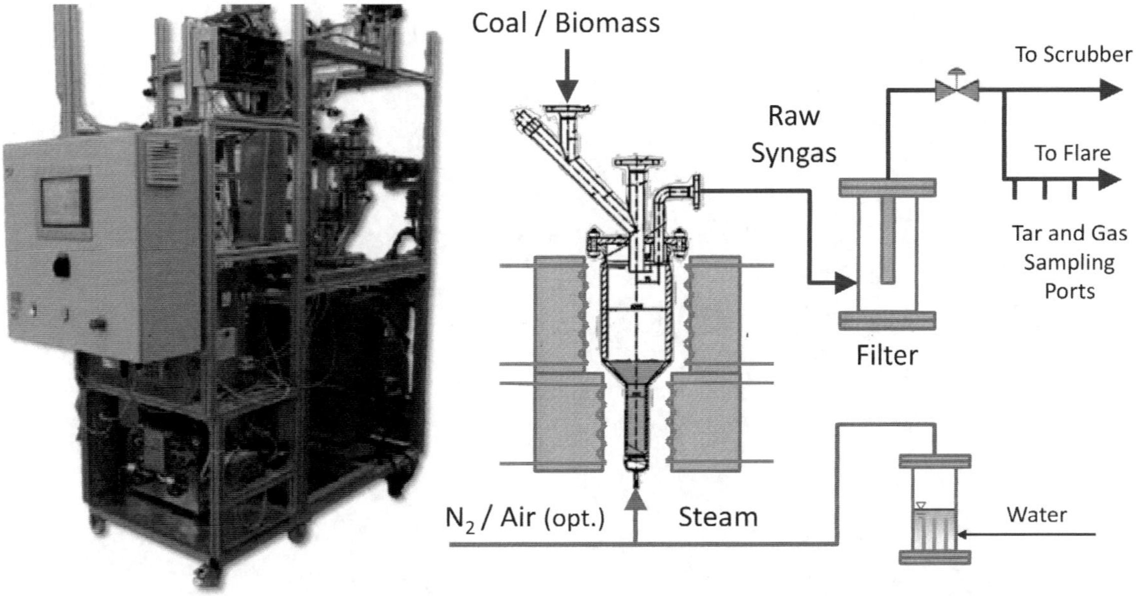

Figure 3: 5 kW bench scale gasifier: setup (left) and function schematics (right)

After the gasifier, the raw syngas is passed on to the bench scale carbonate scrubber (Figure 4) at a pressure of 5.2 bar_a. In the absorber, the raw syngas passes in counter current flow to the solvent over a random packing of 6 mm Raschig rings and exits the column at the top. In contrast to other absorption processes, the solvent is not cooled before entering the absorber (as in amine scrubbing) of the hot potassium carbonate process but it is directly drawn from the reboiler and recycled into the absorber at temperatures only slightly below boiling point (105°C). Therefore, the energy-intensive cooling of the solvent is not needed. Furthermore the regeneration is supported by a partial evaporation of water in the solvent, lowering the partial pressure of CO_2 already at the top inlet of the desorber column. In addition to the absorption of CO_2, also hydrogen sulfide is absorbed and heavier tar components are captured in the solvent flow due to condensation. Therefore, the carbonate scrubbing unit can be considered as bulk removal option for CO_2, H_2S, and heavier tar components.

Table 1: Carbonate scrubbing reactions

$$K_2CO_3 + CO_2 + H_2O \rightleftharpoons 2\,KHCO_3 \quad (1)$$

$$K_2CO_3 + H_2S \rightleftharpoons KHS + KHCO_3 \quad (2)$$

The capture of tar components is mainly based on condensation while the sour gas removal is based on chemisorption with potassium carbonate following the chemical reactions listed in Table 1. The reduced reboiler duty and the omitted solvent cooling are the main advantages of the carbonate scrubbing process compared to other scrubbing processes.

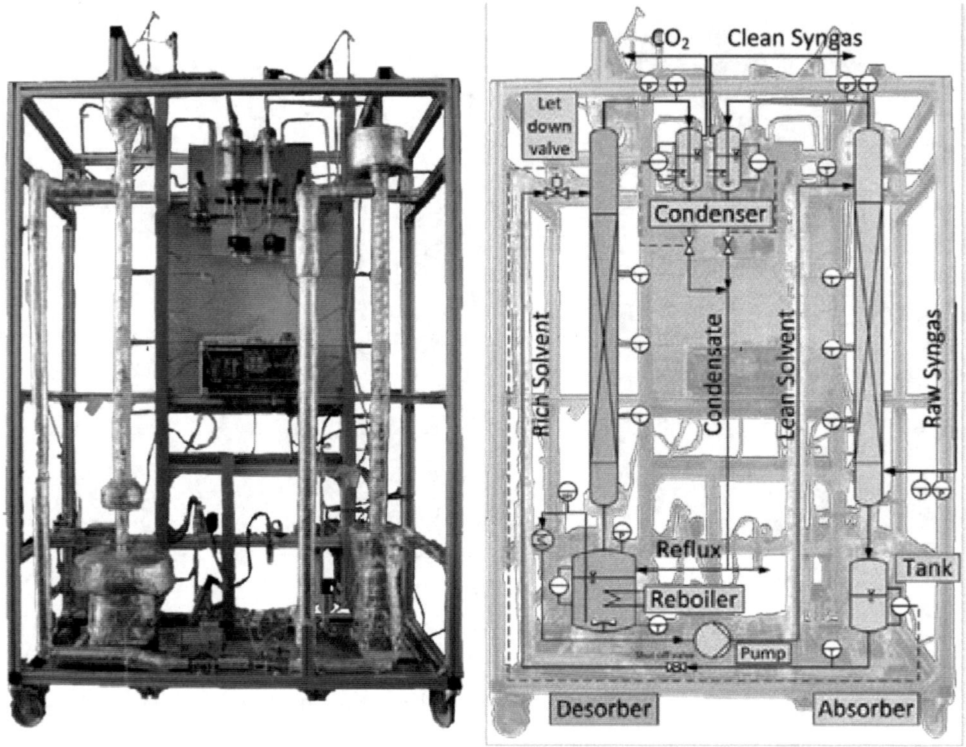

Figure 4: 5 kW bench scale carbonate scrubber: setup (left) and function schematics (right)

100 kW Pre-pilot process chain

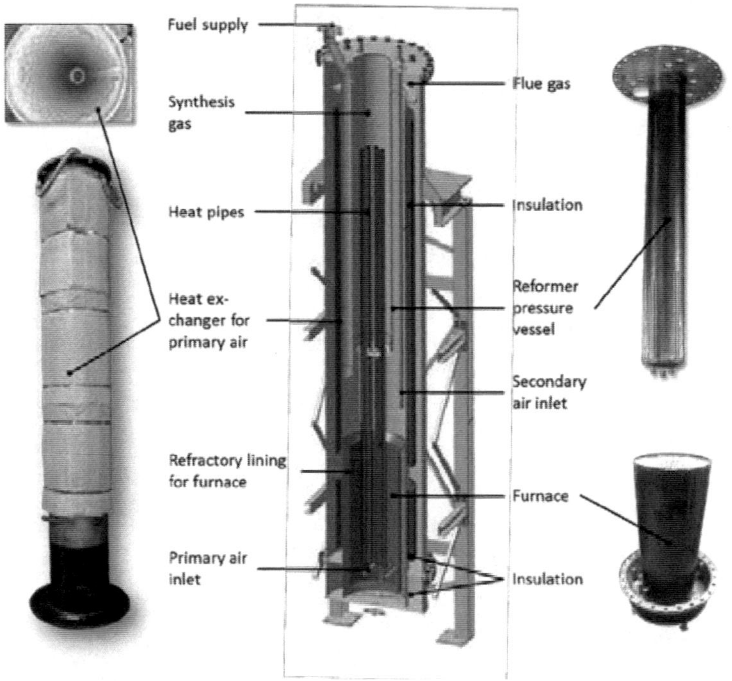

Figure 5: Setup of the 100 kW Heatpipe Reformer (HPR) [3]

Since the 5 kW bench scale test stands are rather prone to heat losses and edge effects (carbonate scrubber columns) the process chain at a 100 kW pre-pilot scale is also investigated. Similar to the bench scale process chain a pressurized, fluidized bed steam gasifier provides the raw syngas. Figure 5 shows the setup of the Heatpipe Reformer. The atmospheric combustion chamber generates the necessary heat which is transported via liquid metal heatpipes to the pressurized gasification reactor at approx. 850°C.

Like in the bench scale process chain, the raw syngas from the Heatpipe Reformer passes over a hot gas filter for the removal of fine coke as well as ash particles. After the filter the syngas enters the absorber column at the bottom inlet. The pre-pilot scale scrubbing unit is shown in Figure 6. The columns (packing height: absorber 3 m and desorber 4 m, diameter: both 100 mm) are packed with 10 mm Raschig rings and have been dimensioned by prior rate based simulations with ASPENplus^{Tm}. The solvent is let down by a throttle valve into a flash drum at the top of the desorber.

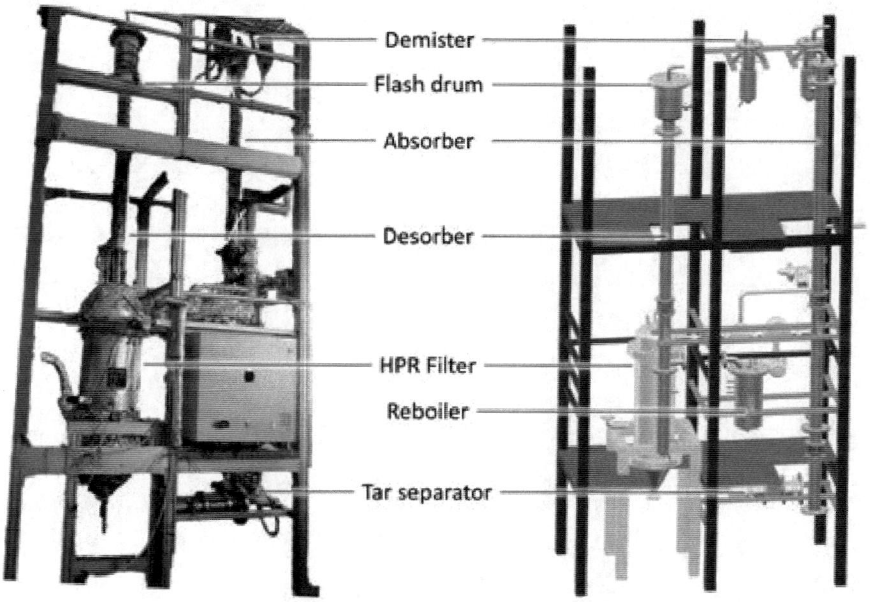

Figure 6: Setup of the 100 kW Benfield scrubber

In contrast to the bench scale scrubber, the reboiler unit of the 100 kW scrubber is not located at the column bottom but is a separate unit which is fed by an additional pump to prevent the enrichment of potassium carbonate in the reboiler. Additionally, a tar separator by means of a coalescence-mesh was integrated in the solvent flow before the main solvent pump. Therefore, a lighter and a heavier tar fraction (compared to the solvent) shall be removed from the solvent flow during regular operation. The total removal of light hydrocarbons is not intended since prior tests have shown that they can be converted in the methanation step [9], [10].

At the top of the absorber column the clean syngas passes a demister and can be either provided for a methanation reactor or disposed in a thermal afterburner. As proposed by Benson and Field [8], [11], [12] a 30 wght.-% potassium carbonate solution shall be used as solvent. Due to reduced solvent volume, the variation of solvent concentration and possible additives is predominantly investigated at the bench scale test rig.

Results

Allothermal steam gasification

In Figure 7 the syngas composition after the gasifier at a stoichiometric steam surplus of $\sigma = 5$ is shown. The relatively high steam surplus causes a syngas which is rich in hydrogen and carbon dioxide.

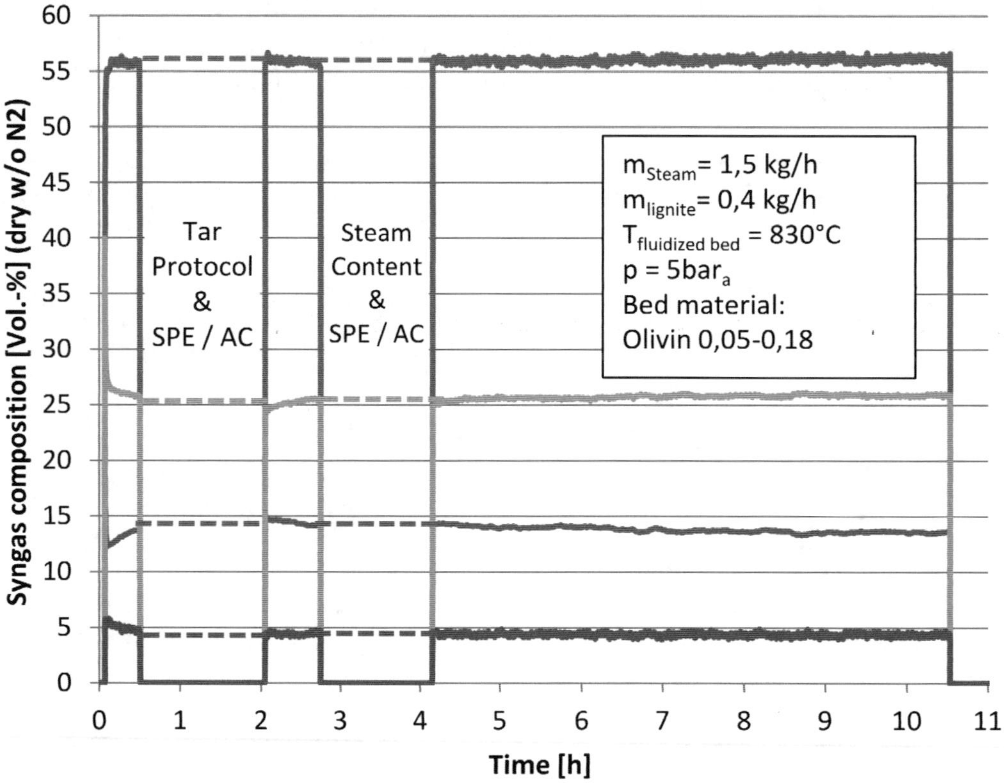

Figure 7: 10-h gasification test at the bench scale gasifier; syngas composition

During the test shown in Figure 7 a steam content measurement and a tar content analysis by means of the tar protocol [13] were performed. The results of the tar protocol measurement are depicted in Figure 8.

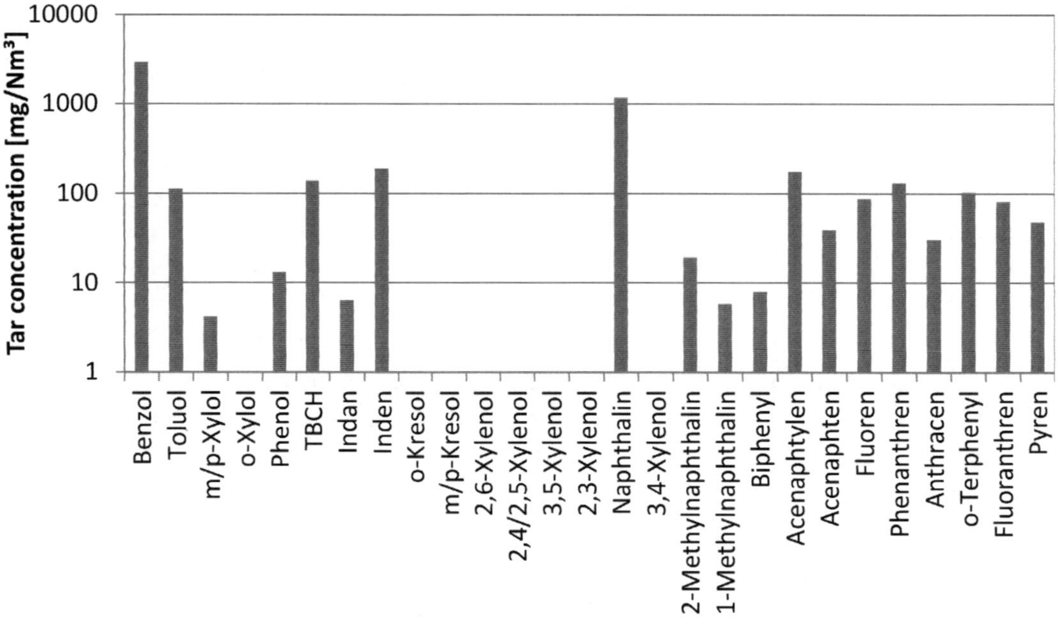

Figure 8: 10 h gasification test, tar concentration in the raw syngas by means of the tar protocol [13]

Integrated syngas cleaning

The influence of the solvent temperature at the absorber inlet as well as the ratio between the solvent and syngas mass flow (liquid-to-gas ratio – L/G) are depicted in Figure 9.

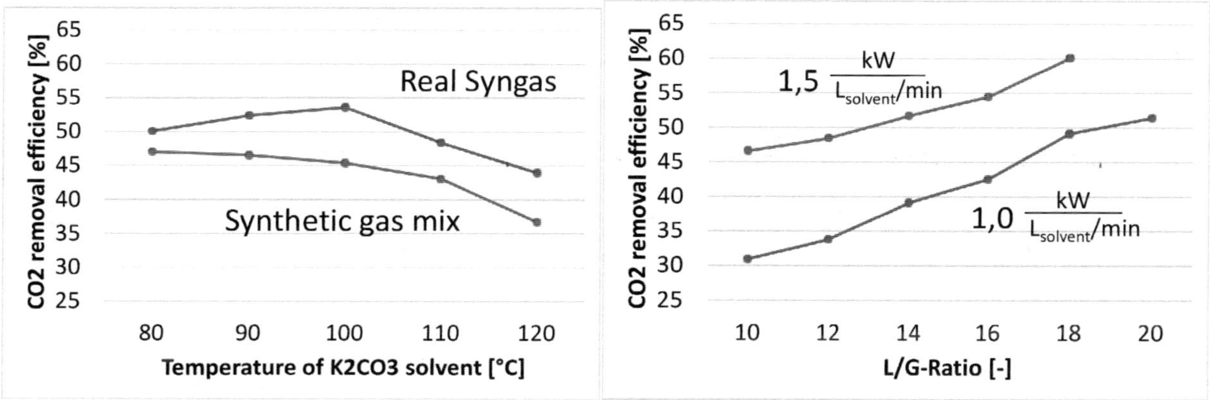

Figure 9: Influence of the solvent temperature (left) and the L/G ratio (right) on the CO2 removal efficiency at the 5 kW bench scale scrubber

At temperatures below 100 °C the CO_2 removal efficiency decreases due to the hindered mass transfer of CO_2 to the solvent and the carbonate reaction rate. At higher temperatures, the solvent loading capacity decreases due to the limited solubility of CO_2.

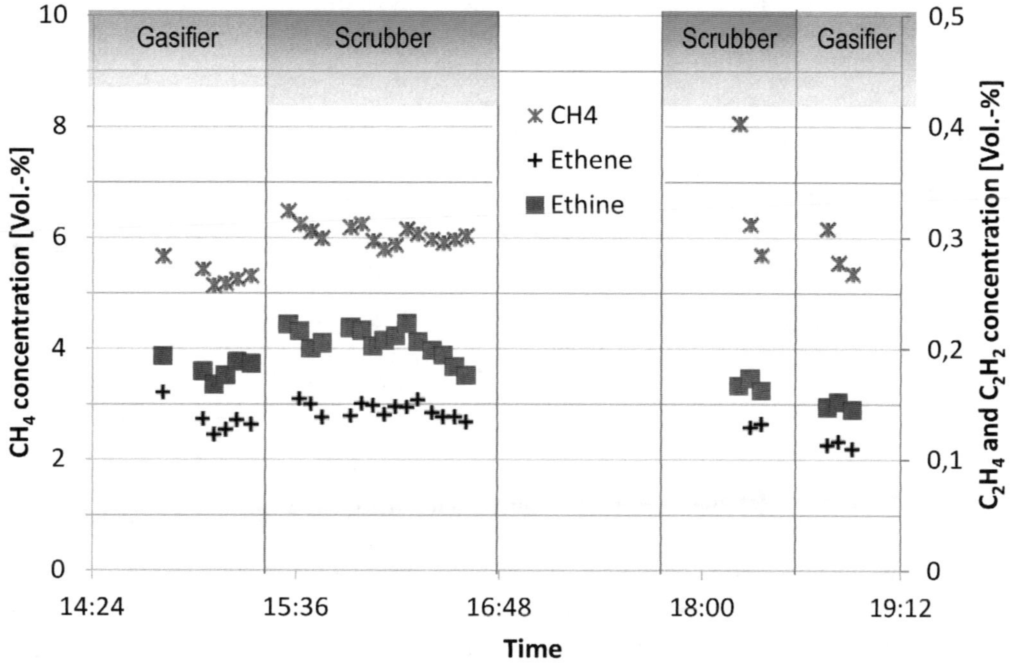

Figure 10: μ-GC measurements after the bench scale gasifier(σ = 5, $T_{gasifier}$ = 830°C, p = 5.2 bar$_a$) and scrubber (L/G = 17, $T_{Solvent}$ = 100°C, p = 5.0 bar$_a$)

Furthermore, a higher removal efficiency of the real syngas could be demonstrated since a slip stream of the raw syngas was used for gas analysis compared to the synthetic tests with bottled gases reducing the syngas volume entering the absorber column. The variation of the solvent flow at constant gas flow proofed an increase in removal efficiency at higher L/G-ratios. An additional increase in the specific reboiler heat duty showed a positive influence on the solvent regeneration and thus an increasing CO_2 removal efficiency.

Compared to the prior simulations, the expected removal efficiency for CO_2 of 90% could not be reached in the bench scale tests. Since the test rig has a relatively small column diameter and the Raschig ring diameter was therefore rather large the scrubber showed dominant edge effects prevented the necessary mass transport in the absorber column. Final tests with the 100 kW pre-pilot test stand are expected to reach the prior simulated removal efficiency.

Besides the direct measurements for the CO2 removal efficiency, also the removal efficiency for other gas components was investigated with a μ-GC. The results shown in Figure 10 demonstrate that light hydrocarbons like ethene and ethine are not influenced by the carbonate scrubbing process and can be measured in constant concentrations in the clean syngas.

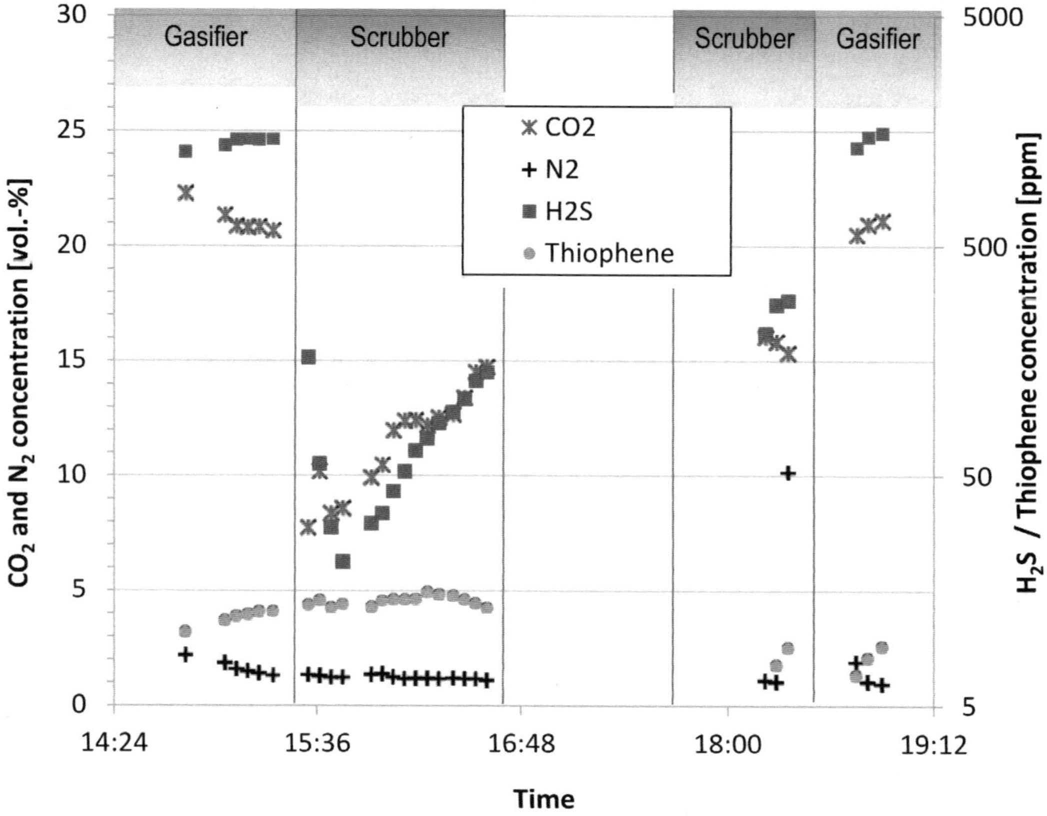

Figure 11: μ-GC measurements after bench scale gasifier and scrubber

The results of the μ-GC measurements in the first half of Figure 11 show the solvent loading process with increasing CO_2 and H_2S content. Since the bench scale scrubber is pressurized and preheated using nitrogen as inert gas, the solvent is stripped to a minimum content of CO_2 and H_2S. In the middle section the product gas after the methanation unit was analyzed showing no content of sulfur species. The final tests demonstrate higher removal efficiencies for hydrogen sulfide than for CO_2 in stationary operation. However the concentration of thiophene was hardly influenced by the carbonate scrubbing unit.

Conclusion

Within the EU research project CO2freeSNG2.0 (RFCR-CT-2013-00008) the process chain for the production of substitute natural gas (SNG) from lignite or biomass for 5 kW und 100 kW fuel input have been constructed. The syngas produced by allothermal steam gasification is treated by means of a hot carbonate scrubbing process and with the same process the bulk share of CO_2 is removed from the syngas stream. First tests could already proof the feasibility of the proposed gas cleaning system which is based on the Benfield process. Besides CO_2 also the bulk share of H_2S was removed from the raw syngas stream while lighter hydrocarbons remained in the gas stream due to higher scrubber temperature. Thus, they are available for conversion in the subsequent methanation step. Organic sulfur components such as thiophene, however, were hardly removed by the carbonate scrubber. Therefore, the proposed process chain demonstrates a bulk syngas cleaning technology for smaller decentralized SNG plants, where conventional syngas cleaning systems are not be suitable due to the high investment costs. In future tests, increased removal efficiency for CO_2 in the 100 kW process chain shall be targeted and the influence of the scrubbing process on tar and sulfur components in the syngas stream will be investigated more thoroughly.

Acknowledgements

The presented test stands and results have been funded by the European Commission and the Research Fund for Coal and Steel (RFCS) as part of the research project CO2freeSNG2.0 (RFCR-CT-2016-00008).

References

[1] J. Karl, *Dezentrale Energiesysteme: Neue Technologien im liberalisierten Energiemarkt.* 2012.

[2] J. Karl, "Biomass heat pipe reformer—design and performance of an indirectly heated steam gasifier," *Biomass Convers. Biorefinery*, vol. 4, no. 1, pp. 1–14, Oct. 2013.

[3] J. M. Leimert, P. Treiber, and J. Karl, "The Heatpipe Reformer with optimized combustor design for enhanced cold gas efficiency," *Fuel Process. Technol.*, vol. 141, pp. 68–73, Jan. 2016.

[4] P. Treiber and J. Karl, "CO2freeSNG2.0," 2013. [Online]. Available: http://www.co2freesng20.eu/. [Accessed: 24-Mar-2014].

[5] C. Baumhakl, J. Karl, T. Kienberger, A. Schweiger, D. Buchholz, S. Bajohr, E. I. Koytsoumpa, and S. Karellas, *Substitute natural gas from coal with internal sequestration of CO 2 (CO2freeSNG) - Final Report.* European Union, 2013.

[6] C. Baumhakl, "Substitute Natural Gas Production with direct Conversion of Higher Hydrocarbons," phdthesis, 2014.

[7] H. E. Benson, J. H. Field, and R. M. Jimeson, "CO2 absorption: Employing hot potassium carbonate solutions," *Chem. Eng. Prog.; (United States)*, vol. 50:7, Jul. 1954.

[8] H. E. Benson and J. H. Field, "Method for separating CO2 and H2S from gas mixtures," US Patent 2,886,405, 1956.

[9] T. Kienberger, C. Zuber, K. Novosel, C. Baumhakl, and J. Karl, "Desulfurization and in situ tar reduction within catalytic methanation of biogenous synthesis gas," *Fuel*, vol. 107, pp. 102–112, May 2013.

[10] C. Baumhakl and J. Karl, "Efficient Bio-SNG Production from Thermal Gasification with Hot-gas-cleaning Technologies," in *21st European Biomass Conference and Exhibition*, 2013.

[11] H. E. Benson and D. H. McCrea, "Removal of acid gases from hot gas mixtures." Google Patents, 10-Jul-1979.

[12] H. E. Benson and J. H. Field, "New data for hot carbonate process," *Pet. Refin.*, vol. 39, no. 4, 1960.

[13] "DIN CEN/TS 15439:2006-08," 2006. [Online]. Available: http://www.din.de/de/mitwirken/normenausschuesse/nmp/din-spec/wdc-beuth:din21:84563381. [Accessed: 01-Apr-2016].

471466 Biocatalytic Polymer Material for Partial Oxidation of Methane to Methanol

Abstract:

An industrial process for the selective partial oxidation of methane under mild conditions would be highly valuable for controlling emissions to the environment and for utilizing vast new sources of natural gas. Methane monooxygenase enzymes perform the difficult activation of methane under nearly ambient conditions, suggesting a biochemical process for methane conversion could be more efficient, less capital intensive, and more easily downscaled compared to current industrial practice. However, the industrial use of enzymes for large-scale applications, especially those involving gas-to-liquid reactions, has been limited by the low enzyme loading and poor mass transfer characteristics of stirred-tank reactors. We have successfully embedded the enzyme particulate methane monooxygenase (pMMO) within a 3D printed hydrogel to create a mechanically robust and gas permeable material that overcomes mass transfer limitations to convert methane to methanol. We studied the kinetics, reactant dependence, and product inhibition of the partial oxidation of methane by pMMO, as these parameters are not well understood and are critical to optimization of our biocatalytic hydrogel material design.

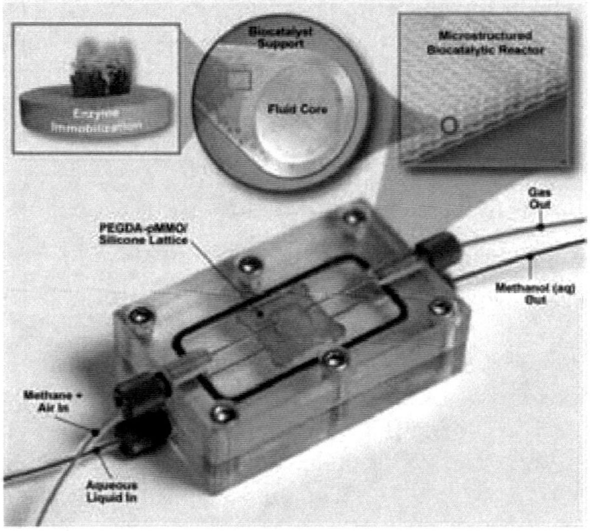

Figure 1: Schematic of enzyme immobilization within hydrogel in a structured polymer scaffold to create a gas-permeable biocatalytic material in a flow-through bioreactor.

Methods: pMMO was immobilized in poly(ethylene glycol) (PEG) hydrogel during crosslinking by UV-initiated photopolymerization. The activity of pMMO was measured by a methanol assay at 45°C using NADH as the reducing agent; methane was injected into the reaction headspace and the methanol produced was quantified by GC-MS. Kinetic studies were conducted by stopped flow analysis with varying pMMO, NADH, dissolved oxygen, and dissolved methane concentrations.

Results and Discussion: Immobilization of pMMO within PEG resulted in a biocatalytic material with nearly 100% physiological enzyme activity. By increasing the surface area to volume ratio of the material using 3D printing, we were able to achieve high volumetric methanol productivity of 2 g/L/hour. Additionally, the hydrogel material can be cured within a poly(dimethyl siloxane) scaffold as part of a flow-through reactor, shown in Figure 1, which is capable of continuous methanol production. Finally, we have determined the dependence of methanol production rate on gas phase reagents, which will facilitate optimization of reactor design parameters such as enzyme loading, membrane thickness, gas-to-liquid volume fractions, surface area to volume ratio, and flow rates.

This work was performed under the auspices of the U.S. Department of Energy by Lawrence Livermore National Laboratory under Contract DE-AC52-07NA27344.

473137 XTL Process with Low Overall GHG Emissions

Abstract:

A patent-pending process that can be used for the production of liquid fuels from biomass/coal based syngas (H_2:CO â ‰¤1) over iron-based catalysts while completely suppressing the formation of CO_2 in the FTS process will be presented. Based on the most optimistic data available from Sasol, the fraction of carbon in the original coal that results in the formation of liquid hydrocarbons is only 16.9 % using iron-based catalysts. The data from research employing this new technology have shown that this fraction can be improved to 51.1 %. This is achieved by co-feeding CO_2 (produced in the upstream processes) such that the CO_2:syngas ratio in the reactor is above 3.33:1 and conducting the FTS at conditions above the critical pressures and temperatures of CO_2. The liquids selectivity moved towards higher hydrocarbons (C_{10}- $C_{22}$with peak content in the range of C_{13}-C_{15}). The $ScCO_2$ â€“ FTS process also takes advantage of the strong solvent yet tunable power of $ScCO_2$ and allows for pressure tunable fractionation of the products. A final but important addition to the $ScCO_2$ â€“FTs process is the addition of product gas recycle capabilities that renders the following two unique advantages. (A) recycle helps in improving the throughput (reduced reactor size). (B)Advantages of $ScCO_2$FTS are further enhanced by the recycle of HC gases (along with the unreacted feed) since HCs like hexane and pentane become supercritical. The results from the various stages of development of this technology will be discussed.

474331 Simulation of Hydrogen and Acetylene Production By Methane Pyrolysis in Thermal Plasma

Abstract:

With the recent shale gas disposal and increase of biogas production, acetylene production from methane could play an important role in near future. Indeed, gas pyrolysis by thermal plasma can produce competitive hydrogen and acetylene for PVC production. Besides, acetylene could be converted to ethylene and ethylene to synfuel which can be an attractive gas to liquid solution to produce transportable fuel [1]. A closer process has been already massively set-up and used in Germany before and after Second World War to produce synthetic rubber and hydrogen for coal hydrogenation process [2]. Recently, we have done many experiments converting coal or liquid hydrocarbons to acetylene by thermal plasma with satisfied performances [3]. However, acetylene is unstable at medium temperature and can convert to tar and oil in about 2 ms and then, the pyrolysis products should be cooled rapidly. Cold gas/liquid hydrocarbons can be used as quenching media to increase the yield of ethylene, i.e. a desired by-product. In order to improve the process efficiency, a deep study has been carried out based on experimental and numerical simulation with methane used as feedstock.

The experimental results show that the methane conversion through both plasma devices is higher than 90% and leads mainly to acetylene and hydrogen with high yields. Also, few-layers graphene have been obtained in most of the cases. This investigation suggests that such a simple technique is easy to operate and suitable for mass production of acetylene and few-layer GNs in a scalable process. To improve and have a better understanding of our lab experimental set-up, a numerical model has been established. The plasma is considered using a MHD model. The reaction kinetics and carbon nanoparticle formation have been considered in a sophisticated model. A parametric study has been made on plasma power and hydrogen/methane ratio. The plasma parameters and the hydrogen ratio could be adjusted to improve the acetylene formation and the graphite nanosheets formation

474574 Enabling C_1 Chemistry through Atmospheric Pressure Corona Discharge (non-thermal) Plasmas

Abstract:

Efficient implementation of C_1 chemistry has been a major goal in industrial chemistry for over a century – currently, the major methane to liquids processes involves reforming to syngas followed by catalytic CO hydrogenation. Our group has developed novel technology using atmospheric pressure corona discharge (non-thermal) plasmas implemented in microreactors. In this work we have demonstrated C_1 coupling to C_2 mixtures (ethylene and/or ethane and excess hydrogen, depending on the gas stream composition) with energetic efficiencies in the order of 70%, as well as to longer hydrocarbons. This presentation will summarize recent developments from our lab in this area.

464739 Process Simulation and Economic Analysis of Producing Liquid Transportation Fuels from Biomass

Abstract:

Although the global energy sector is still driven by fossil fuels, renewable and more environmental friendly energy sources need to be explored due to the insecurity of crude oil supply and the environmental problems associated with fossil fuel consumption. The conversion of biomass to liquid transportation fuels is a promising alternative. A process is developed in ASPEN Plus to simulate the conversion of biomass to produce gasoline, diesel, and kerosene at varying design/operation conditions. Biomass is dried and gasified to generate synthesis gas, which is converted to a mixture of hydrocarbons via Fischer-Tropsch synthesis (FTS). Given the same biomass feedstock, three FTS technologies including conventional FTS, once-through FTS, and supercritical FTS are comparatively studied. Hydrocarbons from FTS are upgraded to liquid transportation fuels which meet all necessary physical property standards through several upgrading technologies (hydrocracking, hydrotreating, isomerization, catalytic reforming, and alkylation). In conventional FTS, the produced $C_1/C_2/C_3$ light gas in the hydrocarbon upgrading process is reformed to produce additional syngas and then recycled back to the FTS reactor along with the fresh syngas to improve the overall carbon conversion rate. For once-through FTS, the $C_1/C_2/C_3$ light gas from hydrocarbon upgrading process and the unreacted syngas from the FTS reactor are directed to a gas turbine for co-production of electrical power. For supercritical FTS, a supercritical fluid is used as the reaction medium to improve/tailor the performance of the FTS reactions.

This study first investigates the product distribution and CO_2 emission of the biomass conversion processes associated with the three FTS technologies. Then heat and power integration is performed for these three biomass conversion process topologies and a given set of operating conditions. Finally, a detailed economic analysis is performed using Aspen Process Economic Analyzer and unit cost functions obtained from literature.

466069 Modeling Plasma Gasification of Biomass with Thermodynamic and Kinetic Approach in Series

Abstract:

Modeling Plasma gasification of biomass with thermodynamic and kinetic approach in series

Babita K. Verma, Rajesh Elangovan, T. Renganathan, S. Pushpavanam*

Department of Chemical Engineering, Indian Institute of Technology Madras,

Chennai 600036, India

*Corresponding author: spush@iitm.ac.in, Tel: +91-44-22574161

Plasma gasification is a promising green technology in which carbon based feed-stocks are converted to synthesis gas. In this process, a plasma arc is used to supply heat to the reactor to sustain the endothermic gasification reactions. Plasma gasification operates at temperatures above 2500 K. Here the reaction rates are very high and this leads to lower residence time as compared to conventional gasification [1]. This drastically reduces the size of gasifiers operating at atmospheric pressure. These advantages motivate the development of accurate mathematical models for plasma gasifiers which can help to analyze the effect of various parameters on the performance.

There are two ways of modeling: Equilibrium and kinetic approach. The purely equilibrium approach does not need information on all reactions which occur at the high temperatures in the reactor here when using the non-stoichiometric approach. The purely kinetic approach requires the consideration of heterogeneous reactions, resistances associated with gas-solid interaction and the formation of radicals. In the present work, the gasifier is modeled as consisting of a high temperature zone followed by a quenching zone where the reactor contents lose heat to the ambient. In the former the reactions are very fast and hence it can be assumed to be at thermodynamic equilibrium. In the quenching zone the reactant mixture cools down and the composition of the gases vary along the reactor length at a finite rate determined by the reaction kinetics. This motivates the reactor to be modeled as being made of two regions namely a plasma gasifier zone (PGZ) which is simulated using a thermodynamic non-stoichiometric approach and a quenching zone (QZ) which is simulated based on a kinetic approach. The schematic representation of the plasma gasifier model [2] is shown in Figure 1

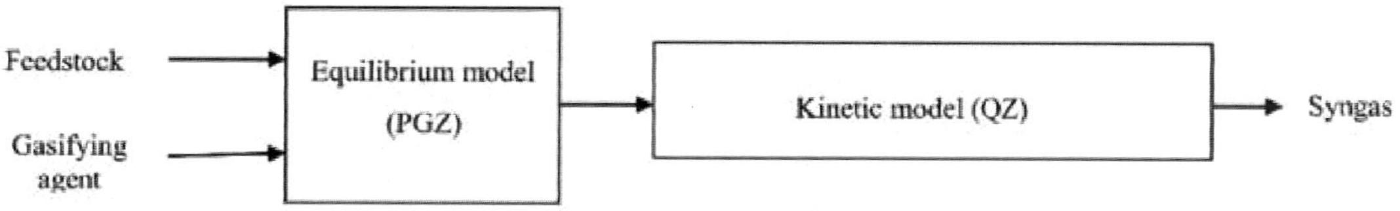

Figure 1. Schematic representation of the plasma gasifier model

The effect of various parameters – gasifier temperature, steam flow rate, air flow rate and feed moisture content – on the performance of the gasifier is studied. Figure 2 shows a typical result i.e. the effect of the gasifier zone temperature on the heat duty and the cold gas efficiency existing in PGZ and QZ region.

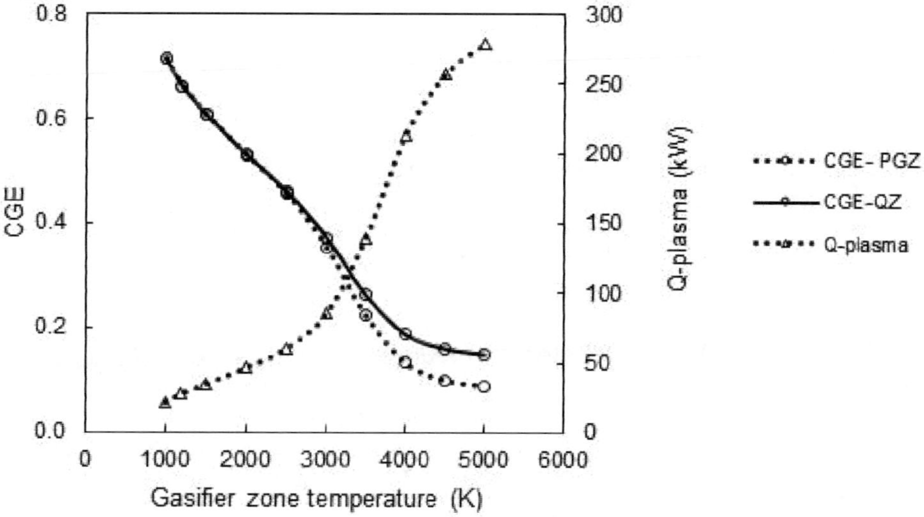

Figure 2: Effect of temperature on heat duty and CGE exiting PGZ and QZ region (Air=13.25kg/hr, Steam=3.90 kg/hr, Moisture=10%)

References:

[1] S. J. Yoon and J. G. Lee, "Syngas Production from Coal through Microwave Plasma Gasification : Influence of Oxygen , Steam , and Coal Particle Size," *Energy & Fuels*, vol. 26, pp. 524–529, 2012.

[2] M. Gorokhovski, E. I. Karpenko, F. C. Lockwood, V. E. Messerle, B. G. Trusov, and a. B. Ustimenko, "Plasma technologies for solid fuels: experiment and theory," *J. Energy Inst.*, vol. 78, no. 4, pp. 157–171, Dec. 2005.

Sustainable Solar Fuel Production by a Mixed Metal Oxide Based Thermochemical H_2O/CO_2 Splitting Cycle

(Principal Author)
Rahul Bhosale
Assistant Professor
Department of Chemical Engineering
College of Engineering, Qatar University, Doha, Qatar
rahul.bhosale@qu.edu.qa

(Other Authors)
Parag N. Sutar
Postdoctoral Fellow
Department of Chemical Engineering
College of Engineering, Qatar University, Doha, Qatar
parag.sutar@qu.edu.qa

Fares A. AlMomani
Associate Professor
Department of Chemical Engineering
College of Engineering, Qatar University, Doha, Qatar
falmomani@qu.edu.qa

Ivo Alxneit
Senior Scientist
Solar Technology Laboratory
Paul Scherrer Institute, Villigen – PSI, Switzerland
ivo.alxneit@psi.ch

Prepared for Presentation at
American Institute of Chemical Engineers
2016 Annual Meeting
San Francisco, California
November 13-18, 2016

Sustainable Solar Fuel Production by a Mixed Metal Oxide Based Thermochemical H_2O/CO_2 Splitting Cycle

Rahul Bhosale
Department of Chemical Engineering
College of Engineering
Qatar University, Doha, Qatar
rahul.bhosale@qu.edu.qa

Parag Sutar[1], Fares AlMomani[1], Ivo Alxneit[2]
[1]Department of Chemical Engineering
College of Engineering
Qatar University, Doha, Qatar

[2]Solar Technology Laboratory
Paul Scherrer Institute, Villigen – PSI, Switzerland

Keywords: Solar Fuel, Mixed Metal Oxide, Thermochemical Cycles, H_2O/CO_2 Splitting, Barium Ferrite

Abstract

This paper reports the synthesis of Ba-ferrite via sol-gel method for the production of solar fuel such as H_2 or syngas via thermochemical H_2O and/or CO_2 splitting cycle. Sol-gel synthesis of Ba-ferrite nanoparticles includes dissolution of metal precursors in ethanol, addition of propylene oxide as a proton scavenger to achieve gel formation, aging of the gel, gel drying, and subsequent calcination in presence of air. The effect of addition of propylene oxide (PO) as a proton scavenger on physico-chemical properties of sol-gel derived Ba-ferrite was investigated in detail. Various characterization techniques such as powder X-ray diffraction (XRD), BET surface area analyzer (BET), and scanning electron microscope (SEM) were utilized in this study.

1. Introduction

Metal oxide based thermochemical production of solar fuels such as solar H_2 or solar syngas (a mixture of H_2 and CO) via a two-step H_2O and CO_2 splitting cycle is considered as one of the most promising new technologies for the storage of solar energy in terms of chemical fuels [1-9]. In past, various metal oxide based thermochemical cycles have been investigated towards H_2O and CO_2 splitting reactions. In recent years, this research has been focused towards mixed ferrite materials.

Different ferrite materials have been investigated which includes: Ni-ferrite, Zn-ferrite, Co-ferrite, Mn-ferrite, Sn-ferrite, Ni-Zn-ferrite, Ni-Mn-ferrite, Ni-Sn-ferrite, and others [10-29]. Although these materials showed promising results in terms of solar fuel production, material stability (grain growth due to sintering) due to thermal cycling is an issue which poses limitation for the ferrite based thermochemical cycles to be a commercial process.

Therefore, in this paper we attempted to dope Ba in the crystal structure of iron oxide to synthesize Ba-ferrite. As an initial step only 10% doping was attempted. To achieve this sol-gel method was employed. Effect of addition of propylene oxide as proton scavenger and aging time on physico-chemical properties of Ba-ferrite and iron oxide mixture was studied. Various analytic techniques such as powder X-ray diffraction (XRD), BET surface area analyzer (BET), and scanning electron microscope (SEM) were utilized in this study.

2. Materials and Methods

Materials: The precursors and chemicals required for the synthesis of Ba-doped ferrite nanoparticles i.e. barium nitrate and iron nitrate, ethanol, and propylene oxide (gelation agent) were purchased from Alfa Aesar and Sigma Aldrich and used without any pre-treatment.

Synthesis of Ba-doped ferrite nanoparticles: The Ba-doped ferrite nanoparticles were derived via sol-gel method. At first, nitrate precursors of barium and iron were dissolved in ethanol (in respective weight ratios) by using a sonic bath. After the dissolution, different amounts of propylene oxide (as a proton scavenger) were added drop-wise and the gel formation was achieved. The Ba-doped ferrite gel was aged for multiple hours at atmospheric conditions. After aging, the gel was dried at 100°C for 2 h using a temperature–controlled hot plate. The dried powder obtained was further calcined at 600°C. The obtained calcined powder was stored in a dry atmosphere and then further characterized to determine the phase composition, crystallite size (CS) and specific surface area (SSA). The process flow diagram for the sol-gel synthesis of Ba-ferrite is shown in Fig. 1.

Characterization of sol-gel derived Ba-doped ferrite nanoparticles: The physico-chemical characterization of the sol-gel derived Ba-doped ferrite was carried out by using different characterization techniques. The phase composition and crystallite size (CS) was determined by using powder X-ray diffractometer. Crystallite size was calculated by using the Scherrer's equation. The specific surface area (SSA) of the sol-gel derived Ba-doped ferrite_was estimated using the BET surface area analyzer via the adsorption/desorption isotherms obtained after degassing the calcined powder at 200°C for 12 h.

3. Results and Discussion

The effect of addition of propylene oxide as a proton scavenger on the gel time, phase composition, CS, SSA and PV of the sol-gel derived Ba-ferrite was studied in detail. At first, the variation in the gel time with respect to the amounts of propylene oxide added was examined and the obtained results are presented in Fig. 1 (10 g of metal precursors + 20 ml of ethanol). As shown in Fig. 2, as the amounts of propylene oxide added increases the time required for gel formation decreases.

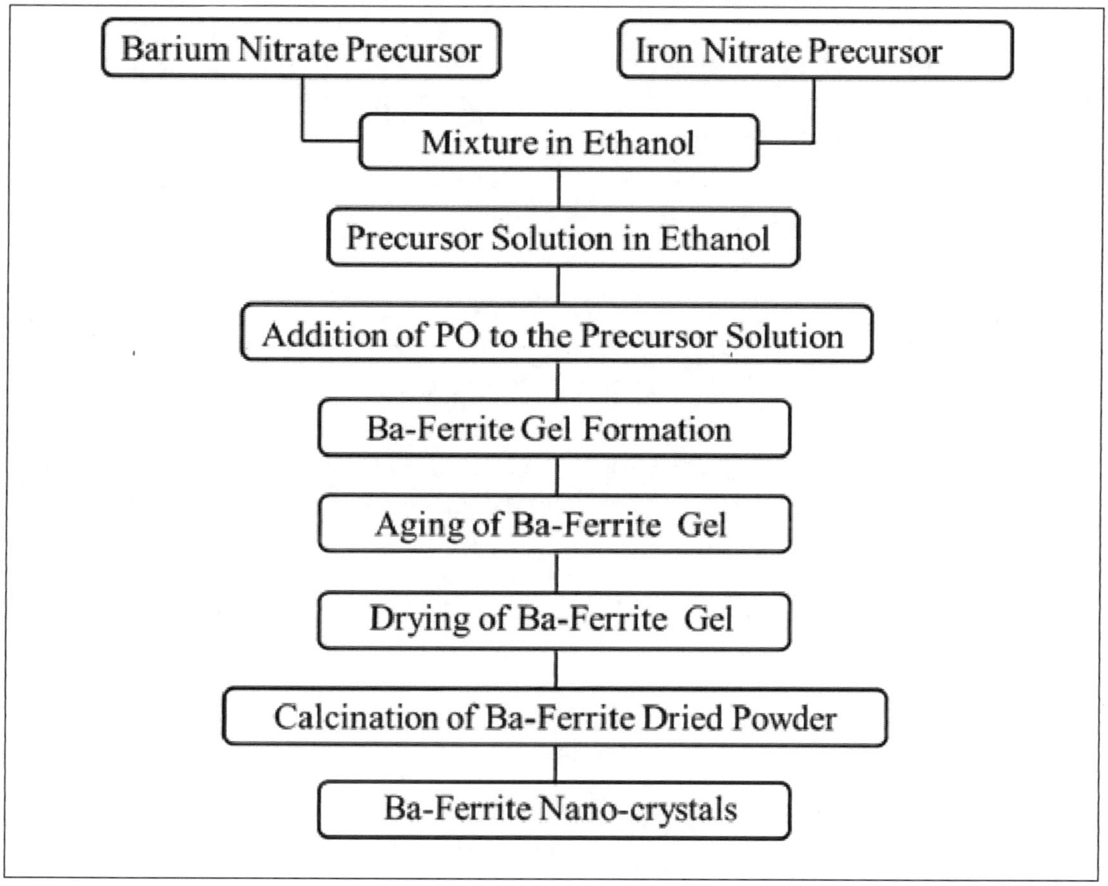

Fig. 1. Process flow diagram of sol-gel synthesis of Ba-ferrite.

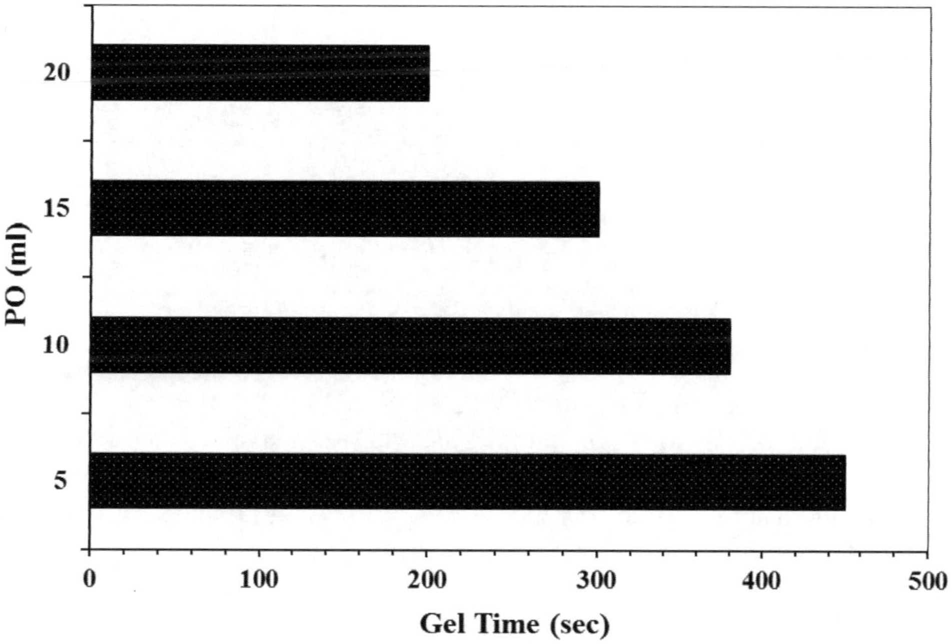

Fig. 2. Effect of PO volume used on gel time of Ba-ferrite.

By adding 5 ml of PO, the gel formation can be achieved in 630 sec. As the amount increase upto 10 ml, the gel time decreases to 565 sec. Furthermore increase in the amount of propylene oxide added induces a quick gel formation. For instance, when 15 and 20 ml of propylene oxide was added the gel was formed in 490 and 400 sec, respectively. Addition of propylene oxide as a proton scavenger helps in freezing the molecular chains and reducing the time required for gel formation.

In addition to the gel time, the phase composition and the CS of the Ba-ferrite gel prepared by adding different amount of propylene oxide and aged for 24 h was also determined. The aged gel was dried at 100°C and further calcined at 600°C in presence of air. The calcined powder was further characterized using a powder X-ray diffractometer. These XRD patterns indicate that due to the addition of different amounts of proton scavenger the phase composition of the Ba-ferrite does not altered. For instance, the XRD of the calcined Ba-ferrite powder is reported in Fig. 3.

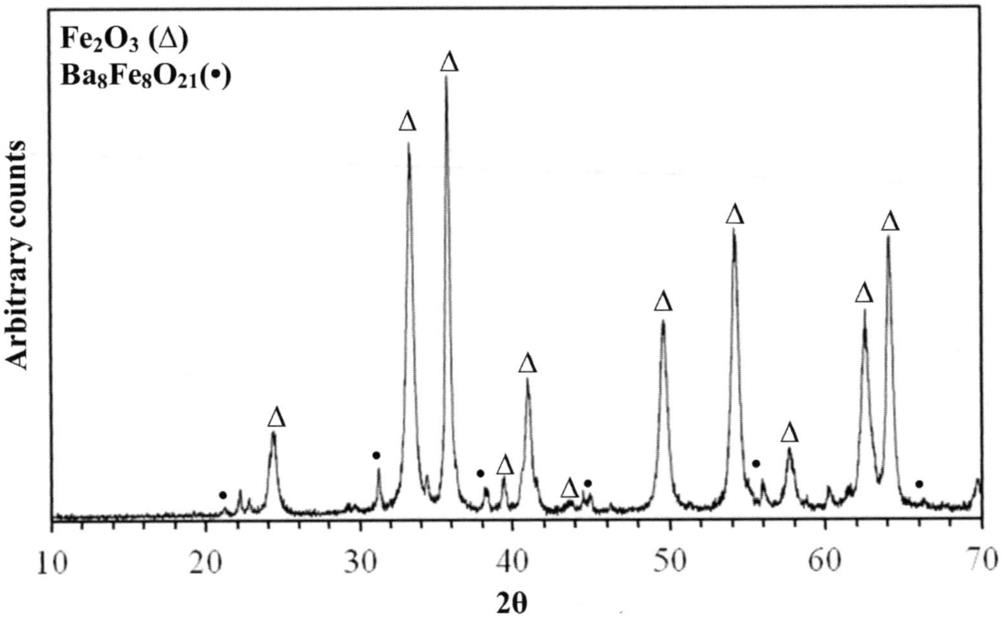

Fig. 3. Powder XRD pattern of Ba-ferrite (10% doping).

Furthermore, by utilizing results obtained via X-ray diffraction, the CS of the Ba-ferrite was calculated by using Scherrer's equation. Negligible variation in the CS of the Ba-ferrite as a function of propylene oxide added as a proton scavenger was observed. For instance, at PO = 5 ml the CS = 22 nm and at PO = 20 ml the CS = 21.45 nm. These results further indicate that the Ba-ferrite nanoparticles are formed.

After analysing the effect of amount of propylene oxide added as a proton scavenger on gel time, phase composition and C_S, the variation in the SSA and P_V was also explored. As shown in Fig. 4, the SSA of the sol-gel derived Ba-ferrite was observed to be increased with the increase in the amount of propylene oxide used. Initially, when 5 ml of propylene oxide was added, the

SSA of the derived Ba-ferrite was observed to be equal to 10 m^2/g. As the amount of propylene oxide added increases to 10, 15, and 20 ml, the SSA was also increased by a factor of 19, 27, and 35 m^2/g, respectively.

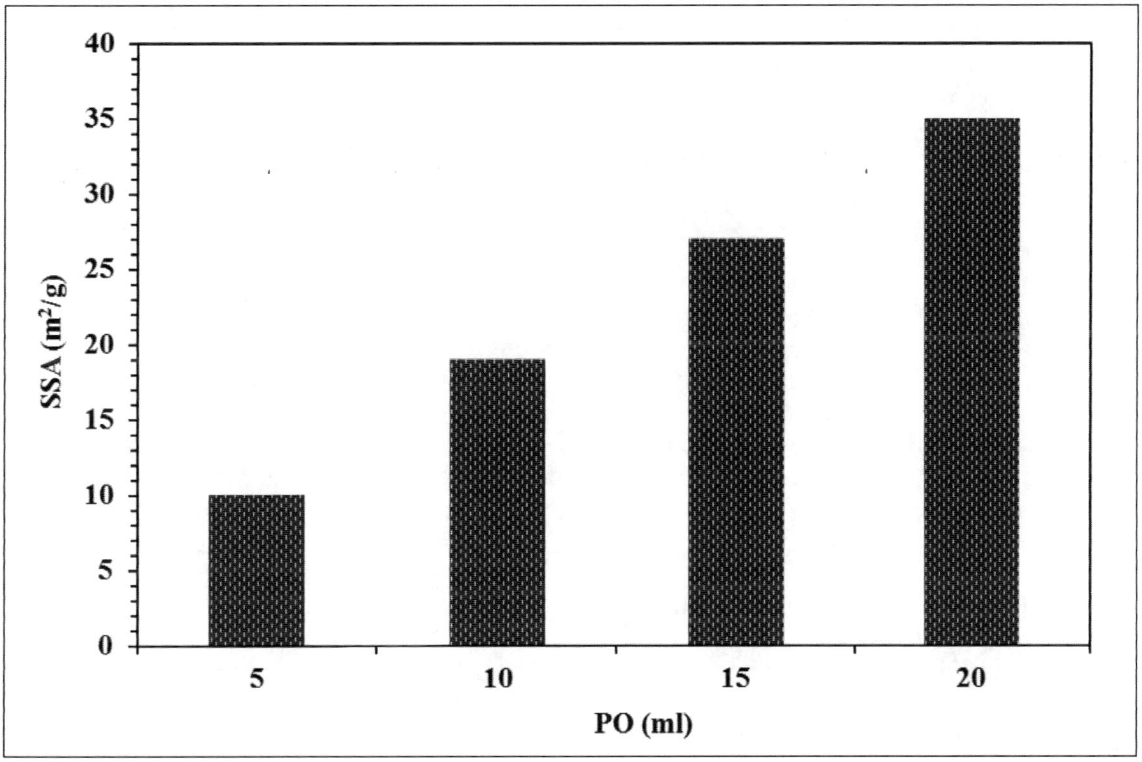

Fig. 4. Effect of PO volume used on SSA of Ba-ferrite.

4. Conclusion

The synthesis of Ba-ferrite (10% doping) was successfully carried out using sol-gel method. The effect of PO addition on physico-chemical properties of Ba-ferrite such as phase composition, gel time, CS, and SSA was studied. It was observed that with the increase in the PO concentration, the gel time decreases and SSA increases. In contrast to this, the phase composition and CS remained unchanged.

5. Acknowledgments

This publication was made possible by the NPRP grant (NPRP8-370-2-154) from the Qatar National Research Fund (a member of Qatar Foundation). The statements made herein are solely the responsibility of author(s).

References

[1] R. R. Bhosale. A. Kumar, F. AlMomani, I. Alxneit, [Sol-gel derived CeO_2–Fe_2O_3 nanoparticles: Synthesis, characterization and solar thermochemical application], *Ceram. Int*, **2016**, in press. doi:10.1016/j.ceramint.2016.01.042.

[2] J. Scheffe, A. Steinfeld, [Oxygen exchange materials for solar thermochemical splitting of H_2O and CO_2 : a review], *Mater. Today*, 2014, 17, 341 – 348.

[3] D. Dardor, R. R. Bhosale, S. Gharbia, A. AlNouss, A. Kumar, F. AlMomani, [Solar carbon production via thermochemical ZnO/Zn carbon dioxide splitting cycle], *J. Emerg. Trends Eng. Appl. Sci*, **2015**, 6, 129 – 135.

[4] D. Dardor, R. R. Bhosale, S. Gharbia, A. Kumar, F. AlMomani, [Solar thermochemical conversion of CO_2 into C via SnO_2/SnO redox cycle: Thermodynamic study], *Int. J. Eng. Res. Appl*, **2015**, 5, 134 – 140.

[5] R. R. Bhosale. A. Kumar, F. AlMomani, U. Ghosh, D. Dardor, Z. Bouabidi, M. Ali, S. Yousefi, A. AlNouss, M. S. Anis, M. H. Usmani, M. H. Ali, R. S. Azzam, A. Banu, [Solar co-production of samarium and syngas via methanothermal reduction of samarium sesquioxide], *Energy Convers. Manage*, **2016**, 112, 413 – 422.

[6] R. R. Bhosale. A. Kumar, F. AlMomani, S. Al-Muhtaseb, R. Gupta, I. Alxneit, [Assessment of $Ce_xZr_yHf_zO_2$ based oxides as potential solar thermochemical CO_2 splitting materials], *Ceram. Int*, **2016**, in press. doi:10.1016/j.ceramint.2016.02.100

[7] A. M. Turkey, [Catalytic oxidation of CO over pure and doped Cu-Ni-O System], *J.Applicable.Chem*, 2016, 5 (1), 219 – 230.

[8] V. K. Mahajan, G. H. Sonawane, [Improved photocatalytic activity of CeO_2 coupling ultrasound for Eosin-Y degradation], *J.Applicable.Chem*, 2015, 4 (5), 1500 – 1506.

[9] B. Suryanarayana, V. Raghavendra, K. ChandraMouli, [Structural, electrical and magnetic properties of $Ni_{0.5}Zn_{0.5}Fe_2O_4$ nanoscale particles by co-precipitation method], *J.Applicable.Chem*, 2015, 4 (4), 1237 – 1242.

[10] C. S. Thatte, M. V. Rathnam, M. S. Kumar, [Synthesis, characterization and application of Chitosan based Schiff base-transition metal complexes (Mn,Cu,Co,Ni)], *J.Applicable.Chem*, 2013, 2 (5), 1192 – 1206.

[11] B. R. Gaikwad, P. P. Khirade, D. V. Kurmude, A. B. Shinde, A. A. Pandit, K. M. Jadhav, [Comparative study of the structural and magnetic properties of magnesium ferrite prepared by ceramic and sol-gel auto combustion technique], *J.Applicable.Chem*, 2013, 2 (5), 1275 – 1280.

[12] M. R. Devi, C. Kannan, [Synthesis of unusual large pore Mg substituted alumino phosphate ($MgAlPO_4$) mesoporous molecular sieve and its catalytic activity], *J.Applicable.Chem*, 2013, 2 (3), 398 – 404.

[13] C. Agrafiotis, C. Pagkoura, A. Zygogianni, G. Karagiannakis, M. Kostoglou, A. G. Konstandopoulos, [Hydrogen production via solar-aided water splitting thermochemical cycles:

Combustion synthesis and preliminary evaluation of spinel redox-pair materials], *Int. J. Hydrogen Energy*, **2012**, 37, 8964 – 8980.

[14] N. Gokon, T. Kodama, N. Imaizumi, J, Umeda, T. Seo, [Ferrite/zirconia-coated foam device prepared by spin coating for solar demonstration of thermochemical water-splitting], *Int. J. Hydrogen Energy*, **2011**, 36, 2014 –2028.

[15] R. R. Bhosale, R. V. Shende, J. A. Puszynski, [H_2 generation from thermochemical water splitting using sol-gel derived Ni-ferrite], *J. Energy Power Eng,* **2010**, 4, 27 – 38.

[16] R. R. Bhosale, R. V. Shende, J. A. Puszynski, [H_2 generation from thermochemical water-splitting using sol-gel synthesized Zn/Sn/Mn-doped Ni-ferrite], *Int. Rev. Chem. Eng*, **2010**, 2, 852 – 862.

[17] F. Fresno, T. Yoshida, N. Gokon, R. Fernandez-Saavedra, T. Kodama, [Comparative study of the activity of nickel ferrites for solar hydrogen production by two-step thermochemical cycles], *Int. J. Hydrogen Energy*, **2010**, 35, 8503 – 8510.

[18] M. Roeb, N. Gathmann, M. Neises, C. Sattler, R. Pitz-Paal, [Thermodynamic analysis of two-step solar water splitting with mixed iron oxides], *Int. J. Energy Res*, **2009**, 33, 893 – 902.

[19] R. R. Bhosale, R. P. Khadka, J. A. Puszynski, R. V. Shende, [H_2 generation from two-step thermochemical water-splitting reaction using sol-gel derived $Sn_xFe_yO_z$], *J. Renewable Sustainable Energy*, 2011, 3, 063104-1 – 063104-12.

[20] R. R. Bhosale, R. V. Shende, J. A. Puszynski, [Thermochemical water-splitting for H_2 generation using sol-gel derived Mn-ferrite in a packed bed reactor], *Int. J. Hydrogen Energy*, **2012**, 37, 2924 – 2934.

[21] F. Fresno, R. Fernandez-Saavedra, M. Gomez-Mancebo, A. Vidal, M. Sanchez, M. Rucandio, A. Quejido, M. Romero, [Solar hydrogen production by two-step thermochemical cycles: Evaluation of the activity of commercial ferrites], *Int. J. Hydrogen Energy*, **2009**, 34, 2918 – 2924.

[22] M. Allendorf, R. Diver, N. Siegel, J. Miller, [Two-step water splitting using mixed-metal ferrites: thermodynamic analysis and characterization of synthesized materials], Energy Fuels, **2008**, 22, 4115 – 4124.

[23] R. R. Bhosale. A. Kumar, L.J.P. van den Broeke, S. Gharbia, D. Dardor, M. Jilani, J. Folady, M. Al-Fakih, M. Tarsad, [Solar hydrogen production via thermochemical iron oxide–iron sulfate water splitting cycle], *Int. J. Hydrogen Energy*, **2015**, 40, 1639 – 1650.

[24] T. Kodama, N. Gokon, R. Yamamoto, [Thermochemical two-step water splitting by ZrO_2-supported $Ni_xFe_{3-x}O_4$ for solar hydrogen production], *Sol. Energy*, **2008**, 82, 73 – 79.

[25] J. R. Scheffe, J. Li, A. W. Weimer, [A spinel ferrite/hercynite water-splitting redox cycle], *Int. J. Hydrogen Energy*, **2010**, 35, 3333 – 3340.

[26] J. E. Miller, M. Allendorf, R. Diver, L. Evans, N. Siegel, J. Stuecker, [Metal oxide composites and structures for ultra-high temperature solar thermochemical cycles], *J. Mater. Sci*, **2008**, 43, 4714 – 4728.

[27] T. Kodama, Y. Kondoh, R. Yamamoto, H. Andou, N. Satou, [Thermochemical two-step water splitting by ZrO_2-supported $Ni_xFe_{3-x}O_4$ for solar hydrogen production], *Sol. Energy*, **2008**, 78, 623 – 631.

[28] R. R. Bhosale. A. Kumar, F. AlMomani, I. Alxneit, [Propylene oxide assisted sol–gel synthesis of zinc ferrite nanoparticles for solar fuel production], *Ceram. Int*, **2016**, 42, 2431 – 2438.

[29] M. Neises, M. Roeb, M. Schmucker, C. Sattler, R. Pitz-Paal, [Kinetic investigations of the hydrogen production step of a thermochemical cycle using mixed iron oxides coated on ceramic substrates], *Int. J. Energy Res*, **2010**, 34, 651 – 661.

[30] S. Lorentzou, A. Zygogianni, K. Tousimi, C. Agrafiotis, A. G. Konstandopoulos, [Advanced synthesis of nanostructured materials for environmental applications], *J. Alloys Compd*, **2009**, 483, 302 – 305.

[31] R. R. Bhosale. A. Kumar, F. AlMomani, [Solar thermochemical hydrogen production via terbium oxide based redox reactions], *Int. J. Photoenergy*, **2016**, vol. 2016, Article ID 9727895, 9 pages, 2016. doi:10.1155/2016/9727895.

[32] S. Lorentzou, A. C. Agrafiotis, A. G. Konstandopoulos, [Aerosol spray pyrolysis synthesis of water-splitting ferrites for solar hydrogen production], *Granular Matter*, **2008**, 10, 113 – 122.

[33] H. Kaneko, T. Miura, A. Fuse, H. Ishihara, S. Taku, H. Fukuzumi, Y. Naganuma, Y. Tamaura, [Rotary-type solar reactor for solar hydrogen production with two-step water splitting process], *Energy Fuels*, **2007**, 21, 2287 – 2293.

Ti-base alloy coking behavior during steam cracking of ethane

Stamatis A. Sarris, Marie-Françoise Reyniers, Kevin M. Van Geem, Guy B. Marin
Ghent University, Laboratory for Chemical Technology, Technologiepark 914, 9052 Gent, Belgium.

Keywords: Ti-base alloy, carburization, steam cracking, pretreatment, cracking, thermal cracking, oxidation, ethane, coke formation, aging, jet stirred reactor, cyclic aging

[*] Corresponding author: Technologiepark 914, 9052 Gent, Belgium;
Kevin.VanGeem@Ugent.be

Introduction

One of the dominant processes to manufacture light olefins is steam cracking of hydrocarbons. The major problem occurring during this process is the coke deposition on the inner wall of the tubular crackers. The formed coke layer reduces the available reactant cross-sectional area causing elevated pressure drop along the reactor. These promote bimolecular over monomolecular reactions, decreasing the olefin selectivity. As a result, more energy is required to maintain the cracking severity and the desired selectivity[1, 2]. At a certain point, either when the metallurgic limits are reached or the pressure drop over the reactor is excessive, the operation is ceased and decoking of the reactor is necessary; typically for 48 h. The latter affects negatively the process finance[3]. In an effort to mitigate coke formation several anti-coking technologies have been evolved, namely 3-D reactor technologies[4-6], application of additives[7-10] and use of advanced surface technologies[11-17].

The present work, testing the cracking behavior of a Ti-base alloy, belongs clearly in the last technologies category. Titanium is used most of the times to enrich the composition of a high temperature alloy in concentrations of 0 to 10 wt %[18] aiming at an increased rigidity. However, to the best of the knowledge of the authors, it has not been used as base material for crackers, therefore no quantitative coking behavior observations are available. In order to evaluate the coking performance of this alloy under ethane steam cracking conditions, an experimental study was performed in a jet-stirred reactor thermogravimetric set-up. As a first step, the optimum pretreatment of Fe-Ni-Cr alloys was applied and compared with a pretreatment increased by 100 K temperature. This, based on the available literature[19-22], aimed at a better oxidation of the surface, thus at an improved coking behavior of the material (see Figure 1). According to that a factor 8 increase in the absolute weight of oxides is expected by increasing the temperature from 1173 K to 1273 K.

The target was partially achieved, the results proved decreased coking rates of the Ti-base alloy after the new treatment in comparison with the standard treatment used for Fe-Ni-Cr alloys, at the expense of the pronounced formation of Carbon oxides. However, the coking performance was still not better than the industrially representative alloy, while the tested Ti-base alloy coupons were partially or fully broken in pieces or slices after application of cyclic aging and cooling down back to ambient conditions. Therefore, the tested Ti-base alloy was not characterized as industrially interesting.

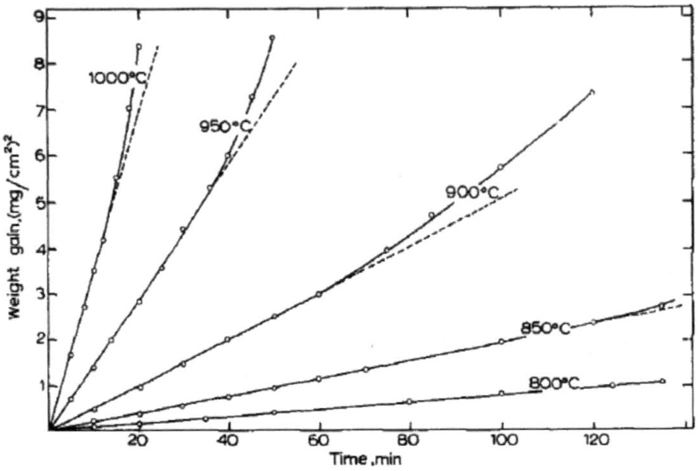

Figure 1. Effect of temperature on the oxides gain over time

Experimental Set-up

A picture of the jet-stirred reactor (JSR) is illustrated in Figure 2. The reactor is entirely made of quartz, including the nozzles and walls, minimizing coke formation on the inner surface. A small flat coupon is placed in the JSR, suspended from the arm of an electro-balance. Typically, the samples are cut by electro-erosion from the internal surface of industrial tubes to dimensions 10 mm x 8 mm x 1 mm. The coupons are polished before assembly to a Ra roughness of 2.6 μm.

The electro-balance can measure the weight difference of the coupons on-line with an accuracy of 1 μg s^{-1}, giving the advantage of actual coking rate values over time. The reactor effluent is quenched to prevent further cracking and coking and its composition is measured online with the aid of two gas chromatographs (GC). Nitrogen is used as internal standard, while the detection limit ranges from methane to naphthalene, satisfactory for Ethane cracking[23].

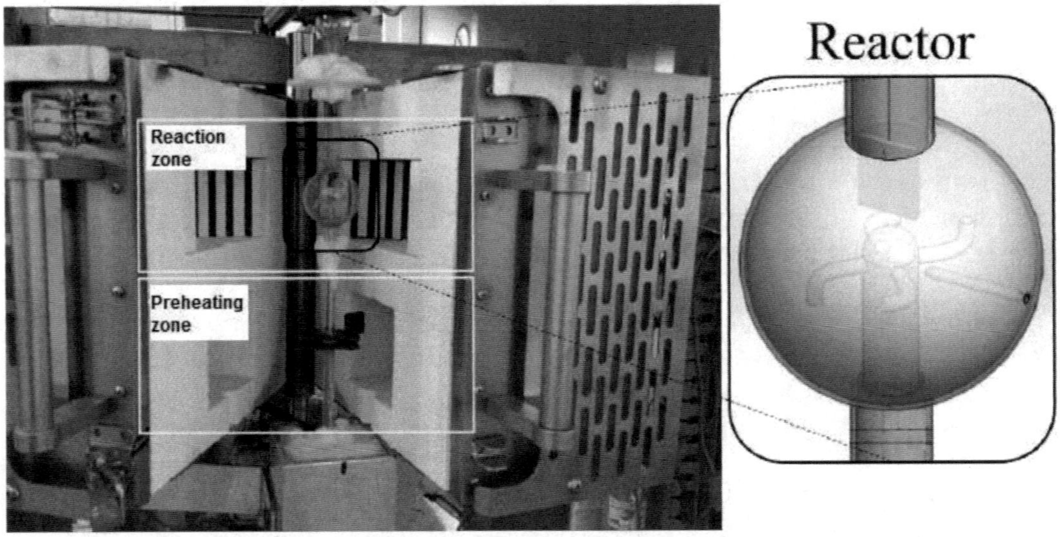

Figure 2. Jet stirred reactor used for coking behavior evaluation during steam cracking

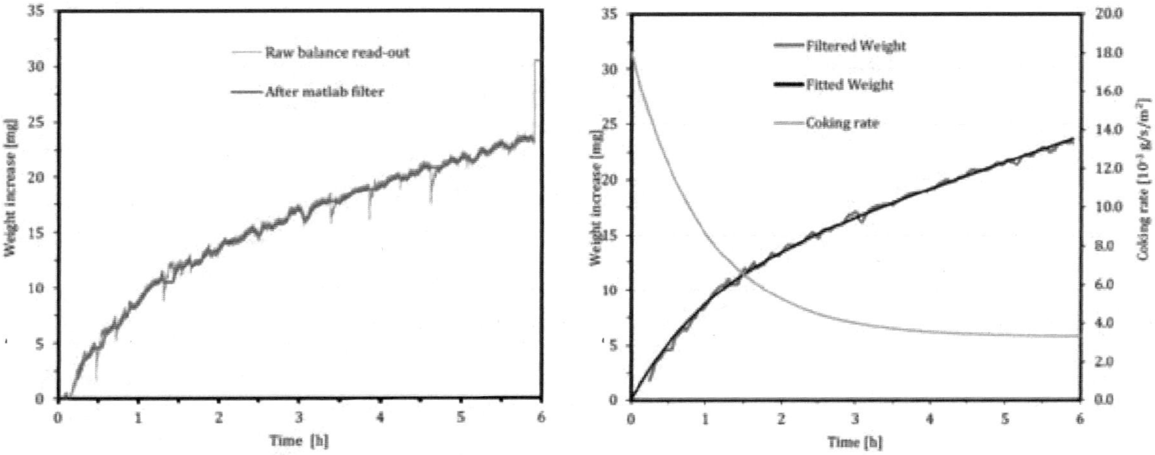

Figure 3. JSR data processing.

A graphical visualization of the data processing as executed by the MATLAB program is shown in Figure 3. In this work, the mean measured value between the initial 15 and 60 minutes is determined as initial coking rate characterizing the catalytic coking mechanism. The asymptotic or pyrolytic coking rate is related to the radical coking mechanism and is reported as the mean measured coking rate between the 5th and 6th hour of cracking.

The samples extracted after the experiment, are ex-situ analyzed by means of SEM and EDX. Similarly to Figure 4, surface and cross sectional analyses are performed aiming at a better understanding of the phenomena occurring during and before cracking.

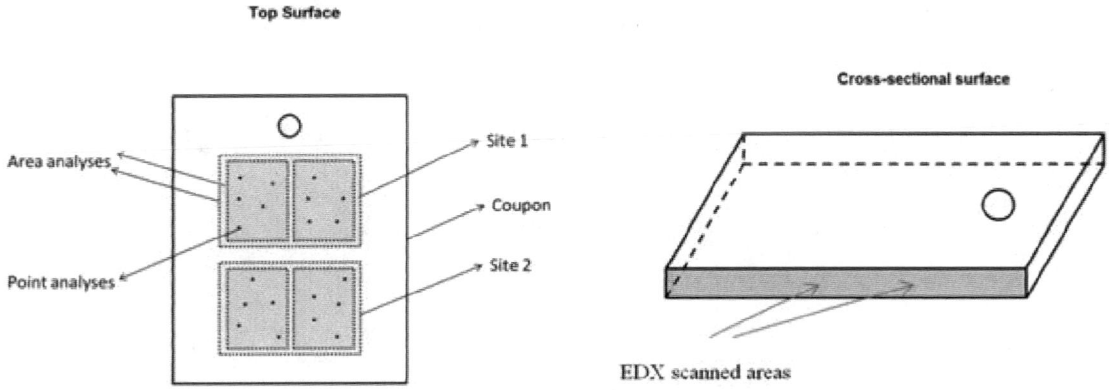

Figure 4. SEM and EDX analysis performed for the JSR samples

Experimental Results

The coking performance of the new Ti-base alloy under ethane steam cracking conditions was evaluated in the JSR set-up of LCT. Initially, the effect of the in-situ pretreatment was evaluated under blank ethane steam cracking conditions.

As mentioned above, a higher temperature pre-oxidation - 1273 K instead of 1173 K - than the typically applied for Fe-Ni-Cr alloys can increase significantly the absolute weight of oxides formed on Titanium[20, 21]. Based on previous work the formation of a uniform and thick oxide layer can improve the anti-coking behavior of a material[11, 12]. However, to the best of the knowledge of the authors, no coking rates results are available for Ti-base alloys.

Table 1 and

Table 2 summarize the experimental conditions and sequences under the two different pretreatments.

Table 1. Conditions of the two different applied pretreatments

Pretreatment	Id	Steps 1	Steps 2	Steps 3
Standard pretreatment	*A*	$6.7 \cdot 10^{-3}$ Nl s^{-1} of Air; 14h; T=1023 K	1:1 $8.15 \cdot 10^{-3}$ Nl s^{-1} of Air:N_2; 30 min; T=from 1023 K to 1173 K	$8.15 \cdot 10^{-3}$ Nl s-1 of Air and $6.67 \cdot 10^{-6}$ kg s^{-1} Steam; 15min ; T=1173 K
High Temperature Preoxidation + Steam Treatment at higher temperature	*B*	$6.7 \cdot 10^{-3}$ Nl s^{-1} of Air; 14h; T=1273 K	1:1 $8.15 \cdot 10^{-3}$ Nl s^{-1} of Air:N_2; 30 min; T=from 1023 K to 1273 K	$8.15 \cdot 10^{-3}$ Nl s^{-1} of Air and $6.67 \cdot 10^{-6}$ kg s^{-1} Steam; 15min ; T=1273 K

Table 2. Experimental sequence during the two performed pretreatments.

	1023 K Air	1023->1173 K N_2+Air	1173 K Air+H_2O	1173 K C_2H_6 +H_2O	1023->1173 K N_2+Air	1173 K Air +H_2O	1173 K C_2H_6 +H_2O	1023->1173 K N_2+Air	1173 K Air +H_2O	1173 -> 273 K He	
A	In situ Pre-oxidation 14 h	mild Pre-oxidation 30 min	Steam Treatment 15 min	1st coking cycle of 6 h	Decoking 30min	Steam Treatment 15 min	2nd-8th coking cycle of [2nd, 3rd and 8th of 6 h, 4-7th of 1 h]	Decoking 30 min	Steam Treatment 15 min	cooling down	time
	1273 K Air	1023->1273 K N_2+Air	1273 K Air+H_2O	1173 °C C_2H_6 +H_2O	1023->1273 K N_2+Air	1273 K Air +H_2O	1173 K C_2H_6 +H_2O	1023->1273 K N_2+Air	1273 K Air +H_2O	1273 -> 273 K He	
B	In situ Pre-oxidation 14 h	High T Pre-oxidation 30 min	Steam Treatment 15 min	1st coking cycle of 6 h	Decoking at higher T 30 min	Steam Treatment 15 min	2nd-8th coking cycle of [2nd, 3rd and 8th of 6 h, 4-7th of 1 h]	Decoking at higher T 30 min	Steam Treatment 15 min	cooling down	time

Table 3 illustrates the most important mass yields during ethane cracking. For a better comparison the effluent results of a typically used material are also included in the presented yields. A factor 10 and 300 increased values of CO, for the typical and higher temperature pretreatment respectively, were measured over the 8 cracking cycles in comparison with the Fe-Ni-Cr alloy. For CO_2, the effect was less pronounced; only after application of the high temperature pre-oxidation (b) an increase of a factor 3 was observed. In addition, the increase of Carbon oxides is followed by an increase of H_2. A minor decrease of CH_4 is also noticed, however it is still within the error margin of measured values. The observations support the idea that Ti and/or Ti-oxides present on the surface have a catalytic effect towards coke gasification. These results already prove that the Ti-base alloy is significantly worse than the reference material.

Table 3. Averaged (over the 8 cracking cycles) mass product yields. Ethane steam cracking: F_{HC}= 29.18 x 10^{-6} kg s^{-1}, δ= 0.33 kg_{H2O} kg^{-1}_{HC}, $T_{reactor}$= 1173.15 K, P= 101.35 kPa, F_{H2O}=9.72 x 10^{-6} kg s^{-1}.

Components [wt%]	Fe-Ni-Cr alloy A	Ti-base alloy A	Ti-base alloy B
H_2 [±0.07]	4.24	4.39	4.45
CO_2 [±0.003]	0.006	0.005	0.019
CO	0.05 ^{1}	0.56 ^{2}	1.87 ^{3}
CH_4 [±0.21]	6.99	6.97	6.91
C_2H_6 [±0.52]	30.13	30.19	30.06
C_2H_4 [±0.25]	49.86	49.61	49.57
C_3H_8 [±0.03]	0.11	0.11	0.11
C_3H_6 [±0.02]	0.74	0.70	0.69
C_2H_2 [±0.05]	1.38	1.46	1.41
1,3-C_4H_6 [±0.07]	2.00	1.93	1.89
Benzene [±0.13]	2.46	2.53	2.53

^{1} [±0.008], ^{2} [±0.07], ^{3} [±0.35]

Nevertheless, coking rates results are also processed and summarized in Table 4. For a qualitative comparison the results of reference material during continuous addition of DMDS are presented as well. Depending on the cracking cycle the Ti-base alloy performs significantly worse than the reference material. The increased temperature pretreatment improves the coking rates of the Ti-base alloy. However, this improved coking behavior still keeps the Ti-base alloy out of competition against the blank reference coking behavior. In terms of anti-coking performance, the Ti-base alloy is placed between the reference during steam cracking with continuous addition of DMDS and the reference with no DMDS.

Additionally, the reference material maintains a rather stable coking behavior over cyclic aging, while the sample of the Ti-base alloy shows increased coking after several cycles.

Table 4. Calculated initial (average 15min-60min) and asymptotic (average 5h-6h) coking rates. Ethane steam cracking: F_{HC}= 29.18 x 10^{-6} kg s^{-1}, δ= 0.33 kg_{H2O} $kg^{-1}{}_{HC}$, $T_{reactor}$= 1173.15 K, P= 101.35 kPa, F_{H2O}=9.72 x 10^{-6} kg s^{-1}.

Alloy Conditions	Fe-Ni-Cr alloy Blank	Ti-base alloy Blank	Ti-base alloy Blank	Fe-Ni-Cr alloy CA
Pretreatment	A	A	B	A
Cracking Temperature (K)	1173	1173	1173	1173
dilution	0.33	0.33	0.33	0.33
CA DMDS (ppmw S per HC)	0	0	0	41
Cc	Coke formed (mg)			
1	1.29	5.48	3.66	7.55
2	1.26	9.01	4.99	8.38
3	1.41	7.43	4.82	8.66
4	0.60	3.57	2.93	4.12
5	0.67	3.03	2.17	4.55
6	0.69	2.97	2.13	4.53
7	0.47	2.95	2.32	4.51
8	1.19	7.69	5.35	12.46
Cc	Initial coking rate [10^{-6} kg/s/m^2]			
1	0.71	1.73	1.23	3.03
2	0.67	3.26	1.81	4.74
3	0.86	3.65	2.59	4.23
4	0.85	3.80	3.11	5.85
5	0.96	3.91	2.76	6.45
6	0.97	3.80	2.70	6.42
7	0.67	3.79	2.79	6.39
8	0.63	3.27	2.46	5.29
Cc	Asymptotic coking rate [10^{-6} kg/s/m^2]			
1	0.22	1.06	0.61	1.53
2	0.22	1.58	0.83	1.43
3	0.23	0.93	0.53	1.61
8	0.21	1.25	0.74	2.47

The most discouraging result obtained by the Ti-base Alloy coupons was their behavior after cooling down to ambient conditions. Most specifically, the coupons broke either in slices or in smaller pieces. In Figure 5, an indication of the thickness decrease after cracking and of the thickness of the slices is given. Clearly, thick oxides are formed during oxidation of Ti, making the coupon more brittle as material towards mechanical stresses. The latter leads to the formation of cracks due to the mechanical stresses evolved during cooling down. An additional example of a cracked coupon is given in Figure 6. For a better understanding of this phenomenon SEM and EDX analysis are performed.

The cross sectional analyses of the two samples treated at the two different temperatures reveal part of the phenomenon occurring in this study (Figure 7). In the standard temperature, 1173 K, 0.1 to 0.2 μm thick oxides are formed and spalling off the coupon from both sides leaving no oxide covering the surface. At 1273 K, thick oxides are formed leading to oxidation of almost the whole coupon volume. Clearly the oxides that are formed are quite brittle. Thus, they are separating from the core of the coupon. In the case of the increased temperature oxidation, the core is too thin, therefore the crack is propagating causing breakage of the coupon in smaller pieces.

Combining the above experimental results with the formation of cracks on the samples' surface can give a fruitful insight on the phenomenon. Based on the increased values of carbon oxides and H_2 differences of the yields, an increased catalytic surface of the Ti-base alloy samples can be assumed. That most probably derived from the presence of the surface cracks. As a result both coke formation and coke gasification were catalyzed.

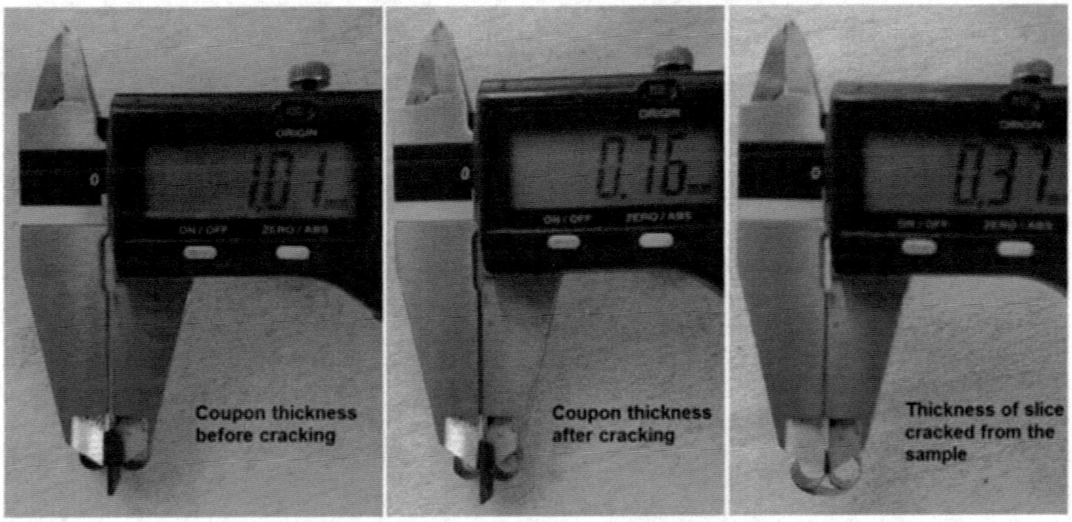

Figure 5. Thickness values of the Ti-base alloy before cracking (left), after cracking (middle) and of a broken slice (right), pretreated at 1173 K.

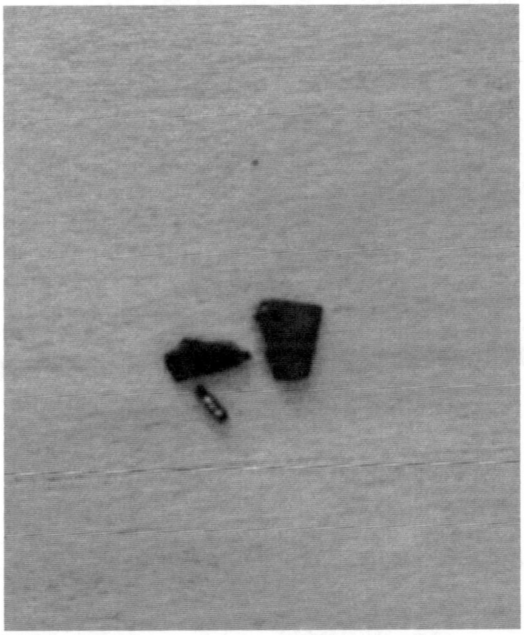

Figure 6. Broken Ti-base alloy coupon after cracking pretreated at 1273 K.

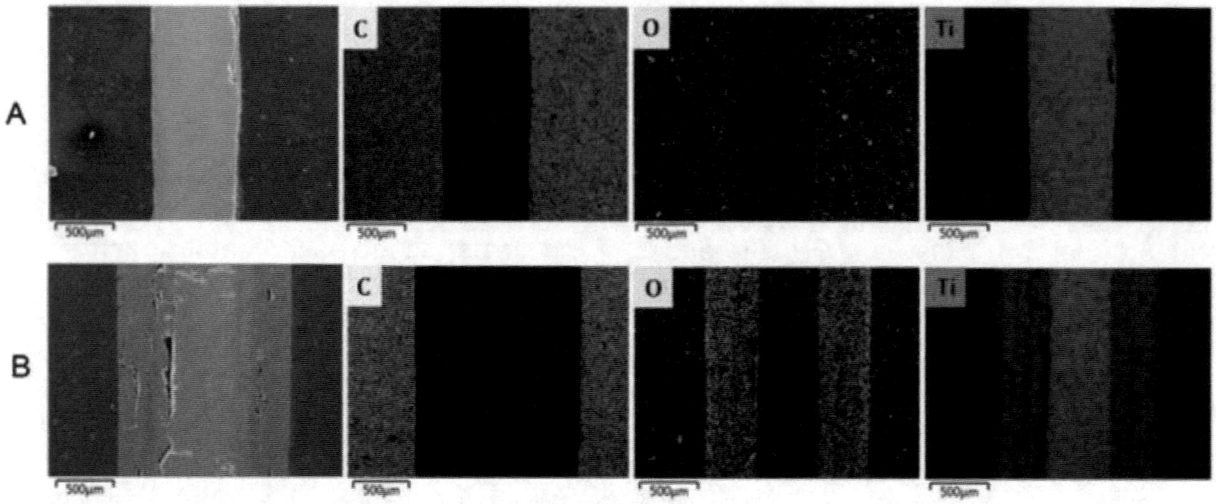

Figure 7. SEM and EDX cross sectional analysis of the two pretreatments. A is referring to temperature of 1173 K and B to 1273 K.

Conclusions

A newly tested Ti-base alloy was evaluated under industrially relevant ethane steam cracking conditions. Outrageously increased values of CO, both for the standard and higher temperature pretreatment, were measured over the 8 cracking cycles in comparison with a typical Fe-Ni-Cr alloy under the same experimental conditions. Notable increase of the CO_2 was observed between the higher temperature pre-oxidation (b) and the typical pre-oxidation used for Fe-Ni-Cr (a) when applied to the Ti-base alloy. No significant difference was observed for the CO_2 values of Ti-base alloy in comparison with the reference material in the standard treatment.

The increase of Carbon oxides is followed by a slight increase of H_2, supporting the idea that Ti and Ti-oxides are catalytic towards coke gasification. The significant differences in terms of gas phase composition can be explained by the increased available surface interacting with the gas phase. Clearly at the points of crack formation coke formation and gasification was catalyzed, affecting the gas phase product distribution.

Additional surface analysis will lead to a better understanding of the crack formation phenomenon observed in these alloys. Judging by these experimental results, the use of the Ti-base alloy in an industrial cracker is considered highly risky and certainly not beneficial in terms of coking or carbon oxides formation.

References

1. Rao, M. R.; Plehiers, P. M.; Froment, G. F., The coupled simulation of heat transfer and reaction in a pyrolysis furnace. *Chemical engineering science* **1988,** 43, (6), 1223-1229.
2. Stefanidis, G.; Merci, B.; Heynderickx, G. J.; Marin, G. B., CFD simulations of steam cracking furnaces using detailed combustion mechanisms. *Computers & chemical engineering* **2006,** 30, (4), 635-649.
3. Heynderickx, G.; Schools, E.; Marin, G., Optimization of the decoking procedure of an ethane cracker with a steam/air mixture. *Industrial & engineering chemistry research* **2006,** 45, (22), 7520-7529.
4. Brown, D. J., Internally finned radiant coils: A valuable tool for improving ethylene plant economics. In *6th EMEA Petrochemicals Technology Conference*, London, UK, 2004.
5. Albano, J.; Sundaram, K.; Maddock, M. *Application of extended surfaces in pyrolysis coils*; New York, NY; American Institute of Chemical Engineers: 1988.

6. Schietekat, C. M.; van Goethem, M. W. M.; Van Geem, K. M.; Marin, G. B., Swirl flow tube reactor technology: An experimental and computational fluid dynamics study. *Chemical Engineering Journal* **2014,** 238, (0), 56-65.
7. Reyniers, M.-F. S. G.; Froment, G. F., Influence of Metal Surface and Sulfur Addition on Coke Deposition in the Thermal Cracking of Hydrocarbons. *Industrial & Engineering Chemistry Research* **1995,** 34, (3), 773-785.
8. Wang, J.; Reyniers, M.-F.; Marin, G. B., Influence of Dimethyl Disulfide on Coke Formation during Steam Cracking of Hydrocarbons. *Industrial & Engineering Chemistry Research* **2007,** 46, (12), 4134-4148.
9. Wang, J.; Reyniers, M.-F.; Marin, G. B., The influence of phosphorus containing compounds on steam cracking of n-hexane. *Journal of Analytical and Applied Pyrolysis* **2006,** 77, (2), 133-148.
10. Wang, J.; Reyniers, M.-F.; Van Geem, K. M.; Marin, G. B., Influence of Silicon and Silicon/Sulfur-Containing Additives on Coke Formation during Steam Cracking of Hydrocarbons. *Industrial & Engineering Chemistry Research* **2008,** 47, (5), 1468-1482.
11. Muñoz Gandarillas, A. E.; Van Geem, K. M.; Reyniers, M.-F.; Marin, G. B., Influence of the Reactor Material Composition on Coke Formation during Ethane Steam Cracking. *Industrial & Engineering Chemistry Research* **2014,** 53, (15), 6358-6371.
12. Muñoz Gandarillas, A. E.; Van Geem, K. M.; Reyniers, M.-F.; Marin, G. B., Coking Resistance of Specialized Coil Materials during Steam Cracking of Sulfur-Free Naphtha. *Industrial & Engineering Chemistry Research* **2014,** 53, (35), 13644-13655.
13. Verdier, G.; Carpentier, F., Consider new materials for ethylene furnace applications: An innovative metallurgy solves maintenance issues. *Hydrocarbon processing* **2011,** 90, (5), 61-62.
14. Wolpert, P.; Ganser, B.; Jakobi, D.; Kirchheiner, R., Using steam; tube with uniform cross-section; pyrolysis; antideposit agents; interior fins; swirling gas flow. In Google Patents: 2005.
15. Jakobi, D.; van Moesdijk, C.; Karduck, P.; von Richthofen, A., Tailor-made materials for high temperature applications: New strategies for radiant coil material development. *CORROSION 2009* **2009**.
16. Jakobi, D.; Karduck, P.; von Richthofen, A. F., The High-Temperature Corrosion Resistance of Spun-Cast Materials for Steam-Cracker Furnaces-A Comparative Study of Alumina-and Chromia-Forming Alloys. *CORROSION 2013* **2013**.
17. Asteman, H.; Hartnagel, W.; Jakobi, D., The Influence of Al Content on the High Temperature Oxidation Properties of State-of-the-Art Cast Ni-base Alloys. *Oxidation of Metals* **2013,** 80, (1-2), 3-12.
18. Petrone, S. S. A.; Mandyam, R. C.; Wysiekierski, A. G., Surface alloyed high temperature alloys. In Google Patents: 2000.
19. Kumar, S.; Sankara Narayanan, T. S. N.; Ganesh Sundara Raman, S.; Seshadri, S. K., Thermal oxidation of Ti6Al4V alloy: Microstructural and electrochemical characterization. *Materials Chemistry and Physics* **2010,** 119, (1–2), 337-346.
20. Kofstad, P.; Anderson, P. B.; Krudtaa, O. J., Oxidation of titanium in the temperature range 800–1200°C. *Journal of the Less Common Metals* **1961,** 3, (2), 89-97.
21. Kofstad, P., High-temperature oxidation of titanium. *Journal of the Less Common Metals* **1967,** 12, (6), 449-464.
22. Kumar, S.; Narayanan, T. S. N. S.; Raman, S. G. S.; Seshadri, S. K., Thermal oxidation of CP-Ti: Evaluation of characteristics and corrosion resistance as a function of treatment time. *Materials Science and Engineering: C* **2009,** 29, (6), 1942-1949.
23. van Goethem, M. W. M.; Barendregt, S.; Grievink, J.; Verheijen, P. J. T.; Dente, M.; Ranzi, E., A kinetic modelling study of ethane cracking for optimal ethylene yield. *Chemical Engineering Research and Design* **2013,** 91, (6), 1106-1110.

469612 Fermentation Design and Gas Transfer Considerations for Biochemical Methane Conversion

Abstract:

Fermentation Design and Gas Transfer Considerations for Biochemical Methane Conversion

Kyle Stone, Matthew Hilliard, Q. Peter He, and Jin Wang

Department of Chemical Engineering, Auburn University, Auburn, AL, 36849, USA

Methane is an essential component of the global carbon cycle and one of the most powerful greenhouse gases (GHGs) with a global warming potential over 20 times that of CO_2. Yet, methane is a rich source of carbon and energy that is utilized for heat, fuel synthesis, and valued chemical production. In addition, there is a rapid expansion of global methane supply predominantly because of increased production of natural gas [1] and biogas [2]. Due to concerns about climate change, abundance of supply, and favorable economic conditions, there has been growing interest in biological processes as a sustainable and efficient way of converting methane to valuable products while reducing GHG emissions.

These biological processes utilize the largely aerobic bacteria called methanotrophs that consume methane as their sole source for carbon and energy. Due to their unique metabolism, methanotrophs can convert methane at ambient temperatures and pressures to many different products. Specifically, methanotrophs have the natural ability to produce methanol, valued proteins, biopolymers, lipids (for biodiesel), organic acids, sucrose, and ectoine [3,4]. With genetic modification, methanotrophs can also produce non-native products such as carotenoids, isoprene, 1,4 butanediol, and farnesene [3,4]. Several of these products can potentially be simultaneously generated with specifically designed processes [4].

However, all processes with methanotrophs face several challenges, such as clean sources of methane, biocatalyst modification, and process design. For process design, the most influential factor is efficient gas-liquid transfer (i.e., methane and/or oxygen gas transfer to the liquid medium where cells reside). The gas-liquid transfer of the poorly soluble methane and oxygen can be strongly affected by reactor design and mass transfer enhancing agents in the liquid medium.

In this work we review recent progress about effective gas transfer for bioprocesses with methanotrophs. One focus will be on the quantification of methane transfer rate and cellular uptake rate. We will also provide a detailed review and discussion of the bioreactor configurations and the influence of these systems on gas-liquid transfer. Finally, we will discuss gas-liquid transfer enhancement with promoting agents. Recommendations will be provided and potential research areas for methane conversion through methanotrophs will be discussed.

[1] International Energy Statistics: Natural gas proved reserves, (2015). http://www.eia.gov/cfapps/ipdbproject/IEDIndex3.cfm?tid=3&pid=3&aid=6.

[2] Biogas Opportunities Roadmap, U.S. Department of Agriculture, U.S. Environmental Protection Agency, U.S. Department of Energy, 2014.

[3] P.J. Strong, S. Xie, W.P. Clarke, Methane as a resource: can the methanotrophs add value?, Environ. Sci. & Technol. (2015).

[4] P.J. Strong, M. Kalyuzhnaya, J. Silverman, W.P. Clarke, A methanotroph-based biorefinery: potential scenarios for generating multiple products from a single fermentation, Bioresour. Technol. (2016).

461790 Integrating Data Sources for Process Sustainability Assessments

Abstract:

To perform a chemical process sustainability assessment requires significant data about chemicals, process design specifications, and operating conditions. The required information includes the identity of the chemicals used, the quantities of the chemicals within the context of the sustainability assessment, physical properties of these chemicals, equipment inventory, as well as health, environmental, and safety properties of the chemicals. Much of this data are currently available to the process engineer either from the process design in the chemical process simulation software or online through chemical property and environmental, health, and safety databases. Examples of these databases include the U.S. Environmental Protection Agency's (USEPA's) Aggregated Computational Toxicology Resource (ACToR), National Institute for Occupational Safety and Health's (NIOSH's) Hazardous Substance Database (HSDB), and National Institute of Standards and Technology's (NIST's) Chemistry Webbook. This presentation will provide methods and procedures for extracting chemical identity and flow information from process design tools (such as chemical process simulators) and chemical property information from the online databases. The presentation will also demonstrate acquisition and compilation of the data for use in the EPA's GREENSCOPE process sustainability analysis tool.

Disclaimer: The views expressed in this presentation are those of the authors and do not necessarily reflect the views or policies of the U.S. Environmental Protection Agency.

457321 How Big Data Can Help Local Community Leaders Optimize Food Security

Abstract:

In 2013, a partnership was formed between the Indy Hunger Network and Dow AgroSciences Hunger Solutions Network. Both Networks were formed to have a measurable impact on hunger in underserved local communities surrounding Indianapolis, IN, USA. In Indianapolis as well as across the US, one in six people are food insecure. One of the common goals for this partnership was to determine the optimal locations for services to meet the real needs of clients of various food-providing community organizations. The IHN identified the needs and requirements, while the HSN brought project management, big data acquisition and organization, and data analytics skills to the table. Since the partnership was formed, several organizations have benefitted from this work.

First, the Gleaners Food Bank in Indianapolis asked the team to help place their next large food pantry. As a result of this work, a large pantry was up and running in 2015 in an underserved area of the city. Next, a group of community organizations needed help in understanding gaps in where senior meals were being served (congregate meal sites). Thirdly, the Women, Infants, and Children Program in the County requested help in locating a new health clinic. Based on the team's modeling, Marion County officials changed the planned location of the next planned clinic to the team's top recommendation.

Finally in 2016, the large Summer Servings program for children, sponsored by the USDA and administered by IHN, was worried that holes existed in their coverage so that many food insecure children were missing the opportunity for a free meal during the summer. The team identified several areas where gaps should be filled by new sites. These big data analytics projects would not have been successful if there had not been the link to the local community to identify needs and identify the client base. Next steps are to find ways to leverage this methodology to other communities and obtain a larger–scale impact.

464708 A Nanomaterial Synthesis Android App: Towards Cleaner Production of Engineered Nanomaterials

Abstract:

Abstract

There are a multitude of synthesis methods for producing engineered nanomaterials (ENMs), some of which can be unreliable and inconsistent. Despite the development of experimental standard protocols and reference materials, inter-laboratory comparisons are rarely conducted for ENMs synthesis and a repository of synthesis information is lacking. Towards more robust nanoinformatics approaches, we have gathered information on available ENMs synthesis methods, developed a platform to search and display them (*i.e.,* an Android app) and will present efforts towards developing schema, knowledge extraction, and ultimately computer aided ENMs synthesis. A particular focus has been placed on enriching ENMs structure activity relationships and applications to energetic materials. Another goal of this effort is to link chemistry and chemical engineering in this emerging and interdisciplinary field.

References

Lewinski, Nastassja A., and Bridget T. McInnes. "Using natural language processing techniques to inform research on nanotechnology." *Beilstein Journal of Nanotechnology,* (2015), 6,1: 1439-1449.

Jensen, Cory D. ChE App Help. AIChE 2015, Salt Lake City, UT, November 8th 2015.

Cinelli, Marco, Stuart R. Coles, Omowunmi Sadik, Barbara Karn, and Kerry Kirwan. "A framework of criteria for the sustainability assessment of nanoproducts." *Journal of Cleaner Production,* (2016).

Thomas, Dennis G., Rohit V. Pappu, and Nathan A. Baker. "NanoParticle Ontology for cancer nanotechnology research." *Journal of Biomedical Informatics,* (2011), 44, 1: 59-74.

466205 A Vision to Create a Framework for the Biopharmaceutical Industry to Approach Maximum Sustainability

Abstract:

A Vision to Create a Framework for the Biopharmaceutical Industry to Maximum Sustainability

In the 80 years since the first commercialised product[1], advances in biotechnology, particularly the development of modern (recombinant DNA) technologies, have enabled the treatment of diseases that were once thought to be untreatable, such as Leukaemia[2–4]. The production of biologics is expanding beyond the US and Europe to developing countries with the emergence of biosimilars and biobetters[5,6]. As the industry matures, the focus is shifting away from breakthroughs towards sustainability. Research on the industry is required to quantify the environmental burden, and diagnose areas for improvement before future manufacturing recommendations are made. Life cycle assessments (LCAs), aided many industries in the past[7–9], are comprehensive environmental impact studies, which consider systems' raw material extraction to their end of life, a tool that aids decision making for future development. However, a framework needs to be developed to ensure resulting recommendations are fully adopted. This paper sets out to analyse the influences on the industry and, from these findings, to present a strategy utilising LCA for the support of continual sustainable development.

There are five major parties that influence biopharmaceutical decision makers: academia, government, non-governmental organizations, the general public and companies' own stakeholders. The level of risk inherent developing a new drug to market is exceptional. The industry is very capital intensive[10]. This poses risk on stakeholders' return. It is in their interest to maximise profit and realising that improving processes reduces cost, sustainable actions are hence chosen. Academics hold some of the fundamentals for the drive to sustainability. Not only they can provide on-going advances in technology for companies to migrate from an unsustainable state to a more sustainable situation[11]; their knowledge and reasoning are cascaded down into education and awareness-raising in our society. The general public have learnt to scrutinize those that do not take the environment into consideration, forcing companies to show leadership in the subject matter to increase social standings to gain competitive advancement[12].

While actions taken towards sustainability might be motived to advance a company economically and socially, it is governments and NGOs policy enforcers and practitioners that ensure industry abides to environmental rules. Laws and regulations have been implemented to reduce pollution due to development and manufacturing waste. However, governmental legislations are often not industry specific and tend to focus on energy intensive industries (EIIs) when it comes to reducing environmental issues. The regulations that have passed are normally set limits that companies and organizations must adhere to. However, EIIs frequently fail to do this[11–13]and on the contrast, the biopharmaceutical industry tends to operate at lower scales, and many individual companies do not reach these governmental limits. Governments have the power to implement industry specific policies to ensure maximum sustainability can be approached by all industries. The biopharmaceutical industry is not currently regarded as an EII but, as it grows, the collective environmental impact can be significant. It will be beneficial to put environmental constraints in place for this developing industry to minimize future burden.

To ensure maximum sustainability is approached, academia, industry and government should work closer together. There are two types of approaches where scientific research can be used to influence policy: (1) data-first – fitting research to governmental body's agenda and (2) co-design and co-production – collaborating with government to solve a common issue[15,16]. Although time consuming, the advantage of co-design is the full adoption of research knowledge into policies. LCA calculates a system's sustainability with regards to selected environmental metrics; economical and social metrics can also be implemented to give full analysis of the industry's status. The comprehensive results can formulate concise and accurate legislation for the single industry. The socioeconomic aspects will prove the relevance and impact of the research, persuading companies that improving environmental sustainability will not come at a cost.

A visionary strategy is to develop a feedback framework by co-design wherein initial analysis informs industry specific legislation based on the specific LCA metrics that has been agreed between research and government. Recommendations and guidelines on biopharmaceutical processes such as the amount of waste per kilogram of product, and carbon emissions can be imposed on the industry as sustainability targets. Then, as new technology arises, new data can be produced which again feeds back to legislators and the resultant recommendations are revised. However, the process for analysing the whole industry is data intensive. There will be implications in gathering accurate results due to the lack of data available. Not only are there few studies on the environmental impact of the industry, but current processes may be subject to non-disclose agreements, which suggests process data is highly sensitive. Partnership between academia, industry and government should be established to ensure the accuracy of acquirable data.

In conclusion, the paper demonstrates the fundamental importance of academia and government in increasing sustainability in the biopharmaceutical industry. Moving from unsustainable to a more sustainable future is a continual process, a feedback framework, indicating sustainability status, developed by the two most influential parties should be implemented to maximise the benefits of new technologies and process methodologies. LCA has transformed industries in the past; co-design and co-production of research knowledge using LCA as a tool is the key to transform legislation to take leadership in continually driving sustainability.

References

1. Quirke V. History of Penicillin. *eLS*. 2001:1-3. doi:10.1016/S0140-6736(80)90580-2.

2. Rader C. Chemically programmed antibodies. *Trends Biotechnol*. 2014;32(4):186-197. doi:10.1016/j.tibtech.2014.02.003.

3. Abegunde D. *Background Paper Essential Medicines for Non-Communicable Diseases (NCDs) Essential*

Medicines for Non-Communicable Diseases (NCDs).; 2011.

4. Projan SJ, Shlaes DM. Antibacterial drug discovery: is it all downhill from here? *Clin Microbiol Infect.* 2004;10 Suppl 4:18-22. doi:10.1111/j.1465-0691.2004.1006.x.

5. Langer ErS. *Report and Survey of Biopharmaceutical Manufacturing Capacity and Production.*; 2015.

6. Wilsdon T, Attridge J, Fiz E, Ginoza S, Mitchell-heggs A. Assessing the value of biopharmaceutical innovation in key therapy areas in middle-income countires. 2015;1.

7. Hunt R, Franklin W. LCA — how it came about. *Int J Life Cycle Assess.* 1996;1(1):4-7. doi:10.1007/BF02978943.

8. Economy GF, Regulations VE. LCA of Automobile Industry. *Phoenix Syst - Process Strateg.* 2009.

9. Klöpffer W. Life cycle assessment: From the beginning to the current state. *Environ Sci Pollut Res Int.* 1997;4(4):223-228. doi:10.1007/BF02986351.

10. DiMasi JA, Feldman L, Seckler A, Wilson A. Trends in Risks Associated With New Drug Development: Success Rates for Investigational Drugs. *Clin Pharmacol & Ther.* 2010;87(3):272-277. doi:10.1038/clpt.2009.295.

11. Scott C. *Sustainability in Bioprocessing.* Vol 9.; 2011.

12. Li W, Yu R. Environmental Responsibility and Biopharmaceutical Companies: Developing a Competitive Strategy. 2011:51-54.

13. Affair VP. Compliance in Manufacturing : A Very Personal Affair. 2013:1-11.

14. Zhou J. Pesticide Manufacturer Fined 12 Million USD In Accordance with China's New Environmental Protection Law | chemical. https://chemlinked.com/news/chemical-news/pesticide-manufacturer-fined-12-million-usd-accordance-chinas-new-environmental-protection-law. Published 2015. Accessed April 28, 2016.

15. US EPA, OECA, OAP I. Civil Cases and Settlements. 2016. https://cfpub.epa.gov/enforcement/cases/. Accessed April 28, 2016.

16. Pohl C. From science to policy through transdisciplinary research. *Environ Sci Policy.* 2007;2. doi:10.1016/j.envsci.2007.06.001.

17. Majone G. Evidence, argument, and persuasion in the policy process. 1992:1-9.

462496 Causal Analysis of the Environmental Distribution of Engineered Nanomaterials Using Bayesian Networks

Abstract:

The increasing use of engineered nanomaterials (ENM) in various products and applications is raising concerns regarding the possible releases of ENMs to the environment and potential environmental and health impact. It is therefore imperative to assess the expected ranges of ENM concentrations in various environmental media (e.g. air, water, soil, sediment) recognizing that ENMs can migrate across environmental phase boundaries. Thus, intermedia transport processes that are governed by various environmental factors are of significant importance when assessing potential ENM exposure levels. However, monitoring of ENMs exposure concentrations in multiple media is a daunting and costly endeavor. In this regard, modeling of ENMs releases and exposure concentrations can inform decision makers as to potential ENM emissions and concentration boundaries. In this regard, mechanistic models are most relevant to describing the distribution of ENMs in the multimedia environment. However, the use of such models for decision analysis is not trivial since multiple simulations may be required to be performed by experts in order to provide the information needed by decision analysis. In order to simplify the approach and make simulation data readily accessible for environmental decision analysis, a data-driven Bayesian Network (BN) model was developed for predicting environmental concentrations of ENMs (Al_2O_3, CeO_2, Cu, SiO_2, TiO_2, and ZnO) over a wide range of environmental scenarios (in different geographical regions and for a range of ENM release scenarios) using data generated from a mechanistic multimedia model of the environmental distribution of ENMs. Using the BN model various environmental scenarios were rapidly evaluated and the impacts of input parameters on resulting ENM concentrations was assessed to provide clear causal relationships. For example, rain scavenging and dry deposition of airborne ENMs were the significant intermedia transport processes leading to the accumulation of ENMs in soil, water and sediment (where there was no direct ENM discharge to water). Through BN approach, it was demonstrated that rapid assessment of the dynamic mass distributions and exposure concentrations of ENMs can be conducted for a broad range of environmental scenarios of interest for direct input to environmental decision analysis.

465016 Flare Efficiency Characterization Using Neural Network and Sigmoid Models

Abstract:

Current EPA regulations have mandated a minimum combustion zone heating value of 270 BTU/scf and NHV_{dil} >=22 Btu/ft^2 for all steam/air/non-assisted flares and 380 Btu/scf for H2/olefin interaction cases while maintaining a high Combustion Efficiency(CE). To achieve the target performance along with satisfying the EPA regulations, it is necessary to understand the influence of various operating parameters and determine their set points. Studying the effect of operating parameters through experiments is both expensive and time intense. It would be more cost-effective to use validated models to guide flare operations. In this study flare data obtained from tests conducted from 1983 to 2014 has been analyzed. The data used in the study has a wide range of exit velocities, heating values and fuel compositions. The purpose of the study is to develop a model that can robustly be used in the industry to achieve the desired CE/Destruction Efficiency(DE) and reduce smoke. Its easier to control assist rates, exit velocities and the vent gas flow rates in an actual flare and so variables like these are used as independent variables in the models built. Multi-variable parameterized sigmoid models and a much complex neural network models were developed for the flare data using various types of fuels like propylene, propane, natural gas, methane and ethylene in this study. Contour plots were developed to determine the desired range of parameters for the independent variables. Both steam and air assist flares were modelled using curve fitting tool box and neural network toolbox in MATLAB. The models show high goodness of fit (R^2)above 0.90. Desirable operating inputs can be set for the incipient smoke point (ISP) and for smokeless flaring (Soot<=$Soot_{ISP}$) with a high CE (>=96.5).

475423 Sustainability Concepts in the Food-Energy-Water Nexus: Chemical Engineering Perspective

Abstract:

The idea of sustainability and/or sustainable development emerged from the World Commission on Environment and Development (WCED), generally known as the Brundtland Commission. It defined sustainable development in "Our Common Future" as that which meets the needs of the present without compromising the ability of future generations to meet their own needs. Environmental sustainability and/or sustainable development "occurs when we maintain or improve the material and social conditions for human health and the environment over time without exceeding the ecological capabilities that support them". As such, sustainability has three underlying bases: (1) economic viability - the cost and business aspects of a project; (2) social concern - human health and social welfare; and (3) natural or ecological issues - depletion of natural capital and environment. Sustainable development has become a major driving initiative in engineering businesses throughout the world today. Our focus is on all aspects of sustainable development including environmental, industrial, business, economical, ethical, and social sustainability as they impact the availability of food, energy, and water. This is important because it ***has been generally acknowledged that the increasing use of non-renewable resources to support an increasing population has created an unsustainable or nearly unsustainable situation where the ability to provide adequate food, energy, and water is in question. Should this rate of use be maintained, two dire consequences will follow: first, the future generations will be unable to maintain a good standard of living, and second, developing countries will have less of an opportunity to bring their living standards to a level comparable to that of the affluent countries. This is important because the availability of food, energy, and water is an indispensable necessity for maintaining a good standard of living. The practice of chemical engineering, perhaps more than any other technical discipline, involves the use of both non-renewable and renewable materials, and energy and exergy (available energy) resources for the production of chemicals, value-added goods, services of commerce, and waste as***

pollution. This paper primarily deals with the ideas of environmental sustainability and/or sustainable development, methodologies to redesign sustainable chemical and allied processes and products development using the LCA approach. These principles are illustrated with a case study for the production of ethanol, an energy source, from molasses, a human food, in a combined agricultural and chemical process that requires substantial amounts of water.

461670 Optimizing Spatio-Temporal Sensor Placement for Nutrient Monitoring: Algorithmic Framework

Abstract:

INFEWS is a new initiative in NSF which deals with Innovations at the Nexus of the Food, Energy and Waters Systems. This project addresses the INFEWS: N/P/H2O solicitation. The food production system generates waste streams that are characterized by high concentrations of organic matter, nitrogen- and phosphorus-containing species in water. Therefore, monitoring nitrogen and phosphorous species is important for water quality requirements for agricultural as well as energy production. Currently these species are monitored via stationary monitoring stations. However nitrogen and phosphorous species move via agricultural run-off to other water systems, and requires portable sensors which can change the positions in real time. This type of dynamic sensing requires novel algorithms which decide sensor locations in real time in the face of inherent uncertainties in the fate and transport of the species. To develop such an algorithmic framework to solve the problem of sensor placement in real time is the objective of this proposal.

This paper presents the first part of the work which is the algorithmic framework for optimal spatial-temporal locations of the sensors in real time. The algorithmic framework is based on a novel algorithm called Better Optimization of Nonlinear Uncertain Systems (BONUS). We present a case study of an agent based model from social sciences for testing these framework. This model is based on behavior of cows in a field where there are grass patches. The cows move in a herd towards the greener pasture and eat grass till it finishes. The grass growth and cow movements are probabilistic. We are using four sensors for object tracking (e.g. cows) and these sensor move spatially and temporally in order to track maximum number of cows. Results show that the optimal movement of sensors allows to track maximum number of cows.

In the future, the algorithmic framework will be coupled with nutrient and fate transport models. A case study of a small watershed will be selected from literature for nutrient monitoring sensor location problem. Optimal spatial-temporal locations will be the result of this new framework for this case study.

Bandwidth Study on Energy Use and Potential Energy Saving Opportunities in Manufacturing Food and Beverages

Presenter:
Caroline Dollinger
Engineer, Energy Analysis
Energetics Incorporated

American Institute of Chemical Engineers
Annual Meeting
San Francisco, CA
November 14, 2016

Collaboration

Caroline Dollinger (Speaker)
Energetics Incorporated
cdollinger@energetics.com

Sabine Brueske
Energetics Incorporated
sbrueske@energetics.com

Joseph Cresko
U.S. Department of Energy
Advanced Manufacturing Office
joe.cresko@ee.doe.gov

Alberta Carpenter
National Renewable Energy Laboratory
Alberta.carpenter@nrel.gov

Energetics Incorporated is an energy consulting company, providing subject matter expertise to private and public sector clients

2

Overview

- An *Energy Bandwidth Study* is currently being performed to analyze energy use and energy saving opportunities in energy intensive manufacturing sectors
- This presentation focuses on the work being done for a *Food and Beverage Manufacturing Energy Bandwidth Study*
- Energy data will broken down by sub-sector, product, and unit operation to identify processes and energy-saving technologies with the greatest opportunities for energy savings

Results presented in this talk are draft data; the work is still ongoing and a final report is expected to be published next year.

3

ENERGETICS A Subsidiary of VSE Corporation

Energy bandwidth studies can be a useful tool for assessing energy savings opportunities in manufacturing

Energy bandwidth studies frame the range (or *bandwidth*) of potential energy savings in manufacturing, and technology opportunities to realize those savings.

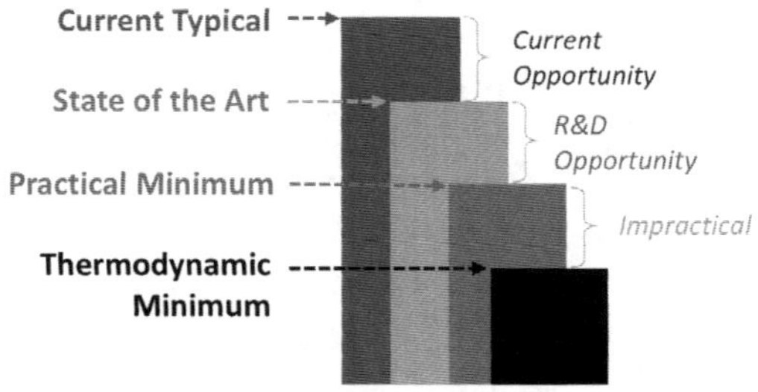

Measures of energy intensity studied:

Current Typical (CT)	State of the Art (SOA)	Practical Minimum (PM)	Thermodynamic Minimum (TM)
Determined from a literature review and stakeholder outreach, based on current typical manufacturing processes in the U.S.	Determined from a literature review and stakeholder outreach, based on the most energy-efficient technologies and practices available worldwide	Modeled based on plausible energy savings from identified R&D technologies under development worldwide	Calculated analytically using a Gibbs free energy approach assuming ideal conditions

4

ENERGETICS A Subsidiary of VSE Corporation

Six new energy bandwidth studies on lightweight materials were recently completed in draft form

- **These studies have previously provide a deep-dive analysis into manufacturing energy use and potential energy savings for:**
  - **Four energy-intensive manufacturing sectors**
  - **Manufacturing of six lightweight materials**
- **This talk will focus primarily on a new bandwidth on food and beverage manufacturing.**
- **Published bandwidths (including this one, once completed) are available on the DOE Advanced Manufacturing Office website**

Summary thumbnail for bandwidth study on petroleum refining

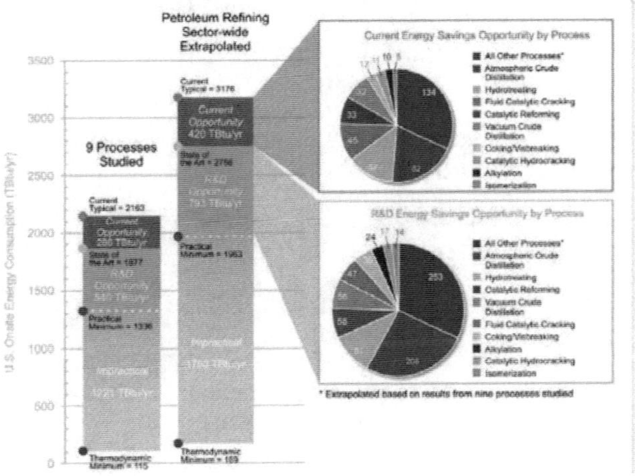

5

First Four Industry Bandwidths – US Results

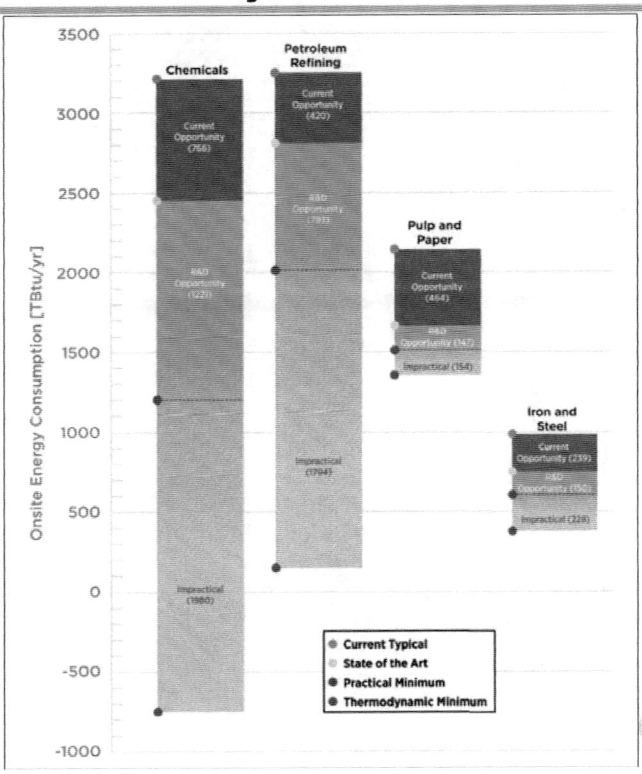

Challenges in Researching Energy Use and Savings in the Sector

- **Highly diversified sector that produces thousands of different products**
- **Processing facilities range from small to large units, normally producing more than one product**
  - **As a result energy intensities will vary from facility to facility and from product to product**
- **Less active in the R&D, with a focus on energy savings technologies, compared to other sectors**
- **Competitive marketplace, where energy efficiency may help to offer a competitive advantage**
- **Largely mechanized processes, difficult to ascertain and source energy data at the sub-process level**

7

ENERGETICS
A Subsidiary of VSE Corporation

Food & Beverage Manufacturing Bandwidth
Study Coverage

Food and Beverage Sub-sectors Studied	Food and Beverage Products Studied
Grain and Oilseed Milling	Wet corn milling- corn oil and by-products Soybean oil and meal Other vegetable oils and meals
Sugar Manufacturing	Beet sugar Cane sugar
Fruit and Vegetable Preserving and Specialty Food Manufacturing	Frozen fruits, vegetables, juice Canned fruits, vegetables Pickled vegetables Dried and dehydrated fruits and vegetables
Dairy Product Manufacturing	Milk Butter Cheese Dry, condensed, evaporated milk Ice cream/frozen desserts
Animal Slaughtering and Processing	Beef, pork, chicken, lamb Animal fat, lard, edible tallow
Beverage Manufacturing	Soft drinks Beer, wine, distilled liquors

8

ENERGETICS
A Subsidiary of VSE Corporation

Food & Beverage Manufacturing Bandwidth
Study Coverage

Food and Beverage Sub-sectors Studied	Energy Intensive Sub-processes
Grain and Oilseed Milling	Crushing Extraction Evaporation Dewatering and drying
Sugar Manufacturing	Dewatering and drying Evaporation Vacuum pans (crystallization)
Fruit and Vegetable Preserving and Specialty Food Manufacturing	Heat sterilization/retorting Cooking/blanching Freezing Concentration Evaporation Drying Canning
Dairy Product Manufacturing	Pasteurization Freezing Concentration
Animal Slaughtering and Processing	Rendering
Beverage Manufacturing	Beer processes (brewhouse, packaging) Wine processes (refrigeration)

Food and Beverage Manufacturing Bandwidth
Scope Description

What's Out- Agriculture Sector NAICS 11

Establishments primarily engaged in growing crops, raising animals, from a farm, ranch, or their natural habitats

111-Crop production

112-Animal production

What's In- Manufacturing Sector NAICS 31

Industries in the Food Manufacturing subsector transform livestock and agricultural products into products for intermediate or final consumption (basically, any food that is processed)

311 – Food manufacturing

What's Out- NAICS 42, 44, 72

424 – Food Wholesalers

445 – Food and Beverage Stores

722 – Food Services and Drinking Places

ENERGETICS
A Subsidiary of VSE Corporation

Food and Beverage Sector
U.S. Energy Consumption Overview

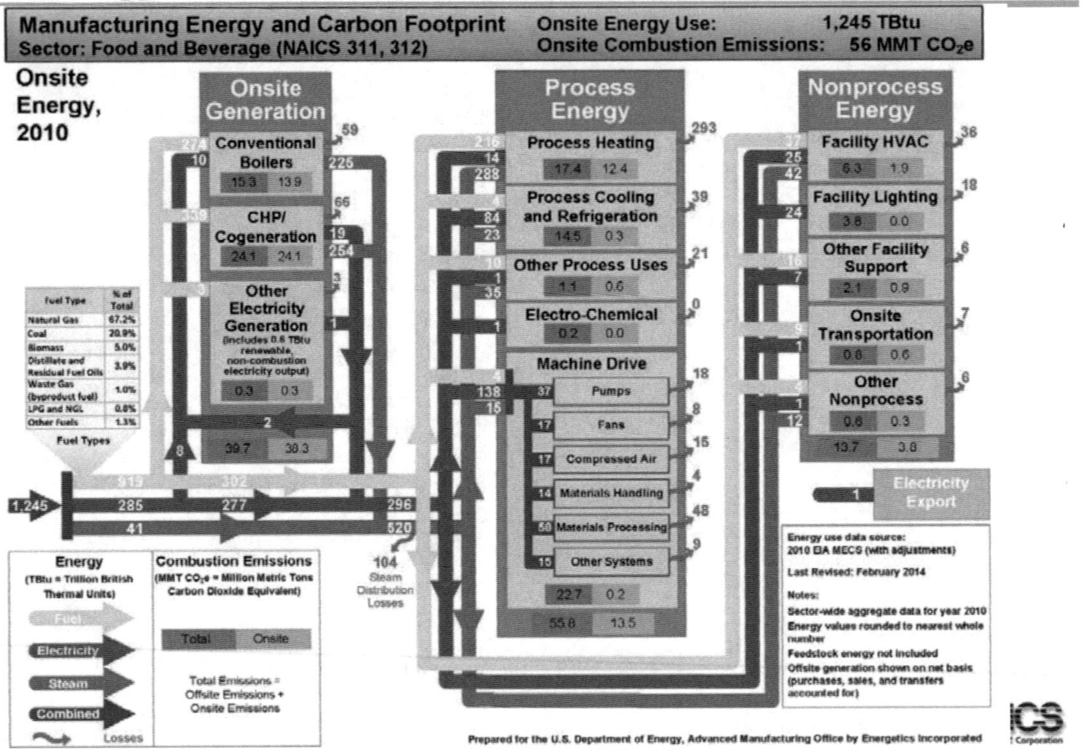

Food and Beverage Bandwidth
U.S. Baseline Energy Consumption, 2010

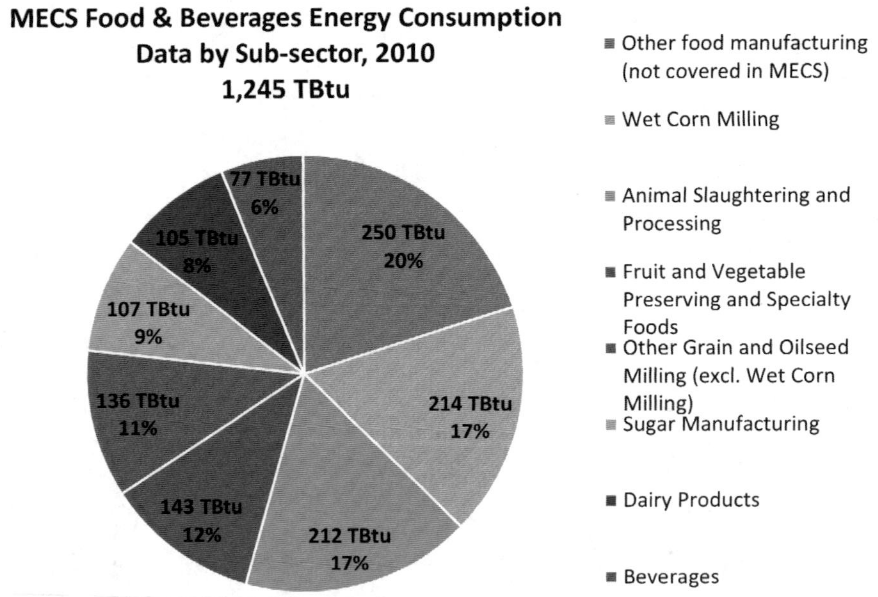

Source: EIA 2014, Manufacturing Energy Consumption Survey (MECS)

12

Food and Beverage Bandwidth
U.S. Production for Sub-sectors Studied, 2010

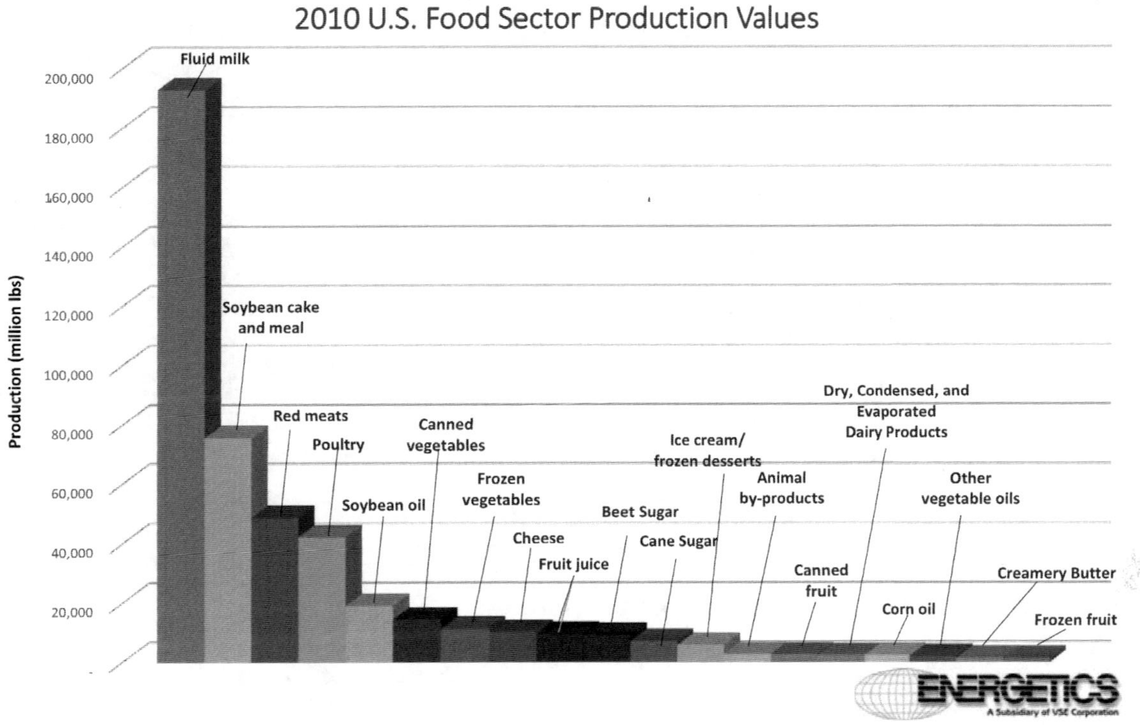

Dairy Products Manufacturing
Energy intensive sub-processes being analyzed

- Pasteurization
- Cooling/refrigeration
- Concentration
- Freezing
- Motors/pumps
- Spray Drying
- Canning

Manufacturing Process	CT (TBtu/year)	% of sub-sector
Pasteurization	20	18.6%
Cooling	16	15.6%
Motors, pumps	9	8.4%

14

Grain and Oilseed Milling

Energy intensive sub-processes being analyzed

Wet Corn Milling

- Starch dewatering and drying
- Gluten thickening and drying
- Gluten feed drying
- Steeping
- Steepwater evaporation
- Fiber and gluten recovery
- Germ dewatering and drying
- Grinding

Other vegetable oils/meals

- Crushing and extraction
- Neutralization
- Oil refining
- Deodorization

15

Fruit and Vegetable Preserving

Energy intensive sub-processes being analyzed

Canned Fruits & Vegetables

- Washing
- Retorting
- Blanching/cooking
- Canning

Frozen Fruits & Vegetables

- Washing
- Peeling
- Blanching/cooking
- Freezing

Other fruit/ vegetable products

- Pasteurization (fruit juice)
- Dehydrating
- Retorting (fruit juice)

16

Animal Slaughtering and Processing

Energy intensive sub-processes being analyzed

Red Meat Products

- Dressing and cutting
- Refrigeration
- Processing (curing, smoking, cooking)
- Packaging
- Splitting
- Scalding
- Singeing
- Blood processing
- Edible rendering

Poultry Products

- Scalding, slaughtering, defeathering
- Evisceration and cooling
- Refrigeration
- Liquid effluent treatment
- Packaging

17

Current Typical (CT) energy intensities and consumption for red meat products, U.S. (2010)

Manufacturing Process	CT (Btu/lb)	TBtu/year (% of sub-sector)
Chilling	468	23 (10.8%)
Processing (curing, smoking, cooking)	320	16 (7.4%)
Edible rendering	4,573	14 (6.5%)
Dressing and cutting	181	9 (4.2%)
Packaging	93	5 (2.1%)
Scalding*	149	3 (1.6%)
Splitting*	117	3 (1.2%)
Singeing*	82	2 (0.9%)

* Pork products

18

ENERGETICS
A Subsidiary of VSE Corporation

Animal Slaughtering and Processing
Example of challenges in data and analysis

- Lack of recent analysis done on energy usage in U.S. manufacturing facilities, must rely on internationally available data
- Most estimates of energy intensity are presented in terms of total energy/product or input (e.g., kWh/tonne), not broken down by sub-process
- Studies done on individual plants may not be applicable to be generalized sector-wide
- Energy data presented in different forms (consumption per head, carcass weight, dressed weight, product weight)
  - Involves accurate conversions to calculate total energy consumption using production values from USDA, other assumed factors
- If data is provided for sub-processes, only provided for a few, not allowing for calculation of balance

19

ENERGETICS
A Subsidiary of VSE Corporation

Current Opportunity and *R&D Opportunity*

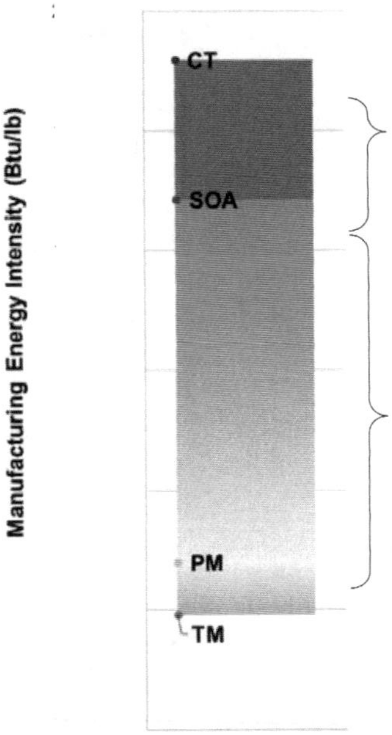

Current Opportunity

Energy savings if the best technologies and practices available were used to upgrade production

R&D Opportunity

Additional energy savings if applied R&D technologies under development worldwide were successfully deployed

<u>SOA Technology Examples:</u>

- Steam system improvements
- Adjustable-speed drives
- Air compression improvements (reduce leaks, efficient vacuum pumps)
- Alternative technologies
- ...

<u>R&D Technology Examples:</u>

- Membrane filtration
- Microwave heating
- Advanced Ohmic Heating
- High pressure processing
- Cost efficient thermal management in motors using nanomaterials
- ...

20

ENERGETICS
A Subsidiary of VSE Corporation

What about the Water part of the Nexus?

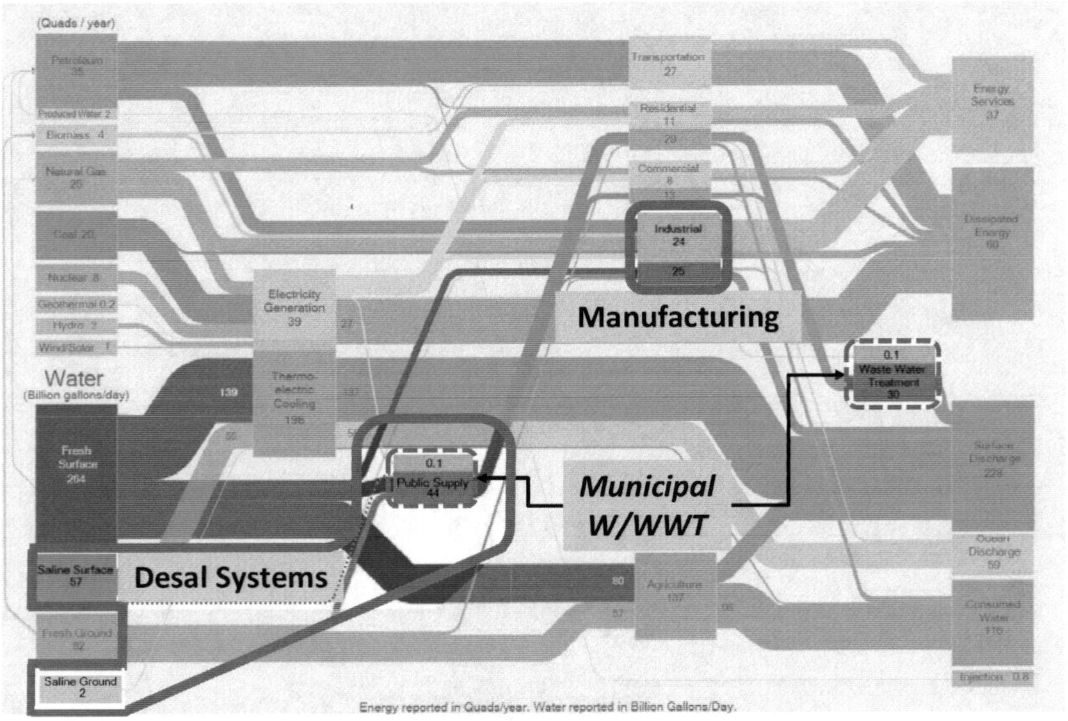

Conclusions

- **An *Energy Bandwidth Study* is being performed to analyze manufacturing energy use and energy saving opportunities for the food and beverage manufacturing industry**
- **Manufacturing energy data are broken down by sub-sector, product, and sub-process to identify processes and energy-saving technologies with the greatest opportunities for energy savings**
- **This type of study has not been done in the past, so data must been gathered carefully; also there is a lack of data in certain areas**
- **While more research is to be done, research to date implies that there may be more current rather than future energy savings opportunity, due to the fact that there are few areas with active R&D in food and beverage, while more traditional areas for energy efficiency savings (steam systems, etc.)**

Results presented in this talk are draft data; studies are currently being peer reviewed. If you are interested in participating, please contact me!

22

ENERGETICS
A Subsidiary of VSE Corporation

Thank you!

Energetics Incorporated

Caroline Dollinger

(CDollinger@energetics.com)

23

468656 Insight-Based Design of Local Integrated Systems for Food, Energy and Water

Abstract:

The importance of Food-Energy-Water nexus (also referred to as Water-Energy-Food nexus) has been widely recognised by national governmental bodies and international organisations such as the United Nations, for its significance to sustainable development and national security. Though these three sectors are inextricably linked, as actions in one sector have impacts in one or both of the others, these areas have too often been considered individually. Thus, there is a need to look at them from a holistic perspective and develop tools that take into consideration their interdependencies. This opens up opportunities for process system engineering research. As such, the aim of this study is to develop a systematic insight-based approach for the design of an integrated local production system particularly to meet local demands for food, energy and water. One of the benefits of the insight-based approach is that it allows decision variables to remain tractable throughout the design process. Unlike mathematical optimisation approaches, intermediate results from insight-based approaches can be analysed, interpreted and used to inform the implications of a given design for the operation of the whole system, thus offering a more practical tool for decision-makers and local planners to use. Insight-based approaches which include techniques such as pinch analysis for heat integration and mass integration have proved particularly useful for the integration of energy and material streams in industrial production processes within eco-industrial parks and industrial symbiosis systems. They have been successfully implemented for enhancing resource efficiency and sustainability of processes involving heat, water and hydrogen, to name a few applications. Although the proposed study also pursues higher resource efficiencies through integration of resources, the design of the local integrated production system has a distinctive emphasis on local resources and demands and on the holistic consideration of all types of agricultural, industrial and municipal processes to take place at the locality of concern. This work will combine existing methods and adapt them to enable the identification of the efficient use of various locally available resources. A net gain indicator will be defined to quantify the avoided cumulative effort that would have been spent for importing resources to satisfy the demand.

Using the net gain indicator, parallel processing and utilisation options within and between different subsystems can be compared and the optimal actions identified. The proposed insight-based approach for the design of local integrated system of food, energy and water will be illustrated through a case study based on an eco-town in the UK.

468765 Recovering Runoff Particulate-Bound Phosphorus Via Fungal Bioextraction

Abstract:

Phosphorus (P) loss is a key issue affecting food-energy-water nexus sustainability. P in fertilizers is critical for ensuring food security for the continuously growing world population. Currently, P is obtained via energy-intensive mining and processing of phosphate rock, which is becoming more scarce and concentrated in a few geographical areas. Excessive fertilizer application due to slow diffusion and heavy fixation of P in soils have led to massive P losses amounting to almost half of all mined P due to agricultural soil erosion and runoff. P-rich runoff streams discharging into surface water bodies can cause eutrophication, toxic algal blooms, and overall quality deterioration of freshwater resources. In this work, we propose capturing and processing runoff sediments using a low-energy and inherently benign fungal bioprocessing method to unlock a previously untappped source of recoverable phosphorus for nutrient recycling and closing the loop of P use. This can potentially reduce the demand for mined sources while at the same time help protect watersheds from the devastating effects of eutrophication.

This paper presents the preliminary findings of our work involving the use of phosphorus-solubilizing fungi (PSF) for extracting and recovering particulate-bound phosphorus in erosion and runoff streams. PSF species can generate organic acids and phosphatase enzymes that can induce mobilization of matrix-associated or surface-adsorbed phosphate ions on inorganic minerals in runoff particulates, and degrade phosphate ester bonds in sedimentary organic phosphorus compounds to facilitate the release of phosphate ions, respectively. Samples of runoff sediments discharging into a eutrophic lake in Kalamazoo, MI were collected and characterized for total P content and P speciation. P extraction experiments on runoff sediment samples were performed using model organic acids and spent culture broth *Aspergillus niger* ATCC 15475. Phosphate solubilization and changes in P speciation of the runoff sediments during treatment were measured using a sequential chemical extraction method and colorimetric analysis. Changes in runoff sediment mineralogy was analyzed using X-ray Diffraction (XRD) spectroscopy. The results

indicate that oxalic and citric acids can extract and solubilize a significant fraction of phosphates bound to the iron, aluminum, and calcium mineral components of the runoff sediments. A kinetic model describing phosphate mobilization via ligand-exchange and induced desorption mechanisms was developed and fitted into the experimental data. We also observed a significant reduction in the organic P fraction of sediments treated with spent fungal culture broth, indicating the possibility of phosphatase enzyme production and activity. Thus, changes in organic P speciation in the sediments were monitored using solution ^{31}P nuclear magnetic resonance (NMR) spectroscopy. The findings demonstrate the feasibility of the proposed technology for P recovery from nonpoint nutrient pollution sources such as runoff sediments. It will help develop an effective, efficient, economical, and environmentally benign system that will benefit the community, agriculture and food production systems, and the environment in terms of resource security and environmental sustainability.

469658 Preliminary Study on Detroit's Urban Food-Energy-Water (FEW) Nexus

Abstract:

With 66% of the worldâ€™s population expected to live in urban areas by 2050, cities around the world will face numerous challenges in sustainable provision of food, energy, and water. These challenges are particularly complex due to the interconnectedness of the food system, energy system, and water system. For example, water is required for food and energy production; energy is needed for food production as well as water and wastewater treatment and distribution; and industrial agriculture and food production lead to oversupply of nitrogen and phosphorus in the urban water system through fertilizer use. Therefore, policy and technology solutions addressing urban food-energy-water (FEW) challenges need to be evaluated through the lens of FEW nexus to identify co-benefits and avoid unintended consequences. This calls for innovative approaches to model the urban FEW nexus as an integrated whole, instead of examining them individually.

Existing studies primarily investigated urban FEW systems individually (e.g., nutrient flows, energy flows, and water flows) or focuses on two of the three FEW systems (e.g., water-energy nexus and food-water nexus). Existing studies that consider all FEW systems primarily aim to understand the impact of one system on the other two, such as how urban water-sector innovations change food and energy flows. There lacks an integrated approach that characterizes the comprehensive interdependence of urban FEW systems.

In this study we developed a systems modeling framework for the urban FEW nexus. We also applied this framework to the Detroit Metropolitan Area (DMA) as a preliminary study. The urban FEW systems modeling framework we proposed is based on material and energy flow analysis (MEFA) which quantifies flows and stocks of a group of substances within a system during a given period of time. This framework includes three steps:

Step 1: Individual FEW systems are modeled as resource flows across major sectors (e.g., water treatment, wastewater treatment, food production) of the urban system. Given the variety of food crops and products, we used nitrogen and phosphorus flows as the currency to represent the food system. Quantities of resource flows (water, energy, nitrogen, and phosphorus) are estimated based on a variety of data such as population, water withdrawal, and wastewater generation that were collected for the DMA case study.

Step 2: We identified and measured interdependence of individual FEW systems. This was done through literature review for common FEW interactions and quantitative estimation of resource flows between individual FEW systems (e.g., the amount of water used for food production and energy production).

Step 3: Each individual FEW system is characterized as a network of resource flows. Based on the interconnections between individual resource flow networks obtained from Step 2, we integrated the individual FEW resource flow networks as a network of four networks (water, energy, nitrogen, and phosphorus). This provides a mathematic representation of the urban FEW nexus, which will be the base of further analysis and modeling.

474593 Innovations at the Nexus of Food, Energy, and Water Systems (INFEWS): A National Science Foundation (NSF) Initiative to Examine Critical Issues in Today's World

Abstract:

At the end of 2015, the National Science Foundation (NSF) put out its first request for proposals for the INFEWS program (Innovations at the Nexus of Food, Energy, and Water Systems). Over the next five years the program plans to fund interdisciplinary projects that fall into four tracks 1. System Modeling, 2. Visualization and Decision Support, 3. Innovative System Solutions, and 4. Education and Workforce Development. The program expects to tackle Food-Energy-Water System (FEWS) research through integration across disciplines, in which agricultural and food systems research is a primary component. To this end NSF has partnered with the United States Department of Agriculture (USDA) to fund related research. This program developed from the Science, Engineering and Education for Sustainability (SEES) program, which in 2015 funded 17 FEWS workshops. These interdisciplinary workshops brought together researchers and other stakeholders from universities, local, state, and federal governments, industry, and non-profits to examine a wide range of FEW questions and identify the research, data and technological needs to solve them. This presentation will cover an overview of the development of the INFEWS program, some pressing FEW needs as identified by the broader research community, and national and international opportunities for future engagement with the program.

474707 The "Root" of the Nexus: Soil-Based Biotechnology for Sustainable Agriculture

Abstract:

Food, energy, and water nexus research by chemical engineers is typically focused on the built environment. For example, innovations in sustainable food production might offer food processing, distribution, or waste recycling technology that minimizes water or energy use. However, recent technological advances in different fields ranging from materials and manufacturing to molecular tools and systems biology are also enabling more sustainable primary production of food in the natural environment. Most of our food is produced in soil, a dynamic natural system of enormous complexity. In particular, the rhizosphere (the region of soil adjacent to plant roots) is a site of steep chemical gradients, plant growth-limiting multiphase transport phenomena, and dynamic inter-kingdom collaboration among members of the root microbiome. Emerging technology is re-engineering the root zone for more sustainable primary food production. Examples of emerging agriculture biotechnology to produce more food with less water and energy include: (i) advanced seed coating technology and nano-enabled agrochemical delivery systems; (ii) plant-growth promoting bacteria and microbial products to improve plant nutrition and enhance moisture retention in soil; and (iii) sensors and sensor networks for precision agriculture. By learning from and working with natural systems, chemical engineers are developing innovative agriculture biotechnology for more sustainable primary food production.

474596 The Role of Waste Streams in the Energy-Water Nexus

Abstract:

The energy-water nexus has helped focus attention on the water used for energy production and energy used in providing water services. However, there has been a lack of attention to the downstream aspects of this nexus; that is, impacts of related waste streams on land, water and air. Primary technology choices (e.g., electricity generation technology, drinking water source) result in differing waste streams which in turn require decisions on the level of treatment and disposition of the related waste products. Each of these choices have related implications for energy use, water use, environmental impact and system vulnerability (e.g., power plant thermal discharge puts operations at risk to drought). Remarks in this panel will endeavor to establish a framework and context for a more holistic treatment of waste streams within the broader energy-water nexus.

474597 Process Development Challenges and Opportunities in Food-Energy-Water (FEW) Nexus

Abstract:

Presentation will start out with a high level view of Food-Energy-Water (FEW) nexus for illustrating complexity of the topic. Global changes in population, energy needs, and water use will drive technological innovations and rapid commercialization needing chemical engineers involved in process development. Connections and interdependencies discussion among these three systems will illustrate how engineering solutions from fields like systems approach, catalysis, biocatalysis, environmental sustainability, etc. will contribute to better understanding the FEW nexus challenges and successfully overcoming them.

Broad overview of current and future process technologies important to FEW nexus selected from various NSF and AIChE workshops and recent developments reported in literature will be presented. Challenges and opportunities for successful transition from concepts to commercialization of such technologies will be discussed with actual examples at various stages of development. Examples will be presented wherever possible to show how process research and innovation, pilot plants, technology transfer, process intensification and micro process engineering as well as product design will contribute to the nexus. Need for collaborative efforts among academics, government labs, and industry across multiple disciplines and countries for local applications for ultimate success in the FEW nexus will be highlighted.

The talk will also cover selected areas where chemical engineers in general and process development engineers in particular can make significant contributions. Critical need for process research, development, and demonstrations that enables new means of adapting to future challenges will be presented. As an example, out of the three main factors nitrogen, phosphorous, and water that limit our ability to produce enough food, nitrogen and water will be covered in detail. This section of the presentation will show how catalytic and waste water treatment technologies will play the role in connection to the FEW nexus.

Main goal of the presentation will be to make audience aware of theoretical and experimental process development challenges and opportunities of the future.

474595 Systems Engineering Methods for Food-Water-Energy Nexus

Abstract:

As the global population grows, consumption of water, energy, and food will also increase, placing stresses on these three sectors, raising the importance of the Food-Water-Energy Nexus (FWEN). This article highlights research challenges and identifies process systems engineering research opportunities to appropriately model and optimize the WEFN. A brief overview of relevant, foundational FWEN research is first presented. We then identify challenges in the multiple scales of the FWEN, ranging from the household scale to the global scale. There are further challenges with appropriate system boundary definitions, and challenges in modeling the decision-making and conflicting objectives of multiple stakeholders in the FWEN. Uncertainties of all kinds appear at all scales of the FWEN and must also be considered. We use two motivating FWEN examples to frame these challenges and propose future avenues and opportunities for research. Possible approaches to the aforementioned challenges are proposed.

459286 Comparative Techno-Economic Analysis of Bio-Hydrocarbon Production from Pyrolysis-Derived Bio-Oil Via Hydroprocessing and Zeolite Cracking

Abstract:

This work examines the comparative techno-economic analysis of two pathways for the upgrading of fast pyrolysis-derived bio-oil into bio-hydrocarbons, *viz.* hydroprocessing and zeolite cracking. The hydroprocessing pathway entails a two-stage hydrodeoxygenation process, followed by a hydrocracking process, which is coupled with a steam reforming process of the aqueous phase of the bio-oil to supplement hydrogen requirements for hydroprocessing. The zeolite upgrading pathway features a single stage regenerator, assumed to be in a full combustion mode and fitted with a catalyst cooler to tackle excess heat due to the accompanying excess coke formation typically associated with this route. The two production pathways were modelled in Aspen Plus® process simulator based on a hypothetical 72 t/day fast pyrolysis plant. Aspen Process Economic Analyser® was used for equipment sizing and cost estimation. The respective minimum fuel selling prices (MFSPs) were estimated by the discounted cash flow method, assuming the plant operates for 20 years at a 10% annual discount rate. The models of the two upgrading pathways were compared in terms of energy efficiency and profitability with regards to their relative MFSPs. Furthermore, sensitivity analysis and uncertainty analysis via Monte-Carlo simulations were conducted for the two upgrading pathways to examine the effect of variations in economic parameters on profitability in a deterministic and probabilistic manner, respectively. The energy efficiencies of the hydroprocessing and the zeolite cracking upgrading routes were 68% and 52%, respectively with corresponding MFSPs of £6.25/GGE and £7.20/GGE. Sensitivity analysis revealed that the MFSPs of both upgrading pathways are mainly sensitive to variations in fuel yield, operating cost, and income tax. Furthermore, uncertainty analysis *via* Monte-Carlo simulations resulted in probability distributions for hydroprocessing and zeolite cracking with expected MFSPs of £6.42/GGE and £8.30/GGE and standard deviations of £1.05 and £1.39, respectively. The results provide evidence to support past research regarding the economic viability of the production

of biofuels from hydroprocessing and zeolite cracking. The results obtained from this work suggest that the upgrading of pyrolysis derived bio-oil *via* hydroprocessing is more techno-economically viable than the zeolite upgrading route.

464269 Economics of Biofuels and Bioproducts from an Integrated Pyrolysis Biorefinery

Abstract:

This presentation describes a novel integrated biorefinery based on fast pyrolysis and subsequent bio-oil upgrading to yield diverse renewable product portfolios. Three portfolios are considered: hydrocarbon biofuels (gasoline and diesel), biobased products (bioasphalt, calcium acetate, and dextrose), and hydrocarbon chemicals (aromatics, olefins, and fuel oil). Based on the economic assumptions and historical market prices of the products, internal rate of returns (IRRs) of the biorefinery (2000 dry metric tons per day) over 30 years are calculated to evaluate the performance of the proposed biorefinery. Given the historical low, average, and high prices of the products, a total of twenty-seven scenarios are explored and the maximum IRR for each scenario is determined. Results indicate that upgrading bio-oil to biobased products generates the maximum IRR of 67.9% if the biobased products can be marketed at a historically high price level (between $257 and $1491 per metric ton) and an IRR of 42.5% at an average price level (between $240 and $950 per metric ton). The capital and operating costs of producing biobased products are $166.5 and $80.6 million. Under some market conditions, the biorefinery should produce hydrocarbon chemicals or biofuels to increase the IRR. IRRs of less than 10% are found under market prices of $820 to $909 per metric ton for biofuels, $222 to $416 per metric ton for biobased products, and $172 to $1356 per metric ton for hydrocarbon chemicals indicating potential financial risks under low market product prices.

470751 A Kinetic Model for Updraft Biomass Gasification By Aspen Plus

Abstract:

Biomass gasification is now widely accepted as a potential and environment-friendly energy resource. In this work, a kinetic model for updraft biomass gasifiers is presented. The model is developed in Aspen Plus V8.6 by solving the detailed gasification kinetics and it approximates the composition of syngas by two steps. The biomass pyrolysis step is accomplished by stoichiometric reactions and the biomass combustion/gasification step is simulated by plugging in heterogeneous and homogeneous reaction kinetic information. Two lab-scale updraft gasifiers experiments available in literature are used to test the kinetic model (K-model) and a widely used minimizing Gibbs Energy model (MGE-model). The syngas composition predicted by K-model agrees well with experimental measurements. K-model is also more accurate than MGE-model especially for the prediction of tar. Moreover, the gasification of new biomass samples at different equivalence ratio (ER), gasification temperature and air/steam ratio are also investigated. Simulation results show that gasification process can be improved if extra heat is available to the gasifier or when it is operated under the desired ER range.

463896 A Multiscale, Multiphysics Modelling Framework for the Processes Involved in Consolidated Bioprocessing

Abstract:

Current biofuel production in the world is primarily from starch or sugar sources, such as corn and sugar cane. This has led to concerns about competition affecting prices as demand for both fuel and food increases. In order to overcome this, focus has turned to lignocellulose resources. The production of biofuels is currently carried out by processes such as separate hydrolysis and fermentation that convert biomass over a series of stages. These have proven to be inefficient and unattractive in terms of cost effectiveness. In order to make biofuels economically viable and competitive with oil-based equivalents, it is necessary to improve the efficiency and cost effectiveness of these processes. One of the ways to achieve this is by consolidated bioprocessing (CBP). CBP involves utilizing microorganisms to produce the cellulases needed to digest the biomass. whilst simultaneously metabolising the sugars produced into desirable products, such as ethanol, all in one reactor. This removes the need for extra reactors, heaters, pumps and other auxiliary equipment. The metabolism of the sugars as they are produced has the potential to reduce product inhibition of the enzymatic hydrolysis step. However, there is not a single organism that is able to both produce the enzymes to hydrolyse the biomass and ferment the sugars to products at a desirable rate and yield. Therefore, research has focused on engineering desirable traits into organisms to create an ideal CBP microorganism, such as modifying *Saccharomyces Cerevisiae* to produce hydrolytic enzymes, or adding fermentative pathways to naturally cellulolytic organisms.

We have developed a multiscale, multiphysics model for the production of valuable fuels and chemicals by consolidated bioprocessing. The thermophile, *Geobacillus Thermoglucosidasius*, has been used as the microorganism for the work. A metabolic reconstruction of the key pathways in a genetically modified strain was carried out. This was then combined with kinetic expressions of intracellular reactions to create a dynamic metabolic flux model that allowed us to visualise fluxes through the organism to products, cell growth and enzyme production. A hydrolysis model describing the breakdown of cellulose into cellobiose using Langmuir adsorption and competitive product inhibition was developed. This, combined with the metabolism model, has allowed us to follow carbon from the biomass through to the end product. A global sensitivity analysis was carried out on model parameters to identify key factors, as well as to locate possible sources of uncertainty in the model. By identifying the most important rate limiting steps we can focus future work on the areas that will have the greatest impact on performance.

465034 Extracting Kinetic Parameters from TGA Datasets - Ensuring Optimal Outcomes for Cellulose Pyrolysis

Abstract:

Thermal gravimetric analysis (TGA) has been used for decades to study the thermal decomposition of whole biomass and its separated component phases, cellulose, hemicellulose and lignin. Extraction of kinetic parameters from TGA datasets, however, has resulted in a wide range of activation energies and associated pre-exponential factors. A proposed global optimization strategy that incorporates a Nelder-Mead simplex algorithm was used to find *optimized* kinetic parameters for the thermal decomposition of crystalline cellulose. Visualization of the objective function topology illustrates the difficulty in locating the global optima. A number of plotting strategies are also illustrated as useful tools for the collective comparison of fit outcomes. A systematic study of the effects of search strategy, objective function choice and sample temperature uncertainty was shown to unify the historical findings.

466394 The Influence of Potassium Content and Heating Rate on Biomass Pyrolysis

Abstract:

A non-isothermal one-dimensional model has been developed to describe biomass pyrolysis at fast heating rate (600-10^4 K s^{-1}), high temperatures (up to 1500°C) and is valid for different biomass particle sizes (< 10mm). The model was developed to estimate the yields of volatile gas and char. The model relies on the concept applied in fast pyrolysis of cellulose through the formation of an intermediate liquid (so called metaplast) which reacts further to char and gas. The kinetics of the fast pyrolysis was described by the Broido-Shafizadeh scheme.

The influence of particle size and shape was included in the model. Cylindrical representation of a biomass particle shape was chosen to be the most suitable in the pyrolysis model. The evolution of devolatilization time required for the complete pyrolysis showed that the particles with a mean diameter < 0.45mm may be considered as thermally thin at high heating rates. The predicted results by one-dimensional model are in agreement with the experimental work, and emphasize a key role of intra-particle heat conduction in biomass particles > 0.45mm.

The potassium influence on the char yield was implemented in the model based on the experimental results in the wire mesh and drop tube reactors with respect to the stronger catalytic effect of potassium on the char yield at low and intermediate heating rates compared to pyrolysis at high heating rates. The heating rate and potassium content affected significantly the char yield as evidenced from the experimental data obtained in the wire mesh and drop tube reactors. Thus, the model including these two parameters provides an acceptable fit of char yield to the experimental data. The present results showed that the proposed kinetic model for the fast biomass pyrolysis is relatively simple and predicts reasonably accurately the char yield of woody and herbaceous biomass particles < 10mm using one fixed set of kinetic parameters valid for woody and herbaceous biomass.

471557 Synthesis and Characterization of Maleic Acid Aluminum Catalyst for Glucose Conversion

Abstract:

Maleic acid (MA) and $AlCl_3$ self-assemble into catalytic complexes (Al-$(MA)_2$-$(OH)_2$(aq)) with improved selectivity for converting glucose to value-added chemicals in water. Use of maleic acid and $AlCl_3$ compared to HCl and $AlCl_3$ shows an increase in selectivity (2.6 to 11) of fructose conversion to hydroxymethylfurfural at 180°C. We report the production of Al-$(MA)_2$-$(OH)_2$(aq) formed by reaction of maleic acid and $AlCl_3$ and confirmed by high pressure liquid chromatography (HPLC) and mass spectrometry. Reaction of 100 mM maleic acid and 100 mM $AlCl_3$ at 100°C for 24 hours shows a modest 25.9% production of Al-$(MA)_2$-$(OH)_2$(aq) with some degradation of maleic acid to malic acid. Increasing beginning concentrations to 500 mM maleic acid and 500 mM $AlCl_3$ shows an increase in selectivity (0.34 to 0.47) for Al-$(MA)_2$-$(OH)_2$(aq) over malic acid production. Storage of the refluxed liquid at 4°C yields formation of precipitate crystals containing 92.6% catalyst complex. Alteration of initial reflux concentrations to 500 mM maleic acid and 250 mM $AlCl_3$ shows an increase in Al-$(MA)_2$-$(OH)_2$(aq) production and purity (18.2% to 20.2%) in Al-$(MA)_2$-$(OH)_2$(aq) present in solution. Forming crystals by precipitation shows a change in morphology of formed crystals and an increase in purity to 94.6% compared to the equal molar reflux liquid. Rinsing crystals with deionized water removes residual maleic acid and malic acid yielding crystals with 99.2% purity catalyst complex. Future work will include a study on x-ray crystallography and a study on complex catalytic conversion of glucose including kinetics and conversion yield.

472913 Catalytic Hydropyrolysis of Milled Lignins Derived from Hardwood, Softwood and Herbaceous Biomass Using Bifunctional Catalysts

Abstract:

Catalytic hydropyrolysis of milled lignins derived from hardwood, softwood and herbaceous biomass using bifunctional catalysts

Yuan Xue[a], Ashokkumar Sharma[b], Xianglan Bai[a], Feng Cheng[c], Catherine E. Brewer[c]

[a] Department of Mechanical Engineering, Iowa State University

[b] Bioeconomy Institute, Iowa State University

[c] Department of Chemical & Material Engineering, New Mexico State University

Lignin is the second most abundant natural polymer after cellulose and accounts for 15-30% of biomass. Lignin is also abundantly produced by paper, pulping industries and biorefineries as a low value by-product. In fact, the phenyl-propane-based lignin polymer can be a potential source of aromatic hydrocarbons. Nevertheless, lignin is also known for its recalcitrant for thermal decomposition and catalytic deoxygenation due to its structural complexity and high char and coke yields. In this work, catalytic hydropyrolysis of lignin and its model compounds was studied using several different types of bifunctional zeolite catalysts at atmospheric pressure. Three milled lignins were derived from

red oak (hardwood), loblolly pine (softwood) and corn stover (herbaceous) to represent the natural lignins in different biomass species. Phenol and guaiacol were used to represent lignin-derived monomers. Catalysts tested are HZSM-5, Ni/ZSM-5, MoO_3/ZSM-5, Fe-HBeta, Ru/Fe-HBeta and Pd/Fe-HBeta, respectively. It was found that the hydrogen environment effectively increased the hydrocarbon yields from the lignins compared to the helium environment for all the catalysts. The catalytic coke with dramatically reduced yields and the aromatic hydrocarbons with an increased selectivity for benzene, toluene and xylene, were observed with the modified HZSM-5 catalysts. The model compound study shows that MoO_3/ZSM-5 and Pd/Fe-HBeta effectively hydrodeoxygenate phenolic monomers, whereas Ru/Fe-HBeta and Ni/ZSM-5 exhibit hydrocracking ability.

460000 Inhibition Effect during Catalyzed Biomass Char Gasification in Steam and Carbon Dioxide and Its Reversal

Abstract:

Gasification is a chemical process in which a gasifying agent reacts with carbonaceous materials to produce gases including high value CO and H_2. Gasification of renewable resources is a promising and environmentally friendly solution to the increasing global demand for energy and chemicals. The process takes place at high temperatures where biomass undergoes pyrolysis to rapidly form the biomass char. The char continuously reacts with the gasifying agent as the rate limiting step in the process. The gas-phase composition influences the gasification rate and is one of many variables that can be used for process improvement. Observations of gasification with low cost gases, steam (H_2O) and carbon dioxide (CO_2), have been reported extensively. The gasification reactivity behavior appears to vary among reports showing either the presence or lack of a synergistic effect for mixtures of these gases. The difficulty of predicting or explaining the optimum gasification rate stems from a lack of understanding the fundamental gasification mechanism in mixtures of steam and CO_2.

This study focused on understanding the mechanism in the evolution of gasification reactivity during conversion of char with CO_2, steam, and their mixtures. Bagasse char as starting material was prepared by pyrolysis of potassium-rich Brazilian bagasse in an inert gas pressurized entrained flow reactor. The char was then exposed to varying gasifying environments in a thermogravimetric analyzer to undergo full conversion. The initial gasification reactivity was higher in steam than in CO_2. However, overall and final reactivities were higher in CO_2 than in steam. Gasification inhibition was observed as a continuously decreasing reactivity with conversion with steam that was not observed with CO_2. On the other hand, CO_2 gasification rate showed a continuous increase with char conversion. The underlying mechanism was explored further by studying the transient behavior under exposure to step changes in model gas compositions of CO_2, steam, H_2, and O_2. It is found that the active sites for both CO_2 and steam gasification are initially the same. However, as the char gasification progresses in steam, the active sites are likely to be blocked by in-situ hydrogen product formation (â€œinhibitionâ€), which does not happen during CO_2 gasification. On the contrary, concentration of active sites is increased as the char gasification progresses in CO_2 due to the increasing K/C ratio. In this study, several methods were demonstrated for the reversal of the inhibition effect. A generalized mechanism, consistent with experimental observations, is proposed. The findings provide a key insight into the source of the elusive synergistic effect during gasification with steam and CO_2.

453986 Acetone Condensation over Ion-Exchanged Hydroxyapatite Catalysts

Abstract:

Biomass can be used to produce platform molecules such as ethanol, acetone, and hydroxymethylfurfural, but further processing is needed to convert renewable feedstocks into useful products. One promising solution is bimolecular coupling of alcohols or aldehydes to form higher-chained commodity chemicals and transportation fuels. Recent reports in literature have shown that hydroxyapatite (HAP; $Ca_5(PO_4)_3OH$) is efficient at catalyzing a variety of coupling reactions, such as aldol condensation, Knoevenagel condensation, and Guerbet coupling; however, the mechanism and site requirements for these reactions remain a point of debate. [1,2]

In this work, we investigate the gas-phase condensation of acetone to mesityl oxide over HAP as a model reaction. Using a post-synthesis ion-exchange method, a series of divalent cation-substituted HAP catalysts were prepared to understand how different elements influence reaction rates. Initial characterization by XRD and BET suggest that the surfaces of exchanged-HAP catalysts are enriched with the introduced dopant while retaining an apatitic structure. To overcome equilibrium limiations, catalysts were tested in a fixed bed reactor using a physical mixture of HAP and Pd/SiO_2 to hydrogenate unsaturated products with H_2. Our results show that the nature of the cation in HAP plays a key role in governing catalytic activity. Isotopic labeling and KIE experiments reveal that enolate formation and C-C coupling are fast and reversible under the range of reaction conditions tested. The rate-limiting step occurs after the formation of diacetone alcohol, which is an intermediate in the reaction. A relationship between key descriptors of HAP structure and microkinetic parameters of acetone condensation will be discussed.

[1] Hanspal, S.; Young, Z. D.; Shou, H.; Davis, R. J. *ACS Catal.* **2015**, 5, 1737â"1746.

[2] Ho, C. R.; Shylesh, S.; Bell, A. T. *ACS Catal.* **2016**, 6, 939â"948.

467494 Cyclic Polyethylene Furanoate As a Monomer from Renewable Resources for Ring Opening Polymerization

Abstract:

Introduction

The replacement of oil based chemicals by renewable resource based chemicals is one of the major issues chemical industry is facing right now. The alternative polymer poly (ethylene furanoate), PEF, is a possible substitute for poly (ethylene terephthalate), PET. Among different process alternatives, PEF can be produced at high polymerization rates and molecular weights via ring opening polymerization of cyclic oligo (ethylene furanoate), cyOEF. On the other hand, the effective production of these cyclic oligomers at high purity is still the major obstacle towards the process industrialization. In this work, different strategies aimed to produce these cyclic molecules from the two monomers Furan dicarboxylic acid (FDCA) and ethylene glycol (EG) are explored. And a model is presented to describe the equilibrium behavior of said cyclic oligomers.

Three pathways towards cyOEF

The three synthetic strategies shown in figure 1 have been investigated. In all cases, high dilutions in organic solvents are required to achieve high yields; the final product is a distribution of cyclic oligomers with different number of repeating units.

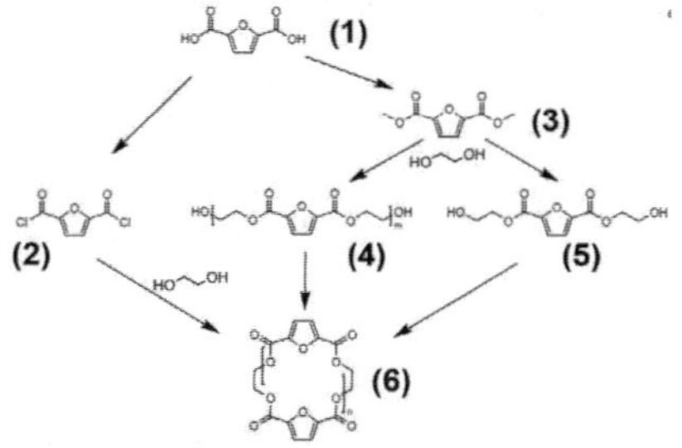

Figure 1 Reaction pathways for the production of cyOEF.

The first (scheme 1-2-6) is a fast cycle formation using the rapid esterification of acid chlorides with alcohols. Moderate selectivity towards cycles of around 50% are in contrast to a complex reaction setup and toxic chemicals involved for the chlorination. The product size distribution is kinetically controlled and consists mainly of C3, with C2 and C4 being the second most abundant species.

The depolymerization (scheme 1-3-4-6) and reactive distillation (scheme 1-3-5-6) are the two more promising routes towards cycles. For depolymerization FDCA and EG are converted to oligomeric chains of around 12 repeating units. This prepolymer is subsequently diluted in a high boiling organic solvent and converted to a mixture of cyclics and linears at 180-200 °C. In case of reactive distillation the FDCA and EG (in excess) are converted to very short chains (1-2 repeating units). The chains are then diluted in a high boiling organic solvent and finally converted to cyOEF by distilling a portion of the solvent together with the excess EG.

Both routes yield a similar distribution of cyclic species. Possible purification routes include precipitation and adsorption processes, which have both been applied to yield cycles with high purity starting from linear species (>99% by HPLC area). The yield of both reactions lies in the range of 60 to 80 % depending on the process conditions. One striking feature of this type of reaction is that it is possible to recycle the solvent and the linear side products in order to increase the yield of the process and furthermore make it more sustainable.

Modeling of the ring chain equilibrium

The equilibrium composition of cyclics and linears in 2-Methylnaphthalene and Dichlorobenzene (DCB) could be modeled using the Jacobson-Stockmayer model (Figure 2). Under high dilutions this equilibrium shifts towards the formation of cyclic oligomers. It is based on computing the probability of ring closure using the assumption of randomly coiled hydrocarbon chains for linear polymers. Two fitting parameters are involved: one is the equilibrium constant for the step growth polymerization, which is in equilibrium with the cyclisation reactions, and the second one is the effective length of the bonds in the polymer chains.

As shown in the figure below, our own results from depolymerization and reactive distillation experiments were fitted using the model. These results suggest that it is indeed possible to use this approach to describe the ring chain equilibrium. The model is currently under refinement in order to describe system-specific effects such as the complete absence of cycles with six repeating units.

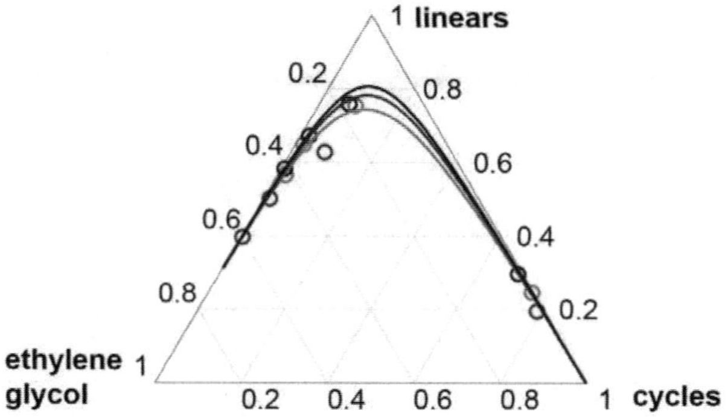

Figure 2 Equilibrium distribution of cyclic oligomers in DCB.

References

[1] Gomes M, Gandini A, *J. Polym. Sci., Part A: Polym. Chem.*, 49, 3759-3768 (2011).

[2] Burch, RR, Lustig, SR, Spinu, M *Macromolecules*, 33, 5053-5064 (2000).

[3] Stockmayer, WH, Jacobson H, *J. Chem. Phys.*, 18, 1600-1606 (1950).

453695 Polymeric Nano-Metal Composite Membranes for Water Remediation

Abstract:

Responsive polymers react to external stimuli like temperature, pH, radiation, ionic strength or chemical composition. These polymers in combination with nanomaterials provide novel hybrid properties such as bactericidal, electrical, hydrophilic, catalytic, selectivity, etc., and they can be self-supported or in a convenient matrix such as a membrane.

These membranes, functionalized with polymers having pH responsiveness (hydrogels), is a promising research field for numerous applications. The functional groups of the hydrogels gain or lose protons based on the pH, collapsing or extending their network, making them swell or un-swell solutions. These nanostructured materials can be exploited to synthesize metal nanoparticles and as ion exchangers. They also prevent the loss of nanomaterials to the environment and improve reactivity due to larger surface areas, expanding its range of applications to selective catalysis, remediation and mining.

The present work describes "green" techniques used to create nanocomposites based on polyvinilidene fluoride (PVDF) hollow fiber (HF) and flat sheet (FS) membranes, both sponge-like and thin. Hydrophobic HF membranes were synthesized with 57.2% porosity and 0.72 µm mean pore size. These membranes were then hydrophilized by a water-based process of cross-linking polyvinylpyrrolidone (PVP) by persulfate (PS). Though, hydrophilization of PVDF membranes using PVP is known in industry, as a post-process has only been implemented at lab-scale.

The PVP cross-linking is confirmed but also a ring-opening is present due to etherification. Hydrophilization was measured by contact angle changes (CA) (76.3° to 60.3°) and surface free energy calculations (40.3 to 49.8 $mJ \cdot m^{-2}$). HF and FS were subsequently functionalized with a poly(acrylic acid) (PAA) hydrogel through free radical polymerizations (thermal or redox) using PS as initiator. This work advanced FS membrane functionalization from lab-scale to full-scale by modifications of the polymerization procedures.

Swelling tests were performed to hydrogels/functionalized membranes finding responsiveness up to 1676% swelling at pH 9.0. Also, mesh size values (ξ) were calculated showing an inverse relationship with membrane permeability, tuning the flux through the membrane. An increase of ξ from 5.89 to 8.33 nm (pH 3.0 to 9.0) represented a 93% permeability decrease (53.9 to 3.4 $L \cdot m^{-2} \cdot h^{-1}$) at pressures between 1.0 and 4.0 bar.

Throughout a double ion exchange of sodium/iron and a subsequent reduction, bimetallic Fe/Pd nanoparticles (60 to 300 nm) were synthesized. The Fe exchanged was almost 500 mg Fe/g PAA. Similarly, it was possible to use the exhausted accelerants of Fe ions of the redox polymerization to synthesize nanoparticles, but having only 9.2 mg Fe/g PAA of loading.

Finally, depending on the Pd concentration in the membrane, the catalytic activity in the reduction of trichloroethylene (TCE) was studied, having enhanced reactivities (up to ten times) in sponge-like membranes compared to previous studies. Using hydrogels alone, similar results to nanoparticles in solution phase were found without nanoparticle loss. 80% of reduction was achieved within 2 hours with chloride production near to stoichiometric values (3/1), demonstrating absence of intermediates.

This research is supported by the National Science Foundation EPSCoR, and by the National Institute of Environmental Health Sciences (P42ES007380). HF membranes were synthesized at the Singapore Membrane Technology Center, NTU (Singapore). Nanostone Water (Oceanside, CA) provided support and collaboration in the development of full-scale flat sheet membranes.

454787 Thermodynamic and Economic Assessment of the Production of Ethylene and Propylene from Bioethanol

Abstract:

Introduction

The increasing worldwide pollution and the environmental problems associated with the petroleum extraction and use, have led to the development of new energy resources and sustainable raw materials [1]. The production of bioethanol from biomass offers a sustainable biotechnology solution at competitive costs with high availability. In Colombia (South America), bioethanol production has increased because the local policies encourage its production, being the third producer in America [2]. However, the focus on bioethanol production has been related to its use as biofuel. Alternatively, there are some high-commercial-value compounds —mostly produced from petroleum— that can also be obtained from renewable sources [3]. Recently, bioethanol has been proposed as a source for producing commodities such as light olefins in the framework of the biorefinery concept [4]. Among these biomaterials, ethylene stands out as one of the most produced chemicals worldwide, having a projected production of 400 million ton in 2020 [1]. In addition, the demand for propylene is growing faster, having a projected production of 100 million ton in 2021[5]. Therefore, this study aims to propose a conceptual design for producing ethylene and propylene from bioethanol samples procured in Colombia, integrating the reaction and separation stages and assessing its technical and economic feasibility. Aspen Plus and Aspen Hysys were used as process simulators.

Methodology

Aspen Plus V7.3 and Aspen Hysys V7.3 (Bedford, USA) were used as process simulators throughout the study. In both, the Soave-Redlich-Kwong (SRK) equation of state was used as thermodynamic model. First, the bioethanol dehydration reaction was analyzed in the RGibbs reactor module in Aspen Plus. Water/ethanol (S/E) feed ratio and reaction temperature were varied in the module in order to assess their effect on products selectivity. Ethylene, propylene and *n*-butene were selected as possible products, and a selectivity response surface was obtained in Matlab R 2014 (Salt Lake City, USA). Aspen Hysys was used to perform the sizing of the process units, heat integration, and process yields. Products purification was modeled in the Distillation Column module and the energetic analysis was performed using Heat Exchanger and Heaters/Coolers modules in Aspen Hysys. Finally, a class-V-order economic evaluation was done to the final flowsheets by determining the capital expenses for all units in the processes [6].

Results and analysis

Figure 1a shows the surface response for ethylene selectivity during bioethanol dehydration simulation, when varying S/E ratio and temperature. The higher the temperature and S/E ratio, the higher the ethylene yield, with a selectivity exceeding 60% beyond S/E = 5:1 and 600 °C. On the other hand, Figure 1b shows a maximum peak of 45% selectivity for propylene at 400 °C, and a slightly positive effect of the S/E ratio. *n*-butene formation also presented a similar trend (Figure 1c), showing a maximum selectivity of 14% at 300 °C. Heavier olefins are formed by ethylene condensation, which is thermodynamically favored at lower temperatures [7]. However, the absence of ethylene at these lower temperatures reduces the amount of this compounds.

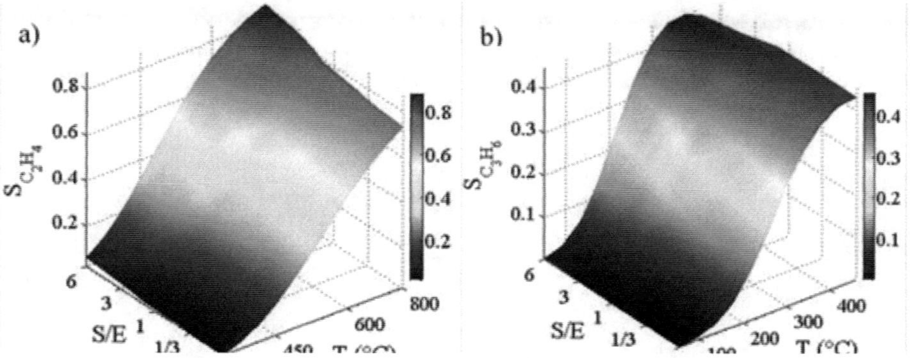

c) d)

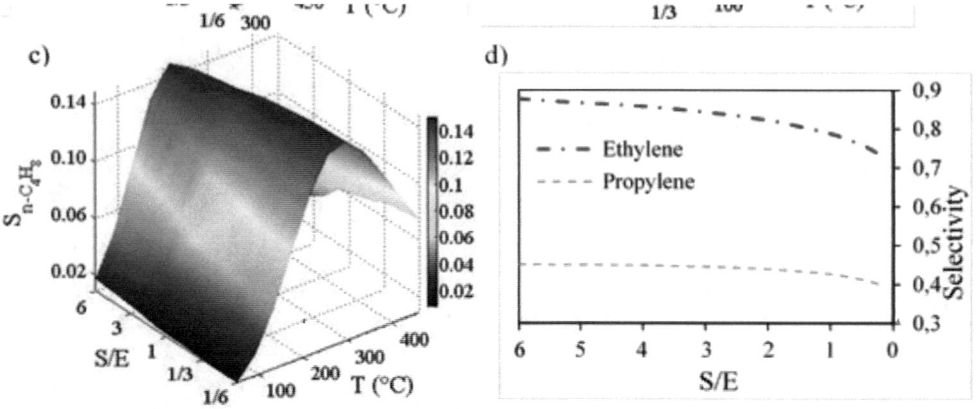

Figure 1. Matlab response surface from Aspen Plus simulated data for the production of (a) ethylene, (b) propylene, and (c) *n*-butene using different water:ethanol ratios (S/E) in the temperature range of 200–800 °C; and (d) the effect of the S/E ratio on the maximum ethylene and propylene yields.

Figure 1d shows the effect of the S/E on the maximum both ethylene and propylene selectivity. The excess of water is slightly positive for the ethylene formation because its acts as a quasi-inert diluent from a thermodynamic point of view [7]. In this way, a bioethanol dehydration reactor feed with a stream of S/E = 5 and operating at 615 °C was selected as the proper design to produce ethylene. In addition, a dehydration reactor operating at 400 °C was selected to produce propylene. These reactors were included in two different processes to analyze the conceptual design of two plants, one producing ethylene and another producing propylene from bioethanol.

Figure 2 shows the designed processes for producing (a) ethylene and (b) propylene from bioethanol. The processes begin with the feed stream of 300,000 ton/year of raw, fermented bioethanol (5 mol% ethanol, S1), which corresponds to the 10% of the national bioethanol production. This stream is heated to 40 °C before its further introduction to a flash separator V-100, which operates at 95 °C and 1 atm. These conditions were established according to the ethanol-water equilibrium to obtain an S/E = 5:1 (16.4 mol% ethanol). After V-100, the product stream rich in ethanol (S3) is heated to either 615 °C or 400 °C (reaction temperatures for ethylene or propylene production, respectively) by a recycle of the flash bottoms stream (S4). S4 is rich in water and is used to the process heat integration. The product streams (S7 in Figure 2a and S6 in Figure 2b) are pressurized to 19 atm in the compressor K-100. The pressurized streams are cooled to 18 °C by fresh water streams. These streams would deliver the energy duty for heating streams (S20 in Figure 2a and S19 in Figure 2b). Subsequently, the cooled streams are used as an inlet in a three-phases flash separator V-101 with the purpose of removing the residual water after the reaction. Finally, the product streams are introduced into distillation columns to separate the olefins.

The ethylene process delivered a stream of 99.7 mol% ethylene with an almost 100% recovery (S16 in Figure 2a), producing 0.60 mol ethylene/mol ethanol feed. In addition, propylene process delivers a stream with 92.0 mol% propylene in S15 in Figure 2b. This corresponds to the 70% of the propylene entering to the distillation tower, producing 0.43 mol propylene/mol ethanol feed.

A class-V economic assessment delivered investments of USD 2 million for producing ethylene and USD 2.23 million for propylene. These costs represent 159 USD/ton-year of ethylene and 327 USD/ton-year for propylene. True [8] reported investment costs of 1300 USD/ton-year for ethylene and 1450 USD/ton-year for propyylene produced from petroleum for a plant constructed in 2016. However, these costs include lumpsum, engineering, procurement, construction and commissioning turnkey contract; which not were included in the current analysis. Nevertheless, this class-V economic assessment shows that olefins production from bioethanol could be a profitable and sustainable alternative to petrochemical processes.

b)

a)

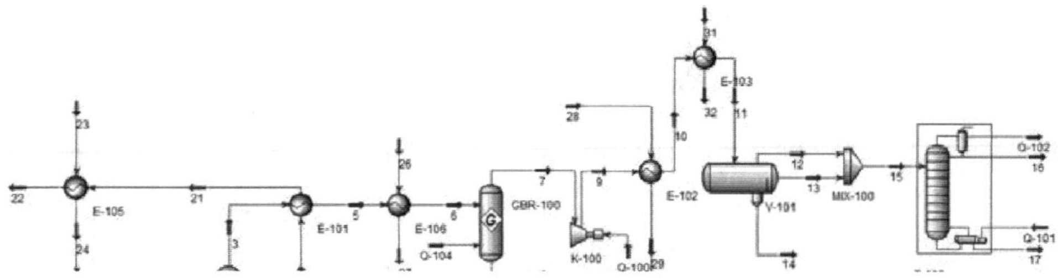

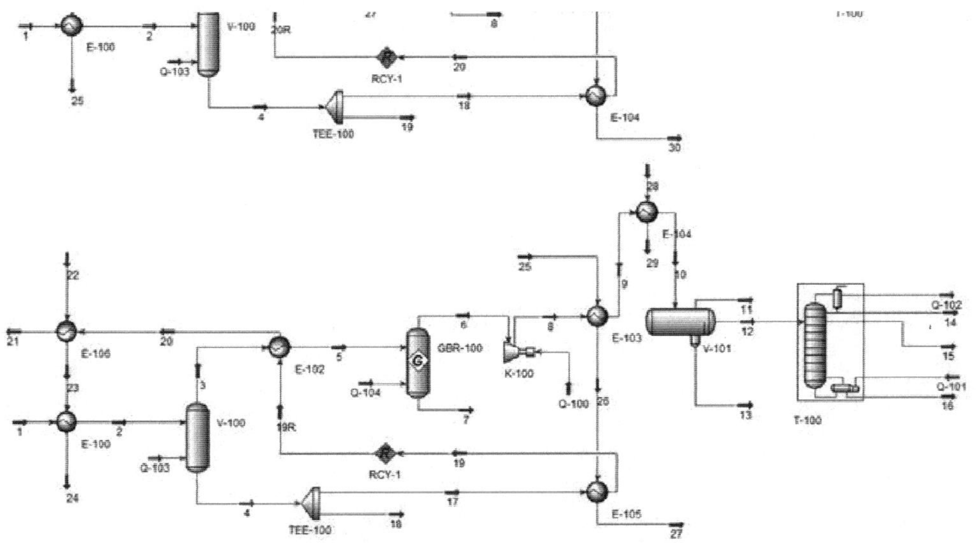

Figure 2. Process flowsheets designed in Aspen Hysys for producing (a) ethylene and (b) propylene from bioethanol. Blue lines represent mass streams and red lines represent energy streams. Equipment notation: E: Heat exchanger, V: Flash separator, K: Compressor, MIX: Mixer, T: Distillation tower, TEE: Splitter, and GBR: Gibbs reactor.

Conclusion

This study evaluated a conceptual design for producing light olefins from bioethanol base on thermodynamic predictions and using commercial software. To the best of our knowledge, it is the first time that reaction and separation stages are integrated and economically evaluated in two different processes, one to produce ethylene and another to produce propylene from bioethanol. Due to water showed a positive effect on the ethanol dehydration reaction, both processes start with a simple flash separator operating at 95 °C and 1 atm to reach the ethanol initial concentration needed in the reactors (S/E=5, 17.6 mol% ethanol). Two different reactors were designed, one operating at 615 °C for producing ethylene and another at 450 °C for producing propylene. The designed ethylene process delivered a stream of 99.7 mol% ethylene with a 60% yield and an investment cost of USD 2 million, equivalent to 159 USD/ton-year of ethylene. On the other hand, propylene production from bioethanol delivers a stream with 92 mol% propylene and 43% yield. Propylene process has an investment cost of USD 2.23 million (327 USD/ton-year propylene). These conceptual analysis show the technical, environmental and economic feasibility of the biorefinery concept development.

References

[1] H. Xin, X. Li, Y. Fang, X. Yi, W. Hu, Y. Chu, et al., J. Catal. 312 (2014) 204–215.

[2] G.P. Ortegón, F.M. Arboleda, L. Candela, K. Tamoh, J. Valdes-Abellan, Sci. Total Environ. 539 (2016) 410–9.

[3] O.J. Sánchez, C.A. Cardona, Bioresour. Technol. 99 (2008) 5270–95.

[4] R. Le Van Mao, T.M. Nguyen, G.P. McLaughlin, Appl. Catal. 48 (1989) 265–277.

[5] Ceresana, Market study: Propylene, (2014). http://www.ceresana.com/en (accessed May 1, 2016).

[6] American Association of Cost Engineers, (2015). http://www.aacei.org/ (accessed November 15, 2015).

[7] T. Lehmann, A. Seidel-Morgenstern, Chem. Eng. J. 242 (2014) 422–432.

[8] W. True, Oil Gas J. 110 (2012).

458322 Design and Economical Evaluation of Polygen Process to Co-Produce Synthetic Natural Gas (SNG), Methanol and Ethylene Glycol

Abstract:

Design and Economical Evaluation of POLYGEN process to co-produce Synthetic Natural Gas (SNG), Methanol and Ethylene Glycol

Bor-Yih Yu and I-Lung Chien*

Department of Chemical Engineering, National Taiwan University, Taipei 106, Taiwan

The steady state design and economical evaluation of poly-generation (POLYGEN) process to co-produce synthetic natural gas (SNG), methanol and ethylene glycol (EG) is studied in this work. POLYGEN has been a widely studied topic recently, in which several different chemicals can be produced parallel at the same time. In POLYGEN process, the relative production rates for different chemicals can be adjusted based on different market demands, daily usages, and also changing political strategies. This character enhances the flexibility of plant-wide operation and also the economic performance. Among the products, SNG holds very similar composition and heat value to typical natural gas, and can be used as a replacement in industrial and home usages. Methanol is one of the most important industrial feedstocks, which can be the basis of various kinds of chemicals. Ethylene glycol is also an important organic chemical, which can be used to synthesize polyesters, antifreezes, solvents, and so on.

The items covered by this work are illustrated in the following. In coal-based POLYGEN processes, syngas is first generated from coal gasification in a CO2-rich atmosphere, and the raw syngas is then passes the treating stage. Syngas treating stages contains two section, namely, sour water-gas-shift-reaction (SWGSR) and acid-gas-removal (AGR). In SWGSR, excess amount of steam is added to adjust the H2/CO ratio. After that it is cooled to low temperature to enter AGR. In AGR, SELEXOL method is used to capture 99.7% of H2S and 92% CO2. Part of the captured CO2 is sent back to gasification section in order to reduce the carbon emission, and the other part is compressed and sequestrated. With proper arrangement in configuration, three streams with qualified composition to be the feedstocks for downstream chemical plants are produced.

For downstream chemical plants, the products produced will meet the industrial standard. The SNG has heat value higher than 965 (BTU/scf), the methanol purity is 99.5 (mol%), and EG purity is 99.8 (mol%). EG is traditionally synthesized through petroleum route, a novel route from coal-derived syngas has been widely considered a potential alternative as the oil shortages continuing. Thus in this work, the production of EG from syngas is adopted. Firstly, CO and H2 in syngas are separated, and are used in different reaction stages. In the first stage, CO is reacted to form dimethyl oxalate (DMO), and in second stage, DMO is further reacted with H2 to form ethylene glycol.

After the design, optimization works are studied based on minimized total annual cost (TAC), and the economical evaluation is performed to investigate the overall economic performance. Besides, some waste heat could be recovered by generating steam, and the steam is used to drive turbine to produce electricity for usage inside the plant-wide process. Thus through this work, a better understanding of POLYGEN process, and the design and operation of coal to ethylene glycol route is reached.

458992 Preparation and Properties of Soybean Oil-Based Composites Containing Natural Fillers

Abstract:

Biobased composites have been prepared, which consist of a vegetable-oil based matrix, with ground nutshell or eggshell as natural, rigid, filler materials.

Crosslinked polymers of acrylated epoxidized soybean oil (AESO) and styrene were first successfully produced across the entire composition range via bulk free radical polymerization. Mechanically, these copolymers varied from hard and rigid at high styrene contents, to soft and rubbery at high AESO contents.

The intermediate 50:50 copolymer was chosen as the matrix for composites containing varying amounts of pecan nutshell flour or powdered eggshells. Composites were produced by pre-blending the fillers into the comonomer mixture before reaction and solidification. Through the AESO and filler contributions, these composites contain substantial fractions of biobased starting materials.

Copolymers and their composites exhibited good structural integrity, and were characterized by flexural testing to fracture, dynamic mechanical analysis, SEM imaging, as well as hardness and impact testing.

Addition of the natural fillers to the matrix resulted in substantial increases in hardness and stiffness, with a nearly six-fold increase in modulus observed at 35 volume% eggshell filler. These increases were accompanied by modest decreases in breaking strength and impact energy, and no observed effect on the glass transition temperature of the matrix. Use of coupling agents [1] to increase adhesion and potentially further modify mechanical properties will be discussed.

General comparison of the filled polymer properties to a baseline of neat polystyrene shows improved or similar individual mechanical characteristics, indicating viability of such biobased composites as a replacement material.

[1] Nachtigall, S.M.B., Cerveira G.S., Rosa S.M.L., Polymer Testing, 26(5), 2007.

460812 Solvent Free Sucrose Esters Production in Reactive Systems Containing Emulsifiers

Abstract:

Surfactants are chemicals used in a wide variety of applications, mainly in the manufacture of consumer goods, industrial cleaners, food-feed, and cosmetic products. In the past, efforts have been made to develop biobased surfactants from renewable raw materials in order to reduce the environmental impact of traditional petroleum-based surfactants. Currently, most biobased surfactants available in the market are lipids-based, but still they contain polar moieties generated from fossil resources (e. g. sulfonic acid, ammonium, or poly-ethoxylated groups) (1). In contrast, in completely biobased surfactants (e. g. acylglycerols, alkyl-poly-glucosides, glycosides of fatty acids) both, the polar and non-polar structures are of renewable origin.

Among biobased surfactants, fatty acid sucrose esters stand out as important additives for the food industry (food ingredient number: E473). These are generally produced by transesterification of sucrose (from sugar cane) with fatty acid methyl esters (from vegetable oils). In addition to their renewable character, sucroesters are high value added, biodegradable and biocompatible surfactants, which can also be used in specialized applications such as cosmetics, clean and care products and pharmaceuticals. Recent estimations (2) indicate that the current market of sucroesters is around than 6000 t/year, with prices around 4 to 20 USD/kg depending on the nature of the product (i.e. lipid chain, degree of substitution, etc).

Sucrose is a polyalcohol with eight hydroxyl groups (See Figure 1), so theoretically, sucrose esters could be monoesters, diesters, triesters, and, so on (up to octaesters). As a result, tuning of reaction selectivity to a specific product is a major challenge. Besides, sucroesters properties and application are heavily determined by the degree of esterification. For example, sucrose monoesters are good water-in-oil emulsifiers, while sucrose octaesters are good oil-in-water emulsifiers. In general, the degree of substitution depends of different process conditions such as temperature, reactants molar ratio, catalyst nature and concentration, and the use of other additives (e. g. solvents, emulsifiers).

In addition to controlling the degree of substitution, a major challenge of the process is the incompatibility between reactants; only 0.1 g of sucrose can be dissolved in 100g of FAME at 140°C (3). This reduced solubility limits the mass transfer and reduces the reaction rate and conversion. A processing alternative to overcome this problem is the use of emulsifiers, which enhance contact and solubility of sucrose in FAMEs (4–7). Some reports indicate that fatty acid soaps (mono- and divalent), sucrose esters (the product itself), and mono- and diacylglycerols can be used as emulsifiers for the reaction. However, it is necessary to study the influence of such emulsifiers (nature and concentration) on the transesterification process (i.e. rate of reaction and esterification degree).

In this work, conversion and selectivity of transesterification reaction to produce sucroesters was experimentally evaluated using three different emulsifiers (potassium palmitate, sucrose palmitate and glyceryl monoester) at different concentrations (up to 20%w). Reactions were carried out in a 150 ml glass-jacketed reactor coupled to a methanol total condenser. The system was homogenized with an Ultraturrax® at 15000 rpm prior to reactions, and maintained under agitation at 600 rpm at constant temperature (110-140 °C) until no further evolution of methanol was observed. FAMEs consumption and sucroesters production were tracked by HPLC analysis of samples. Observed conversion were around 40% and mono substitution increased with the concentration of the emulsifying agents. Higher conversions were obtained when potassium palmitate was used as emulsifier, and reaction times were shorter when emulsifier's concentration was increased. Degree of substitution of the final product was higher when a mixture mono- and diacylglycerols was used as emulsifier and lower when sucrose palmitate was used. Results of this study can be used for preliminary process design in a solvent-free production of sucrose esters with different emulsifying properties.

Figure 1. Emulsifiers used to provide contact in sucrose and fatty methyl estermixtures (Molecular structures taken from ChemSpider)

References

1. HAYES, Douglas G. ***Biobased Surfactants : A Useful Biorefinery Product That Can Be Prepared Using Green Manufacturing***. Knoxville, TN, USA, 2012.

2. OTOMO, Naoya. Basic properties of sucrose fatty acid esters and their applications. In : HAYES, Douglas G., KITAMOTO, Dai, SOLAIMAN, Daniel K. Y. and ASHBY, Richard D. (eds.), ***Biobased Surfactants and Detergents***. Urbana, Illinois : AOCS Press, 2009.

3. ZHAO, Rongrong, CHANG, Zhidong, JIN, Qiong, LI, Wenjun, DONG, Bin and MIAO, Xiaowen. Heterogeneous base catalytic transesterification synthesis of sucrose ester and parallel reaction control. ***International Journal of Food Science & Technology***. 2014. Vol. 49, no. 3, p. 854–860. DOI 10.1111/ijfs.12376.

4. PARKER, Kenneth J., JAMES, K. and HURFORD, J. Sucrose Ester Surfactants — A Solventless Process and the Products Thereof. In : HICKSON, John L. (ed.), ***Sucrochemistry***. Washington, DC : American Chemical Society, 1977. p. 97–114.

5. FITREMANN, Juliette, QUENEAU, Yves, MAITRE, Jean-paul and BOUCHU, Alain. Co-melting of solid sucrose and multivalent cation soaps for solvent-free synthesis of sucrose esters. ***Tetrahedron Letters***. 2007. Vol. 48, p. 4111–4114. DOI 10.1016/j.tetlet.2007.04.015.

6. GALLEYMORE, Harry R, JAMES, Kenneth, JONES, Haydn F and BHARDWAJ, Chaman L. Process for the production of a surfactant containing sucrose esters. 4298730. 1981. United States.

7. RIZZI, George P and TAYLOR, Harry M. Synthesis of higher polyol fatty acid polyesters. 3963699. 1976. United States.

461686 Synthesis and Characterization of Biochar-Based Carbon Supported Metal Nanoparticles

Abstract:

Synthesis and characterization of biochar-based carbon supported metal nanoparticles

Sai Teja Neeli and Hema Ramsurn

Russell School of Chemical Engineering, University of Tulsa, Tulsa, OK 74104

Biochar is a carbon-rich solid formed after biomass has been carbonized. It is also a residue from a number of thermochemical conversion processes whereby biomass is converted to bio-oil and/or syngas. Therefore, this material is environmental-friendly and carbon-neutral. Traditionally, biochar has been used as soil amendments for the healthy growth of plants due to its retention capability of nutrients. This investigation aims to use biochar in a different manner: as a support for metal catalysts to promote green chemistry and engineering. The biochar used in this study is obtained from hydrothermal carbonization of model compounds of biomass namely lignin, cellulose and hemicellulose. This means that high pressure and temperature water is used as a benign solvent to depolymerize the model compounds to biochar. The biochar so produced is chemically activated with 0.1 M HNO_3 solution resulting in activated carbon with defined micropore size distribution, high micropore volumes and large specific surface areas. Metal precursors (like iron nitrate) are then added to the biochar and dried in a convection oven at 110 ^{o}C for one week to obtain metal (iron) impregnated biochar. The dried mixture is thermally-treated in a quartz tubular reactor at 1000 ^{o}C for one hour in a nitrogen flow of 500 ml/min. These synthesized biochar based metal catalysts will then be characterized using a number of techniques including scanning electron microscope (SEM), x-ray powder diffraction (XRD), inductively coupled plasma atomic emission spectroscopy (ICP-AES) and Brunauer-Emmett-Teller (BET)

surface area analyzer. Other metal catalysts such as platinum, gold and nickel can also be impregnated on the biochar in the same manner. It is expected to observe distinct XRD patterns corresponding to graphite for the thermally treated metal-impregnated biochar, due to the conversion of the carbonaceous materials to graphite. SEM images will help not only in sizing the metal nanospheres (about 20 -60 nm) but also in determining how the metal is supported: half embedded or completely enclosed in the porous char matrix. Based on the characterization results, these metal-impregnated biochar catalyst could be tested for a number of reactions. These supports can withstand high temperatures and can even be used for gasification reactions and upgrading bio-oil during hydrogenation reactions. There is also the possibility of using bifunctional metal/acid catalysts on the biochar nanoparticles for dehydration and hydrogenation reactions and for aqueous phase processing to produce selectively targeted alkanes from glucose. The ultimate goal is to use these green catalysts to upgrade the oxygenated biofuels by removing the oxygen through various mechanisms and hence increasing the quality and heating value of the desired product.

461524 U.S. Power System Outlook: A Range of Perspectives from a Suite of Standard Scenarios

Abstract:

U.S. Power System Outlook: A Range of Perspectives from a Suite of Standard Scenarios

Wesley J. Cole and James Richards

Strategic Energy Analysis Center, National Renewable Energy Laboratory, Golden, CO, USA

The U.S. electricity system has undergone significant change over the last decade with declines in coal powered generation and increases in natural gas and renewable energy generation [1]. Many of these changes were largely unexpected. For example, before the shale gas revolution, natural gas prices were expected to be much higher than they are today [2], and solar photovoltaic capital costs have declined much more rapidly that most anticipated (see e.g., [3] vs. [4]). While continued regulation, such as the Clean Power Plan, and expected cost declines in renewable energy costs, may support a continuation of these trends, there is still tremendous uncertainty in terms of what a future power sector will look like. This work aims to identify a range of possible future electric system generation mixes via scenario analyses of some key sensitivities in the power sector. We present scenario results from the National Renewable Energy Laboratory's suite of Standard Scenarios, coupled with future technology cost analysis, to better understand how the power sector might evolve under various futures. Specifically, we consider electricity sector evolution with various natural gas price trajectories, technology costs, and policy drivers. We show how the capacity mix and resulting CO_2 emissions change as different assumptions are applied. The range of outcomes of these possible futures can provide insight into which drivers can have substantial impact on how the system develops.

References

[1] J. Logan, K. B. Medlock III, and W. C. Boyd, "A Review of Sector and Regional Trends in US Electricity Markets: Focus on Natural Gas," National Renewable Energy Laboratory, Golden, CO, NREL/TP-6A50-64652, 2015.

[2] EIA, "Annual Energy Outlook 2010," U.S. DOE Energy Information Administration, Washington, DC, DOE/EIA-0383(2010), 2010.

[3] Black & Veatch, "Cost and Performance Data for Power Generation Technologies," Black & Veatch Corporation, Overland Park, KS, 2012.

[4] NREL, "NREL Annual Technology Baseline (ATB)," 2015. [Online]. Available: http://www.nrel.gov/analysis/data_tech_baseline.html. [Accessed: 21-Apr-2015].

457798 Economics and Dynamic Flexibility of Concentrated Solar Power Technologies

Abstract:

Most techno-economic studies of concentrated solar power (CSP) systems only consider levelized cost of electricity (LCOE) to quantitatively compare economics and performance. This is problematic as LCOE neglects the time varying value of electricity and additional revenue opportunities from providing ancillary services. As consequence, LCOE-centric analyses undervalue the flexibility provided by thermal energy storage and misrepresent the true economics of owning and operating CSP systems [1, 2]. For example, a CSP plant providing 10 MW of regulation capacity (an ancillary service) for all hours of 2015 in the California energy market would have received $500,000 in capacity payments alone. Similarly, shifting 10 MW_e of generation from the average price (30 \$/MWh) to the 1% most extreme prices (97 to 1,621 \$/MWh) yields additional revenues of $400,000/yr. Thus selection of the appropriate economic and performance metrics is paramount when evaluating CSP systems and comparing against other energy technologies.

Beyond economics, cooling system water usage is increasing in importance. CSP systems, in particular, operate in water-sparse areas and thus water conservation is priority if the technology is to be competitive. Cooling systems influence overall system efficiency both due to primary thermal (Rankine cycle) efficiency impacts and parasitic loading associated with cooling system operation, which in turn impact system economics. Most CSP optimization literature, however, focuses only on economic metrics and does not consider water usage.

Compressive analysis of market-based revenue potential and water usage for CSP systems is especially complex because it depends on the selected operating policy. In several studies, revenue estimation is formulated as an optimization problem: manipulate the operational policy (e.g., mass and energy flows) to maximize market revenue subject to physical constraints [3 - 5]. Broadly speaking, CSP operational methods may be classified into two categories: in revenue-focused studies (e.g., [3 - 5]), emphasis is placed on capturing market rules and start-up and shutdown physical restrictions (e.g., minimum up/down times). Simplified correlations (e.g., piecewise linear approximations) are commonly used to model the solar collector and Rankine cycle, and 1 hour time discetizations are considered. In contrast, control focused studies (e.g., [6 - 8]) use much finer time discretizations (e.g., minutes) and consider nonlinear (partial) differential algebraic equation models to capture CSP system physics. These studies neglect market rules and start-up/shutdown limitations, and typically seek to maximize solar energy production or similar metrics. As such, neither of these approaches are capable of analyzing revenue opportunities available at multiple timescales from day-ahead (hour frequency) and real-time markets (minutes frequency).

We present a unified framework that considers market rules, start-up/shutdown restrictions and detailed system physics to analyze multiple CSP performance metrics, especially profits and water usage. The electricity market mathematical model is sufficiently detailed to evaluate revenues from selling/providing both energy and ancillary services, including regulation and (non)-spinning reserve capacity, in the California day-ahead and real-time markets. The CSP physics models include mass and energy balances and nonlinear equipment performance correlations. They are sufficiently complex to resolve temperature profiles in the thermal storage tanks and study the impact of storage dynamics on system performance. Revenue estimation and operational policy determination are formulated as mixed integer nonlinear programs (optimization problems). We present a decomposition strategy that uses physics inspired relaxations to efficiently find approximate solutions that are feasible for all of constraints and locally optimal with respect to a fixed start-up/shut-down schedule. The framework is implemented in Julia/JuMP (an open-source programming language/modeling environment) and leverages large-scale and efficient numerical optimization solvers such as IPOPT and Gurobi.

References:

[1] P. L. Joskow, Comparing the cost of intermittent and dispatchable generation technologies, American Economic Review: Papers & Proceedings 100 (3) (2011) 238-241.

[2] E. Lizarraga-Garcia, A. Ghobeity, M. Totten, A. Mitsos, Optimal operation of a solar-thermal power plant with energy storage and electricity buy-back from grid, Energy 51 (2013) 61-70.

[3] S. H. Madaeni, R. Sioshansi, P. Denholm, How thermal energy storage enhances the economic viability of concentrating solar power, Proceedings of the IEEE 100 (2) (2012) 335-347.

[4] J. L. Gonzalez, I. Dimoulkas, M. Amelin, Operation Planning of a CSP Plant in the Spanish Day-ahead Electricity Market, in: 11th International Conference on the European Energy Market (EEM), 2014, pp. 1-5.

[5] J. Usaola, Operation of concentrating solar power plants with storage in spot electricity markets, IET Renewable Power Generation 6 (1) (2012) 59 - 66.

[6] K. M. Powell, T. F. Edgar, Modeling and control of a solar thermal power plant with thermal energy storage, Chemical Engineering Science 71 (2012) 138 - 145.

[7] K. M. Powell, J. D. Hedengren, T. F. Edgar, Dynamic optimization of a hybrid solar thermal and fossil fuel system, Solar Energy 108 (2014) 210-218.

[8] E. F. Camacho, A. J. Gallego, Optimal operation in solar trough plants: A case study, Solar Energy 95 (2013) 106-117.

460426 Leveraging Storage and Hybridization to Maximize Renewable Utilization

Abstract:

In the pursuit for renewable power generation, a sentiment of competition between renewables and fossil fuels has emerged among much of the population. Renewables are carbon free and sustainable, but lack the reliability of fossil fuels, which can be dispatched on demand. Hybridization is often overlooked as at least an interim solution. By hybridizing a system, the benefits of each type of technology is realized, with the emissions benefits of renewable generation and the reliability of a fossil fuel system. In this work, a hybrid solar thermal and natural gas concept is explored. A central receiver solar concentrating system is used to pre-heat compressed air before it enters a conventional natural gas combined cycle power generation system. High temperature packed-bed thermal energy storage (pre-gas combustion) is used to increase the solar share of the plant. With the plant operating in hybrid mode, the flexibility of the storage and the supplemental natural gas are used in an optimization-based control scheme to maximize the amount of solar energy collected. This configuration demonstrates that by using a fossil fuel with a variable source of renewable energy, a certain synergy is achieved, with the system flexibility being leveraged to maximize the amount of solar energy harvested. The study presented is based on experimental and simulation results and highlights many of the feasibility challenges associated with achieving such a concept at grid scale.

463938 State of Health Estimation Method Design for Energy Storage System of Lithium Ion Battery and Comparative Study : From Cell to Demonstration Sites

Abstract:

State of health Estimation method design for Energy Storage System of Lithium ion battery and comparative study : from cell to demonstration sites

Keonhee Park, Ph.D. candidate
Seoul National University

Recently, Increasing demands of renewable energy and necessity of stable electricity supply are accelerating ESS(Energy storage system) market. ESS, which can stores the power and supplies electricity when it is needed, is possible to increase efficiency of electricity-usage and still has converted from production-consumption paradigm of electricity system to production-storage-consumption paradigm of that so far. ESS is comprehensive system including EMS(Energy Management system), BMS(Battery management system), PCS(Power conversion system). BMS has important role for sustainable ESS operation. Main state variables of BMS are SOC(State of charge) and SOH(State of health). SOC means current residual capacity based on present nominal capacity. And SOH is a prominent indicator to determine degradation of battery. Performance of repeated cycled battery decreased and it should be replaced at an appropriate time for usability. Imprecise SOH can affected SOC accuracy which is important variable for on-line monitoring. However, SOH accurate estimation is challenging due to few obtainable output variables for the battery system and the lack of understanding for electrochemical phenomena in cell. On-line SOH estimation method is needed since a lot of ESS application sites are isolated. This study proposes the modified online SOH estimation method, and compares estimation results of related SOH estimation methods. The results are validated with

experimental cell data and demonstration operation data. This method uses the equivalent circuit model parameters and recursive least square method. The cell experimental data is given by Korea testing certification(KTC) and demonstration data is obtained by 2 sites, which are interconnected with PV power system. The one is installed in micro-grid system in the island and the other is integrated with plant site. From the test results show maximum 2% accuracy in cell data and 5% accuracy in demonstration data. The accuracy of proposed method is dependent on input data quality highly. Also we can estimate the life time of battery system and implement easily without complex change of BMS algorithm.

464641 The Effect of Centralized Energy Storage on an Integrated Distributed Photovoltaics/CHP System for District Power, Heating and Cooling

Abstract:

The rise in availability of Distributed Energy Resources (DERs) has spurred utility companies and grid operators to search for methods to incorporate these resources (i.e., photovoltaics and energy storage) in the energy infrastructure in a manner that keeps the customers connected to the grid without jeopardizing grid stability and reliability. In the past decade, the explosive spread of photovoltaic generation (both rooftop and centralized) has lowered energy demand from conventional generation methods (i.e., power plants) during the afternoon. The result is the famous "duck curve" for conventional electricity demand, with medium demand in the early morning, low demand in the afternoon (when the sun is shining), and a sharp increase to high demand in the evening when the sun sets [1]. While the growth trend for solar photovoltaic generation has important environmental benefits, it can –if not properly managed – lead to an over-generation situation. In this case, demand during the day can be satisfied largely using solar power; however, the increased evening demand, corroborated with a natural drop in solar photovoltaic generation, may require very high power plant ramp rates. Most current power plants are unlikely to be able to withstand such rapid changes in generation rate. Motivated by this, significant research efforts have been expended towards defining strategies for controlling photovoltaic generation and grid-level storage systems. However, there are very few published works exploring the influence of photovoltaics and energy storage on the *design* of future power generation plants.

In our previous work [2], we provided a framework for simultaneous optimization of design and operation strategy for a CHP plant, providing the ability to select equipment sizes as well as plant operation modes to meet the projected electricity, heating, and cooling demand of a residential neighborhood. Based on realistic data collected from Pecan Street Research Inc., our results revealed that in the seasons with low electricity and cooling demand (winter and spring, considering a predominantly cooling climate), rooftop photovoltaic generation may exceed the neighborhood demand, while in the warmer seasons, the cooling capacity can often exceed the system's needs.

Motivated by the above, in this paper, we study the same system operating in an islanded (i.e., grid-disconnected) mode, and consider the impact of a centralized energy storage facility on system design and operations. We discuss an extension of our formulation for simultaneously optimizing the design and operation of the CHP plant and the centralized energy storage to minimize the capital and marginal costs of the CHP plant including transition penalties accrued by the equipment turning on and off. Using the aforementioned data collected from a neighborhood located in Austin, TX, we investigate the benefits of energy storage relative to the level of photovoltaic integration from the neighborhood, as well as identify economic benefits from photovoltaic integration and energy storage in terms of plant savings.

References:

[1] CAISO (2016). *What the duck curve tells us about managing a green grid.* https://www.caiso.com/Documents/FlexibleResourcesHelpRenewables_FastFacts.pdf

[2] Ondeck, A.D., Baldea, M., Edgar, T.F. *Simultaneous Optimization of Design and Operation Strategies for CHP Systems.* Computers & Chemical Engineering. Manuscript in preparation.

468128 Reducing the Cost of Operational Water in Military Water Systems: A Data-Driven Modeling and Optimization Study

Abstract:

Reducing the Cost of Operational Water in Military Water Systems: A Data-Driven Modeling and Optimization Study

Corey James, Michael Webber and Thomas Edgar

Department of Chemical Engineering

The University of Texas at Austin, 1 University Station C0400, Austin, TX 78712

email: cmjames@utexas.edu

Military water systems are tasked to provide safe and clean water to support soldiers, their families and mission related activities. Due to the historic low price and relative abundance of water, there has been little motivation to ensure that water systems minimize water waste after treatment. Currently, water is treated for disinfection as needed to meet demand. The concentrations of common disinfecting agents (chloramines or chlorine) begin to exponentially decay the moment they are placed in the water [1]. A system that treats water without rigorously considering unique demand data and the temperature dependent nature of chloramine decay will inevitably waste water. Waste is created when excess water is treated and placed into the system that exceeds demand. The volume that exceeds demand resides in the system long enough to run out of "lifetime", where the concentration of chlorine in the treated water decays below the legal safe threshold. This water must be flushed, or wasted, in order to ensure a safe water source. Embedded in the water waste is the energy and cost required to pump and disinfect that volume of water. A system that is not optimized to account for chlorine decay could cause thousands of gallons of water wasted on a monthly basis, not to mention the embedded energy and cost.

Approximately 300 million gallons of potable water was wasted in 2013 due to routine maintenance in Texas' 10 largest cities [2]. Water wasted due to decayed chlorine levels is unknown, but likely similar. Lack of optimization of an appropriate objective function in water systems is the main cause of wasted water and energy related to lifetime. Military installations have unique water requirements with regard to emergencies, base resiliency and fires. Volume and pumping constraints within the model are vital to reducing the water and power consumed. Awareness of water shortages, installation of water conserving fixtures/appliances and an increase in stewardship over the past decade created an overall reduction (~13%) in per capita and overall potable water consumption [3]. Unfortunately, the reduction in consumption creates a new unintended consequence for the water system. Water systems are not optimized for the new reduced flow and further waste is created when excess treated water that reaches its disinfectant lifetime is purged. A substantial amount of studies have examined optimizing both residential and commercial water consumption networks, but rigorous studies optimizing water use and related power consumption on military installations are not available .

Results on the optimal production of treated water to minimize waste and energy consumption (pumping costs) while meeting demand and quality constraints utilize data from Fort Carson, Colorado and The Pecan Street Project. The unique demand data and constraints of a typical military system are used to create a time-resolved water profile for an installation, which is then used to examine the fully burdened cost, energy consumption and waste of less than optimal water systems through rigorous multi-tiered optimization and control. Financial, social, economic and environmental impacts are analyzed as part of the fully burdened cost.

References

[1] Peter J. Vikesland, Kenan Ozekin and Richard L. Valentine (2001). "Monochloramine Decay in Model and Distribution System Waters". *Water Resources* 35(7): 1766-1776. <>[2] Amarillo Globe-News, "Cities wasted nearly 300M gallons of water." 4-July-2014.

[3] M.A. Maupin, J.F. Kenny, S.S. Hutson, J.K. Lovelace, N.L. Barber and K.S. Linsey (2014). "Estimated use of water in the United States in 2010". *U.S. Geological Survey Circular* 1405(56).

469581 Energy Demand Management in Process Systems Subject to Time-Varying Electricity Prices: A Decomposition-Based Approach

Abstract:

As the availability and pricing of electricity vary during the day due to intermittency in renewable resources or real-time electricity tariff policies, the chemical industry, especially those with power-intensive processes, needs to rethink operating strategies to maximize economic viability. The implementation of demand response (DR) strategies, which focuses on demand side activities and has been in the center of all pragmatic energy policy decisions, is the key to minimizing production costs and maximizing profits. The economic potential of DR for industrial processes has been acknowledged in numerous studies (see, for example, [1]-[3]).

In developing DR strategies, the target lies in scheduling the production rates, production times and inventory levels at each time step to achieve specific objectives. Therefore, process control-related activities are closely linked to the responsive demand augmentations given that control actions involve real-time manipulation of selected process variables to hold key product quality and production rates near their desired target values. Moreover, as the plant operating schedule varies, process control facilitates the dynamic transition periods between different operating modes. Therefore, it becomes imperative to develop design and operating methodologies that can effectively integrate production scheduling and control functions (e.g., see [4]-[5] for some recent comprehensive reviews of the literature on the integration of scheduling and control).

In a prior work [8], we posed the DR problem through a mixed-integer nonlinear programming (MINLP) formulation where optimal operating policies (modes) were found in response to varying electricity prices and intermittent on-site renewable generation. A key assumption in that work was that the control policies were predetermined and no attempt was made to optimize the transition periods. By relaxing that assumption, the goal of the present work is to offer a new DR formulation for the integration of scheduling and control subject to time-varying electricity prices. The formulation simultaneously takes into account the production scheduling with respect to minimizing production cost, the optimization of controller parameters to realize the objective of reducing the time and lost product during transitions, and the integration of scheduling and control to achieve a more economical operating policy. Instead of formulating one complex and unsolvable MINLP problem, we propose a novel decomposition scheme that generates smaller sub-problems that can often be solved more easily. The proposed strategy is implemented within a receding horizon optimization framework, in which only the first transition is carried out with optimally tuned controller parameters and a subsequent scheduling problem is then reformulated with the new updated storage level corresponding to the optimized transition time. A case study involving a continuous chemical reactor is used to demonstrate the contributions and efficacy of the developed approach.

References:

[1] M. Paulus and F. Borggrefe, "The potential of demand-side management in energy-intensive industries for electricity markets in Germany," *Applied Energy*, **47**, 1492-1503, 2011.

[2] S. Mitra, I. E. Grossmann, J. M. Pinto and N. Arora, "Optimal production planning under time-sensitive electricity prices for continuous power-intensive processes," *Comput. Chem. Eng.*, **38**, 171-184, 2012.

[3] Y. Wang and L. Li, "Time-of-use based electricity demand response for sustainable manufacturing systems," *Energy*, **63**, 233-244, 2013.

[4] S. Engell S and I. Harjunkoski, "Optimal operation: Scheduling, advanced control and their integration," *Comput. Chem. Eng.*, **47**, 121-133, 2012.

[5] M. Baldea and I. Harjunkoski, "Integrated production scheduling and process control: A systematic review," *Comput. Chem. Eng.*, **71**, 377-390, 2014.

[6] C. Tong, A. Palazoglu, N. H. El-Farra and X. Yan, "Energy demand management for process systems through production scheduling and control," *AIChE Journal*, **61**, 3756-3769, 2015.

469073 High Concentration Glucose Production from CELF Pretreated Corn Stover Using Wild Type Clostridium Thermocellum Supplemented with β-Glucosidase

Abstract:

Glucose production from cellulosic biomass can provide a powerful platform on which to build a sustainable fuels industry. Today expensive cellulases are needed to further breakdown pretreated biomass into sugars with high yields. In response, researchers have turn towards biological saccharification of glucan using *C. thermocellum*, a highly cellulolytic anaerobe possessing cellulosomes capable of glucan saccharification at high rates and yields. However, *C.thermocellum* does not produce much β-glucosidase, a necessary enzyme for cellobiose hydrolysis into glucose. Additionally, Co-solvent Enhanced Lignocellulosic Fractionation (CELF) has been shown to be very effective at producing glucan solids from corn stover that are highly susceptible to enzymatic hydrolysis. Thus, the objective of this research is to develop a process by which *C. thermocellum* supplemented with β-glucosidase can achieve high yields of glucose when fed CELF pretreated corn stover solids at high concentrations (>100g/L). Advantages of the process include: 1) reduction of expensive fungal cellulases, 2) no need for genetic engineering, and 3) high glucose concentrations. CELF pretreatment prepared corn stover for biological conversion, and the resulting washed solids were loaded into 50 mL bottles at a 50-100 g/L glucan loading for incubation with *C.thermocellum* supplemented with β-glucosidase at 60°C with a shaking speed at 180rpm. Some samples were also supplemented with sodium azide to suppress cell metabolism and enhance glucose yields. Results will be compared for active and inactive *C.thermocellum* cultures and to results from conventional fungal hydrolysis at high and low loadings at 50°C, pH 5, and 150 rpm.

472224 Switchgrass Solubilization By Mixed Methanogenic Enrichments with Comparison to Pure Cultures of *Clostridium Thermocellum*

Abstract:

To meet anticipated global demand for low carbon liquid transportation fuels, second generation biofuels will likely be produced from lignocellulose, a low cost and abundant plant biomass resource. However, the lignocellulose in plant cell walls is difficult to convert. Increasing lignocellulose conversion will significantly enhance biofuel production and lower capital & operational cost. Microbially-mediated fermentation of lignocellulose is a promising approach. Diverse, mixed microbial enrichment cultures provide an indication of the achievable extent of biomass solubilization. Yet there have been few fundamental studies of grass fermentation under the conditions anticipated for a high-rate, low cost industrial process, including very high solids concentration, short residence times and high temperatures. Motivated by this perspective, we have initiated study of thermophilic switchgrass fermentation using mixed culture anaerobic enrichments at decreasing residence times & increasing grass concentrations.

Inoculum was collected from a two-stage anaerobic digester (Vermont Technical College) fed with manure and food waste. Triplicate semi-continuous, anaerobic digesters were operated at 55°C and 30g/L mid-season switchgrass concentration for 214 days, during which two of them were allowed to come to steady-state sequentially at residence times (RT) of 20 days, 10 days, 5 days and 3.3 days, with the other one running as control at RT = 20 days. This feed rate was maintained by regular replacement of 1/10 of reactor contents to achieve the nominal residence times. Whereas we expected a shift from methanogenesis to acidogenesis as RT decreased to 5 days, surprisingly, even at RT=3.3 days stable methane production was still seen and acids inhibition was not observed, which indicates that the digesters were still healthy and balanced. Total carbohydrate solubilization at RT = 20, 10, 5 and 3.3 days was 69.1%, 62.8%, 54.9% and 50.0% respectively. In addition to these continuous trials, batch studies compared the performance of mixed enrichment cultures to pure cultures of *Clostridium thermocellum*, where solubilization was similar despite

the more limited enzyme diversity of the pure culture species.

16S rRNA analysis was used to characterize microbial populations dynamics. It was found that Firmicutes dominated each of the reactors, representing between 54-96% of the sequence reads, and these populations increased with reduced residence time. In addition, Synergistetes, Euryarchaeota etc. taxa were also well represented and relatively stable. By using bioinformatics tools, species present in the consortia were also matched to closely related cultured species including *Clostridium clariflavum* that have cellulolytic activities. Plate culture isolation further revealed that *Clostridium clariflavumÂ* was one of the major cellulolytic species existing in the mixed culture consortia. Data at feed concentrationsÂ â‰¥Â 75 g/L, approaching industrial conditions, will also be presented.

453000 Substituent Effect of Phenolic Aldehydes Inhibition on Alcoholic Fermentation By *Saccharomyces Cerevisiae*

Abstract:

Phenolic compounds significantly inhibit microbial fermentation of biomass hydrolysates. To understand the quantitative structure-inhibition relationship of phenolic aldehydes on alcoholic fermentation, the effect of eleven different substituted benzaldehydes on the final ethanol yield was examined. The results showed that the degree of phenolic benzaldehydes inhibition was strongly associated to the position of phenolic hydroxyl groups, but not the number of phenolic hydroxyl groups. It was observed that ortho-substituted 2-hydroxybenzaldehyde resulted in 15-20 fold higher inhibition than the meta- or para-substituted analogues of 3- and 4-hydroxybenzaldehydes. By correlating the molecular descriptors to inhibition potency in yeast fermentation, we found a strong relationship between log *P* (octanol/water partition coefficient) of aldehydes and EC_{50} value. The most inhibitory 2-hydroxybenzaldehyde has a highest log *P* and possesses an ortho –OH group capable of forming an intramolecular hydrogen bond, which can potentially increase the cell membrane permeability and their toxicity. The results also indicated that the calculated free energy change between phenolic aldehydes and amino acids can be used to predict their structure-inhibitory activity relationship.

467613 Enzymatic Process for Making Plasticizers from High Oleic Soybean Oil

Abstract:

Epoxidized soybean oil (ESBO) is a chemically modified soybean oil derivative that is used to make plastics pliable as well as for making bio-based polyurethanes for coatings, epoxide-based resins, and high temperature lubricants. While traditional conversion methods use peroxyacetic acid to catalyze the reaction, lipases can also catalyze epoxidation reactions. In this paper we show conversion of soybean oil and high-oleic content soybean oil using immobilized lipase derived from *Candida antarctica* without the need for additional solvents, such as toluene, or added free fatty acids. At 35°C, soybean oil can be epoxidized with >90% yield within 15 hours, while using peroxyacetic acid only obtains 70% yield at 50°C after 10 hours of reaction time as side reactions reduced the selectivity of the epoxidation. Reaction kinetics for individual unsaturated fatty acids will be presented, as well as characterization of the enzymatic ESBO compared to the peroxyacetic acid conversion with both soybean oil and high-oleic content soybean oil. Recyclability of the immobilized enzyme is discussed in the context of potential commercial applications as a continuous process.

473034 Biomass Blending & Densification: Impacts on Biochemical Conversion Performance

Abstract:

A blended feedstock strategy has been proposed to address the high cost of feedstock production and reduce risk to the biomass supply chain. The use of blended and densified feedstocks in biochemical conversion pathways instead of conventionally-ground biomass from a single source addresses several challenges in the current biomass supply chain, including availability, transportation, storage, cost, quality, and supply variability. Blending provides a mechanism to reduce risk to the supply chain and reduce feedstock cost by coupling location-specific resource utilization with biofuel production. Preliminary results suggest blending and densification provide a viable path forward to a low-cost, consistent biomass supply. Recent work has investigated the impacts of blended and densified feedstocks on sugar yields in biochemical conversion pathways. Recent studies have demonstrated that densification does not negatively impact biomass recalcitrance; reactivity was equivalent for ground and pelleted corn stover. Recent studies provide a glimpse into the potential for preprocessing options, like blending and densification, to provide benefits to not only logistics, but downstream conversion. However, the impact blending and preprocessing may have on variability and conversion performance has not been fully vetted. The cumulative impact of densification and blending on biochemical conversion processes will be discussed.

466246 *Tuning Ionic Liquids for Low Cost: Applications in Lignocellulose Deconstruction*

Abstract:

Ionic liquids (ILs) have proven to be highly tunable 'designer solvents' capable of a wide range of exciting chemistries. However, industrial application at large scale is hampered by high solvent cost. This cost is, however, a tunable feature of the solvent itself – provided the ion selection is handled with a careful eye aimed at limiting synthetic complexity. Lowering the solvent cost will increase the attractive opportunities of ILs for bulk processing of lower cost end products – including such applications as biofuels. One of the key challenges in biorefining is the initial separation or deconstruction of lignocellulosic feedstock into separate components. ILs offer unique advantages in this area, due to their unusual thermochemical properties. However, there are serious concerns about the economic viability of their use due to the very high cost of most ionic liquids (> $50/kg). We have overcome this by redesigning the IL based deconstruction process to use low-cost, acidic ILs for lignin dissolution rather than cellulose dissolution, yielding filterable cellulose and a dissolved lignin for precipitation or conversion to high-value chemicals. We have found that processability of the cellulose is high and lignin recoveries near quantitative.

We use a range of 'protic' ILs, the family typically used in IL industrial processes, because their simple acid-base chemistry results in a simple and cheap synthesis, with a cost (< $1/kg) similar to common organic solvents such as acetone or toluene. This presentation will discuss how ionic liquids can be 'tuned' to control cost structure of the final solvent, and what implications this will have for the chemical processes involved. The impact of the solvent on large-scale applications, such as biomass pretreatment, will be discussed, with a focus on performance and process considerations such as how the ILs maintain solvent stability under long-term processing conditions, that they can be recovered and continue to exhibit very good performance after multiple reuses. These properties highlight that the ILs have the flexibility to be useful for a variety of downstream chemical processing techniques, and for use in other applications as well.

471708 Investigation of Lignin Streams Generated during Ionic Liquid Pretreatment of Lignocellulosic Biomass

Abstract:

Lignin is a complex three-dimensional amorphous polymer derived from paracoumaryl, coniferyl, and synapyl alcohols, cross-linked together by different a number of linkages. In plant cell walls, lignin fills the spaces between cellulose and hemicellulose, and it acts like a 'glue' that holds the lignocellulose matrix together. Despite years of research, the lignin structure is still not well understood as a function of feedstock and process. The ability of certain ionic liquids (ILs) to dissolve cellulose and/or lignin enables the possibility of commercial lignin upgrading and valorization to improve the overall biorefinery economics. However, to date depolymerization mechanisms of lignin in ILs is not well understood. In this study, the changes in the chemical composition of lignin after pretreatment with IL were investigated. Three different type of biomass, namely switch grass, eucalyptus and pine were utilized in this study. The relative changes in the common β-O-4, β-β and β-5 inter-unit linkages of lignin during pretreatment process were investigated using solution state two dimensional (2D) nuclear magnetic resonance (NMR). The changes in the size and molecular weight distribution of lignin were analyzed using size exclusion chromatography (SEC) and mass spectrometry.

Swelling and Dissolution of Cellulosic Fibers

Mohammad Ghasemi, Marina Tsianou, Paschalis Alexandridis*

Department of Chemical and Biological Engineering
University at Buffalo, The State University of New York (SUNY)
Buffalo, New York 14260-4200, USA

* E-mail: palexand@buffalo.edu

Abstract

The dissolution of cellulose is a critical step for the efficient utilization of this renewable resource as a starting material for the synthesis of high value-added functional polymers. Important aspects of cellulose dissolution such as solvent-induced decrystallization and chain disentanglement are not well-understood. We have developed a phenomenological model that captures the phenomena governing the dissolution of semicrystalline polymers. This model allows the quantification of two important features, i.e., solvent effectiveness in cellulose (i) decrystallization and (ii) chain disentanglement, the balance of which controls the mechanism and kinetics of cellulose dissolution. The insights obtained from this study improve the understanding of cellulose-solvent interactions underlying decrystallization and disentanglement and their contributions in controlling the kinetics of cellulose swelling and dissolution.

Introduction

The processing of cellulose into desired products typically involves its swelling and/or dissolution.[1] A major barrier to cellulose solubility is the recalcitrance of cellulose, which originates from a partial crystalline structure and extended noncovalent interactions.[2] Only a few solvent systems are effective for direct dissolution of cellulose, and they operate under rather strict conditions of composition and temperature.[3] It is generally thought that the capacity of the solvent to break the intra- and inter-molecular hydrogen bond network of cellulose is a key factor in its dissolution.[4] However, it was recently suggested that, since cellulose has amphiphilic properties, hydrophobic interactions might play a significant role in cellulose solubility.[5]

The homogenous swelling and dissolution of cellulose fibers is shown schematically in Figure 1. When solvent diffuses into the cellulose crystalline domains they are gradually disrupted become amorphous, and the fiber swells.[6] After a certain induction time, cellulose chains gradually become disentangled and move out of their network into the bulk solution.[7] Therefore, the dissolution of cellulose involves two inter-related phenomena: decrystallization and chain disentanglement.[7] If a solvent can destroy the crystalline network of cellulose but is not capable to disentangle the chains, then the crystallinity of cellulose is reduced, however, no dissolution happens, as in the case of 1-N-butyl-3-methylimidazolium chloride ([bmim]Cl) at 70 °C.[8] On the other hand, if a solvent can only disentangle the chains, then only the amorphous regions of cellulose can dissolve.

In order to gain insights into the swelling and dissolution of cellulose, a microscopic description of macroscopic experimental information has been provided by using a newly developed phenomenological model [7] that captures the phenomena governing the dissolution of semicrystalline cellulose. Our model fits well experimental data [8] for swelling and dissolution of cotton fibers in the ionic liquid [bmim]Cl.

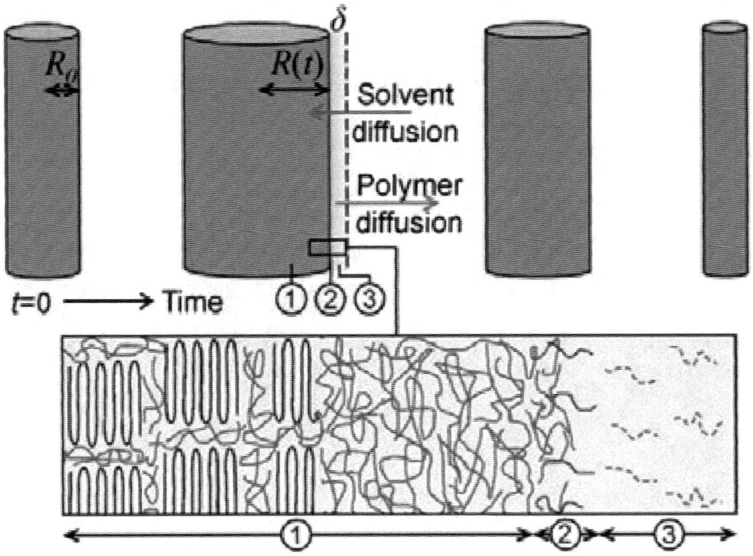

Fig. 1: *Schematic representation of swelling and dissolution of cellulose fiber: (1) fiber interior: crystalline domains (black) and amorphous domains (blue, light shade); (2) fiber-solvent interface at which disentanglement occurs; (3) external diffusion boundary layer at which disentangled cellulose chains (red and dashed) diffuse toward the bulk solution.*[7]

Methods

In the model the cellulose fiber is considered as a long semicrystalline polymer cylinder (consisting of two components; crystalline and amorphous) with a uniform cross section area and with properties that may vary along the cross section.[7] The model describes different steps involved in cellulose dissolution, e.g., solvent diffusion, transformation from crystalline to amorphous domains, specimen swelling, cellulose chain disentanglement, and diffusion of disentangled chains toward the bulk solution, using appropriate transport phenomena and kinetics equations.[7] The ability of solvent molecules to transform the crystal lattice into amorphous and to disentangle cellulose chains is captured by the decrystallization rate constant (k_{dec}) and the disentanglement rate (r_{dis}), respectively. These two are the only parameters fitted to experimental data; all other model parameters are independently obtained.

Results and Discussion

The impact of cellulose decrystallization in the dissolution process is not the same at all decrystallization constant values: at some conditions decrystallization has a strong effect, while at other conditions its effect is negligible (Figure 2a). For example, decreasing k_{dec} from 2.5×10^{-1} to 2.5×10^{-3} s^{-1} has little impact on the total dissolution time (obtained when fraction of dissolved fiber = 1), however, decreasing k_{dec} below 2.5×10^{-3} s^{-1} has a significant impact on the time evolution of fiber dissolution. For $k_{dec} \leq 2.5\times10^{-4}$ s^{-1}, the contribution of solvent molecules to the decrystallization of cellulose is less than their contribution in dissociating the chains, and so the dissolution process is mainly controlled by decrystallization. Because of this, a small change in the k_{dec} value results in a considerable variation in the time evolution of the fiber radius, crystallinity, and solubility, whereas for $k_{dec} \geq 2.5\times10^{-3}$ s^{-1} these fiber characteristics are not affected much by k_{dec}.

When the cellulose-solvent interactions that affect chain disentanglement are stronger than those that influence cellulose decrystallization, the fiber remains crystalline up to almost the end of the dissolution process. This condition happens for $r_{dis} \geq 1\times10^{-9}$ m/s (Figure 2b). Therefore, for $r_{dis} \geq 1\times10^{-9}$ m/s, solvent molecules can disentangle the amorphous domains well, while the decrystallization rate is not high enough to transform the crystalline domains at the fiber surface to amorphous (decrystallization control). However, when $r_{dis} \leq 5\times10^{-10}$ m/s the decrystallization rate of cellulose is relatively greater than the rate of disentanglement, and so the overall kinetics of fiber dissolution are mainly controlled by the chain disentanglement.

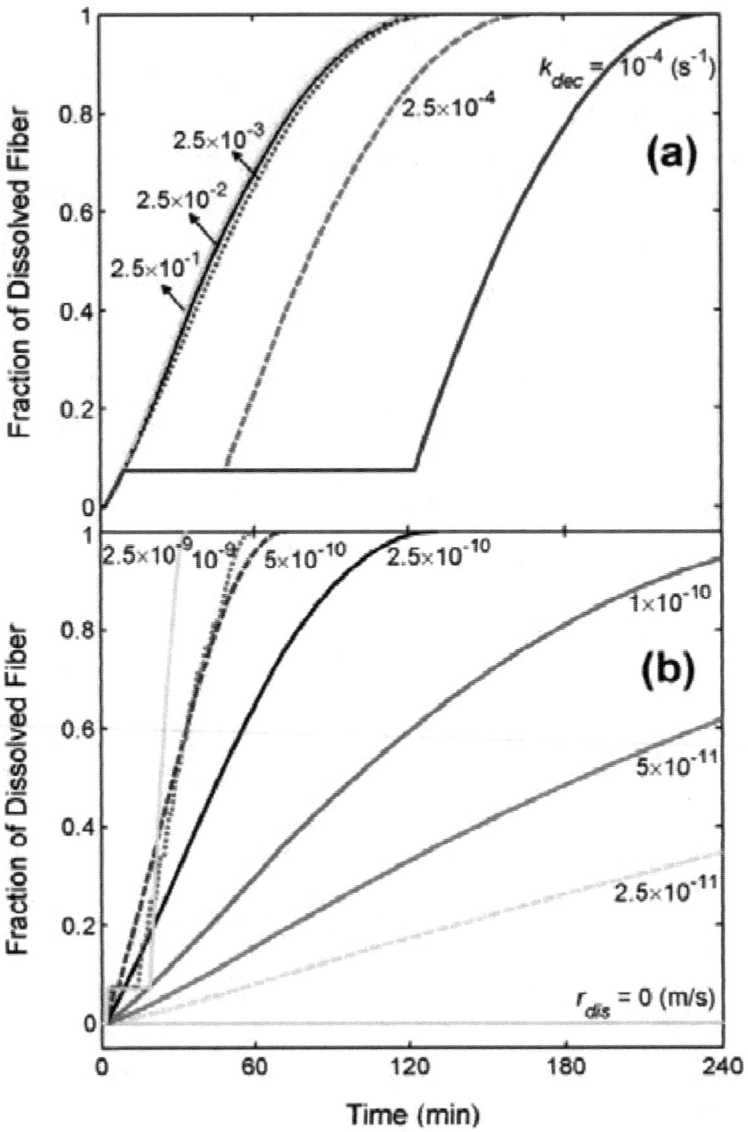

Fig. 2: *Effect of (a) k_{dec} and (b) r_{dis} on the time evolution of the fraction of the fiber that has dissolved.*[7]

The above approach allows us to evaluate the conditions of cellulose dissolution in a variety of systems in the context of the relative contributions of decrystallization and disentanglement. Accordingly, the dissolution conditions can be organized into four regimes: (1) $k_{dec} \leq 2.5\times10^{-4}$ 1/s and $r_{dis} \leq 5\times10^{-10}$ m/s. At these conditions the ability of solvent in both cellulose decrystallization and chain disentanglement is not good, the solubility is limited and the process can be controlled by either decrystallization or disentanglement, depending on the k_{dec} and r_{dis} values. (2) $k_{dec} \leq 2.5\times10^{-4}$ 1/s and $r_{dis} \geq 1\times10^{-9}$ m/s. Here the cellulose-solvent interactions which affect disentanglement are much stronger than interactions that impact decrystallization (decrystallization control).

(3) $k_{dec} \geq 2.5\times10^{-3}$ 1/s and $r_{dis} \leq 5\times10^{-10}$ m/s. At this condition the solvent is more effective in disturbing cellulose intermolecular interactions that result in decrystallization than the interactions that disentangle the chains (disentanglement control). (4) $k_{dec} \geq 2.5\times10^{-3}$ 1/s and $r_{dis} \geq 1\times10^{-9}$ m/s. Here the rates of both decrystallization and disentanglement are very high, the solvent is very aggressive toward cellulose dissolution, and the mechanism of the dissolution process can be controlled by either decrystallization or disentanglement, depending on the k_{dec} and r_{dis} values.

Conclusions

When the solvent effectiveness in the disruption of cellulose crystals is relatively lower than its ability to disentangle the chains (e.g., for LiCl/DMAc[9]), the kinetics of dissolution are controlled by decrystallization. Furthermore, conditions that facilitate cellulose chain disentanglement, in addition to increasing the rate of dissolution, can result in faster decrystallization. The solvent effectiveness in chain disentanglement is the only factor that determines the decrease of the cellulose fiber radius. In cases where the fiber dissolution rate is higher than the decrystallization rate (e.g., in [bmim]Cl at 80 °C[8]), the dissolution of cellulose is mostly controlled by the solvent ability to disentangle the chains. Such knowledge allows the construction of operating condition diagrams for the design of dissolution processes in the biorefinery.

Acknowledgements

We thank the U.S. National Science Foundation (Grant CBET-1159981) for supporting research on cellulose dissolution in our laboratory.

References

(1) Wang, H.; Gurau, G.; Rogers, R. D., Ionic liquid processing of cellulose. *Chem. Soc. Rev.* **2012,** *41* (4), 1519-1537.

(2) Himmel, M. E.; Ding, S.-Y.; Johnson, D. K.; Adney, W. S.; Nimlos, M. R.; Brady, J. W.; Foust, T. D., Biomass recalcitrance: Engineering plants and enzymes for biofuels production. *Science* **2007,** *315* (5813), 804-807.

(3) Liebert, T., Cellulose solvents - remarkable history, bright future. *ACS Symp. Ser.* **2010,** *1033*, 3-54.

(4) Keshk, S. M. S., Effect of different alkaline solutions on crystalline structure of cellulose at different temperatures. *Carbohydr. Polym.* **2015,** *115*, 658-662.

(5) Medronho, B.; Romano, A.; Miguel, M. G.; Stigsson, L.; Lindman, B., Rationalizing cellulose (in)solubility: Reviewing basic physicochemical aspects and role of hydrophobic interactions. *Cellulose* **2012,** *19* (3), 581-587.
(6) El Seoud, O. A., Understanding solvation. *Pure Appl. Chem.* **2009,** *81* (4), 697-707.
(7) Ghasemi, M.; Alexandridis, P.; Tsianou, M., Cellulose dissolution: Insights on the contributions of solvent-induced decrystallization and chain disentanglement. *Cellulose* **2016**.
(8) Jiang, G.; Huang, W.; Wang, B.; Zhang, Y.; Wang, H., The changes of crystalline structure of cellulose during dissolution in 1-butyl-3-methylimidazolium chloride. *Cellulose* **2012,** *19* (3), 679-685.
(9) Ramos, L. A.; Morgado, D. L.; Gessner, F.; Frollini, E.; El Seoud, O. A., A physical organic chemistry approach to dissolution of cellulose: Effects of cellulose mercerization on its properties and on the kinetics of its decrystallization. *Arkivoc* **2011,** *7*, 416-425.

471942 Towards a New Paradigm in the Ionic Liquid Pretreatment for the Production of Lignocellulosic Biofuels: Technoeconomic Insights

Abstract:

Ionic liquids (ILs) have been receiving increasing attention in biomass pretreatment applications (Konda et al., 2014 and 2015) as some ionic liquids are shown to be very efficient in biomass pretreatment. Among several benefits of ionic liquids in general, they are often regarded as *green solvents* due to their negligible volatility. More importantly, given the availability of numerous anions and cations, it is often possible to *tune* their properties to suit specific application(s) by appropriately selecting the combination of anion and cation. Specific to biomass pretreatment, some ionic liquids are shown to be feedstock agnostic (Li et al., 2013) which is a very desirable characteristic especially given the uncertainty with feedstock supply and quality (e.g., seasonal variations). Such a technology can be potentially disruptive and can play a critical role in enabling large-scale cellulosic biorefineries.

To tap into the full potential of ionic liquids, however, it is important that the processes are designed such that they are economically viable as well as environmentally sustainable. These criteria demand for minimal/reduced use of any other chemicals (e.g., co-solvents, water) and energy footprint in the entire process. Furthermore, given the highly integrated nature of these cellulosic biorefineries, these pretreatment technologies need to be developed in such a way that they can be â€˜*plugged-in*â€™ while minimizing any negative impact on the other sections of the process (including downstream). Realizing such a process is often challenging as there exists significant trade-offs amongst these criteria. This emphasizes the need for greater insights into the mass and energy balances of such complex processes. To this end, with the help of virtual biorefinery models, technoeconomic analysis (TEA) offers invaluable guidance to nurture technological advances. Equipped with technoeconomic insights, scientists at Joint BioEnergy Institute (JBEI) are harnessing the tunable nature of ILs, together with the novel process design and operational concepts, to develop advanced pretreatment technologies. Our previous TEA work (Konda at al., 2014) emphasized the importance of high biomass loading, high IL recovery, low IL and water usage to ensure the prospective cellulosic

biorefineries can indeed be economically sustainable. A recent advancement in this direction includes the development of a novel integrated process that is capable of high yields under high-gravity (HG) conditions, while minimizing the use of water and IL. TEA revealed that the continued developments with the proposed integrated process could potentially reduce minimum fuel selling price (MFSP) to below $3/gal.

References:

- Konda et al. (2014). Understanding cost drivers and economic potential of two variants of ionic liquid pretreatment for cellulosic biofuel production. Biotechnology for Biofuels 7: 86.
- Konda et al. (2015). Towards a cost-competitive ionic liquid based pretreatment: Insights from technoeconomic analysis. AIChE Annual Meeting, November 2015, Salt Lake City, UT.
- Li et al. (2013). Comparing the recalcitrance of eucalyptus, pine, and switchgrass using ionic liquid and dilute acid pretreatments." BioEnergy Research 6, no. 1 (2013): 14-23.

474591 Upgrading Corn Stover Pretreatment Wastes into Biodiesel Via Oleaginous *Rhodococci*

Abstract:

The derivation of advance biofuels and specialty oils from agricultural waste streams via oleaginous bacteria is a rapidly developing biotechnology primed for both bioenergy and biochemical sectors. However, finding viable effluents and promising microbes that produce an array of tangible oils for subsequent targeting is a current challenge. Here, we examine and compare alkaline and alkaline-peroxide pretreatments of corn stover as a feedstock for three strains of wild-type *Rhodococci*. In comparison, *R. opacus* DSM 1069 and PD 630 exhibited aggressive degradation of low molecular weight aromatics and glucose monomers compared to *R. jostii*, resulting in heightened viability and improved yields. Interestingly, all three strains yielded oils that could be implemented as a fungible diesel post-transesterification or utilized as a specialty oil for novel biochemical/bio-pharmaceutical applications. Overall, this work presents a novel method for waste stream optimization for bio-refineries.

474502 Biological Design of Lignin Conversion

Abstract:

The utilization of lignin for fungible fuels and chemicals remains one of the most imminent challenges in modern biorefinery. The fundamental challenge in lignin degradation lies in the highly recalcitrant structure that requests redox (oxidation-reduction) reactions to break down the chemical linkages. First, we addressed the challenges by revealing the chemical and biological mechanisms for synergistic lignin degradation via a bacterial and enzymatic system, which significantly improved lignin consumption, cell growth and lipid yield. Chemical analysis suggested that laccase, *R. opacus* cell and Fenton reaction reagents promoted the degradation of different functional groups, elucidating the chemical basis for the synergistic effects. Second, we further improved lignin depolymerization by establishing a new enzyme-mediator system to enhance the electron transfer in enzymatic lignin depolymerization to achieve more than 35% solubilization of kraft lignin without extreme temperature, pH, and pressure. The study has manifested that an efficient enzyme-mediator system can be exploited to establish a bioprocess to solubilize lignin, cleave lignin linkages, modify the structure, and produce substrates amenable to bioconversion, all of which are vital for improving current processing and conversion methods for lignin based biofuels and biomaterials. Third, based on the aforementioned studies, consolidated lignin conversion was engineered in both *Rhodococcus opacus* and *Pseudomonas putida* using systems biology-guided biodesign. Efficient ligninase secretion system has been established through proteomics-guided optimization of promoter/rbs, secretion peptide, and others. The engineered strains have led to significantly increased cell growth and bioproduct yield on lignin substrates. Overall, the biological and chemical design of lignin conversion has opened new avenues for lignin conversion toward various bioproducts.

459510 Selective Conversion of Guaiacol to Substituted Alkylphenols in Supercritical Ethanol over MoO_3

Abstract:

Recently, transition metal phosphides, nitrides, carbides and oxides have been intensively employed in biomass conversion{Ma, 2015 #41}.[1-4] The earth-abundant MoO_3 is regarded as a representative of promising alternative catalysts.[2] However, the bottleneck for the development of HDO technique over MoO_3 lies in the consumption of hydrogen and high temperature. In this work, the conversion of guaiacol to various substituted alkylphenols is investigated using MoO_3 in supercritical ethanol at a low temperature of 280 °C in inert nitrogen. Furthermore, the preliminary reaction pathways are proposed and the molybdenum active state is explored.

The conversion of guaiacol was carried out in a batch reactor at a temperature, in 260-300 °C, for a desired reaction time, 1-5 h over a MoO_3 catalyst. The product sample was injected neat into a GC-MS and a GC-FID instrument for analysis. Several possible reaction intermediates (phenol, catechol, anisole, 4-ethylphenol and 2-isopropylphenol) were employed as reactants to understand the transformation path of guaiacol. XRD, XPS and Raman spectroscopy were used to characterize the catalyst samples.

Fig. 1 exhibits that substituted phenols such as ethylphenols, iso-propylphenols, butylphenols (tert-, iso-) and tert-amylphenol were obtained as the dominating products with a high combined selectivity of 86% and at a conversion of 99% over MoO_3 under 280 °C for 4 h in supercritical ethanol.

To explore the reaction pathways of guaiacol conversion, control experiments with phenol, anisole and catechol as the starting materials under the same reaction condition were carried out. Catechol gave a complete conversion while

phenol and anisole did not. Besides, a similar product distribution compared to guaiacol as the reactant is achieved with catechol. Considering phenol was never found in the guaiacol conversion process, we propose that catechol is formed as the intermediate via demethylation of guaiacol and followed by direct conversion to alkylphenols without the formation of phenol. Subsequently, we employed 4-ethylphenol and 2-isopropylphenol as the reactants and obtained higher alkylphenols as the products. There were no isopropyl, tertiary (iso) butyl or tertiary pentyl derived aliphatic compounds or n-propyl, n-butyl, n-pentyl substituted phenols detected in the liquid and gaseous products of guaiacol conversion. Thus, we speculate that the higher alkylphenols are from consecutive substitution of lower alkylphenols with methyl and ethyl groups produced by ethanol medium, instead of direct substitution with corresponding substituent groups or isomerization with interrelated alkylphenols (Fig. 2).

The Mo (3d) energy region of used MoO_3 is shown in Fig. 3. As can be seen, it contains large amounts of Mo^{5+} (56.4%) with the $3d_{5/2}$ and $3d_{3/2}$ bands located at 230.9 eV and 234.1 eV, respectively. Moreover, molybdenum trioxide (Mo^{6+} state) and molybdenum dioxide (Mo^{4+} state) have been proved to be inactive in control experiments. Therefore, Mo^{5+} was considered to be responsible for the excellent performance of guaiacol conversion.

468094 Enabling Microbial Utilization of Thermally Depolyermized Lignin Monomers

Abstract:

The lignin component of biomass is an attractive source of carbon and energy for the microbial production of fuels and chemicals. This adding of value (*valorization*) to lignin by biological conversion can increase the overall economic viability of biorenewable fuel and chemical production. Thermal depolymerization of biomass through fast pyrolysis shows promise in terms of its ability to recover aromatic monomers from the lignin component of the biomass. Specialized microbes, such as *Pseudomonas putida* KT2440, can utilize many of these aromatic monomers when provided in pure culture [1,2]. However, microbes can be negatively affected by toxic impurities in biomass derived streams, and lignin rich pyrolysis fractions are insoluble in water and therefore inaccessible to microbes in liquid culture medium. Here we addressed these toxicity and insolubility issues by developing and testing an emulsion formula which granted *P. putida* KT2440 access to the phenolic monomers. The optimized emulsion formula contained 18.2 MΩ water, a 70:30 ratio of Tween20® to Span80® surfactant, and the lignin rich pyrolysis fraction. *P. putida* KT2440 was able to use this emulsion formulation as sole carbon source. Additionally, studies with a pure aromatic monomer, p-coumarate, indicated that the emulsion procedure at least partially protects *P. putida* KT2440 from the substrate toxicity of this monomer. To minimize surfactant cost, the surfactant amount was minimized to 7 wt% of the total lignin rich pyrolysis fraction weight. This emulsification method is an important step in enabling the biological valorization of biomass-derived lignin.

Refs: 1. Johnson, C. W.; Beckham, G. T. *Metabolic Engineering* **2015,** 28, 240-247. **2.** Vardon, D. R.; Franden, M. A.; Johnson, C. W.; Karp, E. M.; Guarnieri, M. T.; Linger, J. G.; Salm, M. J.; Strathmann, T. J.; Beckham, G. T. *Energy & Environmental Science* **2015,** 8, (2), 617-628.

472414 Fractionated Lignin Rheology and Structure

Abstract:

Critical technological challenges remain in the development of commercially viable lignin extraction and utilization processes to yield high-value commodity chemicals and novel materials such as aromatics, polymers, adhesives, and flavorings. Processing strategies need to be developed for chemical and enzymatic degradation methods, catalytic upgrading, and purification to make it feasible for commercialization. Designing strategies to process multiphase lignin slurries requires understanding the slurry characteristics, especially quantification of the rheological properties. Fractionating heterogeneous lignin structures yield relatively homogeneous sets that offer simpler processing strategies as compared to inherently heterogeneous whole lignin. The objective of this work was to characterize rheological measurements of fractionated lignin and correlate the rheology to fractionated lignin structures. Organosolv and Kraft lignins were fractionated into five fractions using solvents of increasing polarity. Rheology of lignin fractions dissolved in DMSO was characterized with a rheometer programmed to sweep a range of shear rates for non-Newtonian slurries. ^{13}C nuclear magnetic resonance (NMR) analysis was performed to determine changes in structure and functional group contents of each fractional set. Non-Newtonian behavior increased with increasing lignin concentration within any fractionated group. Although structure varied, there were no obvious rheological differences across solvents or lignin types.

473182 Selective Conversion of Biomass-Derived Lignin to Cyclic Hydrocarbons

Abstract:

A potential value-added stream produced in the DMR process from NREL is the â€œnative-likeâ€ and tractable lignin residues isolated from deacetylation and enzymatic hydrolysis. Nobel metal (i.e.Ru, Rh, Pt) and earth abundant metals (i.e. Cu, Fe, Ni, Zn) immobilized on an acidic carbonaceous material act as efficient catalysts in the hydrodeoxygenation of technical lignins under batch conditions. A synergistic interaction between the metal sites and acid groups within the bifunctional and bimetallic catalyst leads to enhanced catalytic activities for the overall transformation of technical lignins to cyclic hydrocarbons as compared to the individual steps catalyzed by the separate catalytic functionalities. In addition, the selectivity of catalysts will also be addressed in our presentation.

471865 The Value Proposition of an Industry-Defined Smart Manufacturing Platform and Integrated Market Place

Abstract:

The Smart Manufacturing Platform and Market Place facilitates significant productivity gains for manufacturing operations by radically accelerating and lowering the costs of (1) modeling to develop new understandings and opportunities, (2) agile road mapping that can adjust to what was not known and (3) progressive, layer-by-layer gains in trust, confidence, managed risk, human-in-the-loop involvement, technical systems and business outcomes. The SM Platform is designed to be a rich source of validated and certified services, software products and infrastructure choices combined with a rich collaboration environment for new and emerging products and services. It facilitates training, evaluation, testing, development, deployment, scaling and reuse with real-time data to expedite the hard job of working with real data and integrated systems. These capabilities make it possible to significantly improve energy efficiency and energy productivity

Demonstrated in a Department of Energy project involving industry operations, an industry-defined SM Marketplace decouples data and integration services and software products from infrastructure, so that it is possible to compose simple to complex real-time data driven multi vendor solutions and run them to connect operations, businesses and people. The SM Platform brings all of the scalability, cost, per use and security advantages of the cloud provider of choice while still being able to work with multiple vendor products and services. The SM Marketplace is a true market providing access, exposure and evaluation to a wide spectrum of discoverable products, their configurations, and implementation standards and methods. It is an open Market for aligning tools to the needs of the user and empowering the worker with the right tools rather than having to fit the workers to particular tools. As an agile system the SM Platform makes it possible to start small and simple and layer complexity and benefit as understanding is gained.

474496 Manufacturing Systems: From Sustainable to Smart

Abstract:

This talk is organized into two parts. In Part one, I will describe the objective of the "smart" manufacturing and the proposed NNMI Institute for Smart Manufacturing: Advanced Sensors, Controls, Platforms, and Modeling for Manufacturing. Smart manufacturing is an emerging field that combines physical systems with big data and intelligence. It can mean different things to different people. It is also referred to as Advanced Manufacturing, Digital Manufacturing, Cyber Physical Systems (CPS), the Internet of Things (IoT) and in Europe "Industrie 4.0." Smart Manufacturing has the potential to fundamentally change how products are designed, manufactured, supplied, used, remanufactured and eventually retired.

In part two, I will focus on research related to resource efficiency, especially sustainability in manufacturing. Industrial interest in sustainable manufacturing has increased dramatically in recent years, yet the supporting infrastructure and methodologies are still evolving. Ensuring sustainable manufacturing requires an integrated systems approach and spans technical, economic, ecological, and societal issues. Interactions within and across these issues are critical to the fundamental understanding of sustainable manufacturing, because focusing on any single issue could result in suboptimal solutions and the resulting unintended consequences. Industry recognizes that (1) sustainability challenges in manufacturing can only be addressed through multi-disciplinary methodologies and (2) implementing these methodologies require "smart" technologies to enable significant economic benefits for sustainability in general and sustainable manufacturing in particular. It is natural to combine smart and sustainable manufacturing to fully realize efficiency at all levels of manufacturing system.

472630 Manufacturing Sustainable Products: Living Product Challenge and Beyond

Abstract:

Owens Corning is a leader in product transparency and sustainability. By understanding the competitive landscape, Owens Corning is adept to shape the future of industrial sustainability for manufacturers. Indeed manufacturing no longer has to have a negative connotation in regards to the sustainable impact. Shifts in consumer demand now require a higher level of product transparency. Manufacturers must now consider the impact their products have on the end-user; for building materials that means the inhabitants of the building. This work highlights the changes in the marketplace that require a change in historic product disclosures and identify a path forward. We will describe Owens Corningâ€™s experiences participating as a pilot for the Living Product Challenge and identify opportunities for other manufacturers in similar spaces.

As a participant in the pilot program for the living product challenge, Owens Corning helped set a new standard. Living Product Challenge is a new comprehensive product transparency model that extends beyond environmental impact to include manufacturing facility interaction with habitat, worker safety, and responsible supply chain management. The company historically employed multiple third-party agencies to evaluate the product recycled content, VOC emissions, and the safety of chemical ingredients. Under the Living Product Challenge, the fragmented product certification market is consolidated and extended.

Owens Corning Sustainability is on the leading edge and it serves as a model for other manufacturers and hopes to encourage others to join in on the movement.

AIChE 2016 Annual Meeting
Smart and Sustainable Manufacturing

SMART PLANNING & IMPLEMENTATION

"Sustainable Manufacturing Strategy"

DEVELOPING ECONOMY

By

Dr. Hebab Quazi P.E.
The Center for Safety and Sustainability
MARTECH International Inc.

Transformational Strategy

MANUFACTURING ARCHITECURE

MANUFACTURING RESTRUCTURING

Life-Cycle Process
Product/Process Restructuring
Integration
Supply-Chain
Logistics
Efficiency Improvement
Security
Environmental
Policy Realignment
Market Restructuring
Human Capital

Manufacturing Plants Restructured (outside US)

Oil/Gas/Coal
Bangladesh, China, Egypt, India, Mexico, South Korea

Chemical/Petrochemicals
China, Saudi Arabia

Pulp & Paper
Bangladesh, China, El Salvador

Iron/Steel/Mining/Metals Processing
Canada, Chile, China

Power Generation
China, Mexico

Chemicals/Fertilizer Plant Restructured

Existing Product:

Ammonium Nitrate

(Fertilizer for Rice Growing)

Problems:

1. Lost Market Leadership
2. Very Soluble in Irrigating Water
3. High Energy Use
4. High Maintenance
5. Highly Polluting (Air & Water)
6. Losing Money every year

Chemicals/Fertilizer Plant

New Product:

UREA

(Fertilizer for Rice Growing)

Improvements:

1. Enpowers Regaining Market Leadership
2. Reduction in Water Pollution (- 90%)
3. Improved Energy Efficiency (60% +)
4. Production Costs – Made Competitive
5. Identified Export Market (Exporting Plan)
6. Reduced Air Pollution - GHG/Particulates (- 70%)
7. Operation Became Economically Viable
8. Management & Technical Team Trained
9. Received Financing for this "UPGRADE"

SMART PLANNING & IMPLEMENTATION
“Sustainable Manufacturing Strategy”
DEVELOPING ECONOMY

THANK YOU

465848 Strategic and Tactical Control for Sustainable Manufacturing

Abstract:

Sustainable manufacturing has been widely recognized as "the creation of manufactured products through economically-sound processes that minimize negative environmental impacts while conserving energy and natural resources"; it "also enhances employee, community, and product safety" [1]. This definition contains two important aspects: (i) triple-bottom-line based manufacturing, and (ii) continuous rather than one-time adjustment. The former reflects the nature of sustainability, a multi-objective optimization problem, while the latter is a dynamic control problem. In short, sustainable manufacturing is a multi-objective optimal control problem.

In this paper, we introduce a sustainability-oriented hierarchical control methodology for achieving short-to-long-term sustainability goals in large-scale manufacturing. This methodology, by resorting to sustainability fundamentals, system control science, and engineering science, aims at implementing tactical control of individual manufacturing processes, where energy and material efficiencies, waste reduction, product quality, process safety, etc., are all targeted, and realizing strategic control of the manufacturing enterprise to achieve a short-to-long-term sustainability goals, where external information about economic, environmental, and social-related changes will be incorporated. These two layers of control will be coordinated algorithmically. A comprehensive case study on nanocoating manufacturing is illustrated to demonstrate methodological efficacy.

Reference:

[1] EPA: https://archive.epa.gov/sustainablemanufacturing/web/html/

475308 Smart and Sustainable Manufacturing Panel

Abstract:

A panel of invited speakers will conclude the session. The challenge to incorporate sustainable solutions to manufacturing is multifaceted. The purpose of the speaker series is to highlight trends in manufacturing that emerge in light of corporate sustainability initiatives. Companies must consider the return on investment in addition to the sustainable impact of their work. The process, product and lifecycle are all integral players in the success of a sustainable edit to a manufacturing regime. The speakers represent a variety of industries and offer both academic and industry perspectives. The topics of conversation include manufacturing systems, sustainable product development and product life cycle impact.

The panel will focus on opportunities for sustainable manufacturing improvement given todayâ€™s economic climate. These speakers, who are on the forefront of sustainability as it relates to chemical engineering, will offer their thoughts on the role chemical engineers will play in manufacturing of the future. It is imperative that the conversation includes the context for societal and economic influences. The mega-trends of climate change, urbanization and water scarcity each directly affect the ability of manufacturers to conduct business as usual. In light of these issues, they must be addressed in tandem with other initiatives in order to meet the need of tomorrowâ€™s consumer. An interdisciplinary approach is necessary and conversations such as are intended to foster a greater level of critical evaluation in the design and maintenance of manufacturing assets.

469060 Estimating the Life Cycle Impact of Chemicals from Molecular Descriptors and Thermodynamic Properties Via Mixed-Integer Linear Programming

Abstract:

The chemical industry is at present striving to decrease its environmental footprint during the transition towards a more sustainable manufacturing sector. In this context, there is a clear need to develop systematic methods to assess and optimize the environmental impact of chemicals, identify critical hotspots across their supply chains and define the best strategies to implement so as to enhance their sustainability level.

In the recent past, it has become clear that the environmental impact should be assessed considering all the stages in the life cycle of the products/services being assessed. As a result, Life Cycle Assessment (LCA) has become (arguably) the prevalent approach to quantify the environmental burdens of products, finding applications in a wide variety of areas, including the evaluation of the environmental footprint of chemicals from cradle to grave.

One of the main limitations of LCA is that it is highly data intensive, that is, it requires large amounts of data from several echelons of the product supply chain, which are typically owned by different companies that might be reluctant to share this environmental information or even lack the corresponding measurements. This is critical in the chemical industry, where chemicals are produced by complex interrelated chemical networks that exchange energy, mass and water and consume a wide range of intermediate products and feedstocks.

One possible way to simplify the LCA calculations is to resort to Streamlined LCA (SLCA) approaches, which attempt to reduce the upstream and downstream information required in a standard LCA by using proxy data, qualitative models and/or regression equations. These methods are typically tailored to a particular sector, as otherwise the underlying simplifications and assumptions would not hold, thereby leading to poor estimates and large errors.

SLCA methods have been developed for a wide range of production processes, including the building sector (where the size of the bill of materials often makes unpractical the use of standard approaches), urban water treatments and hydroelectric power plants, among others. In the context of chemical processes, SLCA has been applied in oil refining as well as to general chemical processes.

In this work we introduce a novel approach for evaluating the life cycle impact of chemicals that provides better estimates and reduces the complexity of previous models, thereby opening new avenues for their use within the framework of computer aided molecular design tools. In essence, building on the seminal paper by Wernet et Al. (2008)[1] that predicted the life cycle impact of chemicals from their molecular structure using neural networks, we present here an alternative method based on rigorous mixed-integer optimization techniques that uses key thermodynamic properties in addition to molecular descriptors in order to produce more accurate estimates of impact. Our methodology was applied to estimate 9 environmental impact categories of 88 chemicals using 17 structural descriptors and 15 thermodynamic properties. Numerical results show that the life cycle impacts can be predicted with relative errors ranging from 18 to 45% depending on the specific indicator. Furthermore, the inclusion of thermodynamic properties in addition to molecular descriptors improves greatly the quality of the estimate, with relative errors dropping from 22 to 18% when moving from a molecular-based to a combined-based estimation.

1 G. Wernet, S. Hellweg, U. Fischer, S. Papadokonstantakis and K. Hungerbuì˜hler, *Environ. Sci. Technol.*, 2008, **42**, 6717â€“6722.

462465 Optimal Production of Light Olefins from Lignocellulosic Biomass (BTO): Process Synthesis and Global Optimization

Abstract:

The growing interest toward alternative carbonaceous energy sources stems from concerns surrounding energy independency, energy affordability, and the generation of lower carbon energy, as stated in a recent perspective [1]. Specifically, there is a need for the development of sustainable and carbon neutral energy processes that have the potential to reduce global energy related carbon dioxide emissions. Lignocellulosic biomass is a potential feedstock that can serve as an alternative to fossil-fuels and offset the growing energy demand while simultaneously reducing the net carbon emissions due to the absorption of CO_2 during photosynthesis. The U.S. has the potential of sustainably supplying more than one billion tons of biomass annually [2,3]. Biomass has already been shown to be a viable feedstock for the production of liquid transportation fuels at low break-even oil prices [4]. However, the transportation fuel demand is expected to stall in the near future while the chemicals demand is expected to raise by 65% [5]. This increase in the chemicals demand is expected to be the major driver of the crude oil demand growth [5].

This work investigates the optimal production of light olefins from biomass within a process synthesis and global optimization framework. Light olefins are the most demanded class of petrochemicals and command two-thirds of the petrochemical commodities market [6]. These chemicals can be produced from carbonaceous feedstocks through several novel/competing process alternatives [7]. We propose a process superstructure with alternatives regarding (i) biomass conversion, (ii) synthesis gas cleaning, (iii) methanol production and conversion, (iv) olefins purification, (v) light gas recycle, and (vi) hydrogen and oxygen production among others. Simultaneous heat, power, and water integration is included to ensure minimum utility usage. The objective function of the proposed mathematical model maximizes the profit of the refinery. The resulting large-scale mixed integer nonlinear optimization model (MINLP) is solved to global optimality with a branch-and-bound global optimization framework. Several case studies are

investigated to observe the effect of biomass type, plant scale, and topological decisions on the overall plant profitability. Significant reductions in the lifecycle GHG emissions are observed through the utilization of biomass.

[1]: Floudas, C. A.; Niziolek, A. M.; Onel, O.; Matthews, L. R. Multi-scale systems engineering for energy and the environment: Challenges and opportunities. AIChE Journal 2016, 62 (3), 602-623.

[2]: Department of Energy, Biomass as Feedstock for a Bioenergy and Bioproducts Industry: The Technical Feasibility of a Billion-Ton Annual Supply.

[3]: Department of Energy, US Billion-Ton Update: Biomass Supply for a Bioenergy and Bioproducts Industry.

[4]: Baliban, R. C.; Elia, J. A.; Floudas, C. A. Biomass to liquid transportation fuels (BTL) systems: process synthesis and global optimization framework. Energy Environ. Sci. 2013, 6 (1), 267-287.

[5]: ExxonMobil, The Outlook for Energy, A View to 2040

[6]: de Klerk, A. Fischer-Tropsch Refining; Wiley-VCH Verlag & Co. KGaA, 2011.

[7]: Onel, O.; Niziolek, A. M; Floudas, C. A Optimal Production of Light Olefins from Natural Gas via the Methanol Intermediate. Industrial & Engineering Chemistry Research 2016, 55 (11), 3043-3063.

455402 Sub and Supercritical Conversion of Kirchneriella Algal Biomass into Biocrude Oil

Abstract:

Sub and supercritical water technology utilizes wet biomass as feedstock to produce liquid fuels. Water at elevated temperature and pressure exhibit extraordinary solvent properties. Here, water acts as reactant as well as reaction medium in performing various complex chemical reactions. Algal biomass was an interesting feedstock for producing biofuel via sub and supercritical technology due to their high water content. We investigated the potential of microalgae (Kirchneriella sp.) for biofuel production. Several process parameters such as reaction temperature, pressure, time and solid loading were studied. Biocrude oil yield of 45.5% was achieved at 300°C, 30 min, 100 bar and 10% solids loading. Lower biocrude oil was produced at supercritical conditions (374°C and 220 bar) compared to subcritical conditions due to intensive gasification at higher temperatures. The high heating values (HHV) of biocrude oil and bio-char were determined with a bomb calorimeter. The biocrude oils and water soluble organics were Analyzed using gas chromatography mass spectrometry (GC-MS) and calorimetrically. CHNS/O analyzer was used to determine the elemental composition of algae, biocrude and biochar samples.

471483 Selective Aromatic Ring-Opening of Biorefinery Lignin Towards Dicarboxylic Acids: Influence of Reaction Temperature and pH

Abstract:

Lignin is the worldâ€™s only sustainable source of aromatic compounds, which has resulted in a significant quantity of research being conducted on the depolymerization of this complex biopolymer for the production of low molecular weight aromatic compounds. These aromatic products can then be further broken down via selective ring-opening to form dicarboxylic acids.

Dicarboxylic acids are a valuable class of organic compounds which are utilized in numerous commercial industries, including the production of biopolymers, pharmaceuticals, and food products. Like phenolic compounds, dicarboxylic acids are currently dependent on crude oil as a feedstock. However, lignin has the potential to act as a sustainable alternative.

This work investigates the selective production of dicarboxylic acids from several different types of lignins, via thermochemical breakdown. The treatment conditions employed a temperature range of 60 to 250 Â°C, and a reaction mixture pH of 4 to 13. The major products obtained in this work were oxalic, malonic and succinic acids. Yields of up to 10 wt.% oxalic acid were achieved, based on the mass of lignin treated.

467791 Lignin in Ethylene Glycol and Poly(Ethylene Glycol): Fortified Lubricants with Internal Hydrogen Bonding

Abstract:

Lignin, one of the most naturally abundant polymers, has been successfully incorporated into ethylene glycol (EG) and polyethylene glycol (PEG) in this work and fortified lubricating properties were achieved in EG/lignin and PEG/lignin. The molecular interaction between lignin and EG (or PEG) has been revealed as hydrogen bonding, which serves as the dominating factor that determines the thermal, rheological and tribological properties of the mixed systems of EG/lignin and PEG/lignin. The physicochemical properties of the mixed lubricants are tightly related to the state of internal hydrogen bonding (EG-EG, PEG-PEG, EG-lignin, PEG-lignin and lignin-lignin) and are well correlated to their lubrication properties. Generally, larger lignin fraction leads to better lubricating performance in both EG and PEG systems. Lignin liquefaction in PEG has been addressed by catalytic degradation with the presence of sulfuric acid, which was then neutralized by triethanolamine for lubricant development. Lignin in PEG significantly improves the lubricating property at higher pressure conditions, where a wear reduction of 94.6% was observed. Lignin fortified EG and PEG based lubricants show outstanding non-corrosive characteristic to the mostly used metal materials such as aluminum and iron.

469693 Enzyme Selection for Hydrolysis of Lignocellulosic Biomass Coupled with Fermentation

Abstract:

The India - Norway collaborative project EcoLodge aims at developing a new, integrated biotechnological production process for butyl butyrate, a promising supplement/substitute for diesel and jet engine fuels. The feedstock, lignocellulosic biomass is enzymatically hydrolyzed to produce C5 and C6 sugars. Further, the sugars are fermented by the bacteria to produce butanol and butyric acid, which are the building blocks of butyl butyrate. Butanol and butyric acid fermentations are developed and optimized separately to achieve high yield.

Major challenges for fermentation of biomass derived sugars are low product yield and high hydrolysis cost (Sun and Cheng, 2002). Exploitation of all the C5 and C6 sugar polymers locked in the lignocellulosic biomass can provide a solution to the low yield problem. However, the hydrolysis of all the polymers requires an optimal enzyme mixture, which is usually need to be tailor made or carefully selected for the specific substrate (Jørgensen et. al., 2007). The challenge of high enzymatic hydrolysis cost is mainly due to the enzyme since utility costs are low owing to mild operating conditions (pH 4.8 and temperature 45 - 50 °C) (Duff and Murray, 1996). Therefore, this paper suggests an optimal enzyme mixture, which yields the most favorable composition of C5 and C6 sugars necessary for high yield butanol and butyric acid fermentations.

Mathematical models and kinetic parameters for bacterial growth on mixed substrates from the literature are employed for simulation of fermentations with varied sugar compositions (Lendemann and Egli, 1998). Various composition values of C5 and C6 sugar mixtures are taken from the literature and used in the model simulations as well for comparison purposes (Van Dyk and Pletschke, 2012). The simulation results illustrate the effect of sugar composition on the fermentation yield; eventually provide the optimal value of the sugar composition and corresponding enzyme mixtures. The approach enables backward knowledge flow, from fermentation to enzymatic hydrolysis, to investigate

the optimal process conditions. Therefore, it suggests the optimal enzyme mixture after evaluating the possibilities in terms of the cost, composition and the operating conditions.

References

Sun, Y., & Cheng, J. (2002). Hydrolysis of lignocellulosic materials for ethanol production: a review. *Bioresource technology*, *83*(1), 1-11.

Jørgensen, H., Kristensen, J. B., & Felby, C. (2007). Enzymatic conversion of lignocellulose into fermentable sugars: challenges and opportunities.*Biofuels, Bioproducts and Biorefining*, *1*(2), 119-134.

Duff, S. J., & Murray, W. D. (1996). Bioconversion of forest products industry waste cellulosics to fuel ethanol: a review. *Bioresource technology*,*55*(1), 1-33.

Lendenmann, U., & Egli, T. (1998). Kinetic models for the growth of Escherichia coli with mixtures of sugars under carbon-limited conditions.*Biotechnology and bioengineering*, *59*(1), 99-107.

Van Dyk, J. S., & Pletschke, B. I. (2012). A review of lignocellulose bioconversion using enzymatic hydrolysis and synergistic cooperation between enzymesâ€"factors affecting enzymes, conversion and synergy.*Biotechnology advances*, *30*(6), 1458-1480.

466874 Managing Trade-Offs Between Food, Renewable Energy and Ecosystem Services

Abstract:

Competition between renewable energy and food production, driven by increasing food and energy demands, creates incentives to increase agricultural productivity. In the short term, productivity can be increased through the use of intensive farming management practices that use extensive tillage and large quantities of chemical fertilizers. Such practices have been shown to contribute to ecological degradation of farmland via an increase in soil erosion and a decrease in soil carbon, among other impacts. The surrounding watershed is also affected negatively by the nitrate and phosphorus-contaminated runoff from fertilized fields, which degrades the water quality and contributes to the growth of harmful cyanobacteria and other microorganisms. In the long term, this ecological degradation results in reduced agricultural productivity, such that increasing amounts of chemical fertilizers and other technological inputs are required. Farming for maximum productivity without accounting for the ecological impacts thus creates a vicious cycle, in which productivity is maintained by inflicting ecological harm.

Experimental studies have demonstrated that conservation farming practices, with reduced tillage and in some cases reduced amounts of fertilizers, insecticides and herbicides, can mitigate the ecological damage of conventional agriculture. Farming with reduced or no tillage helps to increase and maintain the carbon content in soil, providing an offset for the emissions produced in the agricultural life cycle, and reduces the rate at which soil erodes. Reducing chemical fertilizers or partially replacing them with soil amendments such as agricultural residues reduces the concentration of nitrates and phosphates in contaminated runoff and cuts down on wasted fertilizer as more of the applied fertilizer is taken up by crops. Conservation practices offer myriad ecological benefits but do not necessarily offer similar productivity benefits: the rate at which biomass is produced can be decreased under conservation farming, creating trade-offs between ecological preservation, agricultural production, and farm economics. With demand for both food and energy likely to increase for the foreseeable future, the need for food and bioenergy crops must be balanced against the need for ecological preservation.

The objective of this work is to design a system for the co-production of food and renewable energy with respect to three conflicting criteria: food productivity, energy productivity and net ecosystem service (ESS) supply. Ecosystem services are the goods (food, fuel, materials) and services (climate regulation, water quality regulation, erosion control) provided by nature that are necessary to support human activities, including agriculture. Here we define net ESS supply as the difference between ESS supply, provided by nature, and ESS demand, required by human activity. Climate regulation is a simple example. The net climate regulation supply is the quantity of greenhouse gas emissions generated by human activities, less the quantity of emissions that can be sequestered or otherwise removed from the atmosphere by ecological processes such as biomass growth and sequestration in soil. Net ESS supply provides a quantitative way to measure ecological degradation: a negative net supply means that the demand for ESS is greater than can be supplied by nature, and ecological degradation is the likely consequence. In this work, we will identify options for food and renewable energy co-production that have a net ESS supply greater than or equal to zero, for key services.

The system we consider is comprised of two stages: land use and biomass-to-energy conversion. Land use includes several options for crops both grain and cellulosic, rotation systems and farming management practices, as well as the production of wind and/or solar electricity. Land can also be used to create additional ESS supply through reforestation; however, land used for ESS supply cannot also provide food or energy, thus there is competition between the land uses. Options for biomass-to-energy conversions cover a range of liquid fuels, both ethanol and hydrocarbon, as well as electricity. Grain crops can be sold directly as food products or converted to liquid fuels; the system thus has a variety of configurations in which both food and renewable energy is produced from the same farmland. The land use and biomass-to-energy conversions are modeled at the value chain and equipment scales, respectively, of the process to planet (P2P) multi-scale modeling framework. For this system, the value chain scale may be thought of as a regional scale. The third and largest scale of the P2P framework, the economy scale, is used to capture the life cycle of all inputs to the land use and biomass-to-energy conversion stages. Net supplies of two ESSs are quantified as decision criteria: climate regulation, quantified with carbon dioxide emissions and carbon sequestration, and air quality regulation, quantified with nitrogen oxide emissions and removal. Carbon dioxide emissions (climate regulation demand) is quantified at all scales, equipment, regional and life cycle. Carbon sequestration (climate regulation supply) is quantified at the regional scale only, as the land use options are the only components of the system that provide a variable supply of ESS. Air quality regulation is quantified only at the smaller two scales, with supply again provided only by the land use options and demand created by both the land use options and the biomass-to-energy conversion options. Air quality regulation is not considered at the life cycle scale because the spatial presence of criteria air pollutants and its effects, are either local or regional in nature depending on the pollutant molecule.

To analyze trade-offs between food, energy and ecosystem services, the co-production system is optimized under several constraints. Food and energy production are constrained relative to the maximum production possible, to ensure that all optimal system designs are truly co-production systems. Where possible, net ESS supply is also constrained to be greater than or equal to zero; this is not always achievable for air quality regulation, but net climate regulation supply can reach above zero for many system configurations. Within these constraints, the system is optimized for maximum food production, maximum energy production, and maximum net ESS supply for both services. With the design space thus defined, Pareto curves are found between food production and net ESS supply, and between energy production and net ESS supply, in order to identify sets of compromise system designs. Preliminary results indicate that maximizing either food or energy productivity leads to the worst possible performance for net ESS supply, as expected. However, by either sacrificing some amount of productivity or by using land for both food and energy production and ESS supply, designs can combine adequate net ESS supply with relatively high food and energy productivity.

461799 Efficient Solar Thermal Hydrogen, Electricity and Fresh Water Coproduction Process Synthesis

Abstract:

Global population growth, increase in per capita GDP and change in consumption habits create increasing pressure on food, energy and water (FEW) supply. For the past century, with lower FEW needs and abundant fossil resources, energy, chemical, fertilizer and even fresh water supply via desalination have been predominantly dependent on limited fossil resources. However, concerns regarding limited fossil fuel reserves and increasing greenhouse gas (GHG) emissions from fossil fuels accelerate advancements and implementation of renewable energy conversion processes. Among these alternatives, solar energy conversion processes are particularly prominent because of sunâ€™s tremendous potential [1]. Yet intermittencies and land availability constraints for solar energy collection are the grand challenges for solar energy conversion technologies and demand high conversion efficiency and synergistically integrated energy storage solutions [2-3]. Recently, it has been shown that hydricity concept, synergistic solar thermal hydrogen and electricity coproduction, presents a potential solution for continuous and efficient power supply and also an exciting opportunity to envision and create a sustainable economy to meet all the human needsâ€"namely, food, chemicals, transportation, heating, and electricity [1].

Here, we introduce an integration strategy for the coproduction of fresh water, hydrogen and electricity from concentrated solar thermal energy and present potential impacts of the proposed system on FEW nexus. The integration of solar thermal power production and solar thermal hydrogen production techniques reduces the exergy losses associated with each one of the systems and thus provide an efficient solution. Solar conversion processes are evaluated based on the process *exergy efficiency* that refers to the fraction of incident solar exergy that is directly recovered as the net exergy output, which is defined as the sum of electricity and the hydrogen exergy output.

The proposed efficient solar trigeneration process presents: i) a continuous power and fresh water supply solution that

can achieve high solar conversion efficiencies when combined with the proper hydrogen power cycle and ii) numerous opportunities in chemical industry, transportation sector and in particular in food production via continuous supply of fresh water for irrigation and fertilizer production from coproduced hydrogen. Coproduced products can supply all FEW needs especially in the case of production of fertilizers from the coproduced hydrogen. [1] Gençer E, Mallapragada DS, Marechal F, Tawarmalani M, Agrawal R. Round-the-clock power supply and a sustainable economy via synergistic integration of solar thermal power and hydrogen processes, Proceedings of the National Academy of Sciences (PNAS), 112(52), 15821-15826, 2015.

[2] Gençer E, Al-musleh E, Mallapragada D, Agrawal R. Uninterrupted Renewable Power through Chemical Storage Cycles. Current Opinion in Chemical Engineering, 5, 29-36, 2014.

[3] Gençer E, Agrawal R. A commentary on the US policies for efficient large scale renewable energy storage systems: Focus on carbon storage cycles, Energy Policy, 88, 477-484, 2016.

465212 Food-Energy-Water Nexus: Modeling Energy and GHG Emissions of Water Embodied in U.S. Domestic Food Transfers

Abstract:

Global population growth, climate change, and increasing urbanization are pressurizing already constrained essential food, energy, and water resources. Food, energy and water (FEW) systems are deeply interwoven, interconnected, and necessitate an integrated management approach. Central to understanding the complexity of the FEW nexus is quantitatively mapping and understanding their complex interdependencies. Existing studies have focused mostly on a single or two dimensions of the FEW nexus. There is a critical need for a holistic systems-level approach for studying FEW systems in a joint manner as opposed to studying them in isolation to avoid unintended consequences and encourage optimal outcomes. We integrate concepts of network theory with life cycle information to develop an understanding of the domestic FEW nexus for the United States.

Using publicly available disparate datasets, we develop a weighted and directed network model of interstate food transfers for the United States. The interstate food transfer model is translated into networks of embedded irrigation water, energy, and greenhouse gas (GHG) emission flows utilizing information on water footprints, irrigation datasets, and life cycle inventory databases. Specifically, we focus on four food commodity groups: livestock, cereal grains, meat, and milled grains. We utilize network theory tools and metrics to understand the structure, robustness, and environmental sustainability of these networks. Preliminary results indicate that over 600 million tons of food commodities were transferred across the U.S. along with 230 billion m^3 of virtual irrigation water. Additionally, 450 billion MegaJoules (MJs) of primary energy and 30 billion kg of CO_2-equivalent emissions were embedded in virtual irrigation water transferred across the U.S. It is observed that livestock and meat contributed only 13% by mass but accounted for 60% of embodied water, energy, and GHG transfers compared to grain based commodities. From a network perspective, the unweighted food transfer network is well-connected with majority of states participating in high volume of trade. However, the weighted network structure reveals that the majority of the food flow and embodied resources and emissions are controlled by a few states. Furthermore, meat and milled grain networks are denser compared to livestock and cereal grains networks. A robustness analysis highlights that despite the presence of vulnerable key states, the network is fairly robust to both random and targeted disruptions. These results reveal that while the domestic food transfer network is robust against extreme disruptions on specific nodes, the dense food transfer patterns of resource intensive products have important implications for sustainability of the food network.

465536 Water Footprint of Hydrotreated Renewable Jet Fuel Produced through Rapeseed Rotation with Wheat and Other Crops in North Dakota

Abstract:

Hydrotreated Renewable Jet (HRJ) fuel from oilseed plants has increasingly become important for the aviation sector to address energy security and climate change mitigation. Rapeseed is considered a promising HRJ source for its high quality oil content and the potentially attractive agro-economic benefit to replace the fallow period in wheat/fallow rotations. However, concerns have been raised about the viability of this important change in land use, within which water supplies is one important concern as large scale biofuels production may place pressure on fresh water supplies and water quality. Despite these achievements on analyzing the water impact of production of selected biofuel crops, investigation of rapeseed production for biofuel has been limited. This study develops assessment informed by inputs from multiple models to evaluate the environmental impacts of water utilization and water quality of the different production options for HRJ production from rapeseed cultivation in North Dakota. The biogeochemical based EPIC model is incorporated to simulate the crop growth that influences the hydrological cycle. Output of the land-use management changes from the EPIC is used to run the blue water footprint analysis using the LCA software SimaPro 8.0. Evapotranspiration data from the EPIC model is incorporated to calculate green water footprint during cultivation. The UNESCO Grey Water Footprint Accounting Tier 1 Guideline is followed to estimate the grey water footprint caused by nutrient leaching and runoff. In addition, inputs for multi-modal transportation logistics were obtained from the Alternative Fuel Transportation Optimization Tool (AFTOT) models in collaboration with the US Department of Transportation Volpe National Transportation Systems Center. Co-product allocation methods were employed in scenario analyses, such as energy allocation and market value allocation, therefore producing a robust analysis of water footprint of the rapeseed HRJ life cycle with compiled life cycle inventory.

Our presentation details the methods, assumptions and initial water footprint results for different scenarios of importance in North Dakota, one of the 10 US states studied in this project, when considering the rapeseed HRJ life cycle. Preliminary results indicate that HRJ fuel derived from rapeseed in rotation with wheat has a much larger water footprint when comparing to the conventional jet fuel. In addition, although it is usually neglected, the indirect blue water footprint caused by the energy usage and the material inputs during HRJ production pathway is worth considering due to its important impact on the overall water footprint. Therefore, policy makers should consider highly the water factor when making decisions on transition to large share of biofuels.

468516 NexSym – a Local Nexus Simulation System

Abstract:

Food, energy and water are essential needs to sustain human life. The complex interactions and interdependencies among the infrastructures and decision making processes involved in the provisioning of such needs have recently generated great debate under the umbrella of the Food-Energy-Water (FEW) nexus (also referred to as Water-Energy-Food nexus). As society becomes increasingly aware that resources available to satisfy the needs of a growing population are finite, there is a call to look at the interactions and start creating more efficient solutions to manage the wide range of nexus scenarios, especially under conditions of climate change, increased urbanisation and waste generation. Furthermore, there is awareness of the importance of maintaining ecosystems to achieve sustainable production while expanding land use and using it more intensively to obtain more food, energy and clean water. The nexus may manifest in unique ways in different localities. Similarly, the global conditions will affect each location differently. Therefore, it is urgent to develop holistic analytical tools that could inform decision making for managing the interdependencies of the FEW nexus at the local scale. This is one of the ways chemical engineers are able to contribute with solutions by applying their skills of modelling processes and managing constraints associated with an interconnected system.

In this work, we present the **Local Nexus Simulation System (NexSym)** that integrates models for technological process components involved in the production of food, energy and clean water as well as waste treatment. Furthermore, the techno-ecological interactions can be simulated by integrating models for ecosystem processes, including agricultural production. The tool then allows the user to build a "flowsheet" of the local system with the component of interest (e.g. bioenergy production) and their interacting local system components (e.g. biomass cultivation, water provisioning). This enables examination of the whole picture of FEW nexus interactions at the local scale, which can then be investigated within a global context. Options for managing the nexus from an engineering perspective can then be devised and further evaluated under scenarios of interest to the user. The **NexSym** is expected to have wide application by engineers working across the nexus system components, planners for urban or rural development, policy makers and decision makers in general. The tool will be illustrated with sample case studies featuring interactions between food, energy, water provisioning and waste treatment to demonstrate the usefulness of such a tool in understanding, managing and creating sustainable solutions to the nexus.

472086 Gibbsian Game Theory for Tragedy of the Commons Problems in Food-Energy-Water Sustainability

Abstract:

Sustainability problems in the food-energy-water nexus often involve "tragedy of the commons" (TOC) problems. TOC challenges are often modeled in game theory using n-player Prisoner's Dilemma (NPD) games and other games. However, classical solutions to game theory problems – here called "Nashian games" – have a number of shortcomings, and do not always conform to experimental data. In this talk we treat strategic games as a set of chemical reactions, and we treat choices as types of reactive molecules. Thus, this model provides new solutions to classical games. By using the usual Gibbsian approach to chemical reaction equilibrium, we are able to employ entropic considerations and minimize a type of Gibbs free energy to find game theory equilibria, overcoming the usual shortcomings in game theory. We solve classical games, including the 2-player PD game and the NPD game, using the "Gibbsian game theory" approach, which accounts for "entropic choices". We then show how treating strategic games as a set of chemical reactions can be used to represent decision situations more accurately than typical games, and we solve for the equilibrium of sets of reactions. Then, we apply Gibbsian solutions to NPD games to problems in sustainability. By quantifying concepts like "trust", "power", and "corruption", we explore opportunities for nudging outcomes of contested decisions in sustainability decisions. When overlapping considerations occur, as in the food-energy-water nexus, the NPD games reveal fundamental political conditions that increase the probabilities for when genuine human solutions can advance. In this talk we describe Gibbsian game solutions for NPD problems, and show how they apply to sustainability problems in the food-energy-water nexus.

461657 Water and Energy Systems in Sustainable City Development: Agent-Based Modelling and Resource Technology Optimization

Abstract:

The grand challenges facing human societies are all closely interconnected with the sustainable provisioning of energy, water and material resources for constantly growing and developing populations, as well as the subsequent processing and management of wastes and pollutions [1]. Urban water and energy systems are expanding and require attention and careful planning to minimize their economic and environmental impact both from an operational and investment perspective. However, there is a lack of a sufficiently systematic approach to address these impacts and achieve the optimal strategies either in the context of research or real-world decision making [2].

Facing the challenges of energy and water systems in urban development, this work describes a decision-making platform combining a comprehensive database, agent-based simulation modelling and resource technology network optimization. The current regional demographics, land-use, infrastructure and economic information is used as input for the initial state of the focused urban system incorporating energy, water and other related resources. Detailed spatio-temporal demand data, obtained by simulating a synthetic population using agent-based models, is further used to plan capacity utilization and expansion by supply-side matching on a cost optimal basis, eventually aiming to explore the optimal design and operational strategies for residential, commercial, industrial, and other sectors in urban energy and water systems, and especially their nexus. Moreover, long-term socio-economic scenarios are addressed in the urban system development through a real-time feedback loop.

The outputs depict, with fine spatial and temporal resolutions, an overall energy-water landscape of the studied urban region (e.g., Greater Accra Metropolitan Area in Ghana as a prototype [3]). The material and energy balance of supply and demand from both macro and micro perspectives are obtained, and further used to propose environmental friendly and cost effective sustainable city investment and development strategies. Several representative use cases are built

under different scenarios to maximize utilization of resources, minimize net waste generation, and improve social welfare to engage local stakeholders. This work is to become a core component of the *resilience.io* platform as an open-source, data-driven, integrated systematic tool gathering social, environmental and economic data to inform urban planning, investment and policy-making for city-regions globally.

References

[1] Bringezu, S. and Bleischwitz, R. eds., 2009. *Sustainable resource management: global trends, visions and policies*. Greenleaf Publishing.

[2] Keirstead, J. and Shah, N., 2013. *Urban energy systems: an integrated approach*. Routledge.

[3] Koppelaar, R., Sule, M.N., Kis, Z., Mensah, F., Wang, X., Triantafyllidis, C., van Dam, K.H., Shah, N., 2016. Urban water and sanitation flows in the Greater Accra Metropolitan Area. *Urban Water Journal* (submitted).

465063 Hydrothermal Treatment of CAFO Waste to Recover Nutrients to Prevent Nutrient Runoff and Watershed Pollution

Abstract:

Concentrated animal feeding operations (CAFOs) are an important part of the U.S. economy, providing the country with a substantial low-cost source of consumable meat and dairy products such as beef, pork, poultry, milk, and eggs. However, CAFOs are becoming a greater public/government concern due to the amount of animal manure (waste) generated by such facilities, resulting in both air and water pollution. The United States Department of Agriculture (USDA) estimates that 335 million tons of dry manure waste is annually generated by CAFOs. This waste stream contains significant quantities of nutrients, accounting for nitrogen and potassium totals of approximately 6.17 and 1.85 million metric tons, respectively.

Application of manure to pasture and crop land is the most prevalent CAFO waste management practice. This practice can be beneficial as it improves soil quality by replenishing important nutrients for plant health. However, the amount of applied manure along with nutrient mobility leads to runoff into local watersheds, causing a host of local/regional public health and economic issues. Nutrient runoff (especially P) leads to growth of harmful algal blooms, which has led to the loss of water supply to U.S. cities (due to elevated microcystin levels), loss of aquatic life, and over $100 million in economic damage since 2008.

Due to its distributed nature, nutrient runoff is one of the most prevalent and challenging environmental issues in the U.S. To address this issue, Ohio University (OHIO) is developing a supercritical water (SCW)-based modular process to be operated at CAFO sites in order to recover the nutrients as a solid product. This nutrient-rich solid product may be used as a fertilizer supplement for field application, with significantly reduced potential for runoff into local watersheds. In addition, the process produces a methane-rich gas product, which may be utilized at the CAFO site or upgraded into pipeline-quality natural gas. SCW provides an advantageous media for recovery of nutrients from animal manure; its lower fluid density and hydrogen bonding strength promotes precipitation of solid P components, and provides a gasification medium for undigested carbonaceous species.

To evaluate process potential, OHIO has been conducting both experimental investigations for treating dairy CAFO waste using a semi-continuous bench-scale reactor and process simulations/techno-economic assessments using Aspen Plus™. Major operating parameters being investigated include temperature (250-400°C), pressure (23-25 MPa), manure solids content (2.0-10 wt%), residence time (5-30 min), and oxidizing agent. The liquid and the slurry (liquid and solid contents) phases have been characterized using microwave-assisted acid digestion (AD) and inductively-coupled plasma (ICP) spectroscopy and ion chromatography (IC). The solid nutrient-rich product is collected after each trial and analyzed using AD, ICP, and IC, while the gas product is analyzed using mass spectroscopy. Aspen Plus™ process simulations have been used to develop material and energy balances, heat integration strategies, and process costs. This presentation will review results from this study including nutrient recovery, solid nutrient composition, and product gas volume/composition with operating parameters and treatment costs associated with a 1,000 head dairy CAFO.

470126 A Sustainable Framework for Chemical Manufacturing and for the Integration of Advanced Control with Sustainability Assessment

Abstract:

In recent years, with the progress in understanding, quantifying, and improving sustainability, the incorporation of sustainability into chemical process design, optimization and control has become a research highlight in process systems engineering. Many contributions have been made in sustainable process design and multi-objective optimization. However, an efficient framework for the integration of advanced process control with sustainability assessment tools is still missing. The lacking of such a framework can be attributed to the challenges associated with adding sustainability objectives to process controllers, especially for systems characterized by high-dimensional and nonlinear chemical process models. The objective of this presentation is to provide a systematic framework for the optimization and control of chemical manufacturing systems in order to achieve the most sustainable operating points and life cycle inventory in terms of efficiency, environmental, economic and energy aspects that describe sustainability.

This framework is comprised of four main parts: (i) process and sustainability assessment model construction; (ii) multi-objective optimization formulation; (iii) implementation of the advanced control strategy; and (iv) a life cycle inventory generation. For the model construction part, the process systems model is based on existing mathematical models obtained from the literature, while the system sustainability assessment model can be developed based on the GREENSCOPE [1] sustainability assessment tool indicators in four categories (economic, environmental, efficiency, and energy). Once the models for process and sustainability assessment are established, the multi-objective optimization problem is formulated based on economic and integrated sustainability impact objectives. The resulting multi-objective nonlinear programming problem is solved using optimization algorithms to generate a Pareto-optimal set. Based on a ranking scheme algorithm for decision making, optimal solutions are selected from the Pareto set and sent to the controller as desired set points. The third step then corresponds to the controller implementation to keep the process system at the chosen optimal sustainable operating points. For this step, an advanced biologically inspired optimal control strategy (BIO-CS) [2] is applied to handle the multiple and conflicting objectives. Finally, the fourth step consists of generating a life cycle inventory from the optimized chemical manufacturing stage. Preliminary results associated with the application of the developed approach to a fermentation process example [3] considering performance objectives demonstrated the potential of this framework. In this presentation, we will discuss the implementation of the novel framework to automatically take the selected process to the optimal steady state determined by the multi-objective optimization algorithm according to the decision maker.

References:

1. Lima, F.V.; Li, S.; Mirlekar, G.V.; Sridhar, L.N.; Ruiz-Mercado, G.J. Modeling and advanced control for sustainable process systems In *The analysis, synthesis and design of chemical engineering processes*, Ruiz-Mercado, G.J.; Cabezas, H. (eds.), Eds. Elsevier, 2016.
2. Mirlekar, G.V.; Gebreslassie B.H.; Diwekar U.M.; Lima, F.V.; Design and Implementation of a biomimetic control strategy for chemical processes based on efficient ant colony optimization. *Presented at 2015 AIChE Annual Meeting*, Salt Lake City, Utah, November 2015.
3. Li, S.; Mirlekar, G.; Ruiz-Mercado, G.; Lima, F.V. Development of chemical process design and control for sustainability. *Submitted for Publication.*

466997 Simulation of Work Exchange Systems for Cost-Effective Mechanical Energy Recovery Using Aspen Plus

Abstract:

In many process systems, the operational pressures in different process units could be very different. This renders a need to pressurize/depressurize process streams in different operational stages, where process stream compression and expansion occur. The energy consumption in those operations could be very significant. It is recognized that a great amount of mechanical energy consumed in those operations can be recovered using pressure-driven work exchangers. Note that a work exchanger operates in a hybrid mode (both batch and continuous). Thus, characterization of this type of unit is more sophisticated than that for a process unit operated in a single mode. This kind of characterization is critical to a successful use in process design and operation.

In this work, we introduce a comprehensive steady-state and dynamic simulation technique for characterizing work exchangers. These units will be discussed as a new Aspen Plus tool that can be applied in process simulation to improve the energy efficiency. The developed models are for separately describing the unit operated under isothermal, adiabatic, or polytropic conditions for the process streams in liquid-liquid, gas-gas, or liquid-gas phase. The model-based simulation can facilitate the identification of the bottleneck in mechanical energy recovery and achieve an optimal design of a work exchanger network. Case studies will demonstrate the efficacy of the simulation technique for cost-effective mechanical energy recovery.

471861 An Emergy-Based Sustainability Evaluation and Analysis of Water Resource Usage By the Mining Industry in Colombia

Abstract:

Due to the diversity of weather and soil conditions, Colombia has an abundance of natural resources and a wide range of mining products. These include coal, gold, platinum, nickel, emerald, lime, among others, that are extracted in a smaller scale. In addition, during the last decade, there has been a significant growth in both the volume and value of production of primary goods, particularly in the mining and hydrocarbons that affect positively and significantly the country gross domestic product (GDP).
However, it is unlikely to avoid conflicting objectives, political controversies, economic, environmental, and social challenges around the exploitation of these nonrenewable resources. Moreover, sometimes current mining practices lead to irreversible environmental impacts such as destruction of ecosystems, water scarcity in protected lands, soil, water, and air pollution.
Therefore, there is the need of establishing an evaluation methodology that allows assessing how sustainable is the current management and process mining in Colombia.

Since emergy analysis is a methodological tool for assessing the sustainability of a system or process for environmental decision-making, it will be applied to assess the sustainability of the current mining system in Colombia. This evaluation approach will consider by-products, cost-effectivity, energy efficiency, and environmental impacts through the calculation of emergy-based sustainability indicators. This will allow identifying gaps for improvements and the development of tools to regulate this productive sector in a more sustainable way.

Syngas fermentation for the production of bulk chemicals: early-stage techno-economic and life cycle assessment

Eduardo Almeida Benalcázar, Delft University of Technology, Delft, the Netherlands
Baudine Gevers Deynoot, Delft University of Technology, Delft, the Netherlands
Henk Noorman, DSM Biotechnology Center, Delft, the Netherlands
Patricia Osseweijer, Delft University of Technology, Delft, the Netherlands
John A. Posada Duque [1], Delft University of Technology, Delft, the Netherlands

Introduction

Sustainable production of biofuels and biochemicals through novel routes has been encouraged to reduce dependency on non-renewable resources, cope instability of oil prices and mitigate the growing greenhouse gas emissions. Furthermore, feedstocks such as biomass and other short-cycled carbon sources, *e.g.* CO, CO_2 and CH_4, are sought as possible replacement of petroleum and its derivatives [1]. Currently, commercial-scale production of bio-based chemicals and fuels is mainly done from sugars, starch and oil-containing crops; though ethical and moral debates have appeared since these feedstocks have traditionally been used for food/feed purposes. In contrast, waste and lignocellulosic biomass have no possible competition with food [2].

In general, biomass can be processed mainly by two technologies: i) syngas, and ii) sugar platforms. Syngas production is based on the thermochemical conversion of biomass (gasification), while the sugar platform is based on the enzymatic hydrolysis of cellulose. Syngas is considered as one of the most prominent chemical platforms for biorefineries [3], and is also currently used for heat and power generation, or alternatively converted into fuels and chemicals by catalytic Fischer-Tropsch synthesis (FTS) and fermentation [4].

Acetogenic bacteria are commonly considered as syngas fermentation workhorses since they are able to utilize CO, H_2 and CO_2 as substrates by withdrawing electrons from H_2 and CO. Their carbon fixation into acetyl-CoA follows a metabolic route known as the Wood-Ljungdahl pathway [4]. From this point, they can produce a variety of substances, *e.g.* acetate, ethanol [4], 2,3-butanediol [5] and carboxylic acids [6].

Limited information is available regarding the integrated sustainability assessment (technical, economic and environmental) of syngas production from lignocellulosic biomass and its subsequent fermentation to biochemicals and biofuels. Therefore, the production of ethanol, 2,3-butanediol and hexanoic acid, are here analyzed from the from the perspective of sustainable biorefineries [7]. Thus, techno-economic and life cycle assessments are performed to evaluate the effects of different processing configurations and operating conditions at multiples stages of the supply chain.

Methodology

Process design

The ethanol production scale was defined from an average plant producing second generation bio-ethanol, *i.e.* 20 million gallons per year. In the cases of 2,3-butanediol and hexanoic acid, the production scales were set considering that the input of syngas to the fermentation unit was the same as for ethanol production for comparison.

[1] Corresponding author at: van der Maasweg 9 (Room: C0.360), 2629 HZ, Delft, the Netherlands. E-mail address: J.A.PosadaDuque@tudelft.nl (J.A. Posada)

The syngas derived from four different lignocellulosic materials was considered: pine wood (PW) and corn stover (CS) both produced in the United States (US); and eucalyptus wood (EW) and sugarcane bagasse (SB) both produced in Brazil (BR). In a parallel complementary study [8], we analyzed the syngas production alone, and the resulting compositions from those feedstocks are shown in **Error! Reference source not found.**. In the same study we considered two scenarios for the location of the gasification plant: within the same country where biomass production takes place, and gasification in the Netherlands after biomass import.

Table 1. Composition of syngas fed to fermentation

Component	PW	SB	CS	EW
H_2	60.9	57.0	55.4	56.5
CO	28.1	28.6	30.0	29.0
CO_2	9.9	13.1	13.0	13.2
N_2	1.0	1.2	1.5	1.2
CH_4	593.2 ppmv	379 ppmv	412 ppmv	386 ppmv
NH_3	12.4 ppmv	12 ppmv	13 ppmv	12 ppmv
H_2S	160.6 ppmv	189 ppmv	822 ppmv	-

Fermentation

The stoichiometry and kinetics of the fermentation process reactions, were predicted through a thermodynamic approach as reported in [9,10]. The Gibbs free energy dissipated within the catabolism (see eq. 1 and 2 as examples for ethanol production) was used to define the carbon consumed by microbial growth (see eq. 3 and 4); while the share distribution between main product and the co-product(s) was adjusted according to published experimental results at laboratory scale. The parameters for the Herebrt-Pirt equation were predicted from the maximum substrate consumption rate and common microbial energy needs for maintenance [9,10]. Substrate affinity constants were predicted considering that the concentration of substrate needs to be such as that the energy production through catabolism is higher than 15 kJ/mol of electron donor [10]. All these parameters were adjusted and co-related for the fermentation conditions of temperature, pH and concentration of the main substrates and products.

Ethanol and 2,3-butanediol are produced directly from syngas fermentation [5], while acetic acid and ethanol were here considered as the main by-product respectively. The H_2/CO ratio in which the syngas is fed to the fermentation defines the substrate uptake by microbes, and thus water and CO_2 can either be consumed or produced (see eq. 1 - 4). For both fermentations, the processing conditions used were 37 °C, 1 atm (top pressure) and pH of 5. The resulting concentrations of gases and biomass were obtained from mass balances and calculated by using the mass transfer capacities (MTC) inside the fermentors. The MTC were calculated for each gaseous substrate from superficial gas velocities [11]. The syngas fermentors were designed as bubble columns using main three constraints: a max. liquid volume of 900m^3, max. vessel height of 22m and gas hold-up <15vol.% [11].

Hexanoic acid was assumed to be anaerobically produced from ethanol and acetic acid [8], by a chain elongation reaction (see eq. 5 and 6). The fermentation conditions were set to be 30 °C, pH 5 and low pressure (0.045 bar) to ensure low H_2 concentration at the bottom of the reactor to thermodynamically allow ethanol oxidation. Three key constraints were used for the fermentor design: liquid volume < 2000 m^3, max gas hold-up of 15 vol.% [11] and a gas-liquid mixture height to avoid inhibitory H_2 concentrations.

Ethanol	$6CO + 3H_2O \rightarrow C_2H_5OH + 4CO_2$	(1)
production	$2CO_2 + 6H_2 \rightarrow C_2H_5OH + 3H_2O$	(2)
Microbial	$2CO + 0.5H_2O + 0.25NH_4^+ \rightarrow CH_{1.75}O_{0.50}N_{0.25} + CO_2 + 0.25H^+$	(3)
growth	$2H_2 + CO_2 + 0.25NH_4^+ \rightarrow CH_{1.75}O_{0.50}N_{0.25} + 1.5H_2O + 0.25H^+$	(4)
Ethanol oxidation	$C_2H_5OH + H_2O \rightarrow CH_3COOH + 2H_2$	(5)
Chain elongation	$C_xH_{2x+1}COOH + C_2H_5OH \rightarrow C_{x+2}H_{2(x+2)+1}COOH + H_2O$	(6)

Downstream processing

Different purification routes were designed. Ethanol is continuously removed from the syngas fermentor by the gas excess. The resulting stream is compressed and cooled to condense water and ethanol. Liquids are separated by flashing. The gas stream is treated with monoethanolamine (MEA) to remove non-consumed CO_2. The remaining fractions of CO and H_2 are recycled and mixed with fresh syngas. The pre-concentrated ethanol is pre-heated and distillated. The bottoms stream (containing water, ethanol and acetic acid) is recycled while the top stream is dehydrated by adsorption. The obtained ethanol contains 0.27 wt.% water, while 5.5 % of the product is lost through the DSP.

2,3-Butanediol is continuously removed from the syngas fermentors in the liquid phase with microfiltration for biomass retention, and water is partially removed by reverse osmosis. The permeate is recycled back to the fermenter, while the retentate is distilled to separate water and ethanol. The bottoms stream is flash evaporated to eliminate dissolved solids. 2,3-butanediol is finally condensed and cooled to storage temperature. The final product purity is 99.8 wt.% with an overall recovery of 99.9%.

The DSP for hexanoic acid starts at the chain elongation reactor. The off-gas stream is partially dehydrated by condensation and flashing. The water content, temperature and pressure of the off-gas are adjusted before recycling to the fermenters. Although literature describes different methods for recovery and purification of carboxylic acids [12], only a few authors acknowledge that these products may form an stabilized emulsion with the aqueous phase. In this study, the emulsion is further creamed by microfiltration. The oil-water emulsion is ultimately destabilized by thermal treatment, and the two phases are separated by centrifugation. The inorganic phase is recycled to the chain elongation reactor while the organic is mixed with n-decane to separate the hexanoic acid. The solvent is then recovered by distillation and recycled to the extractor. Finally, the acids mixture is distilled to produce a hexanoic acid ay 99.9 wt.% purity, with an overall recovery of 95.9%.

Economic evaluation

The economic analysis was based on total production costs (capital and operational expenditures (CAPEX, OPEX)) and minimum selling prices (MSP). The MSP is the price that brings the net present value (NPV) to zero at a defined payback time (PBT; 5 years in this case). CAPEX was based on factors related to total equipment purchase costs (EPC) [13] and characteristic size. When needed, the 6/10 scaling factor rule was used for capacity corrections [13]. Prices for all equipment were updated to 2015 by using the Chemical Engineering Plant Cost Index (CEPCI), and the project lifetime was fixed at 15 years.

Aggregation of OPEX was based on [13]: *i.e.* facility-dependent costs (depreciation, maintenance, insurance, local taxes and overhead), process-related costs (raw materials and utilities) and labor. The economic performance was adjusted to the geographical locations by using: i) location factors for CAPEX; ii) specific fuel prices (for utilities); iii) syngas production costs (see **Error! Reference source not found.**) and iv) local economic indexes: interest rate and taxes on revenue. The used values are shown in Table 1.

Table 1. Local economic factors and prices for comparison of the different geographical locations of the processing plants

Country	Location factor	Fuel price, $/GJ	Electricity price, $/MWh	Interest rate, %	Tax on revenue, %
US	1.03	5.075	70.14	1.80	39.0
NL	1.19	12.19	118.0	1.00	25.0
BR	1.01	4.220	134.9	14.25	34.0

Global warming potential (GWP) analysis

The GWP analysis was done according to the guidelines by the ISO 14040 and 14044 [14,15]. A cradle-to-gate plus combustion approach [16] was defined including syngas production, fermentation and DSP stages. Individual impact characterization factors for materials and energy carriers were taken from the *Ecoinvent* V2.2 database, literature and internet references. Factors for electricity production were site-specific. Environmental impact data for syngas production was taken from the complementary work [8].

Results and discussion

Process design

The highest overall mass yield was obtained for 2,3-butanediol from pine wood, which is 13 % higher than for ethanol production, and 33% higher than for hexanoic acid with the same feedstock. An additional advantage of PW is its most compatible syngas for the three products. Additionally, as the CO_2 consumption depends on the H_2/CO ratio, PW the CO_2 consumption is higher from PW), PW leads to a lower dilution of CO and H_2 enhancing the mass transfer capacity and reducing the DSP requirements for CO_2 removal.

Economic evaluation

The direct comparison of MSP, total production costs and global market prices was considered as an indication of the economic potential. In all cases, syngas has the largest contribution to total OPEX (*i.e.*, 42, 34 and 41 %, for ethanol, 2,3-butanediol and hexanoic acid, respectively) followed by facility dependent costs (see Figure 1.a and Figure 1.b)), pH controlling substances and nutrients. Of all cases, the lowest total production costs were for ethanol, followed by 2,3-butanediol and hexanoic acid. For all products, the lowest costs were obtained when PW from the US was used. Although the recovery efficiency and conversion yield of the diol are the highest among the different processes, its DSP is the most resource intensive. This is due to the limited pre-concentration of the diol in the fermentation step. In general, syngas fermentation has the largest contribution to total costs: 90, 75 and 80%, for ethanol, 2,3-butanediol and hexanoic acid, respectively (see **Error! Reference source not found.**). The main reason for this trend is the large equipment and energy requirements to overcome the low CO and H_2 mass transfer capacity. For the hexanoic acid production costs, the chain elongation unit and DSP represent only a minor part of the costs due to significantly lower fermentation volume and easier purification operations. When comparing ethanol production costs through different technologies, data show a great variability with reported values (see **Error! Reference**

source not found.) depending on process configurations and assumptions. The hybrid route is the least cost-competitive option for lignocellulosic ethanol.

The lowest MSP is obtained when production takes place in the US from pine. This is due to the fact that the US has relatively low values for electricity cost, interest rate, location factor, tax on revenue, and fuel prices, plus the lower biomass costs when production is local. On the contrary, BR is the least attractive case, even with the lowest syngas production costs. The obtained ethanol MSP fall above the range of selling prices observed in the global market during the last ten years, meaning that the designed process cannot compete with commercially available ethanol. Due to limited access to 2,3-butanediol prices, it is compared to its equivalent 1,4-butanediol. The MSP of 2,3-butanediol production is within 1,4-butanediol prices range (data no shown) making it a potentially attractive product for the syngas fermentation platform. The MSP of hexanoic acid are compared against jet fuel as the acid could be used as fuel precursor; the acid's MPS's are significantly higher than those for jet fuel.

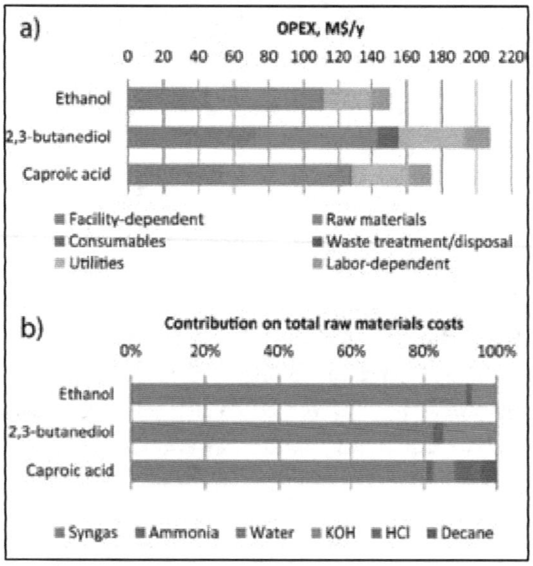

Figure 1. Results of economic analysis for PW in the US via the hybrid process. a) OPEX composition; b) Contribution of raw materials

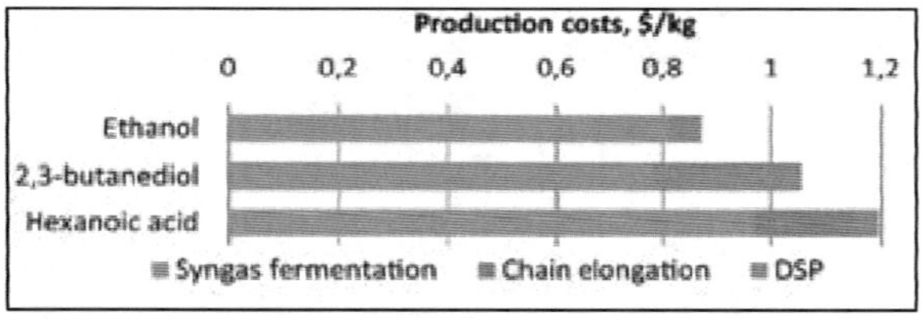

Figure 3. Total production costs of ethanol, 2,3-butanediol and hexanoic acid in the US from PW via the hybrid process

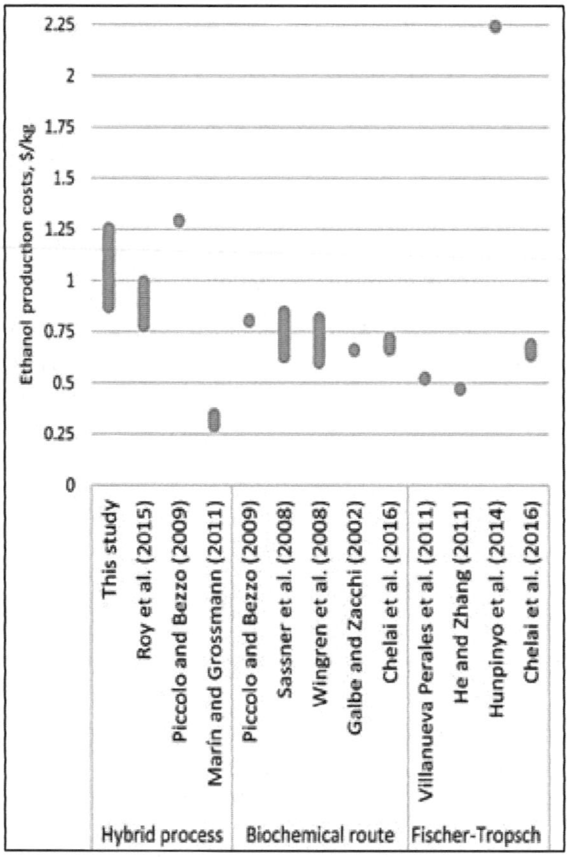

Figure 2. Comparison of different ethanol production costs found in literature. Dots show punctual numbers while lines show ranges. List of references [17-27].

Global warming potential analysis

The main GHG contributors are electricity consumption, syngas production and heat requirements. This tendency is consistent for all scenarios of feedstocks and geographical locations. However, the impacts from electricity are lower in BR than in the US and the NL.

Figure compares the GHG performance of chemicals produced by the hybrid process against alternative production methods by considering other technologies and/or other feedstocks. In some cases similar products were used as benchmark as potential materials to be replaced by the hybrid process. Ethanol produced from lignocellulosic biomass via the hybrid process generally shows higher GWP results, when compared to ethanol derived from first (1G) and second (2G) generation resources (see Figure *4*

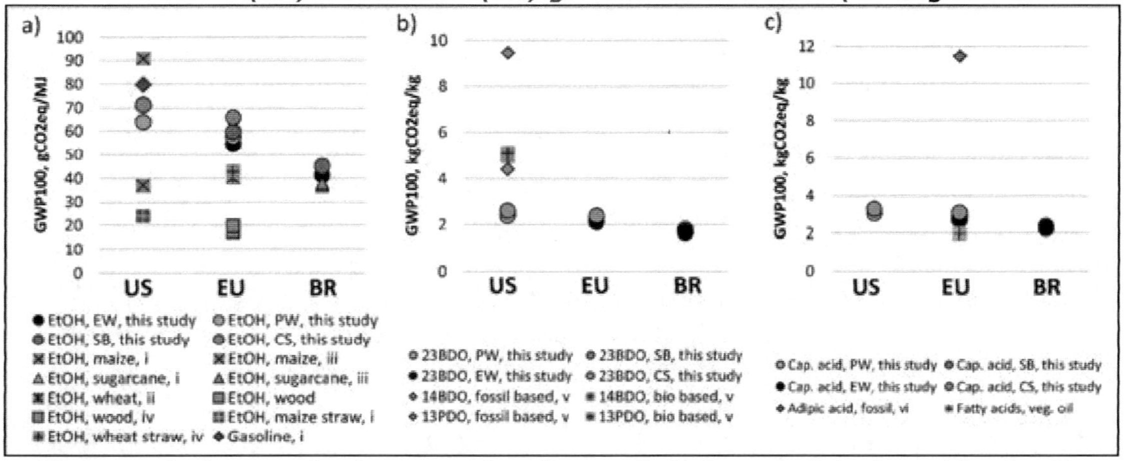

Figure .a). It can also be seen that lignocellulosic materials (wood in the European Union (EU) and maize straw in the US) generate lower GHGs when used as feedstock than 1G raw materials, such as wheat and sugarcane. In general, ethanol performs better than gasoline.

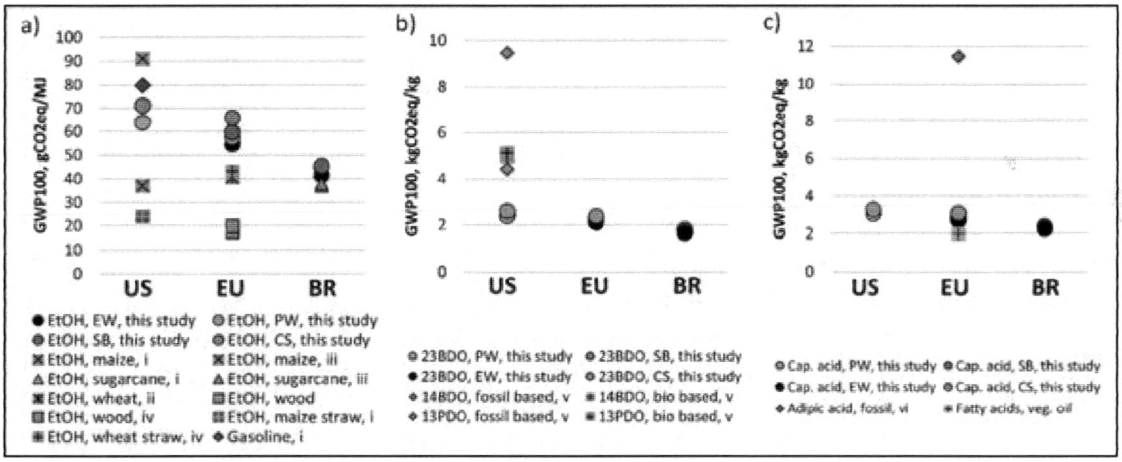

Figure 4. GWP of the hybrid process and comparison with reported data for different routes. a) ethanol, b) 2,3-butanediol, c) hexanoic acid (Literature data adapted from [28])

2,3-butanediol production via the hybrid process on the other hand, resulted in a more attractive option environmentally as well as economically, in comparison to 1,4-butanediol and 1,3-propanediol (both from fossil and bio based raw materials, see Figure

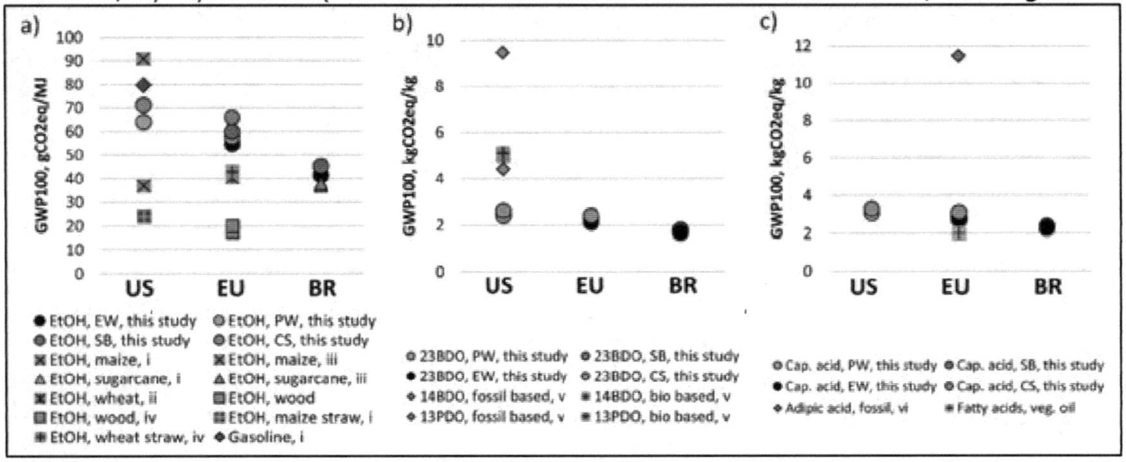

Figure .b); the GHG emissions with the hybrid process represent between 17 – 54 % of what is generated through the production of the two other diols. Hexanoic acid exhibits a similar performance as fatty acids derived from vegetable oils, plus a better performance than adipic acid, a 6-carbon-chain dicarboxylic acid (see Figure .c). Although these last results are attractive, the lack of data about impacts related to 2,3-butanediol and hexanoic acid production through other pathways, brings uncertainty to the analysis performed.

Conclusions

Results of this study suggest that the use of the syngas platform from lignocellulosic biomass, followed by fermentation, has potential for the production of 2,3-butanediol and could even become an interesting alternative for the production of ethanol after future process optimizations.

Application of the hybrid process is especially interesting for the production of bulk chemicals, which require complex separation processes after the fermentation phase, compared to other products that can be separated through phase changes (such as ethanol). The production of 2,3-butanediol has shown better environmental and economic performance compared with fossil and other bio-based technologies for the production of 1,4-butanediol and 1,3-propanediol.

As the fermentation takes economic advantage from the low-cost syngas platform, compared to the expensive sugar-based platform, still further improvements could be obtained by upgrading the gas-liquid mass transfer capacity inside the syngas fermentors, as well as further reduction of the syngas production costs by equipment reduction or the use of cheap feedstocks. Furthermore, the integration of waste heat could be beneficial from an environmental point of view and could possibly reduce production costs.

References

1. Daystar, Jesse; Gonzalez, Ronalds; Reeb, Carter; Venditti, Richard; Treasure, Trevor; Abt, Robert and Kelley, Steve (2014), "Economics, Environmental Impacts, and Supply Chain Analysis of Cellulosic Biomass for Biofuels in the Southern US: Pine, Eucalyptus, Unmanaged Hardwoods, Forest Residues, Switchgrass, and Sweet Sorghum", BioResources, pp. 393–444.
2. Posada, John A. and Osseweijer, Patricia (2016) "Socio-economic and environmental considerations for sustainable supply and fractionation of lignocellulosic biomass in a

biorefinery context", In: Biomass fractionation technologies for a lignocellulosic feedstock based biorefinery, 1st ed. Mussatto SL, p. 674.
3. Moncada, Jonathan; Posada, John A. and Ramírez, Andrea (2015), "Early sustainability assessment for potential configurations of integrated biorefineries. Screening of bio-based derivatives from platform chemicals", Biofuels, Bioproducts and Biorefining, November, pp. 722–748.
4. Phillips, John R.; Klasson, K. T.; Clausen, E. C. and Gaddy, J. L. (1993), "Biological Production of Ethanol from Coal Synthesis Gas", Applied Biochemistry and Biotechnology, p.559.
5. Köpke, Michael; Mihalcea, Christophe; Liew, Fung M.; Tizard, Joseph H.; Ali, Mohammed S.; Conolly, Joshua J.; Al-Sunawi, Bakir and Simpson, Séan D. (2011), "2,3-Butanediol Production by Acetogenic Bacteria, an Alternative Route to Chemical Synthesis, Using Industrial Waste Gas", Applied and Environmental Microbiology, August, pp. 5467–5475.
6. Vasudevan, Divya; Richter, Hanno and Angenent, Largus T. (2014), "Upgrading dilute ethanol from syngas fermentation to n-caproate with reactor microbiomes", Bioresource Technology, January, pp. 378–382.
7. Palmeros-Parada, Mar; Osseweijer, Patricia; Posada, John A. (2016) Sustainable biorefineries, an analysis of practices for incorporating sustainability in biorefinery design. Industrial Crops and Products http://dx.doi.org/10.1016/j.indcrop.2016.08.052
8. Gevers-Deynoot, Baudine; Almeida-Benalcázar, Eduardo F.; Noorman, Henk; Osseweijer, Patricia; Posada, John A. (2016). Sustainability analysis of the whole value chain of bio-syngas production and its potential as chemical platform for syngas fermentation. 2016 AIChE Annual Meeting. November 13-18, 2016. San Francisco, CA
9. Kleerebezem, Robbert and Van Loosdrecht, Mark C. (2010), "A Generalized Method for Thermodynamic State Analysis of Environmental Systems", Critical Reviews in Environmental Science and Technology, January, pp. 1–54.
10. Heijnen, Joseph (2013), "A thermodynamic approach to predict black box model parameters for microbial growth" In: Biothermodynamics, EPFL Press, pp.443-473.
11. Van't Riet, Klaas and Tramper, Johannes (1991), "Basic bioreactor design", Marcel Dekker.
12. Posada, John A; Cardona, Carlos A. (2012). Propionic acid production from raw glycerol using commercial and engineered strains. Industrial & Engineering Chemistry Research 51:2354–2361.
13. Peters Max S. and Timmerhaus, Klaus D. (1991), "Plant design and economics for chemical engineers", 4. ed. McGraw-Hill.
14. ISO, (1997), "ISO 14040 Environmental management - Life cycle assessment - Principles and framework".
15. ISO, (2006), "ISO 14044 Environmental management - Life cycle assessment - Requirements and guidelines".
16. Posada, John A; Brentner, Laura B; Ramirez, Andrea; Patel, Martin K. (2016), "Conceptual design of sustainable integrated microalgae biorefineries: parametric analysis of energy use, greenhouse gas emissions and techno-economics", Algal Research, July, pp. 113-131.
17. Piccolo, Chiara and Bezzo, Fabrizio (2009), "A techno-economic comparison between two technologies for bioethanol production from lignocellulose", Biomass & Bioenergy, March pp. 478–491.

18. Roy, Poritosh; Dutta, Animesh and Deen, Bill (2015), "Greenhouse gas emissions and production cost of ethanol produced from biosyngas fermentation process", Bioresource Technology, September, pp. 85–191.
19. Martín, Mariano and Grossmann, Ignacio E. (2011), "Energy optimization of bioethanol production via gasification of switchgrass", AIChE Journal, December, pp. 3408–3428.
20. Sassner, Per; Galbe, Mats and Zacchi, Guido (2008), "Techno-economic evaluation of bioethanol production from three different lignocellulosic materials", Biomass & Bioenergy, May, pp. 422–430.
21. Wingren, Anders; Galbe, Mats and Zacchi, Guido (2008), "Techno-Economic Evaluation of Producing Ethanol from Softwood: Comparison of SSF and SHF and Identification of Bottlenecks", Biotechnology Progress, September, pp. 1109–1117.
22. Galbe, Mats and Zacchi, Guido (2002), "A review of the production of ethanol from softwood", Applied Microbiology and Biotechnology, September, pp. 618–628.
23. Cheali, Peam; Posada, John A.; Gernaey, Krist V. and Sin, Gürkan (2016), "Economic risk analysis and critical comparison of optimal biorefinery concepts: Economic risk analysis and critical comparison of optimal biorefinery concepts", Biofuels, Bioproducts and Biorefining, May, pp. 435-445.
24. Cheali, Peam; Posada, John A.; Gernaey, Krist V. and Sin, Gürkan (2015), "Upgrading of lignocellulosic biorefinery to value added chemicals: Sustainability and economics of bioethanol-derivatives", Biomass and Bioenergy, April, pp.282–300.
25. Villanueva Perales, A.; Reyes Valle, C.; Ollero P. and Gómez-Barea, A. (2011), "Technoeconomic assessment of ethanol production via thermochemical conversion of biomass by entrained flow gasification", Energy, April, pp. 4097–4108.
26. He, Jie and Zhang, Wennan (2011), "Techno-economic evaluation of thermo-chemical biomass-to-ethanol", Applied Energy, April, pp.1224–1232.
27. Hunpinyo, Piyapong; Cheali, Peam; Narataruksa, Phavanee; Tungkamani, Sabaithip and Chollacoop, Nuwong (2014), "Alternative route of process modification for biofuel production by embedding the Fischer–Tropsch plant in existing stand-alone power plant (10MW) based on biomass gasification – Part I: A conceptual modeling and simulation approach (a case study in Thailand)", Energy Conversion and Management, December, pp.1179–1192.
28. Morales, Marjorie; Quintero, Julián; Conejeros, Raúl and Aroca, Germán (2015) "Life cycle assessment of lignocellulosic bioethanol: Environmental impacts and energy balance", Renewable and Sustainable Energy Reviews, February, pp.1349–1361.

457102 Plasmon-Enhanced Photocatalytic CO_2 Reduction on Nanostructured Composite Electrodes

Abstract:

Using inputs of only sunlight, electricity, carbon dioxide, and water, photocatalytic CO_2 reduction could mitigate the effects of climate change caused by the burning of fossil fuels and produce valuable fuels or chemical feedstocks. However, current CO_2 reduction technologies suffer from high overpotentials and low selectivity, producing a mixture of carbon monoxide, methane, ethylene, formate, methanol, and other products.

Plasmon-assisted photocatalytic CO_2 reduction leads to greater selectivity and lower overpotentials by unlocking unique mechanistic pathways. It has been well-documented that the strongly localized near fields at the surface of plasmonic nanoparticles promote electron-hole pair generation in nearby semiconductors.[1–3] Plasmon decay generates an excited "hot" electron which can be transferred to a surface molecule for direct reduction or injected into an adjacent wide band gap catalyst, effectively limiting carrier recombination through charge separation and expanding the usable portion of the solar spectrum.[4–7] The electron dynamics in an irradiated plasmonic nanoparticle can alter the electronic coupling with surface adsorbed CO_2 and reaction intermediates, thereby changing the binding energy of these species and the catalytic properties of the plasmonic metals. This plasmon-enhanced effect was demonstrated when optically excited surface plasmons at the roughened surface of an Ag electrode in CO_2-containing solutions produced strong enhancement of the cathodic current.[8–9] The wide tunability of the plasmon resonance frequency with shape, size, and material confers fine control over these catalytic mechanisms, allowing for optimization of the photocatalytic performance.

To take advantage of plasmonic catalysis, nano-sized spheres, cubes, and wires of gold, silver, and copper

have been synthesized as these materials exhibit strong plasmonic and electrocatalytic behavior. For the semiconducting component of the photoelectrodes, nanostructured TiO_2 co-catalysts have also been prepared. Transmission electron microscopy (TEM) and scanning electron microscopy (SEM) were used to determine the shape and size distribution of the nanoparticles. X-ray diffraction (XRD) and optical spectroscopy were employed to characterize the electrodes. The CO_2 reduction ability of the plasmonic metal nanoparticles with and without a semiconductor co-catalyst has been tested in a custom glass photoelectrochemical cell with a quartz window that allows for front- and back-illumination of the electrode. The two electrode chambers are separated by a membrane and can achieve a gas-tight seal that allows for precise pressure measurements. Experiments were run both under isolation of the pressurized cell and with a constant CO
spectrometer (GC-MS) that allows detection and quantification of gaseous products. *Ex-situ* high performance liquid chromatography (HPLC) was used to quantify liquid products.

This material is based upon work performed by the Joint Center for Artificial Photosynthesis, a DOE Energy Innovation Hub, supported through the Office of Science of the U.S. Department of Energy under Award Number DE-SC0004993.

The work is supported by the National Science Foundation Graduate Research Fellowship under Grant No. 1106400.

Work at the Molecular Foundry was supported by the Office of Science, Office of Basic Energy Sciences, of the U.S. Department of Energy under Contract No. DE-AC02-05CH11231.

References:

(1) Thomann, I.; Pinaud, B. A.; Chen, Z.; Clemens, B. M.; Jaramillo, T. F.; Brongersma, M. L. *Nano Lett.* **2011**, *11*, 3440–3446.

(2) Ingram, D. B.; Linic, S. *J. Am. Chem. Soc.* **2011**, *133*, 5202–5205.

(3) Gao, H.; Liu, C.; Jeong, H. E.; Yang, P. *ACS Nano* **2012**, *6*, 234–240.

(4) Mukherjee, S.; Zhou, L.; Goodman, A. M.; Large, N.; Ayala-Orozco, C.; Zhang, Y.; Nordlander, P.; Halas, N. J. *J. Am. Chem. Soc.* **2014**, *136*, 64–67.

(5) Mubeen, S.; Lee, J.; Singh, N.; Krämer, S.; Stucky, G. D.; Moskovits, M. *Nat. Nanotechnol.* **2013**, *8*, 247–251.

(6) Naya, S.; Teranishi, M.; Isobe, T.; Tada, H. *Chem. Commun.* **2010**, *46*, 815–817.

(7) Lee, J.; Mubeen, S.; Ji, X.; Stucky, G. D.; Moskovits, M. *Nano Lett.* **2012**, *12*, 5014–5019.

(8) Kostecki, R.; Augustynski, J. *Chem. Phys. Lett.* **1992**, *194*, 386–390.

(9) Kostecki, R.; Augustynski, J. *J. Appl. Phys.* **1995**, *77*, 4701–4705.

459985 Photon, Electron, and Ion Management in Artificial Photosynthesis: Realizing Efficient Renewable Energy to Fuel Conversion

Abstract:

Direct production of fuels from sunlight, air (N_2 or CO_2), and water that can be stored, transported, and later converted into hydrogen or electricity to provide power for transportation and distributed energy generation, have received recent attentions worldwide. This technology could also provide chemicals as synthetic precursors or realize grid-level storage of intermittent solar energy. In artificial photosynthesis based on semiconductor photoelectrochemistry, production of chemical fuels generally requires the coupling of separated electrical charges with electrocatalysts for multi-electron chemical reactions. Meanwhile, the development of integrated, efficient and stable photoelectrochemical (PEC) systems requires the pairing of light-absorbing materials with an optimum bandgap combination. The development of such systems has been hindered in part by the lack of semiconducting materials that can simultaneously provide efficiency and stability in a corrosive aqueous electrolyte, typically strong acid or base. In this talk, I will present our recent progress in the development of efficient and stable PEC devices for H2 production from solar-driven water splitting. First, I will summarize some recent development of photon management in PEC devices and module designs from modeling to experiment. Then, I will discuss methods for heterogeneous interfacial energetic modification on covalence-bond semiconductors for efficient free carriers (electrons/holes) separation. Finally, I will present a novel approach to managing the ionic environment for creating a defect-tolerant condition, which leads to a record-setting solar water-splitting prototype.

462500 Tandem Core-Shell Si-Ta_3N_5 Photoanodes for Photoelectrochemical Water Oxidation

Abstract:

Tantalum nitride (Ta_3N_5
type doping, relatively small bandgap of 2.1 eV, and band positions that straddle the redox potentials of both the hydrogen and oxygen evolution reactions.[1] However, there are other challenges associated with Ta_3N_5 including poor charge transport properties, late photocurrent onset and photodegradation. To address all of these challenges, we fabricated efficient and stable core-shell heterostructured Si-Ta_3N_5 nanowire photoanodes coated with $CoTiO_x$ and NiO oxygen evolution co-catalysts.

Previous device architecture strategies for metal oxide semiconductors with charge transport limitations have included nanostructuring or scaffolding. In this work, a similar approach was applied to Ta_3N_5 to make a nanostructured core-shell Si-Ta_3N_5 device where the Si is a semiconducting, nanowire scaffold and the Ta_3N_5 is coated onto Si as a thin shell through which charge extraction should be efficient. With n-type Si as the scaffold, a heterojunction between Si and Ta_3N_5 is formed enabling a dual-absorber configuration resulting in a 200 mV cathodic shift for photocurrent onset due to the photovoltage gained from Si and therefore improves the overall performance of the Si-Ta_3N_5 device. The Si nanostructures were fabricated by a pseudo-Bosch deep reactive ion etching process, then tantalum oxide (Ta_2O_5) was coated onto Si by atomic layer deposition (ALD), and finally the Ta_2O_5 was nitrided in pure ammonia gas at 900 °C.

To gain a deeper understanding of charge transport within the thin Ta_3N_5 shells and to optimize the electrode

performance, we varied the thickness of the shell between 10 and 70 nm and tested for ferrocyanide oxidation as a facile photoelectrochemical reaction. We found that in a core-shell Si-Ta_3N_5 nanostructured device, photocurrent decreased with increasing Ta_3N_5 thickness, which indicates that charge extraction becomes problematic with thicker Ta_3N_5 films. To verify that decreasing photocurrent was due to poor charge transport in thicker films, we performed absorbed photon to current efficiency (APCE) measurements with Ta_3N_5 deposited on planar Si and quartz substrates. Again, we found that APCE decreased once the films were thicker than 30 nm, demonstrating that the minority carrier diffusion length is likely tens of nanometers.

Finally, to address the problem of Ta_3N_5 photodegradation and to demonstrate photoelectrochemical performance for water oxidation in basic electrolyte with the core-shell system, $CoTiO_x$ and NiO oxygen evolution catalysts were coated onto Si-Ta_3N_5 by dipcoating from a sol-gel and ALD, respectively. In terms of activity, we found that $CoTiO_x$ interfaces well with Ta_3N_5 and yields a high photocurrent density compared to NiO. With NiO we were able to achieve a photocurrent density close to that of $CoTiO_x$ after Fe incorporation into NiO (henceforth called Ni(Fe)O).[2] However, with $CoTiO_x$, the device is not stable; but with only ~1 nm of Ni(Fe)O we see a marked improvement in stability. The device loses 40-50% of its peak photocurrent over a 24 hour chronoamperometry test. Compared to the current Ta_3N_5 reports in the literature, the Ni(Fe)O catalyst affords noteworthy stability.[3-5] Therefore, in this work, we demonstrate a heterostructured Ta_3N_5 system with a tandem core-shell Si-Ta_3N_5 configuration that enabled earlier onset of photocurrent, efficient charge extraction from Ta_3N_5 and relatively stable long term water oxidation with a Ni(Fe)O co-catalyst.

1. Chun, W.-J.; Ishikawa, A.; Fujisawa, H.; Takata, T.; Kondo, J. N.; Hara, M.; Kawai, M.; Matsumoto, Y.; Domen, K. *The Journal of Physical Chemistry B* **2003,** 107, (8), 1798-1803.

2. Trotochaud, L.; Young, S. L.; Ranney, J. K.; Boettcher, S. W. *J Am Chem Soc* **2014,** 136, (18), 6744-6753.

3. Li, Y.; Zhang, L.; Torres-Pardo, A.; González-Calbet, J. M.; Ma, Y.; Oleynikov, P.; Terasaki, O.; Asahina, S.; Shima, M.; Cha, D.; Zhao, L.; Takanabe, K.; Kubota, J.; Domen, K. *Nat Commun* **2013,** 4.

4. Liao, M.; Feng, J.; Luo, W.; Wang, Z.; Zhang, J.; Li, Z.; Yu, T.; Zou, Z. *Advanced Functional Materials* **2012,** 22, (14), 3066-3074.

5. Liu, G.; Fu, P.; Zhou, L.; Yan, P.; Ding, C.; Shi, J.; Li, C. *Chemistry â€" A European Journal* **2015,** 21, (27), 9624-9628.

462505 Heterostructured c-Si/$BiVO_4$ Core-Shell Tandem Photoanode for Unassisted Photoelectrochemical Water Splitting

Abstract:

Photoelectrochemical (PEC) water splitting is a promising way of storing solar energy in hydrogen, which can be used as a carbon-free fuel or a chemical feed stock. In principle, a PEC electrode consists of a single active material capable of handling the key processes: absorption of solar photon energy, charge separation and transport, and electrocatalysis to make and break chemical bonds at the electrode surface. In practice, there is no existing material that can handle all of these functions simultaneously; thus, the multicomponent system that integrate different materials with more specialized functions is a more viable approach.

A tandem-junction PEC device offers a high theoretical solar-to-hydrogen (STH) efficiency due to the additive voltages across two photoabsorbers and the better utilization of the solar spectrum. Modeling STH efficiency as a function of band gap for a tandem device shows that over 20% STH can be achieved with band gaps of 1.2 and 1.8 eV.[1] One promising photoabsorber material combination is c-Si paired with a metal oxide, especially $BiVO_4$. In addition to the appropriate band structure alignment of both semiconductors, c-Si and $BiVO_4$ are low cost, earth abundant, and industrially scalable. However, the key challenges for these materials are the instability of c-Si in aqueous environment and the poor electronic properties of $BiVO_4$.

In this work, we present the fabrication of a heterostructured c-Si/$BiVO_4$ core-shell tandem based device capable of performing spontaneous water splitting without any precious metal. The wafer scaled device architecture was prepared by (1) nanostructuring of the c-Si substrate to serve as both a bottom absorber and a scaffold to improve the charge separation of $BiVO_4$; (2) tuning the c-Si p-n junction depth to maximize the device photovoltage; (3) engineering the

SnO_2 interfacial layer to passivate the c-Si/$BiVO_4$ interface and protect the c-Si p-n junction; (4) conformally coating a $BiVO_4$ thin film on the nanostructured Si; and (5) decorating the c-Si and $BiVO_4$ surfaces with both oxygen and a hydrogen evolution catalysts.

(1) Seitz, L. C.; Chen, Z.; Forman, A. J.; Pinaud, B. A.; Benck, J. D.; Jaramillo, T. F. *ChemSusChem* **2014,** *7* (5), 1372-1385.

462617 The Fabrication of $Cu_2o/g-C_3N_4/WS_2$ Triple-Layered Photocathode for Photoelectrochemical Hydrogen Evolution

Abstract:

The Fabrication of $Cu_2O/g-C_3N_4/WS_2$ Triple-layered Photocathode for Photoelectrochemical Hydrogen Evolution

Xintian Xu, Yuanzhi Zhu, Xiaobin Fan, Guoliang Zhang, Wenchao Peng*

School of Chemical Engineering & Technology, Tianjin University, Tianjin 300072, China.

wenchao.peng@tju.edu.cn

Hydrogen is being vigorously pursued as a future energy carrier due to the fossil energy depletion. Photoelectrochemical (PEC) water splitting is an emerging hydrogen production technology by utilizing the solar light. This technology integrates in the same device both solar energy collection and water electrolysis. Both photo-active semiconductor and catalysts for hydrogen evolution reaction (HER) are necessary in this device. Cu_2O is a direct P-type semiconductor with a narrow band gap (~2.0 eV), which can be activated by the visible light. Moreover, Cu is a kind of earth abundant element, and the synthesis of Cu_2O is facile. Cu_2O is therefore one of the most promising low-cost photoactive semiconductor reported until now. However, the activity of bare Cu_2O is still poor, and it is not stable towards chemical oxidants and light corrosion. Combing with other semiconductors (e.g. TiO_2, ZnO) could improve the activity and stability of Cu_2O effectively.

As we know, Pt and other precious metals are the most efficient catalysts for hydrogen evolution reaction (HER).

However, these metals are rare and expensive to apply. Transition-metal dichalcogenides (TMDs), such as Molybdenum disulfide (MoS_2), tungsten disulphide (WS_2), are emerging HER catalysts that could be used as a substitute for noble metals in HER.

In this study, a novel $Cu_2O/g-C_3N_4/WS_2$ triple-layered photocathode was designed for the PEC hydrogen evolution. Cu_2O was electrodeposited on the FTO glass as the active layer for the light harvesting. A metal-free $g-C_3N_4$ was exfoliated and covered on the surface of Cu_2O as protection layer. $G-C_3N_4$ could also form p-n junction with Cu_2O to increase the activity. $1T-WS_2$was obtained by lithium intercalation and used as HER catalyst for the further modification of the photocathode.

The obtained photocathode was then used for PEC HER test in the Na_2SO_4/H_2SO_4 solution (Î»â‰¥420nm, 100 mW/cm^2). With the irradiation, electron-hole pair will be generated in the p-type Cu_2O, and the excited electron could be then extracted through the $g-C_3N_4$ protective layer to produce hydrogen on the WS_2. The efficient extraction of excited electrons by this triple-layered photocathode leads to photocurrents of up to -6 mA/cm^{-2} at -0.4 V versus RHE (pH 6.0). The corresponding overpotential is -0.23 V (versus RHE) for the H_2 evolution. In sharp contrast, nearly no H_2 evolution happened under dark conditions. This photocathode is facile to synthesis and has great potential for developing low-cost PEC water-splitting device.

473001 *Sustainable System Synthesis and Analysis using a Novel Sustainability Concept*

Abstract:

Sustainable system analysis and synthesis

using a novel sustainability concept

One of the first definitions for sustainability has been introduced by United Nations in 1987, in a report entitled, Our Common Future (also known as the Brundtland report) which define sustainable developments as "...development that meets the needs of the present without compromising the ability of future generations to meet their own needs." Although this definition captured the essence of the word sustainability but it is very vague in order to be applicable for other organizations. Efforts have been made by researchers to quantify the abstract definitions of sustainability and give it an applicable view. Previous attempts include the ecological footprint (1992, 1996), the use of Fisher information theory (2002), the pressure-state-response model of environmental sustainability (2003), the environmental sustainability index (2005), the sustainability assessment by fuzzy evaluation (SAFE) model (2001-2009), and the AIChE Sustainability Index (2007).

In the current work a novel sustainability concept is put forward, which allows a rigorous quantification of sustainability for a variety of system descriptions. The new concept allows societal and personal input to be incorporated in the sustainability assessment process. Rigorous mathematical criteria are provided for its assessment, involving the global solution of finite dimensional optimization problems. In addition, the proposed concept lends itself to the synthesis of sustainable systems. Case studies for 2-dimensional and 3-dimensional dynamical systems are used to illustrate the introduced new concept, and its applicability to both sustainability analysis and synthesis.

473967 Systematic Approach Towards Establishing Thermodynamic Principles of Sustainable Coupled Industrial-Natural Systems (CINS)

Abstract:

Fundamentals of thermodynamics apply to wide variety of systems at multiple scales such as materials and biogeochemical cycles [1, 2] - ranging from quantum to planetary scale. It will not be an exaggeration to say that systems that violate physical laws of thermodynamics cannot exist as stable systems. This raises two very critical questions for the field of sustainability: a) What are the thermodynamic principles of sustainability science and b) How do we approach hypothesizing, establishing and validation of these principles in a scientifically rigorous way? For the first question, I will qualitatively present some plausible theories of thermodynamics in context of sustainability such as Maximum Entropy principle [3] and non-equilibrium thermodynamics [4]. For the second question, I will present a framework that can be adopted to study thermodynamic properties of coupled industrial-natural systems (CINS) and relating it to approaches for testing the applicability of thermodynamic principles for CINS. The quantitative study will focus on the thermodynamic properties of nutrient flows (Nitrogen) in an industrial ecosystems coupled with regional-ecological nitrogen cycle. Following the course of establishment of thermodynamic laws in other fields, it is suggested that quantitative modeling framework to observe thermodynamic properties over a long span may provide clues to answer question 1 whereas more validation approaches may need to be developed for drawing some general principles.

References

1. D Panagiotaras et al., 2015, "Biogeochemical Cycling of Nutrients and Thermodynamic Aspects", J. Thermodynamic Catalysis, 6:2
2. J.J. Vallino and C.K. Algar, 2016, "The Thermodynamics of Marine Biogeochemical Cycles : Lotka

   Revisited", Annu. Rev. Mar. Science , 8:333-356
3. R. E. Ulanowicz and B.M. Hannon, 1987, "Life and the production of entropy", Proc. R. Soc. Lond., B 232, 181-192
4. C.G. Chakrabarti and K. Ghosh, 2009, "Non-equilibrium thermodynamics of ecosystems : Entropic analysis of stability and diversity", Ecological Modelling, 220, 1950-1956

474368 Analysis of United Nations Clean Development Mechanism Carbon Emission Reduction Projects from a Life Cycle Assessment Perspective

Abstract:

We analyze, from a life cycle perspective, the effectiveness of the United Nations Clean Development Mechanism (CDM) emission reduction projects implemented in developing countries. Typically, the carbon offset credit allocated to these projects account for avoided direct emissions of GHGs but do not include indirect emissions that occur during upstream and downstream steps (for both avoided as well as the project emissions). We explore, through a case study approach, how these projects perform when life cycle calcualtions are used. Further, we also explore if these proejcts lead to lower (or higher) health and ecosystem impacts (co-benefits). A small but diverse set of renewable energy generation CDM projects are compared on an equivalent bassis to assess any potential problem shifting issues. Results from this study can help policy makers assess effectivenss of such projects in a more holistic manner and fine tune their emission calculation methodologies.

468174 Evolution and Robustness of the Global Agricultural-Phosphorus Trade Network

Abstract:

Containing one of the three essential nutrients for plant growth, phosphatic fertilizers are an integral input to modern agriculture, and will always have a global demand. As the world population continues to grow through mid-century, a significant increase in food production must follow, potentially furthering our current dependence on synthetic fertilizers. While the global resource-criticality of phosphate rock reserves – from which the majority of these fertilizers are derived – may be uncertain, the trade network upon which many countries depend for access to these fertilizers (and their precursors) is a well-documented complex system, subject to an array of physical, economic, and geopolitical constraints. If we are to collectively achieve global food security by 2030, as per the U.N. Sustainable Development Goals, the global security of agricultural nutrients must also be ensured. To formalize questions of access to these nutrients, network analysis provides an attractive quantitative framework with which to examine features of individual relationships and of the underlying structures of complex systems.

The present work focuses on characterizing the structure and understanding the systemic vulnerabilities of the global agricultural phosphorus trade network (GAPTN). We accomplish this by modeling a directed, weighted global network of annual phosphorus-containing commodity exchanges, including all synthetic phosphatic fertilizers and their chemical precursors. All commodities are further normalized to their weight content of elemental phosphorus. As such, nodes are taken to be countries participating in phosphorus trade, and edges are taken to be the aggregate of the directed trade flows of elemental phosphorus between each pair of countries, resulting in the construction of a global agricultural phosphorus trade network (GAPTN). The primary objectives of this study are to (1) characterize the network topology and evolutionary trends of the GAPTN. (2) Elucidate the sets of nodes and edges that are critical to the overall function of the network. (3) Identify those countries most dependent on the network's functioning, and outline engineering and policy measures such that they may ensure their consistent access to this essential resource.

Edges within the GAPTN display mild heterophily, preferentially connecting nodes of dissimilar degree. Based on statistical testing, the distribution of GAPTN node degrees follows a heavy-tail distribution, as does the distribution of in-, out-, and total-strength for 2014. Such distributions indicate that a small subset of all of the participating countries exchange most of the resources that comprise the system, leaving the network robust to random disruptions and highly vulnerable to inopportune ones. The potential for inopportune disruptions stems from both the price volatility of phosphatic fertilizers with respect to energy prices, and the geopolitical tension that surrounds Western Sahara & Morocco, which owns some 75% of the estimated global phosphate rock reserves. Further confirming the implications of the degree distributions, a robustness analysis of randomized versus targeted removal of nodes from the network indicates that some 80% of the network's overall activity halts when some 5% of the nodes do not participate. Results from node-level analyses reveal a small subset of countries to be particularly vulnerable to supply disruptions, even as the overall network has become more redundant and dense from 1990 to 2014. In light of these findings, both international cooperation on phosphorus management and country-level efforts to close local nutrient cycles can aid in improving food security while also minimizing the ecological impacts associated with fertilizer run-off, especially eutrophication.

471802 A Framework for Considering Synergies Between Nature and Engineering from Process to Planetary Scales

Abstract:

Two major requirements for designing sustainable systems include considering all the systems through the entire life cycle to prevent the shifting of impacts to different scales; and to account for the role played by nature while making decisions in order to stay within the carrying capacity of ecosystems to prevent unintended harm or ecological degradation.

The recently developed Process-to-Planet (P2P) modeling framework [1,2] integrates fundamental engineering models and key processes in the upstream supply chain with macroeconomic systems. It improves upon conventional Sustainable Process Design (SPD) methods by incorporating the entire upstream life cycle while solving a process design problem, thus reducing the chances of shifting of environmental impacts across narrow system boundaries. The Techno-Ecological Synergy (TES) framework [3], on the other hand takes into account the carrying capacity of ecological systems by quantifying the demand and supply of ecosystem services, accounting for all the interactions between technological and ecological systems at multiple scales. Including ecosystems within the boundary of traditional engineering design decisions is expected to identify design solutions that aim to keep the impact of engineering activities within the regenerative capacity of ecosystems, depending on the scale of relevance of ecological systems. While the TES framework is the first of its kind to account for interactions between engineered systems and ecological systems, the framework lacks connections with life cycle models and while the P2P considers a comprehensive analysis boundary, it does not consider ecological systems at any scale.

This work introduces the methodology of "Process-to-Planet Techno-Ecological Synergy" (P2P-TES) for designing sustainable systems. The objective behind developing this framework is to make design decisions at the local scale by capturing the entire upstream life cycle boundary. The Structural Path Analysis algorithm (SPA) embedded within the P2P framework helps to distribute the processes or sectors in the system among the different scales. Inclusion of ecosystem services through TES within the P2P framework will expand the solution space and help to discover optimal solutions which enhance both the environmental and economic profitability of the system. In other words, P2P-TES framework is expected to provide "win-win" solutions to the process design problem compared to one where ecosystem services are not included. In addition, accounting for the imbalance between the demand and supply of ecosystem services will enable making design decisions that keep human activities within the carrying capacity of ecological systems. Ecosystem services that will be included in the framework are carbon sequestration, within the national boundary and air quality regulation service within a regional serviceshed.

P2P-TES framework will be applied to two separate case studies which present their own different set of challenges. Successful application to both of them will prove the robustness of the framework as well expose existent deficiencies.

References

[1] Hanes, Rebecca J., and Bhavik R. Bakshi. "Process to planet: A multiscale modeling framework toward sustainable engineering." AIChE Journal 61.10 (2015): 3332-3352.

[2] Hanes, Rebecca J., and Bhavik R. Bakshi. "Sustainable process design by the process to planet framework." AIChE Journal 61.10 (2015): 3320-3331.

[3] Bakshi, Bhavik R., Guy Ziv, and Michael D. Lepech. "Techno-ecological synergy: A framework for sustainable engineering." Environmental science & technology 49.3 (2015): 1752-1760.

474957 Developing and Piloting Robust Downstream Processes to Utilize Variable Composition and Blended Biomass Feedstocks

Abstract:

Biomass variability can play a key role in end-product quality and downstream conversion process performance and economics. This is a key challenge to the biofuel and bio-based chemical industries in light of the differences in composition and even the identify and utility of biomass crops and residues in different seasons and geographic regions across the US. The Advanced Biofuels [and Bioproducts] Process Demonstration Unit (ABPDU) at the Lawrence Berkeley National Laboratory is working with several National Lab and industrial partners to address these issues. Bench-scale and pre-pilot prototyping of biomass conversion and upgrading processes will be presented using case studies from collaborative projects with the Idaho and Sandia National Laboratories focused on advanced biofuel and chemical product production processes. In addition, results from industrial collaborations and sponsored research at the ABPDU will be used to highlight the importance of robust and semi-universal processes for the conversion of a diverse set of feedstreams to fuels and other bio-based products. The learnings from these bench- and pilot studies will inform larger scale IBR demonstrations focused on multiple feedstocks.

474958 A Review of Minor Feedstock Characteristics that Have a Major Impact on Process Design and Commercial Success

Abstract:

Technology development in the field of lignocellulosics to fuels and chemicals continues, albeit at a reduced rate due to the recent collapse in the global crude price. Key physical and chemical properties of the selected lignocellulosic feedstocks that must be thoroughly understood for successful implementation of first-of-a-kind process technologies will be discussed. A thorough understanding of feedstock properties is not only critical to the design and selection of commercial scale systems but also must be considered during the technology development and scale-up phases. Though focus has historically been on issues related to sugar recovery and downstream conversion process efficiencies, the inorganic fraction composition and general physical properties often introduce unexpected challenges during scale-up and commercialization. When making capital allocation decisions failure to adequately account for feedstock variability and physical properties can result in less than desirable results, sometimes in unanticipated ways. Technology selection practices and investor-related constraints can be in conflict and may extend the project development phases resulting in unnecessary risks and escalated cost.

474955 Innovative Solutions to Feedstock Supply System Challenges Faced by Biorefineries: the U.S. Department of Energy's Feedstock Supply and Logistics Program

Abstract:

Biorefineries face many challenges associated with feedstock supply system establishment and maintenance as well as in-feed handling. Dr. Alison Goss Eng will provide an overview of these challenges as well as potential opportunities for further technology development within the context of the Bioenergy Technologies Office at the US Department of Energy, describing how the Department is working to overcome these challenges with its partners.

468420 Aquathermolysis of Waste Triglycerides in a Continuous Flow Reactor for Jet Fuels Production

Abstract:

The unavoidable dwindling fossil fuel and its attendant climate change in our time have given impetus to researchers in the energy sector to delve into different sustainable energy sources. One of such energy sources is the abundant waste triglycerides whose long chain hydrocarbons can be separated from their short oxygen chain via hydrodeoxygenation, decarbonylation, and decarboxylation in the presence of a heterogeneous catalyst and hydrogen. However, the use of gaseous hydrogen is limited by a diffusion process since it is not readily soluble in liquid (polar) components and requires an excess use of hydrogen. To address these challenges, a single-pot â€˜Formic Acid Assisted Autotheromolysis to Jet Fuel (FAT2JF)â€™ process is studied. The use of aqueous phase (formic acid and water mixture) serves as an *in situ* source of H_2 (hydrogen precursor and hydrogen transfer agent) and organic solvent where the reactions in aqueous media (aquathermolysis) helps in reducing mass transfer resistances that results in significant increase in reaction rates in this FAT2JF process. The phase diagram of water, formic acid and some typical triglycerides will be presented.

The present study on aquathermolysis is focused on developing a metal oxide-based low cost catalyst (e.g. Fe-Ni/activated carbon, Fe-Ni-Co/Al_2O_3) system that is stable in aqueous media. The study is conducted in a packed bed continuous flow tubular reactor in the range of 250 to 400áµ’C under hydrothermal conditions. In order to alleviate coke formation and evaluate the stability of the catalyst, the effects of reactants flowrates, feedstock composition, pressure, temperature, and formic acid concentration are studied. GC-MS and elemental analyzer are used to analyze the product compositions. Thermal gravimetric analysis (TGA), energy dispersive X-ray, SEM, BET method, FT-IR, and XRD are used to characterize both fresh and used catalyst. The proposed novel approach has certain advantages over conventional commercial hydrotreating processes including no handling of hydrogen gas, faster reaction times,

and use of renewable formic acid as hydrogen source.

469513 Rheology of Biomass Slurries to Determine Pumpability for Hydrothermal Liquefaction

Abstract:

The pursuit of renewable fuels and chemicals has led to the development of many different biomass conversion processes. One of these processes, hydrothermal liquefaction (HTL), has the advantage of high energy recovery, no drying step, and utilization of a green aqueous reaction media.[1] However, transport of biomass and water slurries during HTL has proven to be a challenge.[2,3] Typical reactors can only process biomass between 10 and 15 wt% solids because higher solids loadings plug reactor systems. To improve the efficiency of HTL, and other processes reliant on pumping biomass slurries, it is essential to increase solids loading. Increasing solids loading reduces the volume of waste streams and increases process throughput. In an effort to increase biomass solids loadings many people have studied the rheological properties of biomass slurries to improve pumpability.[4] Despite these efforts there has yet to be a strong correlation developed between typical rheological properties, like viscosity, and biomass pumpability.

This work takes a unique approach to studying biomass slurry rheology by combining traditional rheometric measurements, such as viscosity and storage/loss moduli, with a novel flow characterization system. Traditional rheometric measurements are made on a rheometer capable of analysis at temperatures up to 300 °C and 2000 psi, which mimics conditions found in some traditional HTL systems. Novel flow studies performed at Idaho National Laboratory are made on an in-house designed syringe attached to an Instron load frame. This system measures the force required to push biomass through a reducing orifice and can study plugging phenomena in biomass slurry transport systems. These studies have shown how plugging can be related to biomass dewatering, followed by bridging across the flow channel. Pumpability studies have been performed on a wide variety of materials from herbaceous to woody crops and residues. In addition to characterizing pumpability in biomass slurries this work has also focused on improving the pumpability of such slurries through preprocessing techniques including: size reduction, blending, and torrefaction. This work will ultimately lead to improvements in feed handling related to hydrothermal liquefaction and other biomass slurry transport systems.

1 Toor, S. S., Rosendahl, L. & Rudolf, A. Hydrothermal liquefaction of biomass: A review of subcritical water technologies. *Energy* **36**, 2328-2342, doi:10.1016/j.energy.2011.03.013 (2011).

2 Elliott, D. C., Biller, P., Ross, A. B., Schmidt, A. J. & Jones, S. B. Hydrothermal liquefaction of biomass: Developments from batch to continuous process. *Bioresource technology* **178C**, 147-156, doi:10.1016/j.biortech.2014.09.132 (2015).

3 Berglin, E. J., Enderlin, C. W. & Schmidt, A. J. *Review and Assessment of Commercial Vendors/Options for Feeding and Pumping Biomass Slurries for Hydrothermal Liquefaction*. (Pacific Northwest National Laboratory, 2012).

4 Stickel, J. J. *et al.* Rheology measurements of a biomass slurry: an inter-laboratory study. *Rheologica Acta* **48**, 1005-1015, doi:10.1007/s00397-009-0382-8 (2009).

469372 The Upgrading of Biomass-Derived Dimethyl Ether to High-Octane Hydrocarbons: The Effect of Process Conditions on Catalyst Performance

Abstract:

Dimethyl ether (DME) and/or methanol can be converted into C_{5+} branched hydrocarbons at low temperatures and pressures over large-pore acidic zeolites. Recently, we have demonstrated that Cu-modification of the existing commercial catalyst (H-BEA) resulted in a 2-fold increase in total hydrocarbon productivity at 200 °C with co-fed H_2.[1] The Cu/H-BEA catalyst incorporated H_2 with DME into the desired branched hydrocarbons, while maintaining the high C_{5+} selectivity of the parent H-BEA. Here, we compare the performance of Cu/H-BEA and the parent H-BEA at 175, 200 and 225 °C, and at weight-hourly space velocities (WHSV) for DME ranging from 0.43-1.4 h^{-1}. The productivity of C_{5+} products increased with temperature over H-BEA, with no significant changes over Cu/H-BEA due to a decrease in C_{5+} selectivity at elevated temperature, which was observed for both catalysts. Operation at 175 °C resulted in a significant increase in C_{5+} selectivity (75-85% for Cu/H-BEA), but lower productivity than at 200 °C due to lower reactant conversion. Lower space velocities increased reactant conversion over both catalysts, but significantly decreased the C_{5+} productivity over Cu/H-BEA. The Cu/H-BEA catalyst was stable for 100 h at 200 °C after an initial 40 h break-in period. In previous research, we demonstrated that C_4-C_8 olefins produced in the DME homologation reaction can be oligomerized to C_{12}-C_{22} branched hydrocarbons with properties that align with a synthetic paraffinic kerosene product.[2] Following a simple vacuum distillation, the fuel properties of the distillate-range product align well with a jet fuel blend when compared with 5 ASTM methods. This DME-to-hydrocarbons reaction represents a promising route to transform renewable biomass sources into commercially viable, high-performance biofuels.[3]

[1]Schaidle, J. A.; Ruddy, D. A.; Habas, S. E.; Pan, M.; Zhang, G.; Miller, J. T.; Hensley, J. E. *ACS Catal.* **2015**, *5*, 1794.

[2]Behl, M.; Schaidle, J. A.; Christensen, E.; Hensley, J. E. *Energy & Fuels* **2015**, *29*, 6078.

[3]Tan, E. C.; Talmadge, M.; Dutta, A.; Hensley, J.; Snowden-Swan, L. J.; Humbird, D.; Schaidle, J.; Biddy, M. *Biofuels, Bioproducts and Biorefining* **2016**, *10*, 17.

462619 A System-Level Analysis on Biomass Thermal Fractionation and Catalytic Upgrading Processes

Abstract:

Biomass presents a promising source for sustainable liquid fuels production. There are three principle routes for converting biomass into liquid fuels, classified by the initial treatment of the biomass: gasification, hydrolysis, or pyrolysis and liquefaction. The pyrolysis route involves thermal decomposition of the biomass at high temperature (500-800°C), and it has demonstrated high yields. Unfortunately, the wide range of components in the resulting pyrolysis vapor (oil) makes it unstable, corrosive, and challenging to catalytically upgrade. Furthermore, the high content of light (<C6) oxygenates in the pyrolysis are unsuitable for liquid fuels and are therefore of limited utility. Multi-stage torrefaction of biomass with catalytic upgrading presents a promising method of overcoming these challenges. Importantly, the decomposition of the biomass over several stages can reduce the number of product species within each fraction, allowing one to mold the catalytic upgrading strategy to target a subset of the chemical functionalities.

Although staged thermal decomposition of biomass coupled with targeted upgrading has a number of advantages over single stage pyrolysis with hydro-treating, there are many unknowns regarding the best way to design such a system. Importantly, the optimization of the number and conditions of each thermal decomposition stage is a complex problem that cannot be solved without knowing the influence on the final yields and process complexity resulting from the subsequent catalytic upgrading. Furthermore, there are a wide variety of chemistries available to perform upgrading. In addition, the reaction sequence must also be configured to optimize the overall process. In order to improve the impact of such an approach, one must also determine if general practices from this strategy may be applicable to different types of biomass feeds.

Accordingly, the ultimate goal of this work is to design a broad roadmap that can be used to identify the key parameters and trade-offs in a bio-refinery process based on multi-stage torrefaction of biomass with catalytic upgrading. In order to investigate the effect of the number of torrefaction stages and the specific chemistries on the yield of fuel-range grade product and the cost, we first derived a series of process alternatives based on experimental torrefaction yields obtained for a 3-stage torrefaction process (270 °C, 360 °C, and 500 °C). In designing the alternatives, we carefully selected the chemistries that are used to upgrade the thermal decomposition fractions, to maximize the yield of fuel-grade product and to minimize the hydrogen consumption, considering the abundance of various chemical functionalities within the biomass fractionation product. We suggested twelve cases where there is 1, 2, or 3 thermal decomposition stages and where each fraction is upgraded 1, 2, 3, or 4 times. Through the analysis, we found that there exists a trade-off between fuel-range carbon yields versus the number of upgrading steps. In general, increasing the complexity of the fraction upgrading systems increases the ultimate yield of C6+ product, thought there are diminishing returns on the increase in product yield versus the complexity of the catalytic upgrading sequences. In contrast, however, the choice of the number of thermal decomposition stages is found to be not simple, requiring careful consideration of the chemistries available to upgrade different components and the relative abundances of these different components. More thermal decomposition stages allow for more fine-tuning of the upgrading processes but lead to more complex and potentially more expensive processes. Thus, the design of the decomposition conditions and number of stages must be done simultaneously with the design of the fraction upgrading system. Finally, we also suggest some hybrid processes by eliminating several upgrading steps that have small marginal effects on the fuel-range carbon yield and discuss about separations that could enhance the process performance via prevention of unwanted side reactions or protection a functionality from conversion.

461802 Performance of a Bench-Scale Continuous Hydrothermal Carbonization Reactor

Abstract:

Hydrothermal carbonization (HTC, thermal hydrolysis, or wet torrefaction) is a treatment process which converts moist feedstocks into homogenized, carbon rich, and energy dense solid fuel, called hydrochar. One of the main advantages of HTC compared to other thermochemical treatment processes is the use of moisture as reaction medium and reactant, thereby precluding the need for drying prior to treatment. Thermodynamic properties of water change greatly in the subcritical region from 180-280 °C, and as a result, subcritical water behaves as a non-polar solvent and mild acid and base catalyst simultaneously. When subjected to HTC biomass undergoes rapid hydrolysis and other reactions, releasing oxygen-containing volatiles and producing a highly hydrophobic hydrochar.

Although HTC offers a relatively simple and straightforward solution to process diverse biomass feedstocks, the requirements of high pressure and high temperature make the process complex and potentially costly to design and operate. The lab-scale batch process has already been demonstrated in various laboratories around the world, but batch process is not cost-effective for industrial-scale deployment. A continuous process offers a relatively small footprint, energy recovery and efficiency, and notably, economies of scale.

In this study, a bench-scale continuous HTC reactor system has been designed, constructed, commissioned, and operated with model compounds and with whole biomass feedstocks. The maximum temperature and pressure of the reactor are 250 °C and 40 bar. The throughput of the reactor system was maintained at 0.3 L/min with solids loading of ~5%, while the reaction time was maintained at 5 min. The basic reactor configuration is a section of straight pipe, and numerous engineering challenges were encountered during design and commissioning. The primary challenges are related to instrumentation durability, temperature control, and flow control.

Results are presented for conversion of glucose and microcrystalline cellulose. Both solid and liquid products were tested for their physico-chemical properties and compared with the corresponding products from batch process generated in a 2L batch Parr reactor. It was found that temperature and pressure were stable during operation and products are relatively similar to that of batch process.

474623 Spectroscopic and Thermal Characterization of Carbonized Food Wastes

Abstract:

Waste-to-energy conversion technologies may play an important role to improve the efficiency and sustainability of global food supply. In this worked, performed in collaboration with researchers at UNICAMP (Brazil), we used thermal and spectroscopic methods to characterize the carbonized residues that remain after flow-through hydrothermal extraction and hydrolysis. A recent energy analysis of flow-through hydrothermal hydrolysis finds that it may be energetically favorable compared to other pre-treatment and saccharification techniques. Three factory wastes produced by food industries were studied: sugarcane bagasse, green coffee powder, and defatted coffee cake. We foucs here on results obtained for the coffee wastes. Thermogravimetric analysis (TGA) was used to monitor extraction of semi-volatile components after treatment at 150 °C, followed by hydrolysis and carbonization as temperatures were increased to 250 °C. Infrared spectroscopy indicated that the extracted semi-volatile compounds were a combination of valuable lipids and organic acids. At higher temperatures, infrared spectroscopy identified preferential hydrolysis of hemicellulose components, with cellulose and lignin remaining. Raman microscopy gave further detail, identifying the specific compound chlorogenic acid, a valuable compound marketed as a nutraceutical, as one of the semi-volatile compounds to be extracted from the coffee. At higher temperatures, Raman microscopy provided evidence of carbonization of the coffee waste, as suggested by the appearance of graphitic and defect bands previously associated with hydrothermal char. As shown in the Figure, our results suggest temperature-programmed extraction, hydrolysis, and carbonization of coffee wastes to produce biorefinery streams consisting of valuable organic compounds, bioenergy precursors, and carbonized char.

474753 Drying and Pyrolysis of Solid Waste on Spacecraft for Water Recovery and Biochar

Abstract:

Manned spaceflight outside of low-Earth orbit will require significant advances in closing loops within life support systems, especially the recycling of solid and liquid wastes to produce oxygen, food, and fresh water. Pyrolysis is one waste conversion method that has received considerable attention due to its potential to enable improvements in waste reduction, material recovery, and energy production.

Here, we describe the development of a slow pyrolysis reaction system that can transform solid waste and brine from the water treatment system into a nutrient-rich crop growth medium, while recovering water and carbon dioxide. The feedstock for the process followed a standard s recipe used by NASA to represent the average daily solid waste produced by astronauts on manned missions. This recipe included packaging materials (polyethylene, nylon, aluminum foil), clothing and towels (mostly cotton), personal care products (i.e. shampoo, toothpaste, facial tissue, hand wipes, diapers, etc.), food waste, metabolic waste simulant, and miscellaneous materials such as nitrile gloves, duct tape, and cleaning wipes. Solid waste processing consisted of two steps: drying to recover water and pyrolysis at moderate temperatures (400-550°C). Process steps were optimized to recover the most water and produce the most suitable biochar for plant growth applications within the assumed 8 hour total operational time window. Recovered water and biochars were characterized using a variety of physical, chemical and soil nutrient analysis methods. We discuss the challenges of working with this waste feedstock compared to lignocellulosic materials, how this system fits into an overarching life support architecture, and future work needed to adapt this system to a microgravity environment.

474765 Evaluating Physical and Electronic Structure Developments of Loblolly Pine Derived Biochar and Activated Carbon

Abstract:

Biochar is a major byproduct of fast pyrolysis process. Until recently, biochar is considered as an inexpensive solid fuel that is re-consumed for the pyrolysis reactor heating. However, advances in nanotechnologies reveal the potential high-end applications of biochar as electrodes for electrochemical energy storage systems and conductive thin film materials for sensors and electronic devices. The high-end applications of biochar not only contribute to the profitability of overall fast pyrolysis process, but also contribute to the growth of sustainable material market. To attain the objective, it is imperative to understand the complex biomass derived carbon structural development during the carbonization and the activation. In this study, the physical and the electronic structures of loblolly pine (*Pinus Taeda*) derived biochar and activated carbon were systematically analyzed by interpreting X-ray diffraction patterns, electron energy loss spectra, and electrical conductivities. Loblolly pine was carbonized at nine different temperatures (300 , 350 , 400 , 500 , 600 , 700 , 800 , 900 , and 1,000 ,) using quartz tube furnace. Then, four biochar samples carbonized at 300 , 350 , 500 , and 700 were impregnated with NaOH to produce the activated carbon. Carbon sp^2 content was quantitatively determined by the electron energy loss spectroscopy. The sp^2 content of biochar produced at 300 is 58% and the sp^2 content gradually increases to 78% as the carbonization temperature is elevated to 1,000 . The sp^2 contents of activated carbon are 74%, 78%, 79%, and 85% where precursor biochar carbonization temperatures are 300 , 350 , 500 , and 700 . Polyaromatic cluster size and graphitic stacking thickness were determined by analyzing the X-ray diffraction patterns. Four stages of biochar physical structural development were determined. After major biomass components decompose, the average polyaromatic cluster size and the graphitic stacking thickness increase as the carbonization temperature increases. Carbon paste electrode

was produced by mixing biochar and activated carbon powder with polyvinylidene fluoride solution. Biochar electrodes features were closer to insulator that the electrical conductivity was not measurable. Activated carbon electrodes had good electrical conductivities which values are 1.14 S/cm, 1.41 S/cm, 1.3 S/cm, and 1.58 S/cm. The detailed model of physical and electronic structural developments will be discussed.

466364 Characterization of Free Radicals By Electron Spin Resonance Spectroscopy in Biochars from Pyrolysis at High Heating Rates and at High Temperatures

Abstract:

Understanding fast pyrolysis of biomass-derived materials is an important step in optimization of combustion processes. Similar to coal combustion, the fuel burn out is known to be influenced by the yield and reactivity of chars, produced during pyrolysis. The rapid heating of small biomass particles and the short residence time at high temperatures minimize the char yield and increase char reactivity. The differences in chemical composition of organic and inorganic matter between wood and herbaceous biomass affect the operational flexibility of power plants, and increase the complexity of mathematical models that can predict yields, composition and rates of product (char, tar, light gases) formation from fast pyrolysis. The modeling of cross-linking and polymerization reactions in biomass pyrolysis includes the formation of free radicals and their disappearance. Knowledge about these radical reactions is important in order to achieve the high fuel conversion at short residence times. However, little is known about the extent of free radical reactions in pulverized biomass at fast pyrolysis conditions.

The concentration and type of free radicals from the decay (termination stage) of pyrolysis at slow and fast heating rates and at high temperatures (above 1000°C) in biomass char have been studied. A room-temperature electron spin resonance spectroscopy study was conducted on original wood, herbaceous biomass, holocelluloses, lignin and their chars, prepared at high temperatures in a wire mesh reactor, an entrained-flow reactor, and a tubular reactor. The radical concentrations in the chars from the decay stage range up between $7x10^{18}$ and $1.5x10^{19}$ spins g^{-1}.

The results indicated that any differences in the biomass major constituents (cellulose, hemicellulose, lignin) had a minor effect on remaining radical concentrations compared to potassium and silica contents. The higher radical concentrations in the wheat straw chars from the decay stage of pyrolysis in the entrained flow reactor compared to the wood chars were related to the decreased mobility of potassium in the char matrix, leading to the less efficient catalytic effects of potassium on the bond-breaking and radical re-attachments. The high Si levels in the rice husk caused an increase in the char radical concentration compared to the wheat straw because the free radicals were trapped in a char consisting of a molten amorphous silica at heating rates of 10^{3}°C s^{-1}. The experimental electron spin resonance spectroscopy spectra were analyzed by fitting to simulated data in order to identify radical types, based on g-values and line widths. The results show that at high temperatures, mostly aliphatic radicals (g = 2.0026-2.0028) and PAH radicals (g = 2.0027-2.0031) were formed. Free radicals structure from the termination stage was modelled by density functional theory (DTF).

474889 All Biomass Is Local: Paying Farmers More for Cellulosic Biomass Is a Game Changer

Abstract:

Current models of cellulosic biofuel systems require that the delivered price of the cellulosic

biomass feedstock be kept low. Thus the predicted biorefinery size is relatively small, limiting potential economies of scale. However, it is actually the ultimate selling price of the biofuel that largely determines market penetration. We relaxed the constraint of low delivered feedstock price and explored the resulting effects on biofuel price, biofuel volume produced, and global warming impact (GWI). Feedstock price greatly affects the feedstock supply chains that may develop. Increased feedstock price does not affect the final ethanol selling price very much, but higher feedstock prices greatly increase the amount of ethanol produced. Farmers will supply much more cellulosic biomass at higher feedstock prices, leading to shorter transportation distances with reduced transportation costs and enabling larger biorefineries with improved economies of scale, thereby reducing the ethanol selling price.

The cellulosic feedstock supply chain systems were studied as a function of feedstock prices by

determining potential feedstock supply clusters and the maximum capacity of cellulosic biorefineries across the United States. Supply clusters were determined by minimizing costs associated with ethanol production. The analysis is based on county-level cellulosic feedstock production data projected in the *US Billion-Ton Update* report. Each biomass supply cluster is unique in terms of local and regional characteristics (e.g. area, feedstock types), biorefinery capacity, ethanol selling price, and GWI. Very large-scale biorefineries (â‰¥20 000 dry Mg dayâ€"1) may be feasible in some regions.

471184 Impact of Alkaline Deacetylation and Pelleting of Corn Stover at the Depot

Abstract:

Currently, approximately 258 million tons of the Billion Tons (DOE goal) is considered accessible in the existing feedstock supply systems. Additionally, much of this biomass is highly variable in properties important to handling, densification and conversion. An important role of the distributed processing concept is to deliver feedstock to biorefineries with low variability and consistent conversion performance. Achieving this goal would decrease risk to the biorefinery and reduce operating expenses. This work examines the impact of deacetylation at the depot on both the feedstock supply chain and the conversion facility.

Single-pass corn stover was ground to 1 inch using a Wiley Mill and stored at room temperature and humidity. The corn stover was split and deacetylated in a stirred-batch reactor. The recovered solid residuals were collected and compared against raw corn stover using analyses that measured total ash, ash composition and proximate/ultimate. In addition, the chemistry of the deacetylation medium was determined along with the ash and CHNO composition of solids precipitated from the deacetylation medium. In this manner, data was collected to support a full mass balance of the process. The energy to grind and pellet the raw material and alkaline deacetylated material was measured. The alkaline deacetylation resulted in a 13.8% decrease in required grinding energy, and a 23.1% decrease in required pelleting energy. It also resulted in increased carbohydrate content and reduced ash for the biorefinery feedstock. In addition, it was observed that the deacetylated material could be ground at much higher moisture contents. This would allow corn stover to be pelleted, and then dried using a more efficient process. Finally, raw and deacetylated material was subjected to different pretreatment severities with and without disc refining. Key findings were that deacetylation and pelleting were found to: 1) negate the need for disc refining; 2) increase sugar yields by over two-fold with when no pretreatment was applied; 3) require significantly lower severity of pretreatment to achieve the same, or greater, sugar yield; and 4) increase maximum sugar yield to near 100%.

474787 Liquefaction of Lignocellulose

Abstract:

The liquefaction of biomass materials is a process where particulate wood, corn stover, sugarcane bagasse and other forms of lignocellulose feedstocks are converted to an aqueous suspension of solid materials that is suitable for further processing to fermentable sugars and biofuel materials. A number of reports have described the liquefaction of pretreated materials using either cellulose enzyme or an enzyme biomimetic (maleic acid). Our work has evolved to begin to examine the liquefaction of lignocellulosic substrates at 25% solids loading prior to pretreatment, since the ability to obtain pumpable slurries prior to pretreatment has significant benefits in equipment design and operating conditions for the pretreatment step itself. Challenges in obtaining pumpable slurries of biomass materials, either pre- or post-pretreatment include the shear thinning character of the slurry, mixing of the particulate biomass during the liquefaction process, and definition of operating conditions that make efficient use of the biocatalysts required for the liquefaction process. This paper presents an analysis of the insoluble particle characteristics that enable slurries to be obtained, and proposes a model for the liquefaction process that relates particle size, shape, and composition to the liquefaction process.

469070 Technoeconomic Analysis of Pyrolysis and Electrocatalysis Depots

Abstract:

Biomass upgrading depots are small-scale facilities that serve to densify raw feedstocks into transportable and salable commodities with increased energy and economic value. Decentralized systems seek to find the optimum balance of "economies of scale" and "economies of transportation," while creating a venue for value addition to rural communities. Fast pyrolysis is a thermochemical technology that can be deployed in small-scale depots to deconstruct and densify biomass by rapid heating (400-600°C) without oxygen to create bio-oil, biochar, and combustible gas. Bio-oil is a bulk dense and energy dense liquid that is less costly to transport than raw biomass. However, bio-oil is limited as a fuel intermediate because of its poor chemical properties, namely high oxygen content, corrosiveness, and potential to react and form sludge. As a remedy, electrocatalytic hydrogenation and deoxygenation (ECH) has been proposed to partially upgrade bio-oil to create a stable fuel intermediate. ECH works by saturating carbon-carbon and carbon-oxygen double bonds as well as cleaving certain ether linkages on a catalytic cathode. Though ECH is capable of upgrading bio-oil, the cost of doing so must not be prohibitive to its adoption. The economic viability of pyrolysis-ECH depots is a function of capital costs, operating costs, and product values. In this regard, a technoeconomic analysis of pyrolysis-ECH depots will be described to reveal the tradeoffs between depot capacity and the variable costs of transportation. Sensitivities to biomass purchase price, moisture content, electricity source, and electricity cost will be examined. ECH could be the keystone technology in biomass upgrading depots, leading to regionally produced commodities and revenue generation in rural communities.

474887 Electrochemical Bio-Oil Stabilization As an Alternative for Distributed Processing of Biomass

Abstract:

The conversion of biomass to value-added chemicals and liquid fuels is a developing alternative to fossil sources [1]. Bio-oil, which can be obtained from thermal breakdown of forest and agricultural biomass, can be produced near the material source. The distributed production of bio-oil allows the densification of the biomass energy and could reduce the transportation and production costs. However, compared with fossil hydrocarbon sources bio-oil is a chemically unstable micro emulsion that cannot be fractionated [1, 2]. The high concentration of oxygenated compounds (hydroxyaldehydes, hydroxyketones, sugars, carboxylic acids, and phenols) reduces the storage and temperature stability of the bio-oil and results in 50% lower calorific value when compared to hydrocarbon fuels [1, 3, 4]. The typical route to removing oxygen involves a process called hydrodeoxygenation (HDO). Typical HDO conditions (425â—¦C and 2000 psi H2 pressure) promote condensation reactions leading to the formation of coke. A multi-step hydrotreatment process, which include low temperature (140â—¦C) hydroprocessing stabilization under high H2 pressure (1200 psi H2 pressure), reduce the reactivity for HDO [5, 6]. In distributed production mild conditions are desired to reduce capital requirements at remote sites near biomass sources.
In this work an electrochemical process has been investigated as a low temperature alternative for the electrochemical reduction (ECR) and stabilization of bio-oil. The process attempts to reduce site requirements for hydrogen and achieve even milder conditions than used for existing catalytic stabilization. A polymer electrolyte membrane (PEM) configuration allows the processing of the complete bio-oil product without the addition of supporting electrolyte. While work to this point has focused on use of PEM cells employing cation exchange membranes, future work will attempt to use anion exchange membranes where current would be carried by anions of small acid compounds (acetate and formate). This configuration could simultaneously reduce reactivity of oil and also remove small acid molecules that contribute acidic content and consume hydrogen in downstream processes in production of low value gases. Description of the electro-catalytic systems and results obtained with different transition metal catalysts will be presented.

References

[1] D. Mohan, C.U. Pittman, P.H. Steele, Pyrolysis of Wood/Biomass for Bio-oil: A Critical Review, Energy & Fuels, 20 (2006) 848-889.
[2] A.V. Bridgwater, Renewable fuels and chemicals by thermal processing of biomass, Chem. Eng. J., 91 (2003) 87-102.
[3] S. Czernik, A.V. Bridgwater, Overview of Applications of Biomass Fast Pyrolysis Oil, Energy & Fuels, 18 (2004) 590-598.
[4] A. Oasmaa, E. Kuoppala, Y. Solantausta, Fast Pyrolysis of Forestry Residue. 2. Physicochemical Composition of Product Liquid, Energy & Fuels, 17 (2003) 433-443.
[5] D.C. Elliott, Historical Developments in Hydroprocessing Bio-oils, Energy & Fuels, 21 (2007) 1792-1815.
[6] S. Jones, P. Meyer, L. Snowden-Swan, A. Padmaperuma, E. Tan, A. Dutta, J. Jacobson, K. Cafferty, Process design and economics for the conversion of lignocellulosic biomass to hydrocarbon fuels: fast pyrolysis and hydrotreating bio-oil pathway, in, National Renewable Energy Laboratory (NREL), Golden, CO., 2013.

462862 1T-WS_2 on Graphite Foam As a Binder-Free Electrode for Enhanced Hydrogen Evolution

Abstract:

1T-WS_2 on Graphite Foam as a binder-free electrode for Enhanced Hydrogen Evolution

[1]Xiaomeng Guo, [1]Yuanzhi Zhu, [2]Junyi Ji*, [1]Xiaobin Fan, [1]Guoliang Zhang, [1]Fengbao Zhang, [1]Wenchao Peng*

[1]School of Chemical Engineering & Technology, Tianjin University, Tianjin 300072, China,

wenchao.peng@tju.edu.cn

[2]College of Chemical Engineering, Sichuan University, Chengdu, Sichuan 610065, China,

junyiji@scu.edu.cn

Hydrogen is regarded as a clean alternative energy carrier due to its high energy density and environmental friendliness. Electrochemical catalytic hydrogen evolution reaction (HER) is an efficient method for the production of hydrogen. Pt-group metals are the most effective HER catalysts due to their low overpotential and high efficiency. However, these metals are rare and expensive to apply on an industrial scale. To develop highly active and low-cost HER catalysts remains a big challenge. Molybdenum disulfide (MoS_2) and tungsten disulphide (WS_2) are emerging HER catalysts that could be used as a substitute for noble metals in HER.

In this study, a three dimensional graphite foam (GF) was synthesized on the Ni foam using chemical vapor deposition (CVD) method. After the removal of Ni foam with HCl aqueous solution, exfoliated 1T-WS_2 was dipped on the surface of GF to obtain the GF-WS_2 composite. Due to the 3D structure and strong mechanical strength of GF, the composite can be used as a binder-free electrode. The GF-WS_2 electrode was then tested for the HER performance in 0.5 M H_2SO_4 solution, and excellent HER activity was observed with a small overpotential of 0.1 V (vs RHE). Compared to the naked WS_2, the overpotential had been decreased ~0.1 V, and the cathodic current can increase ~10 times at a bias of -0.3V. This activity improvement should be due to the strong chemical and electronic coupling between the GF supports and WS_2 nanosheets. The electrical coupling could afford rapid electron transport from the less-conducting WS_2 nanosheets to the GF with more active sites. The GF-WS_2 is therefore a promising binder-free electrode with great potentials for hydrogen production.

462713 Flame Synthesis of Highly Transparent and Robust Nano-Layers for Enhanced Photo-Electrochemical Water Splitting

Abstract:

Direct chemical storage of solar energy is an attractive concept for off-the-grid renewable power generation. Currently the poor transparency of the thick low-cost catalyst layers required for water oxidation hinders the large-scale fabrication of efficient photo-electrochemical cells for artificial photosynthesis and hydrogen production. Flame synthesis of tailored nanoparticles and direct aerosol deposition is a scalable technology for the production of ultra-porous layers of photo-catalysts and optoelectronic devices[1]. Nevertheless the fragility[2] of this gas-phase self-assembly has limited their application in liquid environments with most studies requiring the in-situ high temperature annealing of the depositing particle to obtain sufficient mechanical stability [3]. This results in a drastic drop of the layer porosity and hinders the fabrication of optimal architecture for electrochemical systems.

Here, we report the engineering of robust high performance photo-electrodes with tunable porosity and composition. We discuss the critical parameters controlling the self-assembly of these nanoparticle layers and present a flexible approach for their mechanical and chemical stabilization. We apply this concept to the fabrication of photo- and electrochemical cells for water splitting demonstrating very high turn-over frequencies, controllable light absorption and efficient electron collection. We envision that this scalable synthesis approach can be readily implemented for the commercial production of low-cost devices for chemical energy storage and renewable fuel production.

References

1. Nasiri, N.; Bo, R.; Wang, F.; Fu, L.; Tricoli, A. *Adv. Mater.* **2015**,27, (29), 4336-4343.

2. Tricoli, A.; Graf, M.; Mayer, F.; Kühne, S.; Hierlemann, A.; Pratsinis, S. E. *Adv. Mater.* **2008**,20, (16), 3005-3010.

3. Tricoli, A.; Wallerand, A. S.; Righettoni, M. *J. Mater. Chem.* **2012,** 22, (28), 14254-14261.

468352 Investigation of the Use of Metal-Organic Frameworks for Combined Water Purification and Catalytic H_2 Production

Abstract:

Lately, a few studies have been reported on the use of metal-organic frameworks (MOFs) for water purification via adsorption of contaminants and/or via photocatalytic degradation. MOFs correspond to 3D crystalline nanomaterials networks consisting of metallic clusters connected via organic ligands. They are characterized by their high porosity and hence have been investigated for applications in water purification, especially for the removal of organics from water. It is interesting to notice that in parallel to those studies, recent work has also been devoted to using MOFs to photocatalytically produce H_2 from water and organic molecules. Up until now, organic contaminants adsorptive removal, photodegradation as well as H_2 evolution have been investigated independently of each other. Hence, an opportunity arises to combine water purification and energy generation (in the form of H_2).

In this work, we have focused on the investigation of two specific MOF families (UiO-66(Zr) and MIL-53) and investigated its adsorptive and photocatalytic properties. The UiO-66(Zr) material was synthesized along with its functionalized counterparts (X-UiO-66(Zr), with X = NH_2, NO_2, Br). MIL-53(M) was synthesized with different metals ions (M = Al, Fe and Cr). The materials were characterized by a number of analytical and spectroscopic techniques in order to determine the effects of functionalization on the band gap and conduction band. In addition, the materials were tested for the adsorption and photocatalytic degradation of organics in water and the simultaneous production of H_2. This work reports on the impact of the abovementioned structural and chemical modifications on the MOFâ€™s electronic properties and their performance as dual-purpose adsorbent/catalyst for water purification and H_2 production.

471808 $Ni(OH)_2$ As a Hole Mediator for Visible Light-Induced Urea Photo-Oxidation

Abstract:

Urea, a major component of human urine, has been proposed as a source of electrolytic hydrogen. We describe here CdS-sensitized TiO_2 assembled with urea electrocatalyst, $Ni(OH)_2$, on fluorine-doped tin oxide (FTO) coated glass ($FTO/TiO_2/CdS/Ni(OH)_2$) used as photoanode to realize simultaneous urea oxidation and water reduction using visible light irradiation. A photoelectrochemical reactor is employed to demonstrate the $Ni(OH)_2$ can serve as hole mediator to drive urea oxidation. The steady-state response of the semiconductor-catalyst electrode in the photoelectrochemical system is investigated, and the photoelectrode transient charge dynamics are studied by employing transient absorption spectroscopy. The obtained carrier dynamics can provide useful information on the competitive processes such as charge transfer and/or recombination within the photoanode.

470119 Study of Photocatalytic Activity of Nano Organic Hybrid Materials (NOHMs) for Photoelectrochemical Reduction of CO_2

Abstract:

A promising opportunity space for novel class of colloidal catalysts has emerged from recent advances in the synthesis of colloidal nanomaterials in which the composition of the inorganic core as well as the organic ligand shell can be precisely tuned. The prospect of tailoring the ligand shell, for optimized interaction with the surrounding fluid and the surface of the colloid, for catalytic conversion is particularly promising for photocatalytic reactions. We have investigated a new class of hybrid colloidal photocatalysts with combined functionality to capture CO_2 and convert it to liquid fuels. The photodriven CO_2-to-fuels reaction has profound technological implications as a prospect to harvest vast solar energy resources, capture CO_2 and store the energy in form of a liquid fuel. Scientifically, the photoreduction of CO_2 invplves a complex interplay of coupled sub-processes including CO_2 transport to and adsorption on the catalyst surface, light absorption, proton-coupled multi-electron transfer, selectivity, and material stability. The photoelectrochemical reduction of CO_2 leads to various â€˜value-addedâ€™ products including methane, olefins, methanol, ethanol, formaldehyde, formic acid, formate, and carbon monoxide. Whereas access to novel colloidal nanoparticle catalysts has enabled advances in fundamental understanding of and control over their catalytic properties, progress towards deploying this know-how by integration with established CO_2 capture technologies (*i.e.*, amine scrubbing) has lagged behind. We embraced this challenge as an opportunity study basic processing-structure-performance relationships of nano organic hybrid materials (NOHMs) applied to the photoelectrochemical reduction of CO_2. We will describe the colloidal synthesis the inorganic nanoparticle core comprised of titania and metal co-catalyst and the CO2-philic ligand shell. We will discuss our recent advances in understanding and controlling the composition, morphology, ligand exchange, optical absorption, and CO_2 chemical and physisorption capacity. We will

relate structure and composition of the catalyst to the photocatalytic performance and will describe reaction products and conversion probed by gas and liquid chromatography.

472234 Nanostructured Silicon Photocathodes for Solar Water Splitting

Abstract:

Photoelectrochemical (PEC) water splitting is a promising route for synthesizing hydrogen using solar energy without direct CO_2 emissions.[1] PEC water splitting devices require highly active hydrogen evolution reaction (HER) catalysts to efficiently produce hydrogen at the photocathode.[2] Platinum and its alloys are the most active HER catalysts but the low availability of platinum may create problems for large scale implementation of solar water splitting.[3] There is a need to replace platinum with a non-precious metal earth-abundant alternative to enable cost effective, scalable devices.[4]

Recently, transition metal phosphides and sulfides have been shown to be among the most active nonprecious HER catalysts on a total electrode activity and turnover frequency (TOF) basis.[5] Though extremely active, the high surface area morphologies of these previously synthesized electrodes present additional challenges for integration onto PEC water splitting electrodes.[6] New methods are required to synthesize HER catalysts with a high density of active sites that also transmit a large portion of incident light.

We discuss our progress in synthesizing highly active transition metal phosphide and sulfide catalysts in nanostructured, low light-absorbing geometries. First, we compare nanostructured silicon semiconductors covered with optically thin catalysts to analogous flat silicon electrodes. These results provide insight into the general design considerations necessary to maximize the device efficiency for any given catalyst-absorber system. Next, we present a device structure which decouples the catalyst nanostructure from that of the silicon absorber. In so doing, we achieve low levels of recombination within the silicon and maximize the current-voltage characteristics of the device. Based on our findings, we discuss the application of our catalyst structure to other semiconductor systems and suggest general strategies for further improving the performance of solar water splitting electrodes.

[1] M. G. Walter, E. L. Warren, J. R. McKone, S. W. Boettcher, Q. X. Mi, E. A. Santori, N. S. Lewis, Chemical Reviews 2010, 110, 6446.

[2] L. C. Seitz, Z. Chen, A. J. Forman, B. A. Pinaud, J. D. Benck, T. F. Jaramillo, ChemSusChem 2014, 7, 1372.

[3] J. Greeley, I. Stephens, A. Bondarenko, T. P. Johansson, H. A. Hansen, T. Jaramillo, J. Rossmeisl, I. Chorkendorff, J. K. Nørskov, Nature Chemistry 2009, 1, 552.

[4] P. C. Vesborg, T. F. Jaramillo, RSC Advances 2012, 2, 7933.

[5] J. Kibsgaard, T. F. Jaramillo, Angew. Chem.-Int. Edit. 2014, 53, 14433.

[6] J. D. Benck, T. R. Hellstern, J. Kibsgaard, P. Chakthranont, T. F. Jaramillo, ACS Catalysis 2014, 4, 3957.

468999 Remarkable Enhancement of Photocatalytic Hydrogen Production in Sensitized and N_2/Ar Plasma Treated, Nanoporous TiO_2 Films

Abstract:

The optical and electronic properties of TiO_2 thin films provide tremendous opportunities in several applications including photocatalysis, photovoltaics and photoconductors for energy production. Despite many attractive features of TiO_2, critical challenges include the innate inability of TiO_2 to absorb visible light and the fast recombination of photoexcited charge carriers. In this study, we prepared ordered nanoporous TiO_2 films co-modified by graphene quantum dot sensitization and nitrogen doping (GQD-N-TiO_2) for hydrogen production from photoelectrochemical water splitting under visible light irradiation. First, cubic ordered nanoporous TiO_2 films were prepared by a surfactant templated sol-gel method. Then, TiO_2 films were treated with N_2/Ar plasma for the incorporation of substitutional N atoms into the lattice of TiO_2. GQDs were prepared by chemically oxidizing carbon nano-onions. The immobilization of GQDs was accomplished by reacting carboxyl groups of GQDs with amine groups of N-TiO_2 developed by the prior immobilization of (3-aminopropyl)triethoxysilane (APTES). Successful immobilization of GQDs onto N-TiO_2 was probed by UV-Vis, FT-IR, zeta potential and contact angle measurements. The nitrogen-doped TiO_2 (N-TiO_2) showed the significant reduction of band-gap (3.5 eV to 2.88 eV) and the unprecedented enhancement (240 times) of photocatalytic hydrogen production visible light illumination. The GQD-immobilized N-TiO_2 film showed about 2 times enhancement compared to N-TiO_2, showing the important role of GQDs. This outstanding enhancement is attributed to the high surface area of nanoporous films and synergistic effects of nitrogen doping and GQD sensitization resulting in enhanced visible light absorption, efficient charge separation and transport.

468774 Bifunctional Porous Materials for Combined CO_2 Capture and Catalytic Conversion

Abstract:

The rapidly-increasing concentration of Greenhouse Gases (GHGs) in the atmosphere has become a critical issue prompting the scientific and engineering community to find ways to efficiently mitigate emissions of GHGs, such as CO_2. Electricity generation using fossil fuels is one of the major contributors of CO_2 emissions and therefore, various end-of-the pipe technologies are currently proposed as carbon management media. A less energy intensive alternative than amine-based solvents are solid adsorbents which are capable of CO_2 capture at low temperatures as part of the Carbon Capture Utilisation and Storage process (CCUS). Another interesting development in CCUS research is the use of captured CO_2 as a chemical feedstock. Although this pathway will not eliminate the need for long term CO_2 storage, carbon utilisation could diversify the range of possibilities

Metal-organic frameworks (MOFs) represent a class of adsorbents with chemical and structural tunability, large porosity as well as high thermal and chemical stability. Owing to these features, MOFs are promising materials for CO_2 capture. In addition, recent research efforts point to the potential use of MOFs as photocatalysts, especially in the context of CO_2 reduction into useful products (e.g. CO and CH_4). Considering these two aspects, we are investigating the application of MOFs as bifunctional absorbent-photocatalyst materials for the combined capture and conversion of CO_2. This approach could potentially enable process intensification and improvement of the overall Carbon Capture Utilisation and Storage process sustainability. A series of MOFs were synthesised and fully characterised by a range of analytical and spectroscopic techniques (i.e. XRD, FT-IR, thermogravimetric analysis, N_2 sorption). These MOFs were tested as CO_2 adsorbents as well as heterogeneous CO_2 reduction photocatalysts in the gas phase using a purpose-built reactor. In the latter case, H_2 and H_2O as sacrificial agents under mild reaction conditions (ambient temperature and pressure). Products were analysed using a GC-MS and the MOFs were characterised after testing to identify any change in structure and/or chemistry. The results were used to identify mechanisms of reactions.

471164 Nature in Engineering: Expanding the Engineering Design Space By Including Ecosystem Goods and Services

Abstract:

Currently, most engineering design methods ignore the role played by nature in supporting engineering systems. Analogous to economic systems, ecological systems also include limits on the stocks of ecosystems goods and services. These can include limits on the availability of ecological resources like fossil fuels, minerals, biomass, fiber and feed, as well as limits on the provisioning of ecosystem services like carbon sequestration, water provisioning, nutrient cycling etc. Ignoring these limits while designing systems have resulted in several unintended consequences, resulting in ecological degradation. For instance, switching farming practices to more intensive agricultural habits to increase crop yield has resulted in harmful algal blooms in Lake Erie due to fertilizer runoff from farms.

In the engineering context, the main focus of conventional design has been to maximize profitability while simultaneously minimizing environmental impacts. Accounting for the role of nature in engineering has been confined to ideas like Biomimicry [1] and Eco-technology [2] such as the use of wetlands for water treatment. However, these methods lack proper integration with concepts of engineering design and such methods do not necessarily identify possible synergies and trade-offs that exist between engineered systems and nature. In addition, all these approaches also neglect the ecological carrying capacity, thus resulting in an increase in disparity between the demand for ecological goods and services and their supply from nature.

The Techno-Ecological Synergy (TES) framework [3] was developed to account for the carrying capacity of ecosystems by quantifying the demand and supply of ecosystem services (ES). Resource inputs for operating engineering systems creates a demand for ecological goods while the emissions, waste and wastewater generated by engineering systems creates a demand for ecosystem services. The supply of ecosystem goods and services is based on the capacity of an ecological system, in the vicinity of an engineering system to provide the necessary ES demanded by the process. The sustainability index for this framework is based on the difference in the supply and demand for ES.

This work integrates the TES framework with concepts of engineering design, resulting in the design of integrated Techno-Ecological systems. The TES design framework accounts for both demand and supply of ecosystem services while designing systems and the objective is to maximize profitability and minimize the disparity between the demand and supply of ES. Thus, the problem focuses on reducing both demand and increasing supply by preservation and restoration of ecological systems, in contrast to conventional approaches of merely minimizing demand.

The presentation conveys the significance of including the role of nature in supporting engineering design. This is indicated by demonstrating that including nature in the design problem expands the solution space of conventional engineering design resulting in the identification of plausible "win-win" scenarios. Design problems are formulated as multi-objective optimization problems, with the objective of maximizing the economic value of the system and maximizing the sustainability index. Trade-offs between these objectives are identified by scalarizing multiple objectives using the epsilon constraint methods. Tradeoff designs are obtained for different ecosystem services, depending on the type of service relevant to the system of interest and the spatial presence of ecological systems.

Four diverse case studies are discussed to demonstrate the significance of including nature in engineering. The first study describes the design of a residential system accounting for ecosystems like trees, lawn and a vegetable garden. While designing a residential systems, technological, ecological and behavioural variables are considered in the design. The objectives are to minimize both the cost compared to basecase and carbon emission. The problem is formulated as a simulation-based Mixed Integer Linear Program (MILP) using EnergyPlus as the simulation engine.

The second study discusses the design of a biosolids network flow system along with the design of ecological systems that supply carbon sequestration ecosystem services, like timberland, forest lands and geological sequestration. Superstructure of possible biosolid flow pathways along with possible pathways for carbon sequestration are identified, and optimal designs of these coupled Techno-Ecological networks are determined by minimizing the network cost and minimizing the difference between carbon sequestration service demand and supply. The entire problem is formulated and solved as a MILP in the Eco-Flow tool. In addition, feedback from ecosystems in terms of biomass from timberland is considered as a feedstock for supplying energy to the wastewater treatment sites.

Third study describes the benefits of including ecosystems in a process engineering design context whereby ecological systems are included in engineering design analogous to unit operations. An integrated design of a biofuel system with a wetland ecosystem is discussed wherein the system is design to maximize provisioning of ecosystem services like water provisioning service and water quality regulation service, while maximizing the net present value.The problem is formulated as an Non Linear Program implemented in GAMS.

The last study focuses on designing agricultural landscapes for producing energy and food, while accounting for services provided by agro-ecosystems. These services include carbon sequestration, water provisioning, and others. The trade-off between monetary objectives and costs are evaluated for various biofuel production alternatives.

Non-inferior pareto frontiers are generated for each ecosystem service for all case studies. Results indicate that including ecological variables in the design results in possible solutions that do not exist in the techno-centric solution space of conventional engineering methods. In addition to this, accounting feedback flows from ecosystems in terms of ecological goods shifts the Pareto frontier to a new range, revealing innovative and novel design solutions. These new solutions obtained from including ecological systems are economically superior compared to conventional solutions; they also have environmental benefits in terms of reduced overshoot.

References

[1] Benyus, Janine M. Biomimicry. New York: William Morrow, 1997.

[2] Straškraba, M. "Ecotechnology as a new means for environmental management." Ecological Engineering 2.4 (1993): 311-331.

[3] Bakshi, Bhavik R., Guy Ziv, and Michael D. Lepech. "Techno-ecological synergy: A framework for sustainable engineering." Environmental science & technology 49.3 (2015): 1752-1760.

466829 Heat-Integrated Work Exchange Network Design

Abstract:

Energy loss in the manufacturing sectors is very high (~50% of the total consumption in 2010, according to a DOE report). While thermal energy efficiency has been greatly improved through adopting effective heat integration technologies over the past decade, the improvement of mechanical energy efficiency is yet to be fully investigated. Mechanical energy, often characterized by process pressure change as the driving force, is more difficult to recover; this is especially true as the pressurization/depressurization involves temperature change, which makes the process non-isothermal. Thus, effective mechanical energy recovery should be coupled with heat integration. Mechanical energy recovery can be realized using work exchangers that are currently used in industrial scale for desalination. The concept and fundamentals of this type of unit, as well as a thermodynamic analysis method for identifying the maximum amount of recoverable mechanical energy from a process system under isothermal condition has been investigated by us previously.

In this paper, a thermodynamic analysis on mechanical energy recovery under non-isothermal condition is described, and a process synthesis methodology for designing a heat-integrated work exchange network is introduced. The methodological efficacy is illustrated by two examples, where both mechanical and thermal energy efficiencies as well as economic feasibility analysis are provided. Furthermore, certain key dynamic performance of the system is analyzed, which is valuable for developing operational strategies for the integrated system.

Reference:

Energetics Incorporated, Manufacturing Energy and Carbon Footprint, Prepared for the Industrial Technologies Program (ITP), DOE, 2010.

468486 Discovery and Assessment of Integrated Waste Biorefinery Paths with Conventional Industry Using an Ontology Engineering Approach

Abstract:

The environmental impacts caused by the extensive use of decreasing fossil fuels has brought tremendous attention towards valorisation of renewable and untapped resources such as urban and industrial wastes. The need to reduce the use of virgin resources gives rise in research areas such as waste Industrial Symbiosis (IS), waste biorefineries, and circular economies. The chemistries or else known as â€œsynthesis pathsâ€, able to extract materials from wastes can be considered abundant.

Unwanted residues are multicomponent materials, containing valuable platform chemicals and numerous added-value specialties. If recovered, they can become standalone feedstock or be also integrated with alike feedstocks to be processed by existing industrial facilities. Such sustainable, retrofitting practice will enlarge process capacity, benefiting from economies of scale, improve LCA performance and cost-effectiveness originating from wasteâ€™s off-market prices.

To achieve such task, a) to appropriately integrate waste resources with virgin feedstocks and b) to identify suitable alternative synthesis paths, producing a diverse gamut of bio-based products, requires the combined use of two well-established research fields: ontology engineering and mathematical optimization. Therefore, it is firstly required a large well-structured database to organize knowledge such as waste properties and valorising technologies characteristics, thus achieving interoperability of information across various manufacturing organizations and processing facilities. Secondly, a screening-level decision support platform is necessary to holistically assess waste synthesis paths.

This work provides a framework that merges an ontological based platform with an optimization process systems tool to first organize and manage tacit knowledge such as chemistries, process models, paths, technology specifications, feedstock characteristics etc. for knowledge sharing across diverse users, and secondly to discover the optimum synthesis path according to a set of environmental, economic, and technical objectives. The semantic algorithm is used to establish layers of interactions and translate process models into networks that are further fathomed using optimization and superstructure technology.

468645 Ionic Liquid Design and Sustainable Process Simulation for Decarbonization of Shale Gas and Gas Separations

Abstract:

Ionic liquids (ILs) have been receiving increasing attention as a potential decarbonization solvent as well as solvent for gas separations, which are usually energy intensive. However, the enormous number of potential ILs that can be synthesized makes it a challenging task to search for the best IL as a decarbonization solvent, for example, for CO2 removal from methane. Also, for gas separations, for example, gas purification processes, ILs can play a major role by replacing the extractive distillation column with a liquid-liquid extractive column, while the recovery column is replaced by an evaporator, thereby saving a lot of energy. The main question is what properties should the ionic liquids and solute (gas)-IL mixtures have to design the optimal separation process? If the IL properties can be matched, then the target process design would be obtained.

In this work, a systematic method is proposed to screen suitable ILs based on a collection of ILs (generated by Computer Aided Molecular Design and collected data); using the COSMO-RS (Conductor-like screening model for real solvents) model and available data in the database, a model for absorption mechanism as well as process simulation-design tools. Several IL properties related to gas absorption, such as the Henryâ€™s constant, the viscosity, gas-solvent solubility data as well as toxicity of ILs are considered for evaluation of targeted decarbonization process and gas separation processes. Furthermore, process simulation is performed to evaluate the new IL-based decarbonization and gas separation technologies. Considering CO2 solubility, CO2/CH4 selectivity, gas solubility, gas/gas selectivities, viscosity and toxicity of ILs, several IL such as [Bmim][NTf2] has been found as a potential solvent from among more than 5000 ILs in the database. Based on reliable experimental data (available for only a fraction of the ILs present in the database), a rigorous thermodynamic model is established by comparing the simulation results with available experimental data. Different process flowsheet options obtained from targeted solvent behavior have been investigated to determine the best IL-based flowsheet. For the decarbonization process, compared with the well-known MDEA (methyldiethanolamine) process for CO2 capture, the single-stage and multi-stage process alternatives with IL reduce the total energy consumption by 42.8%, 66.04%, respectively. For selected gas separation, the potential for energy reduction is even higher â€“ a reduction of more than 70% can be achieved. Net carbon emission as well as eco efficiency indicators are also evaluated to highlight the more sustainable and innovative aspects of the process designs.

455407 Developing Rapid Life Cycle Inventories for Chemical Processes

Abstract:

A product life cycle assessment (LCA) presents an opportunity to consider environmental effects over multiple categories for a system that includes the supply chain and product network (i.e., from cradle to grave). LCA benefits from the flexibility of being able to consider any boundaries for the system. However, this flexibility also means that each system studied can be different. Thus there is a need for rapid methods for developing life cycle inventories. Efforts at the U.S. Environmental Protection Agency Office of Research and Development are focusing on rapid inventory development for processes involved in a product LCA, including top-down and bottom-up methods. The top-down method uses Agency databases of releases to determine an average chemical process inventory. The bottom-up method uses process design methods, available process information, and emission factors to develop an inventory. For this latter method, a step-by-step procedure will be discussed. This presentation will explore the rapid inventory methods, focusing on acetic acid as a case study.

The views expressed in this presentation are those of the authors and do not necessarily reflect the views or policies of the U.S. Environmental Protection Agency.

472750 Sustainable Integration of Refineries and Biorefineries with Forest Plantations to Reduce Emissions

Abstract:

Fossil fuel production typically generates large amounts of greenhouse gas emissions, which are associated to the global warming problem. It is important to note that small increments in the global temperature involve drastic consequences in the world population. For this reason, some authors have stated that it is necessary to limit the greenhouse gas emissions from human activities. Furthermore, several authors have suggested alternatives to address this problem. For instance, biomass has been identified as a high potential renewable energy source for production of biofuels, chemicals and other products using several processing technologies. However, to consider the large scale production of biofuels from biomass, the entire supply chain must be analysed from the economic, environmental and social points of view. It is worth noting that the most of the papers published in the literature present several alternatives for the reduction of the overall environmental impacts during the energy production; however, did not consider a scheme to integrate two or more ways to reduce the global emissions in the supply chain. Therefore, this paper presents a mathematical programming model for the optimal planning for the production of fuels through refineries and biorefineries integrated with the forest plantations focused on decreasing the total greenhouse gas emissions, maximizing the net profit of the system and maximizing the generation of jobs simultaneously. This way, the proposed approach takes into account two ways for the reduction of greenhouse gas emissions, which is given by the production of biofuels in the biorefineries and the forest plantations to capture the CO_2. Additionally, the present approach considers several important issues such as the location of facilities, availability of raw materials, demand of products, inventory of the greenhouse gas emission, as well as the generated jobs for the different activities involved in the supply chain. It is important to mention that the problem is defined by a set of possible oil reservoirs distributed for the extraction of oil to be sent to refineries. Also, the supply chain includes a set of possible harvesting sites to produce different raw materials to be sent to biorefineries. In addition, there is the possibility to produce fuels from oil in the existing refineries, as well as installing necessary refineries to satisfy the fuel demand. On the other hand, it considers the possibility to install new biorefineries to produce biofuels and the option to use existing biorefineries. Furthermore, the proposed model incorporates pollution trading concepts to reduce emissions through a set of forest plantations with different capacities and sizes. It should be noted that the fuel demand, which can be satisfied via fossil fuels or biofuels, is received from several distribution centers. Therefore, the problem consists in determining the topology of the supply chain, the flowrates of products and raw materials, capacities of the refineries, biorefineries and forest plantations in order to obtain the best environmental, social and economic benefits.

The model is formulated as a mixed integer linear programming problem, which involves three objective functions in order to perform the sustainable planning of the production process of fuels in Mexico considering reduction of greenhouse gas emissions, mainly CO_2. In general way, the proposed mathematical formulation is composed of constraints for mass balances, capacity limitations, distribution, carbon emissions and capture as well as objective function equations. The economic objective function considers the maximization of the total profit, which is the sum of the profit from refineries, biorefineries and forest plantations. The profit of refineries and biorefineries includes the revenue from the sale of products, and all the involved costs such as the cost of raw materials for refineries and biorefineries, the operating costs, the capital costs for implementation of new refineries and biorefineries, the transportation costs for raw materials from harvesting places to processing plants, and the transportation costs for products. It also considers a cost for the emissions; for instance, the refineries should pay the forest plantations to reduce their emissions. In case of forest plantations, the profit considers the revenue for the payment that refineries or biorefineries have for their emissions minus the costs of their implementation. The environmental objective is to minimize the total CO_2 emissions, which considers the sum of total emissions from refineries and biorefineries minus the emissions captured by the forest plantations. The total CO_2 emissions from refineries include the ones produced in all the involved activities, including oil extraction, oil refining, transportation of products as well as the emissions obtained from burning petroleum fuels. Whereas the emissions of the biorefineries include the emissions for each of the activities involved in biorefineries such as emissions of CO_2 for biomass transportation, production in biorefineries, emissions produced by the transportation of bioproducts from biorefineries to distribution centers, emissions for the use of bioproducts and the captured emissions of CO_2 by the biomass growth. Furthermore, the captured CO_2 emissions by the eco-industries are given by a unitary factor multiplied by the total number of trees for the forest plantations. Additionally, the model considers a social objective function, which is the maximization of the total generated jobs for the implementation of the proposed scheme. This objective function takes into account the jobs generated in refineries, biorefineries and forest plantations. The jobs of refineries include the ones for processing and installing only in the new refineries, it is important to note that the transportation jobs are not considered because they are given in the transportation network for the existing refineries. On the other hand, the jobs of biorefineries include the ones for biomass plantations, biomass transportation, installing and processing of new biorefineries. With respect to the forest plantations, the considered jobs correspond to the needed people to take care of the plantations.

The results have shown that the relationship between the environmental and the social objectives is inversely proportional, while the behaviour of the economic objective with respect to the environmental objective is different. Also, the results illustrate that petroleum refineries are more economically attractive than the biorefineries, although the biorefineries can satisfy the fuel demand with a minor environmental cost. Additionally, based on the results, it is possible to conclude that the forest plantations can be important sinks for the CO_2 emissions from the fuel production process, which can be afforded by the refineries.

455165 Optimal, Flexible and Feasible Heat Exchanger Network Design Under Severe Uncertain Industrial Operation Conditions

Abstract:

[1]Engineering Department, Crop Science Division of Bayer, Institute, WV, 25112

[2]Dan F. Smith Department of Chemical Engineering, Lamar University, Beaumont, TX 77710

Abstract

Heat Exchanger Network (HEN) synthesis plays an important role in improving the economic performance and energy efficiency of industrial plants, and has been the subject of numerous investigations during the past decades. However, traditionally, most of the research focuses on the design based on determined condition. The operational flexibility and feasibility are less discussed. Chemical process and its production are complex and flexible. Production rate, utility stream quality, process stream temperature, ambient weather condition, and various other uncertain factors, will cause disturbances to the heat exchanger network operation. In this presentation, the authors propose a novel approach for optimal, flexible, and feasible HEN design considering the severe operation uncertainties. It utilizes the Probability Bounds Analysis (PBA) theory to model the uncertain industrial operation conditions, double loop method to sample the uncertainty, and Aspen Energy Analyzer for optimization design. The capability of this approach is demonstrated by a case study which generates a better solution with less cost and good flexibility in uncertain operations than the one generated at a deterministic condition.

461174 From Fossils to Bio-Based Economy: A Revolutionary Transformation in Process Industries

Abstract:

Increasing utilisation and dependence on fossil fuels are attributed to the growing global population and soaring energy demand, and led to the global warming and energy security issues. Employing renewable energy sources such as biomass in new or existing systems and improving the performance of the existing facilities through maximising resource utilisation, enhancing energy efficiency and lowering environmental impact are the keys in addressing these issues. Process industries such as energy, chemical, refinery and manufacturing industries are the major consumers of fossil resources and sources of greenhouse gases emission. The replacement of traditional process design approach by a more holistic approach is exigent for the development of sustainable industrial systems and for effectively reducing the energy and emission intensities in the process industries. The transition from fossil fuel technologies to mature commercialisation of biorefinery may require a couple of decades. Hence, the most strategic solution during this transition is to improve the existing facilities through deployment of decarbonisation system, enhancing the system performance through process integration approach, while carrying out intensive research and investigations into biorefinery technology. This paper provides high level discussion on the global issues on energy and the strategic approach during the transition from fossil era to bio-based economy era. Modern process design approaches using various state-of the-art concepts associated with integrated biorefinery systems, decarbonisation strategies such as CO
mitigating environmental impact and meeting the global demand, will be discussed.

OPTIMAL CO-PRODUCTION OF MARKET BASED POWER GRID SUPPORT AND RENEWABLE FUELS OR CHEMICALS

Robert Weiss, Lotta Kannari, Jari Pennanen, Teemu Sihvonen, Jouni Savolainen*
VTT Technical Research Centre of Finland Ltd, Finland
**Corresponding author: robert.weiss@vtt.fi*

Introduction

This paper addresses the operational profitability and benefits of co-production and dynamics in renewable Power-to-Fuel or Power-to-Chemical processes, when utilized to stabilize the power grid. To reach climate neutrality, the transport sector must switch to climate-neutral or climate-positive fuels or power. This can be done by a proper mix of electric vehicles where the power is based on renewable electricity, hydrogen vehicles where the hydrogen is based on side-product hydrogen or renewable power based processes, or by using in existing vehicle fleet renewable biofuels instead of fossil fuels. For hydrogen-based or hydrocarbon-based chemical industry, similar change needs are obvious. Today, renewable fuels can be produced with biomass based plants or in bio-refineries. If biomass cannot be utilized or there is a lack of some sustainable chemicals like hydrogen in the bio-process, new so called Power-to-X plants [1, 2 and 3] could be used to produce renewable fuels or chemicals of non-biomass origin (X refers to the end product of the plant). These processes are utilizing hydro, solar or wind power via electrolysis for production of the needed additional hydrogen fraction [1]. Optimized day- or week-ahead planning can increase considerably the value of renewable Power-to-X plant operation. Such optimal operation can enhance commercial viability of the Power-to-X plants, and in this way introduce these climate-positive, mid-sized and distributed units into a smart grid. In this paper, we show examples on the value of such optimal dynamic operation together with the value of the dynamics of key unit processes and intermediate storages, while at the same time reaching a high share of sustainable solar and wind power in the end product.

Methodology

Sustainable Power-to-X Concept

Power-to-X (PtX) plant's main products can be typically fuels (H_2, CH_4, CH_3OH, Fischer-Tropsch-products), chemicals (NH_3, light olefins C_2H_4, C_3H_6), or reduction agents (H_2 for metals). Side products are high purity oxygen and low grade waste heat from the electrolysis operation, as well as process steam from some PtX processing units that are highly exotermic, like catalytic methanation. These side products can generally only be economically utilized locally "over-the-fence", and were neglected in this analysis for generality reasons.

For sustainability and avoidance of fossil fuels, the PtX operations should avoid fossil fuel based grid power, and instead utilize high enough content of sustainable renewable power. In our case examples we target to utilize as much wind and solar power as possible which are intermittent. On the other hand, parts of the PtX process may not be as flexible as the intermittent renewable power. Or the demand for the end product itself might even be 24/7, stable, but completely inflexible industrial demand. Finally, for sustainability the PtX plant should participate in power grid stabilization. This operating environment, depicted in Fig. 1, render the production planning a challenging task, necessitating the optimization approach adopted in this paper.

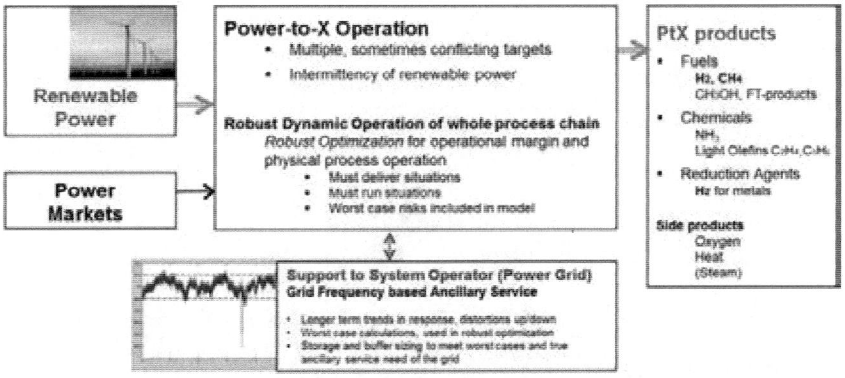

Figure 1 PtX concept and product examples.

Power-to-X Plant Operation and Dynamic Optimization Model

In our analysis, PtX power market contract handling as well as the physical production process is formulated as a dynamic, linear optimization model described in detail in [4], which minimizes the production cost. To our model, we added risk handling related to the power grid ancillary service as well as the utilization of available intermittent solar and wind power as described in the next sub-chapters.

For our analysis, we configured a 4x1.25 MW_e, state-of-the-art commercial PEM electrolyzer system utilizing confidential manufacturer data. As illustrative examples, we present the following example cases in our analysis: ***Case1*** describes on-site electrolytic production of H_2 as metal reduction agent, with a stable but unflexible H_2 demand. ***Case2*** describes Power-to-Gas production of CH_4 as green synthetic natural gas for sustainable mobility, with a flexibility range 70-100% in the catalytic methanation operation. In both cases, we investigate the impact of a 0-24h H_2-buffer storage. The optimization model schematic and case boundaries are depicted in Fig. 2.

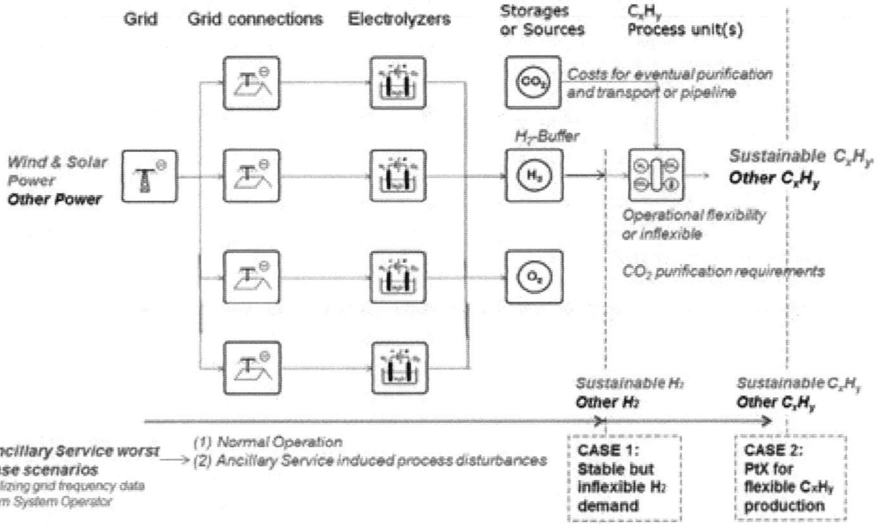

Figure 2 Schematic of PtX scheduling and dynamic optimization model, with the boundaries of example ***Case1*** and ***Case2***.

Markets for Wholesale Power and Ancillary Services

Deregulated electricity markets are generally divided into wholesale energy markets managed by power exchanges, and ancillary service markets managed by Transmission System Operators (TSO). Wholsesale energy is sold and purchased typically on international day-ahead markets for the 24 hours, see e.g. [5] for Nordic System, Baltic countries and Germany, or [6] for Germany, Austria, France and Switzerland. Concerning the ancillary service markets, situation of the Nordic countries compared to Netherlands, Germany and Poland (UCTE area) is well described in [7]. We list some examples on European ancillary service products in Table 1 below.

Table 1 Example on European Primary Frequency Control Schemes

Frequency Containment Reserve - Normal Operation **(FCR-N)** Also known as Primary Frequency Control or PRL	European 50Hz Grids: 49.9 – 50.1Hz or 49.8 – 50.2 Hz control band. Generally a symmetric reserve which is continuously triggered. Asymmetric bidding possible in some markets e.g. Denmark. Day-ahead markets in e.g. Finland, Denmark, and Norway. Week-ahead markets in Germany, Austria, Switzerland and The Netherlands.
Frequency Containment Reserve – Disturbance **(FCR-D)**	Scandinavian 50Hz Grids: 49.9 – 49.5 Hz (only disturbances). Asymmetric, fast. Not often triggered. Day-ahead markets in e.g. Finland, Denmark, and Norway.

In a power system with high penetration of intermittent renewable power production, grid balancing support and ancillary service prices tend to be more and more volatile. As example on this effect, we utilized the hourly markets of the Finnish primary frequency control products FCR-N and FCR-D [8]. In this market, a large number of Combined Heat and Power plants (CHP) are on-line in the winter resulting in high power system intertia and a corresponding low need for additional FCR-N and FCR-D. In contrast, these plants are generally off-line in the summer, with nuclear power plants in maintenace, resulting in low intertia in the system and high prices for hourly additional FCR-N and FCR-D. Examples on this price volatility is shown in Fig. 3 below.

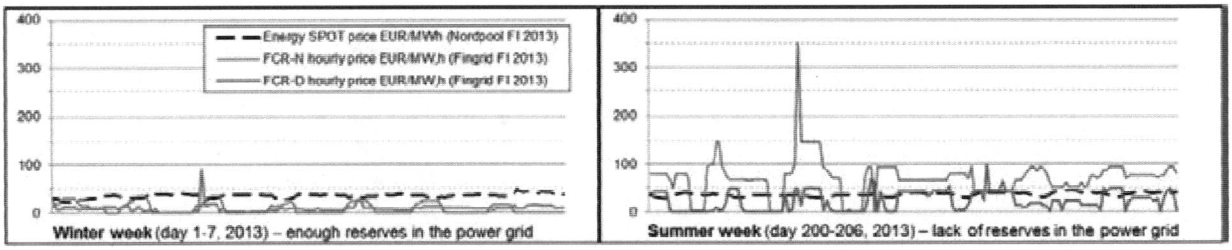

Figure 3 Hourly prices for energy [5] and FCR [8] during winter and summer example weeks.

Uncertainty in Disturbances and Ancillary Service Activations, Robust Optimization Solution

Most ancillary services, including primary frequency control like FCR-N and FCR-D, are capacity options with sometimes a strong uncertainty in the real need for their activation. The power system can have prolonged periods of under- or overfrequency, leading even in symmetric freqency control products (which theoretically should have 0 as net cumulative response) to long cumulative distortions towards large up or down regulation amounts, as depicted in Fig. 4.

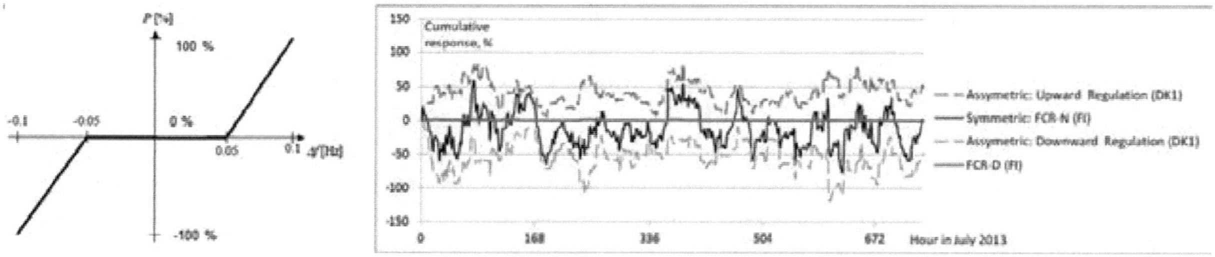

Figure 4 ***Left:*** FCR-N symmetric control scheme in Finland [8]. ***Right:*** Example on the 24 hour cumulative distortions up/down for the response of some primary frequency control schemes.

In our model, we handle such uncertainty with a robust optimization approach, where we solve the conflict between the need for stable unit process conditions versus the ancillary services' need for capacity reservation, fast reactions as well as prolonged activation worst case distortions up/down. Our ancillary service activation risk calculations utilize the *budget uncertainty set* methods from robust optimization theory [10, 11]. Uncertainty sets were defined for combined maximum and duration distributions of the ancillary service control response. These uncertainty set definitions were integrated into the optimization model for the electrolyzer power consumption and H_2 production, H_2 storage as well as H_2 and power balances. This resulted in a min-max bi-level optimization problem which was transformed into a LP model utilizing the duality of the problem. For the Finnish FCR-N and FCR-D schemes, budget uncertainty sets were statistically identified based on the primary frequency control scheme responses and cumulative worst cases identified from TSO frequency data, with a 0.1s resolution for the entire year 2013[8]. These identified uncertainty sets were utilized in our analysis.

Renewable Power and Dimensioning of a PtX Plant

Wind power from wind parks or larger areas can have short storm periods of very high power output still representing only a small energy amount of the entire yearly wind power production, as indicated in Fig 5. If a high annual wind energy share is desired for the PtX production, oversizing the nominal wind power capacity should be considered. Using additional solar power can be an advantage, since wind and solar power production are generally not strongly correlated. Also, in Northern countries like Finland, winds can generally be stronger in autumn and winter while solar production is concentrated to spring and summer. In our example, national wind power measurements from [8] and local solar PV measurements from [9] where re-dimensioned to 133% and 100% nominal capacity of the PtX plant, respectively, as shown in Fig. 5. Together, they resulted in a theoretical minimum curtailment energy loss below 2% of the yearly solar and wind energy available.

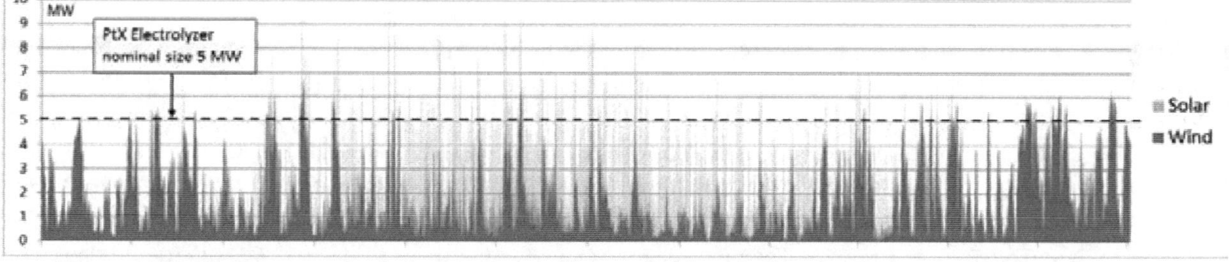

Figure 5 PtX plant nominal size compared to available wind and solar power during one year.

In our optimization study, we assumed wind and solar power to be directly and locally contracted on a yearly base, with only fixed costs but zero marginal costs when available. Forecasting errors of available wind and solar power were assumed to be balanced on average to zero costs on the balancing market. For hours when available wind or solar power where insufficient, the needed additional power could be purchased from the wholesale market, subject to transmission fees and taxes but no transmission congestions. For simplicity, we assumed that surplus wind and solar power not used had to be curtailed at zero prices, not able to reach the market for various congestion reasons.

Results and Discussion

In the analysis below we define the ***part load level sizing*** of the electrolyzer as the ratio of stable H_2 demand divided by the electrolyzer nominal H_2 capacity. ***Net power costs*** are defined as the wholesale power cost + wind and solar power costs – income from FCR-N and FCR-D hourly ancillary service contracts.

CASE 1: PtX with Stable but Inflexible H_2 Demand

Net power costs seem to be clearly lower if the electrolyzer is sized to 50-70% partial load (i.e. oversized) and larger H_2 buffer size decreased clearly the net power costs at all part load levels, as shown in Fig. 6 below. This finding applies both for the whole year analysis, as well as for the selected example weeks displayed with strongly differing pricing in FCR ancillary service (see Fig. 3). The FCR income was very high in the summer weeks, up to 3 US$/kg$H_2$, if large H_2-buffer space was available. For smaller H_2-buffers, our robust optimization restricted strongly the utilization of FCR according to the identified uncertainty sets for FCR control activation and PtX plant dynamics.

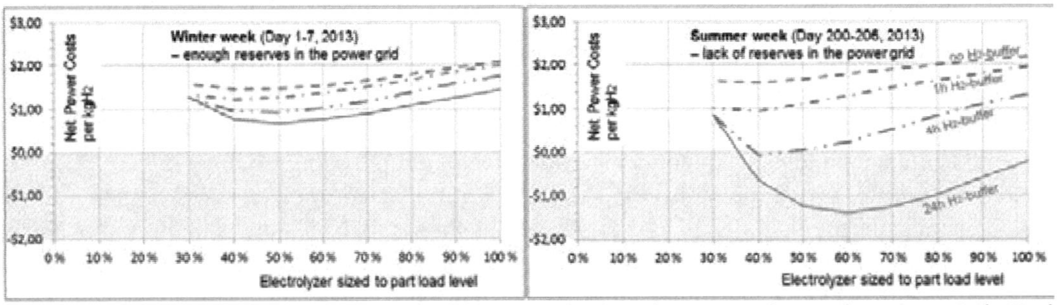

Figure 6 Net power costs for various H_2-buffer sizes during the example weeks. For a 24h H_2-buffer, the FCR income would have led to a net power *income* in the summer week.

Current state-of-the art commercial PEM and AEC electrolyzer technology can also be offered with an overload option, meaning that the electrolyzer may run for a short time substantially over its nominal capacity, but with less efficiency, and must thereafter cool down in the peripheral equipment like cooling circuits and transformers. In our analysis, we evaluated a theoretical +60% overload with a maximum utilization cycle of 0.5h overload + 0.5h cooling, which is quite close to current offerings on the PEM markets. The overload option showed no benefit during the winter example week due to low hourly FCR-N&D prices and absence of heavy peaks in wind power. However, during the summer example week, the overload option could bring substantial savings but only if a large enough H_2 buffer was available storing the overproduction (because of inflexible H_2-demand).

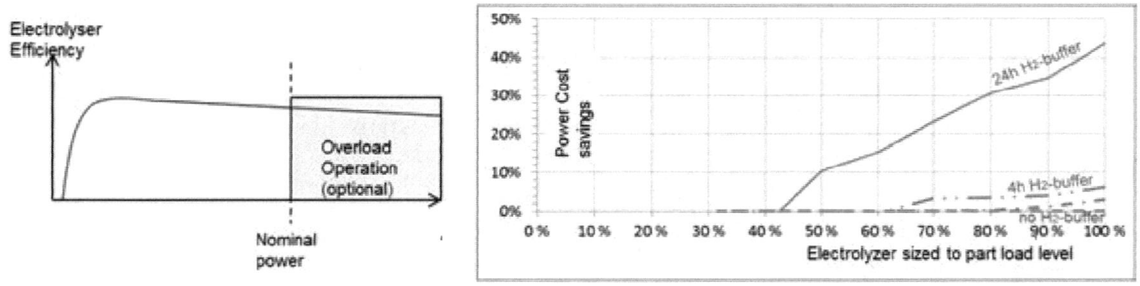

Figure 7 Electrolyzer overload option and power cost savings during the summer example week.

Calculating the hourly dynamic operation over an entire year, we analyzed the situation using a 24h H_2-buffer and 60% overload option, with results displayed in Fig. 8. Almost all available wind and solar power was used if the electrolyzer is sized to 100% i.e. full load. However, the wind and solar power usage decreases if the electrolyzer is sized to part load, which was accelerated below a 70% part load sizing. The reason for this utilization decrease was not the PtX plant flexibility, but the lack of H_2 demand at part load sizing. The share of sustainable power-based H_2 shows an opposite pattern: The share is as low as 50% for full load sizing, but can grow to close to 90% for a 30% part load sizing, being 65% at a 70% part load sizing. The reason for this utilization increase was the increased PtX plant flexibility at part load.

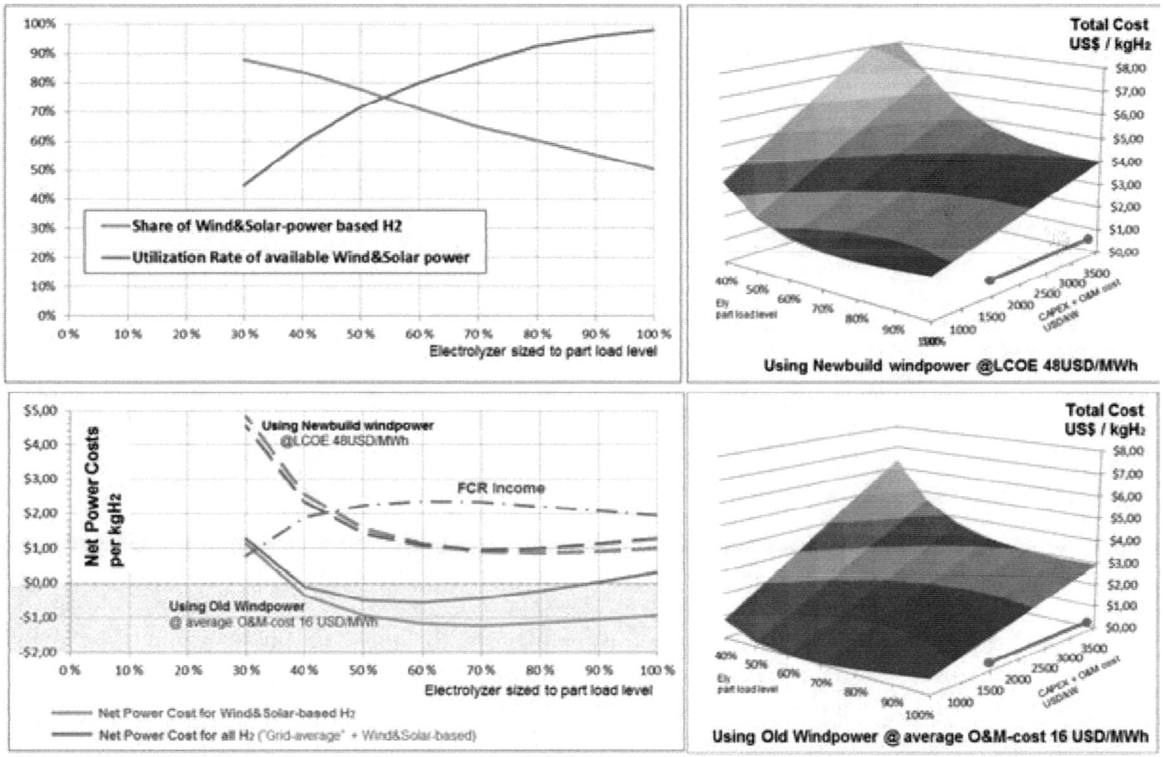

Figure 8 Dynamic operation results for entire year, ***Case 1. Left Up:*** Utilization of Wind&Solar-power, Share of Wind&Solar-based H_2, ***Left Down:*** Net power costs, ***Right:*** Total production costs.

Based on estimates in [3] and confidential information from manufacturers, for current commercial 1-1.5 MW_e AEC or PEM electrolyzer we estimate a fixed cost range between 1500 to 3500 US$/$kW_e$ for CAPEX and 10 year O&M + insurance costs, which is indicated with a red bar in Fig 8. If we use old "stranded" wind power at an O&M cost of 16 US$/MWh, the cost range would result in total H_2-prices 1.3-3.0 US$/kg$H_2$, which are close to or even competitive with steam reforming (avg. cost 1.7 US$/kg). When using new wind power at LCOE of 48US$/MWh, competitiveness can be reached only if electrolyzer CAPEX is clearly below 1000 US$/kWe.

CASE 2: PtX for flexible CH_4 production

Results were generally similar to Case1 and are displayed in Fig 9 below. Under-sizing the methanation unit towards the electrolyzer with a ratio 0.7:1 reduced the utilization of available solar and wind power, but increased the share of solar and wind based CH_4 in the end product gas mixture. The H_2 buffer can be kept at a minimum size of 1-4 hours, and larger buffer does not affect results at all, except if a longer term storage (week or larger) can be added. A more flexible methanation (operation range 10-100% instead of 70-100%) increased by 10% the share of solar and wind based CH_4 in the end product gas mixture, when the methanation unit was sized 1:1 towards the electrolyzer.

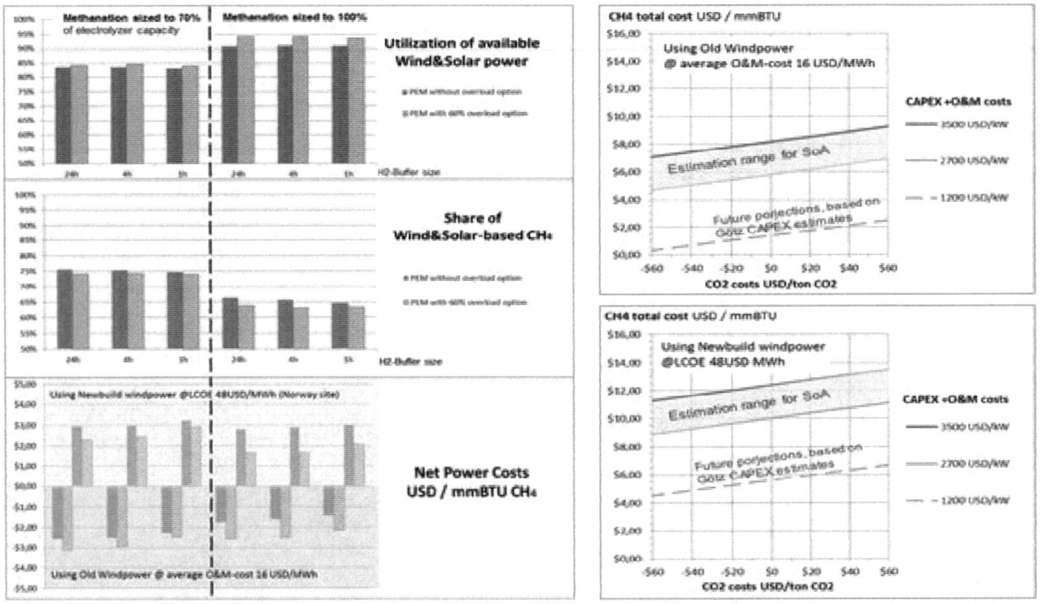

Figure 9 Dynamic operation results for entire year, ***Case 2. Left:*** Utilization of Wind&Solar-power, Share of Wind&Solar-based CH_4, and Net power costs, ***Right:*** Total production costs.

Methanation CAPEX estimation was challenging, since small- and medium scale methanation is not available on the market today. Using the fixed cost estimates for the electrolyzer as presented in previous sub-chapter, and estimates in [3] for methanation, we can see from the results in Fig. 9 that competitive CH_4-prices compared to current US SPOT prices (< 3 US$/mmBTU) seem challenging to reach. Higher CO_2-allowance prices and/or cost efficiency improvements in the electrolyzer and methanation manufacturing, and technology improvements could however change the picture.

Conclusions

PtX example cases for solar- and wind power-based H_2 and CH_4 production were shown for Finnish climate, power market and grid conditions for fast ancillary services. Evidently, concurrent sizing and operational planning is needed to find economical potential solutions and plant configurations. Participation in power grid ancillary services, e.g. primary frequency control, is essential both for plant economics and power system sustainability. Also, state-of-the-art PEM and AEC electrolyzer offer capacity to temporarily overload during primary frequency control price peaks, which is economically beneficial. We demonstrated the ability of robust optimization to handle the activation uncertainty of fast ancillary services, which proved to be an essential part of the optimization solution.

Competitive H_2-prices seem to be reachable for PtX compared to steam reforming, even for an inflexible end customer. In contrary, the sustainable CH_4 case seems to require technological advances or more benefits for the reuse CO_2 and avoidance of fossil fuels. Alternatively, dedicated markets have to be found or created where conventional natural gas cannot be offered, e.g. bio-methane as a sustainable transportation fuel.

References

1. Gahleitner, Gerda (2013), "Hydrogen from renewable electricity: An international review of power-to-gas pilot plants for stationary applications", *International Journal of Hydrogen Energy*, 38(5), February, pp. 2039-2061.
2. Rönsch, Stefan, Schneider, Jens, Matthischke, Steffi, Schlüter, Michael, Götz, Manuel, Lefebvre, Jonathan, Prabhakaran, Praseeth and Siegfried Bajohr (2015), "Review on methanation – From fundamentals to current projects", *Fuel* 166, pp. 276–296.
3. Götz, Manuel, Lefebvre, Jonathan, Mörs, Friedemann, McDaniel Koch, Amy, Graf, Frank, Bajohr, Siegfried, Reimert, Rainer and Thomas Kolb (2015), "Renewable Power-to-Gas: A technological and economic review", *Renewable Energy* 85, January, pp. 1371–1390.
4. Weiss, Robert, Savolainen, Jouni, Peltoniemi, Pasi and Eero Inkeri (2016), "Optimal scheduling of a P2G plant in dynamic Power, Regulation and Gas markets", in: *10th International Renewable Energy Storage* (IRES 2016), Düsseldorf.
5. Nordpool (2016), http://nordpoolspot.com/
6. EPEX (2016), https://www.eex.com/en/products/power/power-spot-market
7. Grande, Ove S., Doorman, Gerard and Bjorn H. Bakken (2008), "Exchange of balancing resources between the Nordic synchronous system and the Netherlands / Germany / Poland." *SINTEF Energy Research*, Technical Report.
8. Fingrid Oyj (2016), Finnish Transmission System Operator, Helsinki, www.fingrid.fi
9. HELEN Ltd (2016), https://www.helen.fi/sahko/kodit/aurinkosahko/suvilahti/
10. Ben-Tal, Aharon, El Ghaoui, Laurent and Arkadi Nemirovski (2009), "Robust Optimization", Princeton University Press.
11. Bertsimas, Dimitris, Brown David B. and Constantine Caramanis (2011), "Theory and Applications of Robust Optimization", *SIAM Review* 53(3):464-501.

453906 Improved Thermal Efficiency Analysis of Biomass Cookstoves

Abstract:

Biomass is an important sustainable energy source, especially in the developing world. One common use of biomass in the developing world is through biomass cookstoves. Biomass cookstoves are used by approximately 2 billion people worldwide on a daily basis. In order for this use to be as sustainable as possible, efforts to reduce wood usage are important to help minimize environmental and health effects associated with pollution and deforestation. Currently the effects of biomass cookstoves on deforestation have been shown to be relatively small, though minimizing wood usage can reduce health, safety and environmental risks.

Because of these risks, the safety and efficiency (both thermal and combustion) of biomass cookstoves has been considered and improved by many researchers through the design of cookstoves and associated auxiliary features such as pot skirts. Once a new cookstove or auxiliary feature has been designed, the design is assessed and optimized. The most common test for assessing cookstoves is the Water Boiling Test (WBT). The WBT consists of three tasks. The first task is to heat water from room temperature to boiling using a "cold" cookstove that hasn't been in use. The second task involves repeating the first task but starting with a "hot" cookstove that was recently used and still has thermal energy in the cookstove structure. The final task directly follows the second task by lowering the feed rate of fuel and keeping the water a few degrees below boiling for 45 minutes.

Measurements of emissions (CO , CO_2, particulate matter, etc.), fuel used, water boiled off, and the time to complete each task are performed during the WBT. Metrics, including thermal efficiency, time to boil, burning rate, firepower, turn-down ratio, and emissions per fuel burned are then calculated from the measurements. Unfortunately, some of these metrics are averages based on endpoint analysis (measurements taken at the beginning and end of the test). This work expands upon the use of the thermal efficiency metric by developing and validating a model that can be used to estimate the average thermal efficiency at any point during the boiling process. This transient average thermal efficiency can be used to gain insight into the attributes of tested cookstoves. Such insights can include the minimal testing time required to determine the endpoint thermal efficiency used for cookstove comparisons, the effects of testing perturbations on thermal efficiency, and the difference in the behavior of thermal efficiency between cold starts and hot starts.

The WBT protocol along with the developed method for estimating thermal efficiency with time was performed using a typical Peruvian cookstove made of two small parallel walls of bricks, modified with both a grate under the fire and pot skirts. The studies using this stove showed that the time required to reach steady-state thermal efficiency was significantly different for cold and hot starts. Additionally, the time and temperature difference required to reach steady state thermal efficiency was significantly lower than required to actually boil the water as required by the WBT. The experimental runs with time-based endpoints were also found to be in good agreement with the estimation shown in this work. This expansion of the WBT protocol should allow for more efficient analysis and optimization of biomass cookstoves.

474844 Pilot-Scale Catalytic Biomass Pyrolysis Studies

Abstract:

Catalytic fast pyrolysis is the direct liquefaction process used by RTI to produce a hydrocarbon-rich biocrude that can subsequently be upgraded by hydrotreating to make refinable intermediates and â€œdrop-inâ€ advanced biofuels. As part of the technology development, the process has been scaled up from bench-scale to 1 ton per day (1TPD) pilot-scale. Evaluation of the process at the pilot-scale is one of the critical steps being taken by RTI to address a number of technical challenges facing catalytic pyrolysis and to collect required processing and engineering data to support the demonstration and commercialization of the technology.

Up to date, several experimental campaigns have been conducted to optimize the catalytic biomass pyrolysis process to achieve high degree of deoxygenation while maximizing the bio-crude production and improving the biocrude thermal stability. A commercially available spray dried, non-zeolite, alumina-based catalyst has been exclusively studied so far. The quality and biocrude yield has been evaluated at different temperatures and other process conditions. Typically, partially deoxygenated biocrude with oxygen contents between 20 and 25 wt% is produced. Gas Chromatography Mass Spectrometry (GC-MS) analysis have shown high amount of oxygenated aromatics in the partially deoxygenated biocrude. Current results from RTIâ€™s 1 ton per day pilot-scale catalytic pyrolysis unit show a biocrude yield up to 19.2 wt% is achievable. This presentation will be used to discuss in detail the results that have been obtained from several studies and highlight the progress that has been made.

Catalytic Pyrolysis of Bio-Oil Model Compounds over La/Ce/Ni Modified HZSM-5 Zeolites

Fengwen Yu, Qiang SONG, Yong NIE, Zhichao XU, Jianbing JI
Zhejiang University of technology, Hangzhou, China

Abstract :

- With the contradiction between supply and demand of petroleum resources and environmental pollution, more and more attention has been paid to the study of clean and renewable energy, especially converting bio-oil into biofuel. Bio-oil catalytic pyrolysis has several advantages, such as low capital cost, easy handling and safe controlling. It has been widely used in improving the quality of bio-oil, for fossil fuels and chemicals production.
- The experimental present a catalytic strategy to upgrade a simulated bio-oil consisting of ethyl acetate, diacetone alcohol, furfural and guaiacol, aiming to study the effect of La/Ce/Ni modified HZSM-5 zeolites on catalytic pyrolysis. The catalysts properties are investigated by BET and NH_3-TPD and the major constituents of the pyrolytic products are carried out by GC/MS. The results have pointed out the yield of organic compounds is increased from 42.26% to 65.05% by Ce-HZSM-5, whilst keeping lower coke formation rates after adding metal cations under the condition of T=450℃, WHSV=4 h^{-1}. The major components of catalytic pyrolysis are aromatic hydrocarbons, ketones, phenols and so on. The optimal selectivity of aromatic reached 46.37% over Ce-HZSM-5. The La-HZSM-5 improved the selectivity of partial deoxidation products from 4.88% to 21.97% .

2016 AIChE Annual Meeting

AIChE

Catalytic pyrolysis of bio-oil model compounds over La/Ce/Ni modified HZSM-5 zeolites

Zhejiang Province Key Laboratory of Biofuel,
College of Chemical Engineering,
Zhejiang University of Technology,
Hangzhou, China

浙江工业大学

Fengwen YU
Qing SONG
Nie YONG
Zhichao XU
Jianbing JI

CONTENTS

Part 1

Background

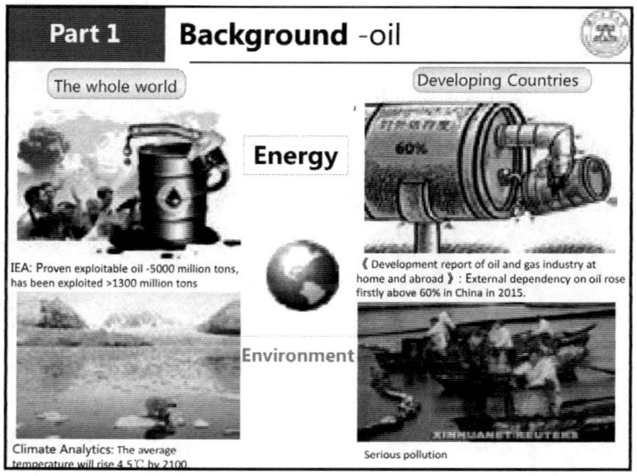
Part 1
Background -oil
The whole world
Developing Countries
Energy
IEA: Proven exploitable oil -5000 million tons, has been exploited >1300 million tons
《Development report of oil and gas industry at home and abroad》: External dependency on oil rose firstly above 60% in China in 2015.
Environment
Climate Analytics: The average temperature will rise 4.5℃ by 2100
Serious pollution

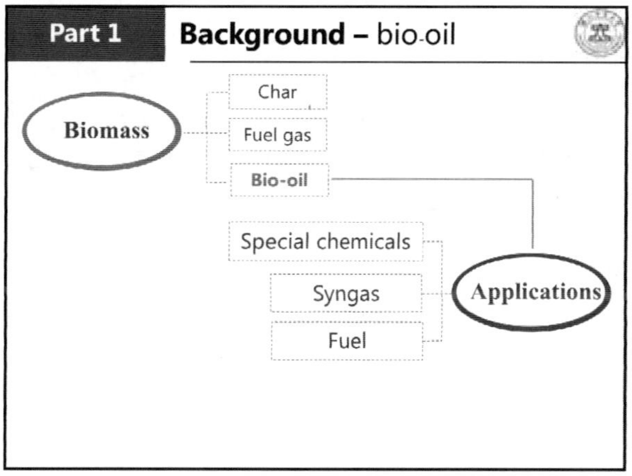
Part 1
Background – bio-oil
Biomass
Char
Fuel gas
Bio-oil
Special chemicals
Syngas
Fuel
Applications

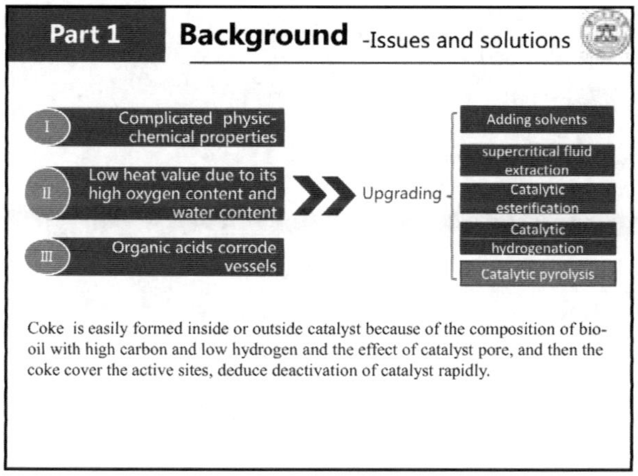
Part 1
Background -Issues and solutions
I Complicated physic-chemical properties
II Low heat value due to its high oxygen content and water content
III Organic acids corrode vessels
Upgrading
Adding solvents
supercritical fluid extraction
Catalytic esterification
Catalytic hydrogenation
Catalytic pyrolysis
Coke is easily formed inside or outside catalyst because of the composition of bio-oil with high carbon and low hydrogen and the effect of catalyst pore, and then the coke cover the active sites, deduce deactivation of catalyst rapidly.

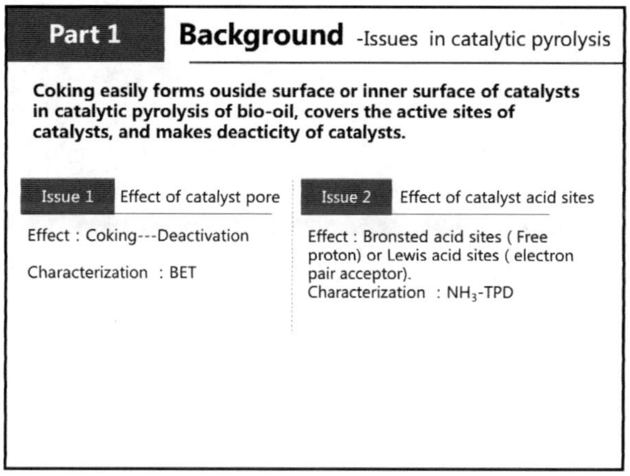
Part 1
Background -Issues in catalytic pyrolysis
Coking easily forms ouside surface or inner surface of catalysts in catalytic pyrolysis of bio-oil, covers the active sites of catalysts, and makes deacticity of catalysts.
Issue 1 Effect of catalyst pore
Effect : Coking---Deactivation
Characterization : BET
Issue 2 Effect of catalyst acid sites
Effect : Bronsted acid sites (Free proton) or Lewis acid sites (electron pair acceptor).
Characterization : NH3-TPD

Part 1 Background –purpose of the research

The key point lies in screening and modifying catalysts in order to obtain high degree of pyrolysis, high effect of deoxidation, and low carbon deposition and coking.

Part 2 Experimental

Part 2 Experimental

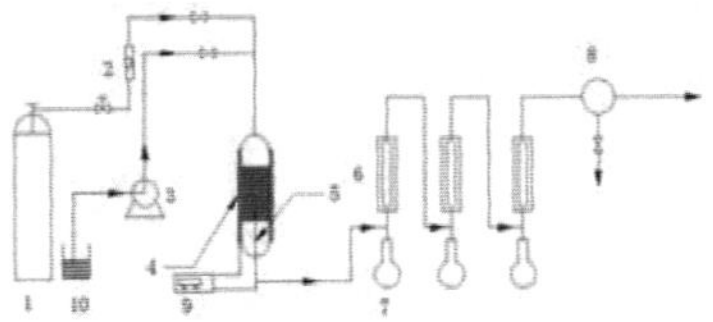

Fig. Experiment flowsheet of catalytic pyrolysis of bio-oil compounds
1-N_2 tank, 2-Rotameter, 3-Peristaltic pump, 4-Heating jacket, 5-Reactor, 6-Condensation pipe, 7-Liquid product receiving bottle,8-Gas buffer bottle, 9-Automatic temperature control unit, 10-Bio-oil (model compounds with Diacetone alcohol, Ethyl acetate, Furfural, Guaiacol)

- Continuous and convenient to control the temperature

Part 2 Preparation of modified HZSM-5 catalysts

Materials

HZSM-5 (Si/Al=50)
$La(NO_3)_3{\cdot}6H_2O$
$Ni(NO_3)_2{\cdot}6H_2O$
$(NH_4)_2HPO_4$

Catalysts

1.0 wt.%La-HZSM-5,
1.0 wt.%Ce-HZSM-5,
1.0 wt.%Ni-HZSM-5

Procedure

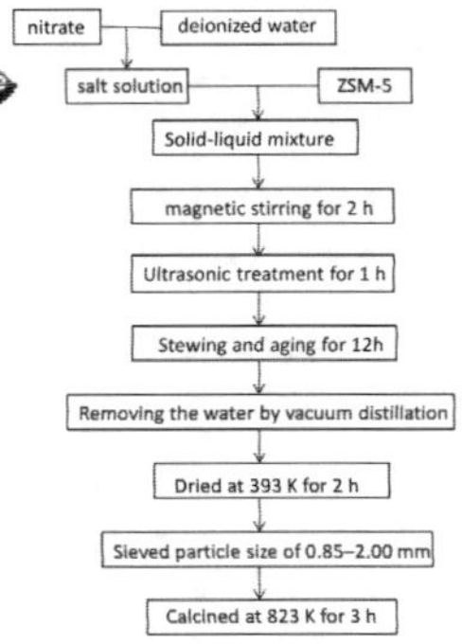

Part 3 Results and discussion

Part 3 Properties of modified HZSM-5 molecular sieve

Table Pore structure parameters of catalysts

Types of Catalysts	$S_{BET}/(m^2 \cdot g^{-1})$	Pore size/nm	Pore volume/ $(ml \cdot g^{-1})$
HZSM-5	308.61	0.77	0.15
La-HZSM-5	300.62	0.79	0.15
Ce-HZSM-5	297.72	0.83	0.15
Ni-HZSM-5	290.42	0.80	0.14

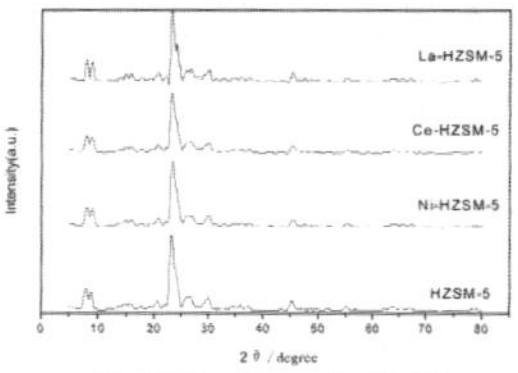

Fig. XRD patterns of La/Ce/Ni modified HZSM-5 zeolites

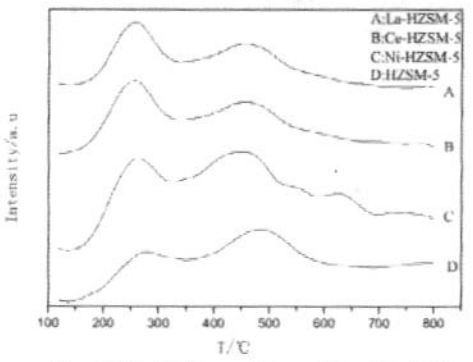

Fig. NH_3-TPD profiles of La/Ce/Ni modified HZSM-5

Part 3 Properties of modified HZSM-5 molecular sieve

Fig. SEM images of La/Ce/Ni modified HZSM-5 zeolites

Part 3 Effect of temperature on model compounds pyrolysis

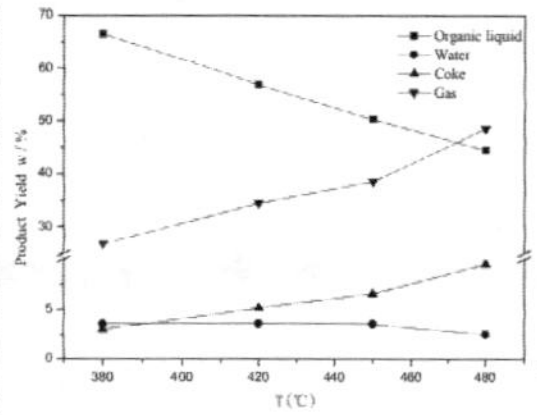

Fig.Products distribution at different temperatures (Feed rate=1.5 ml·min^{-1}, WHSV =4 h^{-1}, no catalyst)

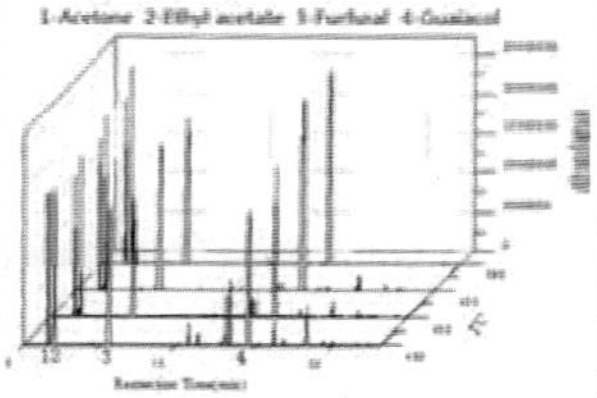

Fig. TIC of organic liquid obtained at different temperatures

- 450℃ is optimal temperature due to higher pyrolysis yield and moderate product distribution.

Part 3 Effect of mdified catalysts on product distribution

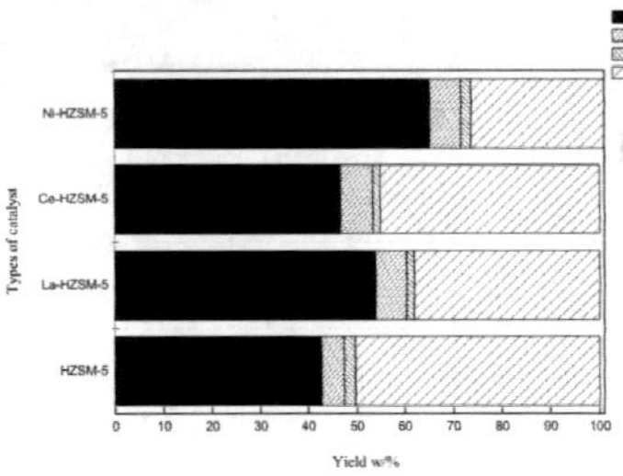

Fig. Products distribution on different catalysts
(At 450℃, feed rate=1.5 ml·min^{-1}, WHSV =4 h^{-1})

- Organic liquid yield increased
  Coking yield decreased
  Gas yield decreased

Part 3 Effect of mdified catalysts on conversions

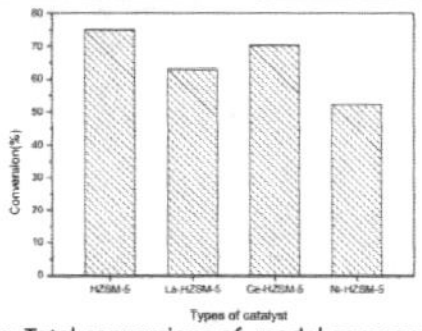

Fig. Total conversions of model components

- Total conversions decreased

- Single conversions decreased except EA over Ce-HZSM-5 and DAA

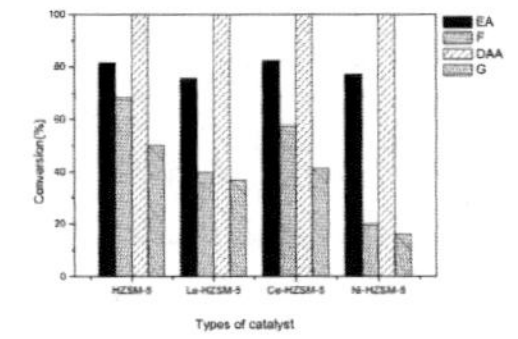

Fig. Single conversions of model component

Part 3 Deoxidizing products over HZSM-5

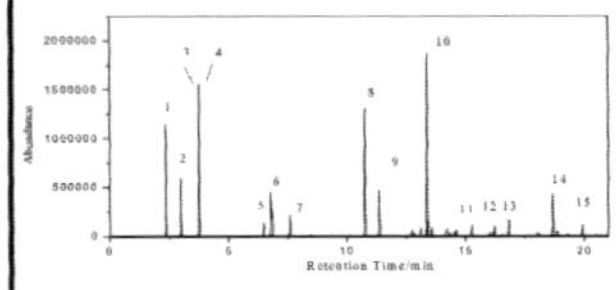

Fig. TIC of the major deoxidizing products of model components

Complete deoxidizing products

Partial deoxidizing products

Table Major deoxidizing products of model components

Number	Retention time/min	Compounds	m/z
1	2.343	Benzene	78
2	2.669	**2-Methyl Furan**	81
3	2.978	**2-Vinyl Furan**	94
4	3.737	Toluene	91
5	6.465	Ethyl benzene	91
6	6.828	M-Xylene	91
7	7.399	P-Xylene	91
8	10.755	**Phenol**	94
9	11.357	**Benzofuran**	118
10	13.374	**O-Phenyl phenol**	108
11	14.193	**M-Cresol**	108
12	15.250	**2-Methyl benzofuran**	131
13	16.817	**Dihydrobenzofuran**	120
14	18.043	Naphthalene	128
15	19.903	**2-Ethyl benzofuran**	131

Part 3 Effect of mdified catalysts on Deoxidizing degree

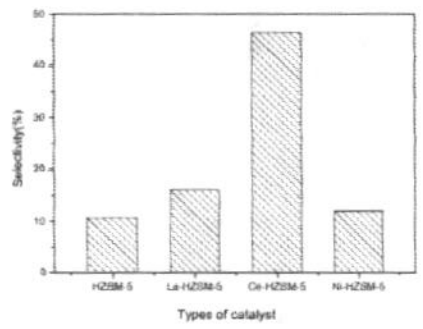

Fig. The selectivity of aromatic on catalysts

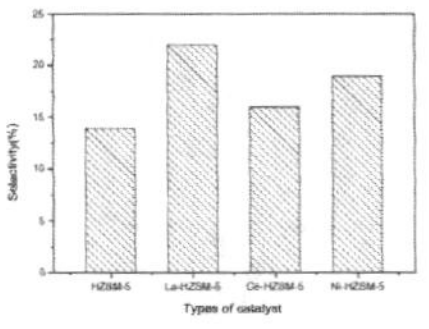

Fig. The selectivity of partial deoxidizing products on catalysts

- Selectivity of aromatic increased, Ce-HZSM-5 was the best.

- Selectivity of partial deoxididation increased, La-HZSM-5 was the best.

Part 4

Conclusion

Part 4 Conclusion

(1) Organic liquid yield increased, coking yield decreased, gas yield decreased over La/Ce/Ni modified HZSM-5 zeolites.

(2) Under the pyrolysis condition of 450℃, WHSV =4 h^{-1}, the selectivity of aromatic increased, Ce-HZSM-5 was the best, 46.37%; the selectivity of partial deoxididation increased, La-HZSM-5 was the best, 21.97%.

(3) Weak acid was increased and strong acid was decreased over La/Ce modified HZSM-5 zeolites, So the anti coking property and the ability of deoxididation were improved.

Part 5

Acknowledgments

Part 5 Acknowledgments

The authors thank the 863 Program (Grant No. 2014AA022103) and Doctoral Fund of Ministry of Education of China (Grant No. 20103317110001).

Part 6 References

Part 6 References

[1] Eoin Butler; Ger Devlin; Dietrich Meier; Kevin McDonnell , A review of recent laboratory research and commercial developments in fast pyrolysis and upgrading [J]. Renewable and Sustainable Energy Reviews, 2011, 15(8): 4171-4186.

[2] Fengwen Yu; Longchao Gao; Weijin Wang; Guodong Zhang; Jianbing Ji. Bio-fuel production from the catalytic pyrolysis of soybean oil over Me-Al-MCM-41 (Me-La, Ni, or Fe) mesoporous materials[J]. Journal of Analytical and Applied Pyrolysis, 2013, 104:325-329.

[3] W T Tsai, M K Lee, Y M Chang. Fast pyrolysis of rice husk: Product yields and compositions[J]. Bioresource Technology, 2007, 98(1): 22-28.

[4] Sathish K. Tanneru, Philip H. Steele. Direct hydrocracking of oxidized bio-oil to hydrocarbons[J]. Fuel, 2015,154: 268—274

[5] A.V. Bridgwater. Review of fast pyrolysis of biomass and product upgrading[J]. Biomass and Bioenergy, 2012, 38:68-94.

[6] YANG Wen-yan, ZENG Yan, LUO Jia, TONG Dong-mei, QING Ren-wei, FAN Yong, HU Chang-wei. Production of bio-oil by direct and catalytic pyrolysis of Nannochloropsis sp. [J]. Journal of Fuel Chemistry and Technology, 2011, 39(9): 664—669.

[7] Supawan Vichaphund ; Duangdao Aht-ong;Viboon Sricharoenchaikul;Duangduen Atong. Production of aromatic compounds from catalytic fast pyrolysis of Jatropha residues using metal/HZSM-5 prepared by ion-exchange and impregnation methods [J]. Renewable Energy, 2015, 79:28-37.

[8] CHANG Sheng, ZHAO Zeng-li , ZHANG Wei, ZHENG An-qing, WU Wen-qiang, LI Hai-bin. Comparison of chemical composition and structure of different kinds of bio-oils[J]. Journal of Fuel Chemistry and Technology, 2011, 39(10): 746—753.

[9] Feiyan Gong, Zhi Yang, Chenggui Hong, Weiwei Huang, Shen Ning, Zhaoxia Zhang, Yong Xu, Quanxin Li. Selective conversion of bio-oil to light olefins: Controlling catalytic cracking for maximum olefins [J]. Bioresource Technology, 2011,102: 9247–9254.

[10] Ji Hongqiang , Zhang Qiang , Dong Xue, Shan Honghong. PERFORMANCE OF METHANOL AROMATIZATION ON NICKEL MODIFIED HZSM-5 CATALYSTS[J]. Petroleum Processing and Petrochemicals, 2015,46(2):58-63.

[11] Shanling Li;Suping Zhang;Zhanyuan Feng;Yongjie Yan. Coke formation in the catalytic cracking of bio-oil model compounds [J]. Environmental Progress and Sustainable Energy , 2015,34(1): 240-247.

[12] S L González-Cortés, S Rugmini, T Xiao, M.L.H. Green, S.M. Rodulfo-Baechler, F.E. Imbert. Deep hydrotreating of different feedstocks over a highly active Al_2O_3-supported NiMoW sulfide catalyst[J]. Applied Catalysis A:General, 2014, 475:270-281.

[13] Minghao Zhou, Yuan Wang, Yanbin Wang, Guomin Xiao. Catalytic conversion of guaiacol to alcohols for bio-oil upgrading[J]. Journal of Energy Chemistry, 2015,24 (4): 425-431.

[14] Wen Chen, Zhongyang Luo, Chunjiang Yu, Yi Yang, Guoxiang Li, Jixiang Zhang. Catalytic conversion of guaiacol in ethanol for bio-oil upgrading to stable oxygenated organics[J]. Fuel Processing Technology, 2014,126: 420-428.

[15] CHEN Jiao-jiao , CHEN Guan-yi , MA Wen-chao , MA Long-long, WANG Tie-jun, ZHANG Qi, LU Wei. Experimental study of aromatics production from catalytic cracking of bio-oil model compounds. Journal of Fuel Chemistry and Technology, 2013, 41(2): 183-188.

[16] Arakshita Majhi;Y.K. Sharma;Rajaram Bal, Babita Behera, Jagdish Kumar. Upgrading of bio-oils over PdO/Al2O3 catalyst and fractionation[J]. Fuel, 2013,107: 131-137.

浙江工業大學
ZHEJIANG UNIVERSITY OF TECHNOLOGY

THANKS!

464724 Process Modeling of Fluidized Bed Biomass-CO_2 Gasification Using Aspen Plus

Abstract:

Biomass can be converted to biofuels using various pathways which are broadly classified as biochemical or thermochemical. Gasification is a common thermochemical process that converts biomass to a gaseous product (syngas) which can be used for power generation, chemical production and as a fuel. The use of air, steam and oxygen as oxidizing agents in this process has been widely studied and reported. Improved CO concentrations have been observed when CO_2 is added as a co-reactive agent in the gasification of biomass.

In this work, a comprehensive model simulating the biomass gasification process using CO_2 in a fluidized-bed gasifier is developed using ASPEN Plus. The model utilizes additional reactors and mechanisms simulating various steps in the gasification process to predict the effect of operating parameters such as temperature and CO_2 flowrate on biomass conversion and syngas composition. A sensitivity analysis of product gas composition and a comprehensive sustainability assessment is conducted to evaluate the economic, environmental, and social performance of the biomass-CO_2 gasification process.

The results from this model are compared to published gasification experimental data obtained on a bench scale fluidized bed gasifier.

451759 Simulation, Heat Integration and Rectisol-Based Decarbonisation for the Production of Synthetic Natural Gas from Biomass Gasification and Landfill Gases

Abstract:

The outline of this research is to develop a systematic heat-integrated gas purification process, followed by methanation for the production of synthetic natural gas or substitute natural gas (SNG). SNG production from biomass is currently considered as an alternative to reduce dependency on high-priced natural gas import and an opportunity to reduce greenhouse gases by applying CO_2 Capture and Sequestration (CCS) technologies. Biomass is thermally converted (e.g. gasification) to produce CO_2, CO, H_2O and H_2 gases that are cleaned to remove CO_2 before being converted to SNG (methane) via a methanation process.

The focus of this project is to develop a comprehensive simulation framework of heat-integrated decarbonized process suitable for the removal of contaminants from biomass-based produced gases to meet SNG production specifications.

In this study, a detailed heat-integrated steady-state simulation framework for biomass thermal conversion, water gas shift (WGS) reaction, gas purification by Rectisol process and SNG production via methanation process has been developed. The main objective was to explore heat recovery opportunities, cleaner operation by reducing contaminants and optimised selectivity with minimal cost. The simulation framework was constructed using Aspen HYSYS (V.2006) process simulator, which is capable to predict the optimised operating conditions of gasification, methanation and Rectisol process at which containments and solvent loss are minimized. Optimization software (SPRINT) is used for data extraction and analysis of heat integration opportunities across the entire process model.

In this project, a biomass (straw) was initially pyrolyzed to mainly produce a gas stream consisting of CO_2, CO, H_2O and H_2 along with tar and char. These streams were then mixed with landfill gases that comprise of methane (CH_4) (40-60%), CO_2 and H_2S as major components that also considered as major greenhouse gases (GHG). Furthermore, these gases are contaminated with sulphur, nitrogen, oxygen, siloxanes, water, halogenated compounds and many other chemicals which require complex gas clean-up process. This mixture was gasified to produce a gas stream consisting mainly of CO_2, CO, H_2O, H_2 and H_2S. The gas stream was then fed to a water gas shift reactor to increase H_2/CO ratio to 3. This gas mixture was cleaned up by a Rectisol process to remove CO_2, H_2S and other containments. The cleaned gas (CO, H_2O, H_2) is finally fed to a methanation process to convert it virtually to pure methane (CH_4), which is the desired final product (i.e. SNG).

The Rectisol process uses refrigerated methanol as a solvent, where 99% removal of CO_2 and H_2S that are particularly present in landfill gases and product gas from biomass gasification was achieved. Two-level refrigeration cycle was developed to maintain the operating temperature of the overall gas purification process. The methanation route was selected for the production of SNG. Methanation is an exothermic reaction and hence, a comprehensive sensitivity analysis was carried out to optimize the process by effective removal of heat of reaction and to achieve maximum methane (CH_4) formation. This analysis showed that four reactors were used to achieve maximum conversion CO to CH_4 which corresponded to CO conversion of 97.8%. After an in-depth heat integration study of the overall process model, 2.5 MW of steam at 320 °C was generated and 0.45 MW process heat recovery was achieved, which resulted in energy savings.

In summary, this project establishes process simulation framework and heat-integrated design of biomass gasification, gas purification and methanation technology. Extensive heat integration strategies have been established based on the following criteria; maximum heat recovery, cleaner and economical production of SNG. The research focused on determining optimal operating conditions for WGS reactor to maintain H_2/CO ratio of 3, for Rectisol Process to maintain 99% removal of CO_2 and H_2S and for methanation process to maximise the conversion rate to SNG. The following conclusions have been drawn from this research:

1. The overall efficiency of SNG production depends significantly on the energy content of the biomass feedstock.
2. Optimizing the refrigeration system for the Rectisol Process plays a vital role in minimizing the total energy requirement of the whole process.
3. Environmental and sustainability factors should always be taken into account prior to selecting the energy conversion process.
4. To meet the SNG specification regarding composition, relative density, calorific value and Wobbe Index; the methanation process must be optimised for maximum conversion of CO to CH_4.
5. Systematic heat integration of the overall process is essential to reduce the utility requirements and provides opportunity for heat and power generation.
6. The cost of electricity from steam generation and cost of carbon capture were estimated to be a function of biomass feedstock cost. It is concluded from this study that biomass feedstock cost is absolutely critical in determining the overall operating cost.

461719 Production of Synthetic Gas through Agricultural Waste Using Fixed BED Gasifier

Abstract:

Gasification is a process that converts organic fuels or fossil fuels (carbonaceous materials) into synthetic gas. It is referred as a producer gas from agricultural waste using gasifier. Biomass can be obtained from living organisms like wood, waste materials and agricultural waste were to be gasified and convert into synthetic gas by gasification process. The results and experiments of solid waste by gasifier is operating based on stoichiometric calculations referred as three different types of agricultural waste like cotton crop waste, tree barks and tree leafs. The material with higher content of the moisture was tree barks waste that was noted as 13.3% whereas cotton crop waste had 12.23% of moisture content and tree leafs waste had 12.81% moisture content. The study showed that cotton crop waste had almost same production of synthetic gas as of tree barks but the productivity was achieved higher in tree barks which had greater quantity of fixed carbon which was noted as 14.13%. The tree leafs also had higher amount of fixed carbon was noted as 12.85% but the ash content in tree leaf was noted as 18.92 %. The composition of synthetic gas from cotton crop waste was 27.52% CO and 32.17% H_2. The composition of synthetic gas from Tree barks was 28.72% CO and 31.88% $H_{2.}$ The composition of synthetic gas from tree leafs were 26.07% CO and 28.12% $H_{2.}$ As examined the increasing fuelâ€™s demand and consumption in a wide country to complete and fulfill the requirement of fuel and energy to achieve the desired targets and goals in Pakistan through the production of alternative fuels with inexpensive way. An innovative process of synthetic gas from solid waste and agricultural biomass is gasification process to generate SYN gas that can be used as replace of natural gas, LPG, power generation and also used for the production of DME (dimethyl ether) to run the mechanical vehicles in the great performance where compared to other fuel resources.

464729 Production of Levulinic Acid in Integrated Biorefineries from Municipal Solid Waste

Abstract:

Demand for material goods is expected to increase due to rising human population and managing municipal solid waste (MSW) resulting after consumption will be challenging. This will require holistic integration of waste processing for efficient resource use, and minimisation of environmental impacts, through pursuit of a circular economy [1]. Integrated waste biorefinery systems are a promising solution for profitable and sustainable waste valorisation [2]. This paper reports an integrated conceptual mechanical biological chemical treatment (MBCT) system in the form of an advanced biorefinery system, for the production of levulinic acid to increase economic margin of municipal solid waste processing by 110-150% [3]. After mechanical separation recovering recyclables, metals (iron, aluminium, copper) and refuse derived fuel (RDF), lignocelluloses from remaining municipal solid waste (MSW) are extracted by supercritical-water. This lignocellulosic stream is then valorised by a chemical process comprising hydrolysis in dilute H_2SO_4 catalyst producing LA as main product and an effluent stream containing some furfural and formic acid (FA). This effluent and the pulping effluent are anaerobically digested into biogas and fertiliser. Produced biogas (6.4 MWh/t), RDF (5.4 MWh/t), char (4.5 MWh/t) are combusted, heat recovered into steam generation in boiler (efficiency: 80%); on-site heat/steam demand is met; balance of steam is expanded into electricity in steam turbines (efficiency: 35%). The economic margin (product prices – annual cost – MSW cost) of the system is 118 Euro/t MSW, with margin of various products, decreasing in the order chemical > metals > fuel > energy > composting, respectively. Value analysis shows diversion of compost like output into chemical eliminates needs for gate fees. This conceptual system demonstrates how the judicious application of process integration combined with economic value and life cycle analyses allows generating competitive and environmentally friendly municipal solid waste biorefinery systems and can contribute to the development of a circular economy.

[1] Satchatippavarn S, Martinez-Hernandez E, Hang MYLP, Leach M, Yang A. Urban biorefinery for waste

processing. Chemical Engineering Research and Design 2015; 107, 81-90.

[2] Sadhukhan J, Ng KS, Martinez-Hernandez E, Biorefineries and Chemical Processes: Design, Integration and Sustainability Analysis. Wiley, Chichester, UK, 2014.

[3] Sadhukhan J, Ng KS, Martinez-Hernandez E, Novel integrated mechanical biological chemical treatment (MBCT) systems for the production of levulinic acid from fraction of municipal solid waste: A comprehensive techno-economic analysis. Bioresource Technology 2016. *In press.*

464135 Synthesis of Processing Paths Fot the Valorization of Specialty Chemicals in Microalgae Biorefineries

Abstract:

Over the last years, biorefineries are gaining ground and numerous studies have been developed with main focus on the exploitation of lignocellulosic feedstock for biofuels and bioproducts. Advances on synthesis, design and integration of such biorefineries are well established and continue to evolve. Third generation biorefineries are currently in the spotlight, especially due to the plethora of chemicals that can be incorporated in the portfolio of products. Microalgae constitute a very promising feedstock with the ability to utilize CO_2 and convert it to specialty chemicals. Thus, challenges arise as to the probe of such an incentive, which does not only include the synthesis and design of a microalga biorefinery, but also its integration with other types of biorefineries.

The study addresses the challenges in the synthesis of microalgae biorefineries by developing a combinatorial approach, in order to simultaneously screen for potential processing paths and formulate the desired product portfolio. The approach enables high-throughput screening with focus on securing the feasibility and profitability of the incentive. The selection of processes and the formulation of the product portfolio are treated as degrees of freedom. The model takes into consideration feedstocks, processes, intermediate fractions and products in a conceptual form and employs economic criteria for the selection of the optimal path. The study takes into account the stages after cultivation and harvesting of the microalgae, where screening paths of candidate technologies and product portfolios is a matter of necessity due to the numerous available options. Competitive chemistries characterizing microalgae as well as the lack of thermodynamic properties regarding involved chemicals add to the complexity of such designs. Extensive knowledge about the use of suitable technologies and required specifications is imperative, thus the need for experimental data becomes indispensable. The methodology deals with these limitations and practicalities, which derive mainly from the lack of industrial applications to draw knowledge from. Therefore, extensive collaboration with practitioners is of vital importance.

For the purpose of this study, the methodology is applied to a real-life case of a microalga biorefinery undertaken by the D-Factory, an FP7 collaborative project, which is relative to the valorization of high-value chemicals from the microalga Dunaliella. This particular microalga is considered a great candidate for mass outdoor cultivation especially because of its extraordinary tolerance to salt stress and potential environmental contaminants. The addressed methodology deals with encountered limitations and constitutes a stepping stone to the development of a systematic approach for the design of integrated microalgae biorefineries. It provides flexibility due to the high profit of specialty chemicals and guidelines of vital importance for future design analysis.

470586 Solvolytic Conversion of Biorefinery Lignin to Value-Added Products

Abstract:

Next generation biorefineries will produce a significant amount of lignin-enriched residues, which are currently just burnt or underutilized. Therefore, new lignin valorization strategies targeting biorefinery lignin streams are important from technoeconomic standpoint. Herein, we show the solvolytic depolymerization of lignin obtained from ionic liquid (IL) pretreatment/saccharification process into low molecular weight compounds. Joint Bioenergy Institute (JBEI) has recently developed one-pot IL pretreatment / saccharification process using biocompatible and bio-derived ILs, which was found to be effective for biomass pretreatment. Residual lignin recovered after one-pot process was subjected to a depolymerization reaction using a Parr reaction with hydrogen donor solvent. The effects of reaction parameters, including reaction temperature (250 ~ 300 °C), time (1 ~ 4 hrs) and catalyst (Ru/C) on depolymerization behavior of lignin were explored. Mass balance analysis shows that approximately 70 % liquid is obtainable from biorefinery lignin. In-depth analysis will be performed to characterize the final products, which will include GC, GPC and NMR. This study will help to provide insights for developing effective depolymerization pathway of biorefinery lignin.

459504 Depolymerization of Kraft Lignin to Value-Added Chemicals over a MoC_{1-X} /Cu-$MgAlO_x$ Catalyst

Abstract:

Different contents of Mo were impregnated on $CuMgAlO_x$ prepared with a co-precipitation method, and the as prepared material was reduced in CH_4/H_2 flow to obtain the MoC_{1-x}/Cu-$MgAlO_x$ catalyst. Lignin conversion experiments were carried out in ethanol with MoC_{1-x}/Cu-$MgAlO_x$ catalyst of different Mo loading at 300 °C for 6 h. The liquid products were identified with a GCâ€“MS instrument and quantified with a GC-FID instrument. Fig. 1 shows the gas chromatography of the liquid product of lignin depolymerization obtained with a 30 % Mo loading catalyst. About 50 kinds of compounds were detected and the products of ethanol self-reactions were excluded by comparison to a blank reaction. 25 compounds from lignin are categorized into five groups, i.e. alcohols, esters, arenes, phenols and benzyl alcohols, of which the detailed yields are given in Fig. 2. When $CuMgAlO_x$ (Mo loading=0) was used as the catalyst, the yield was relatively low with a 584 mg/g lignin total yield. When MoC_{1-x}/Cu-$MgAlO_x$ was used as catalyst, almost all yields increased with the increase of Mo loading. When Mo loading reaches 30 %, all yields reach the maximum values and the yields of alcohols, esters, phenols, arenes and benzyl alcohols are 319, 768, 2, 207 and 125 mg/g lignin, respectively, with the total yield of 1421 mg/g lignin. This yield is lower than that obtained with MoC_{1-x}/AC as a catalyst as we previously reported. However, the selectivity of MoC_{1-x}/Cu-$MgAlO_x$ is different from that of MoC_{1-x}/AC, which obtained higher yields of arenes and benzyl alcohols and lower yields of alcohols, esters and phenols. The difference in selectivity may be caused by the basicity of the $CuMgAlO_x$.

467188 Glycerol Oligomers Synthesized from Etherification of Glycerol with Zeolite-Supported Catalysts

Abstract:

Glycerol oligomers synthesized from etherification of glycerol with zeolite-supported catalysts

Chin-Kai Chen, Bing-Hung Chen*

Department of Chemical Engineering, National Cheng Kung University,

1 University Road, Tainan 70101, Taiwan

e-mail: bkchen@mail.ncku.edu.tw (BHC)

Abstract

In this work, zeolite-supported Ca/Mg/Sr catalysts were synthesized to produce glycerol oligomers from the catalyzed etherification of glycerin which is the abundant byproduct from transesterification of oils and fats for biofuel, using. Both zeolite ZSM-5 and zeolite LTA were chosen as supports impregnated with Ca, Mg, or Sr, to catalyze the etherification reaction of glycerol to glycerol oligomers, mainly the diglycerol and triglycerol, in the Dean-Stark apparatus as a semi-batch reactor. The properties of these zeolite bases catalysts were characterized with XRD, FT-IR, SEM, BET, TGA and TPD. In addition to the heterogeneous zeolite based catalysts, homogeneous catalysts, both acid (sulfuric acid) and base (NaOH), were tested to provide the benchmark. The chemical structure of the product from etherification of glycerin were analyzed with GC-MS, GC-FID and ^{13}C-NMR.It It indicated that diglycerol and triglycerol, either in linear or cyclic form, were dominant products. Moreover, the yield of polyglycerols (diglycerol + triglycerol) could exceed 85% after 6 h of reaction at 260¢XC using Ca-impregnated zeolite ZSM-5. The activation energy of 107.71 kJ/mol of the pseudo-first order reaction confirmed a high energy barrier in glycerin etherification.

Keywords: *Glycerin, Diglycerol, etherification, Zeolite ZSM-5, impregnation.*

Figure 1: The XRD patterns of Zeolite ZSM-5 impregnated with $Ca(NO_3)_2$ solutions

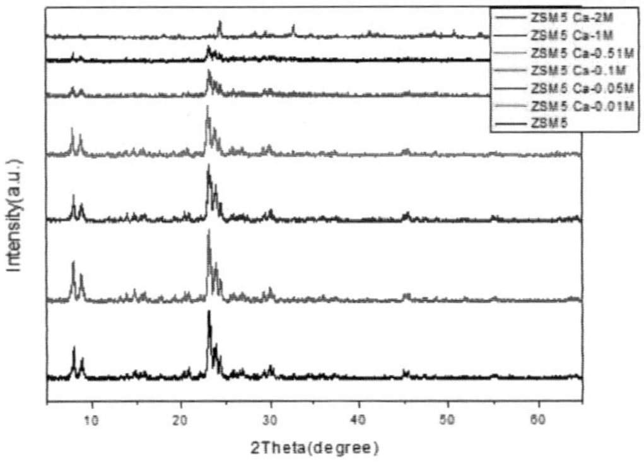

Figure 2: Etherification of glycerin catalyzed by zeolite ZSM-5 impregnated with alkaline earth metals. (Reaction conditions: (1) catalyst loading = 2 wt% of glycerin, (2) reaction time = 8h, (3) reaction temperature = 260¢XC, (4) N_2 flow = 50 sccm).

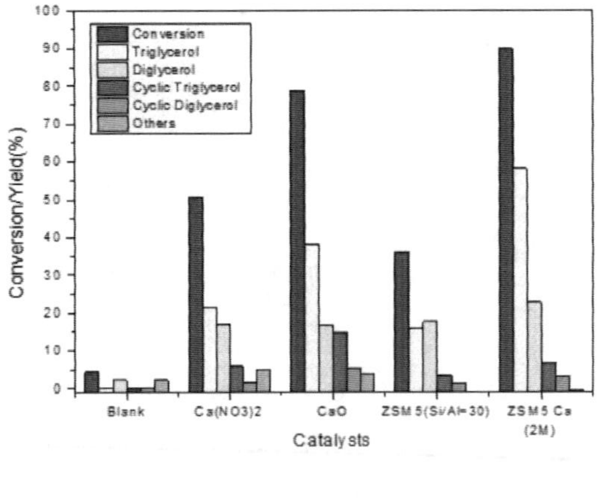

465940 Influence of Pyrolysis Temperature on Characteristics and Aromatics Adsorption Capability of Magnetic Biochars Derived from Rice Straw Pyrolysis Oil Distillation Residue

Abstract:

Influence of pyrolysis temperature on characteristics and aromatics adsorption capability of magnetic biochars derived from rice straw pyrolysis oil distillation residue

Hao Li, Shuqian Xia*, Peisheng Ma

Key Laboratory for Green Chemical Technology of State Education Ministry, Collaborative Innovation Center of Chemical Science and Engineering (Tianjin), School of Chemical Engineering and Technology, Tianjin University, Tianjin, PR China

The bio-oil, derived from rice straw fast pyrolysis, deservedly offers an opportunity as alternatives to replace fossil fuels due to its renewable characteristics. Unfortunately, pyrolysis oil is an unstable product and the physical and chemical properties are changing during storage, which limit its application. Distillation as an effective method for upgrading biomass fast pyrolysis oil has been gained great attention. Although the high-value chemicals or high-quality fuel were obtained from the pyrolysis oil distillation, the distillate yield did not reach 100% and a fraction of solid residue was formed after pyrolysis oil distillation. In this work, the magnetic biochars were synthesized by pyrolyzing the ditillation residue derived from rice straw pyrolysis oil at different temperature of 400, 600 and 800 °C. The effects of pyrolysis temperature on the physicochemical and morphological properties of magnetic biochars were investigated. The experimental results of TGA, FTIR, SEM and XRD indicated that the increase in pyrolysis temperature led to the formation of magnetic biochar with a stable aromatic carbon structure. When the pyrolysis temperature was increased up to 600 °C, the surface area and pore volume of magnetic biochar were enhanced, while a further increase in the pyrolysis temperature lowered the value of surface area and pore volume due to pores widening or blocked pores in the magnetic biochar during pyrolysis. Experimental studies related to the adsorption behaviors of the different magnetic biochars toward various aromatic contaminants (i.e., anisole, phenol and guaiacol), key affecting factors and the underlying mechanisms proposed to explain the adsorption behaviors, have been investigated. The Freundlich adsorption model could describe successfully the adsorption performance and the used magnetic biochars could be easily separated from the aqueous solutions by an outer magnet. An analysis of adsorption mechanisms suggests that the specific surface area and pore volume were the dominant mechanism for adsorption of anisole and phenol on the magnetic biochars while the adsorption of guaiacol on the magnetic biochars were controlled by the ¦Ð-¦Ð electron donor¨Cacceptor (EDA) interaction. Furthermore, compared with phenol and guaiacol, the higher adsorption capacity of anisole can be attributed to the high hydrophobicity (or low polarity) of magnetic biochars.

472208 Evaluation of the Cactus Based-Mucilage As an Alternative Natural Dispersant on Toxicity, Effectiveness, and Surface Tension of Oil-in-Water Emulsion

Abstract:

Cactus based-mucilage has the advantage of being biodegradable and can be obtained through sustainable agriculture. This functional natural surfactant was extracted from the *Opuntia ficus indica* cactus plant, and the extraction process yielded two types of mucilage: the non-gelling extract (NE) and the gelling extract (GE). The properties of the cactus plant-based mucilage were studied on the toxicity, surface tension, droplet size and stability of dispersed crude oil in water emulsions (O/W emulsions). The toxicity of the mucilage and conventional dispersants (Corexit 9500A) were evaluated by a standard EPA toxicity test using Daphnia magna colonies exposed to both NE and GE mucilage extracts in concentrations ranging from 0 to 2000 mg/L for 48 hours. It has shown that mucilage can be classified as practically non-toxic to the Daphnia magna colonies with a LC_{50} above 2000 mg/L. The LC_{50} of Corexit 9500A was below 2 mg/L, which is moderately toxic to the Daphnia magna colonies. Cactus mucilage extracts and conventional dispersants has been evaluated by testing the surface tension and measuring the droplet size of O/W emulsions under different concentrations, including three concentrations of oil (3, 6, 30% v/v) and several dispersant to oil ratios (1:1, 1:3, 1:10). Synthetic seawater and surrogate oil from BP were used. The results of the dispersion obtained using mucilage extracts were compared with using Corexit 9500A. The surface tensions of emulsions with cactus mucilage were similar as with Corexit 9500A. The average droplet size in the systems with cactus mucilage (1.5% of NE has generated 5 μm droplets) were smaller when compared with the systems using the Corexit 9500A (1.5% of Corixit 9500A has formed 6.2 μm droplets). Smaller droplet size implies higher emulsion stability, and may improve the biodegradation of oil. An improved laboratory dispersant testing protocol developed by EPA, baffled flask test, was performed to measure the dispersion effectiveness at DOR 1:20, 1:50, and 1:100. The effectiveness of DOR at 1:20 and 1:50 were around 35%. Cactus mucilage is readily available through sustainable agriculture, it is inexpensive to extract, and has exhibited a long shelf live. Therefore, cactus mucilage can be an alternative technology to mitigate the damage that oil may cause to the aquatic ecosystem and minimize undesired effects associated with the use of synthetic dispersants in oil spills.

462488 Optimization of UF Backwash with Real-Time Control of Coagulant Dosing

Abstract:

Over the past two decades, reverse osmosis (RO) has emerged as the leading technology for seawater and brackish water desalination. However, membrane fouling remains a major challenge for robust RO operation. Hence, effective RO feed pre-treatment is vital to long-term operation of RO desalination plants. Amongst existing RO pre-treatment technologies, ultrafiltration (UF) has proven to be effective in producing consistently high quality filtrate for RO feed. One of the major advantages of UF membrane pre-filtration is the ability to reverse flow through an UF module (a process known as backwash) and thereby removing the foulant layer. After UF backwash, a portion of the foulant may remain on the membrane (i.e., not removed by backwash) as irreversible foulant layer. Irreversible fouling is typically removed through chemical washing which can significantly increase RO feed pretreatment. UF filtration performance can be increased via coagulant dosing of the raw feed water. Studies have also shown that UF backwash effectiveness improves as coagulant dose increases up to a certain threshold, beyond which UF backwash effectiveness no longer improves via additional coagulant dosing. Optimization of coagulant dosing is therefore critical in order to reach both optimal UF and backwash effectiveness while reducing coagulant consumption. However, real-time coagulant dose optimization is a major challenge when feed water quality is temporally variable both in the short term as well as seasonally. Under such conditions coagulant dose optimization via static jar tests is inappropriate. Therefore, in the present work, a novel method is introduced to allow real-time adjustment of coagulant dose so as to ensure effective UF backwash even under conditions of changing water feed quality. In the present method, the change in initial UF filtration resistance (post backwash) is used as a metric of UF backwash effectiveness. The above information is utilized by a coagulant dose controller designed to utilize the above metric to determine whether the coagulant is being underdosed or overdosed, and correspondingly adjust the coagulant dose to the UF feed. The controller was implemented in an integrated UF-RO seawater desalination pilot plant (capable of 18,000 GPD production of fresh water). The controller performance was evaluated, in a field study in California, with respect to its performance, including during a storm event in which raw seawater feed quality rapidly worsened, and its ability to reduce coagulant consumption. Results of the study demonstrated that the present approach enabled system operation that was able to adapt to worsening feed water quality conditions while reducing coagulant dose by up to ~29% relative to operation without real-time coagulant dose optimization.

451292 Performance Analysis of Texas Eagle Ford Shale Oil Hydro Fracturing Produced Water Treatment Process

Abstract:

Performance Analysis of Texas Eagle Ford Shale Oil Hydro Fracturing Produced Water Treatment Process

Chris Skrivanos, Nico Dwarica, Cameron McKay, John Allen Floyd and Mahbub Uddin

Trinity University, San Antonio, Texas

Extended Abstract

With the rise of horizontal drilling, hydraulic fracturing has become increasingly common and it is estimated that 90 percent of new wells drilled in the United States are hydraulically fractured. Hydraulic fracturing involves the injection of a very large quantity of high pressure liquid to fracture the rock formation and pry open the fracture. This fracturing allows natural gas/oil to flow freely from the formation into the well for collection. The components used in the hydraulic fracturing process consist primarily of water, sand, salts, gel, diluted acid, surfactant corrosion inhibitor, scale inhibitor, and other additives to aid in the hydro fracturing process. During the hydro fracturing operation when the pressure used to inject the fracturing fluid into the well is released, some of the fluid returns to the surface; this fluid is called "flowback" water. Once gas production begins at the well, all wastewater emerging from the well is called "produced water". Both types of wastewater – flowback and production phase water – contain potentially harmful constituents. These constituents can be broadly grouped into several principle categories: salts, organic hydrocarbons, metals, chemical additives and some naturally occurring radioactive material. Because of these constituents, shale oil/gas wastewater should not be released into the environment without adequate treatment. The purpose of this paper is to report the performance analysis of a water treatment process of Texas Eagle Ford Shale oil hydro fracturing produced. The treatment process consists of four primary unit operations: Adsorption column, Reverse osmosis, Ion-Exchange and Pump. The produced water is first pumped through the adsorption column, then a second high-pressure pump which then forces the water through the reverse osmosis membrane. The permeate of the membrane then flows through two ion exchange columns and received in a collection bin.

Adsorption is the adhesion of atoms, ions, or molecules from a fluid or dissolved solid to an insoluble surface and is used to remove suspended organics like oil and grease as well as dissolved organics like benzene and toluene. Our adsorption column consists of a cartridge and housing unit rated for operating pressures of 90 psi and temperatures no higher than 125 °F. The cartridge was 10" tall and accepted 10" x 4.5" cartridges. Reverse osmosis filters work on the principle of membrane divergence. An extremely high pressure (800 psi in our process) is applied at the housing entrance that forces the fluid through the fine membrane. Due to the pressure spike, particles that are small enough pass through the fine holes in the filter transfer, while the larger particles are rejected and thus removed from the membrane. In our application, the reverse osmosis membrane rejected the ions and other larger particles like metals, minerals, and organics, allowing only water and smaller particles to remain. The membrane we used is a FilmTec Sea Water Membrane designed to handle increased salinity (greater than 10,000 ppm). The reverse osmosis unit operation used had a flow ratio of permeate to retetante of 73 to 27, which was determined by a series of mass balances across the unit operation. We used a mixed ion exchange in order to remove any remaining positive or negative ions. Our process consists of two ion exchange columns in parallel, both of which were 20" vertical columns housed in separate units. The process uses two pumps. The first pump transferred the water from the tank to the adsorption column at a low pressure. The second pump primed the system to the operating pressure, 800 psi, to transfer the produced water through the reverse osmosis membrane and the ion exchange.

Each water sample analyzed was obtained from one of 4 major positions. Position 1: the untreated water, Position 2: after the activated carbon operation, Position 3: after the reverse osmosis operation and Position 4: after the ion exchange operation. Produced water processed by the system once is defined as first iteration treated water. Produced water processed by the system twice is defined as second iteration treated water. The first iteration sample collection began after flowing untreated produced water through the system for 60 seconds. The second iteration was run identically to the first iteration except produced water that had previously been treated by the system was used instead of untreated produced water. Once collected, samples were labeled according to the iteration, position and the trial number of extraction.

In order to assess the effectiveness of each unit operation and iteration in the wastewater treatment system, a series of experiments were designed to monitor the flow ratio, total dissolved solids present in solution, pH, and change in organic content. Inductively Coupled Plasma Optical Emission Spectrometry (ICP-OES) was implemented as a means to measure the metallic and salt content present in the water at various purification steps. The initial metallic content was determined to be 184.45 ppm, and the initial salt concentration was determined to be 2262.26 ppm for the produced water. A simple pH test using a probe was conducted on samples to monitor the change in acidity of the water as it passed through each unit operation, and the initial pH of the produced water was determined to be 4.25. Gas chromatography measurements were implemented in order to monitor the changing levels of total organic compounds present in the water.

Analysis of Results

First iteration ICP-OES results showed that the initial salt content of the water was 2262.26 ppm and the heavy metal content was 184.45 ppm. The carbon adsorption column removed 99.01 percent of the initial salt content and 98.5 percent of the initial heavy metallic content. The reverse osmosis membrane removed an additional 0.14 percent and 0.16 percent of the initial salt and heavy metal content respectively. After the ion exchange column approximately 100 percent of the total salt content was removed and 99.84 percent of the total heavy metals were removed, leaving a heavy metal concentration of 0.29 ppm and the salt concentration of essentially 0 ppm. Analysis of the second iteration treatment indicated the carbon adsorption column added 0.37 ppm of heavy metal and 2.09 ppm of salt to the water. The reverse osmosis membrane removed 0.29 ppm of heavy metals and added 0.85 ppm of salt content. The ion exchange column removed 0.04 ppm of heavy metals and 1.55 ppm of salts, bringing the final concentration of the second iteration treated water to 0.32 ppm of heavy metals and 1.39 ppm of salts. The first iteration pH results showed that the produced water had an initial pH of 4.25. The carbon adsorption column increased the pH to 8.26, the reverse osmosis membrane lowered the pH to 7.24, and the ion exchange column lowered the pH to the final value of 5.31. Second iteration treated water pH results showed that the carbon adsorption column increased the pH from 5.31 to 8.27. The reverse osmosis membrane decreased the pH to 7.24. The ion exchange column brought the pH to a final value of 6.49. The produced water contained a plethora of unknown organic constituents and it would be unpractical to identify each individual constituent to generate a calibration curve. So instead of measuring the changing concentrations of each constituent, the GC measurements were used to observe the change in the overall organic content in the water, relative to position 1 of the first iteration which is representative of the initial organic content of the produced water. The first iteration GC results showed that the carbon adsorption column removed 48.4 percent of the initial organic content, the reverse osmosis membrane removed an additional 28.7 percent of the initial organic content, and the ion exchange column removed 6.4 percent of the total organic content; the system overall removed 83.5 percent of the initial organic content.

Conclusions and Recommendations

The produced water treated in our process in a single iteration met the EPA standards for acceptable levels of metallic and salt content for reclaimed water, however the pH is at the lower end of the standard. The first iteration treatment reduced the metallic content by 99.84 percent, reduced the salt content by essentially 100 percent, raised the pH from 4.25 to 5.31 and reduced organic content by 83.5 percent. With a second iteration, the treated water pH is within the acceptable range for reclaimed water. The second iteration was shown to be unnecessary since the metallic and salt concentration changed negligibly, and the organic content increased. The pH was raised to acceptable levels, but the addition of a controlled injection pH adjustment unit operation would allow the pH to be raised without the need for a second iteration, making the produced water treatment process more efficient. This change to the process would generate treated water that meets the EPA drinking water standard. Our analysis indicates that the wastewater treatment processes has applications in commercial and residential areas where hydraulic fracturing is common, to treat wastewater that meets EPA standards.

461844 Macroalgae and Chitosan-Macroalgae Biocomposite As Potential Adsorbents of Water-Soluble Hydrocarbons: Effect of pH, Organic Matter and Ionic Strength

Abstract:

Macroalgae and Chitosan-macroalgae biocomposite as potential adsorbents of water-soluble hydrocarbons: Effect of pH, organic matter and ionic strength.

Abstract

Nowadays the recovery of low-molecular aromatic hydrocarbons (HC) released into natural bodies of water continues to be a challenging task. These contaminants bring sever consequences to the environment and to the human health. The oil spill cleanup strategies are primarily design to deal with the heavy fractions accumulated in the water surface. Unfortunately, very limited information is available regarding the treatment of dissolved fractions. Biosorption on macroalgae (Ma) biomass seem to be a potential alternative to overcome disadvantages associated to high-priced costs and complexity of activated carbon production. Three seaweeds of a representative sample of brown, green and red macroalgae (*Macrocystis pyrifera*, *Ulva expansa* and *Acanthophora spicifera*, respectively) were evaluated to remove benzene and toluene from water, which are the most soluble hydrocarbons. Toward this objective, the influences of temperature and ionic strength were also determined. Raw biomasses were characterized by different physical and chemical techniques to assess their potential as biosorbents and the mechanisms involved. The organic and inorganic dissolved fractions were also considered. The characterization techniques mainly identified carbohydrates, lignin and proteins. *M. pyrifera* biomass registered the highest removal capacities: about 112 and 28 $mg \cdot g^{-1}$ for benzene and toluene, respectively, probably attributed to alginate and lignin content. It was found that the hydrocarbons biosorption affinity was not affect up to an ionic strength of 0.6 M, due to the biosorbents pore occlusion by water clusters. For benzene and toluene nonpolar solutes, the biosorption mechanism is an addition of simultaneous interactions between the sorbents and the chemical constituents of the cell wall, presumably by London (dispersion) forces onto â€“OH and â€“NH active sites from polysaccharides and proteins and hydrophobic interactions from lignin and lipid fractions.

Despite these advantageous properties, the physical characteristics (small particle size, low strength and density) of such biomaterials are not viable for a continuous process operation and make biomass difficult to apply. Therefore the development of innovative low-cost techniques of immobilized biosorbents with special attention to increase their effectiveness in the biosorption process is mandatory. The biopolymer-biopolymer interactions between positive and negative macromolecules can enhance the compatibility and stability of biosorbents, and these new structures are called polyelectrolytes complexes (PEC). The linear amino polycationic compound of chitosan (Ch) possess remarkable quantity of amine and hydroxyl groups that interacts with negatively charged substances with excellent chelation behavior and high affinity toward pollutants like metals and dyes. On the other hand pectin (Pe) is a negatively charged polysaccharide composed mostly by esterified D-galacturonic acid residues linked together by Î±-(1-4) chain. Pectine molecule is also used in the food industry for its well-known thickening, stabilization and gelling properties.

Â The main objective of this research was to explore the performance of Ma and PEC complexes of different proportions of Ma/Ch/Pe on biosorption capacity of the two main soluble hydrocarbons (benzene and toluene) in water. The biocomposite synthesis was optimized by application of the factorial design and response surface methodology. Moreover, a through chemical and physical characterization analysis, by textural properties, potentiometric titrations, elemental content, chemical stability, KBr-FT-IR and TGA analysis, were performed to explain the adsorption mechanisms.

458236 Concentrating 17 α -Ethinyl Estradiol (EE2) Using PES Hollow-Fibre Membranes: Experimental Exploration and Molecular Modelling

Abstract:

The 17Î±-ethinyl estradiol (EE2) is a synthetic estrogen derived from 17Î²-estradiol (E2), the major endogenous estrogen in humans. EE2 is widely used as an active ingredient for birth control and other hormonal treatments. EE2 is excreted into the sewage and the environment mainly via urine. Wastewater treatment plants are usually ineffective in removing EE2 from water which results in the release of EE2 into the environment. The concentration of EE2 in surface water bodies are typically at sub-ppb levels. Even at such low concentration, several studies have reported negative environmental effects of EE2 including feminization of aquatic animals. At present, it is not known if EE2, at such low concentration, has any adverse effect on human or other mammals over short and long term exposure. Developing novel strategies for the efficient removal and treatment of EE2 from water has become an imperative in water reclamation and reuse. Here, we report a novel approach whereas a widely used polyethersulfone (PES) membrane was used to concentrate EE2 through adsorption and solvent elution. The concentrated EE2 can then be further treated with high efficiency using approaches such as biodegradation. Molecular mechanisms underlying the interaction between EE2 and PES membrane was elucidated using molecular dynamic simulation.

454263 Natural Organic Matter Fouling Behaviors on Superwetting Nanofiltration Membranes

Abstract:

Natural organic matter fouling behaviors on superwetting nanofiltration membranes

Linglong Shan, Hongwei Fan, Hongxia Guo, Shulan Ji and Guojun Zhang

Beijing Key Laboratory for Green Catalysis and Separation, College of Environmental and Energy Engineering, Beijing University of Technology, Beijing, China. Email: shanlinglong@126.com

In natural water, natural organic matter (NOM) components have been identified as precursors of disinfection by-products. Nanofiltration has been widely recognized as a promising technology to remove NOM from natural water. Currently, the most challenging aspect is the often inevitable occurrence of membrane fouling, which is caused by NOM accumulation on the membrane surface, as in NOM removal in drinking water treatment and production. NOM fouling is considered to be governed by the interface relationship between NOM and the membrane surface in aqueous solution. Much work has been carried out, but the effect of hydrophilicity/-phobicity of NOM and the membrane surface on fouling behavior is still a controversial issue. [1-7] The controversy originates from the unilateral consideration of membrane wettability or foulant hydrophilicity/-phobicity, when in fact both need to be considered. In our study, both superhydrophilic and superhydrophobic membranes with stable structure have been successfully obtained using two facile hybrid approaches (see Fig. 1). The fouling behavior of NOM on superwetting nanofiltration membranes has been extensively investigated using humic acid (HA: model foulant) and real natural water. The extended Derjaguin-Landau-Verwey-Overbeek (XDLVO) approach and nanoindentor scratch tests suggested that the superhydrophilic membrane had the strongest repulsion force to HA due to the highest positive ¦¤G^{TOT} value and the lowest critical load. Excitation emission matrix analyses of natural water also indicated that the superhydrophilic membrane showed resistance to fouling by hydrophobic substances and therefore high removal thereof. Conversely, the superhydrophobic membrane showed resistance to fouling by hydrophilic substances and therefore high removal thereof (see Figs. 2 and 3). Long-term operation suggested that the superhydrophilic membrane had high stability due to its anti-NOM fouling capacity, which is crucial for practical application. Based on the different anti-fouling properties of the studied superwetting membranes, a combination of superhydrophilic and superhydrophobic membranes was examined to further improve the removal of both hydrophobic and hydrophilic pollutants. The R_{UV254} and R_{DOC} could be increased to 83.6% and 73.3%, respectively, by a combination of superhydrophilic and superhydrophobic membranes. In view of the hydrophilic and hydrophobic NOM fouling behaviors on superwetting membrane surfaces, the superwetting strategy can be regarded as a very important approach for obtaining a membrane with both high NOM rejection and good anti-NOM fouling properties.

Keywords: Superhydrophilic, Superhydrophilic, Nanofiltration, NOM, Fouling

References

[1] L. Shan, H. Fan, H. Guo, S. Ji, G. Zhang, Natural organic matter fouling behaviors on superwetting nanofiltration membranes, Water res., 93 (2016) 121-132.

[2] L. Shan, H. Guo, Z. Qin, N. Wang, S. Ji, G. Zhang, Z. Zhang, Covalent crosslinked polyelectrolyte complex membrane with high negative charges towards anti-natural organic matter fouling nanofiltration, RSC Adv., 5 (2015) 11515-11523.

[3] M. Ng, S. Liu, C.W.K. Chow, M. Drikas, R. Amal, M. Lim, Understanding effects of water characteristics on natural organic matter treatability by PACI and a novel PACI-chitosan soagulants, J. Hazard. Mater., 263 (2013) 718-725.

[4] J.A. Nilson, F.A. DiGiano, Influence of NOM composition on nanofiltration, J. Am. Water Works Assoc., 88(5) (1996) 53-66.

[5] H. Yamamura, K. Okimoto, K. Kinura, W. Yoshimasa, Hydrophilic fraction of natural organic matter causing irreversible fouling of microfiltration and ultrafiltration membranes, Water Res., 54 (2014) 123-136.

[6] L. Zhang, G. Chowdhury, C. Feng, T. Matsuura, R. Narbaitz, Effect of surface-modifying macromolecules and membrane morphology on fouling of polyethersulfone ultrafiltration membranes, J. Appl. Polym. Sci., 88 (2003), 3132-3138.

[7] A. Tiraferri, Y. Kang, E.P. Giannelis, M. Elimelech, Superhydrophilic thin-film composite forward osmosis membranes for organic fouling control: fouling behavior and antifouling mechanisms, Environ. Sci. Technol., 46 (2012) 11135-11144.

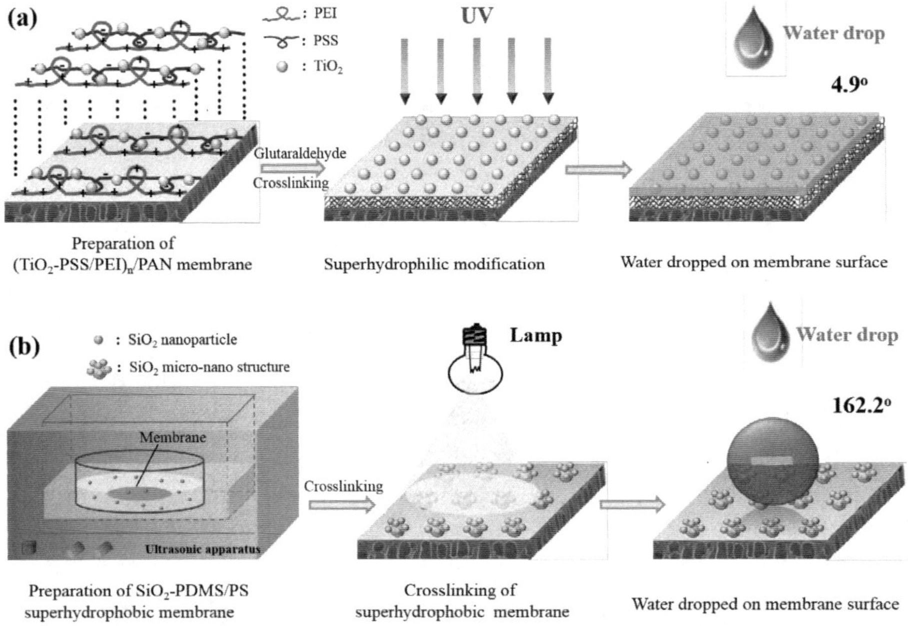

Figure 1. (a) Schematic illustration of the preparation of the superhydrophilic membrane and its surface wettability; (b) schematic illustration of the preparation of the superhydrophobic membrane and its surface wettability.

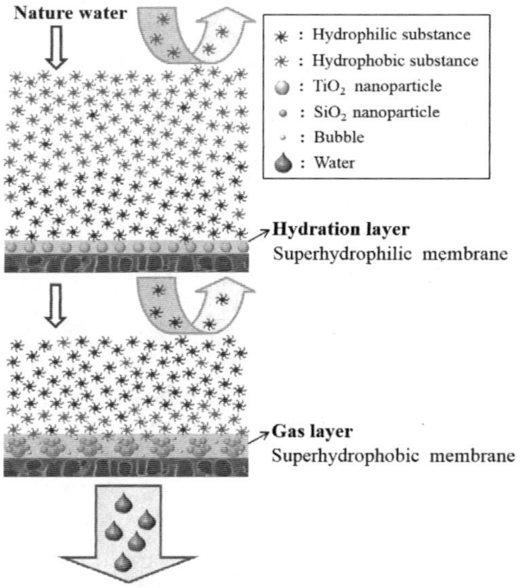

Figure 2. Schematic illustration of different anti-fouling properties of the studied superwetting membranes. Superhydrophilic membrane showed resistance to fouling by hydrophobic substances, while the superhydrophobic membrane showed resistance to fouling by hydrophilic substances.

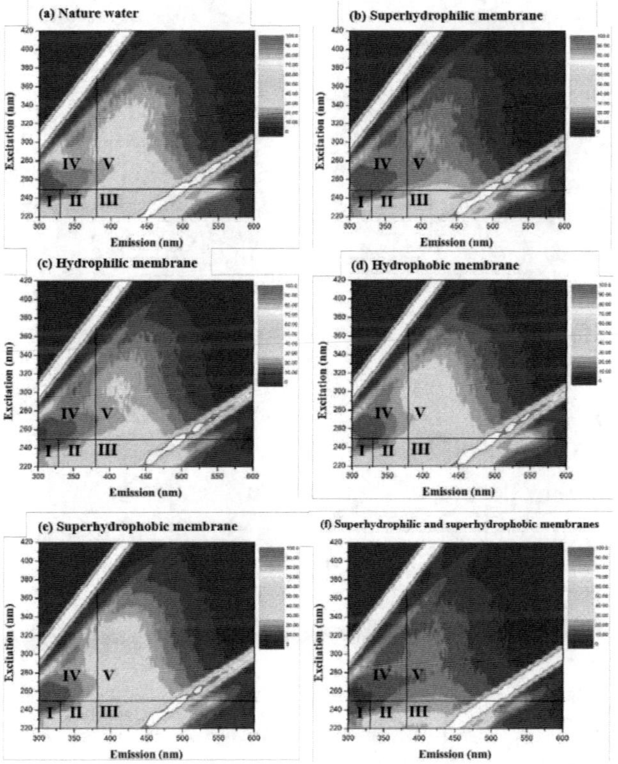

Figure 3. EEMs for (a) natural water; (b) natural water nanofiltrated by superhydrophilic membrane; (c) natural water nanofiltrated by hydrophilic membrane; (d) natural water nanofiltrated by hydrophobic membrane; (e) natural water nanofiltrated by superhydrophobic membrane; (f) natural water nanofiltrated by superhydrophilic membrane-superhydrophobic membrane.

Experimental and Simulation Analysis for Optimization of Solar Desalination System Using Humidification-Dehumidification Method: Open and Close Air Cycle

Nickyar Ghadirinejad [1]***, Farshad Farshchi Tabrizi*** [2]***, Mohammmad Mehdi Zarei*** [3]***, Taher Chegini*** [4]***, Roberto Moreno-Atanasio*** [5]

[1] Chemical Engineering Department, University of Sistan and Baluchestan, Zahedn, Iran
[2] Chemical Engineering Department, Shiraz University, Shiraz, Iran
[3] Chemical Engineering Department, Khazar University, Baku, Azerbaijan
[4] Civil Engineering Department, Khazar University, Baku, Azerbaijan
[5] Chemical Engineering Department, Newcastle University, Callaghan, Australia

Abstract:

Over the last decades, desalination of brackish water and seawater has been studied extensively for producing fresh water. Desalination technologies are divided into two major categories: Thermal and Membrane. Humidification-Dehumidification (HDH) which is the subject of the present research, is one of the thermal technologies. It is inspired by the rain cycle and is based on evaporation and consecutive condensation.

In a series of experiments, the processes of production of fresh water has been investigated under various operational parameters such as temperature and flowrate of cooling water, humidifier intake from water boiler, air flowrate and height of packed bed column. Subsequently, an appropriate mathematical model of the system has been developed and validated using the collected data. Furthermore, based on the developed model a series of simulations has been carried out in ASPEN Plus in order to obtain optimum physical parameters of various parts of the desalination system.

The experimental results showed that under constant heat flux and optimum air flowrate, three parameters increase the production of distilled water the most: increasing flowrate of cooling water, decreasing flowrate of water inlet and increasing temperature. Moreover, although increasing flowrate of cooling water leads to an augmentation in the distilled water flowrate, but this effect is mitigated at high rates.

Furthermore, based on the obtained overall heat transfer coefficients of the humidifier and condenser, the most reliable Height Equivalent to a Theoretical Plate (HETP) model of random packing is found to be Mangers & Ponter. The estimation error of the model in predicting the height of packed bed was less than 4%. Based on the results of this model, closed air system increases the efficiency of the production of distilled water by 15% in comparison to open air. This increase is the result of prevention of wasting of air humidity in the condenser outlet.

Finally, optimum parameters of operational units are obtained by applying a Differential Evolution (DE) strategy. The optimum values of flowrate and temperature of cooling water are 0.0298 m^3/hr, 13 °C, respectively. The best value of flowrate of closed air is found to be 0.008075 m^3/hr and of water intake to be 0.021497 m^3/hr. These values produced fresh water with flowrate of 0.3777 m^3/hr.

Keywords: Solar Desalination, Dehumidification, Humidification, Heat Transfer

1- Introduction:

The rapid growth of industries and population, rivers and lakes pollution problems caused by industrial wastewater and flowing municipal wastewater into freshwater resources have caused a shortage of fresh water [1]. However, it should be noted that the only sources of never-ending water reservoir are oceans; water salinity reduction to overcome the shortage of fresh water, is a more applicable solution [2]. Desalination includes many different methods and required energy for this procedure also encompasses a wide range. The heat produced by fossil fuels and/or the excess heat-released of nuclear reactions, renewable energies such as ocean waves or solar energy, all can supply heat power for water desalinisation plants. Desalination technologies are divided into two categories: Thermal and membrane methods. Water is heated in thermal processes and after evaporation, fresh water is produced via condensation. In membrane methods, thin permeable membranes are used to separate feed into both fresh water flow and more saline flow [3].

In recent years, Humidification-Dehumidification (HDH) system, as a new method has been investigated; through this system, air flows as the intemediate fluid. In fact, this method is a combination of membrane processes and thermal distillation. It operates based on the mass diffusion principles and uses cooling air due to water evaporation and humidification. Fresh water is resulted via dehumidification (condensation of humid air). The obvious benefit of HDH is to be one of the most cost-effective desalination mechanism regarding to terms of low temperature and pressure [4]. The investigated method includes solar collector/s, humidification tower/s (evaporation unit), dehumidification tower (condensation unit) which even to achieve higher efficiency, it could include heat recovery unit as well. In order to increase the efficiency of heat transfer, at first, water is passed through the condenser coil, by heat absorption from hot air, water is pre-heated. Afterward, the pre-heated water is led into a solar heater and an additional heater (auxiliary heater) subsequetly to reach a higher temperature. In the next step, hot water is showered into the humidification tower.

2- Closed-air, open-water (CAOW) water-heated systems:

Alluded humidification tower including random packings/ Bubble Columns/ Wetted-wall Columns [5], showered with hot water, gets totally moist using counter-current air flow from the botton; subsequently, after heat absorption, hot air is guided through the condenser to produce condnesate-water (fresh water). In closed air cycle, after passing the condenser, the air is recycled into the system. HDH method has been studied by Al-Hallaj et al. [6,7], Ben-Bacha et al. [8], Garg et al. [9] ،Nafey et al. [10] ،Nawayseh et al. [11], [12] Muller-Holst.

In order to achieve optimum operational conditions and achieve the best yield, many research projects have been studied [13,14,15,16,17] and various mathematical models have been proposed considering measurement of mass and heat transfer coefficients [13], impacts of input parameters into the HDH system, mass and energy balance equations, Consideration of waste heat during HDH process and finding a solution due to heat recovery [18]. To simulate a system including both a flat-plate water-heater and a double-pass solar air heater, assuming some simplified assumptions, Yildirim and Solmus [19] proposed an optimized mathematical model by using fourth order Runge-Kutta solution method. Cost-estimation of produced fresh water via an elaborated, multiple solar desalination system [20] and via a novel solar HDH system [21] has been investigated.

3-HDH Apparatus:

In the study, the humidifier consisted of emedded random packing (Pall ring type with 25 mm nominal diameter). Water was showered from the top and air wsa flown counter-currently. The coil-condensor was a single 8-meter-long cupper pipe. For circulation of air in side the HDH system, two fans were used, without any direct contact between motors and humid air.

As it shown in figure 1, intake salt water in low temperature (as feed water into the HDH system) entered into the condenser's shell, and after relative pre-heating subsequently entered into the solar heat collector. During experimental research, to achieve higher temperatures an electrically additional heater was used. Heated saline water was showered inside the humidifier. Water flow rate was regulated by passing through some valves and rotameters to reach desired amounts. Air-flow rates adjusted through entire HDH system using fans' regulators. By directly counter-current contact between hot saline water and air in random-packed bed, the air temperature got augmented. Consequently, it tended to embrace greater capacity to absorb more moisture. Passing through the dehumidifier, fresh water was by-passed and collected; outlet air was recycled in closed air system and was blowed from the bottom into the humidifier.

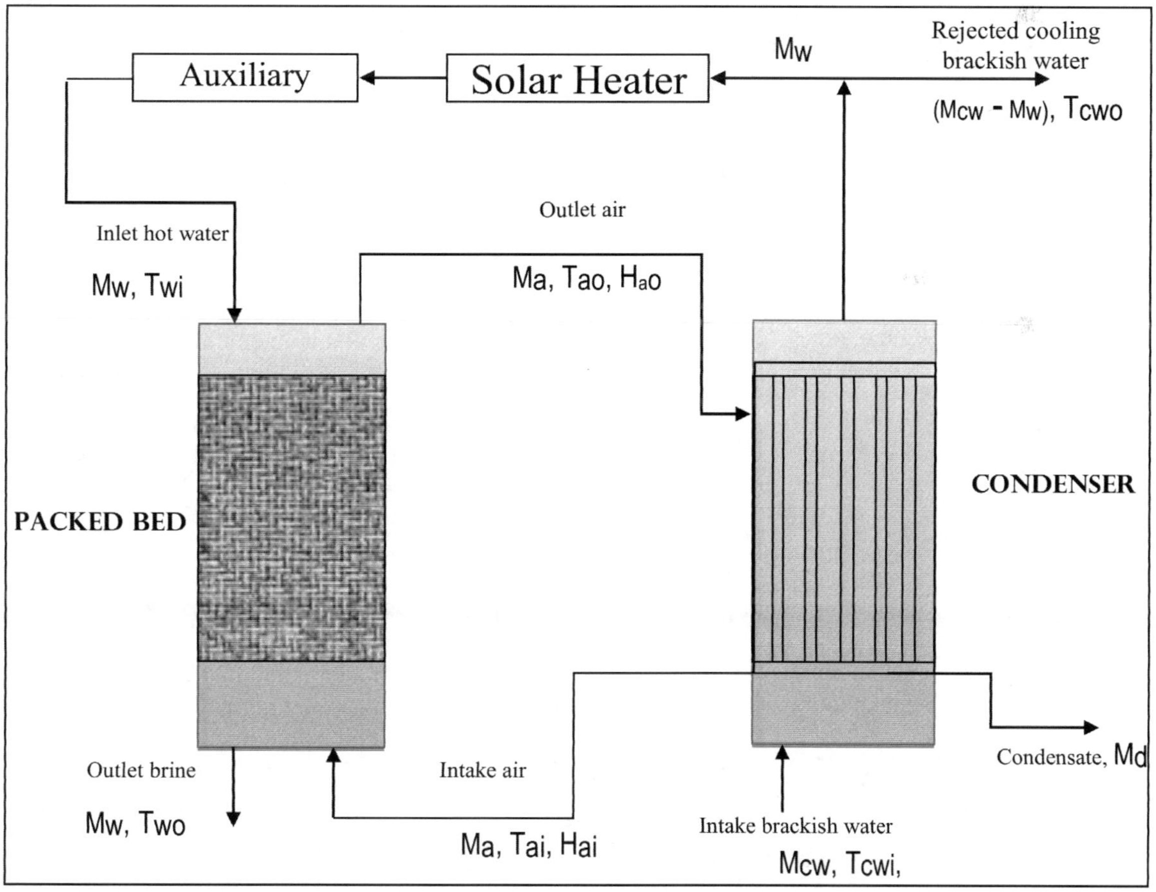

Fig. 1: Layout of HDH system

For experimental data collection, the system was ran continuously for 15 to 20 minutes in the same mode, reaching considered steady-state temperatures and humidity amount in each test; then tests were performed. Each of the experiments were repeated at least three times to ensure their correction.

4- Modeling of humidification-dehumidification system:

In order to determination of diameter and Height Equivalent to a Theoretical Plate (HETP), based on the obtained overall heat transfer coefficients of for the random-packed-bed humidifier and condenser [22], Considering given data about Rasching ring and Pall ring packings [5] following models have been investigated: Shulman et al. [23], Shulman & de Grouff [24], Van Krevelen & Hoftijzer [25,26], Mangers & Ponter [27], Shi & Mersmann [28], Billet and Schultes [29], Onda et al. [30].

The estimation HTEP model of **Mangers & Ponter**:

$$k_L = \frac{(3.22\times10^{8}) Sc_L^{0.5} D_L}{a_s(1-cos\theta)} \left(\frac{\rho_L u_L}{\mu_L}\right)^{\alpha} \left(\frac{\rho_L^{2} g\, h_{pe}}{\mu_L^{2}}\right)^{0.27} \tag{1}$$

As α in the above equation is calculated by:

$$\alpha = 0.49\,[(1-cos\theta)^{0.6} \left(\frac{\rho_L \sigma_L^{3}}{\mu_L^{4} g}\right)^{0.2}]^{0.108} \tag{2}$$

$$k_G = 0.2\,\frac{D_G}{d_c}\,[\frac{\rho_L u_L}{u_L a_p}]^{0.8}\, Sc_G^{\frac{1}{3}} \tag{3}$$

Additionaly, for comparison the results obtained from the mathematical modeling and practical experiments, both physical and chemical operational behaviors of HDH units -at steady state conditions- have been investigated by modeling in ASPEN Plus. Initial feed's conditions to HDH system, constituted 3884.8 ppm salinity, Ca^{2+}, Mg^{2+}, K^{+}, Cl^{-}, HCO_3^-, SO_4^{2-}, NO_3^- Na^{+} aqueous solutions were imported to ASPEN Plus model. Enick's Redlich-Kwong-UNIFAC equations [31] and Pall ring packings were selected for the mentioned processes.

In preliminary calculation, the capacity of existing packing in the dumidifier was assumed 10 HETP. Regrading experimental results, output data from run of each stage in ASPEN Plus were compared.

4-1- Modeling of HDH system in open and close air cycle:

Effects of intake water flow rate and its temperature entering the humidifier, air and cooling water flow rates, and effect of packed-bed height on the yield of desalinated water amounts have been studied by developing mass and energy transfer balances for overally and even on every sinlge unit in the HDH system.

$$Mcw.Cp.Tcwi - (Mcw - Mw).Cp.Tcwo + Q_{Heater} = Mw.Cp.Two + Md.Cp.Ta \qquad (4)$$

Saturated-air, as a main simplifying assumption was considered to solve formulated equations. On the contrary, unlike precedent models, the term of air solubility in outlet saline water, as well as unsteady air and water flow rates recycling the entire HDH system was considered [32].

To improve derived equations and to reach more realistic modeling in the range of $0 \leq S_p \leq 16000$ ppm and $10 \leq T \leq 180$ °C, correlations of seawater density (ρ_{sw}), specific heat (C_p), seawater thermal conductivity (K_{sw}) [32] were applied:

$$\rho_{sw} = 10^3(A_1.F_1 + A_2.F_2 + A_3.F_3 + A_4.F_4) \qquad (5)$$

Which A_n and F_n are specific variables proposed by Isdale and Morris [33].

Evaporation latent heat (λ) (which is T in K degree)

$$\lambda = 2501.89 - 2.407 \times T + 1.1922 \times 10^{-3} \times T^2 - 1.586 \times 10^{-5} \times T^3 \qquad (6)$$

Specific heat coefficient of air (T in K degree)

$$C_{pa} = 1.03409 - 2.84887 \times 10^{-4}.T + 7.816818 \times 10^{-7}.T^2 - 4.970786 \times 10^{-10}.T^3 + 1.077024 \times 10^{-13}.T^4 \qquad (7)$$

To calculate air density with respect to Al-Sahali & Ettouney's proposal [13] for HDH desalination plant, air assumed as an ideal gas.

$$\rho_a = \frac{P.M_{wei}}{R.T} \qquad (8)$$

Thermal conductivity coefficient of air:

$$K_a = -2.276501 \times 10^{-3} + 1.2598485 \times 10^{-4}.T - 1.4815235 \times 10^{-7}.T^2 + 1.73550646 \times 10^{-10}.T^3 - 1.066657 \times 10^{-13}.T^4 + 2.47663035 \times 10^{-17}.T^5 \qquad (9)$$

In order to calculate the amount of dissolved air in the water (dependent on its pressure and temerature) in the humidifier [34] Henry's constants for air solubility in water (K_{sol}) were considered too.

Newton's method was used to solve mentioned non-linear equations, using some most probable intial guesses via experimental data. For optimization of outlet fresh water dependent on Q, Mw, Ma, Tcwi and Mcw values, Differential Evolution (DE) strategy was applied to find optimum values. DE strategy [35], one of genetic algorithm methods, via successive tries to generate correct interpolations was coded in MATLAB software.

In the same method, calculations of open-air cycle in the HDH system were performed by rewriting similar equations as above.

5- Results and discussion:

Keeping constant given thermal energy to the inlet water (including solar heater and additional heater), several sets of experimental data were collected in various conditions. By changing the initial independent parameters, trends of intermediate variables and the final variable (amount of produced fresh water) were evaluated.

5-1 Effect of inlet water flowrate and its temperature on deslination production rate:

At constant thermal energy, inlet water flow rate and its temperature were realized as two dependent vaiables. In this series of tests, parameters of air flow, cooling water flow and packed-bed height were kept constant; and consequently, effects of two mentioned parameters of inlet water entering to the humidifier on the yield of HDH system were investigated (Fig.2). As it is shown, at constant conditions, by increasing the temperature of inlet water, production rate of fresh water increases. It is noticible that during experiments, relative humidty of circulating air was checked continuously and it was near to saturation state (97.33%).

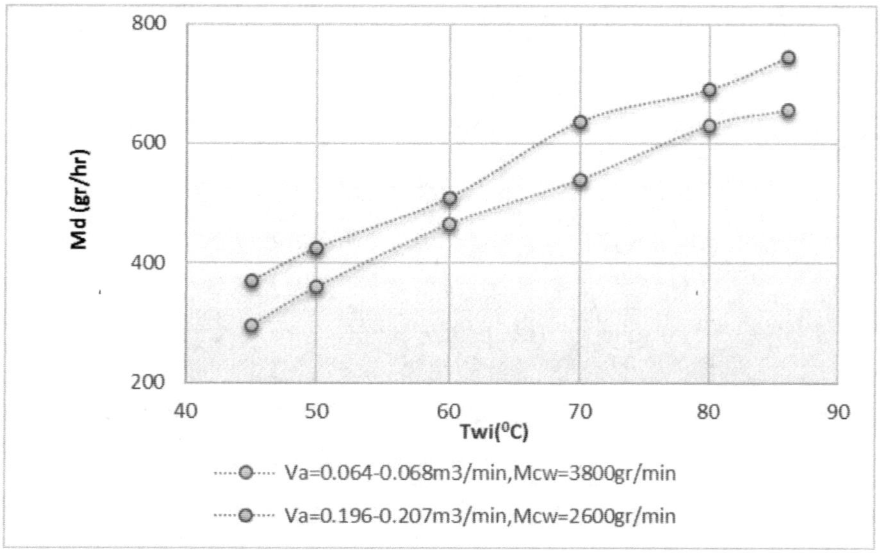

Fig. 2: Effect of temperature of inlet water on produced fresh water

5-2 Survey on air flow rate, considering effects of cooling water flowrate on yield of the HDH system:

As it presented in Figure 3 for three flow rates of circulated-air, the temperature of the exhausted air from the tower is a dependent variable of cooling water flow rate. At constant air flow rate, increasing in water temperature leads to augmentation of air temperature and consequently fresh water production.

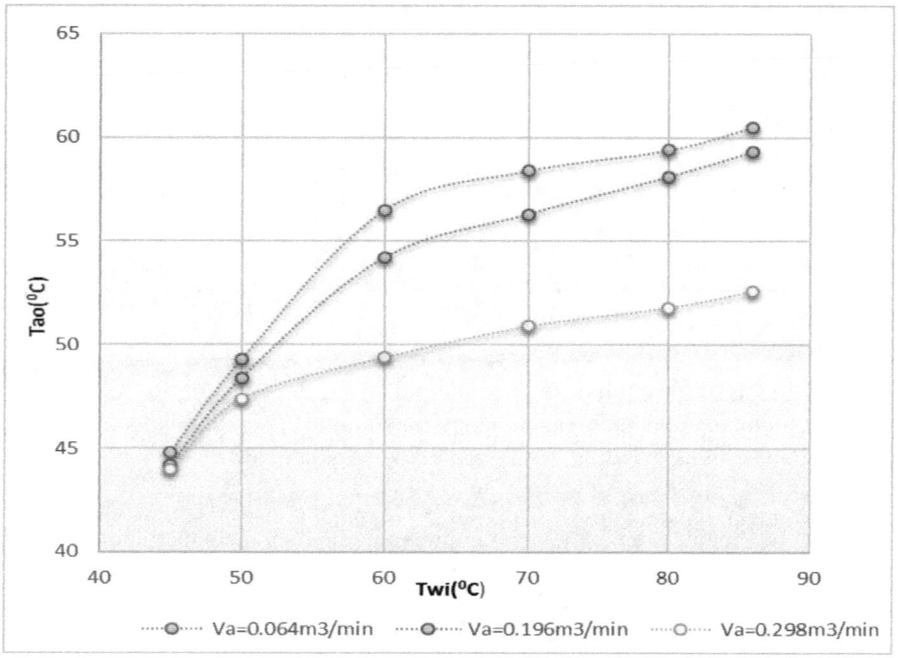

Fig. 3: effects of changes in Twi on Tao

In the present HDH design, to avoid wasting moisture, transfered by exhaust air flow from the condenser, and to spend less energy, close air cycle was contrived. It is noteworthy that the validity of this assumption has been proved by comparing output-results of close and open air models. Experimental tests were conducted in three different ranges (0.064-0.068 $\frac{m^3}{min}$, 0.196-0.207 $\frac{m^3}{min}$ and 0.298-0.310 $\frac{m^3}{min}$). Regarding Fig. 3, 4 and 5, range of 0.196-0.207 $\frac{m^3}{min}$ was determined as the optimal one for the air flow. Obviously, increase of air velocity causes decline of outlet air temperature in the HDH system. Besides, it leads to reduction in air settelment time in the humidifier and finally shrinkage of distilled water.

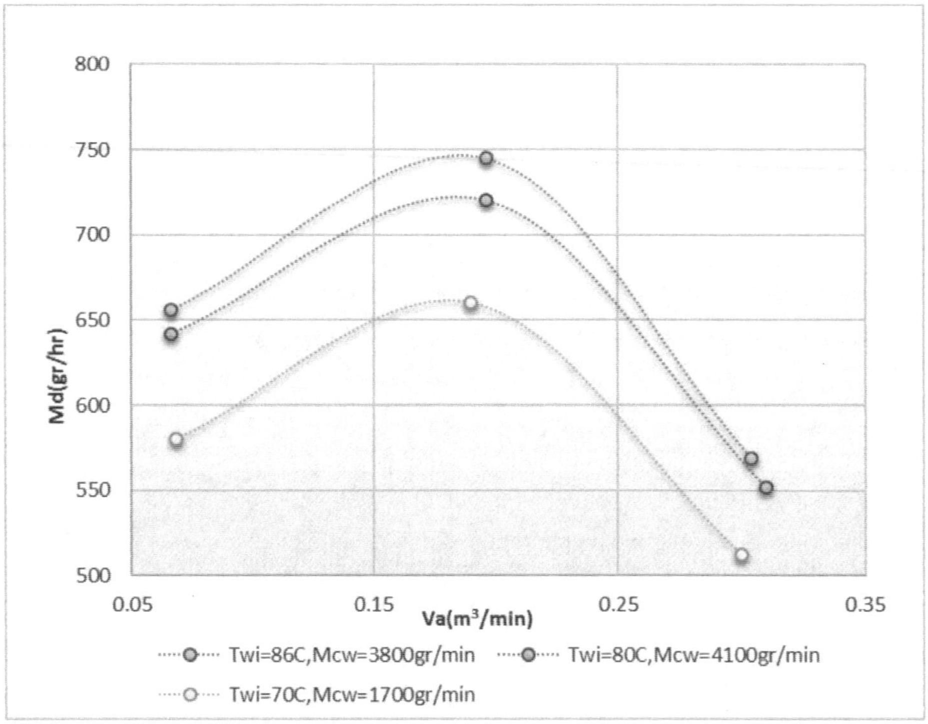

Fig. 4: Trends of M_d in three different air velocities at 70, 80 and 86 °C.

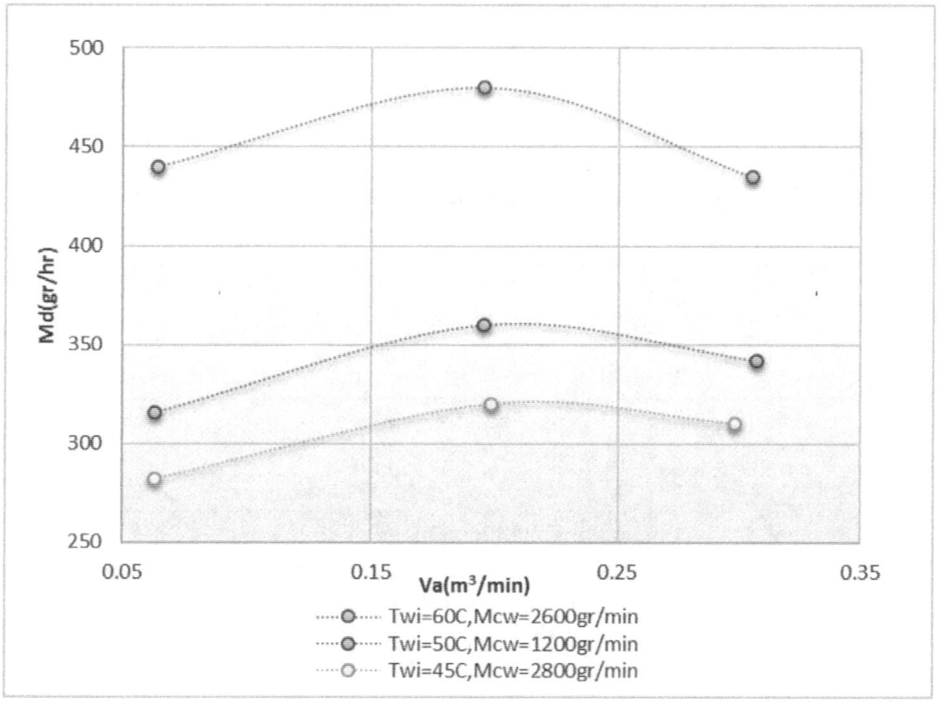

Fig. 5: Trends of M_d in three different air velocities at 45, 50 and 60 °C.

5-1 Distinguishing between results of open and clos air cycle in the investigated HDH system:

This comparibility is shown in table 1. It shows 15% higher productivity of fresh water in close air cycle than production rate in open air cycle. The most probalbe justification for this conlcusion could be waste-heat recovery and lack of moisture loss in the exhaust air from the condenser and recycling to the processes.

Table 1: comparability of 4 samples for open and closed air cycle

Closed cycle				**Open cycle**			
Twi (°C)	Tao (°C)	Tcwo (°C)	Md (m3/hr)	Twi (°C)	Tao (°C)	Tcwo (°C)	Md (m3/hr)
71.6	60.4	32.3	1.70E-04	70.4	56.2	32.1	1.01E-04
91.3	66.5	32.6	1.95E-04	88.4	61.4	31.1	1.50E-04
40.4	41.6	24.9	1.28E-05	37.6	35.1	25.6	1.23E-05
80.2	44.6	33.8	1.71E-04	79.4	39.0	33.2	1.27E-04

5-2 HETP investigation and HDH optimization:

To achieve optimal HETP for humidification tower, several experiments have been executed (Fig. 6). At constant practice conditions, using k_L and k_G equations, the most reliable HETP model has been proposed; Mangers & Ponter's model was determined as the most precise one (Table 2).

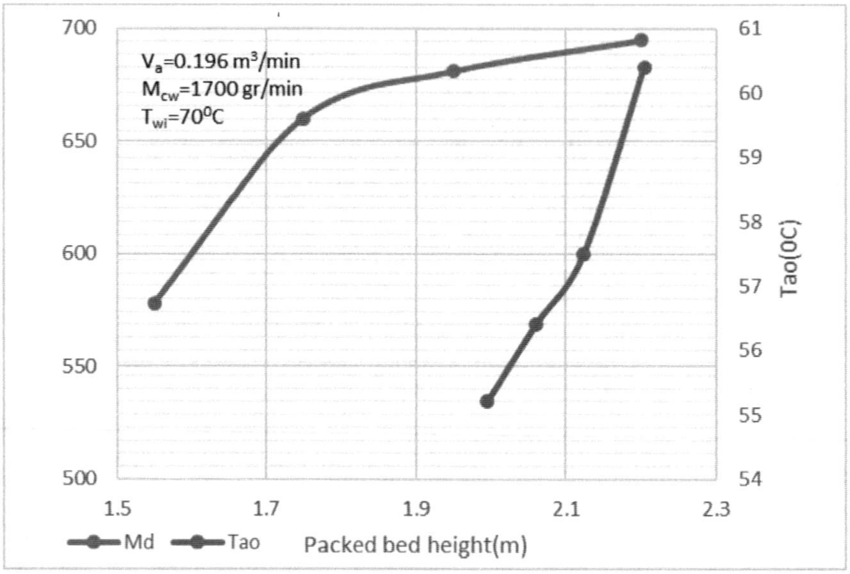

Fig. 6: effects of Packed bed height on M_d $(\frac{gr}{hr})$

Table 2: estimating HETP models

Model	HETP (m)
Shulman et al.	0.042601
Shulman & de Grouff	0.047809
Van Krevelen & Hoftijzer	0.394546
Mangers & Ponter	0.511026
Shi & Mersmann	0.437404
Billet and Schultes	0.407311
Onda et al.	0.660333

Furthermore, due to admit validity of collected data modeling and simulation in ASPEN Plus have been done. At first, 10-equivalent-step was presumed. After every run of functions implemented in ASPEN Plus and comparison between outcome data and experimental values, it was observed that a tower including two or three steps can help us to meet our expected moisture balance (one single run is shown in table 3).

Table 3: ASPEN Plus simulation for 5-equivalent-setps of HETP

Exp. Cond.	Stages	Twi (°C)	Two (°C)	Tao (°C)	Tcwo (°C)	Md (mol/s)	Mw (mol/s)	Md/Mw
Mw=0.00305 (m3/hr) Ma=0.00108 (m3/hr) Tcwi=28(°C) Tai=32.9(°C)	5	82.31	42.23	66.48	32.9	0.04489	0.6095	0.07365
	4	82.17	43.77	65.9	32.9	0.04334	0.6095	0.07112
	3	81.95	46.18	64.92	32.9	0.04089	0.6095	0.06709
	2	81.56	50.26	62.97	32.9	0.03647	0.6095	0.05985
	REAL	80.4	36.2	58.1	30.2			0.06011

According to the Eckert's proposed coefficients [36] and the use of Pall ring packing with 25 mm nominal diameter, the effective height of humidifier was suggested about 170 cm.

Using Differential Evolution (DE) strategy and via differing the number of dependent variables (such as air flow rate, T_{cwi}, M_w and finally M_{cw}) individually and also simultaneously, oprimized model was correlated successfully. Practical values which were resulted optimum M_d=0.3777 $\frac{m^3}{hr}$ were M_{cw}=0.0298 $\frac{m^3}{hr}$ at T_{cwi}=13 °C, wiht M_a=0.008075 $\frac{m^3}{hr}$ and M_w= 0.021497 $\frac{m^3}{hr}$.

6- Conclusion:

One of the most applicable methods for desalination brackish water is humidification-dehumidification system. This thermal method has been studied on a pilot system. After experimental data collecetion due to data validity proof and also finding an optimum fresh water rate, two mathematical models, one on open air cycle and another one on close air cycle have been derived. Furthermore, HETP for the humidifier was studied by comparing experimental data resulted from various tower's height and data from simulation of HDH system in ASPEN Plus; Estimation error of the HETP modelfor Pall Ring packed bed was less than 4%. Yield in closed air system was shown 15% better than open air system.

Convincingly, using of DE strategy, optimal parameters to obtain the choisest yield (0.3777 m^3/hr) have been provided: 0.0298 m^3/hr, 13 °C respectively for flow rate and temperature of cooling water, 0.008075 m^3/hr air flow rate (in closed air system) and 0.021497 m^3/hr for water inake.

Nomenclatures:

D (m)	diameter	$C_{pa}\ (\frac{KJ}{Kg.K})$	Specific heat coefficient of air
G(kg/s)	Gas mass flow rate	$\lambda\ (\frac{KJ}{Kg})$	Evaporation latent heat
L (kg/s)	Liquid mass flow rate	$K_a\ (\frac{W}{m.K})$	Thermal conductivity coefficient of air
M (gr/min)	Mass flow rate	$\mu\ (\frac{kg}{m.s})$	*viscosity*
Ma (kg/s)	Intake air mass flow rate at the bottom of humidification tower	$\rho\ (\frac{Kg}{m^3})$	*density*
Mcw (kg/s)	Mass flow rate of saline water Inlet to cooling tower	$\rho_a\ (\frac{Kg}{m^3})$	Air density
Md (kg/s)	Fresh water (condensed water) mass flow rate	σ	*Thickness*
Mw (kg/s)	Mass flow rate of rejected water from condenser	***Subscripts***	
Q (w)	Heat received /transferred	*a*	*air*
R.H. (%)	Relative Humidity percentages	*ai*	*Inlet air*
T (K)	Temperature	*ao*	*Outlet air*
Tai *(K)*	Intake air temperature at the top of Humidifier	*cw*	*Cooling water*
Tao *(K)*	Outlet air temperature at the top of humidifier	*cwi*	*Inlet Cooling Water*
Tcwi *(K)*	Cooling water inlet temperature	*cwo*	*Outlet Cooling water*
Tcwo *(K)*	Water temperature of rejected water from condenser	*d*	*Distilled water*
Twi *(K)*	Water temperature of inlet water to humidifier	G	*Gas*
Two *(K)*	Water temperature of outlet brine water at the bottom of humidifier	L	*Liquid*
$V_a\ (\frac{m3}{min})$	Volumetric flow rate of air	*w*	*water*
$R\ (\frac{Pa.m^3}{Kmol.K})$	Universal Gas Constant	*wi*	*Inlet water to Humidifier*
M_{wei}	*Molecular weight*	*wo*	*Outlet water from Humidifier*

References:

[1] Malik, M., Tiweri, G.N., Kumer, A., Sodha, M.S., Solar distillation, New York, pergamon, pp. 3-4,1985

[2] Kalogirou, S.A., Solar thermal Collectors and applications, Progress in Energy and Combustion science, Vol. 30, pp. 231-295, 2004.

[3] Watson, I.C., Morin, O.J., Henthorne, L., Desalting handbook for planners, Aqua Resources International Inc., 2003.

[4] Qiblawy, H.M., Banat, F., solar thermal desalination technologies, Desalination, Vol. 220, pp. 633-644, 2008.

[5] Treybal R.E. Mass transfer operations, 3rd ed., NY: McGraw-Hill; 1980.

[6] Al-Hallaj S, Farid MM, Tamimi AR. Solar desalination with humidification–dehumidification cycle: performance of the unit. Desalination 1998;120:273–80.

[7] Farid MM, Al-Hallaj AW. Solar desalination with humidification–dehumidification cycle. Desalination 1996;106:427–9.

[8] Ben-Bacha H, Damak T, Bouzguenda M. Experimental validation of the distillation module of a desalination station using the SMCEC principle. Renewable Energy 2003;28:2335–54.

[9] Garg HP, Adhikari RS, Kumar R. Experimental design and computer simulation of multi-effect humidification (MEH)–dehumidification solar distillation. Desalination 2002;153:81–6.

[10] Nafey AS, Fath HES, El-Helaby SO, Soliman AM. Solar desalination using humidification–dehumidification processes. Part II. An experimental investigation. Energy Conversion Management 2004;45(7–8):1263–77.

[11] Nawayseh NK, Farid MM, Al-Hallaj S, Tamimi AR. Solar desalination based on humidification process. Part I. Evaluating the heat and mass transfer coefficients. Energy Conversion Management 1999;40:1423–39.

[12] Muller-Holst H. Solar thermal desalination using the multiple effect humidification (MEH) method. In: Book chapter, solar desalination for the 21st century; 2007. pp. 215–25.

[13] Al-Sahali M, Ettouney HM. Humidification dehumidification desalination process: Design and performance evaluation, Chemical Engineering Journal, 143, 257–26, 2008.

[14] Farid M.M., Parekh S., Selman J.R., Al-Hallaj S., Solar desalination with humidification–dehumidification cycle: mathematical modeling of the unit. Desalination; 151:153–64, 2002.

[15] Yamali C, Solmus I. Theoretical investigation of a humidificationdehumidification desalination system configured by a double-pass flat plate solar air heater. Desalination 205, 163–177, 2007.

[16] Zamen M, Amidpour M, Soufari SM. Cost optimization of a solar humidification–dehumidification desalination unit using mathematical programming. Desalination 239, 92–99, 2009.

[17] Dai Y.J, Wang RZ, Zhang HF. Parametric analysis to improve the performance of a solar desalination unit with humidification and dehumidification. Desalination; 142:107–18, 2002.

[18] W.F. He, L.N. Xu, D. Han. Parametric analysis of an air-heated humidification-dehumidification (HDH) desalination system with waste heat recovery. Desalination 398 (2016) 30–38.

[19] Cihan Yıldırım,Ismail Solmus. A parametric study on a humidification–dehumidification (HDH) desalination unit powered by solar air and water heaters. Energy Conversion and Management 86 (2014) 568–575.

[20] S.W. Sharshir, Guilong Peng, Nuo Yang, Mohamed A. Eltawil, Mohamed Kamal Ahmed Ali, A.E. Kabeel. A hybrid desalination system using humidification-dehumidification and solar stills integrated with evacuated solar water heater. Energy Conversion and Management 124 (2016) 287–296.

[21] Nabil A.S. Elminshawy, Farooq R. Siddiqui, Mohammad F. Addas. Experimental and analytical study on productivity augmentation of a novel solar humidification–dehumidification (HDH) system. Desalination 365 (2015) 36–45.

[22] Wang GQ, Yuan XG, Yu KT, Review of mass transfer correlation for packed columns. Ind. Eng. Chem. Res. 2005, 44, 8715-8729.

[23]Shulman, HL, Ullrich CF, Proulx AZ, Zimmerman JO, Performance of Packed Columns. 2. Wetted and Effective Interfacial Areas, Gas- and Liquid-Phase Mass Transfer Rates.AIChE J. 1955, 1, 253.

[24]Shulman, H. L.; de Gouff, J. J. Mass Transfer Coefficients and Interfacial Areas for 1-Inch Raschig Rings. Ind. Eng. Chem.1952, 44, 1915.

[25] Van Krevelen, D. W.; Hoftijzer, P. J. Studies of Gas Absorption. 1. Liquid Film Resistance to Gas Absorption in Scrubbers. Recl. Trav. Chim. Pays-Bas. 1947, 66, 49.

[26] Van Krevelen, D. W.; Hoftijzer, P. J. Kinetics of Simultaneous Absorption and Chemical Reaction. Chem. Eng. Prog. 1948, 44, 529.

[27]Mangers, R. J.; Ponter, A. B. Effect of Viscosity on Liquid Film Resistance to Mass Transfer in a Packed Column. Ind. Eng. Chem. Process Des. Dev. 1980, 19, 530.

[28]Shi, M. G.; Mersmann, A. Effective Interfacial Area in Packed Columns. Ger. Chem. Eng. 1985, 8, 87.

[29] Billet, R.; Schultes, M. Advantage in Correlating Packing Column Performance. Inst. Chem. Eng. Symp. Ser. 1992, 128, B129.

[30] Onda, K.; Takeuchi, H.; Okumoto, Y. Mass Transfer Coefficients between Gas and Liquid Phases in Packed Columns. J Chem. Eng. Jpn. 1968, 1, 56.

[31] Eckert, J. S.: Chem. Eng., 82, 70. Apr. 14, 1975.

[32] Mostafa H. Sharqawy, John H. Lienhard, Syed M. Zubair. Thermophysical properties of seawater: a review of existing correlations and data. Journal of Desalination and Water Treatment. 16 (2010) 354–380.

[33] J.D. Isdale and R. Morris, Physical properties of sea water solutions: density, Desalination, 10(4) (1972) 329–339.

[34] Francis L. Smith and Allan H. Harvey (September 2007). "Avoid Common Pitfalls When Using Henry's Law". CEP (Chemical Engineering Progress). ISSN 0360-7275.

[35] Feoktistov, V. (2006). Differential Evolution: In Search of Solutions. Springer. ISBN 978-0-387-36895-5.

[36] H. Bahmanyar, Mass Transfer. Tehran University's Jahad Daneshgahi Publication, 2003.

473748 Process to Planet Framework for Sustainable Design: Systematic Approach for Developing a Multiscale Model and for Multiobjective Optimization

Abstract:

Over the last decade, sustainable process design (SPD) research has utilized life cycle assessment(LCA) methods to incorporate an environmental dimension in process systems engineering(PSE). Conventional SPD involves a â€œbottom up approachâ€ of modeling the system of interest at a smaller scale with detailed process specific fundamental engineering models, while the upstream life cycle of the product is modeled at a larger scale using process based LCA. System boundary selection is a significant challenge for process LCA due toÂ infinite nature of the upstream life cycle network for any process [1]. Process based LCA can lead to significant truncation errors in life cycle resource consumption and emissions, leaving a wide margin for shifting of impacts across narrow system boundaries and leading to illusive optimal design solutions that may in fact be inferior due to incomplete life cycle information [2]. Hybrid LCA(HLCA) methods aim to address the boundary selection problem by combining process level life cycle inventories with environmentally extended input-output analysis(EEIO), which allows for a significant part of the product life cycle to be included in the assessment. The Process to Planet (P2P) framework [3] was developed by introducing the concept of integrated hybrid LCA [4] to SPD. Detailed engineering models of the primary process being designed and other key processes in the supply chain are integrated within a macroeconomic system represented by economic input-output (EIO) model. This ensures solution of process design problems while reducing chances of environmental impacts being neglected due to incomplete life cycle inventories. While bridging the knowledge gap between PSE and LCA, the P2P framework not only improves upon the life cycle assessment component of SPD, but also addresses a major weakness in Hybrid LCA. HLCA is limited to the impact assessment stage of life cycle assessment. Very few studies have dealt with the fourth and final stage of LCA, improvement analysis. This situation is lamentable because inventory scope and impact analysis provide little benefit if not followed by recommendations and decisions that should be adopted to reduce environmental burden. The P2P framework can be visualized as an extension of HLCA method where it enables the incorporation of design variables through process design and allows optimization of the system to improve its environmental footprint. It was successfully demonstrated by application to a corn based ethanol production plant and highly insightful results were obtained that proved the efficacy of the framework [2].

This presentation proposes extensions and modifications that address some of the inadequacies that are inherent in the original P2P framework. While strictly applied to P2P, these extensions can be extrapolated to answer challenges that exist in broad areas of literature, like HLCA. The selection of significant processes that are modeled at the process based life cycle scale and the insignificant processes that are studied at the economy scale is arbitrary. In other words, delineation of the boundary between the different scales of the P2P framework was a subjective task. Previously, this was done based on data availability, accuracy and detail requirements and the complexity of the overall problem which is proportional to the solving time [5,6]. The problem was first mentioned and some methods were proposed for addressing it in a paper by Suh [7]. Lenzen proposed Structural Path Analysis (SPA) involving the decomposition of the Leontief Inverse for addressing this problem [8]. Treloar also used this method to study energy use of individual residential buildings in Australia [9,10]. Lave et al. [11] and Lenzen [12] used SPA to quantify the truncation errors in process-based LCA. A path exchange method using SPA was also developed as an alternative to the method of combining process LCA and IO LCA within integrated hybrid LCA as proposed by Suh. We intend to address this knowledge gap by proposing a path analysis algorithm where the P2P framework superstructure model generation is guided by a hierarchical structural path decomposition method which determines the coverage of the different scales. It is one of the first applications of SPA for solving the boundary selection problem involving the distribution of different upstream processes between multiple scales.

A large number of studies have dealt with multi objective optimization in process design and energy systems involving environmental and economic impacts. A comprehensive list of such literature can be found in Diwekar et al.[13] Reviewing these articles, it becomes apparent that it is necessary to simultaneously work towards both economic profitability and environmental sustainability of the system. Previously, the P2P framework focused only on the environmental dimension of process design. In this work, we propose a multi objective optimization problem, involving the trade-off between an economic objective and an environmental objective function. The novelty in this addendum is that it makes the P2P framework more fruitful for designing any generic process while exercising control over both the economic and environmental scope.

Preliminary results have provided insightful views on the application of SPA algorithm for model generation as well as exposed existing deficiencies to work on. While SPA is successful in mapping the upstream networks of the process to be designed and calculate their contributions to the entire life cycle, the quantification of the quality of the generated model still needs to be addressed. For example, considering two scenarios where an upstream sector is modeled at the IO model scale and process LCA scale respectively, we need to develop parameters that can quantitatively compare between the two cases and rank them objectively. As of now, subjective decisions are being used to build the model, but we intend to develop an algorithm based on error analysis or uncertainty information to determine the model quality. Multiobjective optimization solutions demonstrated that, highly erroneous optimal solutions and counter intuitive results are obtained from practicing conventional SPD approach rather than P2P. In addition to this, it was observed that there is a 59% chance of choosing an inferior solution over the optimal solution when solving the design problem using conventional SPD.

References

[1] Sangwon Suh. Functions, commodities and environmental impacts in an ecological economic model. 48:451-467, 2004.

[2] Rebecca J Hanes and Bhavik R Bakshi. Sustainable process design by the process to planet framework. AIChE Journal, 61(10):3320-3331, 2015.

[3] Rebecca J Hanes and Bhavik R Bakshi. Process to planet: A multiscale modeling framework toward sustainable engineering. AIChE Journal, 61(10):3332-3352, 2015.

[4] Sangwon Suh and Shinichiro Nakamura. Five years in the area of input-output and hybrid lca. The international journal of life cycle assessment, 12(6):351-352, 2007.

[5] H Hondo and S Sakai. Preliminary life cycle inventory analysis (pre-lci) using an economic input-output table. In The Fourth International Conference on EcoBalance, Tsukuba, Japan, pages 181-4, 2000.

[6] G Norris. Selecting and evaluating boundaries for life cycle assessment using economic input-output analysis. In Proceedings from the SETAC 17th Annual Meeting, Washington, DC, pages 17-21, 1996.

[7] Sangwon Suh, Manfred Lenzen, Graham J Treloar, Hiroki Hondo, Arpad Horvath, Gjalt Huppes, Olivier Jolliet, Uwe Klann, Wolfram Krewitt, Yuichi Moriguchi, et al. System boundary selection in life-cycle inventories using hybrid approaches. Environmental Science & Technology, 38(3):657-664, 2004.

[8] Manfred Lenzen. A guide for compiling inventories in hybrid life-cycle assessments: some australian results. Journal of Cleaner Production, 10(6):545-572, 2002.

[9] Graham J Treloar. Extracting embodied energy paths from input-output tables: to-wards an input-output-based hybrid energy analysis method. Economic Systems Re-search, 9(4):375-391, 1997.

[10] GJ Treloar, PED Love, OO Faniran, and U Iyer-Raniga. A hybrid life cycle assessment method for construction. Construction Management & Economics, 18(1):5-9, 2000.

[11] Lester B Lave. Using input-output analysis to estimate economy-wide discharges. Environmental Science & Technology, 29(9):420A-426A, 1995.

[12] Manfred Lenzen. Errors in conventional and input-output based lifecycle inventories. Journal of Industrial Ecology, 4(4):127-148, 2000.

[13] Urmila Diwekar and Yogendra Shastri. Design for environment: a state-of-the-art re-view. Clean Technologies and Environmental Policy, 13(2):227-240, 2011.

462158 Application of a Shale Environmental Footprint Optimization Tool to Enhance Operational Excellence– a Case Study of Selecting Process Fuels with Lower Cost and Less Potential for Impacts to Power Chevron Operations

Abstract:

Unconventional resources represent significant business opportunities; however, their development is subject to much public and environmental scrutiny. Currently there are over 400 local bans and 2 state bans on hydraulic fracturing. Many countries besides the U.S. have placed moratoriums on hydraulic fracturing. The challenge and opportunity that the oil and gas industry is facing is: how to unlock the potential in unconventionals, achieve better economic and operational performance, while still fulfilling our commitment to communities and the environment.

The Shale Environmental Footprint Optimization Tool is a tool developed at Chevron to enable the integration of environmental considerations in process design and decision making related to shale development. It simulates shale operations at the play level, where material and energy flows of in-scope processes are aggregated together and converted to quantitative environmental indicators (i.e. energy consumption, water consumption, and air emissions). It is built with a commercial life cycle assessment (LCA) software (GaBi® Professional) that contains >4000 datasets, and adopts international standards (ISO 14040-10444). The tool also allows sensitivity and uncertainty analysis to better assess and interpret results under varying reservoir and operational conditions.

This tool has been used by business units to better characterize their operations, assess potential environmental risks associated with different decision alternatives, and understand the trade-offs. It also provides them a more holistic view across the value chain, and help them pinpoint major contributors to the total footprint. These insights help business partners prioritize R&D efforts, make informed decisions on mitigation options, and ultimately lead to reduced environmental impact. Externally, the tool can be used address stakeholder concerns and for benchmarking our performance.

In the presentation, we'll show how this tool has been used to assess real world situations (using field data and business scenarios) and generate information to support a critical decision – evaluating and identifying the preferred alternative fuel to power drilling rigs. Optimizing the selection of process fuels could result in significant environmental, social, and economic benefits. One of the scenarios we'll include is to switch from LNG to gas produced in the field for power generation, which may reduce truck traffic on the roads. Reducing trucks is beneficial to safety (MVC, avoid backing/maneuvering incident on well pad), cost (hauling and road repair), land disturbance (less roads and processing units), and emissions (air pollutants & spills). In some locations, it also means improved reliability and fewer burdens on business continuity.

GHG Life Cycle Assessment for the United Arab Emirates Electricity Sector Combining Optimization and Simulation Tools

Paper ID #: 466062

*Alberto Betancourt-Torcat, Mohammed Alkatheri, Ali Almansoori**

Department of Chemical Engineering, The Petroleum Institute, Sas Al Nakhl Campus, Abu Dhabi, P.O. Box 2533, United Arab Emirates.

*Corresponding Author's e-mail: aalmansoori@pi.ac.ae

Keywords: Greenhouse Gas, Life cycle assessment, Electricity sector, Economics, Optimization, Simulation.

Abstract

The United Arab Emirates (UAE) relies heavily on natural gas for power generation, and the demand for gas in a country with one of the highest per capita power consumption rates in the world position the electricity sector as the leading CO_2 emission source. Although the country is very rich in energy resources it is currently facing challenges to meet its domestic gas demand. Thus, the country has been forced to develop its large, but highly sour (high-sulfur content) gas reservoirs. Given the gas high-sulfur content, the sweetening process is particularly very energy and greenhouse gas (GHG) emission intensive compared with conventional gas resources. Additionally, the volume of sweet gas yield is low due to the high proportion of acid gases in the feed gas (e.g., over 35% of H_2S and CO_2 combined). This work quantifies the well-to-grid (WTG) GHG emissions per kilowatt hour (kWh) of domestic gas-based electricity generated in the UAE using mathematical models, simulation tools, and engineering principles. WTG emissions include emissions from sour gas extracting, sour gas processing, sweet gas product transportation, electricity generation, and electricity transmission in the UAE. Two types of gas resources were examined: conventional and ultra-sour gas. The results show that the WTG GHG emissions of domestic gas-based power generation in UAE ranges between 444 to 501 g CO_2 eq./kWh; placing the country in the low-mid range of values compared with the literature 413-587 g CO_2 eq./kWh.

Introduction

Over the last decade, the rapid economic and demographic growth of the United Arab Emirates (UAE) has pushed the country's electricity grid to its limits. Most of the UAE's generation capacity is natural gas-based [1]. Although natural gas is considered to be a relatively clean primary energy source, it is still a fossil fuel source that generates significant greenhouse gas (GHG) emissions. The UAE electricity demand reached a domestic record of 105 billion kWh in 2013, placing the country as one of the highest per capita power consumers in the world [2]. Nuclear and renewable power capacities are currently being added to increase the country's energy mix and reduce GHG emissions [3]. As a result of the steadily growing electricity demand, the UAE plans to increase natural gas production to meet the domestic demand. The majority of the growth will come from the country's largely sour (high-sulfur content) gas reservoirs. The UAE's gas reservoirs have a relatively high sulfur content, which makes it corrosive and difficult to process [4]. The process is particularly very energy intensive and environmentally challenging. Thus, the process is highly GHG-emission intensive, as well vast amounts of sulfur are generated as by-product [5].

There is currently a need to assess the environmental footprints of the UAE electricity sector in terms of its GHG emission life cycle. This could allow benchmarking the UAE's electricity sector against other nations, while finding suitable niches for GHG mitigation strategies. Moreover, since over 98% of the electricity generated in the UAE is gas-based, an analysis on the UAE's natural gas life cycle represents a key factor on determining the overall environmental impact of the electricity sector. The present analysis considers a well-to-grid (WTG) GHG lifecycle assessment; which includes the upstream and downstream operations. The upstream operations includes all activities before fuel is combusted at power plants. Such activities comprise: sour natural gas extraction, sour gas processing, and sweet gas transport to the power plants gate. On the other hand, the downstream operations include the electricity generation and transmission. For the present study, fundamental engineering principles were used to mathematically estimate process specific energy consumption and related GHG emissions; namely for sour gas extraction, sweet gas product transportation, and electricity transmission. Additionally, ProMax® v3.2 [6] was used to simulate the sour gas processing, while performing a parametric analysis to improve the performance of these processes. Also, an optimization mathematical model developed in the General Algebraic Modeling System (GAMS®) [3] was used to illustrate the operation of the UAE power sector, and obtain the average GHG emissions of the country's gas-based power fleet.

Methodology

In the UAE the well-to-grid life cycle starts with the extraction of natural gas from the reservoirs and ends at the electricity transmission. Natural gas is extracted using wells and sent to processing plants where: water, carbon dioxide, sulfur, and heavier hydrocarbons are removed. Then, the resulted sweet gas product is transported through pipelines to the power plants to generate electricity, and transmitted through the electricity grid. The UAE's WTG GHG emissions life cycle can be divided into two inventories: upstream and downstream. The upstream inventory includes all activities before fuel is combusted at a power plant. Such activities comprise: sour natural gas extraction, sour gas processing, and sweet gas product transportation to the power plants gate. On the other hand, the downstream inventory includes electricity generation and transmission. In this study, the GHG emissions associated to gas extraction, sweet gas product transportation, and electricity transmission were estimated based on typical average emission values found in the literature [7]. On the other hand, the GHG emissions generated by the natural gas processing and combustion in power plants were calculated using ProMax® v3.2 (simulation software) [6] and a mathematical model developed in GAMS® (modeling system) [3], respectively. There are two main types of natural gas resources in UAE: 1) ultra/highly-sour natural gas, and 2) conventional natural gas. Therefore, for the present analysis the WTG GHG emissions life cycle was estimated taking into account both natural gas resources. Typical compositions for both types of gas resources are shown in Table 1.

Table 1. Typical compositions for ultra-sour and conventional natural gas resources in UAE.

Main Components	Ultra-sour gas [4]	Conventional gas [8]
H_2S	22	5
CO_2	10	4
C1	54	76
C2	4	8
C3+	5	3
H_2O	5	4

The first stage in the WTG GHG emissions life cycle is the natural gas extraction using a gas well. This operation encompasses various processes such as: rig construction, drilling, well completion, and gas production. Energy is consumed during the drilling process generating GHG emissions. The emissions of natural gas during the extraction stage are considered episodic emissions because they are not part of the daily, steady state of the well operations. For instance, it was assumed that around 20 g CO_2 eq./kWh is emitted to the atmosphere in gas extraction (e.g., mainly point source and fugitive losses) [7]. Moreover, gas production starts after the well is completed. Gas from the wellhead is sent to surface processing, where it is sweetened, dehydrated, and compressed for pipeline transmission. The raw gas from the wellhead is first sent to a three phase gravity separator to separate the gas, light liquids (e.g., mostly sour hydrocarbon liquids C3+), and heavy liquids (mostly sour water). Since no fuel is consumed during this process, it is not considered a significant source of GHG emissions.

After the phase separator, the gas is treated in a gas sweetening unit. In this work, di-glycol amine (DGA) is used as the gas sweetening solvent for ultra-sour natural gas; while methyl-di-ethanol amine (MDEA) is used as sweetening solvent for conventional gas. The units that consume energy in the gas sweetening process are: feed pre-coolers, amine reboilers, circulation pumps, solvent coolers, and acid gas condensers. The amine re-boiler is assumed to be powered by low pressure steam (LPS) (i.e. saturated steam at 3.5 barg), the circulation pump and solvent air cooler are powered by electricity, the feed pre-cooler and acid gas condenser employ cooling water. Later, the wet sweet gas is sent to the dehydration unit; whereas the acid gases (i.e. H_2S and CO_2) are sent to the sulfur recovery unit. In the dehydration unit, water is removed from the sweet natural gas since it promotes corrosion. Tri-ethylene glycol (TEG) is used as liquid desiccant to dehydrate the sweet gas. The glycol reboiler, sweet gas pre-cooler, and pumps are the major energy-consuming units in this process. In the dehydration unit, high pressure steam (HPS) (i.e., saturated steam at 41 barg) is used to power the glycol reboiler. Moreover, electricity is used to drive the pumps, whereas cooling water is used to pre-cool the wet sweet gas. The dry sweet gas from the dehydration unit can be combusted at an onsite power plant to produce electricity. Sometimes the dry sweet gas from the dehydration unit is sent to a natural gas liquid recovery unit to recover ethane and ethylene. However, for simplicity the natural gas liquid recovery unit was not considered in this study.

The liquid hydrocarbons are treated in a liquid sweetening unit using DGA as solvent; while the sour water is treated using a sour water stripper. The sole purpose of both aforementioned treatment processes is the removal of sour components from the hydrocarbon liquids and water. The liquid hydrocarbons sweetening process is similar to the gas sweetening process; thus, involving equivalent energy consuming units. Conversely, mainly two types of units consume energy in the sour water stripper: the water stripper reboiler and pumps.

All acid gases effluent from the gas sweetening unit, liquid sweetening unit, and sour water stripper are collected and sent to the sulfur recovery units (SRU). In these units, the H_2S in the acid gas stream is converted into elemental sulfur. A variety of configurations are available for sulfur recovery. In this study, it was used the "straight-through" Claus process. This process consists of an initial combustion step at high temperature (e.g., above 980 °C), followed by a series of reactions (carried out in a Claus Bed) and condensation steps. Generally two or three such steps are used. Steam is generated in the combustion step; thereby, contributing to meet the steam requirements of the process. The units consuming energy in sulfur recovery are condensers and re-heaters. Condensers and re-heaters are powered by cooling water and HPS, respectively.

The ProMax® v3.2 simulation software was used to calculate all forms of energy required to process conventional and ultra–sour natural gas. However, fundamental engineering principles were employed to calculate the associated GHG emissions. Furthermore, the amount of fuel gas (i.e., dry sweet natural gas) required to produce process steam (e.g., high and low pressure steam needed to process natural gas) was also estimated using the simulator. It is assumed that the fuel gas first combines with air, and then is combusted in a burner. The heat generated from the combustion process is used to boil water and generate high and low pressure steam. From this analysis, it was found that the average emission per slandered of fuel gas is around 66.7 g CO_2 eq./scf. This emission rate is in agreement with values reported by the US Environmental Protection Agency (EPA) [9].

The electricity requirements for the cooling water loop can be estimated using the cooling water flow rate. This can be done calculating the power requirements in pumps and cooling tower fans. For instance, the power requirements in a cooling water pump with a volumetric flow rate (V (m^3/s)) and pressure drop (ΔP (kPa)) can be calculated as follows:

$$Power = \frac{1}{\text{efficiency}} * V * \Delta P \tag{1}$$

For the cooling water loop, it was assumed an overall efficiency of 75% [10], the pressure drop around the water loop was estimated to be 266.7 kPa [10]. Accordingly, the power requirements can be calculated as follows [11]:

$$Fan\ Power = \left(\text{V}\frac{m^3}{s}\right)\left(\frac{60s}{min}\right)\left(\frac{264.172\text{gallon}}{m^3}\right)\left(\frac{0.5ft^2}{\frac{gallon}{min}}\right)\left(\frac{0.041\text{hp}}{\text{ft}^2}\right)\left(\frac{0.7457\text{kw}}{\text{hp}}\right) \tag{2}$$

The emissions of transporting the sweet gas product from process facilities to power plants was assumed to be approximately 30 g CO_2 eq./kWh according to information reported in the literature [7]. Furthermore, a mathematical model developed in GAMS on the UAE power infrastructure [3] was used to calculate the average GHG emissions generated by the gas-based power plants fleet in the country. The simulation results show that in average 370 g CO_2 eq. are generated per kWh of power produced. This value was used to calculate the GHG emissions associated to the electricity requirements involved in the analyzed life cycle. Finally, the GHG emissions associated to electricity transmission was also taken from the literature (i.e., 3.4 g CO_2 eq./kWh) [7].

Results and Discussions

The ProMax® v3.2 simulator was used to calculate the utilities (e.g., high/low pressure steam, cooling water, and electricity) consumed in the processing of conventional and ultra–sour natural gas. Table 2 shows the utility requirements for both types of gas resources per standard cubic feet (scf) of dry sweet gas produced.

Table 2. Utility requirements for conventional and ultra–sour natural processing.

Utility Type	Units	Ultra-sour gas	Conventional gas
Electricity	(Wh/scf)	20.8	4.0
High Pressure Steam (HPS)	(g/scf)	6.1	1.2
Low Pressure Steam (LPS)	(g/scf)	41.5	11.0
Low Pressure Steam (LPS) generated from SRU	(g/scf)	23.7	3.7
Net Required Low Pressure Steam (LPS)	(g/scf)	17.8	7.3
Cooling water	(liter/scf)	1.0	0.3

As shown in the table, the processing of ultra-sour gas requires much more energy than conventional gas per unit of sweet gas yielded. For example, the energy requirements to process ultra-sour gas are roughly four times higher than that for conventional gas. This can be explained by the fact that as the amount of H_2S in the feed gas increases so does the absorption plat utilities demand. This because more acid gases must be removed by the absorber leading to higher recirculating solvent flow rates. Therefore, the power consumed to cool and pump the regenerated solvent back into the absorber, as well as the steam used to regenerate the alkanolamines increases. Both energy forms (i.e., steam and electricity) account for the majority of the utilities consumed in the absorption plant. Likewise, more sulfur must be processed by the sulfur recovery units. Moreover, the sweet gas product recovery rate is much lower for the ultra-sour gas feed. Table 3 shows the recovery percentage for ultra-sour and conventional gas resources.

Table 3. Recovery percentages for ultra-sour and conventional gas.

Feed type	Ultra-sour gas	Conventional gas
Recovery rate	60 %	90 %

As shown in the table, almost 40 % of the ultra-sour gas feed volume is loss due to the high acid gases content; which are removed to obtain the sweet gas product. Conversely, for conventional gas only 10% of the volume is loss. Consequently, more energy is required per unit of sweet gas produced from ultra-sour gas. The electricity requirements for the cooling water loop were estimated using Equations (1)-(2). Accordingly, the GHG emissions corresponding to the electricity requirements were estimated using the gas power generation emission factor. Moreover, the process steam requirements associated to sour gas processing were obtained simulating the gas sweetening plants with ProMax® v3.2 [6]. The GHG emissions values related to HPS and LPS usage, and electricity for processing the natural gas are reported in Table 4. These values are given per standard cubic feet (scf) of dry sweet gas produced.

Table 4. GHG emissions related to conventional and ultra–sour gas processing.

Type of Utility Consumed	Units	Ultra-sour gas	Conventional gas
Electricity	(g CO_2 eq./scf)	7.7	1.5
High Pressure Steam (HPS)	(g CO_2 eq./scf)	0.7	0.4
Light Pressure Steam (LPS)	(g CO_2 eq./scf)	2.7	1.1
Total GHG emissions	(g CO_2 eq./scf)	11.1	3.0

From the table it can be clearly noticed that the GHG emissions associated with ultra-sour gas processing are much higher than those for conventional gas. Additionally, Table 4 shows that more than 50% of the GHG emissions are due to electricity consumption. Using the GAMS model [3], the average amount of natural gas needed to produce one-kilowatt hour of electricity in UAE was estimated. This value was found to be approximately 6.97 scf of gas/kWh. Additionally, all GHG emissions values from natural gas processing were also expressed in terms of per unit of kWh of electricity produced. The GHG emissions involved in natural gas extraction and final product transportation were estimated using factors reported in the literature [7]. The WTG GHG emissions life cycle for UAE domestic gas-based power generation can be estimated as the summation of the GHG emissions from: gas extraction, sour gas processing, sweet gas transportation, electricity generation (i.e., combustion of natural gas in power plants), and electricity transmission. Table 5 shows the single components and total GHG emissions life cycle for electricity produced from natural gas for both type of gas resources.

Table 5. Single emissions and total GHG emissions life cycle for domestic gas-based power generation in the UAE.

Life Cycle Stage	Ultra-sour gas (g CO_2 eq./kWh)	Conventional gas (g CO_2 eq./kWh)
Gas Extraction	20	20
Gas Processing	77.6	20.6
Product Transportation	30	30
Gas Combustion (power plants)	370	370
Electricity Transmission	3.4	3.4
Total GHG	501	444

As shown in the table, ultra-sour gas involves higher GHG emissions than conventional gas; which results from greater energy requirements. It is worth pointing out that most of the GHG emissions originate from the combustion of the sweet natural gas product in the power plants. Additionally, the WTG GHG life cycle analysis shows that for ultra-sour gas the emissions are approximately 13% greater than those involved using conventional gas. Figure 1 shows the percentage share of each stage/activity in the GHG emissions life cycle.

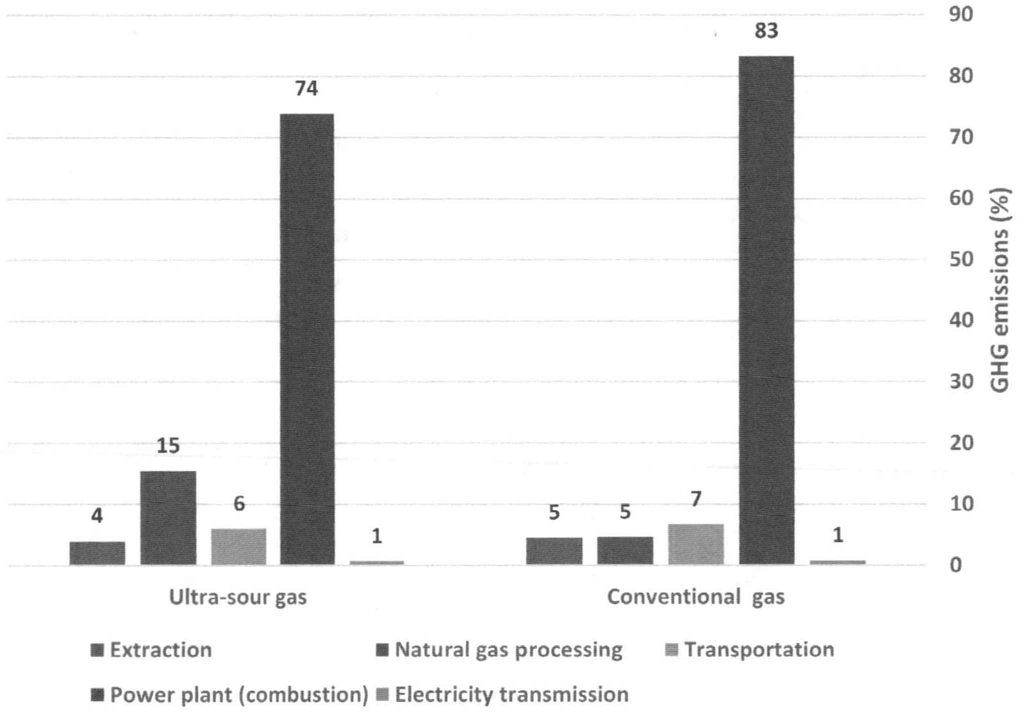

Figure 1. Percentage share per stage/activity in the GHG emissions life cycle for ultra–sour and conventional gas.

As shown in the Figure, sour gas processing and electricity generation (at the power plants) account for more than 88% of the total GHG emissions in the life cycle. Moreover, electricity generation is clearly the leading GHG emission source in the analyzed life cycle. Nevertheless, the GHG emissions generated by the electricity generation operations are rather low compared with other countries. This since the majority of the UAE's current power fleet is composed of modern gas-fired combined cycle plants; which are comparatively more efficient than other gas-based plants. Also, the total WTG GHG emissions of domestic gas-based power generation in UAE places the country in the low range (for conventional gas) and middle range (for ultra-sour gas) of values reported in the literature (e.g., 413-587 g CO_2 eq./kWh) for similar systems worldwide [7, 12, 13].

Conclusion

In this study, the well-to-grid (WTG) GHG emissions lifecycle for domestic gas-base power generation in the UAE is analyzed. The analysis results show that gas-based power generation using ultra-sour gas is more emission intensive than using conventional gas as the power plants' feed. This because much more energy is consumed in ultra-gas processing. Additionally, the yield of sweet gas product is much lower when processing ultra-sour gas compared with conventional gas. Regarding the total emissions of the life cycle, sour gas processing and electricity production account for almost 90 % of the total GHG emissions. Electricity generation is clearly the leading GHG emission source in the analyzed WTG lifecycle.

References

[1] A. Betancourt-Torcat and A. Almansoori, "Design multiperiod optimization model for the electricity sector under uncertainty – A case study of the Emirate of Abu Dhabi," *Energy Conversion and Management,* vol. 100, pp. 177-190, 8// 2015.

[2] Energy Information Administration, "United Arab Emirates " U.S. Energy Information Administration2015.

[3] A. Almansoori and A. Betancourt-Torcat, "Design optimization model for the integration of renewable and nuclear energy in the United Arab Emirates' power system," *Applied Energy,* vol. 148, pp. 234-251, 6/15/ 2015.

[4] Al Hosn Gas. (2015, 22/5/2016). *Abu Dhabi Gas Development Company Ltd. (Al Hosn Gas).* Available: : http://www.alhosngas.com/.

[5] J. Slagle, "Highly Sour Amine Sweetening Process Study for Conditions Common in the Middle East," presented at the The Abu Dhabi International Petroleum Exhibition and Conference, Abu Dhabi, UAE, 2013.

[6] Bryan Research & Engineering Inc., "ProMax," ed. Bryan, Texas, U.S.: Bryan Research & Engineering, 2013.

[7] T. Skone, J. Marriot, J. Littlefield, and B. A. Hamilton, "Life Cycle Assessment of Natural Gas Extraction, Delivery and Electricity Production," National Energy Technology Laboratory, United States of America2012.

[8] P. Pal, A. AbuKashabeh, S. Al-Asheh, and F. Banat, "Role of aqueous methyldiethanolamine (MDEA) as solvent in natural gas sweetening unit and process contaminants with probable reaction pathway," *Journal of Natural Gas Science and Engineering,* vol. 24, pp. 124-131, 5// 2015.

[9] Environmental Protection Agency, "U.S. annual non-baseload CO2 output emission rate," US Environmental Protection Agency, Washington, DC, US2015.

[10] R. Turton, R. C. Bailie, W. B. Whiting, and J. A. Shaeiwitz, *Analysis, Synthesis and Design of Chemical Process*, 3rd ed. Boston, MA, U.S.: Prentice Hall PTR International, 2008.

[11] R. H. Perry and C. H. Chilton, *Chemical Engineers' Handbook*, 5 ed. New York, US: McGraw-Hill, 1973.

[12] L. S. Abrahams, C. Samaras, W. M. Griffin, and H. S. Matthews, "Life Cycle Greenhouse Gas Emissions From U.S. Liquefied Natural Gas Exports: Implications for End Uses," *Environmental Science & Technology,* vol. 49, pp. 3237-3245, 2015/03/03 2015.

[13] World Nuclear Association, "Comparison of Lifecycle Greenhouse Gas Emissions of Various Electricity Generation Sources," World Nuclear Association
London, UK2011.

468547 Data Envelopment Analysis Coupled with Thermodynamic and Life Cycle Assessment Metrics for Solvent Screening: Application to CO_2 Capture

Abstract:

The growing trend towards more sustainable products and processes calls for advanced decision-making tools for the assessment and optimisation of the economic, environmental and social impact of chemicals and chemical processes. The pressure for more sustainable processes is particularly strong in the chemical industry, in which thousands of products are developed and launched at a very fast pace, raising the question (before scaling them up) of whether they are more sustainable than the existing ones or not.

Particularly, in green chemistry applications, engineers and scientists are confronted with the task of screening chemicals (e.g. mainly solvents, additives, etc.) according to sustainability principles. In practice, it is quite hard to find a single chemical/process performing best in the whole range of sustainability metrics, as inherent trade-offs typically arise between conflictive thermodynamic, environmental and safety metrics. There has been a substantial amount of research in the field of multi-criteria assessment of chemical processes.[1-3] In contrast, the problem of screening chemicals according to several criteria has received much less attention to date. In the absence of any rigorous and systematic screening method, practitioners assess them by defining, in either an implicit or explicit manner, subjective weights that are attached to every category of interest (e.g. viscosity, density, solubility, etc.). These weights enable the calculation of aggregated indicators on the basis of which the chemicals can be ranked.[3-5]While it is straightforward to implement, this approach is highly sensitive to the weights defined and provides in turn no insight into the underlying multi-criteria problem.

In this work we propose a new methodology to screen and select chemicals according to the extent to which they adhere to sustainability principles that is based on Data Envelopment Analysis (DEA), a

technique developed in economics for efficiency assessment. Essentially, DEA is a non-parametric analytical tool for quantifying the relative efficiency of a set of decision making units (DMUs) that makes use of linear programming techniques.[6,7]DEA serves two purposes: i) it identifies, among a set of DMUs - each producing multiple outputs from multiple inputs - the efficient ones (i.e. those with the best ratio desired outputs to required inputs); and ii) it establishes efficient targets for the inefficient units that if achieved would make them efficient. DEA has recently found many applications in environmental studies, but to the best of our knowledge has never been applied to assess chemicals in green chemistry applications. In essence, we combine here DEA with a set of thermodynamic and life cycle assessment indicators so as to identify the chemicals that perform best according to technical (economic), environmental and social aspects simultaneously.

The capabilities of the DEA-based approach in the context of green chemistry are demonstrated through the screening of 125 conventional amine-based solvents for CO2 capture according to 10 performance indicators. Our approach eliminates 36% of the solvents for being inefficient, establishing quantitative targets and clear guidelines on how to improve them. The DEA models are capable of identifying the practical downsides of amines as main sources of inefficiency through the calculation of clear improvement targets for properties such as vapour pressure, acute toxicity and several life cycle impacts. The final aim of the DEA-based analysis is to facilitate the selection of more sustainable chemicals in the transition towards a more sustainable chemical industry.

References

1. C. Capello, U. Fischer and K. Hungerbühler, *Green Chem.*, 2007, **9**, 927.
2. G. Wernet, S. Papadokonstantakis, S. Hellweg and K. Hungerbühler, *Green Chem.*, 2009, **11**, 1826.
3. S. Hellweg, U. Fischer, M. Scheringer and K. Hungerbühler, *Green Chem.*, 2004, **6**, 418.
4. M. Tobiszewski, S. Tsakovski, V. Simeonov, J. Namieśnik and F. Pena-Pereira, *Green Chem.*, **17(10)**, 4773-4785.
5. A. I. Papadopoulos, S. Badr, A. Chremos, E. Forte, T. Zarogiannis, P. Seferlis, S. Papadokonstantakis, C. S. Adjiman, A. Galindo and G. Jackson, *Chem. Eng. Trans.*, 2014, **39**, 211–216.
6. W. D. Cook and L. M. Seiford, *Eur. J. Oper. Res.*, 2009, **192**, 1–17.
7. P. Zhou, B. W. Ang and K. L. Poh, *Eur. J. Oper. Res.*, 2008, **189**, 1–18.

470934 Using Multiobjective Optimization and Life Cycle Assessment for the Design of More Sustainable National and International Energy Systems

Abstract:

Using multiobjective optimization and life cycle assessment for the design of more sustainable national and international energy systems

Sabio, N[a]*., Volkart, K[b]., Densing, M[b]., Strachan, N[a].

[a] ***UCL Energy Institute, University College London, 14 Upper Woburn Place, WC1H 0NN***

[b] ***Paul Scherrer Institut, Energy Economics Group, 5232 Villigen, Switzerland***

Nowadays, an increasing energy demand driven by population and economic growth, and the need to reduce green-house gas emissions are the two major axis along which energy systems need to evolve. Not in vain, the design of sustainable energy systems has also been classified as a major challenge in the area of process systems engineering [1,2].

Energy systems models were developed after the first oil crisis using the methods of linear programming for informing policy makers concerned with the importance of energy systems planning and design at the time [3]. These models are still nowadays the main decision-support tools used for informing energy policy.

Although they were initially conceived for analysing energy security and costs, climate regulation has turned environmental factors in the major focus of energy systems analysis. For instance, after the UK committed to reduce its greenhouse gas emissions by 80% in 2050 [4], the Committee of Climate Change produced a set of carbon budgets based on energy systems analysis that outline the pathways and expected changes in the energy system needed to achieve the target [5].

Conflicting objectives lie at the heart of any design problem. Multi-objective optimization (MOO) frameworks were pervasive in PSE since the early days, the combination of multi-objective optimization frameworks combined and life-cycle analysis [6] has gained more recent attention [1,7,8]. Nevertheless, to the extent of our knowledge, this framework has not been applied to energy systems models.

One reason for this might be that multi-criteria analysis is not seen as a popular method in energy systems modelling community [9]. Although the advantages of multi-objective optimization have already been more widely recognised when applied to distributed generation problems involving a variety of stakeholders [10]. However, the challenges for the implementation of life cycle assessment in a whole energy systems model has been subject of more extensive attention in the literature [11].

In the present work we propose a framework for simultaneously optimizing several life cycle environmental impact indicators and the economic performance in energy systems models. For that purpose we use the classical ε-constraint deterministic multi-objective method that a side from being suitable for convex Pareto-sets, it allows to exploit the problem structure showing the different objective spaces without the need to articulate a priori preferences (i.e., weights), and avoiding the numerical problems arising from normalization.

Our results evaluate the effect of simultaneously optimizing several life-cycle environmental impact metrics against the more traditional total system cost optimisation with CO_2 emissions constrained for a newly developed national energy systems model of the UK (UKTM-UCL) [12] and for an international multi-regional energy systems model (GMM) [13].

***Corresponding author: n.sabio@ucl.ac.uk**

References

[1] Grossmann, I.E., Guillén-Gosálbez, G. (2010) Scope for the application of mathematical programming techniques in the synthesis and planning of sustainable processes, ***Computers and chemical engineering***, 34: 1365-1376

[2] Grossmann, I.E.(2012) Advances in mathematical programming models for enterprise-wide optimization, ***Computers and Chemical Engineering,*** 47: 2-18

[3] Stefan Pfenninger, Adam Hawkes, James Keirstead, Energy systems modeling for twenty-first century energy challenges, Renewable and Sustainable Energy Reviews, Volume 33, May 2014, Pages 74-86

[4] Acts of Parliament. ***Climate Change Act 2008***. The Stationary Office Limited, UK

[5] The Committee of Climate Change (CCC) The 5th Carbon Budget: The next step towards a low carbon economy, November 2015

[6] Guinee JB, Goree M, Heijungs R, Huppes G, Kleijn R, de Koning A, van Oers L, Wegener Sleeswijk A, Suh S, Udo de Haes HA, de Bruijn H, van Duin R, Huijbregts MAJ. Handbook on Life Cycle Assessment. Operational Guide to the ISO Standards. Dordrecht: Kluwer Academic Publishers, 2002.

[7] Azapagic A, Clift R. (1999) The application of life cycle assessment to process optimisation. ***Computers and Chemical Engineering***, 10:1509–1526.

[8] N Sabio, C Pozo, G Guillén Gosálbez, L Jiménez, R Karuppiah, Vasudevan, V., Sawaya, N., Farrell, J.T. Multiobjective optimization under uncertainty of the economic and life cycle environmental performance of industrial processes, ***AIChE Journal,*** 60 (6): 2098-2121

[9] Hall, M.H. L., Buckley, A.R. (2016) A review of energy systems models in the UK: Prevalent usage and

categorisation. ***Applied Energy,*** 169: 607-628

[10] Alarcon-Rodriguez, A., Ault, G., Galloway, S. (2010) Multi-objective planning of distributed energy resources: A review of the state-of-the-art. ***Renewable and Sustainable Energy Reviews,*** 14 (5): 1353-1366

[11] Volkart K. (2014).Development of a New Methodology for the Integration of LCA and Energy-Economic System Modelling, ***LCA XIV***, San Francisco, USA

[12]UCL Energy Institute Models Website, Available at: https://www.ucl.ac.uk/energy-models/models/uktm-ucl [10 May 2016]

[13] Turton H., Panos E., Densing M., Volkart K. (2013). Global Multi-regional MARKAL (GMM) model update: Disaggregation to 15 regions and 2010 recalibration, PSI Bericht 13-03

472120 Life Cycle Assessment and Multiobjective Optimization in a Natural Gas Based Petrochemical Complex

Abstract:

In this work, we address the optimization of a natural gas based petrochemical complex using economic and environmental criteria. The mathematical model includes linear and non-linear models for single plants, taking into account main operating variables, intermittent deliveries and inventory variable profiles. The environmental objective is measured with the global warming potential (GWP) and ReCiPe metrics according to the life cycle assessment procedures using indicators at two levels: eighteen midpoint indicators and three endpoint indicators realised in SimaPro (Pre-Consultants, 2015).The resulting mixed integer nonlinear programming (MINLP) models are implemented in GAMS. The bi-criteria MINLP model is solved with the Îµ-constraint method. Mid-point indicators are presented for current conditions. Numerical results show that the petrochemical complex can satisfy product demands while improving the environmental performance by decreasing greenhouse gases emissions in 33 %, from 1,042 to 695.33 kt CO_2-eq/year in the case of GWP minimization.

475481 Corporate Drivers Towards Sustainability: Carrots and Sticks

Abstract:

This presentation will summarize developments that motivate corporations to adopt more sustainable practices and principles ahead of mandatory command and control regulation. Beginning with basic risk management principles, the presentation will review developments that formed the basis for sustainability concepts including litigation, regulation of waste, protection of resources, increased scientific knowledge regarding impacts to the environment and public health, increased population density, greater transparency and conflict. The presentation will continue with a summary of current sustainability principles, voluntary programs and developments, such as guidance from the Securities and Exchange Commission regarding material risk, increased shareholder activism and President Obamaâ€™s Executive Orders, serving to motivate corporations to move forward with revised business models incorporating sustainable decision making, as well as critical developments arising with and after the 2015 Paris Climate Conference which hosted the 21st Conference of the Parties (COP21) and the Sustainable Innovation Forum 2015.

475482 Solarthermal Chemistry – the Path Forward

Abstract:

CSP/PV hybrid farms provide the path forward for large scale solarthermal chemical processing. PV electricity is lower cost compared to CSP as a result of distributed growth that has provided manufacturers and installers valuable experience leading to rapid cost cutting. However, PV can't match CSP in terms of cheap storage. PV can be used to supply parasitic electricity needs on-sun, while CSP can be used for supplying electricity whenever a cloud passes over, as well as into the evening. Directly driving the solarthermal reaction with concentrated sunlight using a hybrid solarthermal/electrical receiver is most efficient since there is no loss in efficiency converting sunlight to electricity. The "sweet spot" for such solarthermal heating is processes operating where fast radiative heating drives the processing and provides for fast chemical reaction kinetics. Hybrid solarthermal/renewable electricity driven reactors are required in order to provide for viable non-intermittent chemical processing. Further, although the overall objective is a totally renewable process to fuels and chemicals, most likely some fossil feedstocks will provide for a path forward. Considerations also need to be made concerning the availability of suitable sunlight and the proximity of required feedstocks. A number of processing challenges need addressed: (1) design of hybrid solarthermal/electric chemical reactors, (2) process control for bumpless transfer between concentrated solar heating and PV/CSP provided electrical resistance heating, (3) improved methods for thermal storage, and (4) process development for appropriate chemical processes, among other challenges. This presentation will attempt to provide for a framework for a path forward using concentrated sunlight and renewable electricity to drive chemical processing. Some example opportunities will be discussed.

475483 The Evolution of Environmental Protection and Associated Engineering Tools: From Environmental Management to Sustainability

Abstract:

The historical evolution of environmental protection is going to be presented from ancient times to the present day. 20th/21st century approaches will be elaborated on, ranging from environmental management to waste reduction, pollution prevention, green chemistry/engineering, and sustainability. Particular emphasis will be given to the concept of sustainability, including the variety of its definitions. Possible interrelations among these different sustainability concepts, and relevant quantifying tools will be discussed. Finally, a novel sustainability concept will be introduced, and tools for its quantification will be discussed.

The Evolution of Environmental Protection

and Associated Engineering Tools:

From Environmental Management to Sustainability

Vasilios I. Manousiouthakis

2016 AIChE Annual Meeting

Abstract

The historical evolution of environmental protection is going to be presented from ancient times to the present day. 20th/21st century approaches will be elaborated on, ranging from environmental management to waste reduction, pollution prevention, green chemistry/engineering, and sustainability. Particular emphasis will be given to the concept of sustainability, including the variety of its definitions. Possible interrelations among these different sustainability concepts, and relevant quantifying tools will be discussed. Finally, a novel sustainability concept will be introduced, and tools for its quantification will be discussed.

461824 Systematic Process Design Strategies for Efficient and Synergistic Integration of Solar Thermal Hydrogen, Electricity and Fresh Water Production Processes

Abstract:

There is an ever-increasing demand for food, energy and water (FEW) due to population growth and change in consumption habits. For the past century, energy, chemicals, fertilizers and even fresh water via desalination or water treatment have been supplied by using fossil fuel resources. However, with decaying fossil fuel reserves and increasing greenhouse gas (GHG) emissions, the grand challenge is to sustainably meet FEW needs without compromising the current level of civilization. Development and implementation of renewable energy conversion processes by systems approach are essential to ensure a smooth transition to a sustainable economy. Among these alternatives, solar energy is particularly prominent because of its tremendous potential [1]. High conversion efficiency and integrated energy storage systems are indispensable to overcome intermittent and dilute nature of harnessing solar energy [2-3]. Recently proposed hydricity concept, synergistic solar thermal hydrogen and electricity coproduction, presents a potential solution for continuous and efficient power supply and also an exciting opportunity to envision and create a sustainable economy to meet all the human needsâ€"namely, food, chemicals, transportation, heating, and electricity [1].

Here, we introduce an integrated process design strategy based on the hydricity concept for the coproduction of fresh water, hydrogen and electricity from concentrated solar thermal energy. Process simulations for the proposed integration are performed in an integrated Matlab and Aspen Plus platform [4]. The operating conditions of various units and topological structure of the integrated process are determined via sensitivity analysis and optimization in Matlab. Detailed process simulations of thermal desalination processes are validated by comparison with performance data from the current operating desalination plants. The resultant desalination processes are further integrated with

solar thermal electricity and hydrogen processes to coproduce electricity, hydrogen and fresh water. Standalone and integrated solar conversion processes are evaluated based on the process energy efficiency and *exergy efficiency*that refers to the fraction of incident solar exergy that is directly recovered as the net exergy output, which is defined as the sum of electricity and the hydrogen exergy output. We have identified exergetically inefficient steps and implemented new process designs to improve the thermal efficiency of standalone solar desalination processes.

The integration of solar thermal power generation, solar thermal hydrogen production and thermal desalination techniques reduces exergy losses associated with each one of the standalone processes and thus provide a synergistic strategy. The implementation of the proposed process synthesis strategy results in efficient solar thermal process designs, which coproduce i) baseload power, ii) fresh water, and iii) hydrogen. Depending on the targeted end-use, hydrogen can be used as an energy career, transportation fuel, feedstock for the production of chemicals, fuels or fertilizers for food production. Coproduced products can supply all FEW needs unveiling a sustainability roadmap.

[1] Gençer E, Mallapragada DS, Marechal F, Tawarmalani M, Agrawal R. Round-the-clock power supply and a sustainable economy via synergistic integration of solar thermal power and hydrogen processes, Proceedings of the National Academy of Sciences (PNAS), 112(52), 15821-15826, 2015.

[2] Gençer E, Al-musleh E, Mallapragada D, Agrawal R. Uninterrupted Renewable Power through Chemical Storage Cycles. Current Opinion in Chemical Engineering, 5, 29-36, 2014.

[3] Gençer E, Agrawal R. A commentary on the US policies for efficient large scale renewable energy storage systems: Focus on carbon storage cycles, Energy Policy, 88, 477-484, 2016.

[4] Gençer E, Tawarmalani M, Agrawal R, Integrated solar thermal hydrogen and power coproduction process for continuous power supply and production of chemicals, Computer Aided Chemical Engineering, 37, 2291-2296, 2015.

472761 Ideas As a Process Intensification Tool with Application to Natural Gas Reforming Based Hydrogen Production

Abstract:

In this work, process intensification is identified as a process synthesis activity aiming at significant improvements over traditional process designs. The infinite-dimensional state-space (IDEAS) conceptual framework is put forward as a systematic tool for process intensification, which can identify fundamental limitations to the attainable performance by networks of technologies under consideration. The employed approach allows for a broad search of the state-space for process intensification opportunities, even by low dimensional IDEAS linear programming approximations that always retain design feasibility. The proposed process intensification methodology is applied to a case study of natural gas reforming based hydrogen production. The obtained results identify intensified process designs featuring hot/cold utility costs that in certain cases can be lower by more than an order of magnitude to those of traditional designs.

464734 Optimal Design of Integrated Upgrading Plant and Utility System for the Oil Sands Industry

Abstract:

Profitability in the oil sands industry historically has been quite unstable and it is significantly affected by the fluctuating crude oil price. Specially, the margin between bitumen production costs and the synthetic crude oil market price is insignificant these days. However, despite the unstable market and low profitability, oil sands production is predicted to be tripled by the year 2020, when taking into account existing projects, projects under development, and announced investments. Hence, it is imperative to reduce the operating costs for both existing and developing upgrading plants to allow for a more sustainable industry with a larger profit margin.

Limited number of works studied the optimization of oil sands upgrading and utility system. Ordorica-Garcia et al. [1] provided a model for the estimation of the energy requirements associated with producing crude oil products from oil sands according to the current commercially used production schemes. In their later work, Ordorica-Garcia et al. [2] presented a model for optimizing energy production for oil sands operations to reach the minimum of the total annual cost of energy consumptions in the oil sands industry. The model determined the optimal combinations of power and hydrogen plants to meet the given energy demands. Betancourt-Torcat et al. [3] incorporated the two previous studies into a more complex model simultaneously searching for the optimal set of oil production schemes and the corresponding energy infrastructures that meet the total production demands. The proposed superstructure involves multiple extraction and upgrading technologies and it is useful for production costs estimation, evaluating future production schemes and energy demand scenarios, and identifying the key parameters that affect the oil sands industry. In a more recent work [4], an energy optimization model for the oil sands upgrading operations was presented to reach the minimum downstream operating costs at the upgrading part in the oil sands industry, and the proposed superstructure of this work only focused on upgrading plants.

In this work, optimal design of integrated upgrading plants and utility systems in the oil sands industry is investigated. For the upgrading plant part, several state-of-the-art applicable upgrading schemes are considered, which include atmospheric and vacuum distillation, LC-fining, delayed coking and hydrotreatment. For the utility system part, a detailed superstructure model is proposed in this work. Compared to previous works [1-4] where the utility systems were just taken account as energy producers and very rough approximations were used for their operational and economic features, this work considers detailed features including process capacity, heating rate, capital cost, and operation and maintenance cost. A more complex polygeneration superstructure is developed in this study for the utility systems which can produce power, hydrogen and steam of varying quality. It includes gasifier (to produce syngas from coke or residue withdrawing from the upgrading plants), syngas cleaning, air separation unit, gas turbine, heat recovery steam generator, gas-fired boiler, and steam turbine. Results of various case studies demonstrate that the proposed work can efficiently address the optimal design of the integrated upgrading plant and utility system, which can lead to significant cost reduction of the oil sands industry operations.

References:

[1] G. Ordorica-Garcia, E. Croiset, P. Douglas, A. Elkamel, and M. Gupta, "Modeling the Energy Demands and Greenhouse Gas Emissions of the Canadian Oil Sands Industry," *Energy & Fuels,* vol. 21, pp. 2098-2111, 2007.

[2] G. Ordorica-Garcia, A. Elkamel, P. L. Douglas, E. Croiset, and M. Gupta, "Energy optimization model with CO2-emission constraints for the Canadian oil sands industry," *Energy and Fuels,* vol. 22, pp. 2660-2670, 2008.

[3] A. Betancourt-Torcat, G. Gutierrez, A. Elkamel, and L. Ricardez-Sandoval, "Integrated Energy Optimization Model for Oil Sands Operations," *Industrial & Engineering Chemistry Research,* vol. 50, pp. 12641-12663, 2011.

[4] J. Charry-Sanchez, A. Betancourt-Torcat, and L. Ricardez-Sandoval, "An optimization energy model for the upgrading processes of Canadian unconventional oil," *Energy,* vol. 68, pp. 629-643, 2014.

452326 Design and Operation of a 10 MWe Supercritical CO_2 Recompression Brayton Power Cycle

Abstract:

Supercritical carbon dioxide (sCO_2) Brayton power cycles are gaining increasing attention as an attractive alternative to conventional Brayton and Rankine cycles using steam. Brayton cycles with sCO_2 as the working fluid have the potential for more compact design, higher thermal efficiencies, and lower cost of electricity. In view of these promising advantages, the U.S. Department of Energy under its Supercritical Transformational Electric Power Program is in the process of designing, with the intent to build, a nominal 10 MWe (net) sCO_2 Brayton cycle test facility to accelerate scale-up, development, and commercial deployment. In this work, steady-state and dynamic process models of a 10 MWe indirect recuperated sCO_2 recompression Brayton cycle have been developed to analyze cycle design and operation for use with power generation applications. The process models are capable of predicting the nominal design point, off-design, and part-load performance.

In this presentation, the 10 MWe sCO_2 recompression Brayton process model, assumptions, and design parameters will be described and the key results for the nominal steady-state operating point will be presented. The additional specifications and methods applied to provide increased model fidelity for generating the rigorous pressure-driven dynamic model will also be discussed, including the use of performance curves for the turbomachinery and flow-dependent pressure drop correlations, volume specifications, and heat capacity calculations for the heat exchangers. The dynamic model predicts time-dependent profiles of key outputs such as power generated and thermal efficiency as a function of compressor inlet temperature and pressure, turbine inlet temperature, bypass recompression fraction, and other key variables for the purpose of designing effective control strategies.

From the dynamic perspective, this presentation will highlight the transient responses of the sCO_2 cycle to reductions in heat input, representing typical off-design and part-load scenarios arising from integration with a load-following energy plant. Based on the open-loop cycle results without controllers, several operational strategies will be analyzed for maximizing cycle efficiency by maintaining turbine inlet temperature during heat input turndown subject to process constraints.

464793 Integrated Thermochemical Process for Optimal Co-Production of Liquid Fuels and Chemicals

Abstract:

Although the current crude oil price is quite low, it is still necessary to develop reliable biorenewable conversion technologies from a long-term perspective due to diminishing petroleum reserves and increasing market requirements. As a very versatile energy resource, biomass can not only be converted to liquid transportation fuels for vehicular use but can also be transformed to aromatics, olefins, lubricants, industrial chemicals, and many other products currently derived from petroleum or natural gas. Indeed, co-producing these value-added chemicals in an integrated biorefinery can incentivize the de-risking of the â€œfront-endâ€ processes that are necessary for liquid fuels production.

An integrated thermochemical conversion based biorefinery plant for coproducing premium quality liquid transportation fuels, olefins, and aromatics is systematically designed and synthesized in this presentation. The central elements contain gasification, fast pyrolysis, hydrocarbon upgrading, methanol synthesis, methanol-to-olefins, methanol-to-propylene, and methanol-to-aromatics. In general, the pretreated biomass is directed to the gasifier for syngas production and to the fast pyrolyzer for crude bio-oil production, respectively. The major portion of the hot syngas from the gasifier is used to provide heat for the fast pyrolyzer and to promote the fluidization. The crude bio-oil is converted to liquid fuels through sequential hydrotreating and downstream separation. The fluidizing gas along with the syngas from the fast pyrolyzer will finally serve as the feedstocks for methanol production. The produced methanol will be converted to olefins and aromatics through alternative routes. In the developed process superstructure, which considers multiple catalysts and reactor types for certain processing units, is formulated as a generalized disjunctive programming model. To reduce the computational complexity of optimization model, a symbolic regression based approach is proposed for generating surrogate models to describe those complex processing units. The bilinear, trilinear, quadrilinear, and concave terms are reformulated for enhancing the optimization model to be solved to global optimality. Water, power and heat integration studies are simultaneously carried out to reduce the utility cost. Trade-offs between process economics and hydrogen source selection are made. The top adjustable factors on overall economic performance are identified through ranking relevant candidates. Through this analysis, the designed integrated process exhibits high flexibility to cope with the uncertainties on market demands and feedstock sources.

465824 Design and Optimization of Integrated Carbon Capture and Conversion with Natural Gas to Produce Syngas

Abstract:

Direct chemical recycling of CO_2 to fuels and chemicals can significantly reduce fossil fuels-related greenhouse gas emission, and create great economic opportunities for energy and chemical sectors worldwide. As much as 2 Gt/yr of fuels and 200 Mt/yr of chemicals can be produced from CO_2 utilization [1]. In this work, we develop and optimize a novel process which takes power plant flue gas and cheap methane sources (e.g., renewable biogas, landfill gas, coalbed methane, CO_2-contaminated natural gas, shale gas) as raw materials and produces syngas (CO and H_2) - the universal precursor for hydrocarbon-based fuels and chemicals. The process is integrated such that it can simultaneously separate CO_2 from power plant flue gas as a source of carbon, and convert CO_2 with methane as a source of hydrogen to produce syngas. The process also utilizes renewable energy such as concentrated solar energy to partially supply the process energy demand.

Specifically, we combine a CO_2-selective zeolite [2] and a dry-reforming catalyst in a sequential configuration to capture and store CO_2 from dilute flue gas, and convert it to syngas using renewables. A special instance of this is when both the zeolite and catalyst are packed inside a single column in a layered configuration. The process is dynamic and operates in a cyclic manner, where each cycle consists of at least two steps. In the first step, the zeolite section separates and intermittently stores CO_2 from flue gas at ambient temperature. In the second step, the zeolite section automatically supplies CO_2 and methane toward the catalyst section for conversion, due to desorption triggered by solar heating and/or the introduction of methane-rich feed. The high temperature for conversion is partially achieved by using concentrated solar energy. The simultaneous desorption and conversion of CO_2 completely eliminates a separate regeneration step, which is typically the most energy-intensive operation in current capture technologies.

The integration of CO_2 adsorption, intermittent storage, desorption using reactor feed, and conversion with methane is largely controlled by the trade-offs between these competing phenomena. Hence, a predictive model of the overall process has been difficult to create – largely because of the time-dependent, distributed and cyclic nature of the process and the complex interaction between the adsorption and conversion sections. In this work, we use a nonlinear, algebraic and partial differential equation (NAPDE)-based model that captures the equilibrium, transport and kinetic behavior of the gases for the dynamic adsorption, desorption and conversion operations. The optimization of the NAPDE model is also challenging. To this end, we have recently developed and applied a data-driven platform for the simulation-based optimization of complex process models, which treats the NAPDE-based model as a black-box system and use it to generate the required data for optimization. We optimize the process for different feed conditions and syngas specifications (H_2:CO ratio). Once optimized, the integrated technology shows significant promise for reducing the cost of CO_2 capture by eliminating the need for a material regeneration step, and reducing the cost of CO_2 conversion by storing solar energy as chemical energy while mitigating GHG emission.

References:

[1] Quadrelli, E. A.; Centi, G.; Duplan, J.-L.; Perathoner, S. Carbon Dioxide Recycling: Emerging Large-Scale Technologies with Industrial Potential. *ChemSusChem*.2011, 4:1194–1215.

[2] Hasan, M. M. F.; First, E. L.; Floudas, C. A. Cost-effective CO_2 capture based on *in silico* screening of zeolites and process optimization. *Physical Chemistry Chemical Physics* 2013, 15 (40):17601–17618.

464736 Optimal Design and Operation of a Semi-Closed Oxy-Combustion Combined Cycle Power Plant

Abstract:

Optimal design and operation of a semi-closed oxy-combustion combined cycle power plant

Holger Teichgraeber[1] Adam Brandt[1]

[1]Department of Energy Resources Engineering, Stanford University, Stanford, California 94305-2220, USA

As gas turbines provide flexibility to energy systems today, in a low carbon future energy system, natural-gas-based power systems with CO_2 capture may take over that role. Hence, we analyse the design and operations of an oxyfuel-based CCS system that is designed to operate flexibly on timescales of order one hour. Such a system can provide an intermediate level of supply flexibility beyond that provided by short response batteries. Such a system could also fill gaps in seasons of low renewable energy availability. (e.g., winter).

Specifically, we present optimal design and hour-to-hour operation of a flexible semi-closed oxy-combustion combined cycle power system. Our method includes co-optimization of the required air separation unit used to produce O_2. Computational optimization is used to maximize net present value, including both operational and design optimization. Operating profit is increased through running the air separation unit at full capacity at times of low electricity prices, storing oxygen and running the oxyfuel-combustion combined cycle at full load in times of high electricity prices. This avoids energy-intensive air separation when the economic penalty is the greatest.

The system consists of a cryogenic air separation unit, oxygen storage facilities, a natural gas combustion turbine, a heat-recovery steam generator, a steam turbine and a CO_2 compression unit. In this integrated system, all components are represented in a modular fashion using energy and mass balances. The air separation unit is modelled for operation at full capacity and partial load based on industrial data. The gas turbine's, steam turbine's and CO_2 compression unit's partial load operation are modelled based on existing literature. In order to model the oxyfuel-combustion gas turbine, the partial load gas turbine model is expanded with pure oxygen combustion and a recirculation fluid cycle. The heat recovery steam generator model is based on previous work [1]. System configuration and the sizing of various components have a substantial impact on operations and system economics.

The system in this work is able to operate flexibly and its overall operation including time-varying electricity prices is optimized. It also includes the air separation unit and an oxygen storage facility within the system boundary. Amongst all system components, the air separation unit is the least flexible and gaseous or liquid oxygen storage would be beneficial in order to provide fast ramping in case of changing electricity prices.

We maximize net present value (NPV) which is a function of design and operational decision variables. A similar optimization approach has been taken and described in previous work [2]. The design optimization is formulated as a mixed integer nonlinear programming problem. It uses the operational optimization as an inner loop in order to find the maximum NPV. This problem is solved using an algorithm that consists of Particle Swarm Optimization (PSO) and Mesh Adaptive Direct Search (MADS). The optimal operation maximizes annual profit using a Sparse Nonlinear OPTimizer (SNOPT) algorithm with multiple starting points due to the nonconvexity of the problem.

[1] Kang, Charles A., Adam R. Brandt, and Louis J. Durlofsky. "Optimizing heat integration in a flexible coal–natural gas power station with CO 2 capture." *International Journal of Greenhouse Gas Control* 31 (2014): 138-152.

[2] Brodrick, Philip G., Charles A. Kang, Adam R. Brandt, and Louis J. Durlofsky. "Optimization of carbon-capture-enabled coal-gas-solar power generation." *Energy* 79 (2015): 149-162.

462461 Natural Gas to Liquids, Olefins, and Aromatics Under Uncertainty in Feedstock and Product Prices

Abstract:

Although natural gas to liquid transportation fuels processes have been investigated for several decades, they were considered economically unfavorable due to high feedstock costs. [1,2] Recent studies have demonstrated the profitability of such refineries can be significantly increased by coproducing valuable chemicals, such as olefins and aromatics, in tandem with liquid fuels. [3,4] Even more recently, new studies have investigated the potential of stand-alone natural gas based chemical refineries, where the main products output are mainly olefins and aromatics, and assessed the technoeconomic and environmental impacts of such processes. [5,6]

The economics of such natural gas facilities is intricately dependent on the complexity of the refinery configuration and the type, quantity, and quality of products produced (i.e., the product slate). The large chemical space available for the manufacture of liquid fuels and chemicals has been expanded even further in recent years due to novel and emerging technologies that can produce liquid fuels, olefins, and aromatics. [3-6] The types of technologies to utilize and the types of products to produce is a tremendously difficult challenge to address, one that is further complicated with the uncertainty surrounding the costs and prices of feedstocks and products, respectively. In this talk, we address these challenges through a robust optimization framework that is incorporated in a large-scale process superstructure that contains several alternatives for the production of fuels, olefins, and aromatics from natural gas. [6-10] We introduce an important production parameter that allows us to investigate the economic effects of natural gas to liquids, olefins, and aromatics refineries. The resulting large-scale, nonconvex, mixed-integer, nonlinear optimization (MINLP) model is solved using our novel global optimization algorithm. [11]

Our comprehensive optimization-based process synthesis framework contains several direct and indirect natural gas conversion technologies and numerous hydrocarbon production alternatives, including Fischer-Tropsch refining and methanol synthesis; each of which is compared to determine the optimal processing pathway. [2-6] Multiple commercial and novel technologies for the production of the olefins and aromatics, such as methanol-to-olefins and toluene alkylation with methanol, are rigorously modeled. [5,6] The key production parameter is introduced as a function of the important products output from the refinery, namely, gasoline, diesel, kerosene, aromatics, and olefins. Robust counterparts are formulated in order to address the uncertainty associated with feedstock and product prices.

Several case studies are presented to investigate the effect of plant capacity and production ratios on the overall profit of the refinery and incorporate uncertainty in the optimal solutions obtained. A discussion on the robust refinery designs that incorporate the effect of uncertainty will follow with emphasis on the major topological decisions as a function of the production parameter. The economic and environmental tradeoffs of the refineries are presented.

[1] Floudas, C. A.; Niziolek, A. M.; Onel, O.; Matthews, L. R. Multi-Scale Systems Engineering for Energy and the Environment: Challenges and Opportunities. AIChE Journal 2016, 62, 602-623.

[2] Baliban, R. C.; Elia, J. A.; Floudas, C. A. Novel natural gas to liquids (GTL) technologies: Process synthesis and global optimization strategies. AIChE Journal 2013, 59, 505–531.

[3] Onel, O.; Niziolek, A. M.; Elia, J. A.; Baliban, R. C.; Floudas, C. A. Biomass and natural gas to liquid transportation fuels and olefins (BGTL+C2_C4): Process Synthesis and Global Optimization. Industrial & Engineering Chemistry Research 2014, 54, 359-385.

[4] Niziolek, A. M.; Onel, O.; Elia, J. A.; Baliban R. C.; Floudas, C. A. Coproduction of Liquid Transportation Fuels and C6_C8 Aromatics from Biomass and Natural Gas. AIChE Journal 2015, 61, 831-856.

[5] Onel, O.; Niziolek, A. M.; Floudas, C. A. Optimal Production of Light Olefins from Natural Gas via the Methanol Intermediate. Industrial & Engineering Chemistry Research 2016, 55, 3043-3063.

[6] Niziolek, A. M.; Onel, O.; Floudas, C. A. Production of Benzene, Toluene, and Xylenes from Natural Gas via Methanol: Process Synthesis and Global Optimization. AIChE Journal 2016, 62 (5), 1531 – 1556.

[7] Li, Z.; Ding, R.; Floudas, C. A. A comparative theoretical and computational study on robust counterpart optimization: I. Robust linear optimization and robust mixed integer linear optimization. Industrial & Engineering Chemistry Research 2011, 50, 10567–10603.

[8] Li, Z.; Tang, Q.; Floudas, C. A. A comparative theoretical and computational study on robust counterpart optimization: II. probabilistic guarantees on constraint satisfaction. Industrial & Engineering Chemistry Research 2012, 51, 6769–6788.

[9] Li, Z.; Floudas, C. A. A comparative theoretical and computational study on robust counterpart optimization: III. improving the quality of robust solutions. Industrial & Engineering Chemistry Research 2014, 33, 13112–13124.

[10] Guzman, Y. A; Matthews, L. R; Floudas, C. A New a priori and a posteriori probabilistic bounds for robust counterpart optimization: I. Unknown probability distributions. Computers & Chemical Engineering 2016, 84, 568-598.

[11] Baliban, R. C.; Elia, J. A.; Misener, R.; Floudas, C. A. Global Optimization of a MINLP Process Synthesis Model for Thermochemical Based Conversion of Hybrid Coal, Biomass, and Natural Gas to Liquid Fuels. Computers and Chemical Engineering 2012, 42, 64-86.

455098 The Sunshot Inititiative: New Opportunities for R&D in Concentrating Solar Power

Abstract:

The SunShot Initiative is a collaborative, national endeavor by the United States Department of Energy's Solar Energy Technologies Office to make electricity generated from the sun cost competitive – on an unsubsidized basis – with traditional forms of electricity generation by the year 2020. Concentrating Solar Power (CSP) constitutes one technological approach to achieving this ambitious goal, generally identified as a levelized cost of electricity (LCOE) $\leq$ \$0.06/kWh$_e$. CSP focuses the sun's rays onto a receiver that converts the photons into thermal energy. Part of this thermal energy can then be transported to storage for use when the sun does not shine, while the remainder is transported to a conventional turbine to generate electricity. The turbine, or heat engine, can realize increased thermodynamic efficiency as the input temperatures increase; however, high temperature CSP systems come with significant materials and cost challenges. Evaluating the interfaces found in the solar receiver, thermal energy storage technologies, heat exchangers, and power blocks will help define the technical challenges and their requisite success metrics. The presentation concludes with remarks, links, and information on DOE's SunShot Initiative so that the audience may stay aware of future Funding Opportunity Announcements.

469740 Non-Stoichiometric Perovskite Oxides As High-Temperature Energy Storage Media and Their Application to Concentrating Solar Power Generation and Hydrogen Production

Abstract:

Although there is a significant abundance of solar resource, the intermittency of the resource compromises its deep integration into the U.S. energy portfolio. However, solar-thermal processes such as concentrating solar power (CSP) are unique in that thermal energy can be efficiently stored in engineered heat-absorbing media. The quantity of heat stored and subsequent power-generation efficiency generally increase as temperature increases, motivating solid particulate storage media for high-temperature systems. However, the storage of only sensible energy limits the potential of both the quantity of stored energy as well as the possible applications of the stored energy. For this reason, energy storage media have been engineered to store both sensible heat and reaction heat, a practice known as thermochemical energy storage (TCES).

Here we present the concepts behind the engineering, synthesis, and characterization of novel doped calcium manganese perovskites ($CaAl_yMn_{1-y}O_{3-\delta}$). These materials exhibit rapid and reversible reduction/oxidation behavior through the introduction of oxygen vacancies in the material structure: $ABO_3 + \Delta H \rightarrow ABO_{3-\delta} + \delta/2\ O_2$. This approach allows the storage of chemical energy through thermal reduction without the kinetic limitations inherent in decomposition reactions. The dopant concentrations of this materials family have been optimized in our laboratories to produce the largest energy storage capacity of a high-temperature perovskite system to date.

Additionally, two applications of the material are presented. First, these materials are applied to a CSP system for electricity generation. In this system, the material is first heated and thermally reduced in the incident solar beam of a solar receiver. The material is then stored in large thermally-insulated bins, after which both the sensible and chemical energy can be utilized by a reaction and heat-transfer process with a working fluid. This high-enthalpy working fluid is passed to a turbine to create power on demand, even when the solar resource is unavailable. Secondly, these materials can operate as thermal-to-chemical energy converters, where the thermal reduction results in a chemically energy-rich species which can be used either fully or in part to facilitate subsequent reactions, such as water-splitting for hydrogen generation. In this embodiment, only the chemical energy is useful, and the sensible energy must be recovered to maximize process efficiencies.

Sandia National Laboratories is a multi-program laboratory managed and operated by Sandia Corporation, a wholly owned subsidiary of Lockheed Martin Corporation, for the U.S. Department of Energy's National Nuclear Security Administration under contract DE-AC04-94AL85000. This work is supported by the U.S. Department of Energy, SunShot Initiative, under Award Number DE-FOA-0000805.

471159 Solar Thermochemical Energy Storage Based on Strontium Carbonate Chemistry

Abstract:

The $SrO/SrCO_3$ chemistry offers the potential for high temperature heat release (~1100-1200°C) upon the carbonation of SrO, which could be a pathway to high efficiency, combined cycle power generation. Building upon previous work on this cycle, this talk will propose a kinetic model for the carbonation reaction derived on experiments using both TGA and packed bed systems. In addition, preliminary technoeconomic and exergy analyses are performed for the cases of CO_2 storage as 1) a compressed gas and 2) ab/adsorbed species in a liquid or solid matrix. Probabilistic methods are used to elucidate the cost drivers and areas for future consideration in evaluation of the cycle feasibility.

452449 Design of Manganese Oxide-Based Particles for High-Temperature Thermochemical Energy Storage

Abstract:

High-temperature thermochemical energy storage shows promise in aiding concentrating solar power plants in meeting variable, grid-scale electricity demand. In this work, manganese oxide-based mixed metal oxide particles have been designed and tested for thermochemical energy storage. Particles are designed for high energy storage capacity, flowability, and physical and chemical stability. We evaluate the effects of Al_2O_3, Fe_2O_3, and ZrO_2 in Mn_2O_3-based spray-dried particles in a TGA between 650°C and 1,200°C over six consecutive redox cycles. Results are compared with thermodynamic predictions from 400-1,400°C under oxidizing and reducing atmospheres. A mixture of 2:1 Fe_2O_3:Mn_2O_3 formed iron manganese oxide spinel (Fe_2MnO_4) on calcination, and demonstrated the highest thermochemical activity despite particle agglomeration and deformation. Conversely, zirconia was an inert support that does not react with manganese oxide. Differences in redox performance between materials with different Fe to Mn ratios have been attributed to ion diffusion and secondary phase formation.

452774 Grid-Scale Thermochemical Energy Storage Using Mixed Metal Oxide Redox Cycles

Abstract:

Thermochemical energy storage (TCES) systems are of great interest in concentrated solar power (CSP) applications: Storing sunlight as chemical energy during the day can enable power generation at night or during cloudy periods, effectively alleviating the inherent intermittency of solar sourced electricity. Metal oxides are among the most attractive TCES materials because they possess high-energy density and high reduction/oxidation temperatures (>1000°C) suitable for driving high-efficiency thermodynamic power cycles. However, limitations involving high-temperature heat transfer, particle handling, sintering, and chemical performance of the metal oxide materials are still problematic in practical application.

A novel, grid-scale CSP/TCES system design that addresses many of the traditional problems associated with metal oxide TCES has been developed at the ANU. This presentation serves as a broad overview of recent work surrounding the designed CSP process, including improvements to the metal oxide active materials, solar reactor design, optical field development, and preliminary techno-economic analyses. Improved chemical performance of the active material is being explored through metal co-doping and the intentional formation of solid solutions. The incorporation of fluidized bed reactors for solar reduction and off-sun oxidation allows for improved heat transfer and enhanced chemical kinetics, and also enables high temperature pneumatic transport of gas-solid reacting flows. The initial solar reactor concept consists of one or several vertical fluidized bed reaction tubes in a beam-up solar cavity receiver, which allows for well-controlled fluidization, high optical efficiency, and minimized convective heat losses. The techno-economic analysis of a 100 MW_{Th} grid-scale facility is employed to predict the levelized cost of electricity.

465618 Kinetic Analysis of High Temperature Thermochemical Energy Storage Based on Calcination–Carbonation Chemical Looping Reactions

Abstract:

This work aims to improve the performance of carbonate storage systems, a form of thermochemical energy storage (TCES) for concentrated solar power (CSP) technologies. TCES has great potential due to its high energy storage densities and inexpensive materials, making it highly competitive with commercial molten salt storage systems. The storage system is based on metal oxide calcination–carbonation chemical looping, a cyclic process for which a broad range of metal oxide candidate materials are considered. Metal oxide produced in the solar-driven endothermic calcination reaction is cycled to the exothermic carbonation reaction, from which thermal energy is released to drive a power cycle.

This work details a comparative thermodynamic and kinetic study of calcination and carbonation reactions, using commercial samples and synthesized mixtures of carbonates such as $CaCO_3$ and $SrCO_3$. Experiments are carried out under a CO_2 atmosphere using thermogravimetric analysis (TGA) and differential scanning calorimetry (DSC) and the materials are selected to attain the desired storage temperature (typically above 565°C), energy density and charging and discharging rates. Chemical and structural cyclability of candidate materials are evaluated as a function of the calcination–carbonation cycle number using several characterization techniques such as scanning electron microscopy (SEM), transmission electron microscopy (TEM), N_2 adsorption-desorption and X-ray diffraction crystallography. We show how different mixtures and synthesis precursors can improve the cyclability of metal oxide–carbonate systems.

465613 Thermodynamic Analysis of $CaCO_3$/CaO Looping for High Temperature Thermochemical Energy Storage

Abstract:

High temperature power cycles, such as the supercritical carbon dioxide (CO_2) cycle, are promising technologies currently being investigated by the international research community for concentrated solar thermal power generation (CSP). These power cycles have the potential to increase the overall efficiency of the power plant compared to state-of-the-art steam Rankine cycles and thereby reduce the levelized cost of electricity (LCOE) of CSP.Â Therefore, novel thermal energy storage materials and processes are required to successfully couple concentrated solar power collectors with high temperature power cycles.

In this context, the $CaCO_3$/CaO thermochemical cycle is an attractive process for thermal energy storage that can provide dispatchable concentrated solar thermal energy at temperatures between 650â€“980Â°C, appropriate to use with high temperature power cycles. Current research efforts focus on material development and solar reactors. However, there is a need for investigation into the field of system integration to support and verify the viability of this concept.

In this work, we evaluate the performance of a $CaCO_3$/CaO system coupled to a supercritical CO_2 power cycle for continuous electricity production. This investigation has identified high-energy demanding subsystems that require minimization to improve the overall efficiency of the process. The effect of the operational conditions (temperature and pressure) in the calcination receiver and carbonation reactor on overall efficiency are also investigated.

470455 Population Ensemble Modeling of Cellulose Dissolution: Insights for Efficient Biomass Processing

Abstract:

A major barrier to the efficient utilization of biomass is the recalcitrance to dissolution of semicrystalline cellulose. The present study addresses the kinetics of swelling and dissolution of cellulose particles in conditions emulating large-scale processing where the particles exhibit a distribution of size and degree of crystallinity. To this end, we have developed a model in which the behavior of a population of particles is obtained from an ensemble of individual cellulose particle dissolution models. The dissolution of individual solid cellulose particles is based on the relevant transport phenomena and kinetics and reveals decrystallization and disentanglement as two important and potentially rate-determinant steps in the process. The average value or the number distribution of any intra-particle property captured by the individual particle model can be determined by simulation of a sufficient number of individual particles such that ensemble averages are independent of the particle numbers and the computed particle distributions are acceptably smooth. Using this population model, various cellulose particle size and crystallinity distributions are analyzed for different dissolution parameters. The findings from this study would be useful for the rational design and optimization of pretreatment processes to reduce the particle size and degree of crystallinity, leading to enhanced biomass utilization.

475618 Renewable Aromatic Hydrocarbon Production from Furfural By a Continuous Dual-Stage Hydrogenation-Cocracking Process

Abstract:

Aromatic hydrocarbon production from furfural by catalytic cracking over zeolite catalysts is significant for realizing biobased aromatics generation, because furfural can be produced from biobased sugars through different transformation routes such as liquid phase conversion and fast pyrolysis. However, due to the high unsaturation degree and high oxygen content, direct cracking of furfural faces low aromatic hydrocarbon yield and coking problems. Therefore, in this work, two improving methods namely hydrogenation pretreatment and methanol-assisted cocracking over HZSM-5 catalyst were combined to convert furfural into aromatic hydrocarbons, and the influence of hydrogenation catalysts was investigated. It was found that in the hydrogenation stage, although the furfural was nearly completely converted over $10Cu/SiO_2$, the furan ring was not efficiently hydrogenated. $5Ni\text{-}5Cu/SiO_2$ and $10Ni/SiO_2$ achieved the further hydrogenation of some furan rings, but over $10Ni/SiO_2$ the primary conversion of furfural was lower and more saturated gaseous hydrocarbons were generated. The different hydrogenation behaviors of furfural affected the subsequent cracking processes. The hydrogenation-cocracking of furfural and methanol over $5Ni\text{-}5Cu/SiO_2$ generated the highest oil phase yield of 26.4%, which was mainly composed of mono-aromatic hydrocarbons such as benzene, toluene, xylenes and trimethylbenzenes. However, because of the insufficient hydrogenation of furan ring over $10Cu/SiO_2$ and the low conversion of aldehyde group as well as some excessive hydrodeoxygenation over $10Ni/SiO_2$ in the hydrogenation stage, the yield and quality of the oil phases in the corresponding following cracking stage both decreased. The comparison of hydrogenation-cocracking process using the optimum hydrogenation catalyst $5Ni\text{-}5Cu/SiO_2$ with other traditional cracking processes including single-stage cocracking of furfural and methanol, hydrogenation-cracking of pure furfural using $5Ni\text{-}5Cu/SiO_2$ as the hydrogenation catalyst and direct cracking of furfural indicated that the hydrogenation-cocracking process was the most favorable technique for the generation of liquid aromatic hydrocarbons and the maintaining of catalyst activity. Finally, based on the experimental results, the hydrogenation-cocracking mechanism of furfural and methanol was proposed to explain the different hydrogenation behaviors of furfural over three catalysts and also the subsequent different cocracking performances.

481269 Biomass Anaeorobic Convertion Technology Development

Abstract:

Base on the demands for environmental protection, energy security and modern agriculture development, with drawing on experiences of vigorously developing anaerobic biomass conversion from developed countries, innovative research on high efficient anaerobic conversion technologies has been conducted by Biomass Bio-chemical Conversion Laboratory, Chinese Academy of Sciences Guangzhou Institute of Energy Conversion.

In recent years, our Laboratory has undertaken more than 30 projects funded by the National Science & Technology Program, National High Technology Research and Development Program, National Natural Science Foundation, National Science & Technology Pillar Program, Provincial and Ministerial Science & Technology Programs and also programs supported by Chinese Academy of Sciences and corporations, focusing on bio-chemical pretreatments of lignocellulosic materials, nutrient balancing regulation base on complementary characteristics of substrates, high concentration co-digestion of multi-feedstock, two-stage continuous desulfuration, treatments of industrial waste (Cassava starch Dreg, Molasses Alcohol Wastewater etc.). Technologies and reactors for breeding/cultivation/pretreatment/anaerobic digestion of energy grassesï¼Œfermentable organics extraction from municipal solid wastes with multiple impurities, two-phase& three-stage anaerobic digestion, combined oil-gas-fertilizer production from food waste have been developed base on our research; integrated technological system for high-efficient, continuous and stable production of biogas from organic wastes has been established. Maximum organic loading rates of 8-10 kgVS/(m^3Â·d), highest solid concentration of 20% as well as stable operation can be achieved within the system; the best performance results of 75% methane content, 89% volatile solid removal rate, 70% energy conversion rate and 45% lignin removal rate have be obtained; the biogas production cost in this system can be lower than 0.5 yuan/m^3. The international leading position of this achievement has been appraised based on the aggregate technology and economic indicators, and it has won the First Award of the Science and Technology Progress of Guangdong Province and Third Award of Energy Science and Technology Progress.
The aim of this report is to systematically introduce our achievement, and we are looking forward to cooperation on relevant research and application with domestic and foreign organizations. This work is funded by national high-tech development programs(No.KFJ-EW-STS-138, KGZD-EW-304-1, 2012AA101802 etc..)

474806 Oxidative Desulfurization of DBT with H_2O_2 Catalysed By TiO_2/Porous Glass

Abstract:

Oxidative desulfurization of DBT with H_2O_2 catalysed by TiO_2/porous glass

Chun Shen[†], Yujun Wang*[,‡], Jianhong Xu[‡], Tianwei Tan[†], Guangsheng Luo*[,‡]

[†]Beijing Key Laboratory of Bioprocess, College of Life Science and Technology, Beijing University of Chemical Technology, Beijing 100029, China

[‡]The State Key Laboratory of Chemical Engineering, Department of Chemical Engineering, Tsinghua University, Beijing 100084, China

Corresponding author: Tel: 86-010-62788568, Fax: 86-010-62788568

Email address: wangyujun@tsinghua.edu.cn and gsluo@tsinghua.edu.cn

Abstract

Aimed at ultra-deep oxidative desulfurization (ODS) of dibenzothiophene (DBT) and 4,6-dimethyldibenzothiophene (4,6-DMDBT) to control air pollution, we specially designed and prepared porous glass supported with TiO_2 nanoparticles acting as an amphiphilic catalyst. Hydrogen peroxide which is considered as the "green" oxidant was used, and for the extreme liquid–liquid phase ratio (usually larger than 1500) reaction system, the pore volume of 0.19 mL/g of the catalyst provides enough space for storage of hydrogen peroxide. The as-prepared catalyst offers a high interfacial surface area of 116.9 m^2/g and enhances the reaction by facilitating the mass transfer. The mono-dispersed TiO_2 exhibited good crystallinity. The mean diameter varied from 2.1 to 7.8 nm with the loading amount increased from 1.27 wt.% to 9.85 wt.%. The catalyst showed high activity and good stability for producing ultra-clean fuels: 100% conversion was obtained within 2 min and the conversion just decreased from 100.0±1.0 % to 94.3±0.6 % after 5 cycles. Overall, this new reusable catalyst provided an alternative for highly efficient ultra-deep desulfurization in a green way.

475573 Hydrolysis of corncob using a modified carbon-based solid acid catalyst

Abstract:

Lignocellulosic biomass is a very important renewable resource that could be used to produce biofuels and various platform chemicals. The conversion of biomass to reducing sugar is an essential intermediate step that can impact the economic benefits of the producing process. The modified carbon-based solid acid is a novel catalyst used in the hydrolysis process. A catalyst material was prepared through the incomplete carbonization and sulfonation of the mixture of glucose and *p*-toluenesulfonic acid, followed by oxidation using 30% H_2O_2 to promote the acidity. The hydrolysis efficiency of the solid acid was significantly increased. The effects of the carbonization temperature, time, and the glucose / *p*-toluenesulfonic acid ratio on catalytic efficiency were investigated. Under the appropriate preparation condition, the catalyst could hydrolyze corncob to produce 78% xylose and 34% glucose. The activity of the catalyst slightly declined after five-cycles application.

474464 Simultaneous Catalytic Conversion of Bio-Oil and Bio-Char for Hydrogen Production

Abstract:

Based on the integrated processes for hydrogen production from biomass, the simultaneous catalytic conversion of bio-oil and bio-char for hydrogen production was investigated in this paper. The potential hydrogen (H_2+CO) yield was increased from 45.70 to 82.19 mmol/g bio-oil with the presence of metals in the bio-char. The effects of metals (K, Ca, Mg, Fe, Zn, and Al) in bio-char on the catalytic conversion of bio-char and bio-oil were also investigated. The catalytic activities of the metals for the conversion of bio-char and bio-oil were followed by the sequence K>Ca>Mg>Fe>Zn>Al. Meanwhile, the alkali and alkaline earth metals (AAEM), including K, Ca, and Mg, were found to play the dominating roles among the metals in bio-char, and the catalytic effect of K was the most obvious. The potential hydrogen yield and tar yields were 143.50 mmol/g bio-oil and 4.96 mg/g bio-oil when adding the metal of 5 wt.% K.

The effects of metals on the biochar and biomass pyrolytic vapor (bio-oil) were investigated. Al had an inhibition on the catalytic activity of biochar, as a result, the soot yield was increased. Naphthalene was the primary tar in reaction temperature was 900 °C, also had a thermal stable in 900 °C. Fe was deactivated by coke and sintering easily.

The integrated process for hydrogen production from biomass was developed

Basing on the gas-solid simultaneous gasification process of the integrated processes of biomass, the effects of metals (K, Ca, Mg, Fe, Zn, and Al) in biochar on the catalytic gasification and catalytic reforming processes were investigated in this paper.

The catalytic reforming of biomass pyrolytic vapor (MPV) over biochar in a fixed bed was investigated. The carbon conversion and potential hydrogen yield of biomass pyrolytic vapor were increased respectively from 68.46% to 86.88% and 42.16% to 82.76% in the presence of biochar. The catalytic reforming activities of different metals were found with the following the sequence: K>Ca>Mg>Fe>Zn>Al. Meanwhile, the alkali and alkaline earth metals (AAEM) like K, Ca, and Mg were found to play the dominating roles in the conversion of MPV, and especially, the catalytic effect of K was the most obvious among the metals in bio-char. The carbon conversion and potential hydrogen yield of pyrolytic vapor were increased by increasing temperature and ratio of steam to pyrolytic vapor.

The effects of metals on the biochar catalytic gasification were also investigated. The catalytic activities of metals on biochar catalytic gasification were in the sequences: K>Ca>Mg>Zn>Fe>Al. AAEM (K, Ca and Mg) and Fe can promote water-gas shift reaction, while Al had an inhibition effect on water-gas shift reaction. The increasing of temperature and steam were benefit for biochar catalytic gasification. When the carbon conversion was more than 90% (S = 20 g/h and T =900 °C), biochar and demineralized biochar need 40 min and 115min respectively.

The effects of metals on the biochar and biomass pyrolytic vapor (bio-oil) simultaneous catalytic conversion were investigated. The potential hydrogen (H_2+CO) yield was increased from 45.70 to 82.19 mmol/g bio-oil with the presence of metals in the biochar. And the tar yield was decreased from 24.88 to 13.84 mg/g bio-oil. The catalytic conversion of biochar and bio-oil were followed by the sequence K>Ca>Mg>Fe>Zn>Al. Meanwhile, the potential of hydrogen and tar yields were 143.50 mmol/g bio-oil and 4.96 mg/g bio-oil when adding the metal of K. The performance of AAEM (K, Ca and Mg) was comparable with Ni in catalytic reforming bio-oil. AAEM can promote water-gas shift and methane reforming reactions, also promote the elimination of soot. Al had an inhibition on the catalytic activity of biochar, as a result, the soot yield was increased. Naphthalene was the primary tar in reaction temperature was 900 °C, also had a thermal stable in 900 °C. Fe was deactivated by coke and sintering easily.

465897 The Pretreatment of Pennisetum Hybrid with Composite Solid Base Catalyst

Abstract:

The Pretreatment of Pennisetum hybrid with Composite Solid Base Catalyst

Zhuang Xinshu, Tan Xuesong, Yu Qiang, Qi Wei, Wang Wen, Wang Qiong, Zhou Gui Xiong,

Yuan Zhenhong*

(Key Laboratory of Renewable Energy, Guangzhou Institute of Energy Conversion, Chinese Academy of Sciences, Guangzhou 510640, China)

Abstract: Pretreatment is a critical step for the downstream enzymatic hydrolysis to producing cellulosic ethanol. Solid base catalyst can be a potential way for lignin removing, xylose releasing and phenols producing. In this paper the effects of the reaction conditions on enzymatic digestibility and total sugar yield of Pennisetum hybrid pretreated in low temperature aqueous environment with composite solid alkaliâ€"CaO/MgO as the catalyst are investigated. The results indicated that 20% catalyst content, 110â„ƒ,80min and 1:20 ratio of solid are the optimal pretreatment conditions, the lignin removal rate is up to 54.18%.The enzymatic digestibility increases from 41.23% to 81.04% with the enzyme loading of 40FPU/g after72h, substrates can be fully hydrolyzed. After pretreatment, the compact structure of the feedstock is damaged, more fibers exposing, the chemical groups such as hydrogen bond in or between are destroyed.

Keywords: Composite Solid Base; Pennisetum hybrid; Enzymolysis; Pretreatment

This work is financially supported by the National Natural Science Foundation of China (51476179) and the Sino-Thai project (51561145015).

Corresponding author: Yuan Zhenhong, 86-20-87057735; yuanzh@ms.giec.ac.cn.

468367 Halophyte Biochar for Water Desalination Brine Concentrate Management

Abstract:

Disposal of concentrate from brackish water desalination is limited by concerns of salt leaching into and contaminating surface and ground waters. These concerns, and measures to prevent leaching, add substantially to concentrate disposal costs with little or no opportunity to recover those costs. One option being explored is the use of concentrate as irrigation water for producing halophyte biomass for forage and other applications. Continuing development of this option requires more knowledge about what happens to the salt in the concentrate once land applied as irrigation water.

The purpose of this project is to determine how much of the salt taken up by the halophyte crops can be protected from leaching by pyrolysis of the biomass, i.e. to determine the potential of sequestering salt in mineral form. Results from this research will be useful for those managing high salinity biomass streams, such as halophyte crop wastes and animal manures, and those seeking local concentrate disposal options with a potential for producing economic and/or environmental returns. The ability to sequester salt in biochar would decrease the negative impacts of land-applying high-salinity materials, including those used in desalination concentrate management systems.

455179 ^{13}c Pathway Analysis for the Role of Formate in Electricity Generation By *Shewanella Oneidensis* MR-1 Using Lactate in Microbial Fuel Cells

Abstract:

Microbial fuel cell (MFC) is a promising technology for wastewater treatment. Basically, the microorganisms sticking on the anode of the MFC utilize various organic compounds from the wastewater as the electron donors to generate electricity, while in the meantime removing the organic compounds in the wastewater. The type of electron donors fed into MFCs affects the electrical performance, and mechanistic understanding of such effects is important to optimize the MFC performance. In this study, we used a model organism in MFCs, *Shewanella oneidensis* MR-1, and ^{13}C pathway analysis to investigate the role of formate in electricity generation and the related microbial metabolism. Our results indicated a synergistic effect of formate and lactate on electricity generation, and extra formate addition on the original lactate resulted in more electrical output than using formate or lactate as a sole electron donor. Based on the ^{13}C tracer analysis, we discovered decoupled cell growth and electricity generation in *S. oneidensis* MR-1 during co-utilization of lactate and formate (i.e., while the lactate was mainly metabolized to support the cell growth, the formate was oxidized to release electrons for higher electricity generation). To our best knowledge, this is the first time that ^{13}C tracer analysis was applied to study microbial metabolism in MFCs and it was demonstrated to be a valuable tool to understand the metabolic pathways affected by electron donors in the selected electrochemically-active microorganisms.

461312 *Advanced Redox-Based Electrochemical Separations for Wastewater Treatment*

Abstract:

Water scarcity is one of the principal challenges for the 21st century, both from a geographical and an economic standpoint. Developing novel, more energetically efficient technologies for wastewater treatment, retreatment and reuse is crucial for long-term water sustainability. One of the principal challenge for water remediation, both in the U.S. and worldwide, is to target contaminants of emerging concern, such as complex organic molecules including pharmaceuticals, chemical precursors, endocrine inhibitors and pesticides.[1] These originate from chemical processes and organic synthesis, the food industry, electronic manufacturing among others. Many of these pollutants disperse in remote locations, such as agricultural locations, in which large-scale purification systems cannot be implemented. Electrochemical methods are particularly attractive for addressing these problems due to their modularity, fast kinetics and integration with renewable energy sources.[2]

We have developed an redox-mediated separation method based on organometallic-functionalized electrodes (e.g. ferrocene or cobaltocene) for the sorption and release of specific ions from the liquid-phase under oxidizing and reducing conditions.[3] Many of the organic contaminants are anionic in nature (e.g. carboxylates, sulfonates or phosphonates), thus selectivity based on chemical interactions with their functional groups is essential in chemical design. Rather than rely solely on charge or size, our immobilized redox-centers targets these compounds through the activation of a specific chemical binding. Moreover, in the charging and discharging steps, electrochemical charge can be stored by ion-selective adsorption onto the porous electrodes, thus dramatically decreasing energetic costs.

This modular system can then be easily implemented at the point-of-exit in chemical manufacturing and wastewater treatment. Here, we present a complete asymmetric redox-system in both batch and flow for (1) water remediation, through the specific separation of anionic and cationic contaminants in water, (2) tandem energy recovery due to pseudocapacitive charge storage and (3) increasing sustainability in chemical processes, through the recovery of expensive synthesis products from complex organic reactions, thus reducing organic waste and decreasing water usage during separation steps.

References

[1] a) Gilliom R.; Hamilton P. *U.S. Geological Survey*. 2014. (USGS fact sheet: http://pubs.usgs.gov/fs/2006/3028/

). b) Doughton, C.G. and Thomas, A. T. (1999). Pharmaceuticals and personal care products in environment: agents of suble change? *EHP* **107** (6), 907-938.

[2] Suss, M. E.; Porada, S.; Sun, X.; Biesheuvel, P. M.; Yoon, J.; Presser, V. *Energy & Environmental Science* **2015**. Zhao, R.; Satpradit, O.; Rijnaarts, H. H.; Biesheuvel, P. M.; van der Wal, A. *Water Res* **2013**, *47*, 1941.

[3] **Su, X.**; Kulik, H; Jamison, T.F.; Hatton, T. A. 2016. Anion-selective redox electrodes: electrochemically-mediated separation with organometallic interfaces. *Advanced Functional Materials.* Advance Article Online. DOI: 10.1002/adfm.201600079

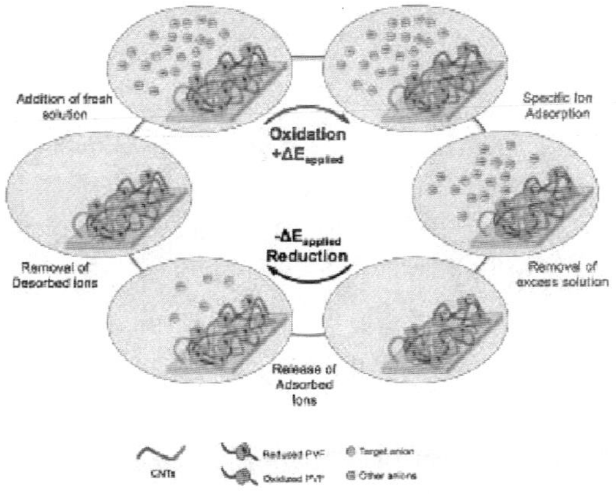

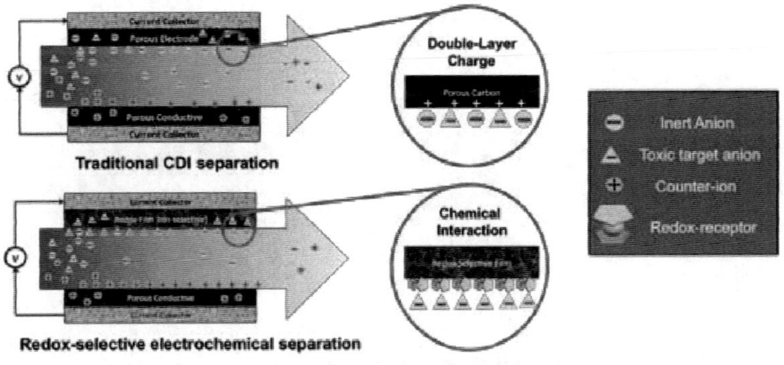

458179 Influence of Elevated Salinity and the Presence of Cytostatic Drugs on Microbial Community in FO Anaerobic Membrane Bioreactors

Abstract:

Wastewater treatment using foward osmotic anaerobic membrane bioreactors (FO-AnMBR) has a number of advantages over traditional wastewater treatment approaches. Previous studies have investigated the overall performance of lab-scale FO-AnMBR in treating synthetic wastewater. During the operation of FO-AnMBR, it is expected that salinity gradually increases, which has also been shown experimentally. In addtion, emerging contaminants such as cytostatic drugs in wastewater would also be accumulated in the FO-AnMBR. Little has been known on whether and how the increased salinity and the presence of cytostatic drugs affect the microbial community in FO-AnMBR. The objective of this study was to investigate the influence of increased salinity and the presence of cytostatic drugs on the structure and function of the microbial community in FO-AnMBR through a metagenomics approach. A lab-scale FO-AnMBR was operated at 25 ^{o}C and fed with synthetic wastewater. Total DNA was extracted from sludge samples taken at pre-determined time points and sequenced using a HiSeq 2500 platform with 250-bp paired-end sequences. Based on the analyses of the sequences, structure and potential functions of the microbial communities were obtained. We found that the increased salinity in FO-AnMBR did not reduce the microbial diversity; however, the relative abunance of methane-producing archaea decreased significantly at high salinity condition, resulting in a reduced methane yield. The general metabolism did not differ significantly under low and high salinity conditions. Intriguingly, genes responsible for cell attachment and biofilm related signalling were enriched at high salinity condition. Transition of the gene clusters for assimilatory and dissimilatory nitrate reduction was also found. The presence of cytostatic drugs only slightly affected the microbial community composition. Biofilm-forming, multidrug resistant and siderophore-producing bacteria were found more dominant after the treatment with drugs.

471970 Batch and Packed-Bed Adsorption Study for the Removal of Dye from Textile Wastewater Using in-House Prepared Silica-Based Nanosorbcats: Kinetics and Computational Modeling

Abstract:

The worldwide shortage of fresh water and the huge competing demands from variety of users stimulates an urgent need for finding alternative cost-effective and environmentally friendly wastewater treatment processes. For instance, synthetic dyes wastewater discharged from the textile and dyestuff industries poses an environmental alert since these organic pollutants are toxic to many aquatic and non-aquatic living organisms. In addition, some of these pollutants are non-biodegradable, and thus, they will exist for a long time in the environment, which may cause a real challenge to the conventional wastewater treatment processes. Accordingly, innovative and environmentally sound techniques are needed. The application of nanoparticle technology as adsorbents and catalysts (nanosorbcats), whether as a standalone or as an enabling technology, in cleaning up wastewater has recently received great attention. This is because of the unique chemical and physical properties of nanoparticles in comparison with their counterparts, which make them superior to the conventional adsorbent/catalysts.

Accordingly, the present study investigated the employment of newly in-house prepared silica-based nanosorbcats functionalized with active species for cleaning up textile wastewater. A facile co-precipitation synthesis route was used to prepare those nanosorbcats, which were characterized by different characterization techniques like XRD, BET, TPD, TPR, HRTEM and zeta potential. The prepared nanosorbcats were employed for the adsorptive removal of cationic and anionic dyes from synthetic and real textile wastewater. Batch adsorption studies were performed for kinetics and equilibrium modeling on the removal of different dyes from wastewater. The operating variables studied were contact time, initial dye concentration, and solution pH. Results showed an optimum pH value, where the maximum removal of dye was achieved. The experimental adsorption data have been modeled using Sips and BET adsorption isotherm models. In addition, a computational modeling on the selected model compounds was carried out to understand the adsorption mechanism on the prepared nanosorbcats. Breakthrough capacities were also investigated in a packed-bed column study. The most successful silica-based nanosorbcats in cleaning up the real textile wastewater in the batch experiments were prepared in extrudate form and used in up-flow packed-bed adsorption experiments. The effects of the influent concentration, flow rate, and bed height were investigated. Breakthrough time and adsorption capacity of the packed-bed increased with increasing bed height, whereas decreased with the increase in influent initial concentration and flow rate. The reliable results of this study will be elaborated and discussed further during the presentation.

463610 Optimization of Operational Parameters for Application of Zirconium/Polyvinyl Alcohol Modified Flat-Sheet Polyvinylidene Fluoride Membrane As Better Adsorptive Material in Treatment of Organic Arsenic Containing Wastewater

Abstract:

Organic arsenic in waters has been a global concern due to its toxicity. In this study, a novel polyvinyl alcohol (PVA)/zirconium modified polyvinylidene fluoride (PVDF) membrane was applied to remove organic arsenic from water. The impregnation of zirconium ions within the modified membrane was attributed to the coordination reactions among the zirconium ions, ether and hydroxyl groups. The synthesized membrane worked better at the acidic conditions and achieved the optimal uptake for both monomethylarsonic (MMA) and dimethylarsinic (DMA) at pH 2.0. The adsorption isotherm study demonstrated that the adsorption of both organic arsenic species was controlled by the mono-layer adsorption process, and the maximum adsorption capacities of 73.04 and 29.78 mg/g for MMA and DMA can be achieved at pH 2.0. The presence of humic acid had a negligible impact on the uptake of organic arsenic, whereas varying impacts on the arsenic adsorption were observed due to the presence of coexisting anions such as fluoride, phosphate, carbonate and silicate. The filtration study was also carried out to evaluate the performance of the modified membrane on the removal of organic arsenic. Based on the XPS analyses, the exchange between chloride ions on the membrane surface and organic arsenic species should be responsible for the removal of both MMA and DMA.

470510 *Smartwaste*- Biomass Conversion into Low Oxygenated Hydrocarbons in Subcritical Water

Abstract:

Sustainable management of food wastes as well as animal manure and other agricultural wastes is a major challenge facing food production worldwide. The dairy and food industries generate approximately 19 and 36 million tons of waste per year, respectively, only in the U.S. The majority of these residues, rich in organic carbon and valuable nutrients, are nowadays disposed in landfills. The current management strategies for manure and food waste are mainly based on biological processes, converting organic wastes into biogas (anaerobic digestion). Limited biogas conversion yield (40-50%) generates a large volume of liquid effluent, containing an aqueous mixture of carbohydrates, proteins, lipids, minerals and nutrients that should be further processed to exploit its economic value. The *SmartWaste* approach is a process integration strategy that uses a series of thermochemical processes to recycle multiple waste streams, producing chemicals, fuels, and nutrients. Furthermore, these fast, selective, and efficient waste-to-fuel processes can be easily integrated into current bio-refineries flow-chart.

Among recycling approaches for high water content waste feedstocks, *SmartWaste* proposed the hydrothermal biomass conversion technology, using both sub- and supercritical water. The suggested approach is based on three-stage processing of waste. First, the pretreatment step focuses on blending different waste streams. Achieving and appropriate feed composition is key to increase the bio-oil production yield in the second stage. In the second step, hydrothermal liquefaction (HTL) of the blended waste stream with subcritical water significantly reduces the oxygen content of the feed to approximately 10% by dewatering and condensation reactions. In the third step, the hydrogenation process, the produced bio-crude oil is further upgraded, by reducing the oxygen content.

The *SmartWaste* process depolymerizes organic waste into smaller structural units. The HTL step is based on fast hydrolysis reactions, followed by dehydration and condensation of sugar, lipids, proteins, and their degradation products, in a single step reaction using subcritical water. This process benefits from the modification in chemical behavior of supercritical water while approaching to the critical point conditions (374 °C and 22.1MPa). For instance, the increase of ionic product (K_W) favors ionic, and acid-base reactions without the addition of a catalyst. Therefore, the optimum processing temperature for this stage is expected to be between 250 and 350 °C. In terms of reaction time, these reactions are usually completed within 20-60 minutes. This represents a big step towards process intensification (reduction of required reactor volume) compared to biochemical processes (enzymes) that require few days.

The main challenges facing residual biomass recycling are the highly complex polymeric composition, together with the high oxygen content, making this feed low in energy content. The flexibility of the hydrothermal technology allows the manipulation of process conditions (reaction time, temperature, pressure) to adapt to different feed compositions and achieve higher bio-oil yields (around 70% on a carbon basis), with higher energy values (oxygen content of ~10% w/w). This oxygen content represents a reduction of 70% with respect to the feed waste. As a consequence of the dehydration and condensation reactions of waste compounds for the generation of bio-oil, an aqueous-soluble carbon phase is also produced. In addition, the solubilized-hydrolyzed amount of solids is not complete, being bio-char being one the products of HTL. To complete the product distribution, a gaseous phase is also generated. The relative amount of any of these four phases depends on reaction conditions, apart from the presence of catalyst.

In this presentation, we will present experimental data for the HTL compartment of the *SmartWaste* project. Carbon mass balances for a wide range of biomass feedstocks will be discussed. This included modeling compounds (polysaccharides, monosaccharides, proteins and long-chain fatty acids) as well as real waste biomass (diary manure digestate collected from anaerobic digestion in upstate NY, and carbohydrate rich food waste from Cornell's dining halls). An example for carbon mass balances is presented in Fig. 1, showing the highest production of bio-crude oil from a mixture of polysaccharides, proteins and lipids at 300 °C with negligible char production. The quality of the bio-crude oil is showed in the Van-Krevelen diagram (Fig. 2), representing atomic rations H/C vs. O/C. HTL significantly reduces the O/C and H/C ratios for all thermal conditions studied. During the HTL process, the quality of the oil intermediate can be highly improved by choosing the adequate biomass feed. For example, the alcoholic groups used for condensation can be synthesized directly in the reactor by lignin hydrolysis, and high pH catalytic media can be achieved by proteins hydrolysis and ammonia release.

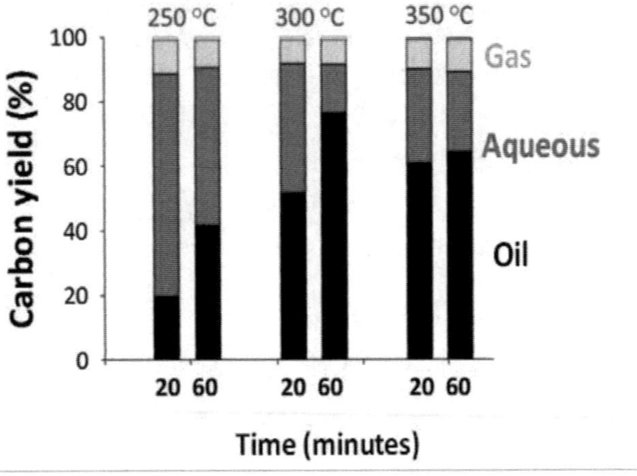

Fig. 1. Carbon mass balance for hydrothermal conversion of ternary modeling mixtures containing polyssacharides, proteins and lipids. The amount of gas phase was estimated as 100-Oil-Aqueous-Solid %.

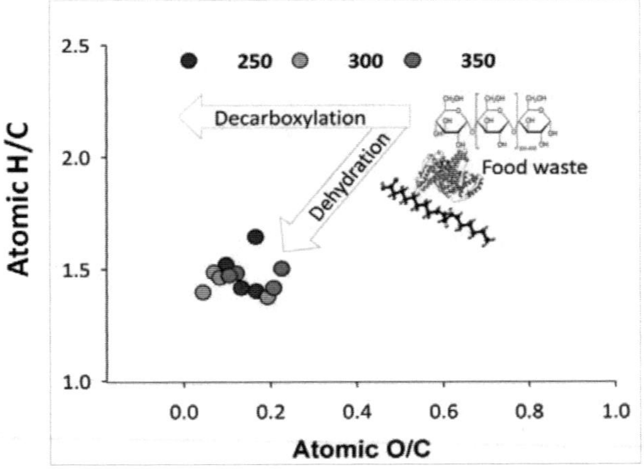

Fig. 2. Van-Krevelen diagram of the produced bio-crude oil upon hydrothermal conversion of ternary modeling mixtures containing polysaccharides, proteins and lipids.

The target product of the *SmartWaste* approach is a stream of alkanes with minimal oxygen content. Further reduction of the bio-oil oxygen content by hydrogenation on supported noble metal catalysts will maximize the energy density of the final products. The production of low polar stream of light straight-chain alkanes (4 to 10 carbons) is achieved by functional reaction routes including hydrolysis, hydrogenation, dehydration, and hydro-deoxygenation steps. Thus, bifunctional acid/redox catalysts are required for the hydrogenation step, where hydrolysis and dehydration processes are carried out on solid acid sites and hydrogenation on active metal sites of the catalyst. The *SmartWaste* produces carbon streams that have similar characteristics to the streams out of an oil refinery. Therefore, future production of biofuels, using small and selective hydrothermal reactors and taking advantage of the available infrastructure for oil refining, without large capital investments, is a reasonable and promising application.

In this research project it has been demonstrated that the hydrothermal liquefaction of waste biomass is an efficient process to convert low value material into enriched fuel and chemicals precursors. The HTL process was able to reduce the oxygen content of the waste streams in 70%. This represents the major oxygen reduction in the whole process. It was also demonstrated that the composition of the biomass reacting is a key factor to consider when designing the process. Adapting the correct combination of sugars, proteins and lipids could increase the carbon oil yields from 20-30% w/w up to 70% w/w.

468527 Catalytic Hydrothermal Gasification of Post Hydrothermal Liquefaction Waste Water for Biogas Production

Abstract:

Hydrothermal liquefaction (HTL) of complex organic materials yields products including bio crude oil, post HTL wastewaster (PHWW). The PHWW contains approximately 100,000 mg/L COD of organics, which are mostly water soluble organic acids and nitrogen-containing aromatics. In order to further assimilate the residual energy and nutrients in PHWW, a sequential catalytic hydrothermal gasification (CHG) was used in this study to produce valuable biogas to maximize the energy recovery from HTL process, reuse the nutrients, and at the same time clean the PHWW. The effect of catalyst selection was investigated using ruthenium (Ru), sodium hydroxide (NaOH), and zeolite, which were used individually and in combination (e.g. Ru plus NaOH) to determine the optimal organic removal rate, gas yield, chemical compositions of gas products, and nutrient fate. Preliminary results demonstrated that a high gasification efficiency of 98% could be achieved in terms of COD removal. Hydrogen production was able to reach up to 28 mmol per g of organics in the PHWW, and methane (CH_4) production up to 21 mmol per g organics in the same condition. The concentration of water soluble short chain organic acids (C1-C4) increased, compared to PHWW feedstock, at lower temperature of 400°C but then decreased and was almost completely removed at higher temperature of 600°C. The concentration of ammonia-nitrogen was raised by 68% in the PHWW after CHG treatments, indicating that a significant denitrogenation of the nitrogen-containing aromatic compounds took place, thus better nutrient recovery was achieved.

463234 Development of an Integrated Process for Sustainable Production of Biodiesel and Plant Protein Isolates from Underutilized Feedstock

Abstract:

â€œFood or Fuelâ€ is an international concern about the use of agricultural land for biofuel production, which results in higher food prices and adversely affects the living standards of poor people. Consequently, it is essential to find compromises that meet the demand for both food and green transportation fuels. Yellow mustard is a Canadian food crop that is a viable source of both oil and protein. At present neither of these valuable components are utilized as most of the seed is now only used as condiment. The proteins have a well-balanced amino acid distribution making them potentially useful as a food ingredient, and the inedible oil, which is high in erucic acid, is desirable in biodiesel due to its superior lubricating properties. In this study, an integrated process was developed to produce food-grade protein isolates, free of solvent contact/residues, and fuel-grade biodiesel from dehulled yellow mustard flour. This process makes novel contributions to the link between energy, food and the environment.

The aqueous extraction of proteins form mustard flour was started by contacting the flour with water at room temperature and alkaline pH 11 followed by centrifugation. The protein was recovered in the form of protein-rich skim fraction with minimal protein damage, while the oil was released as a separate fraction and tied up in a very stable oil-in-water emulsion. The protein-rich skim fraction was further concentrated and purified by membrane processing followed by isoelectric precipitation to produce two high quality food-grade protein isolates: isoelectric precipitated protein isolate (PPI) and acid soluble protein isolate (SPI). The PPI and SPI with desired functionalities had 96.0 and 83.5% protein content on a moisture and oil free basis, respectively.

The underutilized oil-rich emulsion fraction was characterized where its remarkable stability was due to the presence of protein emulsifiers of high molecular weight along with the mixed phospholipid-oleosin layer. The emulsion was successfully destabilized by a novel chemical approach where the organic solvents including isopropyl alcohol (IPA), tetrahydrofuran (THF) or dioxane were used to solubilize the emulsion to produce a single-phase oil-solvent-water miscella. The ternary phase diagrams of THF/oil/water, dioxane/oil/water, and IPA/oil/water were prepared, and the solubility of the oil in these organic solvents in the presence of water was determined. The application of ternary phase diagrams resulted in the development of three-stage destabilization process at which all of the oil was successfully recovered from the emulsion into the miscella. The oil-solvent-water miscella was dehydrated in the next phase by adsorption over zeolite 4A using either batch or continuous fixed-bed systems. The dehydrated miscella was reacted with methanol in a single-phase base-catalyzed transmethylation process with high yields (99.4%) to fatty acid methyl esters (FAME). The resulting FAME satisfied the international standards for use as biodiesel fuel.

Approximately 69% of the protein originally in the flour was recovered as high-quality protein products and over 62% of the oil in the flour was recovered as standard biodiesel through this integrated process. The preliminary economic analysis showed an overall positive margin. The application of this processing approach could help increase Canadian production of both food ingredients and biofuel, without impacting the cost of edible oils, as very large areas of Canada can be readily converted to mustard production.

COMBUSTION OF PYROLYSIS CHAR WITH A FLUIDIZED BED REACTOR FOR GENERATING PROCESS HEAT

Nicholas, R. Schwartz, Paul E. Yelvington*

Mainstream Engineering Corporation, Rockledge, FL

Corresponding Author: nschwartz@mainstream-engr.com

Abstract

Portable biomass fast pyrolysis processes require combustion of byproducts to provide heat to the reactor and to dry feedstocks. Heat integration and byproduct utilization are especially important for small-scale, distributed processes where heat losses become more problematic. Pyrolysis char produced with a 1 ton per day (tpd) pilot reactor from pine woodchips was combusted with a pilot-scale fluidized bed reactor (FBR) to demonstrate the feasibility of process heat generation. The char product was characterized prior to combustion to determine carbon content, moisture, ash content, and heating value. Particle size distributions (PSD) of the char feed and collected ash material were measured. The fluidizing gas flow rate, gas inlet temperature, excess air percentage, and reactor bed temperature were varied to determine the effects on emissions and combustion efficiency. Gaseous emissions were continuously monitored for a range of fluidizing gas velocities and reactor bed temperatures. The effects of excess air on gaseous emissions were determined along with the inlet temperature of the fluidizing gas. The effects of reactor conditions on CO, SO_2, and NO_x were assessed using the FBR with a variable feed rate of 0.3 to 0.4 kg/h. Based on the experimental results, increasing the excess air and reactor bed temperature reduced CO emissions. Increasing the fluidizing gas flow rate decreases the residence time resulting in increased CO. Particulate emissions were determined after each run by measuring the ash blowout from the cyclone separator. Total fine particulates were collected within a baghouse filter after 16 hours of operation. Reactor performance data using the pyrolysis char fuel will be used to scale-up the FBR design to a 7 kg/h system, right-sized for a nominally 1-tpd distributed pyrolysis process.

Introduction

Fast pyrolysis is a process whereby organic material is converted to a liquid bio-oil, char, and combustible non-condensable process gas. The fast pyrolysis process depolymerizes organic material at high temperatures (500°C) in an oxygen-free environment with short vapor residence times, which favors bio-oil production.[1-4] The bio-oil is similar in refinement to a fossil crude oil (higher oxygen and water content), which can be burned directly or further upgraded to higher value products. Pyrolysis is the ideal platform for conversion of organic materials and waste to bio-oil, char, and process gas and has very low emissions compared to incineration. Pyrolysis offers the advantage of processing a wide range of materials at a single set reactor conditions. Byproducts of pyrolysis can be burned for process heat for a near self-sustaining process. Burners tailored to pyrolysis byproducts offer cleaner combustion and lower emissions compared to direct combustion of solid waste material. Bio-oil and char can be stored for later heat or power production with a significant reduction in waste volume.

Previously, a continuous bench-scale unit was optimized over a range of process conditions to maximize the bio-oil yield and quality. The heating values and typical percent yields of the bio-oil, char, and gas for fast pyrolysis are listed in Table 1.

Table 1. Heating Value and Yields for Bench-scale Pyrolysis

Component	Lower Heating Value (MJ/kg)	Yield (wt%), dry
Bio-oil	17.5	65
Char	29.7	15
Non-condensable Gas	7.3	20

Methods

Pyrolysis char collected from the 1-tpd reactor was characterized prior to combustion within a fluidized bed reactor (FBR). Chemical and physical properties of char and ash were characterized by PSD, heating value, moisture content, carbon content, and elemental composition. The PSD was completed following ASTM standard D6913 for the pyrolysis char (shown below in Figure 1).[5] The char product was fine and uniform with 85% by weight is smaller than 0.149 mm and an average particle size of 0.124 mm.

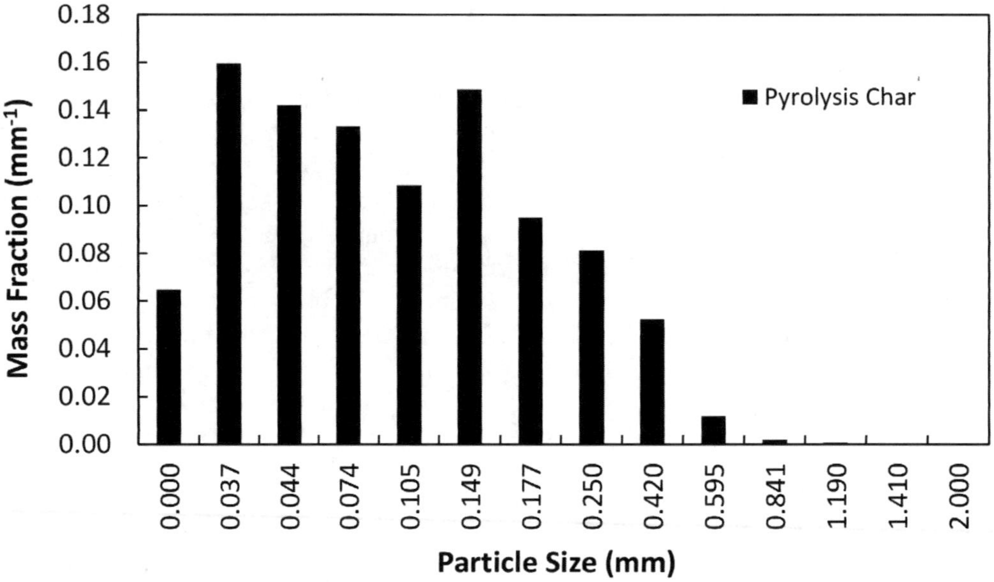

Figure 1. Particle size distribution for pyrolysis char produced by pyrolysis of pine

Heating value, density, PSD, and moisture content are important variables for designing the char combustion FBR. The complete analysis assisted with predicting reactor performance and combustion byproducts. Heating value, moisture, ash, and fixed carbon content for each char material were also determined (results listed in Table 2).

Table 2. Heating Value, Moisture, Ash, and Fixed Carbon Content of Pyrolysis Char

Property	Pyrolysis Char
Heating Value,$_{dry}$ (MJ/kg)	29.7
Moisture (wt%)	6.03
Ash Content (wt%)	5.26
Fixed Carbon (wt%)	82.70

The elemental analysis of the char product determined all detectable constituents of the char and ash. Major constituents of the ash for the char are listed in Table 3. The ash was comprised of mostly CaO, K_2O, Al_2O_3, MgO, and SiO_2, which are beneficial soil additives.[6] One study measured volatile elements produced during pyrolysis and determined most elements are retained within the char.[7]

Table 3. Elemental Composition of Major Ash Constituents

Composition (wt%)	Pyrolysis Char
C	82.70
CaO	7.43
K_2O	3.18
Al_2O_3	2.31
MgO	2.22
SiO_2	1.86
P_2O_5	0.90
Fe_2O_3	0.39
Na_2O	0.26
SO_3	0.25
MnO	0.13

Pyrolysis char produced by the 1-tpd reactor was burned using an automated FBR at the Energy and Environmental Research Center (EERC), which simulates bed chemistry, ash interactions, and emissions under controlled conditions (shown in Figure 2). Emissions and process temperatures were continuously monitored for one hour at steady-state at each set of reactor conditions. Pyrolysis char was fed into the reactor through a single screw feeder. Preheated fluidized gas entered the bottom of the reactor and fluidized the sand and char particles fed into the combustion chamber. External ceramic heaters were used to maintain the combustor and cyclone temperatures. Sand and ash leaving the FBR were collected using an ash cyclone, where the remaining gas exited the system to the gas analyzers.

Figure 2. Fluidized bed char combustor tested at EERC with pyrolysis char

The fluidizing gas temperature, gas velocity, percentage of excess air, and reactor bed temperature were adjusted to determine the effects of process conditions. While burning the pyrolysis char within the FBR, the CO, CO_2, SO_2 and PM emissions were continuously monitored over time for each reactor condition listed in Table 4.

Table 4. Test Matrix for Pyrolysis Char Burning with Fluidized Bed Reactor

Test	Preheat Temp. (°F)	Gas Velocity (ft/s)	Excess Air (%)	Avg. Bed Temp. (°F)
1	400	2.5	25	1,550
2	675	2.5	25	1,550
3	675	2.5	35	1,550
4	675	2.5	15	1,550
5	675	2.5	25	1,650
6	675	2.5	25	1,450
7	675	1.5	25	1,550
8	675	3.5	25	1,550

Results

Flue gas emissions, process temperatures, and flow rates were continuously measured for the FBR for all evaluated reactor conditions. The NO_x, CO, and SO_2 emissions were corrected to a common O_2 content (3% O_2). Average emissions were also reported in terms of a mass-energy basis (lb/MMBtu) for comparison to other fuels. Reactor temperatures were varied from 1,450 °F to 1,650 °F, gas velocity from 1.5 ft/s to 3.5 ft/s, and the excess air from 15% to 35%. Due to the low sulfur content in the pyrolysis char fuel, SO_2 emissions were extremely low and averaged across all tested reactor conditions.

NO_x Emissions

Reactor bed temperature, fluidizing gas velocity, and excess air level all have a significant impact on NO_x emissions. NO_x emissions typically increase with increasing temperature (Figure 3), but some fuels result in decreasing NO_x emissions with increasing temperature. NO_x was highest for the low temperature at 1,450 °F (198 ppm) and lowest for the high temperature at 1,650 °F (178 ppm).

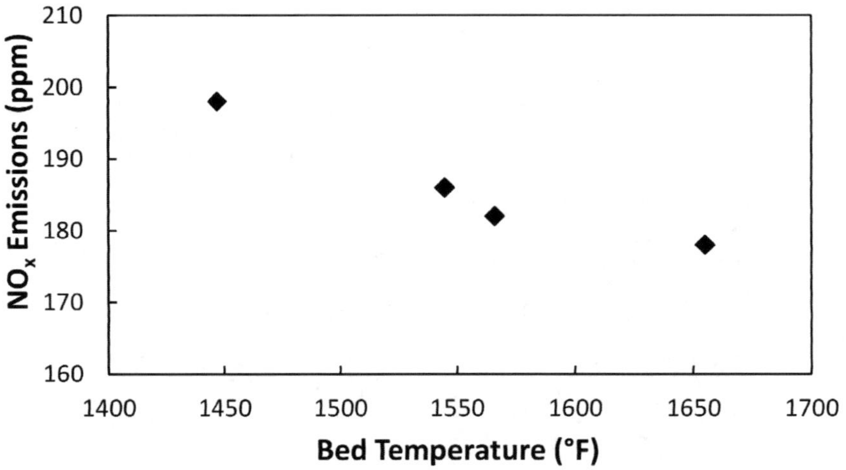

Figure 3. Effect of bed temperature on NO_x emissions

Fluidizing gas bed velocity, which is inversely proportional to gas residence time, has an even greater impact on NO_x emissions (Figure 4). As velocity increased from nominally 1.5–3.5 ft/s, the residence time decreased from approximately 3 to 1 seconds, NO_x increased from 163 to 205 ppm.

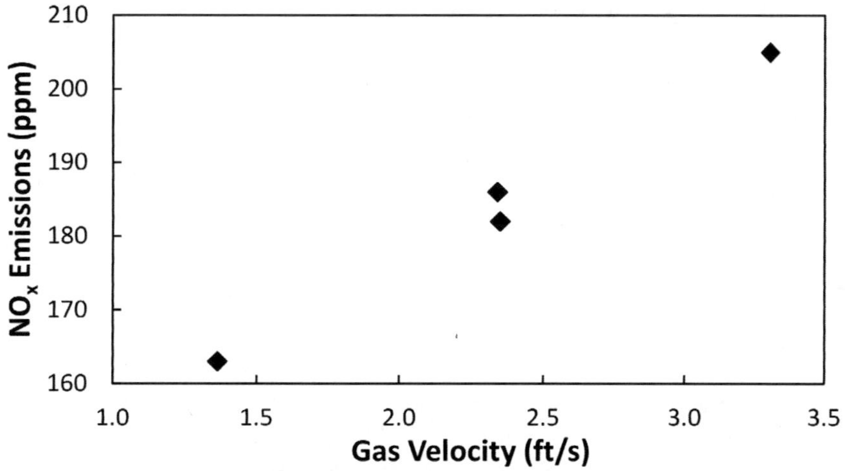

Figure 4. Effect of superficial gas velocity on NO_x emissions

NO_x emissions increased with higher levels of excess air, as expected, increasing from 175 ppm at an excess air level of 11% to 194 ppm at 36% excess air (Figure 5). The NO_x emissions were on average 184 ppm at 26% excess air. As lower excess air improves NO_x emissions, determining optimal reactor conditions will be a trade-off with CO emissions. Average NO_x emissions over all test conditions was 0.331 lb NO_x/MMBtu.

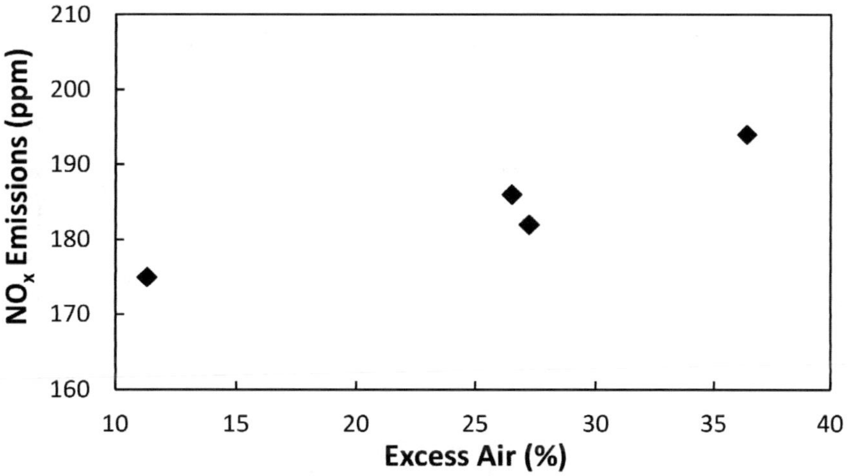

Figure 5. Effect of excess air on NOx emissions

CO Emissions

CO emissions are an indicator of combustion efficiency, with lower CO levels indicating high carbon burnout. As with NO_x, CO emissions can be affected by reactor bed temperature, excess air, and fluidizing gas velocity. Increased reactor temperature typically results in better carbon burnout, which was demonstrated with pyrolysis char (Figure 6). At a high temperature of 1,650 °F, the CO emission was 48 ppm; and at 1,450 °F, the CO emissions averaged 605 ppm.

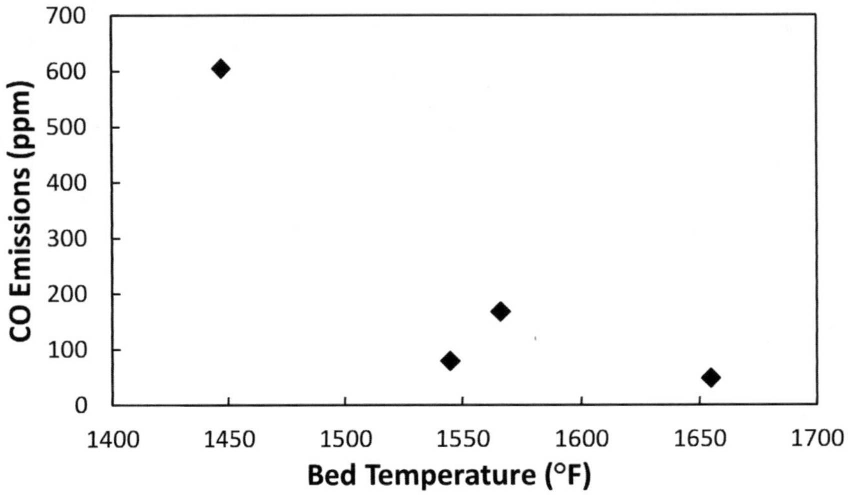

Figure 6. The effect of reactor temperature on CO emissions

The effect of fluidizing gas velocity on CO emissions was more pronounced in comparison to NO_x emissions, as shown in Figure 7. CO averaged 758 ppm at the highest velocity tested (3.3 ft/s), which is significantly higher compared to 83 ppm at 1.4 ft/s and 168 ppm at 2.3 ft/s. There appears to be a specific point at which CO increases sharply with increased gas velocity and the combustion efficiency decreases.

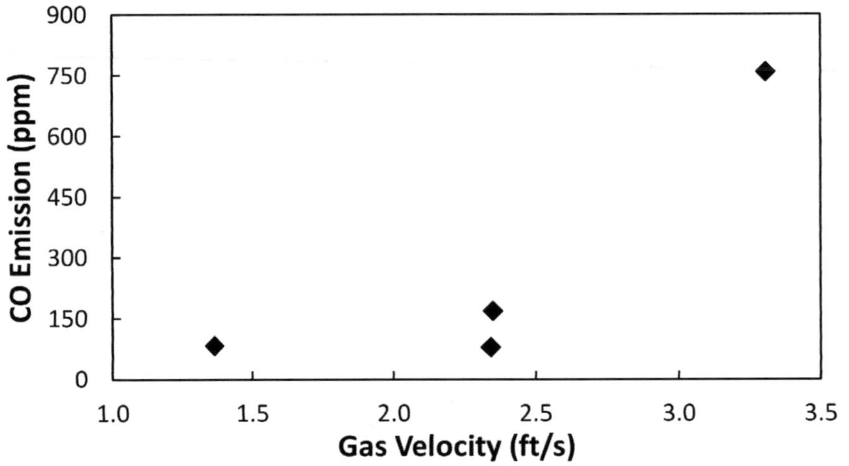

Figure 7. The effect of gas velocity on CO emissions

Excess air had much less of an effect than either reactor temperature or velocity. The CO emissions measured at varying levels of excess air are shown in Figure 8. CO emissions were slightly higher at low excess air (131 ppm) than at high excess air (96 ppm). When the pyrolysis char was combusted with 26% excess air, the CO emissions were 79 ppm with char collected from the reactor bed and 168 ppm with fuel collected after the reactor. Another test that had a lower temperature fluidizing gas velocity with the same level of excess air produced CO emissions of 134 ppm. Over all test conditions, the average CO emissions were 0.224 lb CO/MMbtu.

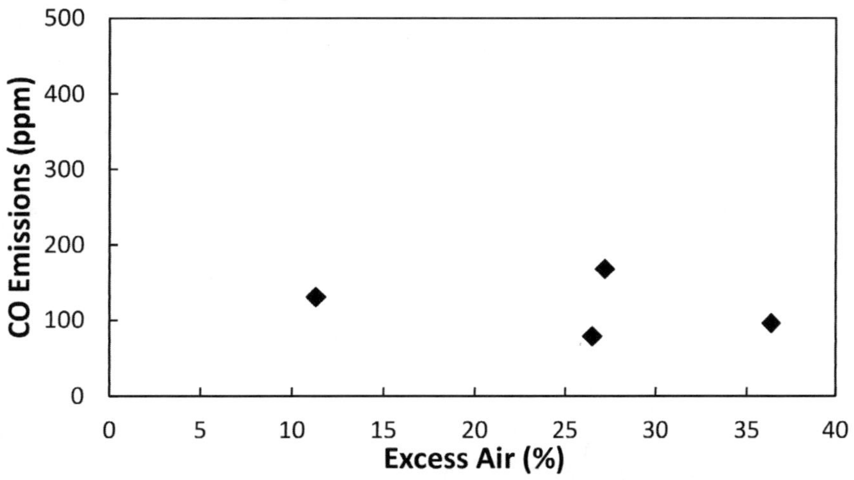

Figure 8. The effect of excess air on CO emissions

Particulate Matter & SO_2 Emissions

The sulfur content of the char was very low, which resulted near-zero SO_2 emissions. Average SO_2 emissions across all reactor conditions was 3.4 ppm (0.020 lb SO_2/MMBtu). Particulate matter (PM) emissions were determined by measuring the total solids captured by the baghouse filter downstream of the FBR over all reactor conditions. The mass balance across the FBR, cyclone, and baghouse filter for ash and sand bed material is detailed in Figure 5. Overall mass balance closure for ash and sand bed material was 85.1%.

Table 5. Mass Balance for Char over FBR, Cyclone, and Baghouse Filter

Total Material In (g)		Total Material Out (g)	
Fuel Feed	6557 g	Cyclone	1200 g
Ash Content	1809 g	Baghouse	17 g
Sand Material	1600 g	Fluidized Bed	1685 g
	3409 g		2902 g

The total amount of exhaust gas was totaled over the 16 hour operating period for an average volumetric flow rate of 1.5 ft^3/min (0.0425 m^3/min). Only as the solids collected by the baghouse filter (total of 17 g) were assumed to contribute to the PM emissions, which was estimated to be 417 mg/dscm. Implementing a baghouse filter into a pyrolysis char burner would capture and eliminate any PM emissions.

Conclusions

Pyrolysis char produced from fast pyrolysis can be combusted for heat generation within a FBR. The effects of reactor temperature, fluidizing gas flow rate, and excess air on NO_x and CO emissions were determined. Low NO_x and CO emissions and high combustion efficiency can be achieved by combusting pyrolysis char at an operating temperature of 1,550 °F or greater within the FBR. Lower fluidizing gas velocities (1.5 ft/s) and excess air (11%) resulted in lower NO_x emissions and improved carbon burnout. CO emissions were strongly dependent on reactor temperature and gas velocity, where higher bed temperatures and lower gas velocities improved combustion efficiency. Changes in excess air had minimal influence on CO emissions across the tested reactor conditions. No agglomeration or fouling was observed during the test, which indicated the pyrolysis char burned cleanly. The fluidized bed pressure drop slowly increased over time, which indicates that intermediate bed draining may be required to maintain pyrolysis char burner conditions. Measured emissions over the tested reactor temperature, flow rates, and excess air provides a starting point for large-scale pyrolysis char FBR operation. Bed depth, diameter, and vapor space height will influence control of reactor conditions and fluidization characteristics. Effects of reactor geometry must be considered along with reactor conditions for the scale-up design for the pyrolysis char burner.

References

1. Bridgwater, A. V., *Fast Pyrolysis of Biomass: A Handbook.* CPL Press: Newbury, UK, 2002; Vol. 2, p 426.
2. Bridgwater, A. V., *Fast Pyrolysis of Biomass: A Handbook.* CPL Press: Newbury, UK, 2005; Vol. 3, p 221.
3. Bridgwater, A. V.; Meier, D.; Radlein, D., Overview of fast pyrolysis of biomass. *Organic Geochemistry* **1999,** *30* (12), 1479-1493.
4. Bridgwater, A. V.; Peacocke, G. V. C., Fast pyrolysis processes for biomass. *Renewable & sustainable energy reviews* **2000,** *4* (1), 1-73.
5. ASTM, Standard Test Methods for Particle-Size Distribution (Gradation) of Soils Using Sieve Analysis. 2009; Vol. ASTM D6913.
6. Heitkotter, J.; Marschner, B., Interactive effects of biochar ageing in soils related to feedstock, pyrolysis temperature, and historic charcoal production. *Geoderma* **2015,** *245-246*, 56-64.
7. Liaw, S. B.; Rahim, M. U.; Wu, H., Trace Elements Release and Particulate Matter Emission during the Combustion of Char and Volatiles from in Situ Biosolid Fast Pyrolysis. *Energy and Fuels* **2016,** *30* (7), 5766-5771.

472362 Small Scale Distributed Ammonia Produced: Analysis of Pilot Plant Runs and Routes to Improve the Economics of Scale

Abstract:

Small Scale Distributed Ammonia Produced: Analysis of Pilot Plant Runs and Routes to Improve the Economics of Scale

Mahdi Malmali[‡], Michael Reese[†], Cory Marquart[†], Kevin Wagner[‡], Eric Buchanan[†]

Alon McCormick[‡] and Edward L. Cussler[‡]

[†] West Central Research and Outreach Center

[‡] Department of Chemical Engineering and Materials Science

University of Minnesota, Minneapolis, MN 55455

Ammonia is one of the most important chemical commodities in the U.S. and will be a key component in helping the world meet the rising demand for food and energy. Ammonia is needed in distributed locations for agriculture (as fertilizer for small grain and corn production), for indirect hydrogen storage[1] (transported as a liquid at moderate pressure to hydrogen stations), or as a liquid fuel[2] (for internal combustion engines or solid oxide fuel cells). Currently, ammonia is manufactured by the Haber-Bosch process, a highly capital- and energy-intensive process requiring large quantities of fossil fuels. Ammonia synthesis is only feasible in massive centralized locations; ammonia consumption thus relies on moving anhydrous ammonia from distant locations to regions where ammonia is needed.

Recently, there has been significant effort to develop scalable technologies for conversion of intermittent energies (e.g., solar, wind) into energy dense carbon-neutral liquid fuels, and ammonia is considered to be a promising option. For the synthesis of carbon-neutral ammonia, hydrogen and nitrogen are delivered from electrolysis of water and pressure swing adsorption of air. Once produced, nitrogen and hydrogen are fed to a catalytic reactor to produce anhydrous ammonia. All these processes are carbon-free and powered by wind turbines or solar arrays.

In this talk, we are going to present a benchmark for the performance of a small-scale ammonia synthesis pilot plant powered by wind energy[3]. The Renewable Hydrogen and Ammonia Pilot Plant located in Morris, MN, is the first local farm-to-coop scale system for ammonia production. The energy used for this plant is stranded, located in areas far from urban centers where ammonia is primarily used as a fertilizer. This plant targets production of small amounts of ammonia for local demand, thus representing distributed production. We will present the analysis of small plant runs, along with some laboratory-scale kinetic studies, to assess the performance of different units, including reaction, phase separation and recycle. A simple but insightful model is developed, in order to understand the performance of the small-scale Haber-Bosch pilot plant and to determine the optimal conditions for operating. This model is successfully employed to predict the performance of each unit, and gives us insight into selecting appropriate unit operating conditions. Such a straightforward model may help us to understand the design of other small-scale processes in the future.

(1) Sørensen, R. Z.; Hummelshøj, J. S.; Klerke, A.; Reves, J. B.; Vegge, T.; Nørskov, J. K.; Christensen, C. H. Indirect, Reversible High-Density Hydrogen Storage in Compact Metal Ammine Salts. *J. Am. Chem. Soc.* **2008**, *130* (27), 8660.

(2) Wojcik, A.; Middleton, H.; Damopoulos, I.; Van herle, J. Ammonia as a Fuel in Solid Oxide Fuel Cells. *J. Power Sources* **2003**, *118* (1-2), 342.

(3) Reese, M.; Marquart, C.; Malmali, M.; Wagner, K.; Buchanan, E.; McCormick, A.; Cussler, E. L. Performance of a Small-Scale Haber Process. *Ind. Eng. Chem. Res.* **2016**, *55* (13), 3742.

468147 A Supply Chain Optimization Framework for Distributed Renewable Ammonia Production

Abstract:

Ammonia is a critically important chemical for global food production which can be either used directly as or processed further into fertilizer. The Haber-Bosch process revolutionized global food production and greatly increased mankind's ability to support a larger population. However, the way ammonia is produced and transported currently is unsustainable. Ammonia is produced using hydrogen obtained from fossil fuel sources, ultimately polluting the environment. It is then transported over long distances by train, pipeline, or even ship: over one-third of ammonia consumed in the United States is imported. It would be more desirable from a sustainability standpoint to have a cleaner, smaller scale, and more local ammonia supply chain. Motivated by this goal, a first of its kind plant has been built in Morris, MN, using the wind-powered electrolysis of water to produce hydrogen without the carbon emissions inherent in fossil fuel gasification. This plant has been demonstrated to produce renewable ammonia with minimal carbon emissions at a scale of 26 t/y. Using data from this plant, it is reasonable to ask how, and at what scale, renewable ammonia plants would be economically feasible with respect to the overall ammonia supply chain.

To address this question, this work aims to create a framework for the analysis of renewable plants in a general region's ammonia supply chain. Candidate sites for renewable plants are placed in towns with nearby wind turbines with maximum plant capacity determined by wind turbine energy output. A supply chain optimization problem is fomulated to decide where and how large to build new renewable plants, as well as how much ammonia should be transported between conventional plants, renewable plants, distribution centers, and demand centers to meet local demand. The problem is a nonlinear program due to economies of scale when scaling up the Haber-Bosch reactor and related separation units. The problem can be set up to minimize economic or environmental objectives. When doing so, a clear tradeoff emerges: a minimal cost supply chain obtains most of its ammonia from a few large centralized plants, while a minimal emissions supply chain obtains its ammonia from many small distributed plants. A multiobjective optimization problem is formulated to analyze these tradeoffs via a Pareto curve of annual cost vs. annual carbon dioxide emissions.

The developed framework is then applied in two separate case studies to two upper midwestern states, Minnesota and Iowa, respectively. Despite the fact that these states are neighbors, they have various key differences that make separate analysis instructive. For example, Iowa has more distribution centers, more pipeline connections, more overall demand, more wind potential, and a more geospatially homogeneous ammonia demand distribution than does Minnesota. Despite these differences, we establish that for both states, a renewable plant scaled up to the order of 100,000 t/y ammonia production is selected for inclusion in the base case optimal supply chain. We then further explore how the differences between the two states affects the solution sensitivity, the impact of a carbon tax on the supply chain, and the improvements required for smaller scale plants to be economically feasible.

NH3 – The Optimal Transportation Fuel
2016 AIChE Annual Meeting

Norm Olson and John Holbrook
Iowa Energy Center
NH3 Fuel Association (https://nh3fuelassociation.org/)

November 8, 2016

Any analysis to help identify an optimal alternative fuel should begin with a logical, comprehensive set of desired characteristics. The following criteria were used to help identify the fuel closest to an "ideal" alternative fuel.

Characteristics of an Ideal Fuel
1. Production flexibility – i.e. the ability to be produced from all primary energy sources (e.g. solar, nuclear, wind, biomass, OTEC, natural gas with CCS, coal with CCS, etc,). Another highly desirable characteristic is that the production of the fuel be a relatively simple process with a wide range of scalability.
2. Affordability. This involves infrastructure costs, production CapX and OpX costs, efficiency and other considerations. The fuel should be cost competitive with gasoline and diesel fuel.
3. Existing storage and delivery infrastructure
4. End use flexibility – suitable for use in spark ignition engines, compression ignition engines, combustion turbines, fuel cells, etc.
5. High engine/fuel cell efficiency potential - the single most important characteristic for high efficiency SI engines is octane rating. A high octane rating allows the use of high engine compression ratios and therefore highly efficient engines. A closely related issue in SI engines is resistance to pre-ignition. High efficiency fuel cell operation is also important.
6. Excellent environmental and health performance
7. Sustainability – renewable resources and widely available, low-cost raw materials
8. Acceptable safety characteristics
9. The ability for any country to produce the fuel and become both power and fuel independent
10. Having additional non-fuel uses of significant benefit to society. E.g.– fertilizer, renewable energy storage, chemical precursor, refrigerant, hydrogen source, household cleaner, etc.

Alternative Fuel Candidates
There are many alternative fuel candidates that partially meet the ideal fuel criteria listed above including but not limited to hydrogen, NH3, drop-in fuels, propane, DME, ethanol, methanol, and butanol. Only hydrogen and NH3 can be produced without using biomass (or a somewhat convoluted, inefficient path involving CO2 capture plus hydrogen). The International Energy Agency (IEA) and the Energy Information Administration (EIA) both estimate the upper limit for biomass-derived alternative fuels at 30% for transportation fuels. This is a very significant issue and would result in continued addiction to petroleum indefinitely.

Algae-based fuels are currently receiving attention as a source of biomass for fuels. In the 1950's algae was viewed as the ultimate food source due to its excellent nutritional value and efficient growth characteristics.

Food vs fuel issues are even more relevant for algae than for corn. Negligible amounts of the type of corn used to produce ethanol can be used for human consumption, this type of corn is used nearly exclusively for livestock feed. If algae can be grown cost-effectively in large quantities, it would make more sense that it be used as a highly efficient, high-quality source of food than as a fuel source. It may prove to be problematic if the future liquid transportation fuel is made exclusively from biomass (although biomass-based fuel additives could be acceptable). In a world where ending the continuing addiction to petroleum is desired, biomass should be used to produce food first, chemicals (as replacements for petrochemicals) second and thirdly as fuel additives.

NH3 (aka "The Other Hydrogen") rises above other alternative fuel candidates based on the criteria listed above. Simply stated, NH3 is the most cost-effective, energy dense means of delivering hydrogen. It has excellent environmental performance, extremely high engine efficiency potential, production flexibility characteristics second to none, universal end use potential, and a decades-long, proven, acceptable safety record. The fact that NH3 also serves as the most cost-effective means of delivering nitrogen fertilizer provides enormous additional benefits, especially to developing countries. NH3 is an efficient, environmentally friendly refrigerant, an excellent household cleaner, is an efficient medium for long-term storage of intermittent renewable energy and a versatile chemical precursor. No other alternative fuel can deliver the comprehensive benefits associated with NH3.

NH3 Production Flexibility

NH3 production is a relatively simple industrial process compared to most petroleum refinery processes. NH3 production facilities are highly scalable and NH3 has multiple, high-value uses including use as a transportation fuel, power fuel, fertilizer, refrigerant, energy storage medium for renewable energy, and a versatile chemical precursor.

NH3 can be competitively priced when produced from natural gas today and has a very small carbon footprint using CO2 sequestration or via thermal cracking co-producing carbon black. It can be produced via nuclear energy and all renewable energy sources when these routes become more cost competitive.

NH3 can be produced using natural gas, coal, biomass, wind, solar, hydro, OTEC, wind, wave, nuclear energy and all other primary energy sources. One of the most promising methods of storing and transporting stranded renewable energy utilizes NH3 as the storage medium.

The two most commonly discussed methods for NH3 production are solid state ammonia synthesis (SSAS) and the Haber-Bosch process both of which are very scalable. Numerous innovative improvements on the Haber-Bosch are currently being developed.

Affordability

NH3 is primarily produced from coal and natural gas at the present time and is normally cost-competitive with gasoline and diesel fuel on a gallon of gasoline equivalent (GGE) basis. The chart below provides characteristic costs for NH3 for a range of natural gas prices and compares these costs to gasoline.

NH3 Cost (GGE) vs Natural Gas Cost

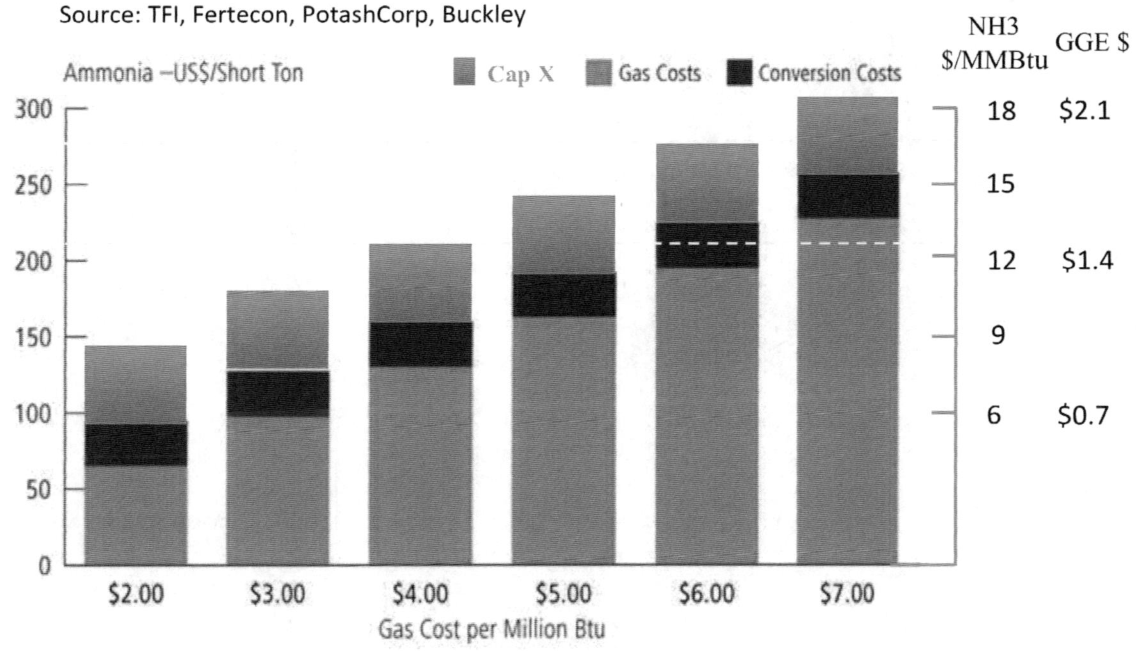

NH3 production facilities, much like ethanol, biodiesel and methanol are very scalable and low-cost relative to petroleum refineries. This makes NH3 "refineries" affordable in most areas of the world. In countries where NH3 fertilizer is crucial to increased food production, a relatively simple NH3 refinery could be used for multiple purposes. NH3 could be used as a transportation fuel, an electrical generation fuel, a refrigerant, a household cleaner, and a chemical precursor in addition to its use as a fertilizer.

Infrastructure costs are low and are similar to propane infrastructure costs. NH3 has an extensive, existing, worldwide infrastructure already in place which will help lower implementation costs.

When NH3 is produced from renewable electricity such as wind or solar PV, assuming an electrical energy cost of 5 cents/kWh (including CapX and OpX costs) and an electrical consumption of 7000 to 9000 kWh/ton, the cost per ton of ammonia produced would be $350 to $450/ton. This corresponds to a gallon of gasoline equivalent (GGE) cost of **$2.45 - $3.15/GGE.**

Storage and Delivery Infrastructure
NH3 is the second most transported chemical in the world and has an extensive, world-wide storage and delivery infrastructure already in place. Infrastructure costs are relatively low and very similar to propane at low storage volumes (<4 ton or ~1500 gallon). At larger storage volumes (30 - 30,000 ton), ammonia can be stored at atmospheric pressure as a refrigerated liquid. It is transported as a refrigerated liquid in ocean-going ships as well. The existing, low-cost storage and delivery infrastructure provides a very valuable advantage to NH3 fuel.

Annual production of NH3 is over 160,000 tonne compared to <100,000 tonne of methanol produced annually. NH3 is the 2nd most transported chemical. Mild steel is the material of choice for pipeline and storage applications. Over 3000 miles of NH3 pipelines have been in place for decades in the U.S. Natural gas pipelines could be modified by replacing compressors with pumps to carry NH3. Atmospheric pressure storage at relatively mild conditions (-28F) compared to LNG (-265F) and liquid hydrogen (-423F). NH3 is less energy intensive to produce, deliver and store per pound of hydrogen delivered than 10kpsi or liquid hydrogen. The volumetric energy density of NH3 is 30% higher than liquid hydrogen. Storage infrastructure capital costs for NH3 can be one to two orders of magnitude lower than hydrogen infrastructure costs.

End Use Flexibility

The use of NH3 in gas turbines, spark ignition engines, compression ignition engines, gas burners and fuel cells has been demonstrated. Although NH3 has a relatively slow flame-spread-speed and high ignition energy, it can easily be partially cracked to produce a mixture of hydrogen and ammonia, which then combusts with characteristics very similar to natural gas (but carbon free!).

NH3 use in diesel engines normally requires the conversion of the engine to spark ignition or dual-fuel use, very similar in concept to dual-fuel diesel/natural gas engine conversions. Advanced compression ignition engines such as those produced by Sturman Industries have demonstrated successful operation using 100% NH3.

Solid oxide fuel cells are very compatible with NH3 as are alkaline fuel cells. Although PEM fuel cells cannot tolerate NH3, there is a commercially available PEM fuel cell that operates on cracked NH3 (i.e. cracked to nitrogen and hydrogen).

Gas burners and gas turbines have both been successfully operated on partially cracked NH3. A federally funded program in Japan has successfully demonstrated the use of NH3 in a gas turbine.

Engine and Fuel Cell Efficiency

High efficiency, affordable cost and low emissions are key metrics associated with optimized internal combustion engines, combustion turbines and fuel cells. The high octane rating and tunable resistance to pre-ignition allow NH3 to be used in the highest compression (and therefore highest efficiency) engines capable of being designed. Just as in real estate the three most important characteristics are location, location, location, the three most important characteristics of a transportation fuel are high octane, high octane, high octane. The highest engine efficiencies will be attained using high octane fuels which also have a high resistance to pre-ignition. Low/zero carbon fuels are extremely beneficial in attaining low emissions and longer lasting engines, due to low carbon deposits.

The most important characteristic of an ideal fuel for spark ignition engines from an efficiency standpoint is a high octane rating. A high octane rating allows the use of high compression ratios and therefore high efficiencies. A related issue is resistance to pre-ignition.

Fuel Octane Rating	RON	MON	(R+M)/2
Gasoline			87
Ammonia	>130[1]		
Ethanol	109	90	99
Methanol	109	89	99
n-Butanol	92	71	83
Hydrogen	>130[2]		
Methane	120	120	120

[1]The minimum ignition energy for ammonia is very high. It is approximately 500 times greater than the energy required for igniting hydrocarbons and 1000 to 10,000 times greater than that required for hydrogen. This makes NH3 much less susceptible to undesirable pre-ignition than hydrogen and other fuels.

[2]Hydrogen does not fit well into the normal definitions of octane number. It has a very high RON and a low MON, so that it has low knock resistance in practice due to its low ignition energy and extremely high flame spread speed.

NH3 is a "tunable" fuel, meaning that depending on the percentage cracked, it can be tuned to perform like propane, natural gas or even hydrogen. This is a very beneficial characteristic allowing NH3 engines to start very easily in cold weather conditions (similar to hydrogen cold start performance) and then be tuned to perform as a very high octane fuel with a very high resistance to pre-ignition in high compression/high temperature conditions, i.e. high efficiency engines. The tunable nature of NH3 also allows continuous performance adjustments for seasonal weathe8r conditions. Gasoline refineries are required to make significant seasonal adjustments to their operations to change from summer to winter blends. In fact, these required seasonal adjustments are often given as a reason for gasoline price increases in the spring and fall. The numerous gasoline specifications required to meet individual state standards has a large impact on the cost of gasoline. NH3 can be tuned to perform optimally for each engine and seasonal condition with firmware onboard the vehicle, providing another huge advantage.

NH3 is an excellent fuel for SOFC's. Intermediate temperature direct ammonia fuel cells have been demonstrated by several research groups.

Environmental and Health Characteristics

NH3 is commonly a naturally occurring chemical and also a common man-made chemical. It is not a known carcinogen, not a greenhouse gas, has an ozone depletion number of zero. NH3 is moderately caustic but is not a toxic chemical. It can be hazardous at fairly low concentrations and can produce base burns (akin to acid burns) when inhaled into lung tissue or when it comes into contact with eye tissue.

With conventional catalytic control systems, NH3 can be used in internal combustion engines with zero measurable criteria pollutants. NH3 fuel cell vehicles would also be zero emission vehicles.

Sustainability, Multiple End Uses, Developing Country Economic Development

NH3 production requires only two widely available raw materials: air (nitrogen) and hydrogen. An iron-based catalyst is used making catalyst availability a non-issue. Any and all primary energy source(s) (e.g. wind, solar, biomass, nuclear energy, OTEC, wave, coal, natural gas, etc.) can be used to produce NH3 and providing an opportunity for unlimited, sustainable production.

Making NH3 the fuel of choice could be a decision that provides developing countries with an opportunity to create a robust, sustainable local economy and greatly enhance the likelihood of developing an export economy as well. Oil refineries and gas-to-liquids facilities normally require multi-billion dollar investments and do not scale down well. NH3 production facilities can be very small and still be economically competitive. The smallest operational NH3 production facility in the world is located in Creston, Iowa. The Creston plant capacity is 100 ton per day which might cost approximately $65 million (~600,000 MMBtu or 5.2 million GGE or 13,000 vehicles at 12,000 miles per year and 30 mpg). For countries without significant fossil fuel infrastructure in place (i.e. natural gas pipelines, utility grid, oil pipelines, etc.), this is the most practical/economical option.

Local energy production creates jobs and energy security for every country in the world, reduces shipping costs. One relatively simple industrial facility producing NH3 can provide fuel for transportation and dispatchable electrical generation, fertilizer for local food production, a high performance long-term alternative energy storage medium (see chart below), an environmentally friendly refrigerant for space cooling and food storage freezers, an environmentally friendly disinfectant/cleaner, and a precursor for many other chemicals.

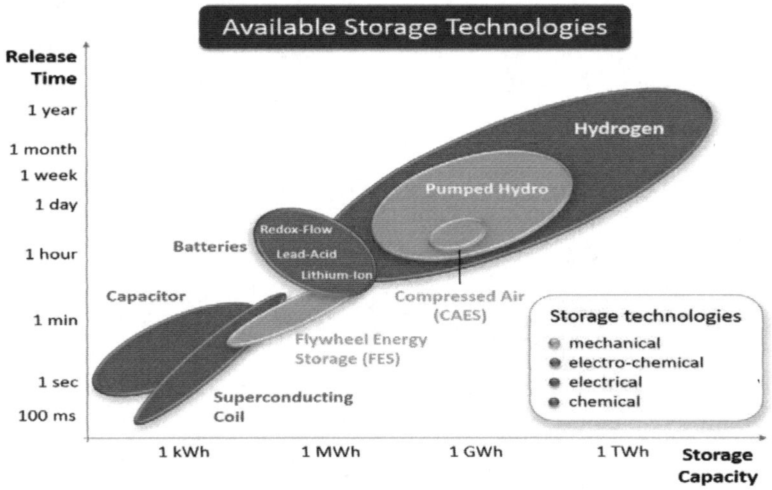

The investment required for a NH3 refinery is relatively low, providing an attractive option for developing countries to drastically improve their standard of living and to spur opportunities for additional economic development. Developing countries open the door to tremendous economic development opportunities by choosing to build their economy around NH3 and the biorefinery model (depicted below) where biomass is used to sustainably produce food, chemicals, fuels and co-products. These countries have an unprecedented opportunity to leap-frog the fossil-fuel addicted OEC's, weighed down by huge existing investments in fossil fuel infrastructure and associated lobbying interests, and to move forward with an affordable, robust, sustainable economy based on renewable NH3.

Biorefinery: Using NH3 to Catalyze the Creation of a Robust Economy via Increased Agricultural Production

Safety

Two highly credible safety studies have been completed to establish the comparative risks of NH3 verses propane and gasoline in transportation fuel applications. The studies: RISO (Europe) and Quest (U.S.) The conclusion of both reports was that NH3 would not present any unacceptable safety risks and in fact would be safer than propane in transportation fuel applications.

An entry-level product design engineer should be capable of designing a NH3 refueling and fuel tank system capable of meeting the world's most stringent safety standards. It is rather a trivial design problem compared to developing an affordable hydrogen infrastructure or an EV with performance equal to liquid fueled vehicles. NH3 fuel systems can be designed to be as safe as they need to be to meet public expectations.

NH3 fertilizer been used in large volumes in an acceptably safe fashion for decades in the Midwest of the United States. There are 800 retail outlets in the state of Iowa alone, many located near residential areas.

NH3 fueling stations be designed to be even safer than the existing fertilizer outlets. Chilled, atmospheric-pressure tanks located in underground containment vaults would provide a very safe design for NH3 fuel. Interlocking refueling hoses where no flow would be allowed until proper connection is verified would provide a re-fueling system that would be safer than existing gasoline re-fueling systems. A typical NH3 fueling station is depicted below.

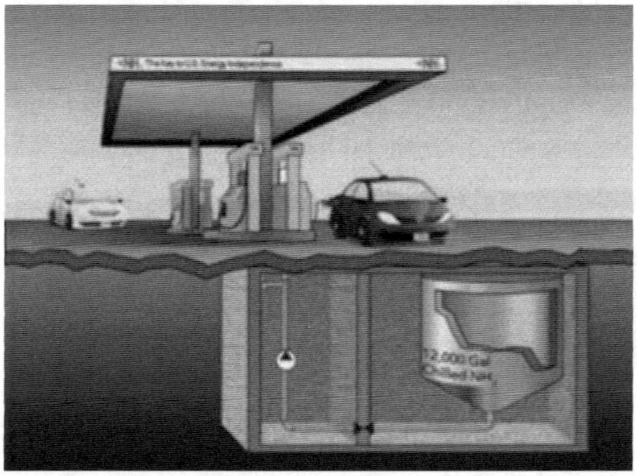

Summary

Unlike some technology areas where "all of the above" has significant advantages, there are tremendous advantages associated with choosing a single, optimized, <u>liquid</u> transportation fuel. The advantages of a <u>logically</u> chosen "ideal fuel" include: 1) low-cost, standardized production facilities; 2) affordable, standardized storage and delivery infrastructure; and 3) optimized, high-efficiency, low-cost, low/zero-emission engines and fuel cells. While electric vehicles will contribute to a reduction in the need for liquid transportation fuels, there will be a substantial need for liquid transportation fuels for the foreseeable future. The cost, efficiency and environmental benefits associated with choosing an optimized liquid transportation fuel are enormous and merit serious consideration.

NH3 most closely meets the criteria for an ideal liquid transportation fuel. It is the most efficient and cost-effective means of delivering hydrogen and has an extensive world-wide delivery system already in place. Additional infrastructure could be added at costs similar to propane infrastructure costs, one of the lowest-cost liquid transportation fuel infrastructure options. Using currently available catalytic controls, NH3 emissions can be even cleaner than hydrogen engine emissions. While NH3 can effectively be produced using biomass, it can also be produced from all other primary energy sources including wind, solar, nuclear, natural gas, OTEC, etc. NH3 can be used in fuel cells, gasoline engines, diesel engines, gas turbines and gas burners. Gasoline engines would require conversions similar to those required for propane. Currently available diesel engines could be modified in a similar fashion to dual-fuel natural gas/diesel fuel engines but NH3 substitution rates could be higher than natural gas substitution rates. Next generation diesel engines have demonstrated the ability to use 100% NH3. The tunable characteristics of NH3 allow its optimal use in almost any combustion application.

Developing countries could create extensive, robust agricultural and energy economies based on NH3 production facilities in a sustainable, environmentally friendly fashion. Competitive energy and agricultural economies would provide the basis for nearly limitless additional economic development opportunities.

Impressive federal programs in Japan and numerous patents field by Toyota combined with significant R&D efforts in the U.S., Canada and Europe are providing exciting momentum for the NH3 fuel movement. It is time to acknowledge the show-stopping deficiencies associated with the usual alternative fuel "suspects" and move forward with the closest thing to an ideal fuel, NH3.

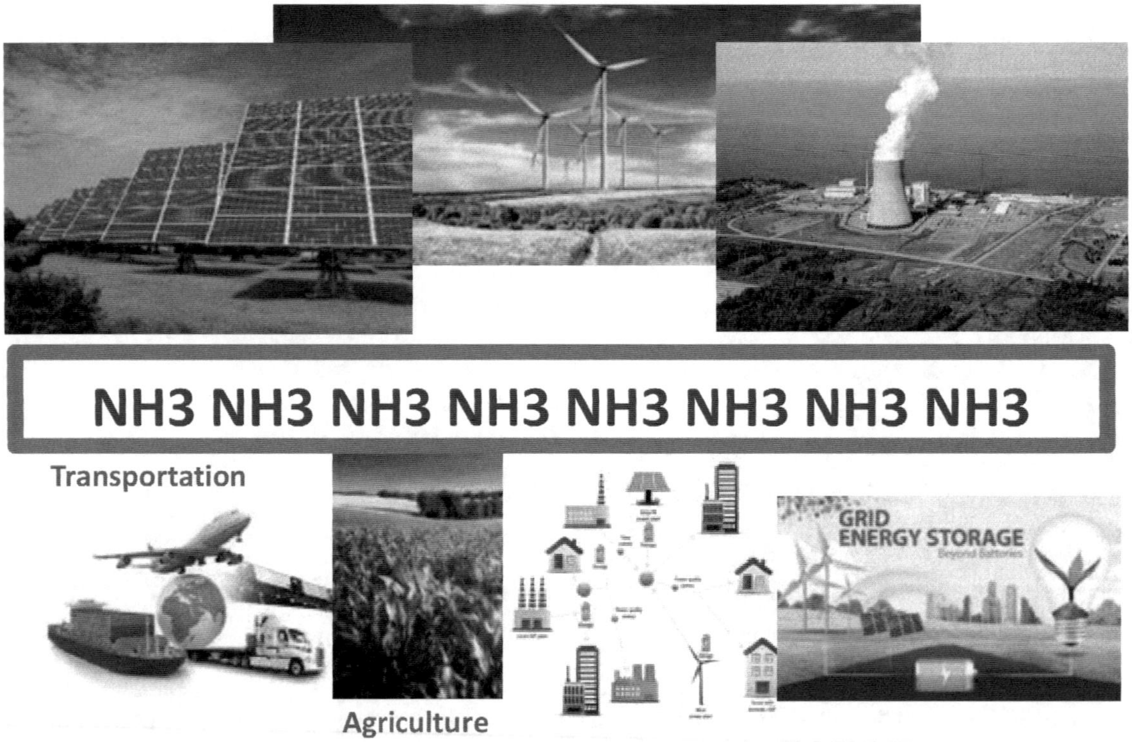

458643 Implementing Robust Vapor-Liquid Equilibrium Calculations in Nonsmooth Multi-Stream Heat Exchanger Models

Abstract:

Optimization of the liquefaction process for natural gas has proven to be a challenging task in particular for two reasons. First, the refrigeration cycles introduce computational loops that cause convergence problems. Second, the refrigerants are mostly multicomponent mixtures with strong interactions in the search for optimal heating profiles to match the cooling profile of the natural gas. In addition, the driving forces in the heat exchangers are very tight, asking for rigorous modeling to make sure the simulation mimics the true behavior of the process.

Both gradient based methods and stochastic search methods have been used to try to optimize processes for Liquefied Natural Gas (LNG) production, where commercial simulators are used to provide the function values in the optimization, however, with limited success beyond the most simple liquefaction processes. Our research groups have been involved in the development of a new model for these processes with the inherent features required to be able to utilize the latest developments in the field of Global Optimization [1].

As an important part of this new approach to LNG optimization, Barton and coworkers at MIT [2] has developed a new non-smooth multi-stream heat exchanger model accounting for phase changes in the process. The model was successfully used to simulate the Poly Refrigerant Integrated Cycle Operations (PRICO) process for liquefied natural gas production. However, it suffered from convergence difficulties when studying more complex single mixed refrigerant (SMR) processes. The primary challenge was flash calculations, which frequently failed even for initial guesses close to the solution. These problems will of course increase even further when moving to Dual Mixed Refrigerant (DMR) processes and the Mixed Fluid Cascade (MFC) process.

In this project, an alternative model structure is studied, replacing the equation-oriented framework with a hybrid solution. Rather than converging the vapor-liquid equilibrium calculations simultaneously with the other model equations, they are included as nested subroutines and solved sequentially.
Nonsmooth Newton-type methods and the implicit function theorem for lexicographically smooth functions are used for computing analytical derivatives in each subroutine [3, 4]. In addition, more robust flash calculations are attempted by developing a nonsmooth version of the "inside-out" algorithm by Boston and Britt [5] that can be incorporated in the existing model framework. The same SMR processes are then simulated in order to investigate whether the alternative framework is more robust.

[1] Floudas, C.A. and Gounaris, C.E., A review of recent advances in global optimization, *Journal of Global Optimization*, vol. 45(1), pp. 3-38, 2009.

[2] Watson, H.A.J., Khan, K.A. and Barton, P.I., Multistream heat exchanger modeling and design, *AIChE Jl.*, vol. 61(10), pp. 3390-3403, 2015.

[3] Facchinei, F., and Pang, J.S., *Finite-dimensional variational inequalities and complementarity problems*, vol. 2. Springer Science & Business Media, 2007.

[4] Khan, K. A., and Barton, P.I., Generalized derivatives for hybrid systems, *Submitted*, 2015.

[5] Boston, J.F. and Britt, H.I., A radically different formulation and solution of the single-stage flash problem, *Computers & Chemical Engineering*, vol. 2(2), pp. 109-122,

465812 Shared and Practical Approach to Conserve Utilities in Eco-Industrial Parks

Abstract:

Eco-Industrial Park (EIP) is a community of manufacturing and service facilities in close proximity. These enterprises collaborate with each other for improving the social, environmental, and economic performance of the whole complex in a sustainable manner by utilizing each other's resources efficiently, and sharing facilities, services, and utilities (Boix et al., 2015; Côté & Cohen-Rosenthal, 1998; Lowe et al., 1996). Examples of such successful industrial symbiosis are Jurong Island (Singapore), Kalundborg (Denmark), and Rotterdam (Netherlands). EIPs such as Jurong Island in Singapore are interested in assessing the potential for inter-plant heat integration among its inhabitants. The aim of such an enterprise wide heat exchanger network system is to exploit the synergistic heating/cooling needs and reduce total utility costs. However, there are many practical challenges in the implementation, construction, and operation of such an enterprise wide network (Chew et al., 2013) that need to be addressed while synthesizing inter-plant heat exchange network (HEN).

(1) The inter-plant heat exchange network synthesis (HENS) methodology should involve all the capital and operating expenses. Since, the exchanger and pipeline costs are comparable in case of long distance transport, inter-plant piping and transport costs cannot be ignored in configuring an EIP-wide HEN.

(2) The independent and diverse profit-making enterprises should be willing to participate and collaborate in such an EIP-wide HEN. Hence, there should be attractive economic benefits and fairness guaranteed to motivate the enterprises for such collaboration.

(3) The exchangers in an EIP-wide direct heat integration must have real physical locations. Furthermore, if a plant hosts a heat exchanger, it must bear the added responsibilities of maintaining and operating that exchanger, and guaranteeing the safety, security, and confidentiality of any guest streams.

(4) Operational disturbances in one plant shouldn't propagate to other plants, as the plants should be able to operate independently.

In this work, we present a novel strategy for EIP-wide HENS that addresses most of the issues highlighted above. We propose a central, shared, mutually agreed location for the HEN. The participating enterprises can appoint a third party logistics provider to manage and operate such a facility. We have considered pumping costs in the operating costs and the pipeline, pump and heat exchanger costs in the capital investment. By optimizing the HEN location, we can minimize the transport and pipeline costs. In our proposed model, we allowed the cooling (heating) of a hot (cold) stream both before entering and after leaving the HEN. To give operational flexibility and control to the parent company, existing in-plant utility heaters/coolers will be retained. Such an arrangement offers several advantages in terms of flexibility, control, cost, safety, security, confidentiality, accountability, operation, and management. We then propose a more practical and rational cost-benefit sharing arrangement. The contribution of an enterprise to capital expenses (CAPEX) and operating expenses (OPEX) is proportional to the utility cost savings it enjoys through the shared HEN. This ensures fairness among the enterprises as each enterprise enjoys an identical rate of return.

The EIP-wide HEN must trade off capital and operating costs versus utility savings. Therefore, we develop a mixed-integer non-linear programming (MINLP) model to maximize the net present value (NPV) of an EIP-wide HEN. The superstructure-based formulations allow sub-streams and non-isothermal mixing in a stage (Huang & Karimi, 2013; Yee & Grossmann, 1990). Since we wish to maximize the net present value of the shared HEN, discounted flows and rate of return has been included in the cost equation. Our model allows arbitrary cost expressions for exchangers with both fixed and variable components and cost multipliers for material and installation considerations. The single HEN location allows the full flexibility of multiple heat exchanges among streams from different plants. We have also improved the efficiency of formulation by adding additional constraints for eliminating redundant or complicated network designs. We have limited the number of sub-streams in a stage, avoided repeat identical matches, and included lower limits on exchanger areas and temperature drops to avoid exchangers with small-duty or small-area. If the HEN looks profitable (NPV > 0), each enterprise will enjoy annual savings. In return, it must share the various costs of the HEN in proportion to its annual savings.

Our proposed model for the case study in Hiete et al. (2012) offers an NPV of \$5.71 million with a utility savings of \$1.74 million/a for the EIP. The piping and pumping costs are major costs in this shared HEN, which restates their importance in EIP-integration. Enterprise E1 contributes 67.4 % to the CAPEX and OPEX, as it enjoys the most savings. In the game theory based formulation in Cheng et al. (2014), enterprise E2 trades streams to E1 and E3, but pays nothing for the exchangers. In contrast, our optimal HEN apportions costs on a more rational and practical basis of utility savings. Since E1 gains the most from the EIP-integration with a utility cost savings of \$126 456/a (\$29 411 /a for E2 and \$71 183 /a) for E3, it contributes the largest share of the costs (E1:55.7 %, E2:13.0 % and E3:31.3 %). Like the case study of direct heat integration in Wang et al. (2015), our HEN also achieves a total utility savings of 95 %, which is the thermodynamic limit. Wang et al. (2015) did not clearly indicate the locations of their heat exchangers (or equivalently the streams being transported). E2 contributes 90.5 % of investment costs, as it supplies the cold streams that would otherwise require expensive hot utilities. Hence, solutions are fair, as companies saving expensive heating utilities will have to contribute more in the investment costs.

Heat integration has been employed over the years in several chemical processes, in particular refineries and petrochemical. As discussed, EIP-wide heat integration poses several unique economic, design, logistics, operations, and safety challenges. But, the utility costs for refineries and petrochemical plants can be huge every year, which is an incentive to reduce these costs in spite of the various challenges. In fact, modern EIPs like Jurong Island in Singapore and Jubail in Saudi Arabia are very much interested in assessing the potential for inter-plant heat integration. The methodology presented here paves the way for assessing and implementing heat integration for such "futuristic" systems.

Boix, M., Montastruc, L., Azzaro-Pantel, C., & Domenech, S. (2015). Optimization methods applied to the design of eco-industrial parks: a literature review. Journal of cleaner production, 87, 303-317.

Cheng, S.-L., Chang, C.-T., & Jiang, D. (2014). A game-theory based optimization strategy to configure inter-plant heat integration schemes. Chemical Engineering Science, 118, 60-73.

Chew, K. H., Klemeš, J. J., Wan Alwi, S. R., & Abdul Manan, Z. (2013). Industrial implementation issues of Total Site Heat Integration. Applied thermal engineering, 61, 17-25.

Côté, R. P., & Cohen-Rosenthal, E. (1998). Designing eco-industrial parks: a synthesis of some experiences. Journal of cleaner production, 6, 181-188.

Hiete, M., Ludwig, J., & Schultmann, F. (2012). Intercompany Energy Integration. Journal of industrial Ecology, 16, 689-698.

Huang, K. F., & Karimi, I. A. (2013). Simultaneous synthesis approaches for cost-effective heat exchanger networks. Chemical Engineering Science, 98, 231-245.

Lowe, E. A., Moran, S. R., Holmes, D. B., & Martin, S. A. (1996). Fieldbook for the Development of Eco-Industrial Parks: Final Report: Indigo Development.

Wang, Y., Chang, C., & Feng, X. (2015). A systematic framework for multi-plants Heat Integration combining Direct and Indirect Heat Integration methods. Energy, 90, Part 1, 56-67.

Yee, T. F., & Grossmann, I. E. (1990). Simultaneous optimization models for heat integration--II. Heat exchanger network synthesis. Computers & chemical engineering, 14, 1165-1184.

462355 Optimal Design and Operation of Integrated Multi-Vector Energy Networks

Abstract:

Integrated multi-vector energy networks are becoming more important because we have a much greater variety of primary resources of energy such as wind, solar, biomass etc. Whereas existing energy systems are dominated by fossil fuels as the primary resource, with electricity and natural gas as the energy vectors, increasing the penetration of renewable energy will change the energy mix and other energy vectors, such as hydrogen, methanol, syngas, ammonia etc. may also play important roles. There are many different pathways to satisfy energy service demands through different energy vectors: for example, hydrogen, biofuels or electricity are considered promising alternatives to petroleum for the transport sector.

The aim of this work is to determine the best design and operation of integrated energy network to obtain the most value from limited available resources while minimising its negative socio-economic and environmental impacts. This endeavour requires the use of mathematical models to provide an accurate representation of the potential technologies and transport infrastructures that may become part of the network, as well as optimisation techniques to determine the most promising designs among the many possible alternatives. This involves selecting the right combinations of conversion and storage technologies, where to locate them, what their capacities should be and how to interconnect them via transport infrastructures. Moreover, the mathematical models need to be sufficiently detailed to account for all of the operational issues at different time scales (e.g. hours, days, seasons, years) and interactions between the different energy vectors. In this conference, we will present a general mixed integer linear programming model that can simultaneously determine the design and operation of any integrated energy networks comprising technologies for conversion and storage and infrastructures for transport of resources (both material and energy). The model takes into account the spatial and temporal distribution of system properties and simultaneously determines the spatial structure, i.e. the location, number, capacity of the technologies and the structure of the transport infrastructures that connect them, as well as the hourly operation of the entire network. It is a dynamic model in order to properly account for the intermittency of renewables and dynamics of energy storage. The model includes detailed and coherent storage inventory balances, thus it is able to track the inventory level of each resource in storage at every hour over the entire planning horizon (e.g. out to 2050).

This presentation will describe the model and discuss a decomposition method that enables this very large, and otherwise intractable, optimisation model to be solved. We will also present case studies wherein the design and operation of an integrated electricity-hydrogen-syngas-natural gas network are determined in order to maximise the value (or minimise the cost or environmental impacts) of meeting the demands for heat, electricity and transport fuels in Great Britain.

464617 Simultaneous Optimization of Design and Operation Strategies for CHP Systems

Abstract:

Combined heat and power (CHP) facilities are a very promising path to reducing CO_2emissions and increasing efficiency in the power generation sector, especially when combined with residential solar photovoltaic (PV) power generation. CHP facilities rely on natural gas, a cleaner fuel than coal, to generate electricity, as well as hot and chilled water to provide heating and cooling. The ability to supply these essential residential utilities in an efficient way on a medium to large scale opens the path for combining district cooling, heating and power generation, and suggests that CHP plants are an appealing choice as an integrated utility supplier for the neighborhood of the future. Yet, there are currently no CHP plants that serve exclusively residential neighborhoods, and published works exploring this possibility are scarce.

In our previous work [1], we demonstrated that a CHP plant with PV integration can in principle meet the demands of a residential neighborhood during the summer months. However, a CHP plant operating in island (i.e., grid-disconnected) mode must be optimally sized to maximize efficiency and to lower the capital and marginal costs, while producing enough electricity, heating, and cooling to meet the neighborhood demands at all times. A majority of the approaches described in literature for the optimal CHP plant sizing tend to select equipment by analyzing the various utility demands (i.e. electricity, steam, and chilled water) individually. Also, they only minimize the capital and O&M costs, overlooking the transitional costs of equipment turning on, by either not allowing the equipment to turn off or ignoring the monetary value of the wear and tear on the equipment. Lastly, many micro-grid papers assume that the main grid will be there to support the system at all times, if the CHP plant is not able to produce the needed energy.

In the paper, we describe a novel simultaneous optimization of design and operation strategies for a CHP plant as a utility producer for a residential neighborhood operating in island (i.e., grid-disconnected) mode, incorporating residential photovoltaics, implemented in a centralized fashion. In order to more accurately predict the operational costs of the CHP plant over a long period of time, the integrated design optimization is combined with an operational-oriented scheduling, forming a two-level problem. The first level consists of the presented integrated design and scheduling optimization, which selects equipment for a plant that must meet hourly residential neighborhood utility demands while minimizing the overall cost of the plant. The second level uses the optimal equipment sizes determined from the first level, and schedules the plant for a year, giving a better result of the operation economics of the CHP plant over a longer period of time. Utilizing data collected by Pecan Street Research Inc., a non-profit smart grid demonstration project headquartered at The University of Texas at Austin, a year of hourly residential heating, cooling, and electricity demand are analyzed and evaluated. These data are then used to create a time-resolved energy demand profile describing residential energy use, which is subsequently serves as a basis for optimizing the size and operation of the CHP plant so that all utility demands are met, while minimizing the capital and marginal costs, including transition penalties.

References:

[1] Ondeck, A.D., Baldea, M., Edgar, T.F. (2015). *Data-Driven Modeling and Optimal Operation of District-level Combined Heat and Power and Photovoltaic Power Generation System*. Applied Energy, 156, pp. 593-606.

462056 Exploiting Dynamic Flexibility to Enable Participation in Multi-Scale Electricity Markets

Abstract:

There is significant interest in re-designing the control architectures for a broad spectrum of energy and manufacturing systems to exploit time-varying electricity prices. For example, the Alcoa Point Comfort Power Plant, which is a utility plant that provides electricity and steam to the adjacent aluminum manufacturing facility, re-optimizes its operations every 15 minutes in response to electricity and natural gas price. These new energy-oriented architectures provide load flexibility to the power grid in exchange for monetary payments. Most academic literature on industrial demand response optimization focuses on scheduling and planning decision layers and considers coarse (e.g., 1 hour) time discretizations with multi-day planning horizons. Emphasis has been placed on optimal planning under time-vary electricity prices, although recent work considers providing reserve capacity and interruptible loads. This emphasis missing flexibility at fast timescales (e.g., minutes to seconds) that may be monetized through participating in real-time energy markets and/or providing regulation capacity and other ancillary services.

We develop a mathematical model for multi-scale electricity markets run by the California Independent System Operator (CAISO). The model incorporates prices from both the day-ahead (1 hour intervals) and real-time markets (5 â€“ 15 minute intervals), and considers both energy and ancillary service (regulation, reserves, etc.) sales. Market rules are expressed as linear algebraic expressions with continuous and optionally discrete variables. The models are general and applicable to both conventional thermal generators and storage devices (e.g., batteries).

Using the market model, we optimize the operation policy of a combined heat and power utility system using real price data for all of 2015. We observe participating in energy markets run at multi-times decreases net operating costs (fuel minus market proceeds) by 14 â€“ 54%, depending on fuel price, without degrading satisfaction of onsite steam and electricity demand. The savings increase if the onsite steam and/or electricity demands are made flexible.

Approximately one-quarter to one-half of the anticipated savings are from regulation capacity sales to the market. In CAISO, the automatic generator control (AGC) system sends new electricity production setpoints to generators providing regulation every 5 seconds that are within a specified flexibility band (units: MW_e). Capacity payments are based on the size of this band. Thus regulation offers a way to monetize the inherent flexibility in manufacturing/energy systems at faster timescales (seconds compared to electricity price variations that occur at slower timescales (minutes to hours). In practice exploiting this flexibility and realizing revenues from electricity markets requires sophisticated controls strategies that manage multiple time-scales and consider uncertainty in market prices.

468112 A Rolling Horizon Scenario-Based Approach for Smart House Management Under Uncertainty

Abstract:

The development of sustainable, efficient and environmental friendly energy supply chains has led to advances in the area of Energy Systems Engineering. In this context, the growing interest in the exploitation of renewable energy sources has led to an increasing use of microgrids, which are decentralized networks able to integrate several heat/power generation systems of low/medium capacity (Hodge et al., 2011; Zhang et al., 2013), with the main purpose of distributing the generated heat/energy locally so as to reduce energy losses and improve the responsiveness to energy demand fluctuations (Kopanos et al., 2013; Tong et al., 2015). These kind of networks usually involve the presence of renewable generators. So, it is imperative that the variations in the behaviour of natural energy sources requires the consideration of uncertainty in microgrids. This variability must be considered in order to ensure the operability and the reliability of the system, since the optimal solution obtained under deterministic conditions can be sub-optimal or even infeasible. Particularly, the operations management of microgrids can be affected by several types of uncertainty, such as energy production and energy demand variations.

This work addresses the presence of uncertainty through a discrete-time two-stage stochastic programming approach, which is combined with a rolling horizon approach (Kopanos et al., 2014) for the simultaneous management of energy supply and demand in microgrids. The proposed Mixed Integer Linear Programming (MILP) formulation contains inequality constraints corresponding to the available heat and energy generation technologies, as well as equality constraints corresponding to heat and energy balance equations that describe flows, production, storage and consumption levels. This mathematical formulation uses a scenario-based stochastic programming approach where the scenarios are associated to internal variations in the duration of the energy consumptions as well as in the overall heat demand. However, the high complexity related to the estimated weather forecast, makes the consideration of all possible external scenarios, computationally intensive. The variability in weather conditions may affect the availability and production capacity of renewable energy generators. Thereby, updating input data (i.e., wind profile, heat and energy demand) is needed to ensure the adequate quality in the obtained results. Hence, an MILP two-stage stochastic programming approach was incorporated within a rolling horizon scheme that periodically updates input data information as the uncertain input parameters are revealed or considered to be known with certainty. The main decisions to be made to maximise the profit of the microgrid includes the execution of consumptions, the amount of heat and energy to be produced or purchased, the heat and energy storage levels and the amount of energy to export to the power grid. The methodology proposed above was tested on a case study and promising results demonstrating the applicability of the methodology for decision making were obtained. This work also forms a basis for future work to address more complex problems.

References

Hodge, B.M., Huang, S., Siirola, J.D., Pekny, J.F. & Reklaitis, G.V. (2011). A multi-paradigm modeling framework for energy systems simulation and analysis. Computers & Chemical Engineering, 35, 1725-1737.

Kopanos, G.M., Georgiadis, M.C. & Pistikopoulos, E.N. (2013). Energy production planning of a network of micro combined heat and power generators. Applied Energy, 102, 1522-1534.

Kopanos, G.M. & Pistikopoulos, E.N. (2014). Reactive scheduling by a multiparametric programming rolling horizon framework: A case of a network of combined heat and power units. Industrial & Engineering Chemistry Research, 53, 4366-4386.

Tong, C.D., Palazoglu, A., El-Farra, N.H. & Yan, X.F. (2015). Energy demand management for process systems through production scheduling and control. AIChE Journal, 61, 3756-3769.

Zhang, D., Shah, N. & Papageorgiou, L.G. (2013). Efficient energy consumption and operation management in a smart building with microgrid. Energy Conversion and Management, 74, 209-222.

450742 Optimization Models for Shale Gas Well Refracture Treatments

Abstract:

Optimization Models for Shale Gas Well

Refracture Treatments

Markus G. Drouven[1], Diego C. Cafaro[2] and Ignacio E. Grossmann[3]

[1,3]Department of Chemical Engineering

Carnegie Mellon University

Pittsburgh, PA 15213, USA

[2]INTEC (UNL – CONICET)

Güemes 3450, 3000 Santa Fe, Argentina

[1]mdrouven@cmu.edu, [2]dcafaro@fiq.unl.edu.ar [3]grossmann@cmu.edu

Abstract

Shale gas wells are well-known for their characteristically steep decline curves. In fact, some shale wells produce more than half of their total estimated ultimate recovery (EUR) within the first year of operation. The initial production peak after well completion is caused by the sudden release of previously trapped hydrocarbons in the pores of the reservoir. The subsequent decline is ultimately driven by pressure depletion and the ultra-low permeability of the shale play. Upstream operators oftentimes struggle with the production decline curves. For one, they are contractually obligated to providing steady gas deliveries to midstream distributors over time – which is difficult to accomplish given the characteristic, dramatic production decline immediately after wells are turned in line. Moreover, as Cafaro and Grossmann [1] suggest and Drouven and Grossmann [2] confirm, operators need to maximize the utilization of production and gathering equipment – such as pipelines and compressors – in order to stay profitable. In reality, however, production and gathering equipment is usually sized based on the initially high production rates. This means that within a matter of months shale gas wells feed into oversized pipelines and compressor stations, and equipment utilization drops. Worse even, in order to satisfy contractual gas delivery agreements, operators see themselves forced to open up new wells continuously to honor their obligations, and hence the process is repeated over and over again.

Refracturing presents a promising strategy for addressing the characteristically steep decline rates of shale gas wells [3]. The core idea behind refracturing is to restimulate the reservoir such that it yields previously untapped hydrocarbons and improves the overall production profile of a well. Whether or not a refracture treatment will reinvigorate a shale gas well depends on a number of factors including the characteristics of the reservoir and the initial completion design. Historically, refracture treatments have been applied predominantly to shale gas wells suffering from low production rates due to known suboptimal initial stimulations and completions. However, Dozier et al. [4] argue that even wells with effective initial treatments have shown significant production improvements when restimulated after an initial period of production and partial reservoir depletion.

The problem addressed in this paper can be stated as follows. We assume that a candidate shale gas well has been identified for refracturing. For this well, a long-term production forecast as well as the production profile after additional refracture treatments at any point in time over the planning horizon is given. A gas price forecast along with expenses for drilling, fracturing, completions and refracturing operations are also available as problem data. The problem is to determine: (a) whether or not the well should be refractured, (b) how often the well should be refractured over its entire lifespan, and (c) when exactly the refracture treatments should be performed. The objective is to maximize either: (a) the estimated ultimate recovery (EUR) of the well, or (b) the net present value (NPV) of the well development project.

First, we present a continuous-time nonlinear programming (NLP) model to determine whether or not a shale gas well should be refractured, and when to schedule the refracture treatment. The proposed NLP model relies on the assumption that the well productivity profile follows a decreasing power function of time. To predict well performance prior to and after a refracture treatment, we propose an effective forecast function that mimics real-life curves. Next, we extend the proposed framework to allow for multiple refracture treatments and present a discrete-time mixed-integer linear programming (MILP). In an attempt to reduce solution times to a minimum, we compare three alternative formulations against each other (big-M formulation, disjunctive formulation using Standard and Compact Hull-Reformulations) and find that the disjunctive models yield the best computational performance. Finally, we apply the proposed MILP model to two case studies to demonstrate how refracturing can increase the expected recovery of a well and improve its profitability by several hundred thousand USD.

References

[1] Cafaro, D. C.; Grossmann, I. E. Strategic Planning, Design, and Development of the Shale Gas Supply Chain Network. *AIChE J.* 2014. doi: 10.1002/aic.14405.

[2] Drouven, M. G.; Grossmann, I. E. Multi-Period Planning, Design and Strategic Models for Long-Term Quality-Sensitive Shale Gas Development. *AIChE J.* 2016 (accepted for publication).

[3] Jacobs T. Renewing Mature Shale Wells Through Refracturing. *SPE News*. 2014, May Issue.

[4] Dozier, G., Elbel, J., Fielder, E., Hoover, R., Lemp, S., Reeves, S., Siebrits, E., Wisler, D., Wolhart, S., Refracturing works. *Oilfield Review*. 2003; 10(8):38-53.

460724 A Composite-Curve-Based Biomass Procurement Planning Approach

Abstract:

A Composite-Curve-Based Biomass Procurement Planning Approach

Wenzhao Wu, Daniel Kurniawan, WenBo Zhu, Christos T. Maravelias*

Dept. of Chemical and Biological Engineering, University of Wisconsin-Madison, Madison, WI 53706

The production of fuels and chemicals from biomass has received considerable attention recently due to environmental concerns[1]. Since the production of biofuels involves relatively expensive feedstock and energy-intensive biomass transportation, any biomass-to-fuels strategy should include an efficient, both in terms of cost and environmental impact, biomass (feedstock) supply chain. Unlike fossil fuels, biomass, as a low-energy density resource, is sparsely distributed. The efficient biomass transportation thus requires biomass procurement planning methods[2]. In many studies, the farms are treated as points without shape or area[3]. This is a reasonable assumption when the transportation distance is so large that the shape and size of the farms can be neglected. In this case, the transportation problem is modeled as a point-to-point (farm-to-refinery) problem. However, the shape and size of the farms cannot be neglected when the refinery is close to the farm, which means that the size of the farm is not significantly smaller than the transportation distances, which in turn means that the error in approximating the real transportation distance with the distance between the center of the farm and the bio-refinery can be quite large. In this case, transportation should be treated as a region-to-point problem. To this end, we discuss a novel approach to biomass procurement planning on a region-to-point basis.

In terms of transportation, we propose a region-to-point modeling approach based on mathematical integration (in a polar coordinate system) over the sourcing region that has unique characteristics such as shape, location, and productivity. The transportation cost is correlated with the amount of biomass ("mass") procured from each farm. The final result is a function $C = f(M)$, where M is the mass, and C is the transportation cost. In other words, the proposed methods generate a function that returns the total cost of procuring M mass, which is graphically represented by a "procurement curve". Both algebraic and numerical solution methods are discussed and demonstrated with examples.

In terms of system-level procurement planning, we develop a composite-curve-based approach that incorporates the regional transportation modeling method, and aims at identifying the biomass procurement plan that minimizes the total procurement cost (including biomass purchasing, harvesting and transportation). The specific steps for the generation of the composite curve using the individual procurement curves, as well as insights into the procurement planning problem are discussed. An analogy to the cold/hot composite curves in the pinch design method for heat exchanger networks[4] is also presented. A case study involving 12 farms (which are approximated as polytopes) and one refinery is presented (see Figure 1A). The corresponding composite curve is shown in Figure 2, and the final procurement strategy is graphically represented in Figure 1B.

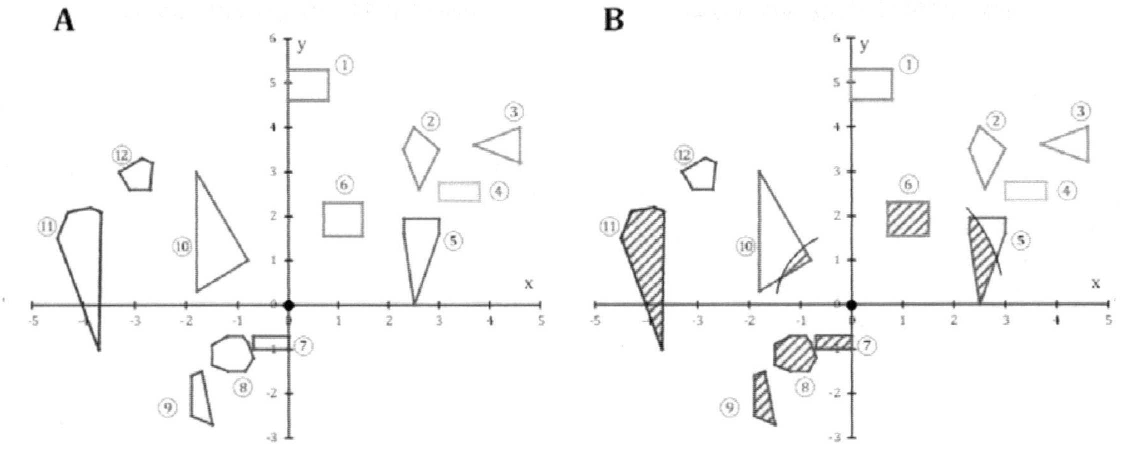

Figure 1. (**A**) Map of farms (the polytopes) surrounding a bio-refinery (the black dot at the origin) for the case study; (**B**) procurement strategy (represented by the green dashed areas) for a 11800 T/year demand. The farm numbers are labeled accordingly.

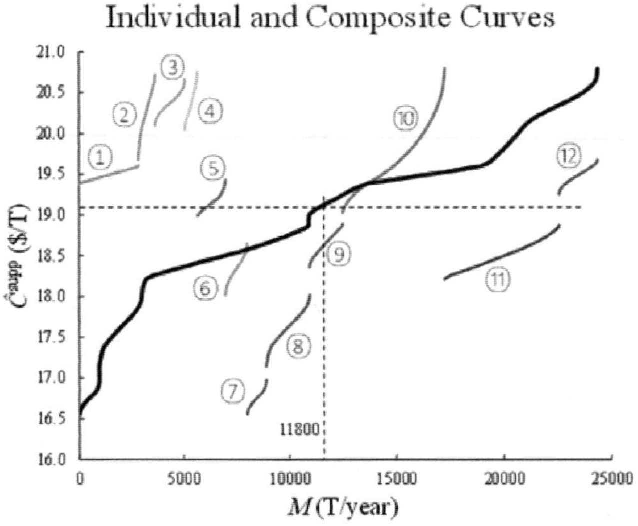

Figure 2. Individual procurement curves and the composite curve for the case study. The composite curve is marked thick. The demand of 11800 T/year and the corresponding per-mass supply cost ($\hat{C}^{supp}$) on the y-axis are marked with dashed lines. The farm numbers are labeled accordingly.

The proposed composite-curve-based method allows us to more accurately calculate transportation distance, and thus transportation costs and GHG emissions due to transportation. It also provides some key insights into the design of biofuel supply chains. In addition, the methods proposed in this work are integrated with mathematical programming to address complicated problems involving, for example, multiple feedstocks, multiple refineries, and multiple periods. We can solve such problems by either generating multiple composite curves, or directly incorporating the $C = f(M)$

function for each farm into a general supply chain optimization model.

References

[1] DOE Bioenergy Technologies Office, 2014. Multi-year program plan, Washington DC, USA: DOE.

[2] DOE/EERE, 2013c. Feedstock supply and logistics: biomass as a commodity, US: Department of Energy, Office of Energy Efficiency & Renewable Energy.

[3] You, F., Tao, L., Graziano, D. & Snyder, S. W., 2012. Optimal Design of Sustainable Cellulosic Biofuel Supply Chains: Multiobjective Optimization Coupled with Life Cycle Assessment and Input-Output Analysis. AIChE Journal, Volume 58, pp. 1157-1180.

[4] Linnhoff, B. & Hindmarsh, E., 1983. The pinch design method for heat exchanger networks. Chemical Engineering Science, 38(5), pp. 745-763.

462162 The Study of Ozone Pollution Superposition Caused By Simultaneous Olefin Plant Startups

Abstract:

It is indisputable that flare emissions from olefin plant turnaround operations generate large amounts of ozone precursors (NOx and VOCs) that will potentially aggravate the ground-level ozone pollution. In an industrial area where chemical and petrochemical plants are heavily populated, flare emissions can come from multiple plants in different spatial locations with different time periods. Thus, it is important to know the synergistic effect on ozone generations caused by flare emissions from simultaneously turnaround operations of multiple olefin plants in an industrial area. In this paper, a systematic methodology to study the ozone pollution superposition caused by simultaneous olefin plant start-ups has been developed. It couples dynamic start-up flaring profiles from multiple olefin plants and regional air-quality modeling together to examine the linearity and nonlinearity of ozone superposition. Based on our case studies, the synergies of ozone pollution due to multiple olefin plant start-ups cannot be simplified as the linear superposition. Meanwhile, the superposition nonlinearity has strong relations with the starting time of plant start-ups operations as well as the difference of plant distances. The general ozone superposition trends under different starting time and plant distance are also disclosed. Based on our study, the ozone superposition gradually approaches the linearity with the starting time of plant start-up operations changes from 00:00 to 10:00. In comparison to plant distances, the ozone superposition at about 32 km can be approximated as the linearity, and such linearity will decrease along with the decrease of plant distance. This study will be helpful to understand the ozone impacts from industrial point sources and identify cost-effective emission control strategies.

470901 Exposome Analysis of Polyaromatic Hydrocarbons

Abstract:

This study demonstrates the use of the INTEGRA platform (coupling multi-source multi-route external exposure modeling with a generic PBTK for internal exposure modeling) for estimating exposure to PAHs under a variety of scenarios of key relevance to REACH and the petroleum industry. The objective of the INTEGRA project is to bring together all information necessary for assessing the source-to-dose continuum over the entire life cycle of substances covering an extensive chemical space through the use of QSARs. The major outcome of INTEGRA is a comprehensive computational platform that integrates multimedia environmental and micro-environmental fate, exposure and internal dose within a dynamic framework in time. The platform allows multimedia interactions across different spatial scales, taking into account environmental releases and related processes at global, regional and local scale, up to the level of personal microenvironment. Coupling seamlessly exposure models with refined computational tools for internal dosimetry transforms risk assessment of environmental chemicals since it allows risk characterization to be based on internal dosimetry metrics. In this way high throughput system data such as the ones generated by Tox21 in vitro testing can be used. This opens the way towards a higher level of assessment that incorporates refined exposure (tissue dosimetry) and toxicity testing (Biological Pathway Altering Dose â€“ BPAD) associated to environmental contamination at different scales.

Exposure to PAHs has became of particular scientific and regulatory interest the last year, especially in view of the potential for petroleum substances to be included in the different REACH processes (notably Evaluation and Authorisation). In order to meet the requirements of REACH, it is of particular importance the capability of models to predict direct (arising from the use of substances) and indirect (e.g. fuel combustion) PAH exposures. INTEGRA fulfils the requirements for carrying out such a comprehensive assessment and goes event beyond, allowing the estimation of internal dose, as well as the assimilation of biomonitoring data. Several scenarios were requested to be evaluated, and the scenarios with the ir results are listed below:

1] Assessing of internal exposure to PAHs (i.e. arising from all sources a consumer is exposed to: environment, food, consumer products) and the relative contribution from different routes (oral-diet; oral-dust, inhalation, dermal). For this scenario, Ïƒtarting from annual emissions of 400 tones B[a]P in air within EU, and for regional emissions of 15 tones, distribution across different media and contribution of different pathways and routes was estimated. Higher intake estimates for children result from the higher bodyweight normalized inhalation rate, amount of soil and dust daily ingested; this is illustrated also in the analysis of various pathways contributions as well as when integrating among routes. Soil and dust ingestion are the dominant exposure pathways for children (uptake rate of 8.5E-03 ng/kg bw/day, followed by diet (uptake rate for both adults and children at 6-7E-04 ng/kg bw/day) and inhalation (uptake rate at 5.7-9.6E-07 ng/kg bw/day). Oral is the dominant route of exposure (uptake rate at 3.2E-03-15.2E-03 ng/kg bw/day) followed by inhalation (uptake rate at 1E-06 ng/kg bw/day) and dermal exposure (uptake rate at 9E-08 ng/kg bw/day). Mean B[a]P blood levels are up to 0.00012 ng/L for adults and 0.0007 for children, while the respective urinary concentration of the major specific metabolite 3-OH-B[a]P is 0.01 ng/L

Internal exposure to PAHs arising from specific use(s)/source(s) was also calculated and split up according to uses and sources. More specifically:

2] Assessment of specific sources such as smoked fish; when accounting for the annual emissions described above, concentration in fish was estimated by the multimedia model equal to 10^{-7} Î¼g/kg. This concentration is compared to the ones identified in the literature from smoked fish analysis, where B[a]P levels in smoked fish range from 0.08 to 4.1 Î¼g/kg (median of 1Î¼g/kg and consumption of 110 grams of fish). Based on the above, intake due to smoked fish consumption dominates among other pathways. The contribution of smoking fish to the total B[a]P content in fish, and consequently the contribution of the consumption of smoked fish to integrated B[a]P exposure (for fish consumers) is 1.4 ng/kg bw/day whereas the corresponding intake rate from soil or dust ingestion are 0.8-2.1E-03 ng/kg bw/day and from inhalation 4E-07 ng/kg bw/day. The respective internal dose of B[a]P upon smoked fish intake results to a peak of 0.4 ng/L of B[a]P in blood, that slowly decreases down to the blood concentration corresponding to the background exposure levels (0.01 ng/L) after almost 72 hours. The peak concentration of urinary 3-OH-B[a]P is 0.01 ng/L, observed after a couple of hours upon smoked fish intake.

3] The contribution of petroleum products (e.g. RAE - Residual Aromatic Extracts/ OLBO - other lubricants based oils) to integrated PAH exposure in Europe was reckoned on the basis of:

(a) contribution via release of PAHs to the environment (production and downstream use sites of a category of petroleum products) leading to indirect human exposure in the vicinity of an industrial plant of petroleum products. In this case inhalation becomes the dominant exposure pathway (uptake of 2.1E-01 ng/kg bw/day)

(b) contribution via consumer use of a category of petroleum products such as lubricants and coatings in rubber boots. Then, the specific consumer use dominates exposure (uptake rate of 6 ng/kg bw/day). In this case, dermal uptake is the dominant route of exposure followed by oral ingestion. However, due to the relatively limited and slow absorption through the skin, the respective peak of internal dose of B[a]P in blood is lower than the one in the smoked fish (0.2 ng/L of B[a]P in blood), but the overall area under the curve is similar, as a result of the slower absorbion into the systemic circulation and concequently the slower elimination.

4] Of particular interest is the estimation of external and internal exposure to PAHs from biomass burning for space heating. In a study evaluating the exposure to PAHs in two different urban locations (namely an urban background and a traffic station), it was identified that exposure to PAHs is greatly affected by biomass emissions. In practice, from the urban background site (where PM precence is dominated by biomass emissions), particles are characterized by lower mean aerodynamic diameter and higher PAH content. Accounting for the differences related to PM HRT deposition and the distribution of PAHs adsorbed on PM of various sizes, it was estimated that 2.5 times higher airborne B[a]P resulted in 6 times higher uptake of PAHs retained by the respiratory system. These findings bring us to the conclusion that PM emitted from wood burning for space heating (mostly in open fireplaces) contribute more to the population exposure to PAHs and to the associated health risk than the PM usually found in this urban environment. This finding brings additional evidence to the long-standing debate about the toxicity of PM emitted from biomass burning. Moreover, biomass burning is always accompanied by incomplete combustion which varies between the different combustion practices (temperatures, appliances and type of biomass used), favoring the formation of PAHs. PAHs, and the most carcinogenic compounds (5- and 6-ring) in particular, including BaP, are adsorbed to finer particles. In practice, this is also reflected in the internal dose as well; although ambient air concentration of B[a]P equivalent concentration was measured equal to 4.5 and 1.7 ng/m^3 for the urban background and the traffic site respectively (for adults), the corresponding B[a]P equivalent levels in blood were estimated equal to 0.38 and 0.06 ng/L respectively.

5] Finally, the time-dynamic nature of the INTEGRA computational platform, allowed also to capture environmental, exposure and internal dose dynamics in high temporal resolution, quantifying the effect of real-life different exposure conditions related to daily activities such as driving, eating smoked fish or operating an open fireplace in actual uptake, internal dose and expected biomarker urinary levels. From this analysis it was indicated that internal exposure to B[a]P (and the other PAHs respectivelly) follows the extensively dynamic pattern of external, however, urinary 3-OH-B[a]P fluctuations are slower. In any case, spot samples of urinary PAHs may not be representative of the overall daily exposure, hence biomonitoerd based exposure assessment of PAHs requires multiple samples per day to be collected.

469013 A New Framework for Radiological Impact Assessment in Life Cycle Assessment

Abstract:

Life Cycle Assessment (LCA) is a technique aimed at understanding and assessing the relations between human activities and environment. Notably, it is extensively used to compare alternative power-generating technologies and waste management approaches from a sustainable point of view. However, in the nuclear industry, the LCA approach is either applied poorly or not at all. The main reason may be found in the lack of a consistent and standardised approach for the assessment of the impact of ionising radiation. In theory such impact is caused by both the direct radionuclide releases to air or waters from operations, and from nuclear waste stored in a final disposal facility. In fact the few methodologies available today neglect the radiological impact arising from solid nuclear waste. This lack of consistency prevents comparative LCA studies to be applied to the nuclear industry and undermines the results of the few published.

The aim of this work is to address this issue by proposing a new framework to be included as a standard procedure in the LCA framework. This framework envisages four steps and assesses the impact of ionising radiation in terms of risk of detrimental effect to human beings. Notably, both air and water releases, and solid waste are included. The fate analysis indeed represents the key step in the methodology, as it aims at quantifying the environmental concentration of a specific pollutant at the receptor location. Given the general framework, two alternative methodologies have been developed and operationalised; they differ in the models underpinning the fate analysis. In one methodology the IAEA generic models [1] coupled with the UK Radioactive Waste Management (RWM) Ltd.'s Post-Closure Safety Assessment [2] is used; whilst, in the other case, a MacKay-type model is adopted to simulate a compartmental multi-media environment. The general approach the two alternative methodologies and the main challenges involving their inclusion in the LCA framework will be discussed at the conference.

For the purposes of demonstrating the proposed framework the UK current approach for Nuclear Spent Fuel management is adopted as a case study. According to the approach proposed by Clift et al. [3], the system is divided into two subsystems: Foreground and Background. 'Real' data from UK nuclear facilities is used for the processes in the foreground system. Along with the proposed methodology, several other impact categories, including Global Warming Potential, will be analysed. Results of the study will show the overall environmental performance and will also allow to identify the sections of the UK spent nuclear fuel reprocessing procedure that have the most radiological and non-radiological impact. Particular emphasis will be given to the comparison between direct radioactive releases and solid nuclear waste. Preliminary results of the LCA study will be presented at the conference.

[1] IAEA, "Generic Models for Use in Assessing the Impact of Discharges of Radioactive Substances to the Environment - Safety Series," Vienna, 2001.

[2] NDA, "Geological Disposal: Generic post-closure safety assessment," 2010.

[3] R. Clift, A. Doig, and G. Finnveden, "The Application of Life Cycle Assessment to Integrated Solid Waste Management," *Process Saf. Environ. Prot.*, vol. 78, no. 4, pp. 279–287, 2000.

465915 Improvement of Dynamic Quantitative Risk Assessment Method: Moving Boundary Conditions.

Abstract:

Improvement of dynamic quantitative risk assessment methodology: moving boundary condition

Wonje Lee, Ph.D. candidate

Seoul National University

To ensure the safety of a process system, engineers use various methods to identify the potential hazards of a system. One of the most popular methods is quantitative risk assessment (QRA) and it can quantifies the risk of a particular process activity. But one of disadvantages of QRA is its isolation and lack of integration with the process operation. A conventional QRA methodology is unable to update risk during the life of process and reflect risk according to the process change because it uses a statistical accident occurrence probability. This was identified as one of the main causes of the BP Texas refinery accident. For this reason, a methodology which is able to calculate the accident occurrence probability dynamically should be studied. Recently, a dynamic QRA methodology which calculate the accident occurrence probability using the remaining time is suggested. The remaining time is calculated by measuring the change amount of a process key variable to upper bound or lower bound. But this methodology has two problems. First, the existing methodology make accident occurrence probability zero when a variable has no change in a dangerous level. If a variable has no change currently but is near boundary condition, it should have a reasonable accident occurrence probability. Second, the existing methodology is unable to calculate the probability proper to the oscillation of a variable. For example, a variable which vibrate with sine function close to upper bound has high probability when it is increasing but when it is decreasing, it has very low probability because of long distance to lower bound. It cannot classify difference between the oscillation and the control of a variable dynamically.

In this study, improved dynamic QRA methodology is suggested being able to consider the current state of a key variable as well as the change amount of it. A imaginary boundary condition is calculated and updated dynamically which has the same distance from current value of a variable to nearer boundary condition and opposite direction. The more a variable is near one of the original boundary condition, the more distance between updated imaginary boundary condition and original boundary condition become narrow. So, it can calculate reasonable probability about small change amount of a variable in dangerous level. This methodology is applied to dynamic simulation data of single compressor system which has recycle line and is verified by comparison to the result of existing methodology.

Session : Sustainable engineering forum – environmental health & safety and sustainability

466860 Safeprops: A Software for Fast and Reliable Estimation of Safety and Environmental Properties for Organic Compounds

Abstract:

We present a new software tool called SAFEPROPS which is able to estimate major safety-related and environmental properties for organic compounds. SAFEPROPS provides accurate, reliable and fast predictions using the Marrero-Gani group contribution (MG-GC) method. It is implemented using Python as the main programming language, while the necessary parameters together with their correlation matrix are obtained from a SQLite database which has been populated using off-line parameter and error estimation routines (Eq. 3-8).

The general form of the MG-GC method is:

$f_i(X) = \Sigma N_jC_j + \Sigma M_kD_k + \Sigma E_lO_l$ (1)

$f(X) = T\theta$ (2)

Robust regression is implemented for the parameter estimation. Here the Cauchy weight is used to place high weights on small residuals and small weights on large residuals:

$\theta^* = \arg\min \Sigma w_i(y_i^{exp}-y_i^{pred})^2$ (3)

$w_i = 1/(1 + (y_i^{exp} - y_i^{pred})^2)$ (4)

Uncertainty of the parameter estimates is based on the asymptotic approximation of the covariance matrix $COV(\theta^*)$:
$COV(\theta^*) = SSE/(n - p) * (J(\theta^*)^T J(\theta^*))^{-1}$ (5)

Following the methodology presented by Frutiger et al. [1] [2], the parameter estimates will be described by a student t-distribution:

$\theta^*_{1-\alpha} = \theta \pm \sqrt{diag(J(\theta^*)COV(\theta^*)J(\theta^*)^T)} * t(n-p, \alpha_t/2)$ (6)

A first guess for the a priori unknown GC factors is provided by applying linear algebra:

$\theta\wedge = (T^{tr}T)^{-1} * T^{tr} * f(X)$ (7)

The SAFEPROPS database is accessed to obtain confidence intervals of the property predictions and the following equation is applied:

$y^{pred}_{1-\alpha} = y^{pred} \pm \sqrt{diag(J(\theta^*)COV(\theta^*)J(\theta^*)^T)} * t(n-p, \alpha_t/2)$ (8)

The software can predict the following safety properties at standard conditions: heat of combustion ($\Delta H_c(T_0)$), the lower and upper flammability limit ($LFL(T_0)$ and $UFL(T_0)$), the auto ignition temperature ($AIT(T_0)$) and the median lethal dose ($LC_{50}(T_0)$). Furthermore, it allows predicting the lower flammability limit at different temperatures (LFL(T)).

The following environmental properties can be estimated: Global Warming Potential (GWP), Ozone Depletion Potential (ODP).

The prediction can be applied to a large variety of chemical compounds including but not limited to hydrocarbons, aromatic compounds, alcohols, acids, esters, ethers, amines, cyclic compounds, hydro-fluorocarbons and chlorine compounds.

The practical application of the software is highlighted in a case study for the screening of organic working fluids for thermodynamic cycles. Safety and environmental constraints should be taken into account as early as possible for the screening of potential working fluid candidates. In this study, a large amount of chemical compounds have been computationally screened. Their performance in the cycle has been analyzed. The corresponding safety-related and environmental properties are then estimated by the new prediction software.

The study shows that the software is a powerful tool for the design of novel organic working fluids, if experimental data is missing.

Finally, a graphical user interface (GUI) is presented which allows the user to search for the compound of interest and obtain the safety and environmental properties including the uncertainty analysis in a tabulated form. An application programming interface (API) is provided to make SAFEPROPS connectible to other modeling environments such as Matlab.

[1] Jérôme Frutiger, Jens Abildskov and Gürkan Sin; A Rigorous Methodology for Development and Uncertainty Analysis of Group Contribution Based Property Models; Abstract from Nineteenth Symposium on Thermophysical Properties, Boulder, United States

[2] Jérôme Frutiger, Camille Marcarie, Jens Abildskov and Gürkan Sin; A Comprehensive Methodology for Development, Parameter Estimation, and Uncertainty Analysis of Group Contribution Based Property Models – An Application to the Heat of Combustion; J. Chem. Eng. Data, 2016, 61 (1), pp 602-613

453168 Enhanced Ethanol Production Via Activated Carbon Addition during Syngas Fermentation

Abstract:

Syngas made by gasification of biomass, coal, pet-coke and waste materials, or present in industrial gas waste streams can be used for production of fuels and chemicals. The biological conversion of syngas components, CO, CO_2 and H_2, to fuels and chemicals at commercial feasibility requires high conversion efficiency, product specificity and operation stability. Acetogenic bacteria convert CO, CO_2 and H_2 via the Wood-Ljungdahl pathway to acetic acid and ethanol as main products. The production of these products relies on efficient transfer of CO, CO_2 and H_2 to the cells at rates that match the bacterium's kinetic capability to process the gas. Cells are inhibited when too much CO accumulates in the fermentation broth, which also decreases H_2 utilization. Batch syngas fermentations using *Clostridium ragsdalei* in a 3-L CSTR with and without activated carbon were compared. Fine powdered activated carbon particles were added to the fermentation broth to alter the gas mass transfer to the bacterium and provide additional surface for growth. Modeled syngas containing 38.0:28.5:28.5:5.0 of $CO:H_2:CO_2:N_2$ was used. Results showed that the addition of activated carbon sustained activity of *C. ragsdalei* in batch fermentations increasing total CO and H_2 uptake six fold compared to no carbon. The increased gas uptake with activated carbon produced 19 g/L ethanol with less than 1 g/L total acetic acid. However, only about 1 g/L ethanol and 5 g/L acetic acid were produced without activated carbon and cell concentrations sharply decreased after 158 h of fermentation with slight syngas consumption measured after 200 h. Alternatively, syngas consumption and cell activity were sustained during the fermentation with activated carbon for over 450 h. The maximum CO and H_2 conversion efficiencies in the fermenter with activated carbon were 88% and 86%, respectively. The operation of syngas fermentation with activated carbon exhibits higher stability, selectivity and energy conservation than any previously reported results, indicating potential for commercial biofuels production and other biological gas conversion processes.

469891 A Discussion of Realistic Feedstocks: Can We Use 2-Year-Old Short Rotation Hybrid Poplar Coppice for Fuels and Chemicals Production Via Biochemical and Thermochemical Conversion?

Abstract:

The economic success of a potential biorefinery is directly related to the use of low cost biomass. Due to the high cost, it is unlikely that a wood-based biorefinery would use only whitewood chips from mature trees in conversion. Using a more heterogeneous raw material is less costly, but efficacy of the conversion using whole short ration coppice (including leaves, bark, branches and white wood) has not been investigated. This study investigated the influence of using whole 2-year-old hybrid poplar coppice on the overall sugar yield via biochemical conversion and the bio-oil production via thermochemical conversion.

By harvesting all aboveground parts of 2-year-old hybrid poplar coppice, the biomass used in this study was comprised of 37% leaves, 9% bark, 12% branches, and 42% whitewood chips (dry weight basis). From this, four biomass samples were prepared and studied: 1) 2-year-old poplar mixture, 2) 2-year-old poplar mixture without leaves, 3) 2-year-old poplar leaves, and 4) 12-year-old poplar wood chips. In biochemical conversion, all poplar samples were steam pretreated at 195ËšC for 5 minutes with 3% $SO_2$and enzymatic hydrolyzed at 5FPU/cellulose enzyme loadings. The sugar recovery and estimated ethanol yield were calculated. In thermochemical conversion, samples were fast pyrolyzed at 500 ËšC, for 1 second in a fluidized bed reactor. The final bio-oil yield and composition were measured.

Results show that converting the 2-year-old hybrid poplar coppice may be challenging, with up to 50% lower sugar yield and close to 30% lower bio-oil yield than 12-year-old poplar wood chips. However, the removal of leaves significantly increased the overall sugar recovery and bio-oil production by 20% and 15%, respectively. Unsurprisingly, the substrate derived from isolated leaves performed poorly in both biochemical and thermochemical conversion. These results show that leaf removal from 2-year-old hybrid poplar coppice prior to harvest may enhance the biorefinery production. Given the yields of both processes, we will evaluate the economics of converting 2-year-old hybrid poplar coppice to sugar and bio-oil via biochemical and thermochemical conversions.

467443 Pyrolytic and Electrochemical Upgrading of Lignins Extracted from Pretreated Biomass to Produce Liquid Fuel Intermediates and Value-Added Products

Abstract:

Pretreatment enhances the effectiveness of enzymatic strategies for depolymerizing the cellulose and hemicellulose fractions of plant biomass. Many different pretreatment strategies have been investigated, including the use of dilute acid, steam explosion, organosolv, alkaline hydrogen peroxide (AHP), and ammonia fiber expansion (AFEX). Extractive ammonia (EA) pretreatment, a more severe version of AFEX, recovers a separate lignin stream from biomass. Upgrading this lignin stream to create fuels and chemicals potentially adds value to decentralized processing depots or centralized biorefineries that practice EA pretreatment. One approach to valorizing this lignin stream is to deconstruct EA lignin using fast pyrolysis to create a lignin-derived bio-oil that is rich in phenolics. Fast pyrolysis involves rapid heating (400-600°C) in the absence of oxygen to create bio-oil, biochar and combustible gas. Upgrading of lignin bio-oil proceeds by electrocatalytic hydrogenation (ECH), preferably using electricity from wind and solar sources. In this study, model compounds that represent lignin bio-oil components were subjected to ECH under mild conditions using ruthenium on activated carbon as a catalytic cathode. To date, several such compounds have been successfully reduced to more stable forms, for example, phenol, guaiacol and syringol have been converted into cyclohexanol at appreciable yields and selectivities. In parallel to model compound studies, catalyst reusability studies have enhanced our understanding of electrocatalyst deactivation and methods for extending catalyst lifetime. By coupling fast pyrolysis and electrocatalysis for lignin deconstruction and upgrading, respectively, a major byproduct of alcohol fermentation can be valorized into fuels and chemical products. Integration of thermochemical, electrochemical and biochemical transformative pathways is paramount for realizing the full potential of plant biomass for displacing petroleum.

466853 Combined Biochemical and Thermochemical Processing of Lignocellulosic Biomass: Techno-Economic Evaluation

Abstract:

The need for environmentally friendly fuels and commodity chemicals has driven research in the production of biofuels and bio-derived chemicals. There is now a broad product portfolio and wide variety of stand-alone technological pathways for biomass conversion systems. In the biochemical pathway of ethanol production alone, there exists multiple conversion technologies available. One such method, the consolidated bioprocessing proposed by Brethauer and Studer [1], seems to offer a viable alternative, as high conversion rates are achieved while combining into one reactor the hydrolysis and fermentation steps. This method of using a microbial consortium for conversion would likely open pathways to other liquid fuels and commodity chemicals as well, as similar work has been done in the production of higher carbon alcohols [2]. But, analogous to an oil refinery, a stand-alone biorefinery must look at coupling heat integration, fuel production, and higher value co-product production. However, work has not yet been widely published to determine feasibility or economic potential for this method of ethanol production. Similarly, scale-up potential and feasibility studies are limited for many of these novel processes.

Moreover, many of these concepts are still in the development stage, which requires the use of process modeling techniques. Modelling of lab-scale based production processes allows for techno-economic analysis and optimization, as estimates of energetic efficiency and scale-up potential of novel conversion processes are lacking. Production feasibility will be highly dependent on the energetic efficiency, environmental sustainability, and economic potential of these bio-based fuels and chemicals, especially in comparison to their fossil-fuel derived counterparts.

This work presents techno-economic process models, using flowsheeting software, of co-production of value added products, liquid fuels, and heat from lignocellulosic feedstock to analyze and compare the annual costs and efficiencies of the different combined biochemical and thermochemical conversion pathways. Furthermore, to contribute to the development and implementation of sustainable biorefineries, a systematic methodology which combines process modeling and optimization approaches is applied and a comparison of the pathways based on economic and energetic efficiencies is then proposed. The results provide a powerful support for the design of biorefinery systems with improved process economics and reduced environmental impact.

468144 Total Site Analysis As a Synthesis Model to Select, Optimize and Integrate Processes in Biorefineries

Abstract:

The paper presents a new approach to apply energy integration in multiple-product plants. The incentives of this work originate in biorefineries. Biorefineries constitute the most promising route for the valorisation of bio-renewables and the replacement of fossil-oil in any application of chemicals and fuels production. Biorefineries are capable to utilize a wide range of biomass sources including agricultural and forestry residues, industrial and municipal wastes as well as aquatic biomass (algae). Several biorefineries have already taken off in USA, EU, China and Brazil; the majority of applications focus on the production of bio-fuels. However, a big number of conventional and novel bio-products â€“ bulk chemicals and specialties, polymers, coatings, resins, pharmaceuticals, enzymes, proteins, fuels etc â€“ can be generated via thermo-processing and bio-processing of biomass. Bio-chemicals production is projected to reach up to 10 % of $1.5 trillion worldwide annual chemicals market [1], while biorefineries are capable to consume up to 84 % less non-renewable sources increasing expectations for GHG emissions reduction from 60 % to 100 % [2].

Biorefineries can be benefited by the integration of multiple products and production lines through the exchange of materials and energy among them. The development of appropriate synergies among the different processes may result in significant reduction up to 90 % in CAPEX [2], 84 % in energy consumption [3] and 58 % in water demands [4]. Provided that numerous products and technologies can get integrated in biorefineries, critical questions rise about which processes portfolio maximizes efficiencies, if collocated and integrated together. Given the high impact of energy cost in sustainability of biorefining, this work focuses on energy integration techniques in upcoming applications.

Energy integration is a powerful tool that estimates the maximum feasible margins of energy savings by exchanging heat between hot (need cooling) and cold (need heating) streams involved in the process (direct integration). In the scope of integrating multiple-process plants (like biorefineries), energy integration takes the form of Total Site Analysis (TSA) [5]. TSA estimates steam utility savings by using excess heat from the individually (directly) integrated processes to cover steam demands of others via steam production-reuse (indirect integration). The graphical tool of Site Sources and Sinks Profiles (SSSPs) is used to estimate steam savings in Total Sites. However, the graphical tool appears impractical and ineffective to address the upcoming industry, since assumes fixed and known processes. In the case of biorefineries, processes are not essentially known and solutions are not profound given the range of candidate technologies that may enter the biorefinery. The biorefinery problem constitutes a synthesis problem, where processes are significant degrees of freedom in integration. Each process features a different energy profile and energy savings will change as far as processes are integrated across different process combinations. In contrast to conventional TSA, heat exchange options in biorefinery cases do not refer to a fixed cluster of processes, but actually expand among all candidates. A potential solution would be the exhaustive use of the graphical tool (SSSPs) to estimate steam savings for all possible combinations and detect the one with the lowest energy cost. In fact, this approach is impractical for large problems (number of combinations explodes) and appears to be inefficient and unsafe to solve the biorefinery problem, as it is proven in this work.

The candidate biorefinery routes from raw materials (biomass) to intermediate and final products are represented in the form of value chain paths. Each path denotes a candidate process, while a set of consecutive paths composes a potential biorefinery plant (Total Site). Alternative biorefinery structures can be developed by choosing each time different processes among competitive paths (they valorise the same chemical), which are branched along the value chain. In the scope of energy integration, the challenge is to systematically address all alternative structures and detect the biorefinery that maximizes steam savings and minimizes energy cost. This means that screening of process combinations and integration (direct and indirect) should simultaneously occur. To address the complexity and multiplicity of the synthesis problem, Total Site Analysis should be upgraded into a synthesis tool by combining thermodynamics with mathematical programming.

The use of SSSPs requires fixed and known processes and, thus, known energy contents of streams that will get integrated. This implies that processes from the value chain should be previously selected and get integrated using SSSPs; accordingly, one process among competitive can be selected and integrated with others each time. Nevertheless, higher savings can be succeeded by sharing upstream chemicals among competitive processes in any way, subsequently changing the capacity and the energy content of involved hot/cold streams in order to improve the heat exchange conditions and secure better synergies. In the courses of a real-life biorefinery problem, this approach resulted up to 8 % higher steam savings compared with savings obtained by using SSSPs. Provided that the graphical tool is not capable to meet with these needs, new comprehensive representations are proposed to aid with the mathematical formulation of Total Site integration and address process capacities as additional continuous variables. The proposed concept could be interpreted as if SSSPs involve moving (variable) curves.

Two new representations, which communicate each other, are introduced to describe the combinatorial synthesis and integration problem. The first, a Biomass Bipartite graph Representation (BBR) is used to translate value chain paths into representations that enable the development of mass balances along the value chain. The second, a Total Site Representation (TSR) uses a properly modified transshipment model that incorporates all heat exchange options (energy balances) related with direct and indirect integration and enable the mathematical formulation of Total Site integration. The mass and energy balances are constructed as an optimization model that is solved using mixed integer linear programming and yields the processes portfolio that minimizes energy cost of the under-construction plant.

The BBR uses three building blocks: (i) product nodes for chemicals (raw, intermediate and final), (ii) process units for paths and (iii) interconnections that connect nodes with units and describe the transportation of chemicals from path to path. The flowrate of each chemical is assigned to the relative interconnection and, thus, mass balances can be recorded for each product node and process unit, given the capacity of raw materials at the beginning of value chain and the conversion rate of each path (given by experimental groups). Accordingly, the flowrates of process inlet interconnections are used to denote the capacity of the associated process. Each path is previously simulated to estimate the normalized heat demands of involved streams per unit mass of inlet chemical. Assuming that mass and energy balances within a process are linearly depended on its capacity, the actual heat content of hot/cold streams can be linearly expressed by the normalized heat demands and the variable flowrates involved in mass balances. This expression secures the connectivity between BBR (selection of paths) and TSR (integration of selected paths).

The TSR incorporates a super-interval cascade, called Total Site Cascade (TSC), which is composed by the temperature intervals of all candidate processes. The process streams contribute to TSC according to flowrates selected by BBR and the modified cascade proceeds with the direct integration of all selected processes simultaneously rather than each separately. As a result, the utility needs of individually integrated processes are estimated by TSC. At the same time, a parallel set of blocks, called utility operators, gathers the utility needs from TSC and proceeds with indirect integration. Specifically, the utility operators (i) translate excess heat from TSC into steam production, (ii) estimate the overlap between generated and demanded steam and (iii) call for external (Total Site) utilities to cover the remaining demands. In fact, the utility operators identify which process are capable to work as steam-source or sink and detect (in combination with BBR) energy complementary process combinations. The energy balances at each internal of TSC and the utility operators are constructed as an optimization model that minimizes energy cost of Total Sites. The combined system of TSC and the utility operators composes the TSR. The model composed by the energy balances of TSR can be also individually used (without BBR) for Total Site integration of a fixed cluster of processes avoiding the graphical tool of SSSPs.

The proposed methodology was applied for the development of a real-life lignocellulosic biorefinery with 9 candidate paths and 11 bio-products yielding in the 5 best processes portfolio that consume 9 % and 14 % lower hot and cold utilities, respectively, compared with the best solution that obtained by exhaustively using SSSPs.

Literature

1. Austin A., Biomass Magazine, 2012, http://biomassmagazine.com/articles/2778/bio-roundtable-feedstock-flexibility-key-in-biochemical-industry-
2. EuropaBio, 2011, http://www.europabio.org/biorefinery-feasibility-study
3. Kokossis A.C., Tsakalova M., Pyrgakis K., 2015, Design of Integrated Biorefineries, Computers & Chemical Engineering, 81, p.p. 40â€“56
4. Nikolakopoulos A., Karagiannakis P., Galanis A., Kokossis A., 2012, A water saving methodology for the efficient development of biorefineries, Computer Aided Chemical Engineering, 30, p.p. 7-10
5. Dhole V.R., Linnhoff B., 1983, Total Site Targets For Fuel, Co-Generation, Emissions, And Cooling, Computers & Chemical Engineering, 1, S101-S109

465112 Long Term Stability and Operational Limits of Lithium Sulphate for Thermal Energy Storage

Abstract:

This work explores the experimental feasibility of lithium sulphate as a thermal energy storage material for concentrated solar power (CSP) technologies. This salt has a crystalline phase change between cubic and monoclinic at temperatures around 576°C. By operating with a minimum cycle temperature, both sensible and latent thermal energy can be utilised. The operating temperatures of the proposed system are appropriate for current CSP technologies based on subcritical steam Rankine or supercritical CO_2 power cycles.

Current phase change materials (PCMs) under investigation utilize solid-liquid transformations which can achieve a higher storage performance than sensible storage materials within a low temperature difference. However, these PCMs suffer from some technical problems such as supercooling, corrosion and volume expansion. Such disadvantages can be overcome when solid-solid PCM is used as the storage material as it is less corrosive and undergoes a much smaller volume change.

In this work, the long term performance and stability of lithium sulphate has been evaluated. This involved long term experiments using 55 g of salt in an air furnace, with characterization of material properties before and after cycling using TGA-DSC, XRD and SEM. From experiment, it was determined that the highest temperature at which the thermal properties and volume of the salt were unchanged after repeated cycles was equivalent to its melting point (859°C). At temperatures lower than 130°C, the material is prone to hydration reducing the performance through significant changes to volume, density and consequently thermal conductivity.

474794 Thermal Properties and Storage Analysis of Nano Enhanced Organic Phase Change Material (NEOPCM) for Thermal Energy Storage Application

Abstract:

The objective of the present work is to model the thermal energy storage unit using Nano Enhance Organic Phase Change Materials (NEOPCM). NEOPCM is capable of absorbing or releasing the large quantity of latent heat during phase change. This paper investigatesNEOPCM with Lauric acid as a base material and nanoparticles (Copper Oxide, Aluminum Oxide)as supporting materials for the preparation of NEOPCM. In this paper, it was studied experimentally the effect of combinations of nanoparticles CuO and Al_2O_3 with different mass fractions to the base PCM, Lauric acid. To ensure the uniform dispersion a field emission scanning electron microscope (FESEM) has been used. In order to determine the changes in the compositional/functional groups during the cycle test, Fourier Transform Infrared (FT-IR) spectroscope is used. Crystalloid phase of NEOPCMs is investigated with an x-ray test conducted by an x-ray diffractometer (XRD). Thermal reliability is carried out by accelerated thermal cycle tester, phase change temperature and latent heat of fusion are measure by Differential Scanning Calorimeter (DSC).Thermal conductivity of NEOPCM is measure by using a KD-2 pro, which is based on hot wire technique.It was also verified experimentally that there is improvement in thermal conductivity and storage capacity of NEOPCM.

Keywords: Thermal energy storage, NEOPCM, Lauric acid, differential scanning calorimeter

474795 Thermal properties and heat transfer analysis of Al_2O_3 as a nano particle dispersed in the base mixture of Therminol VP -1 and Water

Abstract:

This paper presents multi-parameter investigation of Nano HTF with Al_2O_3 as a nano particles dispersed in the base material, Therminol VP-1 and water mixture (50/50 vol.%). In this paper, it was studied the effect of different volume fraction of nanoparticle dispersed in to the base HTF, Therminol VP-1 and water mixture. Nano-fluid sample is characterized by Transmission Electron Microscope (TEM) imaging, particle surface charge (zeta potential), particle aggregate size, light absorbance and PH. Dispersed particle aggregate size measure by particle analyzer based on the dynamic light scattering technique. The results show multiple perspective correlations between nano fluid characteristics and thermal properties. Al_2O_3 properties are very close to desired properties of heat transfer fluid properties and it is cheaper. Its thermal conductivity also higher than those of the conventional heat transfer fluids. It was found experimentally that there is significant improvement in thermal conductivity and considerable lowering of viscosity of Nano HTF which is the main requirement of heat transfer fluid for solar thermal applications.

Key words: Therminol VP-1, NEOHTF, Transmission Electron Microscope (TEM), Al_2O_3.

474519 Conversion of Cheese Whey to Lactobionic Acid

Abstract:

Lactobionic acid (LBA) is a versatile, high-value, low-volume chemical with numerous applications in the food, pharmaceutical, cosmetic, and chemical industries. Microbial production of LBA has the advantages of high selectivity and the ability to use cheese whey, a waste stream from the dairy industry, which is rich in lactose. However, microbial LBA production has not yet been developed as successfully as other organic acids, despite attempts being made to use wild type bacteria, fungi and yeast strains. There is no commercial implementation of microbial LBA production technologies due to the following factors: 1) The most robust strains are pathogenic, which prevents their industrial implementation. 2) The enzyme (cellobiose dehydrogenase or lactose dehydrogenase) which oxidizes lactose to LBA is produced only in the stationary phase. The highest level production of LBA was only achievable under resting cell conditions, which are not usually used in industrial fermentations. 3) The microorganism consumed LBA at the later stage of the fermentation. We report engineering a Neurospora crass strain for cellobionic acid production from cheese whey and solving the problems associated with LBA fermentation using metabolic engineering efforts

468385 Predictive Model and Bioconversion of Mixed Feedstocks

Abstract:

Currently, biorefineries are utilizing single feedstock inputs but in order to reduce feedstock dependency and the related risks, researchers are exploring the option of mixing multiple feedstocks. To understand the implication of mixing feedstocks on sugar yield, we relied on statistical design of experiments to build a predictive model for sugar yield based on varying feedstocks fractions, pretreatment types, and pretreatment reaction parameters. We performed pretreatments at low biomass loading, but for a biorefinery to be economically viable biomass deconstruction has to be performed at higher solid loading. Also, lignocellulosic sugars should be fermentable to produce high energy density molecules such as hydrocarbons.

In this study, three biomass feedstocks: corn stover, switchgrass, and energy cane were chosen based on the least cost formulation model developed by Idaho National Laboratory. These single and mixed feedstocks were pretreated with dilute alkali (DL), dilute acid (DA), and ionic liquid (IL) at severities predetermined by the statistical experimental design and were subsequently, enzymatically hydrolyzed with CTec2® at a loading of 10 mg protein/g glucan. The resulting sugar yields from all treatments, which varied from 25 to 100% w/w untreated biomass, were then fed into SAS JMP® to develop a predictive model that generated a continuous envelope of optimal feedstock mixtures. In order to achieve 90% sugar yield, we have to perform IL pretreatment at 140°C for 120 mins with 12%, 13%, and 75% (w/w) corn stover, switchgrass, and energy cane. More predictions of such kind were generated from this model that can potentially overcome the dependence on a single feedstock while maintaining the overall product yields.

These feedstock/treatment combinations, selected from the model, were tested at higher biomass loading to study the impact on downstream fermentation process to produce ethanol and hydrocarbon, such as bisabolene. At higher biomass loading, mixed feedstock conversion to ethanol performed very similar to single feedstock conversion to ethanol. Rheological characterization was performed on high solid pretreated biomass mixtures to better understand mixing, material handling, and mass transfer issues. This information will be applied in techno-economic analysis of this end-to-end study of deconstruction of optimal biomass mixtures at high solid loading for ethanol and hydrocarbon production.

459583 Two-Phase Kinetics of Enzymatic Depolymerization of Hemicellulose to Soluble Sugars for Liquid Biofuel Production

Abstract:

Hemicellulose is the second most abundant polysaccharide in nature, available not only in its pure form but also as a constituent of all lignocelluloses. The conversion of hemicelluoses to soluble sugars in parallel to its cellulosic counterpart would make liquid biofuels competitive in the market. Of the three key conversion steps (pretreatment of biomass, enzymatic hydrolysis of pretreated biomass to soluble sugars, ethanol from sugars by fermentation), hydrolysis is the slowest and the rate limiting step [1]. Adsorption of the liquid phase enzymes (endo- and exo-xylanases) to the insoluble solid substrate (xylan) is the first step in hemicellulose hydrolysis, followed by enzyme-catalyzed xylan depolymerization reactions in the solid and liquid phases. This work uses a coupled theoretical and experimental framework [2] to quantify the kinetics of product-inhibited solid and liquid phase depolymerization reactions that produce xylose and other reducing (soluble) sugars. We experimentally determine the type of adsorption of the enzyme on the solid substrate, the type of product inhibition in the solid and liquid phases, the transition of two-phase hydrolysis to single in batch reactors. Model-experiment comparisons allow us to quantify the xylobiose inhibition constants in the two phases. Our experiments show that the hydrolysis system transitions from a two-phase (solid-liquid) to a single (liquid) phase process at 11 hours. The enzymatic hydrolysis in both liquid and solid phases is found to be non-competitively inhibited by the product (xylose), with the inhibition being more potent in the liquid phase. Our multi-step reaction kinetic model is simulated using the liquid and solid phase kinetic data and validated with our experimentally measured temporal dynamics of xylose and reducing sugar yields in the liquid phase, and of the solid phase hemicellulose concentrations. We show that the product inhibition in the solid phase increases with the substrate loading, thus lowering the yield of reducing sugar, while the liquid phase product inhibition decreases with substrate loading, thus increasing the xylose yield. Our substrate inhibition studies allow us to recommend 5 mg/ml as the optimum xylan loading, above which substrate inhibition is observed to set in.

References:

[1] Chakraborty, S., Raju, S., and Pal, R., "A multiscale three-zone reactive mixing model for engineering a scale separation in enzymatic hydrolysis of cellulose," *Bioresource Technology*, **173, pp 140-147** (2014)

[2] Dutta, S.K., and Chakraborty, S., "Kinetic analysis of two-phase enzymatic hydrolysis of hemicellulose of xylan type," *Bioresource Technology*, **198, pp 642-650** (2015)

474465 Selective Fermentation of Six Carbons Sugars from Hot-Water Hardwood Extract Hydrolysates

Abstract:

Hot water wood extract hydrolysate contains a mixture of sugars: xylose, glucose, galactose, mannose, arabinose and rhamnose. This sugar mixture presents a challenge in fermentation to ethanol and other chemicals / biofuel. The fermentation efficiency is especially low for xylose. When genetically modified *E. coli* fbr5 was employed to ferment the sugar mixture to ethanol, glucose was the first sugar consumed, and then galactose, arabinose, mannose and rhamnose, xylose was the last to be consumed. While *E. coli* fbr 5 was genetically engineered to ferment Xylose, this outcome is disappointing. Most microorganisms prefer glucose, *E coli* is no exception. Xylose being the dominant sugar species in hardwood hot-water extract hydrolysate and the least preferred sugar for ethanol production present a challenge to the biomass energy industry development. On the other hand, xylose can be employed to produce xylitol, furfural, and a lot more products efficiently. The value of xylose is higher than glucose in the market, and yet it is limiting the efficiency in biomass conversion. Hot water wood extract hydrolysate was subject to fermentation with *S. cerevisiae* and xylose was left unconsumed. Xylose can be recovered from the fermentation broth, while ethanol was produced from glucose.

465914 Effect Temperature and Flow Rate on Lhw Pretreatment SB: Kinetic Study

Abstract:

Assuming xylan hydrolysis to xylose in sugarcane bagasse (SB) by liquid hot water (LHW) and xylose further degradation to furfural were two consecutive first-order reactions, a kinetic model was developed. SB was pretreated by LHW at different temperatures and flow rates, and xylan hydrolysis behavior in SB pretreatment was investigated using the developed kinetic model. Results showed the kinetic model could fit the data well (R^2>0.85). k_S (the rate constant of xylose production) and k_X (the rate constant of xylose degradation) at different temperatures was in accordance with Arrhenius equation. Activation energy (E_a) for xylose production and xylose degradation was 50.39 and 104.66 kJ/mol, respectively, which indicated that xylose degradation was relatively more difficult that xylose production during LHW hydrolysis. The relationship between k_S (or k_X) and flow rate was almost linear. The application of the model in xylan hydrolysis kinetics was successful.

474466 R-HB Production from Glucose By B. Cepacia

Abstract:

Polyhydroxylbutyrate (PHB) is one of the important Polyhydroxyalkanoates (PHAs) that are a promising alternative to conventional, fossil fuel derived plastics. Plastic is placing a burden on the environment by polluting ecosystems, overwhelming landfills, using finite oil resources, and contributing to greenhouse gas emissions. PHAs are polyesters created by microorganisms intracellularly under limited nutrient growth conditions as carbon and energy storage compounds. PHAs are not yet widely accepted and produced because of their high production costs, specifically the carbon source and recovery method, compared to those of petroleum based plastics. In addition, present technology in recovery PHA is not environmentally friendly, leading to whether PHA is a sustainable bioproduct. This presentation shows some of our research efforts in utilizing renewable carbon source. Glucose was employed to produce droxybutyric acid. Butyric acid can be polymerized to produce PHB. The production of butyric acid and polymerization to PHB enables the desired biodegradable plastics to be recyclable.

467000 Process Strategies for High Titers of Lipid Production By Oleaginous Yeasts in Undetoxified Hydrolyzates of Lignocellulosic Biomass

Abstract:

Oleaginous yeasts can accumulate up to 70% of cell biomass as lipid, predominantly as triacylglycerols. Yeast lipid fatty acid profiles have been reported to be similar to that of vegetable oils and consist primarily of oleic, palmitic, stearic and linoleic acids. This capability provides the opportunity to use yeasts to produce bio-based fuels and chemicals from agricultural residues and bioenergy crops, such as corn stover and switchgrass, respectively. Successful cultivation of these yeasts on inhibitory lignocellulosic hydrolyzates is uncertain because little has been published regarding their application to bioconversion of plant biomass. Several novel yeasts producing promising lipid concentrations (>10 g/L) from enzyme hydrolyzates of acid and base pre-treated lignocellulosic biomass were discovered in a prior screening of the ARS Culture Collection (NCAUR, Peoria, IL). Three of the top strains have not previously been reported for the bioconversion of lignocellulose to lipids. Process strategies for maximizing lipid production in one of the more benign hydrolyzates, AFEX-pretreated corn stover, and one of the more toxic hydrolyzates, dilute acid-pretreated switchgrass, respectively, were employed. Hydrolyzates were prepared at up to 20% solids loading to provide over 100 g/L sugars (at ~56:39:5 mass ratio glucose:xylose:arabinose). A two stage process boosting the molar C:N ratio from 60 to well above 400 in undetoxified switchgrass hydrolyzate was optimized with respect to nitrogen source, C:N, carbon loading and pH. Using this process three strains were able to consume acetic acid and nearly all available sugars to accumulate 50-65% of cell biomass as lipid (w/w), to produce 25-30 g/L lipid at 0.12-0.22 g/L/h and 0.13-0.15 g/g (or 39-45% of the theoretical yield) at pH 6 and 7, a performance unprecedented in lignocellulosic hydrolyzates. Depending upon plant biomass yields, we estimated that oleaginous yeasts have the capability of producing ~48 and 190 gal oil per acre from corn stover and switchgrass, respectively. As a frame of reference, ~48 gallons of oil per acre are produced from processing soybeans. The successful application of top-performing lipid-producing yeast in undetoxified lignocellulose hydrolyzates is expected to advance the economic feasibility of high quality biodiesel and jet fuels from renewable biomass, expanding the market potential for lignocellulose-derived fuels beyond ethanol for automobiles to the entire U.S. transportation market.

471952 Evaluation of Alcohol-to-Jet (ATJ) Conversion Technology for Renewable Jet Fuel

Abstract:

Abstract Submission to Be Considered As An Oral Presentation at AIChE 2016 Annual Meeting in the Biochemical Conversion Processes in Forest/Plant Biomass Biorefineries Session

Evaluation of Alcohol-to-Jet (ATJ) Conversion Technology for Renewable Jet Fuel

Scott Geleynse, Manuel Garcia-Perez, Michael Wolcott, Xiao Zhang

Washington State University

The FAA is working to enable the U.S. use of one billion gallons per year of "drop-in" sustainable alternative jet fuels by 2018. Airlines and aerospace manufacturers around the world have also expressed continuously increasing interest in the replacement of jet fuel from fossil fuels to renewable fuels from biomass. Despite the considerable investment in the research of several different conversion strategies, establishing the successful commercial production of biorenewable jet fuel remains a challenge. In this presentation, we will discuss technical aspects of strategies to use alcohols as a platform molecule to generate jet fuel and other products. By using process modeling and techno-economic analysis, we aim to assess the feasibility and readiness of the technology for biorefinery implementation. We hope to provide insight into the various political and economic challenges faced in making and meeting renewable fuel goals.

Presenting author:

Scott Geleynse

Bioproducts Sciences and Engineering Laboratory

School of Chemical Engineering and Bioengineering, Washington State University
2710 Crimson Way, Richland, WA 99354

Contact: scott.geleynse@email.wsu.edu, x.zhang@wsu.edu

473497 Kinetics of the Enzymatic Hydrolysis of Paper Pulp Fibers

Abstract:

Papermaking fibers, particularly waste fibers from recycling operations can be a valuable resource for bioenergy and products such as ethanol, bioplastics or biochemicals. The amount of lignin in such fibers is much lower than native wood or other plant based lignocellulosic resources eliminating the need for pretreatment steps. Moreover, waste fibers are readily available in the waste streams rejected from the papermaking operations to improve paper strength and runnability of the paper machines.

The present work develops a model for the kinetics of the enzymatic hydrolysis of cellulosic pulp fines based on the physico-chemical interactions. The cellulolytic enzymes adsorb to the outer surfaces of the fines after which they diffuse into the interior pores. The fine particles are assumed to be spherical, containing two distinct kinds of pores: one at the macro-scale and another at a micro-scale allowing slow progression of the enzymes into the solid cellulosic matrix. Enzymatic hydrolysis reactions are assumed to occur at the micro-scale within the cellulosic matrix. Convective mass transfer boundary layers reduce the enzyme transport across the external particle surfaces. The main utility of this model is in estimating the optimal reaction conditions and assist in scaling studies of these enzymatic hydrolysis processes.

Keywords: Enzymatic hydrolysis, cellulose, pulp fibers, fines, rejects, paper mill waste solids

472678 Comparing Sugar Titers and Ethanol Yields from Fed Batch Fermentation of Celf and DA Pretreated Poplar Solids at High Glucan Loadings

Abstract:

Biological conversion of lignocellulosic biomass to liquid fuels reduces greenhouse gas emissions and addresses other economic and strategic challenges. However, costs are high due to high enzyme loadings and capital costs, particularly at glucan loadings over 10% needed to realize economically attractive ethanol concentrations. Because solids concentrations above about 10% cannot be readily mixed in batch fermentations, a fed-batch strategy was applied to establish maximum carbohydrate conversions from enzymatic hydrolysis of THF Co-Solvent Enhanced Lignocellulosic Fractionation (CELF) and dilute acid pretreated poplar wood. At 15, 18 and 20% glucan loadings, glucose concentrations of 63, 70, and 75g/l were achieved in only 2 days from CELF pretreated solids while concentrations from hydrolysis of dilute acid solids were only 45, 53, and 60g/l, respectively.

458379 Co-Production of Bioethanol and Bio-Lactic Acid in a Biorefinery Concept: A Comprehensive Study

Abstract:

Biorefineries are seen as a key component for the transition toward a sustainable and low carbon bio-economy. In this research, alternative biorefineries annexed to a typical South African sugar mill are investigated to co-produce ethanol (EtOH), lactic acid (LA) and/or electricity, utilizing bagasse and a component of harvesting residues (brown leaves) as feedstock. Studied scenarios are EtOH production (scenario 1), EtOH form glucose and LA from xylose (scenario 2), EtOH form xylose and LA form glucose (scenario 3) and LA production (scenario 4), all of which are associated with some level of export electricity production. Aspen simulation of the scenarios were developed considering all supplementary units such as evaporation, water treatment, boiler and steam/power generation in addition to the process units. Economic evaluation, energy assessment and Life Cycle Analysis (LCA) of the developed simulations were carried out to determine the viability of each scenario. In order to supply the combined energy demands of both sugar mill and biorefinery, a 35 to 40% bypass of lignocellulose directly to the boiler section was required, to achieve integrated scenarios that were bio-energy self-sufficient, with no supplementation with fossil fuels. Based on the economic analysis, scenario 4 achieved the highest Internal Rate of Return (IRR) of 20.9%, while scenario 1 had the lowest IRR of 10.7%. Since EtOH, LA and enzyme prices strongly affect the profitability of biorefinery scenarios, sensitivity analysis was carried out, which identified scenario 2 as the most economically robust, having the least sensitivity to these price variations.

474486 Thermophiles and Their Thermostable Enzymes in Biofuel Synthesis

Abstract:

Second-generation feedstock, especially nonfood lignocellulosic biomass, is a potential source for biofuel production. Cost-intensive physical, chemical, biological pretreatment operations and slow enzymatic hydrolysis carried out at â‰ ¤50Â°C make the overall process of lignocellulosic conversion into biofuels less economical than available fossil fuels. Thermophiles and their enzymes can play important roles in many kinds of bioprocessing including in conversion of non-food biomass into biofuels. The Homestake gold mine (8000 ft. deep, Lead, SD) offers a unique opportunity for direct exploration of the deep biosphere environment. Using soil/biofilm samples of deep biosphere of the Homestake Gold Mine, compost facility (Rapid City, SD, USA), and Hot Springs State Park (Thermopolis WY, USA), we have isolated several thermophilic cellulose- and xylan-degrading and -fermenting pure cultures belonging to the genera *Brevibacillus*, *Paenibacillus*, *Clostridium*, *Bacillus*, and *Geobacillus*. Unique characteristics of lignocellulose-deconstructing enzymes produced by mine and compost thermophiles include optimum temperatures of >70°C, pH ranges from 4 - 8, and high thermostability (e.g., at 60ºC, 50% of cellulases and xylanases activities were retained in 35 and 23 days of incubation, respectively). Hot Springs State Park thermophiles can grow on various inexpensive regional carbon and energy sources (e.g., prairie cord grass and corn stover), and produce biohydrogen in a single step bioprocessing of biomass. This talk will describe the limitations in exiting lignocellulose conversion technologies and possible ways to overcome those limitations using thermophiles and their enzymes. The influence of high temperatures on various existing lignocellulose conversion processes and those that are under development, including separate hydrolysis and fermentation, simultaneous saccharification and fermentation, and extremophilic consolidated bioprocess will discussed. Integrated decentralized thermophilic biofuel production employing cellulolytic- and fermentative-thermophiles in a single step consolidated process will also be discussed.

463989 Improving the Energy Efficiency of Carbon Capture Processes – Combining Enzyme Accelerated Solvent Systems and Improved Contacting Equipment

Abstract:

In order to achieve the ambitious targets specified by the United Nations Framework Convention on Climate Change in 2011 and reduce GHG emission in the EU by 2020 at least 20% below those in 1990 carbon dioxide (CO_2) emissions have to be reduced significantly. Fossil-fired power plants are identified as major targets for carbon capture processes, since they are responsible for 40 % of the total global CO_2 emissions (Mondal et al., 2012). Efficient post-combustion capture of $CO_2$is one of the most promising options to face this crucial challenge and to achieve at least the specified short-term targets (Rochelle, 2009).

As a result, a wide range of different fluid separation processes, including (reactive) absorption, adsorption, cryogenic distillation and gas permeation, have been intensively studied in recent years (Mathias et al., 2013). Nevertheless, the current state-of-the-art process for CO_2 capture is the reactive absorption in packed columns using amines. Especially primary amines, like Monoethanolamine (MEA), are often used because of the covalent bonding of the CO_2, resulting in high absorption rates (Kenig & Gorak, 2005). However, the recovery of the solvent, which is required for a continuous operation, becomes more challenging the stronger the $CO_2$is bound (Yildirim et al., 2015).

Therefore, the best absorption process will represent an optimal trade-off between fast absorption rates and energy efficient solvent recovery. The latter causes up to 80 % of the total capture costs for the MEA process and results in a significant loss in overall power plant efficiency (Notz et al., 2011). Consequently, solvent recovery represents the bottleneck of the reactive absorption/desorption process, which is practically not considered for large scale industrial application (Mondal et al., 2012). In order to overcome this restriction a lot of research effort is carried out with a focus on more energy efficient solvents. While tertiary amines or potassium carbonate require less energy in the regeneration step, application of these solvents is restricted by their relatively low absorption rate (Kunze et al., 2015).

Process intensification (PI) can help to overcome this restriction in different ways, which are investigated in this study. In order to compensate for the low absorption rates of tertiary amines enzymes are integrated as efficient biocatalysts to the solvent system. Here, especially the enzyme Carbonic Anhydrase (CA) is considered as the fastest catalyst for the hydration of CO_2, which is the prevailing CO_2consuming reaction in tertiary amine systems (Gundersen et al., 2014). A significant enhancement of the absorption performance was observed in previous studies with a wetted wall column set-up, validating the direct influence on the reaction rate (Kunze et al., 2015). However, besides the possible improvement, the application of carbonic anhydrase as a biocatalyst also introduces additional restrictions to the process due to sensitivity towards temperature and pH (Gundersen et al., 2014).

In addition to an intensification of the solvent system, combining high capacity solvents with the efficient biocatalyst, the mass transfer can be further improved by means of intensified contacting devices (ICD), other than making use of conventional columns. The use of a membrane contactor provides a well-defined interfacial area, with packing densities being a magnitude higher as in conventional equipment (Zhao et al., 2016). Due to the fully separated liquid and vapour flow, membrane contactors also allow operational modes different from conventional equipment, which are limited by flooding or loading. The modular nature and operational flexibility make membrane contactors a promising alternative to conventional equipment. However, mass transfer resistance might be increased due to the membrane and the operational window of the membrane contactor is limited as well, such that especially transmembrane pressure differences have to be controlled in order to avoid breakthrough of one of the two phases, which will lead to undesired phase dispersion.

Another ICD that can improve mass transfer in ab- and desorption is the rotating packed bed (RPB) technology. In RPB processes centrifugal forces are exploited to increase acceleration of the liquid. In a counter-currently operated RPB, gas is pushed through the torus-shaped rotating packing from the outside to the middle of the rotor. Liquid is sprayed into the eye of the rotor and flows counter-currently to the gas outwards, driven by the centrifugal forces. RPBs offer mass transfer improvements due to high turbulences and extend the operating window, having the rotational speed as an additional degree of freedom. These advantages result in high capacity combined with compactness (Sudhoff et al., 2015). Yet, application of RPBs is still restricted due to lack of reliable data on the performance efficiency.

While both, ICD and the use of an improved solvent system with CA as catalyst, have the potential to significantly intensify the CO_2capture process, literature studies and research on the combination of both is scarce. In order to investigate if the advantages of the different PI techniques can be exploited and if a combination is feasible, first investigations are performed in order to characterize the operating windows of such intensified processes and to evaluate the potential benefits for an actual implementation. In preliminary work (Kunze et al., 2015), the tertiary amine Methyldiethanolamine (MDEA) has shown promising results to be more energy-efficient compared to MEA and is therefore chosen as solvent (30 wt.-%) for this work. Absorption experiments in a column and the different ICD are performed to determine the operating windows and compare separation efficiency. Subsequently CA is added to the solvent system in order to investigate the potential synergetic intensification of ICD and CA versus the acceleration by CA in a conventional packed column set-up. The results enable the discrimination between the different means for process intensification depending on the determined operating windows and separation performances. Finally, the obtained results are evaluated and benchmarked against the current state-of-the-art process on a theoretical basis. Future work will also investigate the application of the different means for process intensification to the desorption step in order to provide a complete picture for a potential industrial application.

REFERENCES

Gundersen, M.T., Von Solms, N., Woodley, J.M., 2014, Energy Procedia, 63, pp. 624-632.

Kenig, E.Y., Górak, A., 2005, Integrated Chemical Processes: Synthesis, Operation, Analysis, and Control, pp. 265-311.

Kunze, A.-K., Dojchinov, G., Haritos, V.S., Lutze, P., 2015, Applied Energy, 156, pp. 676-685.

Mathias, P.M., Reddy, S., Smith, A., Afshar, K., 2013, Energy Procedia, 37, pp. 1863-1870.

Mondal, M.K., Balsora, H.K., Varshney, P., 2012, Energy, 46 (1), pp. 431-441.

R. Notz, I.Tönnies, H.P. Mangalapally, S. Hoch, H. Hasse, 2011, Int. J. Greenh. Gas Control, 5, 3, 413-421

Rochelle, G.T., 2009, Science, 325 (5948), pp. 1652-1654.

Sudhoff, D., Leimbrink, M., Schleinitz, M., Górak, A., Lutze, P., 2015, Chemical Engineering Research and Design, 94, pp. 72-89.

Zhao, S., Feron, P.H.M., Deng, L., Favre, E., Chabanon, E., Yan, S., Hou, J., Chen, V., Qi, H, 2016, Journal of Membrane Science, 511, pp. 180-206.

Yildirim Ö., Kiss A. A., Hüser N., Leßmann K., Kenig, E.Y., 2012, Chemical Engineering Journal, 213, pp. 371–391.

ACKNOWLEDGEMENT

The research leading to these results has received funding from the European Union Seventh Framework Programme FP7/2007-2013 under grant agreement n° 608535.

464321 Boosting the Multienzymatic Conversion of CO2 through Spatially Separated Immobilization Strategy

Abstract:

Multienzymatic catalysis is considered to be the next generation of biocatalysis. As a central part, rational design and construction of multeinzyme systems have attracted great attention and become a hot research topic nowadays. When designing an efficient multienzyme system, three crucial issues should be taken into consideration: 1) synergistic catalysis among multienzymes; 2) efficient mass transfer of substrates and products; and 3) easy recyclability of multienyzmes. Inspired by the existence form and synergistic catalytic behavior of multienzymes in organisms, we put forward to the strategy of constructing spatially isolated multienzyme systems. In our recent study, a series of spatially isolated multienzyme systems (containing formate dehydrogenase and formaldehyde dehydrogenase) enabled by nanospheres/microcapsules are constructed by combining biomimetic mineralization with other biomimetic platform techniques (*e.g.*, biomimetic adhesion, Pickering emulsion, surface segregation and layer-by-layer (LbL) self-assembly). When utilized for the conversion of CO_2, these systems exhibit desirable catalytic performance and easy recyclability.

Firstly, a spatially isolated multienzyme system enabled by active nanospheres was successfully constructed through the synergy between biomimetic mineralization and biomimetic adhesion, achieving the efficient process intensification. The first enzyme (formate dehydrogenase) was entrapped accompanying the formation of nanospheres through biomimetic titanification. After *in situ* surface functionalization of nanospheres with oligodopa, the second enzyme (formaldehyde dehydrogenase) was immobilized on the surface of active nanospheres through amine-catechol adduct reaction. When utilized for the conversion of CO_2, the as-constructed multienzyme system exhibited a formaldehyde yield of higher than 60.0%. Besides, after evaluating the performance of different-sized multienzyme systems, it was found that smaller sized multienzyme system possessed higher enzymatic activity but lower recyclability.

Secondly, a spatially isolated multienzyme system enabled by nanosphere-stabilized capsules (NSSCs) was constructed through the synergy of biomimetic mineralization and Pickering emulsion, achieving the efficient process intensification. The first enzyme-containing nanospheres were monodispersely spread on the surface of oil droplets through Pickeirng emulsion, maintaining their structural characteristics and integrity. Then, the gel titania, that was produced through sol-gel process of tetrabutyl titanate, was in charge of crosslinking the nanospheres, forming a stable and continuous capsule wall. The second enzyme was immobilized on the surface of capsule wall through amine-catechol adduct reaction between oligodopa and enzyme. During the construction process, the released butanol as a result of the sol-gel process of butyl titanate could be adsorbed on the surface of nanospheres, further inducing the self-assembly of the nanospheres at the oil/water interface. This process played key role on the formation of Pickering emulsions. When utilized for the conversion of CO_2, the as-constructed multienzyme system acquired a formaldehyde yield of higher than 50.0%. Meanwhile, the lower density of the oil core than that of water endowed the multienzyme system with excellent recyclability, especially the formaldehyde yield kept almost unaltered after 10 times recycling.

Thirdly, a spatially isolated multienzyme system enabled by mesoporous hybrid microcapsules was constructed through the synergy of biomimetic mineralization and surface segregation, achieving the efficient process intensification. The first enzyme was entrapped accompanying the formation of the PAH-segregated $CaCO_3$ template. Then, the PAH polymeric network was formed through crosslinking between GA and PAH. Afterwards, this polymeric network was utilized for implementing biomimetic mineralization with *in situ* entrapping the second enzyme. After removing the template, a spatially isolated multienzyme system was acquired. During the construction process, the $CaCO_3$ microsphere served as the dual templates for formation of both capsule lumen and mesopores on the capsule wall, whereas PAH played the following two crucial roles: (1) forming the bulk polymer network to render confined space for biomimetic mineralization and (2) triggering the mineralization of Ti-BALDH to produce titania nanoparticles. When utilized for the conversion of CO_2, the as-constructed multienzyme system exhibited a formaldehyde yield of higher than 70.0%. Meanwhile, the reaction rate was nearly two times higher than the average level. And no catalytic activity was lost after 10 times recycling. Besides, the mesoporous hybrid microcapsules can be utilized in various other applications and exhibited desired performance.

Fourthly, a spatially isolated multienzyme system enabled by mitochondria-like microcapsules was constructed through the synergy of biomimetic mineralization and LbL self-assembly, achieving the efficient process intensification. Inspired by the double membrane structure of mitochondria, we proposed to utilize process and structure biomimicism strategy instead of just process biomimicism strategy for constructing multienzyme systems. The first enzyme was entrapped accompanying the formation of the PSS-doped $CaCO_3$ template. The organic inner membrane was acquired *via* LbL self-assembly of oppositely charged polymers/proteins on the template. The silica template layer was then formed onto the inner membrane through biomimetic silicification with *in situ* entrapping the second enzyme, while the organic-inorganic hybrid outer membrane was acquired *via* biomimetic titanification subsequently. After removing the $CaCO_3$ template and the silica template, a spatially isolated multienzyme system was established. When utilized for the conversion of CO_2, the as-constructed multienzyme system exhibited a formaldehyde yield of higher than 82.0%, with a formaldehyde selectivity of 90.0%. Meanwhile, 70.0% of the initial activity was retained after 10 times recycling.

By using of a series of cases as the evidences, it can be conjectured the spatially separated immobilization strategy can boost the multienzymatic conversion of CO2. This study will shed light on the new strategies for the design and preparation of multicomponent catalysts.

466064 Evaluation of Consumption Energy in CO_2 Absorption Processes with Two-Liquid Phase Separation Type

Abstract:

The global issue of climate change is receiving ever-increasing attention. One of the main focuses of this attention is the reduction of CO_2 emission from activities such as power generation, iron and steels making process. The technology that has the potential to make a significant impact on emission is CO2 capture and storage.

The aqueous of amine CO2 capture proses can remove CO2 from exhaust gas in some industrial process. The major aqueous amine absorbent which is applied in industrial process are primary amines, the secondary amines and the tertiary amines. The aqueous of amine CO2 capture proses are consumed thermal energy for CO2 desorption. A previous study has revealed that a lot of energy for CO2 desorption are CO2 desorption heat, sensible heat of the aqueous for desorption temperature and the heat of vaporization the aqueous.

We suggest new CO_2 absorption processes with two-liquid phase separation type for reducing consumption energy. Absorbing liquid in new process consist of H2O, amines and ether or alcohol additive. In CO2 absorption section, the CO2 react to amine in absorber column. The CO2-rich solution liquid is transferred to the decanter. In the decanter, the CO2-rich solution liquid was separated hydrophilic liquid and hydrophobic liquid. Carbamate and bicarbonate ion exist in hydrophilic liquid. Hydrophobic liquid is only additive. The hydrophilic liquid is transferred to the stripper where comparatively high temperature is used to desorb the CO2. The hydrophobic liquid is transferred to mixer. It is mixed with hydrophobic liquid returned to stripper. The new CO_2 absorption processes can be reduced energy consumption for sensible heat of the aqueous for desorption temperature and the heat of vaporization the aqueous. In this work, we select 2-(Ethyl amino) ethanol, water and diethylene glycol diethyl ether to Absorbing liquid. We examine the CO2 absorption characteristics. We study the effect of the new Absorbing liquid on energy consumption

for process. We determine the comparison of absorbing liquid. (H2O, EAE, diethylene glycol diethyl ether = 10, 30, 60 wt%) We measure the solubility of CO2 on this absorbing liquid. In experiment, we obtained amount of CO2 are measured to be 0.3 (mol-CO2/mol-amine) at absorber temperature 40 degree C. The new absorbing liquid has the enough loading amount of CO2. We find out that the enough performance is obtained as absorbing solution.

Consumption energy of the new process is calculated by process simulator. The consumption energy is calculated to be 2.5 (GJ/t-CO2). This value is lower than that obtained for MEA absorption process. Thus, CO_2 absorption processes with two-liquid phase separation type can be valuable in CO2 capture and separation technology.

466110 The Influence of DissociationÂ Constants on the Kinetics of Carbon Dioxide Absorption in Aqueous Tertiary Amines Solutions Containing Carbonic Anhydrase

Abstract:

American Institute of Chemical Engineers (AIChE) Annual Meeting

The influence of dissociation constants on the kinetics of carbon dioxide absorption in aqueous tertiary amines solutions containing carbonic anhydrase

Bin Liu[a], Xiao Luo[a*], Zhiwu Liang[a,b*], Paitoon Tontiwachwuthikul[a,b*]

[a]Joint International Center for CO_2 Capture and Storage (iCCS), Hunan Provincial Key Laboratory for Cost-effective Utilization of Fossil Fuel Aimed at Reducing CO_2 Emissions, College of Chemistry and Chemical Engineering, Hunan University, Changsha 410082, PR China

[b]Clean Energy Technologies Research Institute (CETRI), Faculty of Engineering and Applied Science, University of Regina, Regina, Saskatchewan, S4S 0A2, Canada

Keywords: Tertiary amines; CO_2 absorption; carbonic anhydrase

* Author for correspondence: Dr. Zhiwu Liang E-mail: zwliang@hnu.edu.cn

Tel: +86-13618481627 and +86-73188573033

ABSTRACT

The chemical absorption of CO_2 into primary amine and secondary amine, such as monoethanolamine (MEA), diethanolamine (DEA), is currently the most widely accepted commercial approach to carbon capture.[1] This solvent is extremely effective in forming stable carbamates when reacted with CO_2, leading to a very efficient removal of CO_2 with a 2:1 reaction stoichiometry.[2,3] However, regeneration of the carbamate species requires a large parasitic supply of energy to release the CO_2 during stripping.[1] This high energy requirement accounts for up to 80% of the operating costs for CO_2 capture and recovery.[4] It is also a highly corrosive substance that readily undergoes degradation in the presence of oxygen, emitting harmful volatile organic compounds.[5]

Considering the high absorption heat and high energy consumption during CO_2 capture process, there are a number of tertiary amines such as methyldiethanolamine (MDEA), dimethylaminoethanol (DMEA), in place of primary and secondary amine for carbon capture. While many of tertiary amines solvents are able to reduce the stripping energy requirement, the reaction rate for CO_2 sorption in the absorber is often slow. This means that it is not possible to recover high purity carbon dioxide within a reasonable column height. Carbonic anhydrase (CA), a naturally occurring enzyme and very efficient catalyst that enhances the reversible reaction of CO_2 to HCO_3^-, was first identified in 1933 in red blood cells and known to catalyze the conversion of CO_2 into bicarbonate (HCO_3^-) at extremely high turnover rates, [6-7] which may be useful for promoting the absorption rates of CO_2 from gas streams when these alternative solvents are used. There is great potential for using CA to hydrate CO_2, especially when it is known that the hydration step is the rate determining step in the CO_2 absorption process. Carbonic anhydrase is not just a single enzyme form, but a broad group of zinc metallo-proteins (enzymes) that exists in three genetically unrelated families of isoforms (¦Á, ¦Â and ¦Ã). Its catalyzed mechanism of CO_2 hydration has been introduced by Lindskog et al. [6-7] and simplified as shown in Figure 1. Penders et al. [8] has been working on CO_2 absorption into amine solutions catalyzed by human carbonic anhydrase in string cell reactor since 2012. This work focuses on the study of effect of dissociation constants (pKa) on the kinetics of carbon dioxide absorption in aqueous tertiary amines solutions containing carbonic anhydrase at various temperature using stopped-flow techniques. The examined alkanolamines were (Figure 2): Triethanolamine (TEA), Methyldiethanolamine (MDEA), Dimethylethanolamine (DMEA), 2-(Diethylamino)ethanol (DEEA), 1-Diethylamino-2-propanol (1DEA-2P), 3-Diethylamino-1-propanol (3DEA-1P) and 2-(Dimethylamino)isobutanol (2DEA-2M-1P). The values of pKa of different amine was determined using conventional pH method as shown in Figure 3 and its final results was presented a function of temperature in Figure 4.

It can be seen from the results (Figure 5) that DMEA with a higher pKa value than MDEA and TEA hence showed a higher reactivity with carbon dioxide in the absence of enzyme. It can also been found that the reaction rate constant (k_0) in amine-CO_2-H_2O system was significantly enhanced in the present of BCA, in which the enzyme activity is higher in TEA solutions. This is because Enzymes often display a pH-dependent changes in activity, in characterizing the pH dependence of the enzyme activity, the experimenter often observes a bell-shaped curve in plot of activity versus pH or s-shaped curves in plots of activity versus pH (either falling from optimal activity or rising to optimal activity) with an inflection point at some pH value. This result would provide a useful information to understand the reaction activity of CO_2 absorption into various amines system.

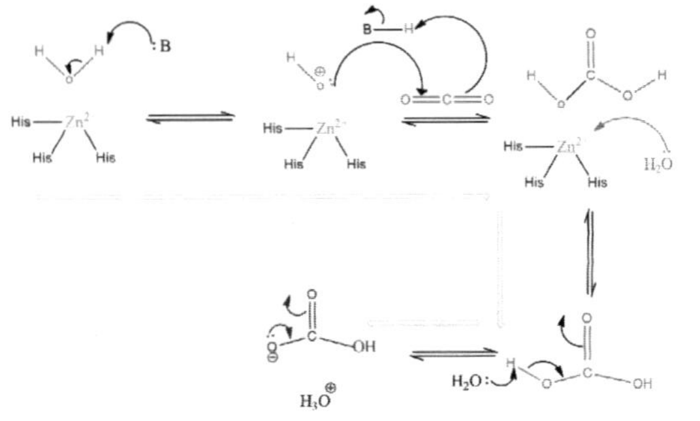

Figure 1. The catalytic Mechanism of enzyme in aqueous solutions

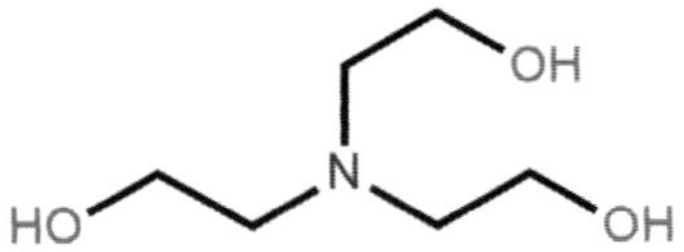

TEA

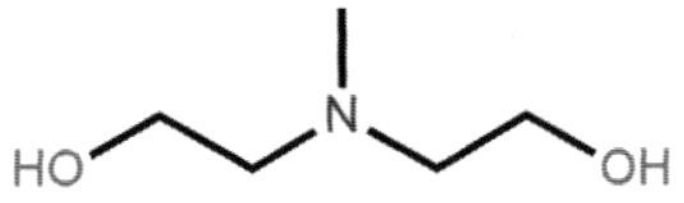

MDEA

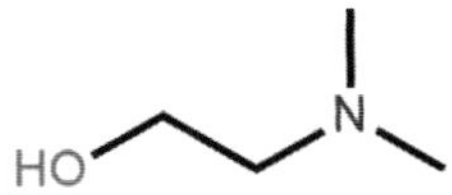

DMEA

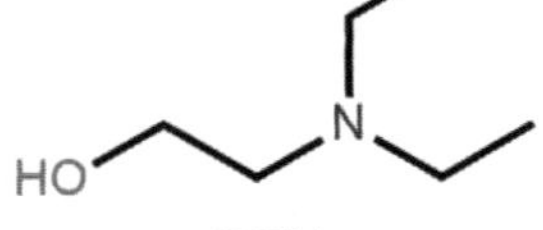

DEEA

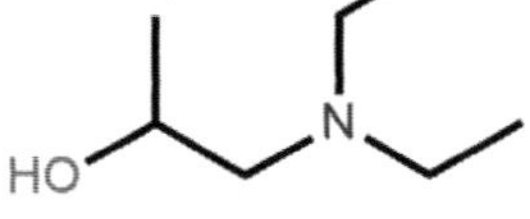

1DEA-2P

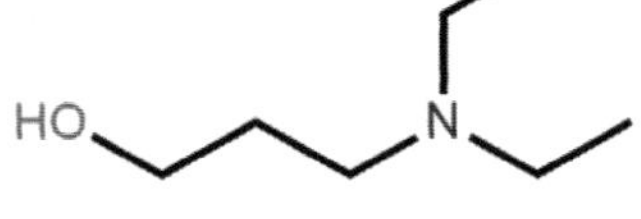

3DEA-1P

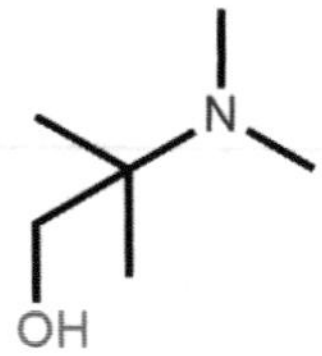

2DEA-2M-1P

Figure 2. The chemical structure of amines studied in this work

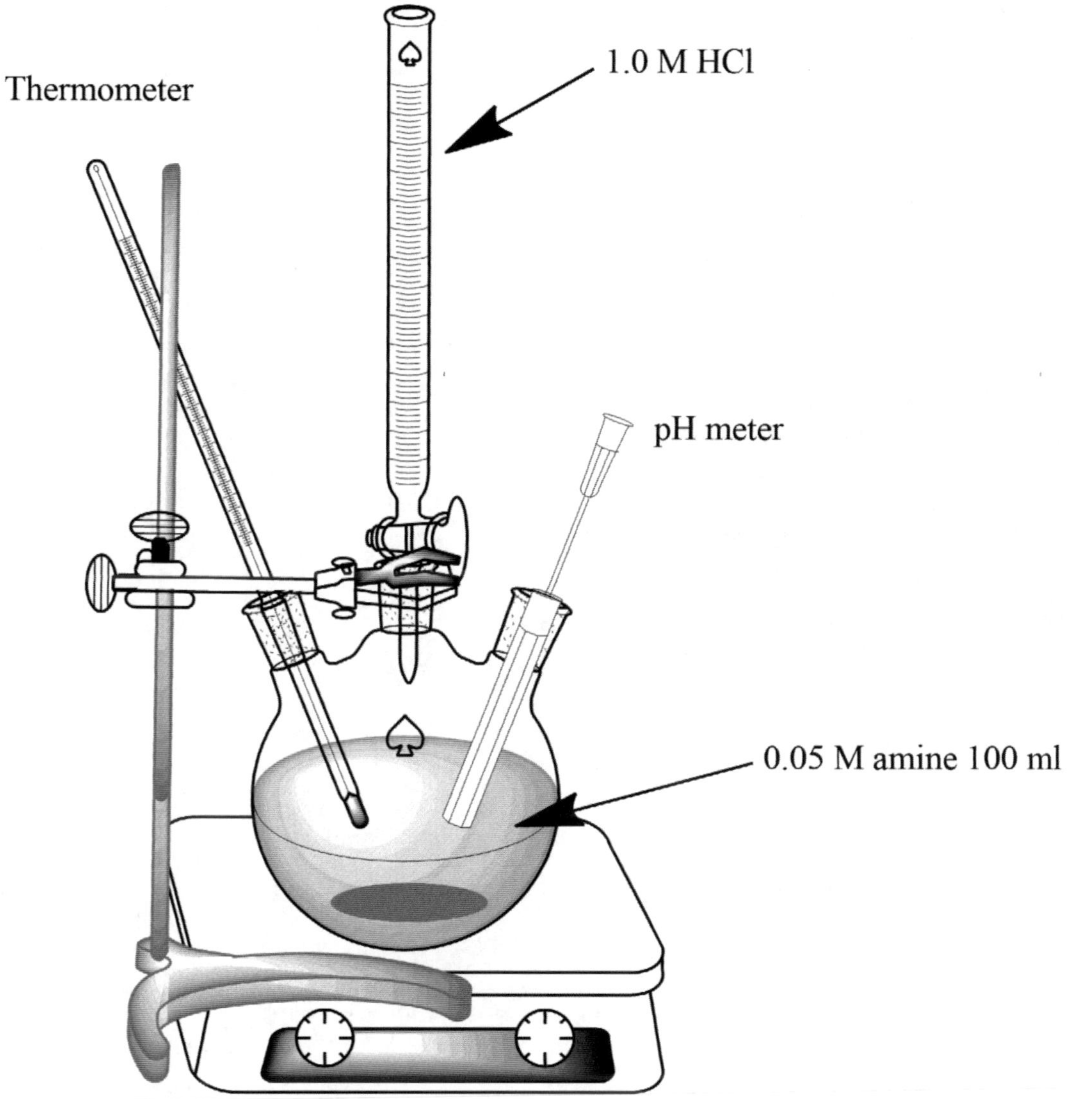

Figure 3. String cell for determination of amine pKa

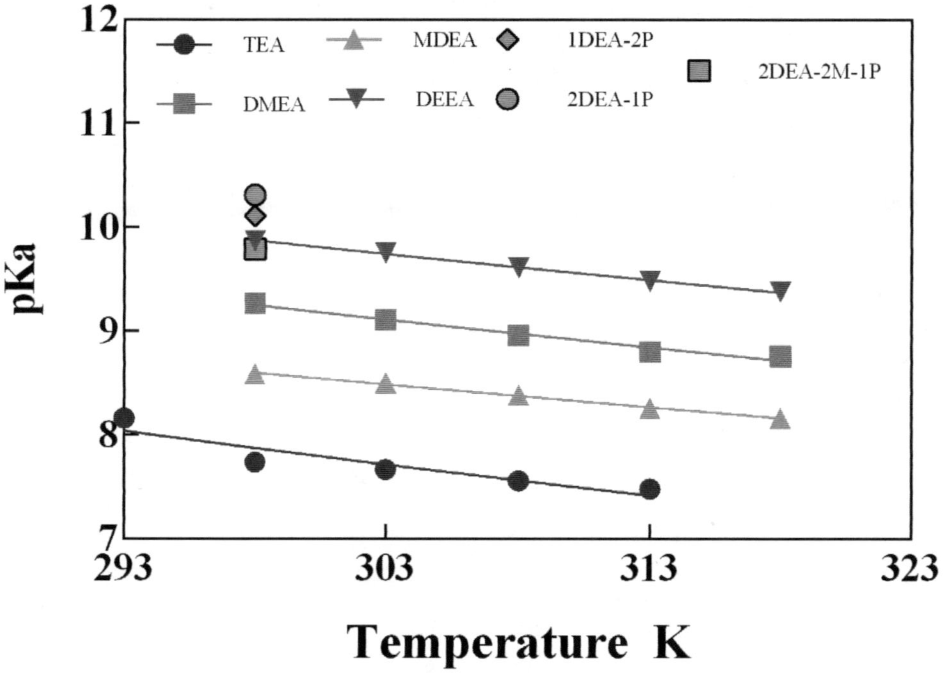

Figure 4. The dissociation constants show as a function of temperature

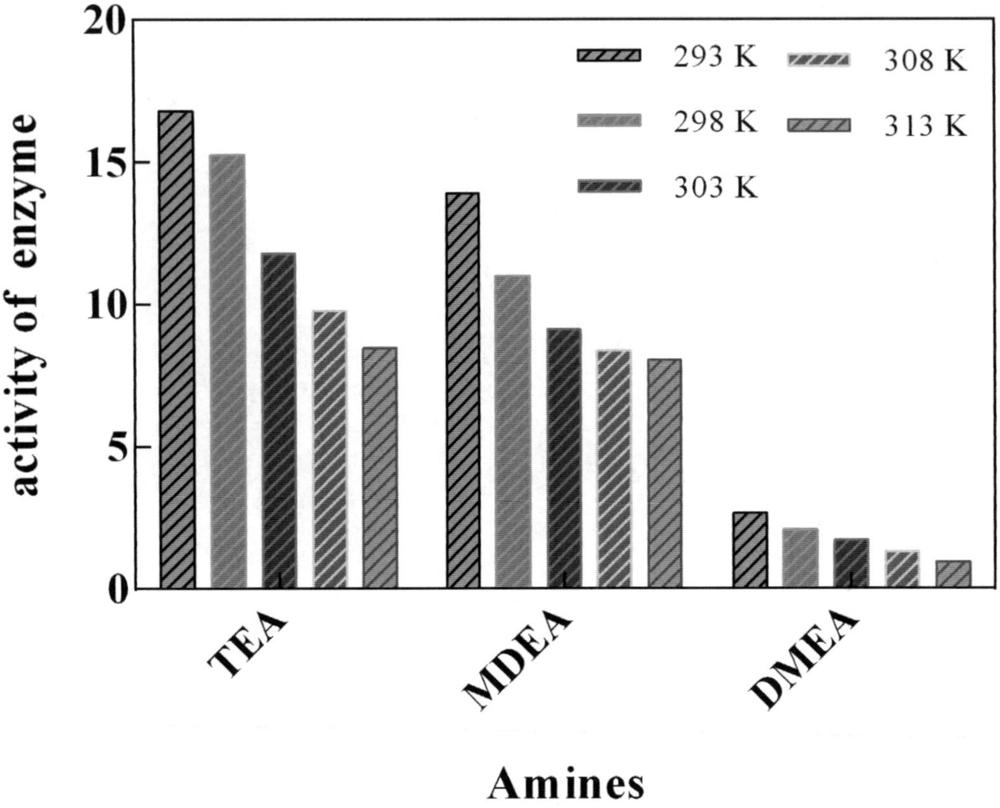

Figure 5. Comparison of enzymes activity in different aqueous amine solutions

Acknowledgment: The financial supports from the National Natural Science Foundation of China (Nos. 21476064 and 21406057), National Key Technology R&D Program (Nos. 2012BAC26B01 and2014BAC18B04), Innovative Research Team Development Plan-Ministry of Education of China (No. IRT1238), and China¡¯s State ¡°Project 985¡± in Hunan University-Novel Technology Research and Development for CO_2 Capture as well as Hunan University to the Joint International Center for CO_2 Capture and Storage (ICCS) is gratefully acknowledged.

References

1. Tontiwachwuthikul P, Idem R, Gelowitz D, Liang ZH, Supap T, Chan CW, et al. Recent progress and new development of post-combustion carbon-capture technology using reactive solvents. Carbon Management. 2011; 2: 261-3.

2. McCann N, Phan D, Wang X, Conway W, Burns R, Attalla M, et al. Kinetics and mechanism of carbamate formation from CO2 (aq), carbonate species, and monoethanolamine in aqueous solution. The Journal of Physical Chemistry A. 2009; 113: 5022-9.

3. Yu C-H, Huang C-H, Tan C-S. A Review of CO2 Capture by Absorption and Adsorption. Aerosol and Air Quality Research. 2012; 12: 745-69.

4. Liang ZH, Sanpasertparnich T, Tontiwachwuthikul P, Gelowitz D, Idem R. Part 1: Design, modeling and simulation of post-combustion CO2 capture systems using reactive solvents. Carbon Management. 2011; 2:265-88.

5. Reynolds AJ, Verheyen TV, Adeloju SB, Meuleman E, Feron P. Towards commercial scale postcombustion capture of CO2 with monoethanolamine solvent: key considerations for solvent management and

6. Lindskog S. Structure and mechanism of carbonic anhydrase. Pharmacology & therapeutics. 1997; 74: 1-20.

7. Lindskog S, Silverman DN. The catalytic mechanism of mammalian carbonic anhydrases. The Carbonic Anhydrases: Springer; 2000. p. 175-95.

8. Penders-van Elk NJMC, Derks PWJ, Fradette S, Versteeg GF. Kinetics of absorption of carbon dioxide in aqueous MDEA solutions with carbonic anhydrase at 298 K. International Journal of Greenhouse Gas Control. 2012; 9: 385-92.

A study of porous support amine-RTILs binary system for CO_2 capture

Min Xiao [a], *Helei Liu* [a,b]*, *Raphael Idem* [a,b], *Paitoon Tontiwachwuthikul* [a,b], *and Zhiwu Liang* [a,b]*

[a] *Joint International Center for CO_2 Capture and Storage (iCCS), Provincial Hunan Key Laboratory for Cost-effective Utilization of Fossil Fuel Aimed at Reducing CO_2 Emissions, College of Chemistry and Chemical Engineering, Hunan University, Changsha 410082, PR China*

[b] *Clean Energy Technologies Research Institute (CETRI), Faculty of Engineering and Applied Science*

University of Regina, Regina, Saskatchewan, S4S 0A2, Canada

Abstract

In this work, room temperature ionic liquids (RTILs) 1-butyl-3-methylimidazolium tetrafluoroborate ($[BMIM]BF_4$) and 1-butyl-3-ethylimidazolium tetrafluoroborate ($[BEIM]BF_4$) are synthesized. The cation structure of ionic liquids is confirmed by NMR spectroscopy. The thermal decomposition temperature of synthesized ionic liquids is measured and both ionic liquids show good thermal stability. The viscosity of ionic liquids is determined over the temperature range of 298 K to 353 K. A mixed system made by blending amine and ionic liquid is proposed for CO_2 capture. Investigation of the performance of this binary system on a porous support medium for CO_2 absorption shows that amine is able to greatly intensify the ionic liquids' ability to absorb CO_2. Furthermore, when blending tertiary amine and ionic liquids to absorb CO_2, it was found that only physical absorption takes place and that the viscosity of the absorbent increases slightly after CO_2 absorption. Based on the results of this work, the binary system of tertiary amine and RTILs is proven to be an alternative absorbent for CO_2 capture.

Keywords: *amine-RTILs; CO_2 capture; low viscosity; physical absorption*

* Author for correspondence:
Email: zwliang@hnu.edu.cn (Z. Liang) & lhl0925@hotmail.com (H. Liu)

1. Introduction

Global warming, which brings extreme weather, has become a serious issue and is more and more drawing people's attention. As one of many greenhouse gases, carbon dioxide is mainly blamed for the rise in global temperatures. Large amounts of CO_2 are directly released to the atmosphere through the burning of fossil fuels to meet the global demand for energy. There is an urgent need to reduce CO_2 emissions to protect the environment. Meanwhile, CO_2 is also a potential resource which can be applied to Enhanced Oil Recovery (EOR) or reformed with methane to produce synthesis gas etc. Thus, much research has been done on separating CO_2 from combustion exhaust gases. Although absorption of CO_2 with alcohol amine has been widely studied for years and proven to be able to capture CO_2 efficiently, its wide application is limited by a number of remaining problems such as corrosivity, volatility, degradation, and especially high energy consumption. The amount of energy required for regenerating rich amine solvents accounts for as much as 70-80% of the total operating cost of the system. It is thus of primary importance to reduce this energy consumption in order to bring down the cost of CO_2 capture.

A new kind of compound named room temperature ionic liquids (RTILs) has aroused people's interest because of their many unique properties. RTILs usually exist as liquid at room temperature with low vapor pressure, something that commends them as environmental friendly. Besides, RTILs show good thermal/chemical stability and tunable physicochemical characteristics, properties that have allowed them to be successfully used as green solvents in reactions and as catalysts for organic reactions [1]. There is also some work focusing on using RTILs for CO_2 capture. However, using pure RTILs for CO_2 capture as an answer to the world's CO_2 problem still has a very long way to go. When physical absorption dominates CO_2 capture with RTILs, the procedure is usually performed under extreme conditions, such as a very high pressure (90 bar or higher), and with a very long time being required to reach equilibrium (up to 24 h) [2]. To intensify CO_2 capture performance, a kind of ionic liquids named task-specific ionic liquids was developed. From the work of Bates et al. [3], the molar uptake of task-specific ionic liquids, which was modified with amino group, approached 0.5 mol CO_2•mol amine^{-1} during 3 h. However, high viscosity appears to be a new problem making those particular ionic liquids challenging for application in CO_2 capture [4]. The viscosity of ionic liquids is commonly high and will dramatically increase after CO_2 absorption, especially for task-specific ionic liquids [2].

To address these problems, a kind of low viscosity amine-IL mixed system is proposed for CO_2 capture in this work. Imidazole based ILs 1-butyl-3-methylimidazolium tetrafluoroborate ($[BMIM]BF_4$) and 1-butyl-3-ethylimidazolium tetrafluoroborate ($[BEIM]BF_4$) were chosen for their relatively low viscosity compared with other ILs. Conventional amines, Monoethanolamine (MEA) and Methyldiethanolamine (MDEA) were selected as the amines for blending with the ILs. The structures and abbreviation of amines and cation of ILs are shown in **Figure 1**.

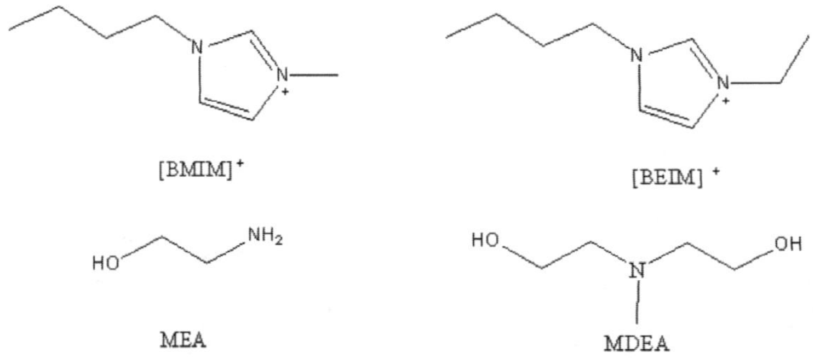

Figure 1 The structure and abbreviation of conventional amines, $[BMIM]^+$ and $[BEIM]^+$

2. Materials and experiment

2.1. *Materials*

CO_2 with purity of 99.9% was obtained from Changsha RiZhen Gas Co., Ltd. MDEA with purity of 98% was from Aladdin Industrial Corporation. MEA with purity of 99% and n-Butyl bromide (≥ 98%) was obtained from Sinopharm Chemical Reagent Co., Ltd. 1-Methylimidazole with purity of 99% and 1-Ethylimidazole with purity of 98% were supplied by Shanghai Macklin Biochemical Co. Ltd.

2.2. *Experiment*

[BMIM]BF_4 and [BEIM]BF_4 were synthesized on basis of imidazole. The general preparation was similar to the work of Harris et al [5]. Complete anion exchange was confirmed when no bromide could be detected with silver nitrate. After evaporating any unreacted components and water, the RTILs were stored in a vacuum drier. A ^{13}C NMR Spectrometer (Bruker 400MHZ Advance) was employed to characterize the cation structure of the RTILs and the absorbents after CO_2 absorption. Thermogravimetric Analysis (Shimadzu DTG-60) was introduced to evaluate the water content and thermal stability of the RTILs. Then the dynamic viscosities (Brookfield Viscometer DV2T) of pure RTILs and the amine-RTIL binary systems were measured.

The CO_2 absorption experiment process is briefly described below. Approximate 0.8 mL absorbent was stabilized into a porous filter candle which was then placed into a gas-washing bottle. The gas-washing bottle was immersed into a water bath to control the temperature. The pure CO_2 flow rate was fixed at 50 $mL{\bullet}min^{-1}$ by a mass flow controller, and fed into a conical flask with concentrated sulfuric acid to get rid of all trace amounts of water. Then, the CO_2 was fed to the gas washing bottle continuously. The amount of absorbed CO_2 was determined by detecting weight change using an electronic analytical balance with accuracy of 1 mg. When the weight of the absorbent stopped increasing, equilibrium was considered to be achieved.

3. Results and discussion

3.1. *Thermal stability of RTILs*

To evaluate the thermal stability of [BMIM]BF_4 and [BEIM]BF_4, Thermogravimetric Analysis (TGA) was introduced. The results for pure [BMIM]BF_4 and [BEIM]BF_4 are shown in **Figure 2**. Water content of the RTILs was confirmed to be less than 1% and thermal decomposition temperatures (10% mass loss) of [BMIM]BF_4 and [BEIM]BF_4 were found to be 635.65 K and 646.96 K, respectively. Results showed that both RTILs have good thermal stability. Because the side chain of $[BEIM]^+$ was one carbon longer than $[BMIM]^+$, it showed better thermal stability.

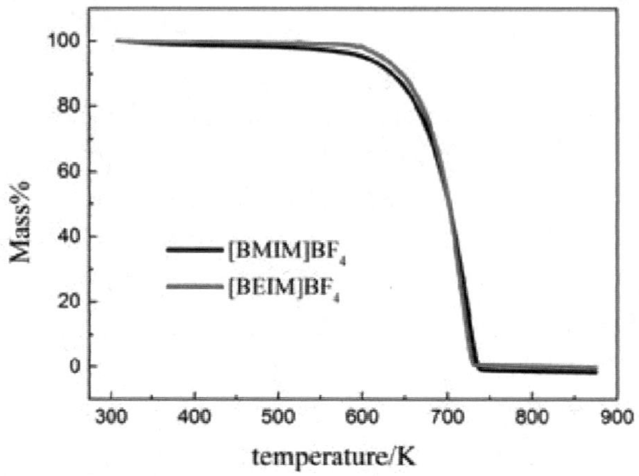

Figure 2 Thermogravimetric Analysis on $[BMIM]BF_4$ and $[BEIM]BF_4$

3.2. *Viscosity of RTILs*

The viscosities of $[BMIM]BF_4$ and $[BEIM]BF_4$ were measured over temperature ranges from 298 K to 353 K. It was found that $[BMIM]BF_4$ shows viscosity of 101.2 cP at 293 K, which agreed well with Harris's work with a deviation of 2.6% [5]. The viscosity of $[BEIM]BF_4$ is slightly higher than $[BMIM]BF_4$ due to its longer carbon chain. An extended Arrhenius equation (1) is applied to correct viscosity with temperature:

$$\ln\eta = \frac{E_a}{RT} + \ln\eta_\infty \cdots\cdots\cdots\cdots(1)$$

The R-square of viscosity fitting for $[BMIM]BF_4$ and $[BEIM]BF_4$ is 0.997 and 0.998, respectively. The equation is able to correct the experiment data well. The summaries of E_a and η_0 are listed in **Table 1**.

Table 1 Summaries of E_a and η_0 for pure $[BMIM]BF_4$ and $[BEIM]BF_4$

	characteristic parameter	
	$E_a/(kJ\bullet mol^{-1})$	$\eta_0/(\bullet 10^2 cP)$
$[BMIM]BF_4$, Harris's work [5]	33.53	0.14
$[BMIM]BF_4$, this work	34.15	0.0079
$[BEIM]BF_4$, this work	32.62	0.015

3.3. *CO_2 absorption*

In this work, the CO_2 absorption by RTILs and some binary systems was studied at the temperature of 298 K under atmospheric pressure. The absorption curves of the pure RTILs are shown in **Figure 3**. CO_2 equilibrium solubility of $[BMIM]BF_4$ and $[BEIM]BF_4$ at 298 K was 0.039 and 0.049 mol $CO_2\cdot$mol ILs^{-1}, respectively. Equilibrium was reached within 10 minutes and $[BEIM]BF_4$ showed higher equilibrium CO_2 solubility than $[BMIM]BF_4$.

However, both RTILs revealed low physical CO_2 absorption without the amines. To improve CO_2 absorption, 30% mass fraction of MEA and MDEA was blended with the RTILs, respectively. It was found that CO_2 absorption could be greatly improved by these added amines and the CO_2 equilibrium could be achieved within 120 minutes. Because the amines dominate the absorption of CO_2, the result was expressed using the units of mol CO_2•mol amine^{-1}. Four CO_2 absorption curves are shown in **Figure 4**. The equilibrium CO_2 loadings of MEA-[BMIM]BF_4 and MEA-[BEIM]BF_4 at 298 K were 0.494 and 0.515 mol CO_2•mol amine^{-1}, respectively. Chemical reaction between MEA and CO_2 was confirmed by the detection of a new resonance at 165.08 ppm using ^{13}C NMR technology. Two MEA molecules combined with one molecule of CO_2 to form carbamate without the presence of water. The CO_2 absorption of MDEA-[BMIM]BF_4 and MDEA-[BEIM]BF_4 was accomplished within 80 minutes and the equilibrium CO_2 loadings at 298 K were 0.214 and 0.334 mol CO_2•mol amine^{-1}, respectively, which was relatively lower than the MEA-RTILs systems. It has to be pointed out that for the MDEA-RTILs systems, no additional peak could be detected after CO_2 absorption. This means that only physical absorption exists in this mixed system. The CO_2 molecule is stabilized in the absorbent by van der Waals force, which is much weaker than the force of chemical bonding. Consequently, regenerating the absorbent requires less heat compared with the system where chemical reaction happens. Furthermore, amine-[BEIM]BF_4 performs better than amine-[BMIM]BF_4 on CO_2 absorption. It can be concluded that RTILs with higher CO_2 capacity are able to enhance the amine absorption of CO_2 as well.

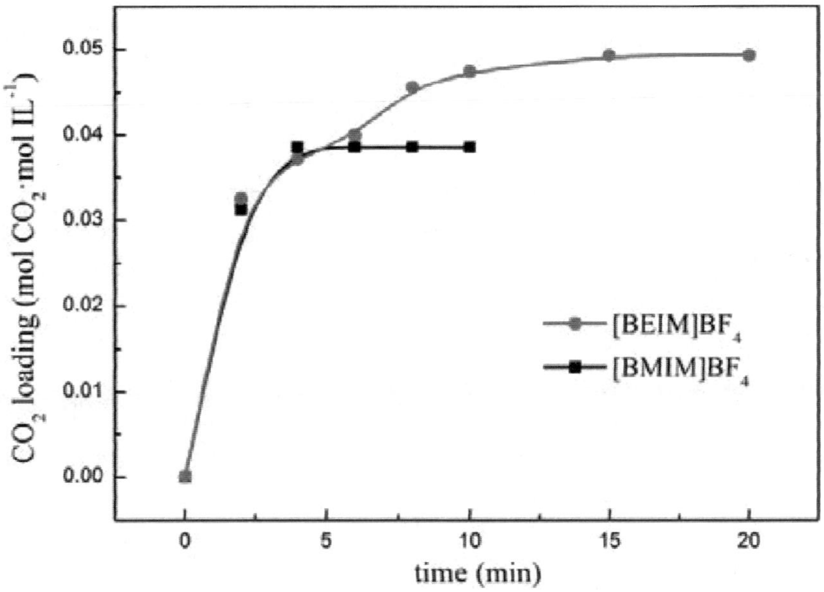

Figure 3 The CO_2 loading of pure RTILs versus time at 298 K

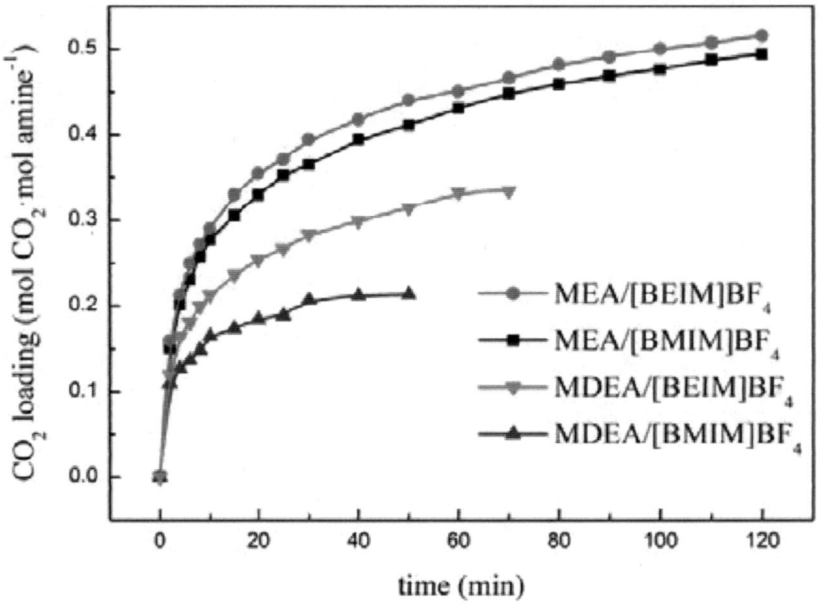

Figure 4 The CO_2 loading of mixed system with amine mass fraction of 30% at 298 K

3.4. Viscosity of mixed system before and after CO_2 absorption

To further study the viscosities of the mixed MDEA-RTILs systems, approximately 1 mL of each absorbent was exposed to a pure CO_2 atmosphere for 180 minutes. It has to be pointed out that during the CO_2 absorption using amine-RTILs mixed system, MEA was found to precipitate out in the form of carbamate after reacting with CO_2. The viscosity of MEA-RTILs system was thus unable to be measured. MDEA was fully dissolved into the RTILs before and after CO_2 absorption. Thus the viscosities of the MDEA-RTIL systems at 298 K were measured and are summarized in **Table 2**. Results show that the viscosity of MDEA-[BMIM]BF_4 increased from 49.46 cP to 63.40 cP after CO_2 absorption while for MDEA-[BEIM]BF_4, it increased from 49.79 cP to 77.77 cP. It can be concluded that the viscosity of both absorbents increased slightly after CO_2 absorption. In the case of [BEIM]BF_4, it showed higher viscosity than [BMIM]BF_4, the MDEA-[BEIM]BF_4 system revealed higher viscosity than MDEA-[BMIM]BF_4 before and after CO_2 absorption as well.

Table 2 The viscosity of mixed MDEA-RTILs system before and after CO_2 absorption at 298 K

	Before CO_2 absorption	After CO_2 absorption
	η(cP)	η(cP)
MDEA/[BMIM]BF_4	49.46 ± 1.02	63.40 ± 1.23
MDEA/[BEIM]BF_4	49.79 ± 1.02	77.77 ± 1.61

4. Conclusion

In this work, the thermal decomposition temperature and viscosity of ionic liquids [BMIM]BF_4 and [BEIM]BF_4 was measured. Both ionic liquids showed good thermal stability. [BEIM]BF_4 was found to be more stable than [BMIM]BF_4 due to its longer carbon chain. Meanwhile, the viscosity of [BEIM]BF_4 was slightly higher than that of [BMIM]BF_4. The absorption of CO_2 by amine-RITLs mixed systems was investigated as well. The viscosity of MDEA-RTILs was measured before and after CO_2 absorption. Results show that MDEA-[BEIM]BF_4 proves to be a potential binary system for CO_2 capture. The mixed system is thermally stable, able to capture CO_2 quickly and shows relatively low viscosity even after CO_2 absorption. Although MEA-RTILs show higher CO_2 capacity than MDEA-RTILs, it has to be pointed out that only physical absorption occurs in MDEA-RTILs systems, which means the mixed system can readily be regenerated by heating to strip out CO_2. Consequently, MDEA-[BEIM]BF_4 is considered to be an energy saving absorbent.

Acknowledgements

Financial support from the National Natural Science Foundation of China (Nos. 21476064, 21376067 and U1362112), National Key Technology R&D Program (Nos. 2012BAC26B01 and2014BAC18B04), Innovative Research Team Development Plan-Ministry of Education of China (No. IRT1238), Shanxi Yanchang Petroleum (Group) Co., LTD, Specialized Research Fund for the Doctoral Program of Higher Education (No. 20130161110025),and China's State 'Project 985' in Hunan University-Novel Technology Research and Development for CO_2 Capture as well as Hunan University to the Joint International Center for CO_2 Capture and Storage (iCCS) is gratefully acknowledged.

References

1. Joan F. Brennecke, Maginn EJ. Ionic Liquids: Innovative Fluids for Chemical Processing. *AIChE Journal.* 2001;47(11):2384-2389.
2. Zhang Y, Zhang S, Lu X, Zhou Q, Fan W, Zhang X. Dual amino-functionalised phosphonium ionic liquids for CO2 capture. *Chemistry.* 2009;15(12):3003-3011.
3. Eleanor D. Bates, Rebecca D. Mayton, Ioanna Ntai, James H. Davis. CO_2 Capture by a Task-Specific Ionic Liquid. *J. Am. Chem. Soc.* 2002;124(6):926-927.
4. Seo S, Quiroz-Guzman M, DeSilva MA, et al. Chemically tunable ionic liquids with aprotic heterocyclic anion (AHA) for CO_2 capture. *The journal of physical chemistry. B.* May 29 2014;118(21):5740-5751.
5. Harris KR, Kanakubo M, Woolf LA. Temperature and Pressure Dependence of the Viscosity of the Ionic Liquid 1-Butyl-3-methylimidazolium Tetrafluoroborate: Viscosity and Density Relationships in Ionic Liquids. *J. Chem. Eng. Data.* 2007;52:2425-2430.

COMPARATIVE STUDY ON LOW AND HIGH PRESSURE CO2 ADSORPTION CAPACITY OF ORGANIC MATERIALS

Mert Atilhan and Ruh Ullah*
Chemical Engineering Department, Qatar University, Doha

Santiago Aparicio
Department of Chemistry, University of Burgos, Burgos, Spain

Hasmukh Patel
Northwestern University, Evanston, USA

*Cafer T. Yavuz**
Department of Chemistry, KAIST, Daejeon, Republic of Korea

Abstract

Application of solid sorbents for CO_2 capture and separation has been under consideration since last decades owing to the promising feature of repeatable use, low cost of regeneration and thermo physical stability. However, the capturing capacity, selectivity and the overall performance of adsorbents at different temperatures and pressures still require further investigation to make these materials capable of replacing the mono-ethanol system. Here in we report synthesis, characterization and application of covalent organic polymers. Capturing capacity of COPs at low temperature and pressure has been experimented for two COPs materials (namely COP-9, COP-10) for CO_2 and N_2 adsorption. Materials were also characterized with BET and TGA.

1. INTRODUCTION

Till date cost of mono-ethanol amine regeneration and reduction in the electricity production are still among the major challenges of the plants using fossil fuels as an ultimate source for power generation. The demand of more power production due to the technological development for the improvement of life style is increasing world-wide which in turn enhances the usage of coal and subsequently rising CO_2 emission and concentration in the environment [1]. The renewal energy sources such as solar power, wind power, nuclear power and bio fuel power are still in the research and initial stages and consequently the world will have to depend mostly on fossil fuels for next few decades. It has been estimated that on average a single power plant using fossil fuels as power production source emit at least 3.5 million tons of CO_2 per year, whereas the number of such power generation industries increases globally. Cement industries and transportation, which are the second and third major sources of CO_2 emission and environment pollution are also multiplying every year and accordingly enhancing the concentration of CO_2 in atmosphere. Research has shown that the content of CO_2 in atmosphere may exceed 500 ppm by 2020 if not handle properly [2, 3]. This alarming increase in CO_2 emission has severely threatened life on the earth and issues like global warming, severe weather conditions spread of very dangerous disease such as respiratory illnesses and asthma are all due to the excessive and uncontrollable emission of CO_2 [4-9]. The new energy production plants using fossil fuels however now changes to oxy-combustion fuel system and integrated gasification combine cycle system where CO_2 can be effectively captured using advanced technologies as well as MEA[4, 10]. Solid sorbents possess various advantages such as simple regeneration, repeated applications, low manufacturing cost, no extra-cost on electricity production, environmental and economic viability, tunable capturing capacity and changeable selectivity. Solid sorbents mainly have applications in pre-combustion system like IGCC, where, CO_2 can be separated from syngas after water shift in the process[10]. However, this requires the adsorbents to be physically strong which can hold high pressure and high temperature unlike the post-combustion process where CO_2 can be separated at atmospheric pressure. In addition to the thermal and physical stability, solid sorbent needs to be extremely selective for separation of CO_2 from the stream, which have H_2 as fuels, should have high surface area and large pore volume and better affinity for CO_2. Various solid sorbents such as activated carbon, metal organic framework, zeolites, SBA-15, nanoclays, and many organic polymers have been investigated both for low and high-pressure CO_2 adsorption and separation. Attempts have been made to increase the CO_2 capturing capacity of solid materials by inserting various amine function groups within the porous structure of adsorbents[11]. As an example SBA-15 has been impregnated with different function groups and tested for low and high pressure CO_2 adsorption. Results indicated that amine modification have very limited effect on the capturing capacity at high pressure. Here in we report synthesis and characterization of covalent organic polymers, which possess internal nitrogen functionality and can be assumed to be the more favorable compounds as long as the affinity of CO_2 [12-15]. We prepare two COPs materials using different building block monomers and linked with two different linkers to investigate the effect of different functionality on the CO_2 and N_2 adsorption capacity at low and high pressure under different temperatures.

2. EXPERIMENTAL

2.1. Materials and Synthesis procedure

N,N-Diisopropylethylamine, 1,3,5-benzene tricarbonyltrichloride, 2,5-Dimethylpiperazine, cyanuric chloride, 2-imidazolidinone, dimethylacetamide, and tetrahydrofuran (THF), purchased from Sigma-Aldric, however, dioxane and ethanol were obtained from TCI, Japan. All the chemicals were used without any further modification and/or dilution. COP-9 and COP-10 were prepared through the following procedures. In a typical procedure N,N-Diisopropylethylamine (1.9 mL, 10.8 mmol) was added to trans-2,5-Dimethylpiperazine (0.48 g, 4.2 mmol) dissolved in THF at 0-5 ^{0}C. Cyanuric chloride (0.5 g, 2.7 mmol) dissolved in THF was added dropwise to the above solution with continuous stirring at 0-5 °C in N_2 environment. The solution was stirred at 0-5 °C for 1 h. Subsequently, the temperature of the reaction mixture increased to 25-30 °C and stirred for 2 h, followed by heating at 68-70 °C for 21 h. The precipitate was washed with THF, dioxane and ethanol (three times each). Finally, the precipitate was dried at room temperature for 2 h and 80-90 ^{0}C for 5 h under vacuum and labeled as COP-9. N,N-Diisopropylethylamine (2.1 mL, 12 mmol) was added to 2-imidazolidinone (0.4 g, 4.65 mmol) dissolved in Dimethylacetamide (DMAc, 35 mL) at 25 ^{0}C. 1,3,5-Benzenetricarbonyl trichloride (0.8 g, 3 mmol) dissolved in DMAc (25 mL) was added dropwise to the above solution with continuous stirring at 25 ^{0}C. The solution was stirred at 25 °C for 2 h. Subsequently, the temperature of the reaction mixture increased to 80 ^{0}C and stirred for 16 h. The reaction mixture cooled to room temperature and 250 mL of de-ionized water added which resulted precipitate. The precipitate was washed with water and ethanol (three times each) and it was dried at room temperature for 2 h and 80-90 ^{0}C for 5 h under vacuum and labeled as COP-10.

2.2. Characterization and Adsorption Setup

Thermal stability of COP-9 and COP-10 was conducted with Perkin Elmer Pyris 6 (Thermogravimetric Analysis) TGA model 4000 system. TGA system was properly balanced and calibrated according to the purging gas (N_2), crucible and sample holder stem. TGA machine was equipped with chiller from Huber model Pilot One to automatically maintain the temperature of the system and N_2 was used as the heating environment, where the materials were heated directly from 30 °C to 800 °C at the rate of 1° per minute. Physical parameters such as pore volume and surface area of materials were estimated using BET analyzer Model micrometrics ASAP 2420. The solid materials were degased for at least 12 hours prior to the BET measurement to avoid any blockage of pore due to the water.

PCT-Pro was used for the low-pressure adsorption process, where each pressure step was 0.1 bars up to the maximum of 1 bar. Here only CO_2 and N_2 were calculated at three different temperatures of 298 K, 308 K and 323 K [16]. Two gas cylinders (high purity grade CO_2, and N_2) were connected to gas dosing system (GDS) via external pump to pressurize the required gas into the measuring cell. The maximum set pressure (1 bar) was applied stepwise by increasing the pressure gradually from 0 bar up to 1 bar using each step of 0.1 bar. After reaching the maximum set pressure points, pressure of the system was gradually reduced exactly with the same measurement points starting from the high pressure and reaching to the last evacuation point to complete the desorption process. The system is fully automated and the pressure goes to next higher point after completing the previous measurement point.

3. RESULTS AND DISCUSSIONS

Thermal gravimetric analysis of both covalent organic polymers COP-9 and COP-10 (shown in figurer 1) reveals that COP-9 is thermally more stable than COP-10, since; the former is almost stable up to 400 °C. TGA further indicates that COP-10 is thermally degraded in three steps, while COP-9 decomposes in two steps. The initial degradation of COP-10 starts at around 75 °C which can be associated to the evaporation of stored water within the structure, 2nd degradation of COP-10 starts at around 200 °C, while the third one starts at 400 °C. Contrary, COP-9 is about completely stable up to 400 °C however, beyond this point it decomposed quickly and in another 100 °C increase in temperature, almost 50 % of the material losses. TGA analysis confirmed that COP-9 is much stable than COP-10, however, it degrades gradually while a sharp degradation in COP-9 can be observed after 400 °C.

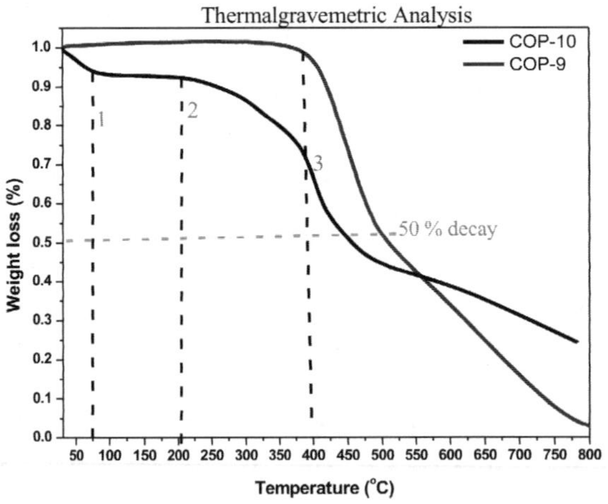

Figure 1. Thermal-gravimetric analysis of COP-9 and COP-10

Physical properties and proposed chemical structure of COPs are given in table 1 below which shows that COP-9 has smaller pore volume than COP-10, however, the former has larger surface area and lower tapped bulk density. It has been assumed that low pressure gases uptake by solid materials is mostly governed by the physical parameters such as pore volume and surface area[17]. On the other hand adsorption capacity at high pressure does not necessarily depend upon the porosity of the structure, but most often varies with affinity of the materials. Thus the material with lower porosity and lower surface area may have higher affinity at high pressure and therefore will capture more CO_2 at elevated pressure as compared to those adsorbents which have larger physical parameters [18, 19].

Table 1. Physical Parameters and structure of covalent organic polymers (COPs)

Sample	**Structure**	**Surface area (m^2/g)**		**Pore volume (cm^3/g)**	**Density (cm^3/g)**	
		BET	**Langmuir**		**Bulk**	**Tapped**
COP-9	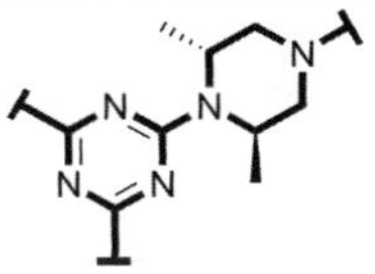	146.4	204.3	0.244	0.181	0.202
COP-10	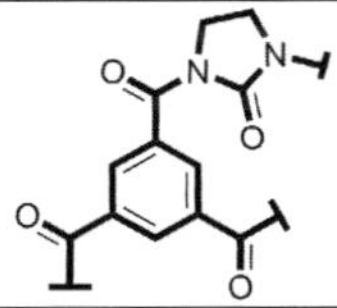	84.5	114.6	0.634	0.129	0.258

Figure 2, shows the low pressure CO_2 and N_2 adsorption of COP-9 (c, d) and COP-10 (a, b), which clearly indicates that due to the larger pore volume of COP-10 it captures comparatively more CO_2 at lower pressure of 1 bar. Xiang et al reported that COP-1 possess the highest surface area captured more N_2 and CH_4; however, COP-2 with larger pore volume captured more CO_2 and H_2 than all other organic polymers[20]. On the other hand COP-10 uptakes very low quantity of N_2 at 1 bar and three different temperatures as compared to COP-9. It must be noted that selectivity of COP-10 for CO_2:N_2 at low pressure is much higher than that of COP-9. Additionally as given in table 2 selectivity of COP-10 for CO_2:N_2:H_2 is also better than that of COP-9 at 308 K, while at all other temperatures selectivity of both polymers is similar.

Additionally, the selectivity of COP-10 at atmospheric pressure is increasing with rising temperature and becomes infinity at temperature above than 308 K. As shown in figure 2b COP-10 is totally inert for N_2 at higher temperatures, although it captures a bit smaller quantity of N_2 at standard temperature and pressure. Figure 2c and 2d indicates that COP-9 uptake less CO_2 than COP-10, but its N_2 uptake is comparatively larger than that of COP-10. Table 2 indicates that the overall performance of COP-10 at atmospheric pressure and different temperatures is comparatively better than COP-9, since the former adsorb more CO_2 and less N_2 which makes this material favorable for the pre-combustion CO_2 capture and separation.

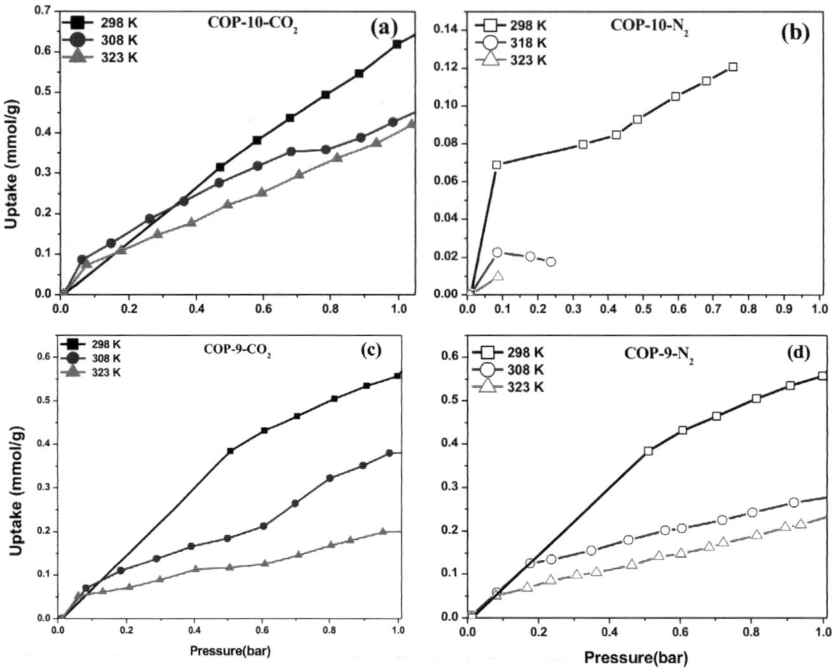

Figure 2. CO_2 adsorption of COP-9 (c & d) and COP-10 (a & b) at maximum of 1 bar and three different temperatures obtained with PCT-Pro.

It can be concluded that two covalently bonded organic polymers COP-9 and COP-10 were prepared from two different core building block monomers and linked with different linking agents. Both COPs material were tested for low and high pressure CO_2 and N_2 adsorption at three different temperatures. It was found that COP-10 has almost three times larger pore volume than COP-9 and therefore captures slightly larger CO_2 than COP-9 at atmospheric pressure and room temperature. Additionally, COP-10 uptakes significantly lower quantity of N_2 at atmospheric pressure and room temperature while it becomes almost N_2 phobic at high temperature and lower pressure, thus resulting in very good selectivity under these conditions.

In addition to the effect of pore volume and surface area on the gases uptake capacity of solid sorbents, affinity of the gas molecules toward the attractive sites also contribute to the adsorption capacity of materials. It can be assumed that, COP-9 may have more interacting sites than COP-10 since the core monomer of COP-9 has three nitrogen nodes whereas the core monomer of COP-10 has less nitrogen species. On the other hand COP-10 may have more oxygen OH functionality that may have less affinity for CO_2 than the N functionality. Thus it can be argued that, although COP-9 has lower pore volume but it has more and stronger attractive sites for CO_2 than COP-10, therefore, it captures comparatively more CO_2.

ACKNOWLEDGMENTS

This paper was made possible by the support of Qatar National Research Fund, National Priorities Research Program grant (NPRP# 5-499-1-088) and (NPRP# 6-330-2-140). The statements made herein are solely the responsibility of the authors.

REFERENCES

1. Aaron, D. and C. Tsouris, *Separation of CO2 from Flue Gas: A Review.* Separation Science and Technology, 2005. **40**(1-3): p. 321-348.
2. Boot-Handford, M.E., et al., *Carbon capture and storage update.* Energy & Environmental Science, 2014. **7**(1): p. 130-189.
3. Chu, S. and A. Majumdar, *Opportunities and challenges for a sustainable energy future.* Nature, 2012. **488**(7411): p. 294-303.
4. Ahmed, U., et al., *IGCC process intensification for simultaneous power generation and CO2 capture.* Chemical Engineering and Processing: Process Intensification, 2016. **101**: p. 72-86.
5. Awadallah-F, A. and S. Al-Muhtaseb, *Carbon dioxide sequestration and methane removal from exhaust gases using resorcinol–formaldehyde activated carbon xerogel.* Adsorption, 2013. **19**(5): p. 967-977.
6. Berger, A.H. and A.S. Bhown, *Selection of Optimal Solid Sorbents for CO2 Capture Based on Gas Phase CO2 composition.* Energy Procedia, 2014. **63**(0): p. 2092-2099.
7. Cavka, J.H., et al., *High Pressure Adsorption of CO2 and CH4 on Zr-MOFs.* Industrial & Engineering Chemistry Research, 2014. **53**(40): p. 15500-15507.
8. Chen, J., L.S. Loo, and K. Wang, *High-Pressure CO2 Adsorption on a Polymer-Derived Carbon Molecular Sieve.* Journal of Chemical & Engineering Data, 2008. **53**(1): p. 2-4.
9. Gonzalez-Salazar, M.A., et al., *Comparison of Current and Advanced Post-Combustion CO2 Capture Technologies for Power Plant Applications.* Energy Procedia, 2012. **23**: p. 3-14.
10. Ekström, C., et al., *Techno-Economic Evaluations and Benchmarking of Pre-combustion CO2 Capture and Oxy-fuel Processes Developed in the European ENCAP Project.* Energy Procedia, 2009. **1**(1): p. 4233-4240.
11. Hedin, N., et al., *Adsorbents for the post-combustion capture of CO2 using rapid temperature swing or vacuum swing adsorption.* Applied Energy, 2013. **104**(0): p. 418-433.
12. Ullah, R., et al., *Insights of CO2 adsorption performance of amine impregnated mesoporous silica (SBA-15) at wide range pressure and temperature conditions.* International Journal of Greenhouse Gas Control, 2015. **43**: p. 22-32.
13. McKeown, N.B. and P.M. Budd, *Polymers of intrinsic microporosity (PIMs): organic materials for membrane separations, heterogeneous catalysis and hydrogen storage.* Chemical Society Reviews, 2006. **35**(8): p. 675-683.
14. Modak, A. and A. Bhaumik, *Porous carbon derived via KOH activation of a hypercrosslinked porous organic polymer for efficient CO2, CH4, H2 adsorptions and high CO2/N2 selectivity.* Journal of Solid State Chemistry, 2015. **232**: p. 157-162.
15. Plaza, M.G., et al., *CO2 capture by adsorption with nitrogen enriched carbons.* Fuel, 2007. **86**(14): p. 2204-2212.
16. Deniz, E., et al., *A combined computational and experimental study of high pressure and supercritical CO2 adsorption on Basolite MOFs.* Microporous and Mesoporous Materials, 2013. **175**: p. 34-42.
17. Belmabkhout, Y., V. Guillerm, and M. Eddaoudi, *Low concentration CO2 capture using physical adsorbents: Are metal–organic frameworks becoming the new benchmark materials?* Chemical Engineering Journal, 2016. **296**: p. 386-397.
18. Veneman, R., et al., *Adsorption of CO2 and H2O on supported amine sorbents.* Energy Procedia, 2014. **63**(0): p. 2336-2345.
19. Zulfiqar, S., M.I. Sarwar, and C.T. Yavuz, *Melamine based porous organic amide polymers for CO2 capture.* RSC Advances, 2014. **4**(94): p. 52263-52269.

20. Xiang, Z., et al., *Covalent-organic polymers for carbon dioxide capture.* Journal of Materials Chemistry, 2012. **22**(42): p. 22663-22669.
21. Mason, C.R., et al., *Enhancement of CO2 Affinity in a Polymer of Intrinsic Microporosity by Amine Modification.* Macromolecules, 2014. **47**(3): p. 1021-1029.
22. McKeown, N.B. and P.M. Budd, *Exploitation of Intrinsic Microporosity in Polymer-Based Materials.* Macromolecules, 2010. **43**(12): p. 5163-5176.
23. Ullah, R., et al., *Synthesis, characterization and evaluation of porous polybenzimidazole materials for CO2 adsorption at high pressures.* Adsorption, 2016: p. 1-14.
24. Patel, H.A., et al., *Highly Stable Nanoporous Sulfur-Bridged Covalent Organic Polymers for Carbon Dioxide Removal.* Advanced Functional Materials, 2013. **23**(18): p. 2270-2276.
25. Patel, H.A., et al., *High capacity carbon dioxide adsorption by inexpensive covalent organic polymers.* Journal of Materials Chemistry, 2012. **22**(17): p. 8431-8437.

466434 Carbonic Anhydrase Enhanced Carbon Capture: Kinetic Measurements and Pilot Plant Trials

Abstract:

Carbonic Anhydrase Enhanced Carbon Capture: Kinetic Measurements and Pilot Plant Trials

Arne Gladis[1], Maria T Gundersen[2], Philip L Fosbøl[1], John M Woodley[2], Nicolas von Solms[1]*

1) Center for Energy Resources Engineering, Department of Chemical and Biochemical Engineering, Technical University Denmark, Kgs. Lynby 2800

2) CAPEC-PROCESS Center, Department of Chemical and Biochemical Engineering, Technical University Denmark, Kgs. Lyngby 2800

Keywords:

Wetted wall column, carbonic anhydrase, carbon capture, enzyme kinetics, amine, carbonate, absorption, carbon dioxide

Abstract:

In this study the effect of carbonic anhydrase addition on the absorption of CO_2 was investigated in a wetted wall column apparatus. Four different solvents: MEA (a primary amine), AMP (a sterically hindered primary amine), MDEA (a tertiary amine) and K_2CO_3 a carbonate salt solution were tested in concentrations from 5 to 50 wt%. Necessary mass transfer parameters such as liquid side mass transfer coefficient and solvent and enzyme reaction rates

were determined in a temperature range from 298 to 328 K and benchmarked to a 30 wt% MEA solution.

The study reveals that the addition of the enzyme carbonic anhydrase (CA) dramatically increases the liquid side mass transfer coefficient for 30 wt% MDEA and 15 wt% K_2CO_3. 30 wt% AMP has a moderate increase whereas 30 wt% MEA was unchanged. The results confirm that bicarbonate forming solvent which do not produce carbamate benefit from CA. The results reveal the impact of temperature in relation to CA. A temperature increase resulted in lower liquid side mass transfer rate for 30 wt% MDEA and 15 wt% K_2CO_3 but in higher rate for 30 wt% AMP. The overall first order enzyme reaction rate (s^{-1}) was linearly dependent on enzyme concentration for 30 wt% MDEA and 15 wt% K_2CO_3 at 313 K. The derived enzymatic reaction rate constant k_{enz} (m^3 kg^{-1} s^{-1}) for 15 wt% K_2CO_3 at 313 K was about 9 times higher than for 30 wt% MDEA and 10 times higher than for 30 wt% AMP. Temperature and concentration did not observably influence the enzymatic rate constant in the concentration range of 5 to 15 wt% K_2CO_3. The higher solvent concentration only led to a slightly higher reaction rate. A solution with 20 wt% K_2CO_3 had almost 3 times higher enzyme reaction rate compared to 15 wt% at 298 K and increased with temperature to almost 5 times faster at 328 K. The enzymatic reaction rate for MDEA decreased with both temperature and solvent concentration from 15 to 30 wt%. An increase to 50 wt% resulted in a decrease in reaction rate due to less water present.

Pilot plant campaigns were carried out for different solvents and conditions and the results were successfully modelled using intrinsic data obtained from the wetted-wall column experiments

460384 Gasification Reactivity of Model Refuse Derived Fuel (RDF) Char and Its Components in CO_2 and Steam

Abstract:

The discarded portion of Municipal Solid Waste (MSW) is subjected to pre-processing steps (e.g., glass and metal separation) to produce Refuse Derived Fuel (RDF). For this study, a model RDF composition was established based on the EPA data and pyrolysis results of individual components in different categories, e.g., plastics, paper and food waste. Pyrolysis studies were carried out in a thermogravimetric analyzer (TGA) at atmospheric pressure and at temperatures between 25 â€“ 800 Â°C with a heating rate of 20 K/min. Gasification of pyrolyzed feed was conducted at 800 Â°C using pure CO_2 and steam as the gasification reagents.

Inorganic materials inherently present in the feed have an effect on the gasification reactivity. Alkali and alkaline earth metals were catalytically beneficial, while other inorganics, such as silicon and phosphorous, had an inhibiting effect. The three major inorganics affecting the gasification reactivity in the model RDF feed were potassium from orange peel, calcium from paper components and silicon from paperboard and rubber. Gasification behavior of the model RDF char in CO_2 showed a synergistic effect, which is attributed to the redistribution of inorganic species during the gasification process. Potassium present in the orange peel char catalyzes the gasification of other RDF char components due to its mobility at high temperatures. This observation was verified by comparing the sum of individual gasification reactivities with mixed RDF char composite reactivity. Additionally, EDX of the chars confirmed redistribution of potassium during gasification.

The effect of the inorganics was also evident when steam was used as the gasifying agent. However, the transient reactivity profiles in steam were different as compared to those during CO_2 gasification. Tissue paper (calcium-rich), orange peel (potassium-rich), rubber (silicon-rich) and model RDF were gasified in steam and the reactivity trends were compared to those observed during CO_2 gasification. Both the type and amount of inorganics present had an effect on the gasification reactivity in different gasifying mediums. In addition to the effect of the inorganics, steam gasification reactivity profiles were affected by hydrogen inhibition. Various RDF components rich in different inorganic species thus exhibited varying reactivity trends during steam gasification.

460629 Catalytic Hydrotreatment of Pyrolysis Oil Model Compounds in a Batch Reactor

Abstract:

Bio-oil generated from the pyrolysis of lignocellulosic feedstocks is a complex mixture of highly oxygenated hydrocarbons (>30wt% oxygen), and is poorly miscible with conventional fossil fuels. Therefore, catalytic hydrodeoxygenation (HDO) is necessary to lower its oxygen content and to produce a compatible â€œdrop-inâ€ transportation fuel. We conducted batch HDO reactor studies using representative phenolic compounds found in pyrolysis oil including anisole, m-cresol and phenol, both individually and in mixtures. These tests were conducted at pressures (50 bar) and temperatures (250 to 350°C) close to industrial conditions to obtain high conversion of the initial reactants.Â Experiments were conducted with finely powdered catalysts in a well-mixed Parr reactor to reduce mass transfer effects, and samples were collected regularly until completion. Instead of traditional sulfided HDO catalysts, the studies employed platinum on alumina (Pt/Al_2O_3) and palladium over carbon (Pd/C). Utilizing reaction mechanisms and pathways proposed in our previous studies, the data from the batch reactor experiments was used to fit the kinetic rate constants and Arrhenius parameters for the primary reactants and intermediate compounds. The analytical results show that many of the reactants proceeded through similar intermediates; therefore, a hybrid, lumped-parameter model was created and compared to the data. Higher temperatures led to shorter intermediate lifetimes, and more secondary reactions. While platinum on alumina was primarily focus, tests run with palladium on carbon exhibited less rearrangement of methyl groups. The reaction model developed may then be used to predict product yields, hydrogen consumption, energy requirements, and greenhouse gas (GHG) emissions. This capability is necessary for a more accurate Life Cycle Assessment (LCA) of this biofuel route, and to better assess the economic viability and sustainability of the process. Catalyst coking at longer test times was an issue, and thermal desorption tests followed by mass spectroscopy were performed to determine the rate and extent of the coking process.

467202 Supercritical Fluid Processing of Biomass-Derived Pyrolysis Oil

Abstract:

Liquid hydrocarbon fuels are essential to our current modes of transportation due to their high energy density. However, new technologies are needed to satisfy the demand for these fuels using sustainable resources. Biofuels are an attractive renewable substitute for petroleum-based fuels. The challenge is that woody biomass, although abundant enough to supply a meaningful fraction of the nationâ€™s fuel, is recalcitrant to many chemical processes. Fast pyrolysis to convert biomass into a crude bio-oil has gained recent attention as the basis of a distributed biofuels production capability. Despite promise, the disadvantages of fast pyrolysis are that the resulting bio-oil has low energy density, high water content, acidic pH, and poor storage stability. Vigorous research efforts have been mounted in the past few years to upgrade and stabilize pyrolysis bio-oil. Unfortunately, bio-oil upgrading has been plagued by unacceptable hydrogen consumption, catalyst failure, and inability to process whole bio-oil.

A particularly attractive scenario is to efficiently separate out multiple fractions of bio-oil so that each portion can be treated with a process specifically tailored to the functionalities present in that phase. Bio-oil can be selectively condensed to create multiple fractions, or separated during post-processing. Efficient extractive post-processing is an appealing approach due to the potential for tailoring an extraction for specific components, and facilitating in-situ upgrading where hydrotreating/cracking can be performed during the extraction. Supercritical fluid processing is especially promising for this application due to the ease of solvent recovery, good solvency for organic chemicals and the flexibility of tailoring the solvent for desired chemical fractions. Results of supercritical fluid processing of biomass-derived pyrolysis oil will be presented. This includes characterization of the fractions obtained during various processing conditions as well as conversion strategies tailored to these specific fractions.

462128 Commodity Chemicals from Lignocellulosic Biomass: Conversion of Furfural into 1,5-Pentanediol

Abstract:

Lignocellulosic biomass is a low cost, abundant, and renewable feedstock that can be used to produce commodity chemicals. α,ω-diols like 1,4-butanediol (BDO) and 1,6-hexanediol (HDO), are commodity chemicals produced at high volume (annual market of 2,500,000 ton and 138,000 ton, respectively) and high value ($2,200/ton and $4,400/ton, respectively). In this presentation we will discuss the production of 1,5-pentanediol (PDO) from furfural.

Furfural, is one of the most important platform chemicals produced from biomass. It is commercially produced from the hemicellulose portion of biomass by hydration to xylose and subsequent dehydration (Xing et al., 2011). Furfural can then undergo hydrodeoxygenation to produce PDO with tetrahydrofurfuryl alcohol (THFA) as a precursor. The furfural is first hydrogenated to THFA (Nakagawa and Tomishige, 2010; Nakagawa et al., 2012). The resulting THFA can then be converted into PDO by a Rh-ReOx/C catalyst at high selectivity (above 97.2% selectivity). However, the bottlenecks for this process are: (1) the production of PDO at low conversion; (2) the high catalyst cost resulting from using an expensive noble bimetallic RhRe catalyst; and (3) the high capital cost of reaction and separation systems resulting from low feedstock loading in the liquid phase.

In this work, we propose a novel multi-step catalytic process to convert furfural into PDO at high feedstock loading (up to 50 wt% in water) using non-precious metal catalysts. We also design a separation subsystem for the recovery of the final PDO product from process water. We have developed an experimentally based process simulation model (mass and energy balances) to determine the economic potential of PDO production. It is shown that this strategy leads to much lower catalyst costs and a higher overall PDO yield (over 80%) comparing to one-step THFA hydrogenolysis (Chia et al., 2011) using a noble metal catalyst. If we consider a plant processing 135 ton per day furfural (equivalent to 1,000 dry metric tonne per day white birch) feedstock and producing 38,520 ton per year PDO, the integrated strategy leads to a production cost of $1,600 per ton of PDO. We estimate that this cost is over 35% lower than the cost of production of PDO via a noble metal catalyst. Finally, we perform sensitivity analysis to identify the major economic drivers of the process and suggest future research directions.

References

1. Chia, M., Pagán-Torres, Y. J., Hibbitts, D., Tan, Q., Pham, H. N., Datye, A. K., Neurock, M., Davis, R. J., Dumesic, J. A. (2011). Selective Hydrogenolysis of Polyols and Cyclic Ethers over Bifunctional Surface Sites on Rhodium–Rhenium Catalysts. J. Am. Chem. Soc., 133 (32), 12675.
2. Nakagawa, Y., Nakazawa, H., Watanabe, H., Tomishige, K. (2012), Total Hydrogenation of Furfural over a Silica-Supported Nickel Catalyst Prepared by the Reduction of a Nickel Nitrate Precursor. ChemCatChem, 4: 1791–1797.
3. Nakagawa, Y., Tomishige, K. (2010). Total hydrogenation of furan derivatives over silica-supported Ni–Pd alloy catalyst. Catalysis Communications, 12 (3), 154-156.
4. Xing R., Qi W., Huber G. W., (2011). Production of furfural and carboxylic acids from waste aqueous hemicellulose solutions from the pulp and paper and cellulosic ethanol industries, Energy Environ. Sci., 4, 2193.

466868 Cyclization and Dehydration of Aldaric Acids to 2,5-Furandicarboxylic Acid

Abstract:

Aldaric acids (2,3,4,5-tetrahydroxy-1,6-hexanedioic acids) are formed by catalytic oxidation of glucose or by oxidation of uronic acids, the monomers of pectin and alginate biopolymers. Aldaric acids are emerging as important biorenewable intermediates for use in consumer goods and as platforms for further conversion. The direct cyclization and dehydration of glucaric acid and galactaric acid and their esters to form 2,5-furandicarboxylic acid (FDCA) has been carried out. The production of FDCA is attracting a great deal of attention, as it is a monomer of polyethylenefurandicarboxylate (PEF) polymers, which exhibit superior barrier properties to PETE packaging materials. Aldaric acid cyclization to FDCA takes place in an anhydrous environment in the presence of strong acids; we have found that under some conditions the presence of an ionic liquid solvent (1-butyl-3-methylimidazolium with various anions) enhances the yield of FDCA. An optimized set of reagent compositions and reaction conditions have been identified to give FDCA yields approaching 60% of theoretical from galactaric acid. Undesired reaction pathways include lactonization of the aldaric acids under anhydrous conditions, and total dehydration to form carbonaceous residues. A method to recover pure FDCA directly from the reaction mixture has been developed, along with recycling of the ionic liquid and acids. This new route to FDCA avoids the use of difficult-to-handle intermediates such as hydroxymethylfurfural (HMF), and appears promising given the recent advancements in commercialization of the aldaric acid precursors.

472905 Thermo-Catalytic Conversion of Biomass Derived Lipids to Fuels and Chemicals over HZSM-5

Abstract:

In this study, thermo-catalytic cracking of triglyceride was carried out in a continuous pyrolysis system with feed injected through an atomizer. This allowed introduction of micron-sized droplets of oil that could be rapidly vaporized inside the hot reactor. With this novel design, we were able to achieve short vapor residence times (optimum Ï„$_{vap}$ was 20s in order to achieve high liquid product yeilds and triglyceride conversion) without use of carrier gas, which would significantly reduce the overall cost of pyrolysis. Effects of reaction temperature (450<T_{rxn}<500 Â°C) and catalyst loading (HZSM-5; 0-5g) on conversion, product yields and composition were investigated. At T_{rxn} =500 Â°C and without catalyst, nearly quantitative conversion of feed was achieved and the yield of pyrolysis liquids was as high as 89% (relative to feed mass). Under these conditions, the identified products consisted of 37% hydrocarbons (C_5-C_{17}), 35% long-chain fatty acids (C_{16}-C_{18}, but primarily oleic acid) and 10% short-chain fatty acids (C_6-C_{12}). At T_{rxn} =500 °C using 5g of HZSM-5, the yield of liquid products was 65%. The identified liquid products contained 73% aromatics (C_6-C_{12}, but primarily benzene, toluene and xylene), 5% aliphatic and only 4% long-chain fatty acids. Overall, our results demonstrate the feasibility of producing liquid products at high yields, including a wide range of fuels (gasoline and jet) and enriched oleic acid (for oleochemicals production) with integration of conventional downstream separations, such as distillation.

469768 An Ontology Supported Integration Framework for Models and Data in Biorefining

Abstract:

Process modelling and simulation are vital tools to plan, evaluate, assess and develop different alternatives in the biorefinery design for production of value added chemicals, materials, fuels, and energy from biomass. However, modelling for biorefining processes poses a number of technical challenges due to the complexity of the characteristics of biomass feedstock, and associated processing technologies evolving multiple pathways, including thermochemical-, catalytic-, biochemical- and hybrid-conversion routes. Due to the complexity and variety of these processes, numerous software tools have been developed to obtain the best-of-breed software for each application area, to fulfil the requirements of particular engineering tasks. During the process of developing a new model, the best suited modelling tools for different parts of the process are employed. However, to understand process design as a continuous work process from an integrated perspective requires the use of tools from a diverse range of sources and disciplines simultaneously. With the current increasing number of Computer Aided Chemical Engineering (CAPE) tools, the emergence of interoperability between heterogeneous modelling tools and process simulators was demonstrated by Braunschweig et al. (2000), which has ultimately led to the CAPE-OPEN initiative.

The CAPE-OPEN has conceptualised and developed a standard to define a set of interface specification as a method pertaining to the interoperability of the mainstream flowsheet applications (Pons 2010). The interoperability is facilitated by a standard middleware service, which enables the use of a single set of common interface as a communication tool (Braunschweig et al. 2004; Morales-Rodríguez et al. 2008). Having the common interface increases the potential reusability of existing models without having to invest in the software development, which further reduce the developing time, efforts, and the need for expertise (van Baten & Pons 2014). However, to fully exploit the potential reusability of existing models, the task to become aware of and have access to all of those models from a variety of sources is a challenge, yet required in order to identify the most suitable model to achieve the best solution for a particular engineering problem. As a consequence, Yang et al. (2008) indicated the potential misuse of the models that can further lead to inefficient or even wrong solution to the problem.

This paper is intended to address the challenges that previous research, CAPE-OPEN framework has faced and hence to minimise user intervention in selecting a model for the purpose of model integration. To enable automated discovery process in order to reduce the reliance of user intervention, the semantic representation is used to describe the models and data in terms of their input and output parameters, preconditions, and functionality (Koo & Cecelja 2015). Additionally, the Semantic Web Services (SWS) are employed to create a semantic profile of models and data representing biorefining processes. To this end, the domain ontology assures unified description of models by guiding the model registration and hence instantiation of ontology through ontology parsing, whereas the SWS assures the process of model discovery, its invocation and consequently integration. To demonstrate proposed approach, the framework is built for the models and data residing in web-based semantic repository in the domain of biorefining. In the domain ontology, the models are characterised in terms of functionality, modelling methods, model granularities, required software environment, etc. and their input and output parameters are characterised in terms of type (i.e. material, utility, economic and emissions), compositions and associated parameters, etc. The classification and characterisation of parameters used in matching process are not limited to numerical values (i.e. value for the flow rate), but also includes descriptive (i.e. type of material, energy) or even composite (i.e. range of flow rate with minimum and maximum values) types. The options of the discovered models to the requesting model, the input-output (I/O) matching technique is employed to identify the models that matches to a certain degree of request, which permits for a high degree of flexibility by allowing a partial matching rather than a full (exact) matching between I/O parameters. The discovered models from matching process are then ranked according to a set of preferences and the best option that satisfies the matching criteria is proposed to the user. To verify the performance of model discovery and selection, models representing biochemical- and thermochemical conversion processes are used to demonstrate as a case study with an aim to evaluate the best suited model whilst avoiding wrong design solution.

References:

Braunschweig, B.L. et al., 2000. Process modeling : The promise of open software architectures. Chemical engineering progress, 96(9), pp.65–76.

Braunschweig, B. et al., 2004. CAPE web services: The COGents way. Computer Aided Chemical Engineering, 18(C), pp.1021–1026.

L. Koo, F. Cecelja, "Model Integration Using Ontology Input-Output Matching," Computer Aided Chemical Engineering, 2015, volume 37, pp. 2567-2572.

Morales-Rodríguez, R. et al., 2008. Use of CAPE-OPEN standards in the interoperability between modelling tools (MoT) and process simulators (Simulis® Thermodynamics and ProSimPlus). Chemical Engineering Research and Design, 86(7), pp.823–833.

Pons, M., 2010. How to make use of CAPE-OPEN? In 2010 AIChE Annual Meeting, 10AIChE.

van Baten, J. & Pons, M., 2014. CAPE-OPEN: Interoperability in Industrial Flowsheet Simulation Software. Chemie Ingenieur Technik, 86(7), pp.1052–1064. Available at: http://doi.wiley.com/10.1002/cite.201400009.

Yang, a. et al., 2008. A multi-agent system to facilitate component-based process modeling and design. Computers and Chemical Engineering, 32(10), pp.2290 2305.

466202 A Biorefinery Boutique Made in Switzerland

Abstract:

Many challenges pave the road from an oil-based to a renewable economy. In the chemical industry, a substantial amount of work needs to be undertaken for renewable-based processes to compete with their petrochemical counterparts. Nowadays, the biggest hurdle is to evaluate which biorefinery concepts are environmental friendly and economic competitive at an early stage. Therefore, an innovative exploration was undertaken to create a biorefinery boutique for Switzerland and Europe for different future industrial scenarios.

Traditionally, there are two approaches to find biorefinery concepts. The first is to use overly simplified models, which can be analyzed according to their production portfolio and evaluated based on their profitability. In the second, detailed process simulations are used, which give detailed technological insight into each technology, but cannot be connected with each other due to their high complexity. This work employs a novel technique which connects detailed model biorefinery technologies in a network and finds tailor made biorefinery concepts for multiple future business scenarios.

Besides state of the art bioprocessing technologies, the biorefinery boutique incorporates life cycle assessment to indicate environmental issues and an economic assessment to project future profitability.

460997 New Short-Cut Tools for Early-Stage Investment Evaluation of Biorefineries

Abstract:

New short-cut tools for early-stage investment evaluation of biorefineries

Biorefineries offer a promising alternative to fossil-based processing industries and have met rapid development during the last years. Biorefinery processes employ state-of-the-art technologies and thus, represent high risk business decisions. Researchers undertaking the development of a biorefinery, often have to estimate capital and manufacturing costs based on minimum information, in order to decide wisely on project continuity and justify further funding for their research. Most R&D engineers rely on literature information to estimate the costs as historical cost data are proprietary information and are seldom announced. They also draw on various costing techniques and heuristics, which were developed for the needs of the petrochemical and the Oil & Gas Industry and require high level of process detail, which is not available at the early-stages of process conception.

We have evaluated some of these methods with means of both deterministic and statistical comparison from reported literature estimates and commercial biorefineries' cost data. As the majority were published during the 1970's-80's and were derived from petrochemical processes, they do not account for the technological progress often met within the state-of-the-art biorefinery processes and report discrepancies in their results. Fig. 1 illustrates the aforementioned discrepancies for dry corn mill ethanol biorefineries. We have been collecting capital cost data for commercial corn-to-ethanol plants (marked as circles). Only three out of the six cost methods fall within the 90% Confidence Intervals of the best regression line (marked as triangles), while the capital estimation reported in literature reference greatly underestimates the plant's final cost (marked as x). Our work reports similar figures for chemical and thermochemical biorefineries, as well CITATION Tsa16 \l 2057 [1].

Therefore, we decided to attempt new capital and manufacturing cost estimation methods to meet the needs of non-experienced cost estimators which undertake the design of first-of-a-kind biorefineries: the methods require information available at the start of the process conception. The first capital cost estimation method proposes new investment factors that draw on historical investment costs along with probabilistic estimation of the total investment. The second capital cost estimation method proposes cost estimation relationships for chemical, biochemical and thermochemical biorefineries, as well as simple techniques to define the uncertainty of the estimate. Finally, we have developed a new rapid manufacturing cost estimation methodology with uncertainty: it relies on factors requiring minimal user input, i.e. raw materials cost calculation, for determining the net production cost of the process, employing probabilistic means of determination.

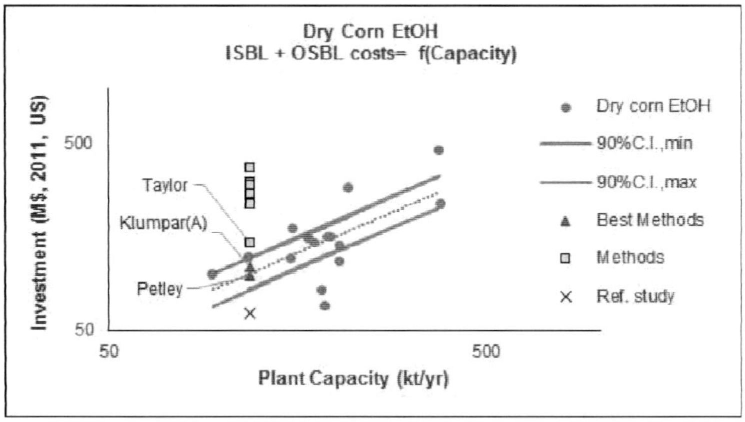

Fig. 1. Dry corn mill bioethanol production plants Investment Costs (M$, 2011,US) vs Plant Capacity (kt/yr)

Acknowledgements

The European Commission is gratefully acknowledged for funding the Renewable Systems Engineering project (RENESENG - FP7 Marie Curie project).

BIBLIOGRAPHY

[1] M. Tsagkari, J.-L. Couturier, A. Kokossis and J.-L. Dubois, "Early-stage capital cost estimation of biorefinery processes: a comparative study of heuristic techniques," *ChemSusChem*, 2016 *(under review)*.

466913 Global Sensitivity Analysis of Economic Assessment of Early Stage Process Design: The Case of the Glycerol Biorefinery

Abstract:

Nowadays it has become progressively important to achieve a sustainable process performance in order to keep a viable and competitive advantage in the international markets. Thus, the development of all-inclusive/holistic/comprehensive and systematic methods to accomplish this goal is the topic of this study. To this end, a multi-level framework for early-stage design and screening of process options is proposed for techno-economic and environmental sustainability analysis through risk assessment. The alternatives composing the design space are assessed by following the framework's steps, targeting to (i) quantify the economic risk; (ii) quantify the potential environmental risk under uncertainties by using monetary valuation of environmental impacts techniques; (iii) measure the concepts' eco-efficiency; and, finally (iv) facilitate the decision-making procedure by interpreting the trade-offs through a qualitative and quantitative sustainability risk assessment matrix. The framework application is highlighted through the critical assessment and screening of early stage glycerol-based biorefinery design concepts. It was found that the best potential options for glycerol valorisation is are obtained through the production of either: (i) lactic acid; (ii) succinic acid; or, (iii) 1,2-propanediol. To reduce the potential economic risk, a multi-product biorefinery is suggested, whose strategy is based on its capability of switching between the production of lactic acid and succinic acid.

The risk assessment tool described above provide valuable information on how uncertainties (technical versus external such as feed/product prices) impact the decision of optimal process concept and how to better mitigate these impacts by improving technology development. In this contribution, we take one step further and ask the question, a given a certain economic risk associated with a conceptual process alternative, what is the contribution of input uncertainties((e.g. feed composition, yield, feedstock prices, product prices, etc)) to the risk? If risk is defined as variance (deviation from an expected value), what is the individual variance attributed to each input uncertainty? And how we can use this information to further minimize economic risk and improve technology advancement? To answer these questions, we use sensitivity analysis methods including local methods such as one-factor-at-a-time (OAT) as well as global methods such as variance decomposition to compute first order as well as total sensitivity indices.
We implemented algorithms in Matlab (R2015a) based on Monte Carlo sampling using Latin Hypercube Sampling with Iman Conover correlation control (Sin et al 2009) and sampling based on conditional probability of dependent variables proposed by Kucherenko et al 2012 to take into account the correlation structure of input uncertainties (especially price correlation matrix). The economic model used in this work is the discounted cash-flow rate of return (DCFR). The OAT sensitivity analysis has shown that deviations in the product's and feedstock prices, total production cost, fixed capital investment as well as discount rate, among others, have a high impact on the project's profitability (quantified with NPV and MSPA global sensitivity analysis of NPV indicated two major factors contributing to its variance (risk): uncertainty in fixed capital cost estimation (82%), and, uncertainty in product prices (14%). The NPV meta-model showed that improving the accuracy of fixed capital cost estimation (5% *std*) leads to a 95% reduction in the NPV's uncertainty. This global sensitivity analysis confirms the process engineering expectation that more detailed capital cost estimation supported through pilot scale studies or more rigorous model analysis is expected to decrease the economic risk and better judge the transferability of technology to the market. Global sensitivity analysis is an important and complementary tool to study and decompose impact of uncertainties to economic assessment of conceptual process design studies.

Keywords: techno-economic assessment, uncertainty analysis, glycerol-based biorefinery concepts, risk-based decision making, global sensitivity analysis.

References:

Kucherenko, S., Tarantola, S., & Annoni, P. (2012). Estimation of global sensitivity indices for models with dependent variables. *Computer Physics Communications*, *183*(4), 937-946.

Sin, G., Gernaey, K. V., & Lantz, A. E. (2009). Good modeling practice for PAT applications: Propagation of input uncertainty and sensitivity analysis. *Biotechnology progress*, *25*(4), 1043-1053.

Cheali, P., Gargalo, C., Gernaey, K. V., & Sin, G. (2015). A Framework for Sustainable Design of Algal Biorefineries: Economic Aspects and Life Cycle Analysis. In *Algal Biorefineries* (pp. 511-535). Springer International Publishing.

460678 General Bio-Separation Superstructure Optimization Framework

Abstract:

General Bio-Separation Superstructure Optimization Framework

Wenzhao Wu, Kirti Yenkie, Christos T. Maravelias*

Dept. of Chemical and Biological Engineering, University of Wisconsin-Madison, Madison, WI 53706

Recent advances in synthetic biology and metabolic engineering have enabled the production of a range of chemicals using engineered microorganisms[1,2]. However, despite the intriguing potential, the biological production of high-value chemicals is challenging because it is likely to have low titer and substantial amounts of excreted byproducts, and the purity specifications for high-value chemicals, as opposed to fuels, are rather strict, which means that separation costs are likely to represent a large fraction of the total production cost (more than 70% of total cost[3]). Thus, the efficient synthesis of bio-separation processes becomes a critical task. Although this synthesis problem has been studied for various chemicals in the past, these studies were mostly performed on a case-by-case basis. There has been limited research towards the development of systematic methods for bio-separations, applicable to all chemical targets. Accordingly, the goal of the present work is to develop a general bio-separation superstructure optimization framework, aiming to provide guidance on the preliminary synthesis of separation networks.

Based on general bio-separation principles and insights obtained from industrial separation processes for specific products, we first identify four separation stages: (1) cell treatment, (2) product phase isolation, (3) concentration and purification, and (4) final refinement. Specifically, cell treatment starts with the harvesting of cells (if the product is intracellular), which can be performed using solid-liquid separation technologies. Cell harvesting is followed by cell disruption to release the intracellular products. The first stage is skipped if the product is extracellular, and the separation starts from the second stage - product phase isolation, where the product-rich phase is isolated. In the concentration and purification stage, we remove large amount of water along with other impurities, by utilizing the differences between the product and the other components on volatility, molecular size, diffusivity, solubility in solvents, etc. In the final refinement stage, we further remove trace impurities and perform refined operations to satisfy special product specifications, such as colorlessness, complete dryness, and crystal form.

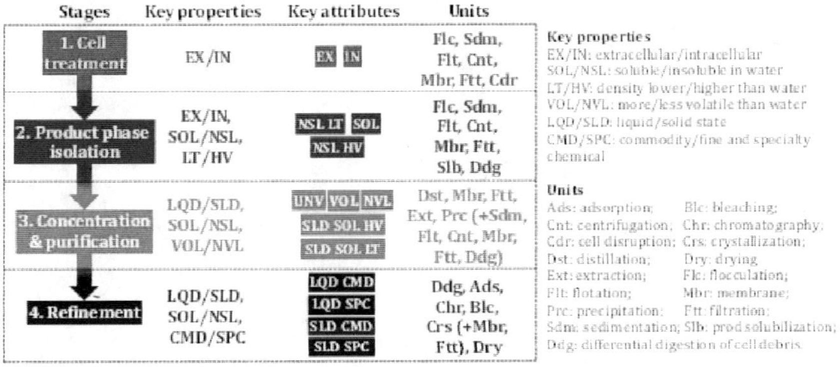

Figure 1. Stage-wise analysis of bio-separation processes, including key product properties, attributes, and units for each stage.

Next, we identify key product properties and key attributes that affect the separation processes in each stage, as shown in Figure 1. For example, in Stage 2, the different attributes imply different locations of the product (e.g., "NSL LT" indicates that the product is located at the top, while "SOL" indicates that the product is dissolved in water and evenly distributed in the system), and hence different separation processes. Major units in each stage are also identified.

Then for each stage, we systematically implement a set of connectivity rules[4] to develop attribute-specific superstructures. They are subsequently combined to generate a stage-specific superstructure. Finally, all the four stage-specific superstructures are integrated into a general superstructure (shown in Figure 2) that accounts for all types of bio-chemical products.

We further develop a superstructure reduction method to solve product-specific instances, based on product type, unit availability and suitability, case-specific considerations, and final product specifications. The reduced superstructure for an example case is shown in Figure 2, where an EX NSL LT NVL LQD CMD product (see abbreviations in Figure 1) is required to be completely colorless in its final product form, and all units in the general superstructure are available except for filtration.

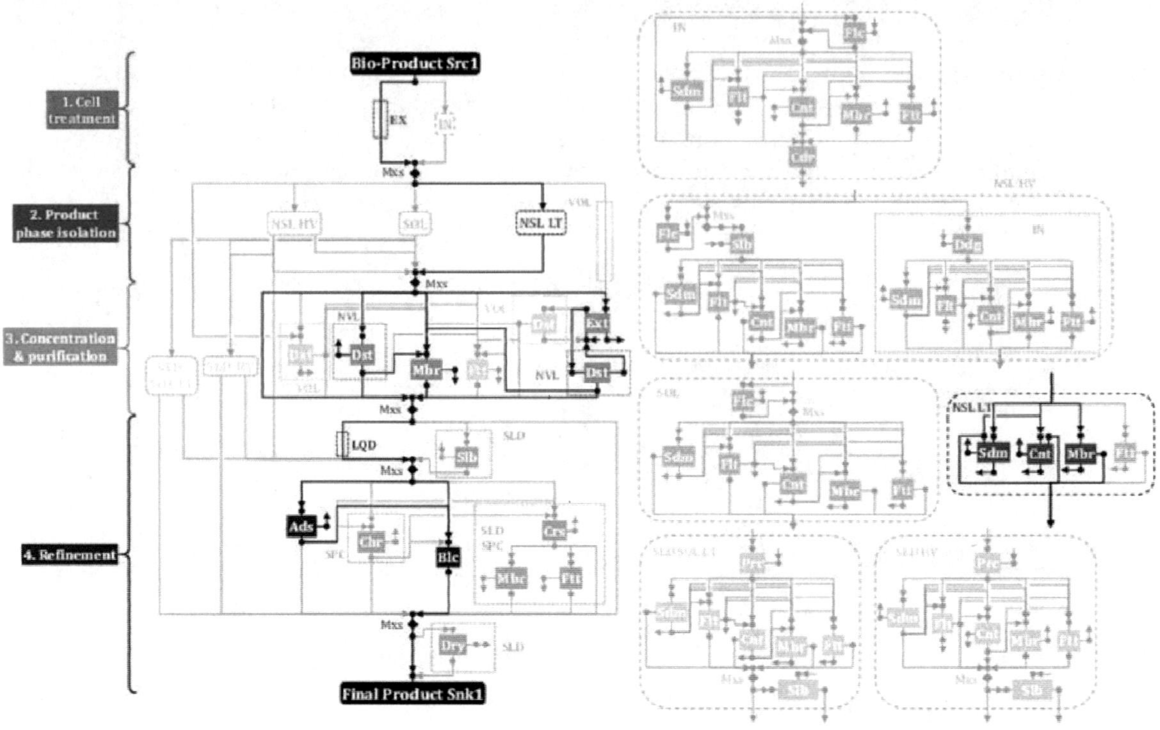

Figure 2. The general bio-separation superstructure (including the "dimmed" parts), and the reduced superstructure (excluding the dimmed parts) for an example case, where an EX NSL LT NVL LQD CMD product is required to be completely colorless in its final product form, and all units in the general superstructure are available except for filtration.

Finally, a general MINLP optimization model, including short-cut unit models (e.g., Fenske-Underwood equations for distillation unit models) for all types of separation units considered in the framework, is formulated. In the modeling of a specific reduced superstructure, only models of relevant units (which exist in the reduced superstructure) are included.

The proposed framework is applied to three case studies. Each case considers different set of product types, unit availability and suitability, case-specific considerations, and final product specifications. The common objective is to minimize total cost of purifying a dilute product stream (5 wt% concentration) to a product stream with at least 90 wt% purity.

References

[1] Gavrilescu, M. & Chisti, Y., 2005. Biotechnology- a sustainable alternative for chemical industry. *Biotechnology advances,* Volume 23, pp. 471-499.

[2] Bornscheuer, U. T. & Nielsen, A. T., 2015. Editorial overview: chemical biotechnology: interdisciplinary concepts for modern biotechnological production of biochemicals and biofuels. *Current Opinion in Biotechnology,* Volume 35, pp. 133-134.

[3] Kiss, A. A., Grievink, J. & Rito-Palomares, M., 2015. A systems engineering perspective on process integration in industrial biotechnology. *J Chem Technol Biotechnol,* Volume 90, pp. 349-355.

[4] Wu, W., Henao, C. & Maravelias, C., 2016. A Superstructure Representation, Generation and Modeling Framework for Chemical Process Synthesis. *AIChE J.*

459851 Technoeconomic Study of AB Synthesis for Biobutanol Production

Abstract:

Over the last 15 years there has been an important effort towards climate change mitigation. Among the different actions, one of the more challenging is reducing emissions in the transport sector. Ethanol and biodiesel, from different biomass sources, represent the typical alternatives to substitute crude based gasoline and diesel nowadays. However, there are other options such as FT fuels (Martín & Grossmann, 2011), glycerol ethers (Martín & Grossmann, 2014) or biobutanol. Biobutanol is considered an advanced fuel. It has similar properties to gasoline, it can be directly used in gasoline engines without modification, and it has higher energy density and boiling point than ethanol reducing evaporation losses. Moreover, it is less hydroscopic than ethanol (Durre, 2007).

The ethical issues involving first generation biofuels have led to focus on obtaining biofuels from agricultural residues. Corn being the major crop in the world, results in the fact that corn stover is widely available. However, in the case of biobutanol, the yield has been low due to the fact that together with biobutanol, ethanol and acetone are typically obtained. Thus, this work is divided in two parts: the first part is devoted to the experimental evaluation of the lignocellulosic biomass degradation. In the second one, a technoeconomic study of a new production process is described. The novelty of this work relies on the scale up of AB fermentations, that does not produce ethanol, (Qureshi et al., 1999). In addition, in-situ recovery techniques such as evaporation are also included at the process to make it economically competitive with conventional fuels and bioethanol.

In the first stage of the work, comprehensive evaluation of the experimental conditions to break down switchgrass and corn stover from Castille and Leon region (Spain) has been performed. The study considers both pretreatment and hydrolysis. With regards to pretreatment, several physio-chemical techniques such as acid (H2SO4), alkali (using KOH or NH4OH), ultrasounds processing, surfactants or organic solvents are tested. With regard to hydrolysis, various enzyme cocktails were used. It turns out that corn stover reaches the highest yield, 677.00 mg of sugars by gram of biomass, when it is pretreated using a diluted sulfuric acid solution 0.5% v/v, autoclaving the mixture during 90 min at 121 ºC with no sonication stage. The hydrolysis method is set up at 50 ºC and 150 rpm with a solid loading of 2.5 % w/v and a ratio of enzymes/biomass of 400 and 440 µl/g of NS50013 (cellulose complex) and NS50010 (β-glucosidase cocktail) respectively. The sugar recovery obtained is competitive with current best available technologies i.e. (Sundaram and Muthykumarappan, 2016).

In a second stage, we synthesize a process for the production of biobutanol from corn stover including the use of novel technologies such as pervaporation to improve the yield from biomass and reducing the by-products by means of AB fermentation. The processes consists of four sections, pretreatment, hydrolysis, fermentation-pervaporation system and product purification. We use the experimental results to determine the yield of the pretreatment and hydrolysis. A hybrid simulation using EXCEL to develop black box kind of models for the experimentally validated units and ChemCAD has been performed to evaluate the technoeconomic feasibility of this process.

The results show an investment of 186 MM€ and a butanol production cost of 1.09 €/kg for a 27 kt/y plant. The yield to butanol is 0.21 kg/kg of biomass resulting, altogether, in a fuel closer to become competitive versus second generation bioethanol and fossil based fuels in terms of yield and cost.

References

Durre, Biobutanol: an attractive biofuel Biotech. J., 2 (2007), pp. 1525–15234

M.S. Madihah, A.B. Ariff, K.M. Sahaid, A.A. Suraini, M.I.A. Karim (2001) Direct fermentation of gelatinized sago starch to acetone–butanol–ethanol by Clostridium acetobutylicum. World J. Microbiol. Biotechnol. 17 (2001), pp. 567–576

Martín, M., Grossmann, I.E. (2011) Process optimization of FT- Diesel production from biomass. Accepted Ind. Eng. Chem Res, 50 (23),13485–13499

Martín, M.; Grossmann, I.E. (2014) Simultaneous dynamic optimization and heat integration for the co-production of diesel substitutes: Biodiesel (FAME & FAEE) and glycerol ethers from algae oil. Ind. Eng. Chem. Res. 53, 11371-11383

Qureshi, N., Hans P. Blaschek. Production of Acetone Butanol Ethanol (ABE) by a Hyper-Producing mutant strain of Clostridium beijerinckii BA101 and recovery by Pervaporation. Biotechnol.Prog. 1999, 15, 594-602.

Sundaram, V., Muthukumarappan. K., Influence of AFEXTM pretreated corn stover and switchgrass blending on the compaction characteristics and sugar yields of the pellets. Industrial Crops and Products 83 (2016) 537–544.

465757 Effect of Market and Technical Parameter Uncertainties on the Optimal Design of Integrated Biorefineries

Abstract:

Effect of Market and Technical Parameter Uncertainties on the Optimal Design of Integrated Biorefineries

As research is exploring possibilities for developing renewable bio-based fuels and chemicals, there arises a need for development of strategies which can design sustainable value chains that can be scaled up efficiently and provide tangible net environmental benefits. Recent ventures in biofuel production have been fraught with corporate failures. A driving reason for these unsuccessful ventures, in part is governed by the lack of proper planning in designing plants and supply networks. Modeling strategic and operational level decision processes of an enterprise which produces renewable-based fuels and chemicals, using engineering and financial tools, can provide valuable insight into the inter-play of technology and process interactions for different product streams.

In this study we present the development of a comprehensive decision support tool, within which the business value of a multiproduct biorefinery is optimized by considering all types of uncertainties including uncertainties at strategic and operational level. A stochastic linear model is first developed to optimize production capacity of the plant for the desired planning horizon and then process simulation coupled with stochastic optimization algorithm is employed to optimize the operating condition of the plant under uncertainty. The proposed methodology is based on an iterative framework that utilizes systems-based optimization in conjunction with detailed mechanistic modelling and simulation of the process. Market uncertainties are taken into account at the strategic planning level, and uncertainties related to parameters characterizing the processing technologies are addressed in operational level optimization. Monte-Carlo based simulation and global sensitivity analysis are utilized to identify the most critical parameters and optimize the operating conditions of the plant accordingly. Additionaly, risk management strategies are introduced to the framework for explicit treatment of strategic and operational risks. The other key feature of our approach is the development of

dynamic data exchange strategy embedded as part of the process simulation to incorporate the complex kinetics of bio-reactions in the simulation model. The final results of the proposed framework include strategic capacity plan, optimal expected NPV, and optimal operating conditions of the process.

461131 Resource Recovery from Municipal Solid Wastes: A Case Study in the UK

Abstract:

Increasing waste generation in the UK has called for a sustainable municipal waste management strategy. It is imminent to reduce the volume of wastes going into landfill to avoid permanent resource loss, health impacts and environmental pollution. Prevention of waste generation is the most preferred option, however this is not easy to achieve in practice. Resource recovery from wastes by numerous treatment technologies is the second preferred option for not only to eliminate landfilling, but also to close the loop for a circular economy. Systematic techno-economic analysis and environmental impact assessments of various mechanical biological treatment (MBT) configurations for recovering metals, fuel and energy as resources from municipal solid wastes (MSW) are essential for selecting the most sustainable option. The systems under consideration include MBT systems with aerobic and anaerobic digestion (non-thermal), processes as well as incineration (thermal) processes. In the present study, MBT plant with aerobic digestion (windrows and in-vessel compositing); MBT plant with aerobic digestion and combined heat and power (CHP), with different degree of RDF and metal separation, with and without RDF incineration; and MSW incineration and CHP have been investigated. The MBT system with anaerobic digestion coupled with biogas and RDF CHP has shown promising economic performance giving an additional benefit of 113.5 Euro/t from the electricity generation from RDF incineration and CHP, that is almost 18 times increase compared to the case without electricity generation from RDF incineration and CHP, at an additional capital investment cost of 18 Euro/t. MSW valorisation through recovering metals such as aluminium, zinc and copper has the potential of generating additional revenue of 152 Euro/t. It has been identified that recovered aluminium is the main contributor to the economic value while recovered copper has the highest avoided environmental impact. Metals containing in the residues from mechanical separation worth 1.8 Euro per tonne of MSW could be lost if they are not recovered. Higher economic incentives and environmental impact savings for MSW treatment can thus be realised through effective resource recovery strategies, demonstrated in this work.

471007 The Case of Making Fuel Oil from Sanitary Sewage

Abstract:

What if we make fuel oil from our sanitary sewage? The ramifications of this question posed a decade ago turned out in various research areas such as resource recovery, wastewater treatment, chemical reaction engineering, catalysis, bioprocessing, separation processes, modeling-simulation, systems integration, and supply chain. The enhancement and extraction of oil from microbial consortia of activated sludge (AS) generated by wastewater treatment facilities have been studied with the potential as viable platform for organic waste-to-fuels and organic waste-to-chemicals transformation. The fundamental principles of this technology emulate the principles of single cell oil technology and resource recovery from wastewater treatment. The technology was established in 2005 through a project on biodiesel and value-added chemicals extraction from wastewater sludges [1]. Significant amounts of AS microbial oil were extracted and trans-esterified to biodiesel [2], but the enhancement of oil content is needed to make the technology economically feasible [3]. The oil content of AS was then enhanced using the principles of single cell oil technology in which high carbon to nitrogen ratio (mole C/mole N) triggers generation of structural lipids (phospholipids, sphingolipids, etc.) and reserve lipids (mainly TAGs) [4]. With sugars such as glucose and xylose as model carbon sources, researchers established that activated sludge oil can be enhanced and that the microbial process has similar microbial kinetics to that of a single cell oil accumulation [5, 6]. The commercialization of the aerobic lipid accumulation by AS, however, is challenged by the relatively high cost of the typical sugar substrates glucose and xylose. Several studies explored alternative carbon sources for microbial lipid accumulation and suggested the use of short chain fatty acids (SCFAs) such as acetic acid, propionic acid and butyric acid as substrates [7, 8]. These SCFAs are considered cost-effective because they are intermediate compounds in anaerobic digestion, a common wastewater treatment process [9]. Recent studies indicated that the microbial consortia of AS feeding on acetic acid (an SCFA) can accumulate lipids [10, 11]. This talk assesses the maturity of the technology in terms of the findings and the pertinent questions yet to be answered. On-going studies are also presented in order to set the immediate directions of some active research paths.

[1] Zappi ME, French WT, Hernandez R, Dufreche ST, Sparks DJ. Production of biodiesel and other valuable chemicals from wastewater treatment plant sludges Mississippi State University, US Patent, 2009.

[2] Dufreche S, Hernandez R, French T, Sparks D, Zappi M, Alley E. Journal of the American Oil Chemists' Society 2007;84(2):181-187.

[3] Mondala A, Liang K, Toghiani H, Hernandez R, French T. Bioresource Technol. 2009;100(3):1203-1210.

[4] Papanikolaou S, Aggelis G. European Journal of Lipid Science and Technology 2011;113(8):1031-1051.

[5] Mondala A, Hernandez R, Holmes W, French T, McFarland L, Sparks D, Haque M. AICHE J. 2013;59(11):4036-4044.

[6] Mondala AH, Hernandez R, French T, McFarland L, Santo Domingo JW, Meckes M, Ryu H, Iker B. AICHE J. 2012;58(4):1279-1290.

[7] Fei Q, Chang HN, Shang L, Choi J-d-r, Kim N, Kang J. Bioresource Technol. 2011;102(3):2695-2701.

[8] Fontanille P, Kumar V, Christophe G, Nouaille R, Larroche C. Bioresource Technol. 2012;114:443-449.

[9] Chang HN, Kim N-J, Kang J, Jeong CM. Biotechnol. Bioprocess Eng. 2010;15(1):1-10.

[10] Fortela DL, Hernandez R, French WT, Zappi M, Revellame E, Holmes W, Mondala A. Renew. Energ. 2016;96, Part A:11-19.

[11] Fortela DL, Hernandez R, Zappi M, WT French, Bajpai R, Chistoserdov A, Revellame E, Holmes W. J. Bioprocess. Biotech. 2016;6(4):275.

473515 Synthetic Diesel Production through Catalytic Pyrolysis of Biomass-Waste Tire Mixtures

Abstract:

The continuance of the modern world luxurious life-style, threatened by imminent depletion of cheap crude oil reserves, demands both the identification of alternative, sustainable, dense, cheap energy resources and means for effective waste reuse and disposal. This is of particular importance in the context of Jet fuels, for which energy demands are significant. The pyrolysis of waste tire rubber (WTR) results in the production of a pyrolytic oil (PO) in many ways similar to the crude exploited from fossil resources. PO has become a promising product since it addresses all of the aforementioned demands simultaneously. The yields and composition of this PO are strong functions of the production conditions and feedstock. In this work, the reaction conditions for PO synthesis using pulverized WTR alone and a mixture of WTR with sugarcane bagasse (30 and 70% w/w respectively) were experimentally assessed. Slow heating rates of approximately 5.5ºC/min were applied to 500 g of the feedstock in a batch reactor at a nitrogen flow rate of 2 L min^{-1} with and without MgO derived catalysts. Liquid yields were obtained in the range from 50 to 56% for WTR alone and 47 to 50% for WTR+SCB, with catalyst having a negligible effect on the conversion to PO. Cetane number of PO was estimated using NIR spectra, with average values of 43 for bulk oils of both studied feedstock. For WTR pyrolysis, gas yield was observed to be close to 6% for experiments run without catalyst with a two-fold yield increase if catalyst was present. In contrast, WTR+SCB yields approximately 20% gas, with no definitive catalyst effect on gaseous phase. Pyrolytic processes require energy, which may partly come from the fuel gasses obtained during the process itself; therefore, a characterization of the gas phase comes convenient. In the experiments run within the context of this work, the obtained gas of WTR pyrolysis was composed mainly of CH_4, with a peak measured composition of over 96% CH_4. SCB addition resulted in a 75% reduction of CH_4 and a 70% increase in CO_2 concentration.

471932 Conversion of Municipal Solid Waste to Methyl Ketone Using Ionic Liquid Based Process

Abstract:

Municipal solid waste (MSW) has the potential to serve as a feedstock for the production of advanced biofuels and biochemicals but has not been extensively studied to date in terms of conversion efficiency and hydrolysate quality. In this study, we evaluated the conversion of MSW blends to methyl ketones (MKs) using ionic liquid (IL)-based hydrolysates followed by fermentation using an engineered E. coli strain. The hydrolysates were produced from a one-pot IL process (which included both pretreatment and enzymatic saccharification) using different types of MSW blends and renewable, choline-based ILs. For the range of MSW blends and ILs tested, similar trends of OD600, sugar consumption, and MK production of hydrolysates were observed. Comparable or even greater amounts of medium-chain (C11-C17) MKs were produced using IL-based hydrolysates compared to the simple sugar controls with the same amount of sugars at the starting point. The results showed that this engineered E. coli strain can tolerate IL-based hydrolysates and that the neutralized ILs are acting as the extra carbon source for MK production.

462491 Upgrading of Hydrothermal Liquefaction Biocrude Oil Converted from Wet Biowaste

Abstract:

Wet biowaste is an increasingly studied feedstock for producing renewable bioenergy, reusing nutrients and improving the environment. Each year, the United States produces 79 million dry tons of wet biowastes from food processing and livestock manure and could be recovered as renewable energy. One of the limitations for wet biowaste to energy conversion is the high water content (80-99%).

Hydrothermal liquefaction (HTL) is a technology to reduce complex organic materials into biocrude oil and an aqueous fraction containing original nutrients and wastewater. Research has shown that HTL can convert 30-70% of different volatile solids into biocrude oil with heating values between 32-38 MJ/kg. HTL biocrude oil converted from different types of wet biowastes contains impurities: 10-20% oxygen, 3-7% nitrogen, and up to 20% of moisture. It is necessary to separate and/or upgrade the HTL biocrude oil for transportation fuel application. Distillation can reduce the oxygen content in the biocrude oil to 5% and increase the heating value to 41-45 MJ/kg. We hypothesize that with appropriate separation techniques, such as distillation, the HTL biocrude oil can be separated into different fractions, and some fractions can be used as transportation fuel including aviation fuel and diesel.

Our initial results suggest that distillation of HTL biocrude oil converted from swine manure (SW) and food processing waste (FPW) is very promising. Analyses demonstrate that the distillates from SW- and FPW-derived biocrude, respectively, contain 15 wt.% and 45-50 wt.% fractions that have heating values similar to those of petroleum aviation fuel and diesel without any further upgrading. In addition, an aviation biofuel blend prepared with 10 wt.% HTL distillates with the highest heating values (46-47 MJ/kg) and 90 wt.% petroleum jet fuel (or bio-kerosene BK10), exhibited the same properties including density, viscosity (6.3% lower), total sulfur content, copper corrosion performance, and net heat of combustion (1.30% lower) as those of commercial petroleum jet fuel. Although the distillation significantly improved the fuel's properties, more improvement is needed for use in current engines. For instance, the BK 10 blend had four times the acidity and a 20% lower flash point than those of petroleum jet fuel. We also found that oxygenates and nitrogen-containing compounds in the distillates from SW-derived biocrude led to excessive gum in BK10 blend. Thus, additional upgradation is needed.

Two strategies were investigated to make the HTL distillates converted from SW and FPW more compatible with petroleum transportation fuels. First, esterification was conducted to reduce the acidity in FPW-derived distillates. An orthogonal experimental design was operated to examine the effect of reaction temperature, reaction time, the catalyst loading (H_2SO_4), and the molar ratio of methanol on the esterification yield and acidity of the distillates. After esterification, the acidity of the distillates was reduced from 27.5 mg KOH/g to 0.8-7.3 with a yield of about 88 wt.%. The highest esterification yield was achieved at 70 °C for a 0.5 h reaction time with 2 wt.% catalyst loading under a 1:9 molar ratio of methanol; and the lowest acidity was realized at 50 °C for a 2h reaction time with 2 wt.% catalyst under a 1:15 molar ratio of methanol. The orthogonal test also indicates that the reaction time impacted the esterification yield the most, while the reaction temperature affected the acidity the most.

On the other hand, catalytic deoxygenation and denitrogenation under sub-/supercritical water are investigated to reduce the oxygen and nitrogen contents in the SW-derived distillates to reduce gum contents for BK10 blend. Preliminary data showed that supercritical water is a promising reaction media to inhibit the char formation and promote deoxygenation/denitrogenation for SW-derived distillates. Different types of catalysts (*e.g.,* Raney-Ni), reaction temperature, reaction time, and reaction media (e.g., supercritical water) are examined. The physicochemical properties of the upgraded distillates are analyzed to elucidate the upgrading mechanism. Elemental compositions, chemical compositions, heating values, flash point, viscosity, gum, and ash contents of the upgraded distillates are characterized and compared to those without upgrading.

471372 Evaluation of Optimum Cell Disruption for Microbial Lipid Extraction Using High Pressure CO2

Abstract:

The economy of lipid extraction from non-food sources of triglycerides (e.g. microbial oil) depends on lipid extraction and feedstock cost. Lipid can be extracted either from dried biomass (needs drying of the biomass up to 95%) or from wet biomass (consists mostly moisture). The microbial lipid extraction from wet biomass has advantages over dried biomass due to the elimination of energy intensive drying process. To improve lipid recovery from wet biomass, the cell wall of the microbes needs to be disrupted prior to lipid extraction for releasing the intracellular lipid from microbes that is usually difficult to extract using traditional solvent extraction process. The cell disruption is found to be affected by applied gas pressure, temperature, exposure time, depressurization rate, agitation, etc. This research will investigate the effect of different factors that affect the microbial cell disruption using high pressure CO_2. A fractional factorial design of experiment will be used to evaluate the controlling factors in microbial cell disruption. The number of cells before and after the treatment will be used as a response variable. After evaluating the main factors, a central composite design will be used to find the optimum cell disruption using a response surface methodology.

475190 Alternative Process Design for Addressing the Engineering Grand Challenges

Abstract:

The U.S. National Academies has examined engineering Grand Challenges [1-3], and a number of the challenges can be addressed through the perspective of alternative process design. Alternative process design connects chemicals with their production routes; the combination affects the sustainability of various human and Earth systems. Sustainability intends to satisfy the environmental, economic, and social needs of future generations without compromising current needs. Among the challenges are the need to develop alternative fuels and feedstocks for the chemical industry, to approach systematic problems with analyses such as life cycle assessment, and to enable sustainability through education [1]. Another report emphasized the need to harness biological understanding; microstructured products and their processing; sustainably extracting resources; and preventing, controlling and treating wastes [2]. A third report considered general engineering challenges, beyond chemicals [3]. It extols advances in solar energy, carbon sequestration, clean water, urban infrastructure, health information and medicines, the brain, nuclear aspects, cyberspace and virtual reality, personal learning, and tools for scientific discovery. This presentation will review these and point to accomplishments and continuing needs.

[1] National Research Council (2005). Sustainability in the Chemical Industry: Grand Challenges and Research Needs. National Academy Press, Washington, DC.

[2] National Research Council (1988). Frontiers in Chemical Engineering: Research Needs and Opportunities. National Academy Press, Washington, DC.

[3] National Academy of Engineering (2008). Grand Challenges for Engineering. National Academy Press, Washington, DC.

477179 Shock Electrodialysis and Other Electrochemical Water Purification Methods

Abstract:

Large-scale seawater desalination is essentially a solved problem, despite its inherent cost, as reverse osmosis plants approach the thermodynamic efficiency limit. In contrast, grand challenges remain in how to efficiently treat lower salinity feedwater, e.g. from brackish sources, municipal waste, or industrial streams, in order to reduce salt, disinfect, remove toxins, and/or separate valuable chemicals with geographical flexibility and smaller, portable systems. Existing electrochemical methods, such as electrodialysis and capacitive deionization, can address some, but not all, of these needs. This talk will introduce a new process called "shock electrodialysis", which exploits deionization shock waves in charged porous media for continuous and simultaneous deionization, separations, and disinfection. The first shock ED prototype achieves 99.99% salt removal and up to 80% water recovery, although material optimization is still required to improve the energy efficiency and boost the flow rate.

476195 Battery Storage   the New Renewable Force

Abstract:

Battery storage for solar and wind is critical in providing sustainable renewable energy, consistently and reliably from it source to the intended user. The discussion will focus on R&D efforts being undertaken by Cal Clean Earth (CCE) in improving storage capacities and efficiency. Emphasis during the panel discussion will be placed on Lead Acid, Lithium Iron Phosphate and others. Additionally, we will look at productivity aspects of these systems, e.g. standalone against hybrids. For the success of these projects, funding is essential, thus will share approaches taken through Power Purchasing Agreements (PPA), Property Assessed Clean Energy (PACE) and also how rebates/offsets are processed through Self-Generation Incentive Programs (SGIP).

459515 Review on Rheological Studies of Sludge from Different Sections of Municipal Wastewater Treatment Plants for Improved Process Performance

Abstract:

Wastewater treatment process generates large quantities of sludge from different operational steps. The rheological behaviour of sewage sludge at different steps of wastewater treatment plant (WWTP) is useful for the design, operation, optimization of polymer dosing, hydrodynamics and mixing in anaerobic digesters, and for efficient sludge management. Rheological characterization of sludge is also known to be an essential tool to optimize flow, mixing and other process parameters in WWTPs. In this review article, the impacts of different parameters like solid concentration, temperature, pH, floc particle size, mixing ratio, ageing and dose of sludge conditioning agents on the rheological behaviour of sludge from different parts of WWTPs were analysed. Raw primary sludge, excess activated sludge, thickened excess activated sludge, mixture of raw primary and thickened excess activated sludge (mixed sludge), digested sludge, and biosolid rheological and flow behaviour under the influence of different operating parameters and the impacts on process performance were reviewed in this study. Solid concentration, temperature, primary to thickened excess activated sludge mixing ratio, and sludge age were found to affect the viscosity, yield stress, flow index and flow consistency of mixed sludge., viscosity also showed a decreasing trend with decreasing total solid concentration and percentage of thickened excess activated sludge in the mixture. In contrast, yield stress and viscosity generally showed reduction with increasing temperature. The effect of dose of conditioning agents on the rheological behaviour of thickening of excess activated sludge and digested sludge was also addressed in this review. The relationship between activation energy and viscosity for varying temperature has also been discussed. The rheology of digested sludge affects the flow hydrodynamics, dewaterability and the polymer consumption in WWTPs. The rheological characteristics of digested sludge are highly dependent on changes in total solid concentration, temperature and polymer dose. The rheological characteristics of digested sludge during polymer conditioning and

flocculation process was significantly affected by temperature and solid concentration; hence, polymer dose can be reduced by operating the dewatering process at optimum temperature condition and varying the polymer dose as a function of the total solid concentration and viscosity of the digested sludge. Volatile solid content of digested sludge fed to centrifuge has inverse relationship with total solid content of biosolid or biosolid capture rate. Hence, improving the volatile solid and COD removal/reduction by enhancing the performance of anaerobic digesters or by using other digested sludge conditioning methods will significantly improve the flow behaviour of digested sludge and dewatering performance. Investigating the relationship between rheological and physico-chemical characteristics of sludge can also serve as a tool to optimize essential process parameters. The applicability and practical significance of various rheological models such as Bingham, Power Law (Ostwald), Herschel-Bulkley, Casson, Sisko, Careau, and Cross models to represent experimental rheological characteristic of different sludge types was also investigated in this review.

460652 Carbon-Based Rods with Anomalous Water Expulsion at High Humidity

Abstract: Three water adsorption-desorption mechanisms are common in inorganic materials: chemisorption, which can lead to modification of the first coordination sphere; simple adsorption, which is reversible; and condensation, which is irreversible. Regardless of the sorption mechanism, all known materials exhibit an isotherm in which the quantity of water adsorbed increases with an increase in relative humidity. Here, we show a unique water adsorption phenomenon of carbon-based rods (PNNL-Carbon Rods; PCR). The rods adsorb water at low humidity and spontaneously expel about half the adsorbed water at high relative humidity. The water expulsion is attributed to the interfacial forces between confined rod surfaces. At wide rod spacing, a monolayer of water can form on the surface of the rods, which subsequently leads to condensation in the confined space between adjacent rods. As the relative humidity increases, adjacent rods (confining surfaces) in the bundles are drawn closer together via capillary forces. At high relative humidity, the size of the confining surfaces decreas to a critical distance, and water condensed inside the confined space is released as a vapour. The process is reversible, as indicated by over three cycles of water sorption-desorption. The rods synthesised under different reaction conditions (eg., reaction temperature, amount of $FeCl_3$ catalyst used, and $FeCl_3$ source) have different morphology, size and exhibited different RH for the onset of evaporation and amount of water expelled (Figure 1). PCR were synthesized at relatively low temperature compared to temperatures used for carbon nanotube synthesis (200 °C vs 600 °C, respectively). We will present the physical, thermodynamic and spectroscopic evidence for this behavior and discuss its potential applications.

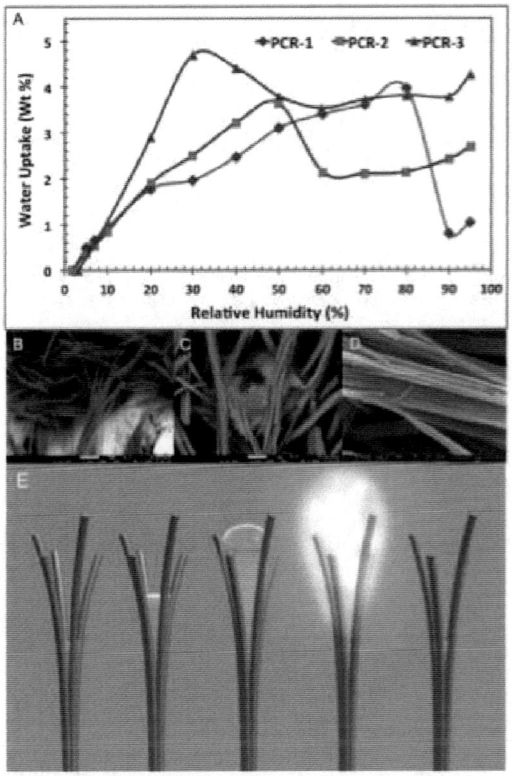

Figure 1: (A) Water sorption isotherms of carbon-based rods synthesized using 0.1 M (PCR-1; synthesized at 220 °C) and 0.2 M $FeCl_3$ (PCR-2; at 220 °C, PCR-3; at 230 °C), (B-D) SEM image of rods. (E) Schematic of proposed water expulsion mechanism.

- Abstract Text

465679 An MINLP Model for Sustainable Water Management in Macroscopic Systems Under Economic Penalty Scenarios

Abstract:

In a sustainable system, the amount of water available from the natural sources in a watershed would be sufficient to meet the demands of the users without compromising their future availability. Furthermore, the amount and type of pollutants discharged over the water bodies would be completely degraded by the natural flora and fauna present in the aquatic ecosystems. However, this is not a common scenario.

Optimal resources distribution and wastewater treatment in a system (which involves a complex network of usage and disposal of water) cannot always be achieved just by imposing environmental regulations. In most cases, a compromise between the economic and environmental aspects has to be taken into account. This is not a simple task, because the needed investment for water treatment sometimes becomes prohibitive, especially when the users (e.g. domestic, industrial, agricultural) have many economic concerns regarding budgets, profitability of their products, market share, etc. An efficient strategy is still needed to mitigate harmful consequences on the environment.

Literature reports several efforts to deal with pollution over water bodies from a trading and penalty perspectives. These approaches are based on an ideal scenario, where charging environmental violators with fines and exchanging pollution credits among various sources achieve lower global costs while environmental constraints are met. However, none of those approaches has considered a macroscopic system which also takes into account the sustainability of water resources within a watershed and its surroundings.

This paper proposes a model to account for the sustainability of a macroscopic system through resources management, including the evaluation of alternative technologies for fresh water consumption reduction, and a distributed treatment system to treat wastewater from different sources; however, our approach also considers the possibility of penalty scenarios due to the violations of environmental regulations. The aim is to assess the implications of the penalties over the costs and the environmental goals of the problem.

Our approach results in an MINLP multi-period model which has been solved through the GAMS® modeling environment. A case study with different scenarios shows the scope of the proposed approach and the results show the significance of the results; the solution of the optimization problem allows finding the minimum investment to meet the desired environmental goals with respect to freshwater consumption and pollution over water bodies. This investment consists of the number, size and location of rainwater storage devices, the number, size, type and location of treatment plants in the distributed wastewater treatment system, the total amount of recycled wastewater and the total amount of fines charged to the users for violation of environmental constraints regarding pollution over water bodies.

469744 Nitrogen Doped Magnetic Carbon Nanocomposits Synthesized from Melamine with Enhanced Cr(VI) Removal

Abstract:

With the growth in various industries, such as metal cleaning, electroplating, pigment, mining and leather processing, heavy metals (Cr, Cd, Hg, Pd and As) have become a widespread pollutant in the surface water and groundwater. Cr(VI), a typical heavy metal ions, is a commonly identified contaminant because of its high toxicity and mobility. Recently, magnetic carbon nanocomposites with large specific surface area were introduced to remove the heavy metal in the wastewater and displayed the excellent efficiency. In this work, we successfully synthesized the novel nitrogen doped carbon composites with higher magnetism and specific surface areas using melamine as carbon and nitrogen source through a simple pyrolysis method. After that, we utilized the as-synthesized materials to treat the Cr(VI) solution, and found that the excellent performance were observed (removal capacity: 29.46 mg/g, pH at 7; 2001 mg/g, pH at 1), which can be attributed to the nitrogen dopant and the iron nanoparticles on the surface. The nitrogen dopant on the carbon surface can increase its negative charge density and then enhance the ability for the adsorption of Cr(VI) ions. These results demonstrated that nitrogen doped magnetic carbon nanocomposites are the promising materials for the environmental remediation.

470707 Techno-Economic Assessment of Membrane Distillation for Shale Gas Produced Water Treatment

Abstract:

Unconventional shale gas is a promising energy resource with major economic benefits but is accompanied by a host of environmental challenges. In particular, the rapidly developing shale gas industry faces a critical challenge of managing vast quantities of high salinity wastewater generated in the process of hydraulic fracturing. Membrane distillation (MD) is an emerging desalination technology for treating high salinity wastewaters including the produced water from shale gas activities. MD has lower operating temperature (30-90^{0}C) and pressure compared to conventional desalination technologies. Moreover, MD can treat wastewaters with up to 350,000 mg/Liter total dissolved solids (TDS) with up to 99% salt rejection where most pure thermal processes or pressure driven membrane processes are not applicable. While MD offers several advantages, techno-economic assessment (TEA) is necessary for evaluating the economic feasibility of MD for shale gas produced water treatment. To date, there has been little emphasis on evaluating the performance of MD technology for treating produced water and to best of our knowledge TEA of MD technology for produced water is not available in the existing literature.

We present a detailed TEA for a hypothetical MD plant for treating shale gas produced water based on a combination of experimental results, ASPEN Plus process model, and best available engineering knowledge. The analysis is carried out for a hypothetical 0.5 million gallons per day (MGD) MD plant that concentrates produced water from 10% (100,000 mg/Liter) TDS to 30% salinity. Preliminary results reveal that thermal energy cost for MD operation contributes the most to total cost of treating produced water in a MD plant. We also perform sensitivity analysis to identify input parameters that have the highest impact on total cost of purified water using MD technology. The results of sensitivity analysis reveal that plant capacity and feed TDS level have a significant impact on the cost of treating produced water. Subsequently, we evaluate the economic potential of MD for treating shale gas produced water under the scenario of utilizing waste heat for meeting the thermal energy requirements of the MD process. Specifically, we focus on utilizing available waste heat from existing natural gas (NG) compressor stations (CS) to meet the heat requirements of the MD process. Our prior work has developed quantitative estimates of waste heat from existing NG CS in the U.S. This information is coupled with experimental results on MD performance to propose a systems-level integration of MD with available waste heat at NG CS. Subsequently, we quantify the amount of produced water that can be treated using MD driven by available waste heat using ASPEN Plus simulation and experimentally measured flux rates. The results of this study reveal that utilizing waste heat as the energy source for MD reduces the total water cost significantly. A comparison of MD with existing desalination technologies including reverse osmosis (RO), multi-stage flash (MSF), and multi-effect distillation (MED) will be discussed. We also compare our findings with the business-as-usual scenario where produced water is transported and disposed in class II injection wells to highlight the potential and limitations of the MD technology for produced water treatment.

470025 Perchlorate Destruction from Ion-Exchange Regenerants By Catalytic Reduction Process Increases Brine Recovery

Abstract:

Perchlorate salts are generated as a byproduct from rocket fuels and other explosives. Ingestion of perchlorate inhibits normal function of the thyroid gland and contributes to hormone imbalance. Recently, the USEPA determined that perchlorate must be regulated as a water contaminant under the Safe Drinking Water Act (SDWA). One method of removing perchlorate from drinking water and potable water sources is through selective ion-exchange resin. Because perchlorate ion binds very tightly to the strong base anion exchange resin, a solution of having extremely high salt concentration, typically between 7 â€" 12%, is required for regeneration of the resin. Furthermore, disposal of high concentration of perchlorate containing regenerant is problematic. Accordingly, there is a need for an improved method of removing perchlorate from water, including brine recovery.

Using a nanofiltration (NF) process, perchlorate is separated from the brine. More than 95% perchlorate was separated and retained in the reject. Permeate primarily containing brine was recycled back to the Ion exchange process to be used in the next stage of regeneration. Low volume of the NF reject containing high concentration of perchlorate ions is mixed with a reducing agent in a reactor where the perchlorate ions and the reducing agents undergo an oxidation-reduction (Redox) reaction. During the Redox reaction, perchlorate ions are reduced to chloride ions and the reducing agent is oxidized. The oxidized reducing agent is separated from the water which is regenerated by means of a second reducing agent, and reused in the Redox process. Redox reaction kinetic was significantly improved in presence of catalyst. Various reducing agents including Titanium (III), Manganese (II), and Iron (II) at different temperatures were investigated. The impacts of pH on the reduction kinetic were examined. All reactions were conducted with and without the catalyst. The efficiency of TiO_2 and MnO_2as the catalyst was compared.

Using an ion Chromatography (IC), perchlorate was measured. Results reveal that within the first two hours of

reaction, more than 98% perchlorate was converted to chloride ion in presence of a catalyst. Increasing temperature increased prechlorate destruction efficiency. The process, and the impacts of various process parameters on the perchlorate destruction efficiency will be presented and discussed.

475639 High Temperature Solar Thermochemical Energy Storage Materials

Abstract:

High temperature solar thermochemical energy storage (STCES) has promise to provide a low cost and high temperature storage solution for solar thermal power.Â However, the current status of STCES is at an early stage of development and substantial research and development is required to realize practical implementation. Two thermochemical energy storage systems with high theoretical energy densities at high temperatures have been investigated. In both the cases, we have been able to develop stable energy storage materials with energy densities in excess of 1200 MJ m^{-3} that can operate at temperatures above 1100 â„ƒ in both cases.

The first system pertains to thermochemical energy storage via decomposition and carbonation of strontium carbonate/strontium oxide in the temperature range of 1150-1235Â°C. Porous structures of pure $SrO/SrCO_3$ were studied via a fractional factorial experimental study to verify the possibility of synthesizing stable undoped $SrO/SrCO_3$ structures using the sacrificial pore formation with graphite (SPFG) method. Results of the experimental study indicate that the SPFG method is not suitable for fabricating porous structures of pure SrO or $SrCO_3$ for high temperature thermochemical energy storage. Cyclic volumetric energy density (including only the chemical reaction contribution) calculated from thermogravimetric measurements decreases substantially to values below 600 MJ m^{-3} upon repeated cycling, demonstrating that the SPFG method not viable. Further investigations revealed that doping $SrO/SrCO_3$ with MgO and BaO improved the cyclic stability significantly. Mixtures of SrO-MgO and SrO-BaO in the molar ratio of 2:3 appear to achieve cyclic stability at 860 ± 5 MJ m^{-3} (500 ± 3 kJ kg^{-1}) and 1408 ± 15 MJ m^{-3} (1001 ± 11 kJ kg^{-1}) respectively over 15 cycles. Pure BaO was also tested for thermochemical energy storage and is found to be more stable with a higher energy density of 1690 ± 24 MJ m^{-3} (1070 ± 15 kJ kg^{-1}).

A second study was performed on transition metal oxide based thermochemical energy storage systems. The purpose of the investigation was to validate the use of magnesium oxide (MgO) as a sintering inhibitor/porous structure stabilizer for high temperature solar thermochemical energy storage (STCES) using reversible redox reactions. Solid solutions of reactive metal oxides in MgO prepared using the solid state reaction method (SSR) were investigated for STCES applications. A new concept of using steam and carbon dioxide to lower the partial pressure of oxygen during thermal reduction was tested and validated. 40 and 33 mole % MnO in MgO, 20 mole % CuO in MgO and 40 mole % Fe_2O_3 in MgO were investigated for high temperature thermochemical energy storage via redox reactions. A composition of 40 mole % MnO in MgO appeared to be the most promising material among those tested here for high temperature thermochemical energy storage. The material can release heat at a temperature greater than 1200°C and has an energy storage capacity of 64212 $kJkg^{-1}$ (122236 MJm^{-3}). The material appeared stable over 20 cycles in a thermogravimetric analyzer and 12 cycles in a tube furnace bench scale reactor and is expected to maintain stability over repeated cycling.

466067 Techno-Economic Assessment of the Polygeneration of Liquid Fuels and Electricity By Solar Hybridised Gasification of Coal

Abstract:

Solar gasification offers the potential to convert a greater portion of the fuel input to syngas than is possible with a conventional gasifier, resulting in a higher production of Fischer-Tropsch liquids (FTL), by displacing the carbonaceous fuel that is otherwise burnt in O_2 to provide process heat. In this hybrid gasification process, concentrated solar thermal energy is used when it is available, while autothermal gasification is used when the incoming solar insolation is insufficient. Thus, the intermittency of the solar resource is accommodated through the use of intermediate storage of syngas and turning down the liquid fuel production rate. This paper presents a techno-economic evaluation of a coal-to-liquids process integrated with a solar hybridised, oxygen blown, atmospheric pressure vortex-flow gasifier (SCTL) located at Daggett, USA. The influence of the solar multiple (SM) and syngas storage capacity (hours) on the plant feasibility was assessed. Here, the SM is the ratio of the heliostat field area to the required heliostat field area that can generate a net annual peak solar thermal energy matching the energy required by the gasification process.

The annual specific FTL output and the specific total energetic output (FTL and electricity) of the SCTL plant relative to the conventional coal-to-liquid (CTL) were found to increase by 12% and 15%, respectively, as the SM was increased to 3. On the other hand, the annual reduction in mine-to-tank (MTT) CO_2 emission from the SCTL relative to the CTL was reduced by 17%. The utilization factor of the FT and power generation plants in the SCTL was significantly influenced by the syngas storage capacity. The utilization factor for the SCTL with no syngas storage was calculated to be 74% compared with 93% for the SCTL with 4 hours syngas storage capacity.

To produce 1500 barrels per day of FTL, the levelised cost of fuel (LCOF) for the CTL was estimated to be USD1.2/L_{2012}. The LCOF for the SCTL with no syngas storage was increased from USD1.3/L_{2012} to USD1.4/L_{2012}, and from USD1.4/L_{2012} to USD1.5/L_{2012} with 4 hours syngas storage capacity, as the SM was increased from 0.5 to 3. Despite the fact that the outputs from the SCTL were higher than that of the CTL, this could not overcome the high cost of the solar components and the syngas storage system. Furthermore, the LCOF for the SCTL is sensitive to several parameters identified in the sensitivity analysis, such as the total plant cost, the assumed cost for carbon capture and sequestration, and the operation and maintenance costs.

452450 Solar Metallothermic Production of Rare Earth Elements

Abstract:

Rare earth elements (REEs) have many applications in metal alloys, catalysts, petroleum refining, glass additives, polishing compounds, and ceramics; they also underpin many critical clean energy technologies such as PV devices, batteries, high-performance magnets, and phosphors for lighting and electronics. The demand for rare earth elements is expected to grow in the coming years, especially in uses related to the energy sector. At present, the extractive processes for REEs and rare earth oxides (REOs) are driven by fossil fuel based energy sources with very high greenhouse gas impacts and resource depletion potential. The separation and reduction stages of the REO/REE production process have the highest energy and material consumption requirements.

Cerium (Ce), lanthanum (La), neodymium (Nd), and praseodymium (Pd) REEs and REOs accounts for 54, 25, 4, and 3% (86% combined) of the global REE supply and usage. Ce, La, Nd, and Pr metals are commercially produced by electrolytic processes where the anhydrous chlorides of the metals are dissolved in molten NaCl, KCl, $CaCl_2$, or other salts. The electrolysis of these rare earth chlorides must occur at approximately 50°C above the melting point of these metals, which are 795, 920, 1021, and 931°C for cerium, lanthanum, neodymium, and praseodymium, respectively.

Energy consumption and environmental concerns, coupled with the fact that the current electrolytic production method must occur at high temperature, provides significant motivation for investigating solar thermal processing as an alternative to conventional REE production. Here, the potential for solar metallothermic production of rare earth elements, especially Ce, La, Nd, and Pr, is critically evaluated using thermodynamic analysis.

473021 Manganese As a Redox Reactant for Sustainable Thermochemical Ammonia Synthesis

Abstract:

Fixed nitrogen, as ammonia (NH_3), is the main ingredient of most fertilizers and its availability is the most common yield-limiting factor in industrial agricultures. Responsible for up to 50% of crop yields, ammonia fertilizers provide the nitrogen necessary to sustain over half of the global population. Annual global ammonia production, currently greater than 140 million metric tons, will need to expand along with the expected increase in demand for fertilizer for food production. Catalytic ammonia synthesis via the Haber-Bosch process, operates at some of the most intense process conditions for chemical production, with temperatures as high as 500°C and pressures reaching 300 atmospheres. The Haber-Bosch process consumes 0.6 tons of natural gas and emits 2.3 tons of carbon dioxide per ton of ammonia produced.

A chemical loop is experimentally demonstrated here using manganese as a cycled redox reactant for ammonia production from nitrogen, water, and methane at atmospheric pressure. Manganese is used to reduce the nitrogen which is then reacted with steam to form ammonia in the presence of a sodium promotor. The corresponding manganese oxide is reduced by a dilute methane stream also producing syngas. The role of sodium hydroxide in the promotion of ammonia liberation is investigated. The thermodynamic predictions of the system are discussed, and the reaction kinetics are evaluated.

The thermochemical cycle appears a promising method to sustainably produce ammonia at near-ambient pressures. Separating the ammonia synthesis into nitrogen reduction and subsequent protonation steps introduces an additional degree of freedom for optimizing the redox reactant and process conditions. Reduced carbon dioxide emissions, and valuable syngas coproduct are advantageous for this novel approach going forward.

448616 CO_2/Brine/Rock Interactions in Lower Tuscaloosa Formation

Abstract:

Saline aquifers are the largest potential continental geologic CO_2 sequestration resource. Understanding of potential geochemically-induced changes to the porosity and permeability of host CO_2 storage and sealing formation rock will improve our ability to predict CO_2 plume dynamics, storage capacity, and long-term reservoir behavior.

Experiments exploring geochemical interactions of CO_2/brine/rock on saline formations under CO_2 sequestration conditions were conducted in a static system. Chemical interactions in cores samples from the Lower Tuscaloosa formation from Jackson County, Mississippi with exposure CO_2 saturated brine under sequestration conditions were studied through six months of batch exposure. The experimental conditions to which the core samples of Lower Tuscaloosa Massive Sand and Selma chalk were exposed to was a temperature of 85 °C, pressure of 23.8 MPa (3,500 psig), while immersed in a model brine representative of Tuscaloosa Basin that was mixed with CO_2. Computed tomography (CT), X-Ray diffraction (XRD), Scanning Electron Microscopy (SEM), brine chemistry, and petrography analyses were performed before and after the exposure. Permeability measurements from the sandstone core sample before and after exposure showed a permeability reduction. No significant change of the permeability measurements was noticed for the core sample obtained from Selma chalk after it was exposed to CO_2/brine for six months. These results have implications for performance of the storage interval, and the integrity of the seal in a CO_2 storage setting.

In addition, a numerical core scale model was also developed to simulate reactive transport with porosity and permeability change of the sandstone from the Lower Tuscaloosa formation and Selma Chalk rock from the sealing formation above the Lower Tuscaloosa formation. The model predicted a 5.2% permeability decrease in the sandstone after 180 days of exposure to CO_2-saturated brine, which was close to laboratory-measured permeability results. For the Selma Chalk sample, there was a very small porosity decrease in the interior of the sample after 180 days of exposure. The model predicted no significant change of permeability for the Selma Chalk sample after 180 days of exposure, which was consistent with laboratory-measured permeability results.

455391 Biofuel Production Via Thermochemical Conversion of Algae and Optimizing Coproduct Utilization

Abstract:

Hydrothermal liquefaction (HTL) is a method for thermochemical conversion of wet algae that can produce energy dense biocrude oil as well as gaseous, aqueous and solid by products. In this study, three different algal species such as Galdieria, Kirchneriella and Micractenium were used as feedstock for the biofuel production. Process parameters like reaction temperature, time and solids loading were optimized to achieve maximum biocrude oil yield. A bench top 250 mL batch reactor was used for the HTL experiments and dichloromethane was used as a solvent for extraction. The maximum biocrude oil yield of 45.5% was obtained for Kirchneriella species at reaction conditions of 300 C, 90 bar, 30 min and 10% solids loading. Since the maximum biocrude oil yield achieved was less than 50% and to make this process economically viable, other co-products need to be utilized which adds value to the algal biorefinery. The aqueous phase will be rich in carbohydrates and soluble proteins and can be recycled to the cultivation facility. This nutrient rich aqueous phase upon required dilution was successfully used to cultivate algae. The biochar produced during the hydrothermal liquefaction has significant amount of phosphorous and nitrogen. These components were extracted as phosphates and ammoniacal nitrogen and were used as macronutrient supplement to grow algae. The growth experiments were investigated using a tissue Culture Roller drum apparatus with 16 mm borosilicate glass tube photo reactor and confirmed using a Microplate Assay at 40â°C and 2-3 % COâ,, in an incubator. The remaining biochar was dried and used to produce porous carbon. Various analytical instruments such as gas chromatography mass spectrometry (GC-MS), CHNS/O analyzer, spectrophotometer, semi micro calorimeter and inductively coupled plasma optical emission spectrometry (ICP-OES) were used to analyze algal biomass and different products. In conclusion, HTL along with coproduct development has the potential to make algal biofuels sustainable.

456357 Hydrothermal Pretreatment of Rice Husk with Biodiesel Waste As the Reaction Medium

Abstract:

Agricultural waste is becoming more popular as cellulosic biomass for bioethanol production because it is not competing with human food supply. Rice is the largest agriculture product of Japan including in the small district in Oasa, Kita Hiroshima, Hiroshima Prefecture. As the consequence, rice husk is produced annually in large amount in this region. Rice husk is mostly remains unutilized and it contains cellulose. Therefore, it has the potential to be converted into bioethanol. Hydrothermal process is known as environmental friendly pretreatment technique to convert cellulosic biomass into various products including bioethanol since it is using hot compressed water solely as the medium. However, studies shown that catalyst addition improve the yield of products. Alkali catalyst like sodium hydroxide is often used in hydrothermal pretreatment since it is low cost chemical and easily obtained. Nonetheless, the reduction of catalyst cost will be beneficial to make bioethanol more economically competitive. Interestingly, in district of Oasa, Kita Hiroshima Japan, biodiesel has been produced from vegetable cooking oil waste and being utilized for diesel engine vehicles. This biodiesel production generates biodiesel waste that has very high pH (alkaline) and treatment is required to neutralized it before discarding it into the environment. In this research, the possibility of using the alkali characteristic in biodiesel waste to replace commercial alkali catalyst for hydrothermal pretreatment of rice husk is studied. The study was conducted using autoclave reactor made from stainless steel. The experiments were carried out at reaction temperatures of 150, 200 and 250 °C, reaction times of 30 min, and the biodiesel waste addition varied 0, 10, 20, 50 % v/v. To measure the effectiveness of pretreatment, enzymatic hydrolysis was employed after pretreatment to obtain glucose yield. The effect of biodiesel waste addition on rice husk conversion to glucose by hydrothermal pretreatment at various temperature was elaborated.

Thermal Resistance Network Model for Heat Pipe – PCM Based Cool Storage System

Sean Hoenig[1*], CJ Pan[2], Chien-Hua Chen[1], and Richard Bonner III[1]

[1]Advanced Cooling Technologies, Inc., Lancaster, PA 17601, USA

Concept Overview

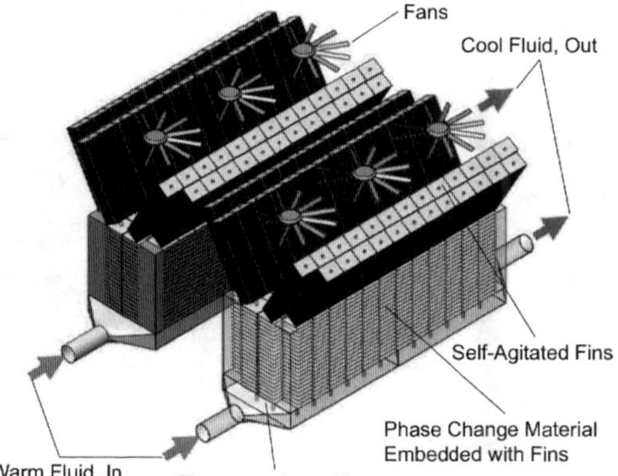

Figure 1. Schematic of proposed cool storage system

Motivation

- For the ARPA-e ARID Program, ACT proposed a supplemental cool storage system that utilizes an integrated heat pipe-phase change material (PCM) heat exchanger to transfer heat from a low-grade steam loop into a salt-hydrate PCM reservoir. During an operational cycle, the stored heat is rejected at night to the cooler ambient air, which solidifies the PCM for repeated use in the cooling system.
- To model the thermal performance and obtain a practical capital cost analysis of the system (<$150/kW$_{th}$), a transient numerical based thermal resistance network (TRN) was developed.
- A CFD based model is impractical to complete for a detailed parametric study and cost analysis. An Excel-based analytical design tool can demonstrate the optimization of design components, thermal capabilities, and overall capital cost to represent the technology's value proposition.

Model Configuration

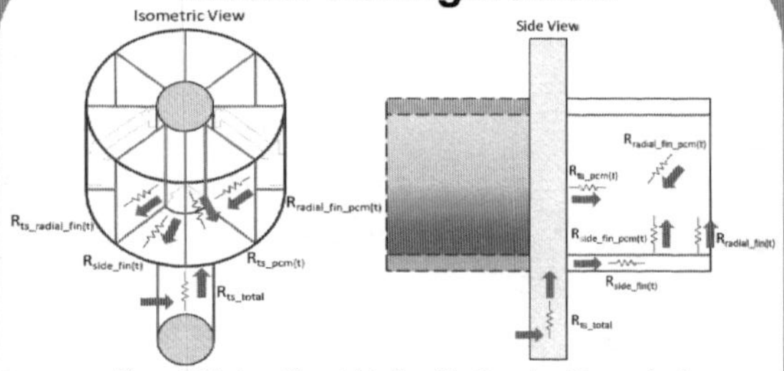

Figures 2 & 3. Isometric and side view of the thermal resistance network

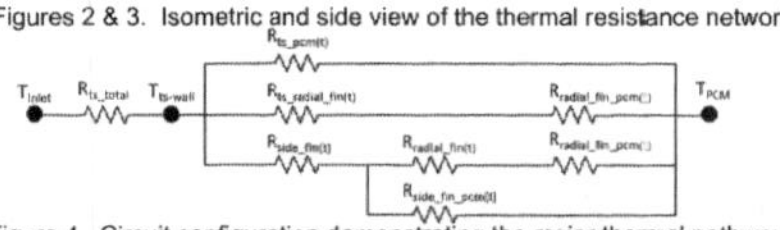

Figure 4. Circuit configuration demonstrating the major thermal pathways

Calculation Methods

- Based on Figure 4, a combined series-parallel resistance calculation is used to derive the changing melt front propagation with each time step. An energy balance based on the accumulated latent heat of PCM and heat into the system through the overall network is used to accomplish this task.

Accumulated | In

$$q = \frac{d(H\rho V)}{dt} = \frac{H\rho A_s \Delta L_c}{\Delta t} = \frac{T_i - T_m}{R_{eq}}$$

$$q_{total} = q_1 + q_2 + q_3$$

$$\Delta L_{c,1} = \frac{\frac{T_i - T_m}{R_{eq,1}}\Delta t}{H\rho A_{s,1}} \qquad \Delta L_{c,2} = \frac{\frac{T_i - T_m}{R_{eq,2}}\Delta t}{H\rho A_{s,2}} \qquad \Delta L_{c,3} = \frac{\frac{T_i - T_m}{R_{eq,3}}\Delta t}{H\rho A_{s,3}}$$

- As the PCM melts, a liquid PCM boundary layer develops, which is the major transient resistance pathway to be analyzed.
- A fin efficiency analytical method is used to account for the changing temperature gradients along the various fin structures as the PCM melts.

Results

- To validate the use of this model, comparisons with ANSYS Fluent were done.
- Figure 5 comparison was completed at experimental conditions and the melt time results match well.

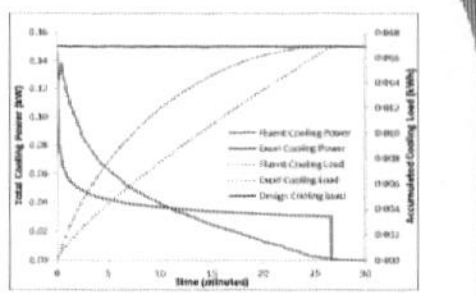

Figure 5. Model comparison with ANSYS Fluent

- Parametric studies determine the optimum trends in design (e.g. thermosyphon diameter, fin spacing).
- Larger diameter thermosyphons are beneficial due to the insignificant price increase and larger achievable heat flux, resulting in lower melt times.

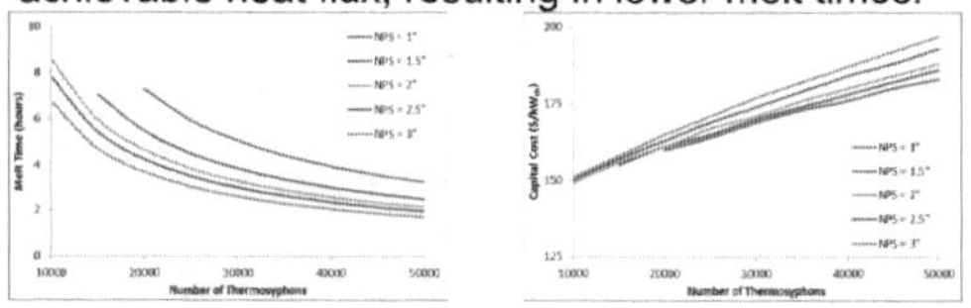

Figures 6 & 7. Initial parametric study to demonstrate thermosyphon sizing optimization

Future Work

- A numerical Excel-based thermal resistance network can demonstrate the thermal performance and capital cost of this technology.
- Further progression of the aforementioned high fidelity methods will ensure precise simulation of the PCM melting/solidifying process.
- Extensive parametric studies will demonstrate a holistic understanding of optimum system design.

Acknowledgements

This project is supported by ARPA-e Award DE-AR0000582.

*Corresponding author: Sean Hoenig, R&D Engineer
E-mail: Sean.Hoenig@1-ACT.com

The author would like to thank his coworkers at [1]ACT and the [2]Lehigh University team for their efforts in support of this work.

459853 Life Cycle Carbon Footprint of Renewable Electricity Generation from Aspen Forest Intensive Harvest in Wisconsin, USA

Abstract:

The Renewable Portfolio Standard mandates increased production of electricity from renewable sources. There are large quantities of timber net growth in excess of removals in the Great Lakes region. The excess timber wood can be a promising feedstock to address the forest products industry while also providing renewable feedstock for energy production. In this study, we aim to evaluate the life cycle greenhouse gas (GHG) emission of renewable electricity produced from cogeneration of aspen biomass in Wisconsin. The typical harvest age for aspen is 40 years. We propose that one way to provide biomass for electricity generation is to intensively harvest aspen by shortening rotation cycles. First, soil carbon (0-60cm depth) was measured from 4 age classes of aspen, namely, 10yr, 20yr, 30yr, and 40-60yr. CO_2 emissions due to intensive aspen harvest were estimated based on differences in soil C levels of various age classes. Next, the optimal biomass supply chain will be determined by minimizing the economic, environmental and social cost of shipping biomass from forest to the cogeneration facility. Finally, a life cycle assessment model is used to quantify the cradle-to-grave GHG emission of renewable electricity, including soil C change, wood harvest, trans-loading and transportation, and combustion in fluidized bed boiler. The life cycle GHG emissions of renewable electricity will be compared with that of Wisconsin grid mix.

460758 Modeling Differential Catalytic Hydrotreatment of Surrogate Compounds for Pyrolysis Oil

Abstract:

The final product distribution of catalytic hydrotreatment of pyrolysis oil depends strongly on processing parameters, especially temperature, hydrogen to oil ratio, and catalyst choice. In this study, we have investigated the hydrotreatment of individual surrogate compounds (anisole, m-cresol, guaiacol, and furfural) in a continuous reactor maintained at 7 bar ranging from 250-350°C, over 3 different catalyst types: acidic (ZSM-5), hydrogenation (Pd/C), and bifunctional (Pt/Al_2O_3). Reaction pathways were studied for each temperature and catalyst; higher temperatures led to greater conversions and more cracking, cracking and rearrangement of methyl groups was seen with ZSM-5, ring saturation dominated with Pd/C, and Pt/Al_2O_3 showed a combination of these functionalities. These tests aim for low conversions to study differential rate kinetics that can be used to model the process. This model uses a lumped parameter approach that groups reaction pathways together and sorts compounds by functional groups.

461080 Pretreatment of Black Liquor to Improve the Recovery Boiler Operation

Abstract:

This work presents a possible method to decrease the alkali related problems in a recovery boiler. This will improve the performance and make a better utilization of the renewable feedstock. The new process design is done by a thermal pretreatment in order to minimize entrained and evaporated alkali.

Black liquor is an intermediate stream that is formed during the production of kraft paper pulp where wood is used as raw material. It contains most of the lignin from the wood but also some parts of the hemicelluloses and celluloses. In addition, it also contains the ingredients of the cooking liquor used to deliberate the different wood components. This makes the black liquor highly rich in alkali and sulfur. It is not uncommon with a portion of 20 wt% for sodium and 7 wt% for sulfur on dry material basis. In order for the pulp mill to close its material balances and to produce necessary steam and electricity, the black liquor is combusted in a recovery boiler. Before the combustion, the black liquor is concentrated to about 80% before it is fed into the boiler through the liquor guns.

During combustion, there are three distinct phases for each individual droplet. These are: drying, pyrolysis and combustion. Due to the high concentration of alkali metals (sodium and potassium), the amount of these that are entrained in the gas flow or just evaporated due to the high temperature is rather large. This creates problems in the boiler due to deposition of material on the heat exchangers, which might cause plugging and thus decrease the heat transfer but also by blocking the passage and create an unwanted pressure drop. Another problem is the corrosive combination between alkali and chlorine (chlorine compounds are present in the original wood material and ends up in the black liquor) which are corrosive and thus detrimental for especially the super heater tubes inside the boiler. Most of the sodium reacts rapidly and forms sodium carbonate or sodium sulfur oxides such as sodium sulfates or sodium thiosulfates. Thereafter, the pyrolysed droplet starts to combust but as a design feature of the recovery boiler, the

droplet should contain some carbon and not be completely burned out when it reaches lower part of the boiler in order to reduce the sulfur compounds to sulfides.

The thermal pretreatment has been performed in our laboratory on black liquor from two Swedish kraft pulp mills at different temperatures and heating rates. In combination to laboratory test for the amount of sodium leaving the black liquor during pyrolysis, we have also performed thermodynamic modelling of the lower part of the recovery boiler to investigate if all sodium would stay in the melt or if it is evaporated in this last stage. It has been previously showed that the sulfur oxide compounds in the black liquor will mainly form hydrogen sulfide in oxygen lean environment at temperatures below 600°C which is the condition for mild pyrolysis but the equilibrium also shows that it will completely form sodium sulfide at high temperatures (above 800°C) which is the condition in the smelt bed.

Within this project we have showed that the thermal pretreatment is a possible way to minimize the sodium loss due to entrainment and evaporation during the free fall period and that the equilibrium calculations indicates that this higher amount will stay in the molten phase and thus increase the direct recovery of sodium from the boiler. This could give higher efficiencies and a lower cost for boiler operation due to lower need for soot blowing steam, possible higher temperatures in the steam cycle and less maintenance.

461121 Characterization of Physical Properties of Macroalgal Pyrolysis Products for Process Simulation

Abstract:

Pyrolysis is a thermochemical conversion of carbon based material to lower molecular products. This conversion is performed at elevated temperatures ranging from 450°C to 500°C in the absence of oxygen and the products are solid carbon-rich char, organic liquids called bio-crude, and hydrocarbon gasses. A major goal of pyrolysis in the field of biofuel production is the maximization of bio-crude yield through appropriate optimization of process parameters. Since the number of components in macroalgal bio-crude is vast and mechanisms of product generation and distribution from macroalgal pyrolysis are not very well understood, accurate and representative simulation of such process is challenging.

Nevertheless, this study presents an approach to characterize the liquid product mixture, bio-crude, for simulation of macroalgae pyrolysis using Aspen Plus. Primarily, bio-crude was modeled in order to simulate the conversion accurately and in time efficient manner. Original liquid product mixture contained around 200 compounds (detected experimentally by GC/MS analysis) in water soluble and water non-soluble organic phases. In this study this number was reduced to 21 representative compounds. In its databases Aspen Plus contains information about physical properties of vast number of compounds. Often, simulations of thermochemical conversions of biomaterials require definition of properties for components which are not found in the databases. This study demonstrates the methodology of reducing compositional experimental data based on chemical structure and boiling point range similarities and translating into components which are already contained in the Aspen databases. This is, in essence, a crucial step of simulating a complex thermochemical conversion.

462358 Experimental Investigation on Expanded Granular Sludge Bed Reactor to Maximize the Methane Gas Production in the Biogas

Abstract:

Experimental Investigation on Expanded Granular Sludge Bed Reactor to Maximize the Methane Gas share in The Biogas Production

Haider Al-Rubaye, Joseph D. Smith, Ph.D.

Chemical & Biochemical Engineering Department, Missouri University of Science and Technology, Rolla, USA

Abstract

The increase in the global temperature, the greenhouse gas emissions, air pollution and the instability in the fossil fuel prices played an important role in making the world more aware of the renewable energy developments, By most definitions alternative energy(have low carbon emissions, a distinction which separates it from renewable energy which may or may not have significant environmental impact.

The Anaerobic Digestion Process, is one of the most contributed renewable energy sources as it generates biogas that could be used for heating and powering purposes, it could be used also in all equipment designed for the natural gas with minimal adjustment to account for the lower BTU content of the methane gas, upgrading the biogas sometimes is preferred in order to increase the composition of the methane gas in it. In general, the biogas consists of methane gas about 50-70% and carbon dioxide 30-50%.

There are many ways to upgrade the biogas and increase the methane share in it. One of the methods is a biochemical process, which basically introduce the hydrogen gas into the process in order to react with the carbon dioxide. In this way, methane increases about 20-30% while the carbon dioxide would decrease about 10-25%.

The experiments would be conducted in an expanded granular sludge bed reactor which treats the industrial and municipal waste water to generate biogas, sensitivity analysis will be performed to maximize the composition of the methane in the biogas. Moreover, the effect of the hydraulic retention time and the substrate concentrations will also be investigated.

462833 Modelling of Exponential and Stationary Phases in Microalgae Growth Using a Population Balance Equation

Abstract:

Being the fastest growing biomass on earth [1], makes microalgae an interesting bio-product [2] source and, depending on the environmental conditions, it is possible to further enhance the productivity. In the past few years, numerous works have focused on developing models able to predict the growth [3, 4], taking under consideration parameters such as temperature, light intensity, nutrients concentration, etc. However, all of these models do not consider the growth as a consequence of the cell life cycle. In other words, the reactor is considered as black box, that under certain conditions (input) gives a correspondent growth (output). Differently, aim of this work is to consider the whole life cycle of the cells, which means that the cell will first grow, then get mature and ready for division and, finally, give birth to daughter cells.

Looking at the downstream process, we know that the harvesting is one of the most energy consuming steps. Knowing that some microalgae may have a wide size range [5], we can use the model to predict which is the most suitable moment to collect the cells. On one side we have that larger cells will require less energy for their collection (assuming that the cell density doesnâ€™t change). Also, the moment in which harvesting is actuated, may affect the next batch. For example, if the collected population is old, the initial growth (in the subsequent batch) will be particularly good because the cells will be ready for reproduction and, at the same time, the amount of useful components (e.g. lipids) inside the cell will be much higher than the one present in younger cells.

The population balance equation has been used in numerous and diverse applications in science and engineering (dispersed phases and microbial populations to mention some). Also, applications related to microalgae can be found in literature [6]. The equation can allow to monitor the cell population over the time in different conditions. In this work, the objective is to simulate the exponential and the stationary phases of microalgae growth and evaluate the size

distribution.

The work has been experimentally validated. Inspired to open ponds, which are among the most common type of photobioreactors, the experiment was run using some open roof tanks (approximately 10 L) and the turbulence was maintained using some water pumps. Temperature was kept about constant using a temperature controller, connected to a heater placed inside the tank. Nutrients were introduced into the system at the beginning of the experiment, while lights were turned on during all the experiment The concentration of cells as well as the size distribution have been monitored daily. In few words, data related to mass concentration, number concentration and size distribution over the time were collected. The developed model can predict all of the mentioned parameters dynamically in the same experimental conditions.

References

[1] R. P. John, G. Anisha, K. M. Nampoothiri and A. Pandey, "Micro and macroalgal biomass: a renewable source for bioethanol," *Bioresour. Technol.*, vol. 102, pp. 186-193, 2011.

[2] M. A. Borowitzka, "High-value products from microalgaeâ€"their development and commercialisation," *J. Appl. Phycol.*, vol. 25, pp. 743-756, 2013.

[3] Q. BÃ©chet, A. Shilton and B. Guieysse, "Modeling the effects of light and temperature on algae growth: State of the art and critical assessment for productivity prediction during outdoor cultivation," *Biotechnol. Adv.*, vol. 31, pp. 1648-1663, 2013.

[4] H. Chang, Y. Huang, Q. Fu, Q. Liao and X. Zhu, "Kinetic characteristics and modeling of microalgae Chlorella vulgaris growth and CO 2 biofixation considering the coupled effects of light intensity and dissolved inorganic carbon," *Bioresour. Technol.*, vol. 206, pp. 231-238, 2016.

[5] K. Miklasz, *Physical Constraints on the Size and Shape of Microalgae*, 2012.

[6] E. Pahija, Y. Zhang, M. Wang, Y. Zhu and C. W. Hui, "Microalgae growth determination using modified breakage equation model," *Computer Aided Chemical Engineering*, vol. 37, pp. 389-394, 2015.

463767 Novel Process of Biofuel Production from Acid Oil and Hydrous Bioethanol Using Ion-Exchange Resin; II. Introducing Life Cycle Perspectives in Optimization of Operating Conditions

Abstract:

This research is separately reported in two presentations (Parts I and II) in sessions that best suit for the content. The two presentations therefore share parts of introduction section.

Liquid biofuels offer attractive advantages over other forms of biomass energy resources in terms of high energy density and ease of storage and transport. Bioethanol is useful as a gasoline additive (up to 10%), with production worldwide in 2014 of 94 million kL. Biodiesel, or fatty acid methyl ester (FAME), is also used as a fuel alternative (100%) to petroleum diesel, with worldwide production in 2014 at 30 million kL. However, there has been no significant increase in the production of these fuels in the past few years. The reason is that there is no profitability for producers owing to high production cost and no benefits for consumers because of unstable product quality and lower fuel economy and power. Moreover, emergence of next-generation automobiles (e.g., battery-powered and fuel cell electric vehicles) will lower the demand for gasoline. The water content of ethanol used as the gasoline additive should be less than 0.5 wt% and the dewatering process requires a large energy input. Thus, the necessity of anhydrous ethanol as gasoline additive fades and new technology which can use hydrous ethanol is desired. On the other hand, the diesel fuel continues to be used for heavy duty on-road commercial transportation, non-road construction, agricultural machines, railroads, and various types of ocean vessels. If new technology to economically produce high quality biodiesel can be developed, the necessity of biodiesel utilization will increase.

In Part I of this research[1], biodiesel is produced in the form of fatty acid ethyl ester (FAEE) using hydrous ethanol by esterification of free fatty acid (FFA). The FAEE has higher cetane number and fuel economy and power than the

traditional fatty acid methyl ester (FAME) and FAEEâ€™s low temperature fluidity is also improved[2]. However, the conventional technologies for esterification of FFA never use hydrous ethanol because the water in the reaction solution must be completely removed to obtain a high conversion. We have reported to produce FAME at more than 95% conversion by esterification of acid oil using a cation-exchange resin catalyst without dewatering and adding excess alcohol[3]. This means that the catalytic activity of the resin is not inhibited by water. Thus, using the cation-exchange resin as solid catalyst the batch and continuous esterification experiments with hydrous ethanol are performed. The maximum water content in hydrous ethanol to obtain high conversion is clarified.

The experimental study exhibited how the rate of esterification reaction slows down as the water content becomes higher. Lower reaction rate leads to an increase in energy consumption, thus contributes negatively from a life cycle perspective by inducing more environmental impacts associated with energy. It will also lead to higher frequency of regeneration of resin, which requires additional loading. On the other hand, increased water content means lower duty for distillation of ethanol. To allow a system-wide optimization, in Part II of this study, modeling of the system with a scope that contains all the relevant product life cycles is performed. Using the data acquired from the experiments with continuous configuration, the abovementioned trade-off is highlighted using a case study with a Japanese example. The systems are modeled so that is possible to conduct a comparative LCA. For the bioethanol, the base case is set at the situation where fermented liquor is distilled and dehydrated to 99.5% to replace gasoline, whereas in the evaluated case, it is used to produce FAEE. For biodiesel, in the base case, FFA and methanol are subject to esterification reaction whereas in the evaluated case, FFA and hydrous ethanol at concentration of 90%-99.5% are assumed to be used for esterification reaction. The inventory data for the continuous production are obtained from the pilot scale experiments performed in this study. The details of the experiments are described in Part I[1]of this report. Electricity consumption is continuously measured by built-in Watt-Hour meter. Molasses from cane sugar mills are assumed as the source of bioethanol. Yeast is used for fermentation to convert sugars to ethanol. The yeasts are recycled after the fermentation. Economic allocation is applied to the surplus yeast after fermentation as it is valuable as a fertilizer. In Japan, most of the cane sugar mills produce only the sugar as product, and the molasses are collected by a private company as raw material for other product. Heat and power required for sugar extraction is usually entirely supplied from bagasse boiler attached to the sugar mill. The bagasse is usually not enough if ethanol fermentation and distillation is required. Therefore, combustion of fuel oil in the bagasse boiler is assumed as the source of heat and power for bioethanol production, in the case study. The factors that influence the trade-off in life cycle greenhouse gas emissions and cost of operation over the water content in the bioethanol used for FAEE production is discussed.

Reference

1) N. Shibasaki-Kitakawa et al., Submitted for presentation at 2016 AIChE Annual Meeting

2) N. Shibasaki-Kitakawa et al., Kagaku Kogaku Ronbunshu, 42 (2016), 30-36.

3) N. Shibasaki-Kitakawa et al., Fuel, 139 (2015), 11-17.

464779 Advances in the Pyrolysis of Municipal Solid Waste Mixtures

Abstract:

Municipal solid waste (MSW) generation is one of the most critical tasks in environmental management and assuring public health of cities all over the world. Its storage in landfills is currently leading to many concerns, not only due to the scarcity of land available but mostly for the environmental issues related to it. According to the U.S. Environmental Protection Agency, landfill emissions contributed 18% to total methane emissions from 1990 to 2013 in the US. Therefore, alternative solutions for MSW disposal should be investigated. Thermochemical conversion offers an interesting option to address this issue, by providing a useful product (biofuel) with the possibility of reduce greenhouse gases (GHG) emissions. In a recent study by our research group we found that the production of alternative liquid transportation fuels using MSW can generate less GHG emissions compared to landfill storage even when considering mixtures of organic and plastic waste. However, little research has been reported about the effects of important operation conditions for this process.

In the present work a series of different operation conditions will be tested on a micro-pyrolysis scale coupled to a gas chromatograph-mass spectrophotometer. The objective will be to obtain processing conditions that increase bio-oil production. First, different mixture compositions of organic and polyolefin plastic waste will be tested, over the range of plastic waste blends up to 20% (wt.) in organic MSW (paper, cardboard, demolition wood). Pyrolysis temperatures from 500-600ËšC will bc investigated to determine its effect on pyrolysis bio-oil composition and quality (oxygen content, molecular weight distribution). These findings will provide useful data to aid in the design and simulate economic and environmental effects of waste-to-fuels facilities, which appear as a promising alternative to solve waste management tasks by more sustainable means.

462697 Development of Methanol Synthesis Process Using Self-Heat Recuperation

Abstract:

Many methanol synthesis processes have reactant recycle system with product separation to achieve high conversion ratio of the whole process, as well as other chemical processes. For this product separation from reactant recycle stream, a gas-liquid separator has often been used and distillation column has often been used for product purification. However, this gas-liquid separator and distillation column are well known as energy consuming processes. In fact, a huge amount of coolant is required for the cooler located between reactor and the gas-liquid separator in the methanol synthesis process. Furthermore, there are two distillation columns are normally located in the methanol synthesis process to purify the crude methanol and these columns consumed a large amount of energy. In addition, some of the exothermic reaction heat are utilized for the preheating of the reactant stream, leading to reduction of recovered heat amount by steam generation.

Recently, authors have developed self-heat recuperation technology. In this technology, each process unit is divided into functions to analyze the process energy balance and not only the latent heat but also the sensible heat of the process stream can be circulated without any heat addition, leading to huge reduction of energy requirements in several chemical processes.

In this research, we investigated the feasibility of applying self-heat recuperation technology to a methanol synthesis process using industrial conditions and data and developed an innovative process for methanol synthesis process from the energy saving point of view. By installing the self-heat recuperation technology to the methanol synthesis process, the energy consumption of the process can be greatly reduced. In addition, all of the exothermic reaction heat of the methanol synthesis can be recovered by steam generation.

464428 The Impact of Fugitive Particulate Matter on the Environmental Health and Sustainability of Dry and Arid Regions

Abstract:

The developing arid and dry regions (i.e. North Africa, Middle East, Central Asia, and Australia) are experiencing a large wave of urbanization and industrialization, adversely affecting the area's air quality. In 2008, outdoor air pollution contributed primarily to 7.3% of the total deaths in the UAE (Gibson *et al.*, 2013). Lately, particulate matter (PM) pollution in these regions had become a topic of increasing concern (Nasser *et al.*, 2015; Tsiouri *et al.*, 2014), giving the recorded health hazards associated with its exposure. According to the air quality guidelines set by the World Health Organization, annual mean concentrations of $PM_{2.5}$ and PM_{10} should not exceed 10 μg m^{-3} and 20 μg m^{-3}, respectively (WHO, 2005). Nevertheless, annual PM levels estimated in several Middle Eastern cities between the years 2010-2012 have greatly exceeded these limits (WHO, 2014). Fugitive particulate matter (fPM) are unrestrained particles that escape to the atmosphere when applying a mechanical force on an exposed surface (e.g. emissions caused by wind shear, material transfer process, or other mechanical forces from non-point sources). fPM is also a substantial source of airborne pollution in these regions. However, accurate determination of fPM emissions and their impact on environmental health, humans' well-being, and overall sustainability of these regions has been an ongoing challenge for the air quality research community because of the induced health effects and the large uncertainty on their determination. Despite the efforts invested in developing emission inventories for standard air pollutants, it has been observed that (i) comprehensive spatially and temporally resolved emission inventories for fugitive PM sources in the developing arid regions are non-existent, (ii) the knowledge on fugitive (non-point) sources in the existing inventories is sparse (Pouliot *et al.*, 2012; Winiwarter *et al.*, 2009), and requires extensive attention.

In this study we present a holistic approach towards the assessment of the fPM impact to health and sustainability. We have selected the fPM emissions from loose Calcisols (i.e. soils with substantial accumulation of secondary carbonates) which are a characteristic soil for developing areas; usually very close to, or within residential areas. Multiple campaigns have been conducted to monitor fPM a) emissions from anthropogenic activities e.g. construction, b) building infiltration, and c) composition and size. We developed new emission factors for fPM from anthropogenic activities, assembled high resolution emission inventories, and finally assessed the impact of fPM on local air quality, environmental health, and sustainability of the involved society.

Acknowledgements

This publication was made possible by the support of the Qatar National Research Fund (a member of The Qatar Foundation) [NPRP 7 - 674 - 2 - 252] and the Ministry of Public Health [Contract No SCH/PA/18/2015] State of Qatar. We would like to extend our thanks to our research teams, owe to technical limitations not all contributing researchers have been included here. The statements made herein are solely the responsibility of the authors.

References

Gibson, J. M., J. Thomsen, F. Launay, E. Harder and N. DeFelice, 2013: Deaths and medical visits attributable to environmental pollution in the United Arab Emirates. *PLoS ONE*, **8**

Nasser, Z., P. Salameh, W. Nasser, L. A. Abbas, E. Elias and A. Leveque, 2015: Outdoor particulate matter (PM) and associated cardiovascular diseases in the middle east. *International Journal of Occupational Medicine and Environmental Health*, **28**:641-661.

Pouliot, G., T. Pierce, H. Denier van der Gon, M. Schaap, M. Moran and U. Nopmongcol, 2012: Comparing emission inventories and model-ready emission datasets between Europe and North America for the AQMEII project. *Atmospheric Environment*, **53**:4-14.

Tsiouri, V., K. E. Kakosimos and P. Kumar, 2014: Concentrations, sources and exposure risks associated with particulate matter in the Middle East area-a review. *Air Qual Atmos Health*:1-14.

WHO. (2005). WHO Air Quality Guidelines for Particulate Matter, Ozone, Nitrogen Dioxide and Sulfur Dioxide - Global Update 2005. In W. H. Organization (Ed.). http://www.who.int/phe/health_topics/outdoorair/outdoorair_aqg/en/.

WHO. (2014). Ambient (outdoor) air pollution in cities database 2014. http://www.who.int/phe/health_topics/outdoorair/databases/cities/en/.

Winiwarter, W., T. A. J. Kuhlbusch, M. Viana and R. Hitzenberger, 2009: Quality considerations of European PM emission inventories. *Atmospheric Environment*, **43**:3819-3828.

464865 Viologen-Catalyzed Carbohydrate Oxidation for Biofuel Cell Applications

Abstract:

Biomass has potential as a renewable source of electricity, but currently faces a number of challenges that inhibit its suitability for widespread use. One method of converting carbohydrates to electricity is with a biofuel cell. However, such fuel cells have been plagued by low reaction rates and incomplete conversion of the carbohydrate. Viologens are stable, non-biological catalysts that appear to enable nearly complete carbohydrate conversion. The process consists of a homogeneous reaction between viologen catalyst and the carbohydrate in solution, coupled with a second reaction (electrochemical or with oxygen) for re-oxidation of the catalyst. The purpose of this study is to investigate and quantitatively describe the reaction between the catalyst and the carbohydrate. To do so, experiments were performed as a function of temperature for a variety of carbohydrates obtainable from cellulose and hemicellulose at different molar ratios of viologen to carbohydrate. The rate of the reaction was tracked by spectroscopic measurement of the viologen concentration as a function of time. In addition, the extent of carbohydrate conversion was determined from the reaction products with use of 13C NMR. The reaction was observed to proceed rapidly for all of the biofuels tested; the most rapid rates measured were for the carbohydrates with fewer carbons. Of particular note is the observation of carbohydrate conversion >80% complete (i.e., ~20 of 24 electrons released from glucose). Finally, a comprehensive kinetic model was developed for the oxidation of carbohydrates by a viologen catalyst under anaerobic conditions. The results of this study support the feasibility of using viologen-catalyzed carbohydrate oxidation in a biofuel cell for power generation, and provide quantitative data needed for initial reactor design.

465067 Multi-Objective Optimization of Integrated Aspen Plus Unsteady-State Batch and Fed-Batch Fermentation and *in Situ* Gas Stripping Simulations

Abstract:

The batch Acetone-Butanol-Ethanol (ABE) fermentation process produces low final ABE concentrations, yields and productivity as a result of substrate and product inhibition to the microorganisms. Approaches to reduce product toxicity to the microorganisms include integrated fermentation and *in situ* product recovery techniques, such as gas stripping, adsorption, pervaporation, etc. Fed-batch fermentation reduces substrate inhibition by supplying substrate as it is consumed to keep the substrate concentration in the reactor below the inhibitory levels to the microorganisms.

Process simulation can predict the most informative experiments to conduct for process design, control and optimization. Previous Aspen Plus simulations and analyses of the ABE fermentation represented the batch and fed-batch processes with steady-state stoichiometric reactors with fixed product yields relative to key inputs, such as substrate (glucose) concentration. These steady-state representations of the batch and fed-batch fermentation processes are inadequate as these processes are inherently unsteady-state; the fermentation environment changes due time-dependent concentrations of cells, substrate, intermediates and products. This simulation approach results in Aspen Plus steady-state simulations that decouple the ABE fermentation kinetics from the fermentation environment and may not be representative of the integrated batch and fed-batch ABE fermentation and *in situ*gas stripping. The operating conditions of the fermentation must be directly linked to the fermentation kinetics in order to optimize productivity and control product profiles in the fermentation process.

To this end, this study will use a multi-objective optimization strategy to maximize the ABE productivity, yield and concentration by manipulating the operation conditions of integrated batch and fed-batch fermentations and *in situ* gas stripping processes. The integrated batch and fed-batch ABE fermentation process will be simulated as unsteady-state processes using a Fortran user kinetics subroutine linked to the batch reactor (RBatch) in Aspen Plus. This research can serve as a basis to evaluate the economic viability and technical feasibility of bioprocesses as a decision-support tool for researchers, investors and policy makers in the long run.

468409 Characterization Based Design of Ionic Liquids for CO_2 Absorption

Abstract:

Chemical engineering design problems have evolved in recent years from large scale bulk processes to targeted small batch approaches. As a result, product design is reaching beyond the traditional solutions to classic chemical engineering problems such as extraction, purification, removal of sulfur from fuel, CO_2 capture and sequestration, energy storage and biomass pre-treatment. Ionic liquids can provide environmentally benign solutions that can be tailored to specific process requirements. Due to the unique structure of ionic liquids, the cation, anion and length of the alkyl chain can be varied to create a molecule with specific physical properties. It has been estimated that over 10^{14} unique molecules exist at room temperature, so experimental trial and error methods will not be effective. More advanced methods must be created determine the ideal combination for a specific set of process requirements.

The purpose of this work is to determine the ideal ionic liquid to capture CO_2, a question which has been the topic of many experimental laboratory studies. However, only a small percentage of the potential ionic liquids have been synthesized and tested for the solubility of CO_2. This innovative approach decouples the solution from time-consuming laboratory study and streamlines the search for the most promising candidates that should be synthesized.

A characterization based group contribution method is combined with density functional theory to determine the ionic liquid that can most effectively absorb CO_2. Infrared spectra data contains descriptor data that can be used to estimate properties of ionic liquids, but does not exist for all ionic liquids. Density functional theory is used to create IR data based on a training set of experimental data. Principal component analysis and partial least square techniques will be employed to reveal important features and patterns in the molecular architecture. A characterization based group contribution method is used to estimate the properties of the new structures. The reverse design of potential ionic liquid molecules will be completed by an exhaustive search of combinations with various cation, anions and lengths of alkyl chains until a candidate molecule is found that provides the highest solubility of CO_2.

467962 Simultaneous Optimization for Heat-Exchanger Network Synthesis of Natural Gas Combined Cycle Power Plant Integrated with Post-Combustion CO_2 Capture and Compression

Abstract

In this work the thermal integration of natural gas combined cycle (NGCC) power plant integrated with post-combustion carbon capture (PPC) and CO_2 compression is proposed. With the objective of found a heat exchanger network (HEN) that decrease the energy impact of the PCC process, which decreases the power of the turbines as it requires a large amount of steam for its operation and also it uses part of the power generated by the NGCC power plant. This study consists of a 453 MW_e NGCC power plant with an MEA-based PCC process which are simulated in ASPEN PLUS® to get all the process inventory that includes stream flows, composition, temperature, pressure, equipment heat, work, and also the identification of hot and cold process streams. Using the model of simultaneous optimization of thermal integration SYNHEAT scheduled in GAMS™ it was obtained as a result the decrease in the use of steam in the striper reboiler to 65.43 Kg/s with a HEN that decrease also the use of hot water to heat the input of natural gas producing a 126.571 MW_e power output of the steam turbines increasing the thermal efficiency to 50.6996%, due to the use of the heat of CO_2 compressor train streams in the HEN.

- Abstract Text

468973 Overview on Evaluation of Sustainable Biofuel Conversion Processes

Abstract:

This paper reviews the existing methodologies on how to assess the overall sustainability of a biofuel production process, including approaches on how to quantify and compare the key indicators. It is always highly desirable to develop a more sustainable biofuel production process, for better economy, environment, and society. A holistic sustainability analysis for the production of biofuel plays a critical role on the improvement of the overall life cycle sustainability. Once a conceptual conversion process is developed, the next step is to evaluate its overall sustainability, based on a process sustainability evaluation methodology using various indicators in the areas of environment, efficiency, energy, and economics. One outcome of the sustainability evaluation is to allow researchers to identify process areas in need of sustainability improvement, as well as the challenges and opportunities for achieving the best possible sustainability target.

468418 The Role of 'Free Radical Based Mechanisms' in Wood-Feeding Termites for the Deconstruction of Lignocellulosic Biomass

Abstract:

Wood-feeding termites are very efficient in deconstructing the plant cell wall, leading to more than ~90% of cellulose and hemicellulose solubilization in a span of 24 hours. The key to this efficient natural process is the *mechanism of selective lignin modification* in the termite gut as shown by previous studies, which characterized lignin subunits in termite chewed wood particles and feces. These studies have led to an important scientific question- what is the mechanism that helps termite to cleave selective lignin bonds, enough to avoid lignin recalcitrance for saccharifying enzymes?

In the present research, we are answering this question by designing a speculated *free-radical driven system* that can participate in lignin bond cleavages in such a manner that it would avoid lignin recalcitrance and subsequently allow bigger molecules like-enzymes to enter the matrix and catalyze the saccharifying reactions. The free radicals are small, highly reactive and unstable molecules like hydroxyl, superoxide etc., which play an important role in termite based pretreatment. These radicals can be generated and utilized in many enzymatic and chemical reactions. Therefore, this proposed free-radical system could be targeted as a mimic for designing a novel pretreatment process.

469935 Life-Cycle Assessment of Biofuels and Chemicals from Unconventional Methane Feedstocks

Abstract:

This is the result of an ongoing collaboration with a team of researchers in an ARPA-E project to investigate the feasibility of utilizing stranded, small-scale sources of natural gas to produce liquid transportation fuels and value added chemicals. In addition to technical and economic considerations, environmental impacts are being studied through a cradle-to-gate life-cycle assessment approach. Here, we present preliminary findings for the fuel life cycle, involving methane procurement, conversion to lipids in a bioreactor, and further downstream processing. We have defined several potential scenarios for using methane derived from stranded gas, associated gas, and landfill gas. The requirements for procuring each gas stream are taken into account, along with fugitive emissions and considerations of alternate usage for each gas stream. Bioreactor input requirements, along with outputs of products and other material streams, were evaluated in consultation with LanzaTech process engineers. Environmental impacts being considered are global warming potential and fossil energy demand. Initial results reveal that major differences emerge between potential gas feedstock scenarios based upon alternate fate of the gas (flaring vs. utility generation, for instance). Estimated fugitive emissions rates also play a large role in determining the embodied global warming potential of feedstock gases and will need to be thoroughly documented in each commercial setting to verify optimal performance. Large improvements to the overall environmental performance of the system are gained when next-generation LanzaTech bioreactor technology is utilized, in comparison to conventional loop or bubble column reactors, due to improved mass transfer properties and lower gas compression requirements to facilitate the methane fermentation reaction inside the bioreactor.

469948 Climate Change Impacts of Bioethanol Production from Willow

Abstract:

Abstract:

The use of cellulosic biofuels is projected to increase in order to offset greenhouse gas (GHG) emissions from the combustion of non-renewable fuels. Hot water extraction (HWE) is a potential pretreatment step for multi-pathways biorefinery to yield natural, wood-based chemicals and materials. A cradle-to-grave life cycle assessment (LCA) was performed to assess the GHG emissions associated with the production of 1 MJ of ethanol by fermentation of sugar derived from HWE of willow biomass. The system boundary includes cultivation and harvesting of willow, transportation from the field to the biorefinery, hot-water extraction, fermentation and separation of ethanol, as well as the transportation of ethanol and its combustion. The system boundary was expanded to include co-products and underground carbon sequestered. Two conversion scenarios were considered. In the first scenario, electricity and heat required for the processes are cogenerated on site by the combustion of a fraction of the extracted willow biomass. In the second, electricity for the processes is supplied exclusively by the national grid. A sensitivity analysis and Monte Carlo analyses were performed for each scenario. Preliminary results show that GHG emissions for scenario one are -36.9 g CO_{2eq} MJ^{-1}. For scenario two, GHG emissions increase to 117.9 g CO_{2eq} MJ^{-1}. The amount of yeast used and the yield of willow were the most sensitive variable parameters. The fermentation and distillation of ethanol accounts for up to 70% of the total greenhouse gas emissions. The source of heat and electricity is the main driver in reducing greenhouse gas emissions when gasoline is replaced by ethanol produced from sugars that are extracted by HWE of willow.

470418 CO_2 and Water Adsorption on Seeded Growth of Mg-MOF-74 on Carbon-Nanotubes

Abstract:

Increased atmospheric CO_2 concentrations result in increased demand for novel and efficient CO_2 control technologies. Metal Organic Frameworks (MOFs) have recently attracted research interests due to their tailorable porous structure and potential application as novel adsorbents. The presence of open metal sites in some MOFs fortifies their adsorption sites, which is favorable for CO_2 capture. In this research, the adsorption performance of Mg-MOF-74 is enhanced by incorporating MOF on a membrane surface. Mg-MOF-74 was synthesized by solvothermal method and seeded on the surface of plasma treated carbon nanotube bucky paper which has demonstrated a superior CO_2 adsorption. The plasma treatment was conducted in either O_2/Ar or NH_3 gas flows. All samples were characterized using XRD, XPS, SEM, and CO_2 volumetric adsorption. XRD results show identical pattern for Mg-MOF-74 grown on CNT and Mg-MOF-74 alone, confirming the intact crystalline structure of MOF. XPS measurements illustrates that plasma treatment with O_2/Ar can yield samples with higher wettability which is favorable for further surface modification and increases affinity towards water vapor by anchoring hydroxyl functional group on the surface. The incorporation of MOF on CNT increased the CO_2 adsorption performance of MOF. The highest CO_2 adsorption capacity of 5.6, and 34.7 mmol/g was obtained for Mg-MOF-74, and Mg-MOF-74-CNT, respectively. Maximum water adsorption capacity of 9.0 and 20.6 mmol H_2O/g were obtained for Mg-MOF-74 and Mg-MOF-74-CNT, respectively. Comparing CO_2 and H_2O isotherms for prepared samples demonstrates that generally the seeded growth Mg-MOF-74 on CNT-BP performed better than Mg-MOF-74.

470328 Optimization of Biomass Processing through Physical Fractionation

Abstract:

The nascent cellulosic biofuels industry is struggling due to challenging economics derived from high cost of cellulosic sugars relative to starch and sucrose (and recently the substantial drop in the price of petroleum) with contributions arising from the high capital and operating costs associated with large centralized lignocellulosic biorefineries. Well-recognized, critical challenges include the low bulk density represents an important challenge for the logistics of feedstock transport and storage, the substantial within- and between-feedstock heterogeneity of herbaceous feedstocks such as corn stover and switchgrass that can contribute significant challenges to processing and conversion, and the need for high-value co-products or co-products targeting diverse market sectors to both add value and to improve the economic robustness to fluctuations in both global and local market forces.

Biomass heterogeneity has important impacts on a number of important processing variables. These include differing responses of the biomass to comminution that result in wide particle size distributions, substantial differences in particle hygroscopicity and drainability, a diversity of compositions (including lignin and ash), and a wide range of responses to chemical pretreatments and hydrolysis. This work will, work will assess physical fractionation approaches for corn stover and energy sorghum that mimic industrially relevant fractionations utilizing either manual classification that mimics in-field classification that could be achieved during harvest and combinations of comminution and sieving that could be performed either during harvest or at decentralized processing depots. Employing these approaches in this work, we demonstrate that physical fractionation can yield biomass fractions with diverse properties that include wide ranges for lignin content and hygroscopicity that impact feedstock processability (*i.e.*, response to pretreatment and enzymatic hydrolysis).

471527 Kinetics Modeling of Fed-Batch Lipid Accumulation of Activated Sludge

Abstract:

Recent studies have demonstrated the capability of activated sludge (AS) microbial consortia to accumulate lipid droplets under fed-batch cultivation on acetic acid as carbon source. The experiments were conducted in laboratory-scale bioreactors. The prediction of system performance at a large-scale implementation of the technology requires the mathematical modeling of the laboratory-scale bioprocess, so kinetic models must be fitted to the experimental data. This study models the kinetics of the AS microbial culture during the fed-batch cultivation.

Monod-type kinetic models were regressed to the experimental data using various numerical methods implemented in MATLAB. Initial data-fitting via unconstrained nonlinear optimization of least-square-error suffered from instability of solution and long computation time. Then the Markov Chain Monte Carlo (MCMC) method was implemented resulting to improved regression. That is, the model parametric constants converged while minimizing the least-square-error at a faster rate. This work describes the various components of the problem-solving strategy that enabled the kinetics modeling.

472988 Pyrolytic Fractionation of Algal Biomass: Catalytic Conversion of the Resulting Vapors to Fuels and Chemicals

Abstract:

Our group recently developed a novel pyrolytic fractionation method for converting oleaginous biomass (microalgae and/ or oil seeds) to fuels and products. We have observed that the biopolymer components (protein, starch and lipids) volatilize over narrow and distinct temperature regions. This allows recovery of oil and gas separately from each fraction. Through pyrolytic fractionation, volatile N-containing compounds and carbohydrate products can be removed from the biomass by first heating to ~320 °C and holding temperature steady until most protein and starch are thermally degraded or stabilized via polymerization to char. The biomass remaining after thermal removal of protein and starch (first pyrolysis fraction) can be further processed (at temperatures > 450°C) to recover low-N bio-oil from lipids, again as a separate fraction. We are investigating the production of higher value aromatics and nitriles from each of these fractions (320 °C and 450°C) by upgrading the generated vapors using an H-ZSM-5 catalyst bed. Our recent findings on this project will be presented and discussed

471255 Biomethane Potential Determinants of Different Agro Industrial Substrates

Abstract:

Anaerobic digestion is a promising technology offering the opportunity for conversion of biomass into biogas, a renewable fuel that can be used for energy generation in internal combustion engines, microturbines and boilers. Throughout the years this technology has been applied for the valorization of waste streams including but not limited to manure, olive mill waste and municipal sludge. While the method is offering a promising pathway for generation of biogas, it may suffer by low efficiency, inhibition by product or substrate accumulation and lack of available enzymes. The literature is rich with studies describing the biomethane potential of comparable (according to the presented analysis) substrates where the efficiency and the recovered yields showed significant variation. For example Pitk et al. (2013) studied the biomethane potential of different slaughterhouse wastes. The sterilized mass (i.e. bone and meat meal) used in this study achieved a production of 721 mLCH4gVSadded. In contrast Wu et al. (2009) reported a yield of only 381 mLCH4gVSadded while assessing bone and meat meal. Furthermore, Ebner et al. (studied the digestion of cattle manure and reported a biomethane potential of 238 mLCH4gVSadded which is at least 30% higher to this described by Amon et al. (2007). Bayr et al. (2012) assessed the biomethane production of fat and reported yields of upto 403 mLCH4gVS FM (FM = 935g/Kg, 100%VS) corresponding to approximately 430 mLCH4/gVSadded. In contrast to this, Martínez EJ et al. (2016) reported yields higher than 800 mLCH4/gVSadded. Despite the fact that all these values are well within the theoretical limits, the reason of the variability is not easily distinguished. While the variations can be the outcome of differences in the influent substrate and the presence of inhibitors, other technical issues and lack of standardization cannot be excluded.

In this work we investigated the anaerobic digestion of four substrates (animal fat, wheat straw, sterilized slaughterhouse wastes and glucose). The different substrates selected were representative of the most common feedstock used in industrial digesters.

The substrates were assessed in batch vials with the reaction time set to 30 days under mesophilic conditions. The

loading rates applied were controlled to range between 10 and 20 kgVS/m3, while the inoculum to substrate ratio between 2 and 0.5. Furthermore, the experiments were divided into two stages with two complete sets of vials inoculated with different inoculums sourced from reactor operating under different loading rates and feed composition regimes. The last experiment targeted the recovery of data from where information related to the importance of the initial inoculum can be extracted.

According to the results the source of inoculum appears as the most important factor for the AD process. The inoculum from the reactors treating the agro industrial waste (Inoc A), was able to offer a robust process when fat was used as substrate with only the 0.5 inoculum to substrate ratio failing. On the other hand, when the inoculum recovered from the reactors treating manure (Inoc B) was used only digestion where the lowest loading rate applied was successful. In this case the ultimate yield was at least 25% lower compared to this achieved by Inoc A. The same is true for the case of the sterilized mass where the daily productivities of the batches with Inoc A were up to 100% higher than these offered by Inoc B. On the other hand when sugars were digested Inoc B trsulted in higher yields and productivities compared to Inoc A. The second important parameter is the initial inoculum to substrate (I/S) ratio. A low I/S ratio can delay the biomethane production and in some cases inhibit the overall process which can be easily described by the limited number of available cells in the inoculum and the detrimental effect that an inhibitor can have in an unbalanced biological process. The least important factor among the ones assessed is the initial organic load per batch. Loadings up to the 20 kgVS/m3 can be sustained by AD systems treating slowly hydrolyzed substrates. On the other hand, addition of 20kgVS/m3 of simple sugars into AD systems can inhibit the process through intermediate product accumulation, manifesting the inherent problem of the process and the different rates of the biochemical reactions involved.

Our analysis underlined the critical role that inoculum (both in terms of its source and in terms of its relative concentration in the reactor) plays on the determination of the overall methane production efficiency in anaerobic digestion systems. Further work will investigate the microbial and biological profile of the system to determine optimal operational regimes as a function of the performance determinants analyzed above.

Amon, T., Amon, B., Kryvoruchko, V., Zollitsch, W., Mayer, K., Gruber, L. 2007. Biogas production from maize and dairy cattle manureâ€"Influence of biomass composition on the methane yield. Agriculture, Ecosystems & Environment, 118(1â€"4), 173-182.

Bayr, S., Rantanen, M., Kaparaju, P., Rintala, J. 2012. Mesophilic and thermophilic anaerobic co-digestion of rendering plant and slaughterhouse wastes. Bioresource Technology, 104(0), 28-36.

Ebner, J.H., Labatut, R.A., Lodge, J.S., Williamson, A.A., Trabold, T.A. Anaerobic co-digestion of commercial food waste and dairy manure: Characterizing biochemical parameters and synergistic effects. Waste Management.

Martínez EJ, Gil MV2, ernandez C1, Rosas JG, X., G. 2016. Anaerobic Codigestion of Sludge: Addition of Butcher's Fat Waste as a Cosubstrate for Increasing Biogas Production. PLoS One(http://dx.doi.org/10.1371/journal.pone.0153139).

Pitk, P., Kaparaju, P., Palatsi, J., Affes, R., Vilu, R. 2013. Co-digestion of sewage sludge and sterilized solid slaughterhouse waste: Methane production efficiency and process limitations. Bioresource Technology, 134(0), 227-232.

Wu, G., Healy, M.G., Zhan, X. 2009. Effect of the solid content on anaerobic digestion of meat and bone meal. Bioresource Technology, 100(19), 4326-4331.

472961 Flash Pyrolysis and Fractional Pyrolysis of Oleaginous Biomass in a Fluidized-Bed Reactor

Abstract:

In this study, flash pyrolysis of oleaginous biomass was investigated to produce liquid fuels. Milled soy flakes were used and can be considered a surrogate for algal biomass due to similar compositional, proximate, and ultimate characteristics. A fluidized-bed flash pyrolysis was designed and constructed to facilitate rapid heat transfer to the biomass along with short vapor residence times to minimize secondary degradation of bio-oils and achieve high yields of bio-oil. In addition, a dry-ice packed bed condenser was used to recover the generated vapors. The novel condenser provides a large area of cold surfaces that allow dilute vapors to impinge, condense and coalesce effectively. Single-step pyrolysis experiments were performed at several temperatures over the range 250-610°C with a vapor residence time between 0.3-0.6s. Biooil yields were highest at 550°C (~70% g/g-feed) with only a 12% loss as uncondensed vapors. Moreover, 90% of the feedstock lipids were recovered in the bio-oil fraction. In addition, three-step fractional flash pyrolysis experiments were performed to assess the possibility of producing higher quality bio-oils since a large fraction of protein and carbohydrates degrade at lower temperatures (310-350°C). A low temperature pyrolysis step (250ºC) was first performed and was followed by pyrolysis of the residues at higher temperatures (310 and 530ºC). The bio-oil from last step (530ºC) resulted in higher quality bio-oil with low water- and nitrogen- content, and high heating value of 33 MJ/kg.

473026 Life Cycle Assessment of Renewable Hydrotreated Fuels from Castor

Abstract:

Hydrotreated renewable transportation fuels (HRTF), i.e., green diesel and bio-jet fuel, have become promising options for reduction of fosil fuels consumption and mitigation of climate change [1]. These biofuels have to be produced, however, in a sustainable way. The life-cycle assessment (LCA) have shown to be an useful and accurate tool to evaluate CO_2production and energy and water consumption in the Biojet Fuel Production [2].

LCA allows the evaluation of the environmental impact of the production of hydroprocessed renewable fuel against conventional fuels, the main objective of this alternative is to reduce the environmental impact. The limits and objectives of the LCA were defined in order to measure CO_2emissions during cultivation, transport and industrial production, everything applied at the conditions of Oaxaca State. This research is part of the research to demonstrate that hydrotreated renewable fuels are a real option to substitute conventional hydrocarbon-based fuels in Mexico.

Thus this work evaluates the life cycle assessment of green diesel and bio jet fuel from castor plant, which has a large growth in rural areas along Mexico. The LCA methodology has been used in this work to quantify GHG emissions. No land change has been considered, due the nature of the crop, which is a wild and not eatable plant. The presentation details the methods, assumptions and LCA results for the production of renewable fuels in Oaxaca State. Preliminary results indicate that HRTF are capable of being producedin large scale to supply a significant fraction of energy demand of the country, reducing at the same time the GHG emissions. These results contributed to a favorable carbon footprint result for castor HRTF, yielding savings of greenhouse gas emissions compared to fossil fuels.

References

[1] J.Han et al, "Life-cycle analysis of bio-based aviation fuels", Bioresource Technology, pp 1-10, 2013

[2] D. B. Agusdinata et al, "Life Cycle Assessment of Potential Biojet Fuel Production in the United States", Enviromental Science & Technology, pp 1-11, 2011.

473262 Enhanced Design of Hybrid Distillation Processes for Separation of Biomass Derivatives

Abstract:

Global warming is the current major environmental issue that modern society is facing due to the world's overdependence on the use of mineral resources as raw materials. Therefore, extensive studies of renewable resources as feedstock to substitute fossils have been carried out. Therefore, the interest of biomass, which is the only renewable resource of fixed carbon, has increased considerably in recent times. Biomass can not only produce conventional hydrocarbon liquid transportation, but also petrochemical products. Several value-added platform chemicals with a range of potential uses from biomass were investigated based on the potential markets, derivatives and the complexity of the synthetic pathways. On the other hand, biomass production on an industrial scale is limited due to the expensive feedstock, low synthetic yield and lack of detailed process design. Several promising hybrid separation processes for recovering valuable biomass derivatives from inexpensive raw materials have been investigated. This work proposes a systematic procedure of enhanced design of hybrid distillation process for separation of biomass derivatives. Initially, enhanced separation process configurations were assessed based on a literature survey, equilibrium distribution and a separation feasibility study. Several configurations were chosen for detailed design and optimization on an industrial scale and compared in terms of the energy requirement, total annual cost (TAC) and total annual CO_2 emission (TAE). All distillation columns were designed with a load of approximately 85% of the load at the flooding point to prevent flooding in the columns. Some preliminary simulations were conducted to determine the initial structures for the extractor and all distillation columns. A rigorous simulation using the simulator, Aspen Plus V8.6 was then conducted to identify the optimal design in terms of minimizing the TAC. The capital cost, total annual operating cost (TOC) and TAC were calculated based on Guthrie's modular method. The results showed that the innovative hybrid distillation configuration could save up the energy requirement and TAC significantly compared to the conventional sequence.

474714 A Chemical Engineer's Guide to Contemporary Problems of Climate, Energy and Environment

Abstract:

From artists to engineers, how does an instructor best prepare his or her students for issues they will face related to climate change, energy supply, and the nexus of these two issues? Future public policy decisions may very well be in the hands of those without a strong scientific background. Meanwhile, as many engineering undergraduates are developing the technological background to tackle these challenges, they may miss out on many aspects of a broader education that will put future solutions in context. Topics regarding energy and the environment are much broader and more complex than can be covered in the traditional one hour lecture period. Over the past year, the authors have been working on a general education book entitled, "How Science is Done: Contemporary Problems of Climate, Energy and Environment". Although targeted to a general (i.e. non-scientific) audience, the book is relevant to providing the future chemical engineering workforce a broader understanding of how technology will impact society and the environment. In the book, we have also worked to incorporate a number of detailed discussions on the physics of how the earth's climate is determined and regulated, the thermodynamics and mechanics of engines, energy conversion and conservation, and the mass flows of carbon throughout the environment. A number of these examples are relevant to the basic chemical engineering curriculum. Even more of these examples can be used to engage first-year engineering students in the impact engineering has on the society and the environment. Several classroom examples will be shared in this session, with a focus on how they might be utilized within the existing chemical engineering curriculum. A longer term "green engineering" project that incorporates hands-on campus-wide sustainability initiatives into an advanced undergraduate elective will also be discussed. To "engineer a sustainable future", our undergraduates will need a broad contextual understanding of energy usage, energy supply, carbon emissions, and environmental ramifications of technology.

475634 Residential Wood Pellet Heating System Optimization Using Process Dynamic Simulation

Abstract:

As a growing part in biomass energy utilization, the use of wood pellet boilers for residential space heating have been increased over the past decades in EU and North America. A typical wood pellet heating system has three major components: the boiler, thermal storage tank, and heat consumption unit. Considerable effort has been applied to the design of the boiler to build high efficiency and low emissions (HELE) boilers. However, currently there is no systematic process optimization in terms of the system efficiency, especially the optimization of the operation of the thermal storage tank with the heating system.

In this study, a 25 kW nominal output wood pellet boiler was installed and monitored through the past two heating seasons (from February 2015 to May 2016). Two 120 gallon tanks were connected in parallel with the boiler as the thermal storage unit. Field data were collected for boiler thermal efficiency calculation. Meanwhile, several runs of CTM-039 stack sampling were performed to evaluate the emissions from the boiler. In the end, a process dynamic simulation using VMGSim^{TM} was built for the residential wood pellet heating system. The boiler simulation is based on wood combustion modeling of the boiler and tuned with field data.

The whole system was optimized for a better combustion and a higher system efficiency from the following aspects. Firstly, optimize the combustion efficiency through control of excess air ratio, pellet quality, etc. Secondly, optimize boiler thermal efficiency by running the boiler at maximum load and employing more efficient cleaning mechanisms. Thirdly, optimize of the tank sizing with respect to boiler nominal output and heat consumption from the building for a higher tank efficiency and better use of water stratification in the tank. And lastly, a sensitivity analysis was performed for the validation of the model.

This work can substantially improve future residential space heating system designs based on wood pellet heating technology.

475766 Introduction and Study on Electron-Beam Assisted Stabilization Process for Cost-Effective Manufacture of Carbon Fibers: Process to Bring about Wide Application of Lightweight Structural Composites for Sustainable Environment

Abstract:

To lower energy consumption and carbon emission in transportation is essential choice for future sustainable environment. Employing a lightweight structural composite with high mechanical properties in the structure of transportations can be thought as one of the approaches. As a reinforcing material in the composite, a carbon fiber (CF) is a promising candidate due to its high strength and modulus per unit weight. Even though the CFs have been commercialized and available in the market, the application of CFs is still limited to specific transportation such as aircraft parts due to their high manufacture cost. The wide application to common transportations such as personal vehicles will be realized by decreasing the manufacture cost of CFs.

The manufacturing process of CFs derived from polyacrylonitrile (PAN, the most common precursor polymer of CF) can generally be divided into three processing steps: 1) spinning and stretching, 2) stabilization (200~300 Â°C) and 3) carbonization (â‰¥1000 Â°C). A number of experimental parameters associated in each step have been extensively optimized for the energy and time effective manufacture of the CFs. Among these three steps, the stabilization is the most crucial step as a determinant of not only high performance but also of processing cost due to lengthy heat treatment. The conventional stabilization is carried out by passing the fibers through a series of convection ovens with gradual temperature gradient (200~300 °C) for two to three hours, which is the most time and energy consuming step. As one of the methods to shorten such lengthy step with expensive energy consumption, electron beam irradiation on PAN precursor fibers has been employed prior to short thermal treatment. In conclusion, with short treatment of electron beam (200 sec), the irradiated fibers were successfully stabilized with short time of thermal treatment (30 min). In the presentation, we briefly overview the environmental impacts if the CF composites are commonly applied to transportation parts and approaches towards low manufacture cost of CF for each of manufacturing step. In particular, as one of the specific suggestions to the problem and study, the electron beam treatment for energy-effective stabilization will be explained in detail by presenting several characterizations of irradiated PAN fibers.

475416 Life Cycle Assessment of Pharmaceutical Manufacturing Processes

Abstract:

Continuous processing, as a form of process intensification, is one of the keys of green engineering research and development in the pharmaceutical industry. It has the potential to reduce solvent use and cost of production as well as increase production quality and operational safety. The goal of this work was to identify environmental hotspots, quantify and compare the differences in environmental burdens between batch and continuous manufacturing processes. A comparative cradle-to-gate life cycle assessment between the batch and continuous manufacturing processes of 4-D-Erythronolactone at pilot plant scale is presented as a case study. The assessment begins with defining the scope, the system boundaries, and functional unit. The system boundaries include from raw material extraction, transportation, the manufacturing process, equipment cleaning, facility requirements (lighting and heat, ventilation and air-conditioning) and waste incineration. Two other waste handling options (wastewater treatment and solvent recovery) are investigated as they are common operations employed by the pharmaceutical industry to treat waste. As part of a modular approach to obtain life cycle inventory data, each manufacturing system was part of a theoretical production campaign to fulfil the functional unit. A variety of methods such as kinetic and reactor modelling, input-allocation inventory models is used to fill data gaps.

For the case study, environmental hotspots of the batch system were centered on the production of inputs, washing activities and waste disposal. The adoption of continuous processing from batch manufacturing produced an overall reduction in impacts. With continuous processing, less solvents were used in washing and it translated to reduced impacts from washing and waste disposal. However, more energy was required for continuous processing due to the use of less concentrated starting materials. This case study provided insights into the life cycle impacts of batch and continuous manufacturing processes and helped develop a modular approach to compare the two manufacturing systems.

475872 Advancing Building Materials Product Transparency with the Living Product Challenge

Abstract:

General Concept, Background, Motivation

Product transparency has become a mega-trend in the building materials industry. Consumers now demand that increased scrutiny be placed on the materials they interact with on a daily basis. To that end, ingredients once considered proprietary to product manufacturers must now be disclosed.

Method

The Living Product Challenge is a comprehensive 3rd party certification that goes beyond single attribute programs to encompass all the life cycle stages of a product as well as the social impacts, in addition to environmental and financial impact. The International Living Future Institute's Living Product Challenge challenges the status quo; it encourages manufacturers, architects and builders to consciously increase the level of disclosure they offer the public through their structured product certification process. The work described in this project outlines Owens Corning's journey as a pilot participant in the Challenge to successfully gain certification for their unbounded loosefill fiberglass insulation as the world's first Living Product. The process caused Owens Corning to thoroughly examine the way it conducts business, manufactures products and quantify the positive impact its operations have the environment.

Results and Interpretation

The successful facility and document audit lead to imperative certification for Owens Corning's leading fiberglass insulation products. This test-and-learn project served as a case study for the Owens Corning internally for use as a model for documenting the sustainable impact of other products within the company's portfolio. The intent was to encourage other manufacturers to join Owens Corning in increasing their commitment to product sustainability.

The process, product and lifecycle are all integral players in the success of a sustainable edit to a manufacturing regime. The Living Product Challenge directly addresses marketplace demand for comprehensive product transparency while leveraging single attribute certifications to document the sustainability gains of industry leading companies like Owens Corning.

475935 Process Design, Modification, and Sustainability Assessment of Coal and Biomass Co-Fired Plants for Generation of Transportation Fuel: A Case Study in Kentucky

Abstract:

As demand for in-country generated transportation fuel increases, methods that utilize region-based resources to produce biofuels are sought. Such is the case in Kentucky, United States, where the biofuels blend rate in transportation fuels is likely to increase from 10% to 25% by 2022. Consequently, a total of 775 million tons of transportation fuel are demanded to keep the state from importing 90% of its fuel in 2022. This challenge can be met by investing the implementation of coal and biomass co-fired plants in Kentucky to produce various fuels, including diesel, kerosene, and gasoline. To analyze this possibility, process design and modification of coal and biomass co-fired plants was conducted for energy and material efficiency improvement and waste reduction. Process simulation was performed, with the evaluation of process sustainability using the IChemE Sustainability Metrix System. The results of this case study will be presented.

475948 Synthesis, Properties, and Processing of Melt-Spinnable Acrylonitrile Copolymers

Abstract:

Carbon fiber (CF) which is one of outstanding structural materials have received much attention for application such as composite materials, textiles, microelectrodes, and catalyst because of its superior mechanical properties, heat resistance, low thermal expansion, chemical resistance, and so on. CF has been developed from several kind of precursors such as polyacrylonitrile (PAN), pitch, and rayon. In particular, because CF derived from PAN shows especially high mechanical properties and its manufacturing process has been well studied, CF commercial makers has mainly used PAN as a CF precursor. Until now, CF and its composites have been only applied in high tech fields such as aerospace because of the cost problem. Recently, those application to automobile is a rising issue because those properties, light weight and high strength, can make increase of performance and fuel-efficiency. However, reducing the cost of CF is a precondition for these application. In addition, because the automobile application is not required for very high mechanical properties, a little decrease of the properties is allowed. In this regards, many research institutes have found the method for reducing manufacturing cost. There have been several approaches such as using low cost precursors and changing manufacturing procedure. In particular, traditional PAN precursors is prepared by the wet spinning which required expensive organic solvent recovery process and waste water treatment process. In this research, we have synthesized acrylonitrile copolymers containing functional components for reducing manufacturing cost by using melt spinning process which can spin fibers with a high speed than wet spinning and do not require solvent and water handling. In conclusion, we could prepare acrylonitrile fibers by melt spinning and stabilize the fibers by using other treatment with thermal method. In the presentation, the thermal and rheological properties of the synthesized polymers will be mainly discussed in detail.

473927 Refineries in a Carbon Contrained Future - the Potential for Facilty-Wide CO2 Capture at Plant Utilities

Abstract:

What can refineries do in a carbon constrained future? Obviously refineries will produce products to satisfy the product demand slates â€“ primarily liquid transportation fuels. However, refineries can also remove the CO2 emissions associated with their operations. We will explore a CO2 capture concept comprised of a utilities plant that and produces hydrogen for combustion in the refinery furnaces and CO2 to send to storage outside the refinery boundary. This option could be the least-cost pathway for refineries to remove their CO2 emissions.

CO_2-Utilization in Chemicals Production Based on Evidence from Start-Ups and Literature

Abstract:

Greenhouse gas emissions and the resulting anthropogenic climate change create tremendous economic, ecologic and social challenges for societies today and in the future. To mitigate these effects, world leaders agreed to limit global warming to two-degrees compared to preindustrial times, which translates to an atmospheric CO_2-concentration of 450 ppm [1]. To reach this goal, global annual greenhouse gas emissions have to decrease by 40 to 70 percent until 2050 and to near-zero until 2100. Major reductions are suggested to be generated from energy efficiency measures, renewable energies and carbon capture and storage [2]. One additional emissions reduction pathway is CO_2 utilization, which takes CO_2 from point sources or from air and utilizes it in large-scale applications. This way CO_2 becomes a process agent or even a carbon source, which leads to the substitution of fossil fuels and lower overall greenhouse gas emissions [3].

Whether or not emission reductions from CO_2 utilization can be realized, largely depends on the rapid implementation in large-scale industrial processes and markets. While the political pressure and support is high, most current CO_2 consuming reaction pathways have not yet passed lab-stage technology levels and industrial feasibility is mostly not yet proven [4]. Many projects described in literature are in the stage of catalyst and reaction optimisation. In many cases evaluations of early-stage base chemicals, intermediates, polymer or fuel reaction pathways have been carried out in literature (e.g. [6], [7], [8], [9]), often described as "techno-economic" assessment. However it remains unclear what indicators such a techno-economic assessment should include or which procedure it should follow. While critical success factors for large-scale implementation lie within the fields of technologic, economic but also environmental feasibility, very few studies manage to evaluate technologies in all three fields. Additionally assessments in high level of detail remain challenging, since many processing parameters are not defined at an early stage of technology. A multidimensional evaluation with short-cut indicators seems to be beneficial to clarify potentials and challenges with proposed CO_2 utilization technologies compared to current market benchmarks.

In this contribution we first outline the approaches of techno-economic assessments performed in chemistry-related start-ups, gathered by qualitative interviews. Furthermore we analyse current techno-economic assessment approaches in the literature. Based on this evidence we present an integrated approach for rapid and systematic evaluation of chemical CO_2 utilization technologies in the three fields of technologic, economic and environmental feasibility. The method leads to recommendations for further catalysis research and process development in the prior domains and facilitates decision making for start-ups, funding agencies and strategic investors. Additionally we show first results of its application in scientific literature of selected CO_2 utilization reactions.

This contribution is funded by EIT Climate KIC's project "enabling CO_2 reuse" $EnCO_2re$.

[1] International Energy Agency (2012): World Energy Outlook Special Report. Redrawing the Energy Map. Paris. Available online http://www.worldenergyoutlook.org/media/weowebsite/2013/energyclimatemap/RedrawingEnergyClimateMap.pdf.

[2] IPCC (2013): Climate Change 2014 Mitigation of Climate Change. Geneva. Available online http://www.ipcc.ch/report/ar5/wg3/.

[3] P. Styring, E.A. Quadrelli, K. Amstrong (2015): Carbon Dioxide Utilisation: Closing the Carbon Cycle. 1. edition. Amsterdam, Oxford: Elsevier.

[4] A. Boxin-Dumitriu, M. Perez Fortes, E. Tzimas, T. Sveen (2013): Carbon Capture and Utilisation Workshop Background and proceedings. Luxembourg. Available online http://publications.jrc.ec.europa.eu/repository/handle/111111111/30232.

[5] H. Audus, H. Oonk (1997): An assessment procedure for chemical utilisation schemes intended to reduce CO2 emissions to atmosphere. Energy Conversion and Management, 38, S409-S414.

[6] K. Mueller, W. Arlt (2014): Shortcut Evaluation of Chemical Carbon Dioxide Utilization Processes. Chemical Engineering & Technology, 37 (9), 1612-1615.

[7] M. Lehner, M. Ellersdorfer, R. Treimer, P. Moser, V. Theodoridou, H. Biedermann (2012): Carbon Capture and Utilization (CCU) – Verfahrenswege und deren Bewertung. BHM Berg- und Hüttenmännische Monatshefte, 157 (2), 63-69.

[8] A. Otto, T. Grube, S. Schiebahn, D. Stolten (2015): Closing the loop: captured CO2 as a feedstock in the chemical industry. Energy & Environmental Science, 8 (11), 3283-3297.l

[9] I. Dimitriou, P. Garcia-Gutierrez, R. H. Elder, R. M. Cuellar-Franca, A. Azapagic, R. W. K. Allen (2015): Carbon dioxide utilisation for production of transport fuels: process and economic analysis. Energy & Environmental Science, 8 (6), 1775-1789.

460305 Assessing the Potential of CO_2 Utilization with an Integrated Framework for Producing Power and Chemicals

Abstract:

Energy is an indispensable part of our daily lives. The last decade of development has resulted in a sharp rise in energy demand. Reducing CO_2 emissions urgently is a global issue, and the stakes are increasing with the passage of time. The major focus so far has been on sequestering the captured CO_2, but this has its own problems and concerns. Thus, the idea of "using" or "converting" CO_2 rather than "disposing" seems more attractive! The direct use of CO_2 being limited, we need ways to convert CO_2 into something useful such as valuable chemicals and fuels. Carbon dioxide utilization (CDU) is seen as a strategic measure to mitigate CO_2 emissions. In CDU, CO_2 is viewed as a carbon source for producing chemicals and fuels instead of treating CO_2 as a mere waste. However, a quantitative assessment of the global impact of CDU is a crucial yet unaddressed issue.

In this work, we propose the concept of CO_2 neutral chemicals and power production. We propose a network of reactions that begins with oxy-combustion of methane (the energy source) and ends with production of chemicals that have significant global demands. Dimethyl ether and methanol are considered as potential alternatives for conventional fuels. The reactions in the network are categorized as combustion reactions, hydrogen (H_2) producing reactions, CO_2 consuming reactions and secondary reactions. CO_2 consuming reactions use any CO_2 produced within the system. Secondary reactions produce any other reactants required to achieve CO_2 neutrality. The superstructure of all possible reactions is solved as an optimization problem to obtain the reaction network that achieves CO_2 neutral chemicals and electricity production. The process is entirely self-sufficient in its thermal and electrical energy needs with a net surplus of electricity, and completely utilizes all CO_2 produced.

The only caveat in achieving CO_2 neutrality mentioned above is the availability of hydrogen, as hydrogen produced internally is not sufficient to achieve both CO_2 neutrality and meet global demands for the chosen chemicals. Hydrogen can either be produced within the proposed framework (i.e. fossil based), or obtained from electrolysis of water using renewable energy resources. Thus, based on the source of hydrogen, we present three different scenarios. Firstly, we assumed availability of unlimited renewable energy to produce hydrogen by electrolysis of water. In this scenario, the estimated CO_2 avoidance was 59% as compared to the 2013 emission level. In the second scenario, we assumed the availability of renewable energy at current global levels. CO_2 avoidance dropped dramatically to 6% in this scenario. Finally, in the third scenario, we assumed zero external hydrogen and the avoidance of CO_2 emissions reduced to a mere 1%. Our analysis quantitatively establishes the fact that availability of renewable hydrogen will play a pivotal role in determining the future of CDU. Note that our framework does not advocate centralized production of bulk chemicals in one place. We also tested the idea of producing individual chemicals within our framework, but it obviously results in lower CO_2 avoidance.

Our work has taken a process systems viewpoint to show that utilizing CO_2 to produce alternate fuels along with bulk chemicals has the potential to combat CO_2 emissions at the global scale. Our framework allows us to obtain an upper bound for the global impact of CO_2 utilization by taking into account the maximum current and projected global demands for all chosen chemicals. Market demand of chemicals derived from CO_2 will have a significant impact on CDU. With the rising concerns about CCS, CDU may well be the next strategy for consideration in the long run. However, availability of abundant renewable hydrogen will be necessary to realize the full potential of CDU.

463458 CO2 Utilization for Methanol Production As an Intermediate for Formaldehyde: Technoeconomic and Sustainability Assessment of the Value Chain in the Case of Sweden

Abstract:

To enable CO_2 re-utilization, a study in West Sweden aimed to identify partnerships between companies producing CO_2 and companies that may reuse CO_2 in their industrial processes. The study was initiated by a methodology for location, quantification and characterization of CO_2sources providing the first test-bed with large replication potential to other regions in Europe. The test-bed currently focuses on technologies mature enough to be implemented in a short-term timescale (5 to 10 years). On this basis methanol production has been prioritized as one of the technologies where emphasis should be given (e.g., some of the others being algae production, horticulture production, power to gas, concrete curing, mineral carbonation, etc.).

Methanol is one of the most common platform chemicals (i.e., a precursor of MTBE, DME, formaldehyde, acetic acid, ethanol, olefins, etc.) and is traditionally produced from synthesis gas (syngas). Syngas is mainly produced by the reforming of natural gas; however, gasification of solid (e.g., wood residues, coal) and liquid (e.g., black liquor from pulp and paper industries, refinery by-products) fuels is also gaining attention. The global annual use of fossil-derived syngas is approximately 6 EJ, 10% of which is used for methanol production. Currently, the methanol synthesis plants operate in a stable, high-throughput production mode at low CO_2/CO ratios in feed (e.g., preferred by copper catalysts up to concentrations of 10-12% CO_2). It is an issue of on-going research to develop catalysts and processes that can utilize high-CO_2content feeds in a techno-economically viable operating mode (generally high pressures, low temperatures are preferred for increasing the methanol yield).

Depending on the source of CO_2, hydrogen, the other methanol synthesis reactant, may be already present in the gas mixture (e.g., syngas from gasification, steel work off-gases), may be available from a different industrial source or provided from water hydrolysis using external, cheap energy sources (e.g., surplus energy from the grid). Hybrid solutions have also been proposed, e.g., where methanol is produced by combining biomass based synthesis gas with steel plant off-gases typically having very diverse composition (e.g., 3-65% H_2, 5-60% CO, 0-20% CH_4, 2-25% CO_2, 3-55% N_2) and LHV values (e.g., 3-18 MJ/Nm^3).

This study focuses on the perspective of utilizing available industrial CO_2 sources to produce methanol as an intermediate for the subsequence synthesis of formaldehyde, which is of particular industrial interest for chemical production companies in West Sweden. It compares 6 process alternatives of different technology readiness level (TRL), including synthesis routes through the reversed-water-gas-shift reaction, formic acid, organo-carbamates, methyl formate, dimethyl carbonate and an upcoming indium oxide based technology. Rigorous process modelling is used to provide the necessary data for the technoeconomic (i.e., considering CAPEX and TRL) and sustainability (i.e., considering LCA and hazard assessment). The potential impact of impurities is discussed for the whole value chain from the CO_2 and H_2 sources to methanol and the subsequent formaldehyde production. The assessment considers uncertainties in the information for the CO_2 and H_2sources, the process modelling, and various future scenarios in the development of the energy technology sector.

474798 Life Cycle Emissions Assessment of a Renewable Fuel Process: Impact of Catalyst Performance on the Net GHG Emissions of Methanol Production By Direct Electrocatalytic Reduction of CO2

Abstract:

The ability to synthesize liquid hydrocarbon fuels from renewable energy and CO2 would provide a stable, energy-dense storage medium for solar energy, and provide a non-fossil, low-emissions route to liquid fuels. Research is now underway to address technical challenges such as reducing the energy cost of CO2 capture and increasing the efficiency of CO2 reduction catalysts. However, the full system-wide emissions balance of such a process has received little attention. Although the CO2 that is incorporated into a product is diverted from emission to the atmosphere, other aspects of the utilization process may produce greenhouse gas (GHG) emissions that negate some or all of this benefit. To determine whether a CO2-to-fuel process would provide a net emissions benefit requires a life cycle assessment (LCA) that extends across the full life cycle of the utilization process.

We present a GHG LCA for a CO2-to-fuel process to assess the net emissions benefit of such a process, and to evaluate the importance of catalyst performance and other system parameters in achieving net emissions reductions. As an example of a CO2-to-fuel process, we examine the low-temperature, six-electron reduction of CO2 to methanol ("deCO2rr"). We consider a scenario in which CO2 is supplied from a concentrated point source (coal plant flue gas), and process energy for the electrocatalytic reduction is provided by a low-emissions source (wind generated electricity). We evaluate the total GHG emissions from producing both the methanol fuel and the electricity output from the power plant, and compare to the emissions when producing these from conventional processes.

The analysis projects that the deCO2rr pathway could produce methanol and electricity with lower GHG intensity than conventional processes, with a sufficiently efficient CO2 reduction catalyst (e.g. 40% faradaic efficiency, 40% selectivity for methanol). The maximum emissions reduction potential of the deCO2rr pathway, using a near-ideal catalyst, is a 60% reduction relative to the conventional processes for methanol and electricity production. When the faradaic efficiency of the CO2 reduction catalyst is above 30%, further improvements in faradaic efficiency have little impact on the full system emissions of the deCO2rr route. This analysis demonstrates the opportunity for full system LCA to complement fundamental research and guide technology strategy for sustainable outcomes.

470127 Integration of Sustainability Assessment in the Early Design Phases of Amine Based Post-Combustion CO_2 Capture

Abstract:

Climate change is one of the biggest environmental and engineering challenges of the 21st century. Post-combustion CO_2 capture (PCC) is one of the most mature technologies to be deployed in an attempt to maintain atmospheric CO_2 levels below 450 ppmv. Controlling CO_2 levels would be necessary to keep the average global temperature rise below 2 °C compared to pre-industrial levels to avoid irreversible changes as a result of climate change. Given the environmental concerns driving process development, it is important to fully study the impacts of deployment of such technologies beforehand to avoid replacing one problem with potentially more severe ones.

Monoethanolamine (MEA) is currently the industry standard for amine solvents in PCC applications. However, there is increasing pressure to develop alternative solvents that avoid some of the drawbacks encountered in the MEA process without posing additional health and environmental risks. A new solvent would have to reduce the energy penalty associated mainly with solvent regeneration as well as circumvent problems caused by solvent degradation while maintaining a high reactivity and capacity for CO_2 absorption. Progress in development of more effective solvents and processes has been slow due to the need for extensive experimental research and pilot plant campaigns. There is a vast number of combinations of potential solvent candidates, including solvent blends, and process configurations; the experimental evaluation of each option is expensive and time-consuming.

To help speed up this effort, computer-aided design methods are used to guide the search for improved solvents and process schemes as well as to focus the experimental effort. The early integration of sustainability assessment in the process and solvent design is crucial to avoid costly choices and impacts that will be difficult to mitigate in later design and operation stages. Sustainability assessment is defined here as a combined hazard and life cycle assessment. The

hazard assessment evaluates potential damage in accidental scenarios, while the life cycle assessment presents the environmental impact of normal process operation. A rigorous hazard and life cycle assessment is first conducted for the well-studied MEA process and compared to amine-based CO_2 capture processes that have been proposed in literature. The findings of this assessment are used to guide the development of a robust sustainability assessment tool to be integrated into computer aided molecular design (CAMD) applications for initial solvent screening.

Typical solvent design applications use properties such as solubility parameters, pKa, vapour pressure, density and viscosity among others as proxy indicators for the technical feasibility of a solvent and its performance in a capture process. Such properties could represent a solventâ€™s capacity for CO_2 capture, expected fugitive losses, pump work and heating loads for solvent regeneration. All these factors are also important as indicators for the life cycle assessment. QSAR models for estimating physical properties are largely available. The hazard assessment requires the estimation of additional properties such as solvent flammability, toxicity and persistence in the environment. In a smaller extent, the use of shortcut group contribution methods also allows for the integration of the impacts of solvent production to the life cycle assessment. Severe data gaps however exist for the prediction of a solventâ€™s potential to degrade and the impact of solvent degradation. Solvent degradation can lead to many operational problems such as reduced heat transfer efficiency, foaming, corrosion and the formation of hazardous degradation products in addition to loss of CO_2 absorption capacity. Therefore, degradation potential is an important indicator for a solventâ€™s process, hazard and life cycle performance and should be included as early as possible.

A database of hazard and environmentally related properties is created comprising 80,000 chemical structures. Physical and chemical properties are collected mainly from available material safety datasheets. This database facilitates the automation of the online connection of the CAMD optimization routine and the complimentary sustainability assessment module. The connection can be established where an algorithm combines the proposed molecular groups into all potential structure formations and runs the proposed structures through the internal database for hazard related properties. If all proposed molecules are found the mean of the properties is taken, otherwise QSAR models are used wherever possible to predict the missing properties Experimental results are used in combination with literature data to develop a QSAR model for predicting solvent degradation rates. The model is developed using Principal Component Regression (PCR) and Multi Non-Linear Regression (MNLR) to correlate degradation rates with molecular structure, thermodynamic properties and operating conditions.

In conclusion, this work introduces a framework for the integration of the sustainability assessment at early stages of solvent and process design for post combustion CO_2 capture. It shows the translation of rigorous hazard and life cycle assessment into indicators that can be easily integrated in a CAMD routine. It presents a database of hazard and environmental properties, which facilitates the online connection to an optimization routine. Finally, it proposes a QSAR model for the determination of solvent degradation rates under the PCC conditions, which enables the addition of an important and currently missing solvent feature at molecular design stages for a more comprehensive assessment.

472144 Optimal Design and Techno-Economic Analysis of CO_2 Mineralization Process Using Aqueous NaOH

Abstract:

CO_2 mineralization is one of the prospective CO_2 utilization methods, which can achieve both economic benefit and CO_2 reduction by converting CO_2 to more stable and valuable mineral form. As a potential CO_2 mineralization process, sodium bicarbonate ($NaHCO_3$) production using aqueous sodium hydroxide (NaOH) was suggested earlier by Skyonic Corporation. The overall process is composed of chlor-alkali process, CO_2 absorption process, and product recovery process. One feature of this mineralization process is that a variety of product choices are possible. In addition to $NaHCO_3$, one can produce hydrochloric acid, sodium hypochlorite, and hypochlorous acid from NaOH, Cl_2 and H_2 which are intermediate products of the chlor-alkali process. In this study, optimal design of the CO_2 mineralization process using NaOH is mainly examined. A process design achieving maximum CO_2 utilization is considered as a base case in which all of NaOH produced is consumed to convert CO_2. The base case is designed as a reference to the Skyonicâ€™s process using AspenPlus^{TM}. Additionally, techno-economic analysis is carried out to discuss the feasibilities of the base case in terms of economics and CO_2 reduction in comparison with a conventional $NaHCO_3$ production process (â€˜Solvayâ€™ process). Finally, optimal process designs leading to maximum profit are suggested among several possible process designs with different production portfolios. Particularly, various scenarios of the market demand and product prices are considered in finding the optimal design.

463256 New Conceptual Design Models for Carbon Capture and Utilization Via Carbonate-Based Construction Materials

Abstract:

Carbon capture and sequestration by amine-based absorption technology is too costly for widespread deployment and also suffers from safety and environmental concerns about the ultimate fate of the sequestered fluid. We prefer carbon capture and utilization as carbonate-based cementitious material [1]. This CO_2 utilization strategy exhibits two major advantages; it turns a waste into a useful product and obtains the collateral benefit of reducing CO_2 emissions from cement kilns. The overall cost to produce carbonate containing structural materials will be highly sensitive to process design, so the problem of analyzing the technical and commercial feasibility of carbonate-based supplementary cementitious material becomes an exercise in creative conceptual design and process intensification. Point sources of CO_2 exist at power generation facilities as low pressure, relatively dilute gaseous mixtures. Geologic brines are a reasonable candidate for calcium and magnesium sources to minimize limestone calcination. Therefore, rational conceptual design in this system requires a framework for collectively considering three-phase contact, rate-based chemical reaction, and multicomponent, electrolytic mass transport. There are methods in the literature for dealing with these effects in binary combinations, but the combined effect of concentrated multicomponent reaction and diffusion among electrolytes occurring in conjunction with both absorption and precipitation processes does not find a complete treatment in the literature. Here, we present a framework for considering these effects collectively and consistently using the Maxwell-Stefan description of diffusion [2]. We present some preliminary results and address the remaining challenges that exist in applying these methods to the production of carbonate containing cementitious material to capture and utilize CO_2.

References

[1] Constantz, Brent R., Andrew Youngs, and Terence Hollands. Reduced-carbon Footprint Concrete Compositions. Calera Corporation, assignee. Patent WO 2010048457 A1. 29 Apr. 2010.

[2] Taylor, R., & Krishna, R. (1993). *Multicomponent mass transfer* (Vol. 2). John Wiley & Sons.

472620 Renewable N-Containing Chemicals Production By Ammonization of Biomass Derived Lipid over Heterogeneous Catalyst

Abstract:

N-containing chemicals are widely used in the pharmaceutical and polymer industries. Nitrilation of fatty acids by direct reaction with ammonia is mostly carried out in a batch reactor, in which the acid reactant is in liquid phase and the nitrile remains in the reactor. The process needs a few hours of reaction at high temperature, during which undesirable side reactions are likely to occur, especially with the unsaturated fatty acids (e.g. oleic acid). Alternatively, ammonization of fatty acid can be carried out in vapor phase and at a significantly shorter residence time to decrease the side reactions and improve the nitrile yield.

We are investigating the reaction of fatty acids with ammonia in the vapor phase and presence of a heterogeneous solid catalyst bed. The feed is continuously introduced into the reactor through an atomizer. Our previous work has shown that this approach allows the introduction of micron-sized droplets of feed that are rapidly vaporized inside the reactor. The fatty acid vapors can then easily react with NH_3 in the presence of metal oxide catalysts to produce fatty amide as intermediate products. Subsequently, the fatty amide dehydrates and produces fatty nitrile. We will present the effect of reaction temperature, vapor residence time and catalyst on conversion, yield and selectivity of nitriles formation.

467828 Catalytic Conversion of 5-Hydroxymethylfurfural into Tetrahydrofuran Dimethanol in Aqueous Media Using a Continuous-Flow Reactor

Abstract:

AIChE meeting 2016

Catalytic Conversion of 5-Hydroxymethylfurfural into tetrahydrofuran–dimethanol in aqueous media using a continuous-flow reactor

Sérgio Lima, Klaus Hellgardt, David Chadwick

Chemical Engineering Department, Imperial College London , London SW7 2AZ, United Kingdom

During the coming decades, the world is being challenged to reach its goals to replace a major part of its use of oil based feedstock for the production of new commodity chemicals and fuels. In this context, a particular attention is now given to the introduction of renewable carbon (e.g. biomass) in the chemical industry. Among the large variety of biomass available on earth, lignocellulosic biomass, a non-edible resource, has the potential to provide a direct access to valuable chemicals, ideally with similar or improved properties as compared to fossil-derived chemicals. Its high oxygen content, low accessibility to (bio)catalysts and its polyfunctionality makes the transposition of current catalytic processes to lignocellulosic biomass rather challenging. Tetrahydrofuran–dimethanol (THFDM) is a key derivative for the production of a wide range of non-petroleum-derived chemicals. In particular it can be used for the production of 1,6-hexanediol (1,6-HD; Scheme 1). 1,6-HD can be used to produce caprolactam, the monomer for nylon-6, a widely used synthetic polymer and is extensively used in the production of polyesters for polyurethane elastomers, coatings, adhesives, and polymeric plasticizers.[1,2]

In this work we wish to report on the heterogeneous catalysed continuous synthesis of tetrahydrofuran–dimethanol (THFDM) from 5-hydroxymethylfurfural (5-HMF) in water using a continuous-flow reactor. Raney-Cu and Raney-Ni were found to be effective catalysts for this reaction, using water as solvent in a two-steps process. The highest yield of THFDM (higher than 90 %) was achieved by first hydrogenation of 5-HMF to 2,5-furandimethanol (FURDM) in the presence of Raney-Cu follow by hydrogenation of FURDM to THFDM under 90 bar H_2 at 90 °C. Results indicate that large-scale production of low-cost platform and commodity chemicals for the polymer industries from carbohydrate biomass can be achieved in continuous flow processes.

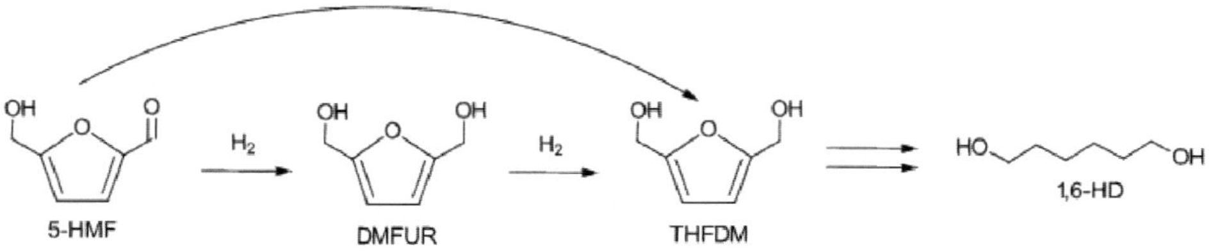

Scheme 1

1. Buntara T., Noel S., Phua P. H., Cabrera I. M., de Vries J. G., Heeres H. J., *Angew. Chem. Int. Ed.* **2011**, 50, 7083-7087.

2. Figueiredo F. C. A., Jorda E., Carvalho W. A., *Appl. Catal. A* **2008**, 351, 259-266.

472847 Catalytic Co-Hydropyrolysis of Biomass and Plastics for Improved Product Quality

Abstract:

Catalytic co-hydropyrolysis of biomass and plastics for improved product quality

Yuan Xue[a], Xianglan Bai[a]

[a] Department of Mechanical Engineering, Iowa State University

Every year, a tremendous amount of Municipal Solid Waste (MSW) is generated in the US. In 2013, Americans produced 254 million tons of trash and only recycled 1/3 of it. The environmental pollution and land usage from the improper disposal of MSW has become an increasingly urgent issue. Among MSW, the major compositions of waste plastics are polyethylene (PE), polypropylene (PP), polystyrene (PS) and polyethylene terephthalate (PET), which are petroleum-derived synthetic polymers, rich in carbon and hydrogen. In the present work, catalytic co-pyrolysis of the plastics and biomass is investigated by co-pyrolyzing red oak with PE, PP, PS or PS using HZSM-5 zeolite catalysts. The work is conducted in a tandem micropyrolysis both in-situ and ex-situ under a helium and hydrogen environment. It was found that in-situ catalytic co-hydropyrolysis significantly improved the product distribution for all cases compared to the helium environment, by dramatically increasing the yield of aromatic hydrocarbons and reducing the formation of solid residues. The yield of alkanes also increased at the expense of the decreasing yield of alkenes for co-hydropyrolysis of red oak with PE and PP. For the co-pyrolysis of red oak with PS and PET, the yield of CO increased, whereas the yield of CO_2 decreased in hydropyrolysis compared to pyrolysis under a helium environment.

For the ex-situ catalytic co-pyrolysis, hydropyrolysis was less effective as there were fewer changes in the product distribution compared to pyrolysis under a helium environment. The study suggests that the external hydrogen plays an important role in promoting a series of reactions involving the biomass and plastics over the zeolite catalyst, which include cracking, deoxygenation and aromatization.

464230 Pt Catalysts for Efficient Aerobic Oxidation of Glucose to Glucaric Acid in Water

Abstract:

Pt Catalysts for Efficient Aerobic Oxidation of Glucose to Glucaric Acid in Water

Jechan Lee, **Basudeb Saha**, and Dionisios G. Vlachos

Catalysis Center for Energy Innovation & Department of Chemical & Biomolecular Engineering, University of Delaware

Non-food biomass is an appealing and sustainable starting material for renewable liquid fuels and chemicals that can minimize our dependence on petroleum and mitigate concerns on greenhouse gas emissions. One such chemical is glucaric acid which has use in detergents and polymeric products. Glucaric acid can be upgraded to adipic acid (AA) for the production of nylon 6,6, polyurethanes and adipic esters. The current manufacturing process of AA (>2.3 million tons per year) from petroleum-based KA oil[1-3] employs corrosive nitric acid and generates greenhouse gas, N_2O.[4,5] Therefore, development of sustainable manufacturing processes for glucaric acid for use in detergents and for upgrading to AA is a viable strategy.[6] However, efficient aerobic oxidation of glucose to glucaric acid has proven to be challenge because of slow and difficult oxidation of gluconic acid, an intermediate oxidation product of glucose.[7]

We studied aerobic oxidation of glucose to glucaric acid using activated carbon, SiO_2 and Al_2O_3 supported Pt catalysts in water. The Pt/C catalyst exhibits the highest activity of these in glucose conversion and glucaric acid yield. Experiments are conducted systemically in a wide range of pH, O_2 pressure, temperatures and glucose/Pt to optimize the reaction conditions. The reactions in base-free and mild basic conditions give best yield to glucaric acid as compared to acidic and highly basic (pH 13.2) media. A maximum 74% glucaric acid yield at initial pH of 7.2, 80 °C, 13.2 bar O_2 and glucose/Pt molar ratio of 54 was achieved, which is the highest reported to date. C–C bond scission of glucaric acid in basic pH via retro-aldol results in low carbon chain carboxylic acids. Higher temperatures and higher Pt loadings cause degradation of glucaric acid, resulting in lower yields. Importantly, the Pt/C catalyst is stable after five consecutive catalytic cycles. Recyclability and characterization studies reveal that the catalyst is stable after five cycles with no sign of Pt leaching into the solution.

Mechanistic studies of the oxidation process entails rapid conversion of the aldehyde group of glucose and much slower oxidation of the $-CH_2OH$ group of gluconic acid, limiting the overall process efficiency. In order to rationalize the observed phenomenon, we have estimated the free energy of adsorption of various compounds. Kinetically, the transformation of an alcohol group $-CH_2OH$ to an acid requires first dehydrogenation to –CHO followed by oxidation to –COOH.[8-10] Upon dehydrogenation, the chemistry becomes facile. These fundamental findings are consistent with the general knowledge that aldehydes are much more reactive than alcohols. We propose that the oxidation of gluconic acid is slow due to the (oxidative) dehydrogenation of $-CH_2OH$ to –CHO being slow; this dehydrogenation step, which is important for activating the molecule, is unnecessary in oxidation of glucose to gluconic acid. We believe high pressure of O_2 facilitates oxidation of gluconic acid to glucaric acid in our study. This explanation is in line with previous work where the formation of glucaric acid is reported at high pressure of O_2.[11]

References:

1. K. Weissermel and H.-J. Arpe, *Industrial Organic Chemistry*, Wiley-VCH, Weinheim, 2003.

2. S. V. d. Vyver and Y. Román-Leshkov, *Catal. Sci. Technol.*, 2013, **3**, 1465-1479.

3. R. Beerthuis, G. Rothenberg and N. R. Shiju, *Green Chem.*, 2015, **17**, 1341-1361.

4. R. A. Reimer, C. S. Slaten, M. Seapan, M. W. Lower and P. E. Tomlinson, *Environ. Prog.*, 1994, **13**, 134-137.

5. C. Aellig, C. Girard and I. Hermans, *Angew. Chem. Int. Ed.*, 2011, **50**, 12355-12360.

6. T. R. Boussie, E. L. Dias, Z. M. Fresco, V. J. Murphy, J. Shoemaker, R. Archer and H. Jiang, *US Pat.,*2010/0317823 A1, 2010.

7. X. Jin, M. Zhao, J. Shen, W. Yan, L. He, P. S. Thapa, S. Ren, B. Subramaniam and R. V. Chaudhari, *J. Catal.*, 2015, **330**, 323-329.

8. B. N. Zope, D. D. Hibbitts, M. Neurock and R. J. Davis, *Science*, 2010, **330**, 74-78.

9. S. E. Davis, B. N. Zope and R. J. Davis, *Green Chem.*, 2012, **14**, 143-147.

10. S. E. Davis, A. D. Benavidez, R. W. Gosselink, J. H. Bitter, K. P. d. Jong, A. K. Datye and R. J. Davis, *J. Mol. Catal. A-Chem.*, 2014, **388-389**, 123-132.

11. V. J. Murphy, J. Shoemaker, G. Zhu, R. Archer, F. Salem and E. L. Dias, *US Pat.,* 2011/0306790 A1, 2011.

454177 Synthesis of Methyl Levulinate from Glucose Using Solid Acid Catalyst

Abstract:

1. Introduction Levulinic acid or alkyl levulinate is one of the important chemicals, which can be an intermediate of many types of chemical products. The production method of levulinic acid is, for example, treating saccharide with mineral acids. However, the recovery of levulinic acid after using homogeneous catalyst is generally difficult, and there are few reports about recovery or further purification of produced levulinic acid. In this study, the production method of methyl levulinate using solid acid catalyst was examined. Furthermore, the purification of methyl levulinate and the reusability of recovered catalyst and methanol was examined as well.

2. Experimental method Glucose was used as a saccharide material. 4 types of zeolites consist of different SiO_2/Al_2O_3 ratio (Name: Cation type: SiO_2/Al_2O_3 = HS-320: NH_4^+: 6, HS-341: H^+: 7, HS-500: K^+: 5.5, HS-690: H^+: 200) were used as solid acid catalysts after the calcination under air at 550°C for 6 h. A certain amount of glucose, catalyst, and methanol were sealed in a batch reactor, and then plunged into preheated oil bath. After the reaction, a reactor was cooled in a water bath, and then products were recovered by vacuum filtration. A recovered catalyst was again calcined under the same condition for reuse. A filtrate was distilled at 80°C and then at 120°C for the recovery of methanol and the purification of the product.

3. Results and discussion

3.1 Conversion of glucose into methyl levulinate using zeolite 0.2 g of glucose, 0.1-1.0 g of zeolite, and 10 mL of methanol were reacted at 80-160°C for 1 h. The conversion pathways of glucose into methyl levulinate consist of the isomerization of glucose to fructose, dehydration of fructose to 5-hydroxymethylfurfural, hydration of 5-hydroxymethylfurfural to levulinic acid, and at any time intermediates form dehydration products of methyl-intermediates. When 0.1 g of HS-320 was used at 120°C, fructose was obtained at the yield of 42mol% and the selectivity of 78%. This yield was close to glucose-fructose equilibrium which indicated that HS-320 was an effective catalyst for glucose isomerization. Lewis acidity of zeolite is expressed by the cleavage of Si-Al bonds, therefore, Lewis acidity of zeolite is considered to be stronger as its SiO_2/Al_2O_3 ratio is closer to 1. Isomerization of glucose to fructose is generally promoted by Lewis acid, therefore, HS-320 whose SiO_2/Al_2O_3 ratio was the closest to 1 among examined catalysts resulted in the best yield of fructose. The best yield of methyl levulinate among the examined condition was 54mol% which was presented by 0.5 g of HS-341 via 4 h of the reaction at 160°C.

3.2 Conversion of glucose into methyl levulinate using several types of zeolites To increase the yield of methyl levulinate, the use of several types of zeolites was examined. Firstly, sequential reaction of isomerization using HS-320 and conversion using HS-341 was examined ("2-steps synthesis"). As the 1st step, 0.2 g of glucose was treated with 0.1 g of HS-320 and 10 mL of methanol at 120°C for 1 h, and then as the 2nd step, HS-320 was exchanged with 0.5 g of HS-341 and treated at 160°C for 0.5-4 h. Secondly, simultaneous reaction of isomerization using HS-320 and conversion using HS-341 was examined ("one-pot synthesis"). 0.2 g of glucose was treated with 0.5 g of blended catalyst and 10 mL of methanol at 160°C for 0.5-4 h. The blended catalyst consisted of 0.1 g of HS-320 and 0.4 g of HS-341. Through "2-step synthesis", the yield of methyl levulinate was lower than that obtained using only HS-341 at any reaction time. This is probably because of the material loss at the 1st step as byproducts. On the other hand, the yield was slightly increased by "one-pot synthesis", which resulted in the yield of 58mol% after 4 h of the reaction.

3.3 Recovery and reuse of catalyst

During the reaction at 160°C for 4 h, the catalyst HS-341 changed its color from white into brown. This was because undesirable byproduct of humin was adsorbed on the surface of the catalyst. The brown catalyst turned back into white after re-calcination, however, the yield of methyl levulinate changed from 54mol% to 45, 41, 34mol% after 1st, 2nd, 3rd time of reuse, respectively. For the reuse of the catalyst, therefore, it was implied that the further examination of the calcination condition which could only remove humins was indispensable.

3.4 Purification of methyl levulinate and reusability of recovered methanol Filtrate obtained after the reaction at 160°C for 4 h with HS-341 contained 1wt% of methyl levulinate. Through the simple distillations at 80°C and 120°C, colorless transparent liquid which was considered as methanol was obtained, and the concentration of methyl levulinate increased to 33wt%. The obtained clear liquid through the distillation at 80°C was then used as a reaction medium. 0.2 g of glucose was treated with 0.5 g of HS-341 and 10 mL of obtained liquid at 160°C for 4 h. As the result, 51mol% of glucose was converted into methyl levulinate. The yield decreased by 3mol% comparing with the yield using pure methanol, however, the decrease was not considerable. Therefore, it could be concluded that some amounts of methanol was reusable in this process.

4. Conclusions Glucose was converted into methyl levulinate using solid acid catalyst at the yield of 58mol%. The possibility of reuse of the catalyst was indicated. The purification of the product was indicated to be possible through the simple distillation, and the reusability of methanol was confirmed.

465690 Processing Carbonaceous Feedstock Using Ionic Liquids for Sustainable Carbon Engineering

Abstract:

Energy security is an important issue for the sustainable development of the global community. The United States faces major energy security challenges due to the high energy demand from residential, commercial, industrial, and transportation sectors, which spur the development of low energy and clean processes that produce fuels and chemicals from domestically available carbon-based feedstocks. Using lignocellulosic biomass to produce biofuels for transportation has emerged as a focus of interest in recent years. Biomass provides a renewable carbon-based source to reduce GHG emissions through the capture of CO_2 during photosynthesis, and decrease the reliance on petroleum for the production of both fuels and chemicals. However, biomass conversions to fuels and chemicals also suffer from its' own set of problems: the distributed nature of biomass resources, low energy density, high cost of transporting sufficient biomass to allow refineries to take advantage of economies of scale.

One promising alternative is to develop a hybrid feedstock system. This hybrid energy system by combining biomass with other carbonaceous feedstocks in a single conversion platform would reduce supply fluctuations, mitigate GHG emissions, and be more politically palatable. However, current research has mostly focused on thermochemical conversion of single type of feedstock at high temperature and elevated pressure, which requires high energy consumption. Developing low energy, environmental benign, feedstock agnostic technologies, is critically important to address concerns of energy security and process economics. Ionic Liquids (ILs) as a class of novel environmental benign solvents are receiving increased attentions. Many researches have focused on ILs' applications in lignocellulosic biomass dissolution for biofuels production, while the work on blending biomass with other feedstock sources to reduce the supply cost and improve the conversion efficiency is limited. This works looks into the feasibility of processing woody biomass mixture with other carbonaceous feedstocks using IL solvents for production of clean energy and high valuable materials. Various types of ILs and feedstocks, and process conditions are investigated. The IL processed carbonaceous materials are further characterized for their compositional, physiochemical and rheological properties changes to advance process optimization and development towards an economic feasible carbon engineering technology by employing wide range of regional feedstocks.

469354 A New Lignin Based Polymer

Abstract:

Lignin, the most abundant renewable aromatic polymer on earth, is usually treated as a waste stream in the biorefinery industry. Developing new applications for lignin as valuable products or materials is a well-recognized challenge. Here, we present a template free method for synthesizing lignin-resorcinolâ€“formaldehyde porous gel with high lignin loading, by utilizing lignin with low molecular weight and less condensation. Up to 75% lignin loading was obtained by blending lignin.

471770 Process Modeling and Techno-Economic Assessment of the Conversion of Biorefinery Lignin to Dicarboxylic Acids

Abstract:

The development of commercially viable, integrated biorefineries remains a difficult challenge. A key barrier is the lack of a strong value proposition for the lignin waste stream. Most approaches towards lignin valorization have focused on the depolymerization of lignin to phenolic compounds, but a class of chemicals â€“dicarboxylic acidsâ€“ have largely been overlooked as a potential lignin product. Dicarboxylic acids are important and highly valuable industrial chemicals and intermediates used across the biopolymer, pharmaceutical, and food additives industries. Though the oxidative aromatic cleavage of lignin is a well-known reaction during the delignification of wood pulp, it is only recently that the selective production of dicarboxylic acids from biorefinery lignin has been demonstrated as a mild, environmentally friendly process. Development of this new pathway and its potential impact on biorefineries remains to be explored.

In this study, we will present several process models for the oxidative conversion of biorefinery lignin to dicarboxylic acid products, developed in Aspen Plus. We will evaluate and compare their technical and economic feasibilities, and present sensitivity analysis on key process parameters. Process and cost bottlenecks will be identified and discussed as potential areas for future research and improvement.

472014 Integration of Renewable Jet Fuel Production with the Pulp Industry through Alcohol Conversion

Abstract:

Abstract Submission to Be Considered As An Oral Presentation at AIChE 2016 Annual Meeting in the Chemical Conversion Processes in Forest/Plant Biorefineries Session

Integration of Renewable Jet Fuel Production with the Pulp Industry through Alcohol Conversion

Scott Geleynse, Senthil Subramaniam, Xiao Zhang

Washington State University

Among increasing interest in the production of alternative fuels from biorenewable resources, considerable research has provided numerous routes to renewable jet fuels. Along with these new technologies, there is need for design concepts to integrate them with existing industries in order to enable new facilities and ease the transition toward the new renewable biorefinery industry. In this presentation, we explore one such idea in which the alcohols-to-jet conversion technology is implemented into an established kraft pulp mill. By utilizing a large portion of a pulp mill's processes, a biorefinery can be constructed on already established supply lines and facilities with significantly reduced capital investment. This may not only be a possibility to enable to generation of new renewable jet fuels on a large scale, but an option for pulp mills to benefit by reinventing their process and enter entirely new markets.

Presenting author:

Scott Geleynse

Bioproducts Sciences and Engineering Laboratory

School of Chemical Engineering and Bioengineering, Washington State University
2710 Crimson Way, Richland, WA 99354

Contact: scott.geleynse@email.wsu.edu, x.zhang@wsu.edu

472054 Novel Solvent Deconstruction of Woody Biomass and Isolation of High Purity Lignin

Abstract:

Abstract Submission to Be Considered As An Oral Presentation at AIChE 2016 Annual Meeting in the Chemical Conversion Processes in Forest/Plant Biomass Biorefineries Session

Novel Solvent Deconstruction of Woody Biomass and Isolation of High Purity Lignin

Karissa Garcia, Ruoshui Ma, Scott Geleynse, Xiao Zhang
Washington State University

Lignin is the most abundant renewable aromatic polymer on territorial earth present in all high plants. Emerging interests in commercialization of lignocellulosic biofuel process injected new enthusiasm to lignin valorization. Due to the main objective of optimizing cellulosic fiber and/or carbohydrate recovery, most of current lignin isolated from either paper making or biorefineries are often difficult to purify and/or has been indiscriminately modified depending on the process conditions. It appears that one strategy has been overlooked is to devise a new method and/or identify a new solvent to extract lignin from plant biomass with a high purity and minimized structural heterogeneity.

Herein, we will report a new method of applying Deep Eutectic Solvents (DES) for extracting lignin from woody biomass with high yield and high purity. Four different DES were evaluated for treatment of hardwood and softwood. It was found that these DES treatments can selectively extract a significant amount of lignin from wood with high yields: 78% from hardwood and 58% from softwood. The extracted lignin has high purity (95%) with unique structural properties. The mechanism of DES cleavage of ether bonds between phenylpropane units was investigated.

The results from this study demonstrate that DES is a promising solvent for wood delignification and the production of a new source of lignin with promising potential applications.

Presenting author:

Karissa Garcia

School of Chemical Engineering and Bioengineering

Washington State University

Contact: karissa.garcia@wsu.edu, x.zhang@wsu.edu

456861 Low Temperature Conversion of Biomass into Activated Biochar for Water Treatment in Resource-Constrained Settings

Abstract:

Access to clean water is a major challenge in many parts of the world, especially in remote rural areas such as rain forest communities in Africa. New approaches are needed to ensure adequate water supply and quality, as they are inextricably intertwined with food security and public health. In this presentation, we will focus on the synthesis, characterization, and application of activated biochar for water purification. Unlike conventional gasification reactor systems designed to operate at large scales, not much has been investigated in the past about small-scale batch pyrolysis. Nevertheless, batch systems offer advantages such as increased char yield and net exothermicity. Various types of cellulosic biomass (eg: sawdust, coconut shells) can be used as carbon precursors via pyrolysis. We will describe a process where catalytic activating agents are used in batch setups to lower the pyrolysis temperature and simultaneously activate the carbon produced to BET surface areas in excess of 500 m2/g . A systematic study of process parameters and characterization of the resulting activated carbons is used to optimize the chemistry and minimize contaminants from the produced filter-carbon.

468039 Small-Scale Multiple Effect Distillation (MED) for Brackish Groundwater Desalination

Abstract:

Multiple effect distillation (MED) is the oldest thermal desalination process and has a typical plant capacity of 600 to 300,000 m^3/day. Because MED units do not use membranes, the capital cost, system complexity, and maintenance costs are usually lower than those for reverse osmosis (RO) systems. The purpose of this study was to engineer a small unit that could be operated in the field using renewable energy-sourced heat, namely, heat from the pyrolysis of waste biomass.

A small-scale MED unit was designed and fabricated at NMSU to produce approximately 0.5 m^3/day of very pure water (<10 ppm total dissolved solids (TDS)) from brackish groundwater. The unit is being used to study scaling behavior as a function of water chemistry and identify optimum operating conditions for each water. The fabricated MED unit consists of two falling film evaporators with horizontal smooth copper tubes in a parallel feed arrangement; the flow rates are approximately 60 kg/hr. brackish feed water and 20 kg/hr. of distilled water based on brine concentration factor of 1.5. The feedwater is sourced from the Brackish Groundwater National Desalination Research Facility (BGNDRF) in Alamogordo, NM and has a TDS concentration of 1,000-5,000 ppm. The MED operates under partial vacuum (0.2 bar absolute) and temperatures of 60-70°C. The MED performance was improved by adding a heat exchanger, water heater, and pump to bring the feedwater temperature closer to its boiling point prior to be sprayed into the effects. The aluminum parts in the MED effects were anodized to prevent corrosion. Remote thermocouples were used to monitor the temperature at the surface of copper tubes to study the effects of dry spots. The temperatures and pressures selected in this design are meant to be compatible with local brackish groundwater chemistry in terms of salt solubility to limit scaling on the heat exchanger surfaces.

472093 Bench Scale Investigation of the Biosand Filter Media As the Fluoride, Turbidity, and *E. coli* Removal

Abstract:

Biosand filter (BSF) is a simple and inexpensive facility that can be built as a community-based water treatment technique using locally-available materials. It has been proven that BSF can remove turbidity, *E. coli*, arsenic in water efficiently by modifying the filtration media. Drinking water fluoridation is prevalent in the world. High fluoride intake would cause dental fluorosis and other toxic health effects, especially for children. Therefore, drinking water de-fluoridation is necessary to safeguard public health. There is little information on the use of BSF for fluoride removal. For instance, some previous works reported that aluminum oxide-coated pumice (AOCP) was effective for fluoride removal and had better removal efficiency compared to activated alumina. Other research works found that pumice media BSF can provide similar water quality as conventional sand BSF but 30% more filtered water. Based on these findings, we have investigated the optimization of using AOCP surfaces to serve as media for BSF to clean contaminant water. Preliminary results have shown that AOCP can remove 93% of fluoride in surface water. In our setting, the traditional sand layer has been replaced by AOCP surfaces. We have found that filtration rate, contaminants removal efficacy are prime parameters that enhance BSF performance. It is expected that our findings will be able to predict the behavior of BSFs and their implementation in low-income communities or as point of use (PoU) systems.

463624 Capacitive Deionization of Brackish Waters Using Biomass-Derived Carbons

Abstract:

Irrigated agriculture water represents the bulk of demand for water in most of the countries suffering from acute water scarcity in the Near East and North Africa, as well as Mexico, Pakistan, South Africa, and large parts of China and India. Greenhouses are emerging solution. Current testing in Mozambique and Sierra Leone has demonstrated significantly reduced water use. Yet even a few gallons per day per greenhouse is challenging because many local water sources are brackish. Capacitive deionization (CDI) is a sustainable, energy efficient, and cost effective technology for removal of low or moderate salinity typical of brackish water as compared to technologies that remove the water instead. Indigenous production of the carbon for the electrodes holds the key to such technology being adopted as a source of irrigation water, simultaneously facilitating the use of such using renewable energy sources. Biochars have recently been explored as electrode materials in CD, and can be easily produced from local plant-based agricultural waste materials.

The goal of this study is thus to identify biomass process conditions to enable the best performance as electrode material in CDI. Studies on CDI so far have been targeted towards achieving the best adsorption capacity, which directs towards the use of highly processed and tailored carbons. This is mostly achieved through chemical activation processes using toxic or corrosive and expensive materials which require deionized water for subsequent washing of the carbon, for achieving better CDI performance. Hence, with the effort directed towards making the process viable for use locally in regions with acute water scarcity, a change in the approach to tailor the electrode material for use in CDI is proposed. This approach is to determine the best possible material utilization by ensuring maximum possible contact between the liquid and the biochar surface (wettability) and by generating the least Ohmic loss thus resulting in highest electrical efficiency.

472455 The Use of Electrodeionization for High Water Recovery in Brackish Water Treatment

Abstract:

Currently reverse osmosis dominates desalination because of low cost, reliability, and choices in the market place. However, there are many new desalination opportunities, especially in brackish water applications, where the size and scale make it possible to consider other technology. One of these technologies is electrodeionization. Electrodeionization is a technique in which we use wafers to remove ions to low concentration through an electricially driven process. Because the cost is related to the amount of ions being moved this technology can be cost competitive at low to medium ion concentration and low to medium scale. The purpose of this presentation is to explore places where electrodeionization can be economically feasible and show the fundamental scientific challenges with the technology. This work will specifically focus on the high water recovery possible with electrodeionization with experimental results to show high current efficiencies at concentration concentrations of greater than 10% salt.

466903 Total Phosphorus Removal from Sewage in Microbial Electrochemically Assisted Septic Tanks

Abstract:

Background

Subsurface sewage treatment systems (SSTS), commonly known as septic systems, typically consist of a septic tank and an absorption field (or drainfield or leachfield), to treat sewage from households. The septic tank as pretreatment step decreases total suspended solids (TSS), biochemical/chemical oxygen demands (BOD/COD), and pathogen levels from sewage influent, while the drainfield or soil absorption field captures nutrients (nitrogen and phosphorus) from the tank effluent so that the water leaving SSTS would have minimal environmental impact. By modifying the conventional septic tank through incorporating electrode materials to make use of microbial electrochemical process, this study assessed this novel design in terms of sewage treatment performance. Since phosphorus becomes increasingly threatening to water environment, it was the goal of this novel design to remove it from liquid phase and potentially capture it at electrode surface.

Methods

The experiments were conducted with 1-L scale simulated septic tanks at four different applied voltage levels (0.50 V, 0.63 V, 0.75 V, and 0.88 V, selection based on previous preliminary results) with control of without applied voltage. Tanks were set up and operated at two temperatures (15 and 25 °C), a hydraulic retention time of 8.3 days, two cycles of feeding and withdrawal during each HRT. The treatment conditions of the reactors are listed in REF _Ref450506281 \h Table 1
08D0C9EA79F9BACE118C8200AA004BA90B02000000080000000E0000005F005200650066003400350030003500300036003200380031000000
. ANOVA and pairwise multiple comparisons based on Bonferroni test were conducted for water characteristics with OriginPro Version 8 at significance level of $\alpha=0.05$.

Table SEQ Table * ARABIC 1. Treatments of simulated novel septic tanks for 171-day operation

	Treatment*		
Reactor #	Temp, oC	Voltage, V	Chlortetracycline (CTC)
Reactor #1	25	0	No
Reactor #2	25	0.50	No

Reactor #3	25	0.63	No
Reactor #4	25	0.75	No
Reactor #5	25	0.88	No
Reactor #6	15	0	No
Reactor #7	15	0	1 mg-CTC/L
Reactor #8	15	0.50	No
Reactor #9	15	0.63	No
Reactor #10	15	0.75	No
Reactor #11	15	0.88	No
Reactor #12	15	0.75	1 mg-CTC/L
Reactor #13	15	0.88	1 mg-CTC/L

*Water characteristics data of reactors #6 and #7 were pooled for data analysis, and the same procedure was applied to reactors #10 and #12, and reactors #11 and #13, for the reason that CTC addition was not detected to affect other water characteristics.

Results and Discussion

For total phosphorus, results suggested that septic tanks partially removed it from influent (REF _Ref450507954 \h Table 2 08D0C9EA79F9BACE118C8200AA004BA90B02000000080000000E0000005F0052006500660034003500300035003000370039003500340000000), and the novel design of microbial electrochemical further decreased its concentration (p=0.0001) with an influence from temperature (p=0.0003), while the interaction of voltage and temperature was not significant (p=0.2723). Those performance differences in total phosphorus removal were further elucidated by pairwise multiple comparisons given in REF _Ref450508201 \h Table 3 08D0C9EA79F9BACE118C8200AA004BA90B02000000080000000E0000005F0052006500660034003500300035003000380032003000310000000 . The results concluded that at 25 °C, 0.63 V or higher voltage at the selected electrode material is practically effective to remove most total phosphorus (>89.7%) from sewage, resulting in an effluent TP concentration less than 0.82 mg/L over 171-day operation; and that at 15 °C, phosphorus removal efficiency was smaller and it required 0.88 V to remove over 90% of total phosphorus from sewage.

Table SEQ Table * ARABIC 2. Total phosphorus concentrations in influent and effluents of different reactors

	Ave	SD	TP removal
	mg/L	mg/L	%
Influent	7.95	1.39	-
Control, 25 °C	6.98	1.72	12.2
0.50 V, 25 °C	1.81	3.67	77.2
0.63 V, 25 °C	0.82	2.92	89.7
0.75 V, 25 °C	0.14	2.30	98.3

0.88 V, 25 °C	0.11	2.31	98.7
Control, 15 °C	7.36	1.60	7.45
0.50 V, 15 °C	6.30	3.84	20.7
0.63 V, 15 °C	4.31	3.90	45.8
0.75 V, 15 °C	2.26	2.48	71.6
0.88 V, 15 °C	0.49	2.36	93.9

Table SEQ Table * ARABIC 3. Pairwise multiple comparisons results for total phosphorus levels in effluents

Comparisons (Bonferroni Test)	Significance at alpha=0.05 or lower
0.50 V, 25 °C vs. Control, 25 °C	*
0.63 V, 25 °C vs. Control, 25 °C	*
0.63 V, 25 °C vs. 0.50 V, 25 °C	NS
0.75 V, 25 °C vs. Control, 25 °C	*
0.75 V, 25 °C vs. 0.50 V, 25 °C	*
0.75 V, 25 °C vs. 0.63 V, 25 °C	NS
0.88 V, 25 °C vs. Control, 25 °C	*
0.88 V, 25 °C vs. 0.50 V, 25 °C	*
0.88 V, 25 °C vs. 0.63 V, 25 °C	NS
0.88 V, 25 °C vs. 0.75 V, 25 °C	NS
Control, 15 °C vs. Control, 25 °C	NS
Control, 15 °C vs. 0.50 V, 25 °C	*
Control, 15 °C vs. 0.63 V, 25 °C	*
Control, 15 °C vs. 0.75 V, 25 °C	*
Control, 15 °C vs. 0.88 V, 25 °C	*

0.50 V, 15 ºC vs. Control, 25 ºC	NS
0.50 V, 15 ºC vs. 0.50 V, 25 ºC	*
0.50 V, 15 ºC vs. 0.63 V, 25 ºC	*
0.50 V, 15 ºC vs. 0.75 V, 25 ºC	*
0.50 V, 15 ºC vs. 0.88 V, 25 ºC	*
0.50 V, 15 ºC vs. Control, 15 ºC	NS
0.63 V, 15 ºC vs. Control, 25 ºC	*
0.63 V, 15 ºC vs. 0.50 V, 25 ºC	*
0.63 V, 15 ºC vs. 0.63 V, 25 ºC	*
0.63 V, 15 ºC vs. 0.75 V, 25 ºC	*
0.63 V, 15 ºC vs. 0.88 V, 25 ºC	*
0.63 V, 15 ºC vs. Control, 15 ºC	*
0.63 V, 15 ºC vs. 0.50 V, 15 ºC	*
0.75 V, 15 ºC vs. Control, 25 ºC	*
0.75 V, 15 ºC vs. 0.50 V, 25 ºC	NS
0.75 V, 15 ºC vs. 0.63 V, 25 ºC	NS
0.75 V, 15 ºC vs. 0.75 V, 25 ºC	*
0.75 V, 15 ºC vs. 0.88 V, 25 ºC	*
0.75 V, 15 ºC vs. Control, 15 ºC	*
0.75 V, 15 ºC vs. 0.50 V, 15 ºC	*
0.75 V, 15 ºC vs. 0.63 V, 15 ºC	*
0.88 V, 15 ºC vs. Control, 25 ºC	*
0.88 V, 15 ºC vs. 0.50 V, 25 ºC	NS

0.88 V, 15 ºC vs. 0.63 V, 25 ºC	NS
0.88 V, 15 ºC vs. 0.75 V, 25 ºC	NS
0.88 V, 15 ºC vs. 0.88 V, 25 ºC	NS
0.88 V, 15 ºC vs. Control, 15 ºC	*
0.88 V, 15 ºC vs. 0.50 V, 15 ºC	*
0.88 V, 15 ºC vs. 0.63 V, 15 ºC	*
0.88 V, 15 ºC vs. 0.75 V, 15 ºC	*

* indicates significant difference, p-values less than 0.05; NS indicates non-significant difference, p-values equal or larger than 0.05.

Sludge solids were collected from simulated septic tanks (reactors at 25 ºC, REF _Ref450512244 \h Figure 1 08D0C9EA79F9BACE118C8200AA004BA90B02000000080000000E0000005F005200650066003400350030003500310032003200340034000000) and quantified for mixed-liquor (ML) total suspended solids (TSS) and total solids (TS), sludge volume index (SVI), and composition. After log transformation, both MLTS and MLTSS displayed a linear increase along with voltage level. At 0.63 V where TP removal achieved close to 90%, the MLTSS and MLTS increased by 2.53 and and 1.17 folds, respectively. However, the corresponding SVI of the sludge was only 52.5% of the control reactor, indicating the formation of denser sludge, which suggested a smaller increase of sludge volume (85.3%) than the weight increase. The composition analysis based on energy dispersive X-ray spectroscopy (EDS) revealed the concentration and amount of phosphorus in sludge samples (REF _Ref450512615 \h Table 4 08D0C9EA79F9BACE118C8200AA004BA90B02000000080000000E0000005F005200650066003400350030003500310032003600310035000000). More phosphorus was captured in sludge sediment when voltage was increased, but the phosphorus balance, assumed to be captured on electrode surface, was higher at voltages levels of 0.50 and 0.63 V than 0.75 and 0.88 V.

Figure SEQ Figure * ARABIC 1. Mixed-liquor total suspended solids (A), mixed-liquor total solids (B), and sludge volume index (C) of the sludge from reactors with different voltage treatments at 25 ºC.

Table SEQ Table * ARABIC 4. Phosphorus balance in different reactors.

	Amt of eff P	Amt P removed	Amt P in Sed	P of balance (on electrodes)
	mg	mg	mg	mg
Inf	178	-	-	-
Control, 25 ºC	143	35	13	21
0.50 V, 25 ºC	37	141	26	114
0.63 V, 25 ºC	17	161	37	125
0.75 V, 25 ºC	3	175	92	83
0.88 V, 25 ºC	2	176	92	83

Conclusions

These results confirmed the phosphorus retention capability of the novel septic tank, suggesting a lower burden on the soil absorption for nutrient removal. The amount of sludge solids was also increased, which may require a more frequent tank cleaning.

466262 On the Systematic Integration of Different Generation Biorefineries

Abstract:

As the world strives to create a more sustainable environment, the biorefinery community fights to minimize the environmental impact of bioprocesses and build a green economy. The use of biomass as raw material for production of energy carriers and chemicals has emerged as a promising alternative to fossil resources for mitigating climate change and enhancing energy security. Second generation biorefineries (2G) came up as the natural evolution of the first generation (1G) bioprocesses to preserve food resources, while, recently, seaweed and microalgae processes have also started to hatch as the third generation (3G) biorefieneries. As different generation biorefineries treat different feedstock, their integration can reduce energy consumption and result in a more economical design with lower environmental impacts than the stand-alone.

To date, first and second generation processes have been integrated by the adoption of grassroots design methodologies [1-4], while, in reality, their coupling is far more complex. The upgrade of first generation plants constitutes a retrofit problem, constrained to make the best use of the invested equipment of the existing plant. Second and third generation upgrades account for new technology and investment very flexible to adjust so that the upgrades are best, but the portfolios of products need also to be investigated. Apart from integrating energy, there is also significant potential to exchange (by-) products. For example, bagasse can be used as feedstock in the second generation process, and, similarly, lignin can be used as fuel for the co-generation unit, while wastewater could be used for algae cultivation. Moreover, the processes may share common sections, like the downstream processes. Finally, the co-generation (CHP) unit, after modifications, can support both plants. Thus, the integration of first and second generation biorefineries is a combined grassroots–retrofit design problem, much larger and complex than most conventional applications, that hides inner trade-offs and proves the deficiency of present-day practices in proposing realistic. The size and the complexity of the problem impose the need for a systematic approach

in the form of mathematical models.

This work studies the different symbiotic options in the integration of different generation biorefineries, scoping for the development of a new systematic methodology, which can combine different process systems engineering tools for the integration of grassroots-retrofitted problems. Detailed designs for sugarcane (1G), bagasse (2G), and seaweed (3G) ethanol production are developed in Aspen Plus®, validated by the industrial partner (CIMV Process™) and laboratory experiments, and the potential of cultivating microalgae in wastewater is also investigated. The detailed models feed the mathematical models, developed in GAMS®. The processes are integrated following a mixed mathematical model. The model applies the methodology of retrofit integration on first generation [5] and grassroots integration on second and third generation technology [6, 7], following the transhipment analysis. Preliminary results indicate a significant potential in internal trade-offs savings, both in operating and capital cost.

Acknowledgment: The financial support from Marie Curie European Research Program, RENESENG (FP-607415) is gratefully acknowledged.

References

[1] R. Palacios-Bereche, A. Ensinas, M.Modesto, and S. Nebra, 2012, Ethanol production by enzymatic hydrolysis from sugarcane biomass–the integration with the conventional process. In Proceedings of the Ecos 2012–25th international conference on efficiency, cost, optimization, simulation and environmental impact of energy systems (pp. 26-29).

[2] M. Dias, O. Cavalett, R. Maciel Filho, and A. Bonomi, 2014, Integrated first and second generation ethanol production from sugarcane, Chemical Engineering Transactions, 37, 445-450.

[3] P.R. Lennartsson, P. Erlandsson, and M.J. Taherzadeh, 2014, Integration of the first and second generation bioethanol processes and the importance of by-products, Bioresource technology, 165, 3-8.

[4] J. Moncada, J. A.Tamayo, and C.A. Cardona, (2014). Integrating first, second, and third generation biorefineries: Incorporating microalgae into the sugarcane biorefinery. Chemical Engineering Science, 118, 126-140.

[5] R. Smith, M. Jobson, and L. Chen, 2010, Recent development in the retrofit of heat exchanger networks, Applied Thermal Engineering, 30(16), 2281-2289.

[6] A. Kokossis and A. Yang, 2010, On the use of systems technologies and a systematic approach for the synthesis and the design of future biorefineries, Computers & Chemical Engineering, 34, 1397-1405.

[7] A. Kokossis, M. Tsakalova, and K. Pyrgakis, 2015, Design of integrated biorefineries, Computers and Chemical Engineering, 81, 40–56.

470769 Use of Multi-Objective Optimization for Selecting Optimally Integrated Biorefinery Processes

Abstract:

Biorefineries are integrated processing facilities that convert biomass into value added products and energy after many physical, chemical and thermal conversion steps, aiming the maximization of the economic value of biomass while minimizing the waste streams. Recently, the attention has been focusing on developing feasible integrated biorefinery concepts as well as on the identification of the most promising bio-based chemicals. This is due to rising needs for renewable feedstocks as alternatives to petroleum and high environmental impacts of petroleum-derived products. The design of a biorefinery brings a significant number of challenges due to the large number of processes and products available. Hence, it requires the use of advanced process synthesis and optimization methods to ensure that these systems are economically viable, energy efficient, have minimum environmental impacts and realize the best use of the biomass resource.

In this study, a systematic process design methodology which incorporates process optimization and energy integration is applied to assess the thermo-economic performance, the environmental impact and the energy requirement for several processing pathway options. A superstructure of process models in a lignocellulosic biorefinery converting the feedstock in sugars (biochemical conversion) and syngas (thermochemical conversion) platforms for different bio-based fuels and valuable chemicals is developed. The heat recovery is represented by characterizing the process units energy requirement using pinch analysis.

Multi-objective optimization is implemented, using different objective functions such as economic and environmental criteria that are simultaneously considered to show the trade-offs between these conflicting objectives. Different performance targets are established to compare the alternative designs, to increase the utilization of biogenic carbon and to understand the best combination of products and the synergies between them. The overall design problem is formulated as mixed integer linear programming (MILP) optimization problem and epsilon-constraint method is combined with integer cut constraints (ICC) algorithm to systematically generate the list of competing options in a Pareto front. The results of the proposed approach provide a set of non-dominated solutions (i.e., Pareto front), and each solution shows different biorefinery configuration. Finally, these trade-off solutions can be used as a decision support for engineers and decision-makers to identify the best biorefinery pathways.

Keywords: Integrated Biorefinery Systems, Process Design, Process Integration, Multi-Objective Optimization

465681 Process Models to Assess the Co-Processing of Gasoil and Bio-Based Feedstocks in Hydroprocessing Unit Operations

Abstract:

As the bioenergy industry focuses on the near-term development of conversion pathways from biomass feedstocks to infrastructure-compatible fuel blendstocks, a major area of interest is the integration of partially upgraded, biomass-derived intermediates into existing refinery upgrading unit operations. Development of analysis capabilities and process models are needed to assess these new ideas for refinery integration strategies on the basis of potential economic viability, impact on quality of refinery fuel products and benefits versus ill effects to reliability of petroleum refinery operations. One major area of focus for the modeling effort, enabled by funding from DOEâ€™s Bioenergy Technologies Office (BETO), is the development of hydroprocessing models in Aspen Plus and Aspen HYSYS to assess the co-processing of fossil-derived gasoils and bio-based feedstocks. One of the case studies explored through this initiative is the co-hydrotreating of gasoil and triglycerides.

This presentation will focus on (1) an overview of the analysis approaches and process models developed under this effort, (2) the benefits for utilizing the tools for assessing co-processing scenarios for biomass-derived intermediates, (3) insights gathered from model development and analysis for gasoil and triglyceride co-processing and (4) future work for this effort in assessing blending of oxygenated compounds.

469498 Design and Economic Analysis of a Macroalgae-to-Butanol Process Via a Thermochemical Route

Abstract:

Design and Economic Analysis of a Macroalgae-to-Butanol Process via a Thermochemical Route

Chinedu O. Okoli[1], Thomas A. Adams II[1][*], Boris Brigljević[2] and Jay J. Liu[2]

[1]Department of Chemical Engineering, McMaster University, 1280 Main Street West, Hamilton, Ontario, L8S 4L7

[2]Department of Chemical Engineering, Pukyong National University, 365 Sinseon-ro, Nam-gu, Busan 608-739,Republic of Korea

-

Biofuels produced from macroalgae are termed "third generation" biofuels and are becoming of increasing interest in the renewable fuels community. The advantages macroalgae have over first and second generation biofuel feedstocks include fast growth rates, which enable up to 4 - 6 harvest cycles per year, and the ability to be grown in the sea thus eliminating issues relating to land use and irrigation water [1]. Despite these advantages, question marks around their economic and environmental potential remain because of uncertainties around optimal macroalgae cultivation and harvest methods as well as the technologies to convert them to fuel. The objective of this presentation is to address the question of macroalgae conversion to fuel.

One such fuel receiving increased attention is biobutanol, as it is a better alternative to bioethanol as a gasoline replacement in current automobile engines and fuel pipeline networks [2]. Two major routes exist for biobutanol production; the biochemical route which primarily proceeds via fermentation of biomass feedstock to butanol, and the thermochemical route which proceeds via gasification of biomass feedstock to syngas and the conversion of syngas over catalysts to butanol. This work assesses a first-of-its-kind process for butanol production from macroalgae through a thermochemical pathway.

The design and simulation of different configurations was performed in Aspen Plus. In addition, the potential of the different configurations were assessed using economic and environmental metrics such as the minimum butanol selling price (MBSP), and cost of CO_2 equivalent emissions (CO_2e) avoided under different market scenarios. The resulting values of the metrics were then compared amongst the different configurations and also against standard literature references of similar processes. Finally, the impact that variations in key parameters have on the metrics were assessed using sensitivity analysis.

The results showed that the lowest MBSP was obtained with configurations which import natural gas and electricity as utility sources alongside the macroalgae feedstock. However these options perform poorly when the cost of CO_2e avoided is factored in. On the other hand, macroalgae-only configurations provide the best potential for cost of CO_2e avoided but perform poorly for the MBSP metric. Finally, evaluation of the sensitivity analyses results show that gasoline price changes have a high impact on the South Korean based plant configurations as shown in Fig. 1 [3].

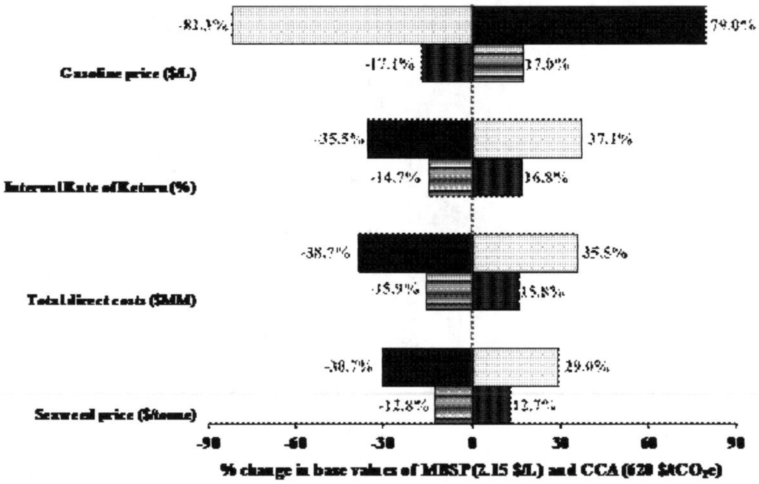

Figure 1: Sensitivity of key parameters from their base case values on MBSP (bottom bars with hatched fill) and CCA (top bars with solid fill) of the South Korea - self-sufficient plant scenario.

REFERENCES:

[1] G. Roesijadi, S. B. Jones, L. J. Snowden-Swan, and Y. Zhu, “Macroalgae as a biomass feedstock: a preliminary analysis,” Pacific Northwest National Laboratory, Richland, 2010.

[2] M. Kumar and K. Gayen, “Developments in biobutanol production: New insights,” *Appl. Energy*, vol. 88, no. 6, pp. 1999–2012, Jun. 2011.

[3] C. Okoli, T. A. Adams, 2016. "Design and economic analysis of a macroalgae-to-butanol process via a thermochemical route". Manuscript submitted for publication, 2016.

[*] Corresponding author: Email: tadams@mcmaster.ca; Ph: (905) 525-9140 x24782

462493 Robust Optimization of Biomass and Natural Gas to Liquid Transportation Fuel Refineries: Process Synthesis Under Uncertainty in Feedstock and Product Prices

Abstract:

Biomass and natural gas are two important feedstocks for the future production of liquid transportation fuels in the United States [1]. Biomass, as a renewable feedstock, can provide significant greenhouse gas (GHG) reductions for refineries, and the increased abundance of natural gas has significantly lowered its purchase price. Through process synthesis techniques and utilization of deterministic global optimization algorithms, optimum refinery topologies can be found that utilize both biomass and natural gas as feedstocks [2-7]. There are significant benefits to hybrid refineries in which both biomass and natural gas are converted to fuels, often through synthesis gas as an intermediate before upgrading through methanol or Fischer-Tropsch synthesis routes. These hybrid systems can offset the potentially high cost of biomass through low-cost natural gas, while providing significant GHG emission reductions based on the utilization of biomass.

In most process synthesis approaches, model parameters are assumed to be known and retain static parameter values. In reality, this is often not the case for parameters such as prices for feedstocks and products, among others; uncertain parameter realizations can have drastic impacts on the objective function values of the optimal solution, or even on overall model feasibility. Robust optimization is a framework for incorporating uncertainty which allows the optimization of large-scale process synthesis models which have uncertain parameters participating linearly in model constraints [8,9]. By imposing uncertainty sets for the parameters onto the model constraints, optimal solutions are found that ensure feasibility with known probabilities of constraint violation. Recent advances in robust optimization theory have greatly reduced the conservatism of robust solutions and allow the determination of competitive refinery topologies at known levels of risk through an iterative method utilizing *a priori* and *a posteriori* probabilistic bounds

[10-13].

The process synthesis superstructure for converting biomass, specifically hardwood or switchgrass, and natural gas to liquid transportation fuels is a non-convex, nonlinear mixed-integer optimization problem, and is solved to global optimality using a rigorous branch-and-bound algorithm [14]. Robust solutions will be presented when robust optimization model counterparts are incorporated into the process synthesis superstructure for a refinery with uncertain price parameters. These parameters appear in the objective function, allowing probabilistic guarantees on the level of profit to be provided. The impact of uncertainty on the objective function value, feedstock utilization, product distribution, and investment costs will be discussed. Case studies will be conducted using the box, interval + ellipsoidal, and interval + polyhedral uncertainty sets, and an iterative method will be utilized in order to provide high quality robust solutions at low probabilities of constraint violation.

[1] Floudas, C. A.; Niziolek, A. M.; Onel, O.; Matthews, L. R. Multi-scale systems engineering for energy and the environment: Challenges and opportunities. AIChE Journal 2016, 62 (3), 602-623.

[2] Baliban, R. C.; Elia, J. A.; Floudas, C. A. Biomass to liquid transportation fuels (BTL) systems: process synthesis and global optimization framework. Energy Environ. Sci. 2013, 6 (1), 267-287.

[3] Baliban, R. C.; Elia, J. A.; Floudas, C. A. Biomass and Natural Gas to Liquid Transportation Fuels: Process Synthesis, Global Optimization, and Topology Analysis. Industrial & Engineering Chemistry Research 2013, 52 (9), 3381-3406.

[4] Baliban, R. C.; Elia, J. A.; Floudas, C. A.; Gurau, B.; Weingarten, M. B.; Klotz, S. D. Hardwood Biomass to Gasoline, Diesel, and Jet Fuel: 1. Process Synthesis and Global Optimization of a Thermochemical Refinery. Energy & Fuels 2013, 27 (8), 4302-4324.

[5] Onel, O.; Niziolek, A. M.; Elia, J. A.; Baliban, R. C.; Floudas, C. A. Biomass and Natural Gas to Liquid Transportation Fuels and Olefins (BGTL+C2_C4): Process Synthesis and Global Optimization. Industrial & Engineering Chemistry Research 2015, 54 (1), 359-385.

[6] Niziolek, A. M; Onel, O.; Hasan, M. M. F.; Floudas, C. A Municipal solid waste to liquid transportation fuels - Part II: Process synthesis and global optimization strategies. Computers & Chemical Engineering 2015, 74 (0), 184-203.

[7] Matthews, L. R; Niziolek, A. M; Onel, O.; Pinnaduwage, N.; Floudas, C. A Biomass to Liquid Transportation Fuels via Biological and Thermochemical Conversion: Process Synthesis and Global Optimization Strategies. Industrial & Engineering Chemistry Research 2016, 55 (12), 3203-3225.

[8] Li, Z.; Ding, R.; Floudas, C. A. A Comparative Theoretical and Computational Study on Robust Counterpart Optimization: I. Robust Linear Optimization and Robust Mixed Integer Linear Optimization. Industrial & Engineering Chemistry Research 2011, 50, 10567-10603.

[9] Li, Z.; Tang, Q.; Floudas, C. A. A Comparative Theoretical and Computational Study on Robust Counterpart Optimization: II. Probabilistic Guarantees on Constraint Satisfaction. Industrial & Engineering Chemistry Research 2012, 51 (19), 6769-6788.

[10] Li, Z.; Floudas, C. A. A Comparative Theoretical and Computational Study on Robust Counterpart Optimization: III. Improving the Quality of Robust Solutions. Industrial & Engineering Chemistry Research 2014, 53 (33), 13112-13124.

[11] Guzman, Y. A; Matthews, L. R; Floudas, C. A New a priori and a posteriori probabilistic bounds for robust counterpart optimization: I. Unknown probability distributions. Computers & Chemical Engineering 2016, 84, 568-598.

[12] Guzman, Y. A; Matthews, L. R; Floudas, C. A New a priori and a posteriori probabilistic bounds for robust

counterpart optimization: II. A priori bounds for known symmetric and asymmetric probability distributions. 2016, In Preparation.

[13] Guzman, Y. A; Matthews, L. R; Floudas, C. A New a priori and a posteriori probabilistic bounds for robust counterpart optimization: III. A posteriori bounds for known probability distributions. 2016, In Preparation.

[14] Baliban, R. C; Elia, J. A; Misener, R.; Floudas, C. A. Global optimization of a MINLP process synthesis model for thermochemical based conversion of hybrid coal, biomass, and natural gas to liquid fuels. Computers & Chemical Engineering 2012, 42, 64-86.

465132 Dynamic Modeling of a Continuous Acetone-Butanol-Ethanol (ABE) Fermentation Process

Abstract:

Continuous acetone-butanol-ethanol (ABE) fermentation has drawn interest for large-scale production of biobutanol due to biobutanolâ€™s superior fuel properties over bio-ethanol and biodiesel [1]. Compared to batch fermentation, continuous ABE fermentation enables better control over process parameters and the possibility of achieving higher biobutanol productivity owing to the continuous product removal [1,2]. Dynamic modeling and analysis of integrated ABE fermentation and separation processes offer opportunities for optimal process design and operation. Dynamic models have been reported for fermentation processes integrated with product recovery units such as hollow fiber membrane, liquid-liquid extraction, and pervaporation [3]. In this work, we present a dynamic model for an integrated continuous ABE fermentation process that consists of a continuous fermentation unit [4], a tangential filtration unit, a membrane vapor extraction (MVE) unit [5], and a recycle stream. MVE is a novel product recovery technique that separates the vaporized solvent products from fermentation via solvent transport through a microporous, omniphobic membrane followed by isothermal absorption into a nonvolatile solvent [5]. The MVE unit enables continuous product removal with minimal energy consumption. An extensive dynamic sensitivity analysis has been performed to systematically investigate the effects of various kinetic and process parameters on the butanol productivity. The analysis provides guidelines for future process integration and optimization.

[1] P. Dürre. Biobutanol: An attractive biofuel. *Biotechnology Journal*. 2007; 2:1525.

[2] E. M. Green. Fermentative production of butanol- The industrial perspective. *Current Opinion in Biotechnology*.2011; 22:337.

[3] R. Mayank, A. Ranjan, and V. S. Moholkar. Mathematical models of ABE fermentation: Review and analysis.

Critical Reviews in Biotechnology. 2013; 33: 419.

[4] E. A. Buehler and A. Mesbah. Kinetic study of acetone-butanol-ethanol fermentation in continuous culture. *PLoS One*. 2016; Submitted.

[5] D. E. Liu, C. Cerretani, R. Tellez, A. P. Scheer, S. Sciamanna, P. F. Bryan, C. J. Radke, and J. M. Prausnitz. Analysis of countercurrent membrane vapor extraction of a dilute aqueous biosolute. *AIChE Journal*. 2015; 61: 2795.

471285 Dynamic Data Reconciliation and Model Validation of a CO_2 Capture Process Using Pilot Plant Data

Abstract:

The post-combustion CO_2 capture systems would be subjected to fluctuating flue gas flow rate and composition as the power plants follow the load. In addition, future CO_2 capture plants can vary the extent of CO_2 capture in response to the electricity price. Therefore optimal servo control and disturbance rejection characteristics are highly desired while designing the control system for CO_2 capture processes. One essential element for designing the control system is a high fidelity dynamic model that is validated with the experimental data . However, there is a lack of validated dynamic models for solvent-based CO_2 capture processes. With this incentive, a rigorous dynamic model of a MEA-based CO_2 capture process is developed and validated using steady-state and dynamic data obtained from the US DOEâ€™s National Carbon Capture Center (NCCC) in Wilsonville, Alabama.

Rigorous thermodynamic and transport models with quantified uncertainties are first developed and validated using large number of experimental data. A novel approach is developed to obtain a rigorous mass transfer model by considering the wetted wall column and packed column data simultaneously while regressing the parameters of the interfacial area model, mass transfer coefficient models, and diffusivity models together. In addition, rigorous kinetic model, holdup model, and pressure drop model are developed. The steady-state model is validated using large number of data collected from NCCC.

One issue with the limited dynamic data that are available in the open literature is that they span a limited range of operating conditions and often only contain a single step change. In this work, the dynamic test runs were conducted by designing a test protocol that helps to reflect the process nonlinearity and ensures persistence of excitation

approximately. The dynamic data include solvent composition and loading data that were manually collected every 5 minutes and analyzed in the NCCC laboratory. The dynamic data collected from the NCCC pilot plant were found to have significant measurement noise and some inaccurate measurements. In addition, there were a number of key variables that were unmeasured. To handle these issues, a dynamic data reconciliation (DDR) framework is developed to ensure closure of material and energy balances prior to model validation. As the rate-based models are not currently supported in Aspen Plus Dynamics, the software platform of our choice, an approximate method that matches the results from the rate-based Aspen Plus model closely is developed and implemented using scripts in Aspen Plus Dynamics. It is observed that the results from the dynamic model match very well with the experimental data that span more than three days. The dynamic model is then leveraged to perform a number of transient studies by simulating expected disturbances and setpoint tracking problems.

463128 Rigorous Modeling of a MEA-Based CO_2 Capture Process with Uncertainty Quantification and Validation Using Large Scale Pilot Plant Data

Abstract:

The US DOE's Carbon Capture Simulation Initiative (CCSI) has a strong focus on the development of state of the art process models to accelerate the development and commercialization of post-combustion carbon capture system technologies. One of CCSI's goals is the development of a methodology for a "gold standard" process model that will serve as a definitive reference for benchmarking the performance of steady-state and dynamic CO_2 capture systems. Although this methodology is intended to be applicable for various solvent types and for a wide range of operating conditions at different scales, an aqueous monoethanolamine (MEA) system is treated as a baseline in this work due to its widespread use as the industrial standard. This work is focused on developing the "gold standard" process model in Aspen Plus® and validating the model with pilot plant data.

The MEA model in this work is developed through a compilation of submodels, including physical property, reaction kinetics, mass transfer, and hydraulic models. The model parameters are regressed with experimental data that span the temperature, pressure, and composition ranges of interest. For each submodel, a Bayesian inference methodology is used to quantify the parametric uncertainty.

This model is validated with high quality steady-state data (23 total trials) collected from the National Carbon Capture Center (NCCC) in Wilsonville, Alabama. While most of the steady-state test runs reported in the literature are focused on narrow operating range, the key manipulated variables such as solvent flowrate and reboiler steam flowrate and disturbance variables such as the flowrate and CO_2 concentration of the inlet flue gas were varied widely in these test runs. Moreover, the number of beds, and thus packing height, and the presence of intercooling in the absorber are variable in the test runs. For each of the 23 cases, uncertainty in the key variables of interest, including percentage of CO_2 capture in the absorber and required reboiler duty in the regenerator, are determined by propagating parametric uncertainty through the process model. This rigorous methodology for obtaining uncertainty ranges on key operating variables is not only novel, but also provides insight into the effects of the operating variables on output uncertainty and helps to identify process submodels that have significant impact on output uncertainty. The insights gained through this developed methodology can play a key role in reducing scale up uncertainty of CO_2 capture processes.

471729 Immobilizing Metal Organic Frameworks on Active Substrates for CO_2 adsorption

Abstract:

Scaling up gas separation processes requires novel approaches for manufacturing of adsorbent materials into scalable contactors such as monolithic structures that offer a cost-effective platform for practical applications in comparison to traditional packing systems. In this study, several metal-organic frameworks (MOFs) are supported on a variety of commercial and synthesized monolithic substrates and their dynamic adsorption behavior are assessed in a rapid pressure swing adsorption (RPSA) system for post-combustion CO_2 capture. To gain control over crystal nucleation and growth, various bottom-up growth techniques are employed and optimized with respect to uniformity, thickness, and adsorption characteristics of the MOFs films. To further improve the equilibrium capacity, post-functionalization of composite adsorbents are carried out using several aminopolymers and aminosilanes. In addition, long-term stability of the structured adsorbents are evaluated by performing cyclic operation. Furthermore, to assess the influence of water on adsorption behavior of coated monoliths, both dry and humid runs are performed under real conditions. Finally, the influence of SO_x/NO_x impurities are determined by exposing the structured adsorbents to a simulated flue gas containing SO_2 and NO.

467239 Incorporation of 1-Butyl-3-Methylimidazolium Hexafluorophosphate into ZIF-8: Elucidation of Interactions and Their Consequences on Gas Separation Performance

Abstract:

Incorporation of 1-butyl-3-methylimidazolium hexafluorophosphate into ZIF-8: Elucidation of Interactions and Their Consequences on Gas Separation Performance

Fatma Pelin Kinik, Seda Keskin,* and Alper Uzun*

Chemical and Biological Engineering Department, Koç University, Istanbul 34450, Turkey

Metal organic frameworks (MOFs) have been popular recently in adsorption-based gas separation applications due to their configurable pores, large surface areas, robust framework structures and high chemical and thermal stabilities.[1] Lately, ionic liquids (ILs) have been used to enhance the gas separation performances of MOFs. ILs are liquid salts at room temperature which consist of organic cations and inorganic or organic anions.[2] Because of their favorable physical and chemical properties such as non-flammability, negligible volatility, high thermal stability and remarkable gas solubility (especially towards CO_2), ILs have been used in gas adsorption and separation processes.[3] In this work, an IL was used to improve gas adsorption and separation properties of a MOF. The influence of IL confinement into MOF structure was examined for CO_2/CH_4, CO_2/N_2 and CH_4/N_2 gas separations. ZIF-8 (2-methylimidazole zinc salt) was selected as the MOF as it is a thermally and chemically stable zeolite-type material with large pore size of 11.6 Å and small pore size of 3.4 Å.[4] Because of having high surface area, significant porosity, water resistance and commercial availability, ZIF-8 has been commonly used in gas separation applications.[5] As IL, 1-butyl-3-methylimidazolium hexafluorophosphate, [BMIM][PF_6], was selected because of its hydrophobic nature, low cost, and notable affinity for CO_2.[6]

Crystalline structures, thermal stabilities and interionic interactions were investigated by combining various characterization techniques. XRD and SEM results showed that the crystalline structure of the MOF is preserved upon the incorporation of IL. Moreover, XRF and BET measurements indicated that impregnation of the IL was successfully achieved at an IL loading of 24 wt.%. TGA and FTIR results proved that there is an interaction between IL and MOF. Adsorption measurements of CO_2, CH_4, and N_2 gases in ZIF-8 and [BMIM][PF_6]/ZIF-8 were performed using the volumetric method at room temperature up to 35 bar and ideal selectivities were calculated for CO_2/CH_4, CO_2/N_2, and CH_4/N_2 separations. It was shown that the CO_2 selectivities of IL/MOF samples at low pressures significantly increased compared to the pristine ZIF-8, while CH_4/N_2 selectivity of IL/MOF samples remained lower than the one of pristine ZIF-8. There is a considerable improvement in the CO_2/N_2 selectivity of IL-incorporated MOF, which is almost three times of pristine ZIF-8 up to 0.3 bar and two times of pristine ZIF-8 up to 1 bar. Also IL-incorporated MOF has approximately two times more CO_2/CH_4 selectivity than pristine ZIF-8 up to 1 bar. As a conclusion, IL-modified ZIF-8 is a promising material for gas separation applications, especially for CO_2 capturing processes such as natural gas sweetening and removing CO_2 from power plant flue gas. Tunable physicochemical properties of ILs and MOFs offer great potential for optimizing these structures for better gas separation performance.

Acknowledgement

This work is supported by the Scientific and Technological Research Council of Turkey (TUBITAK) under 1001-Scientific and Technological Research Projects Funding Program (project number: 114R093) and by Koç University Seed Fund Program. Financial support provided by the KUTEM (Koç University TUPRAS Energy Center) is gratefully acknowledged. S.K. acknowledges TUBA-GEBIP and A.U. acknowledges the support by the Science Academy of Turkey under the BAGEP Award Program. Authors thank Koç University Surface Science and Technology Center (KUYTAM) for providing help with the sample characterization.

*Corresponding authors: skeskin@ku.edu.tr and auzun@ku.edu.tr

References:

[1] Li, J.-R.; Kuppler, R. J.; Zhou, H.-C., Selective Gas Adsorption and Separation in Metal-Organic Frameworks. *Chemical Society Reviews* **2009,** 38, 1477-1504.

[2] Marsh, K.; Boxall, J.; Lichtenthaler, R., Room Temperature Ionic Liquids and Their Mixtures—a Review. *Fluid Phase Equilibria* **2004,** 219, 93-98.

[3] Hasib-ur-Rahman, M.; Siaj, M.; Larachi, F., Ionic Liquids for CO_2 Capture—Development and Progress. *Chemical Engineering and Processing: Process Intensification* **2010,** 49, 313-322.

[4] M. Zhu; Venna, S. R.; Jasinski, J. B.; Carreon, M. A., Room-Temperature Synthesis of ZIF-8: The Coexistence of Zno Nanoneedles. *Chemistry of Materials* **2011,** 23, 3590-3592.

[5] Venna, S. R.; Carreon, M. A., Highly Permeable Zeolite Imidazolate Framework-8 Membranes for CO_2/CH_4 Separation. *Journal of the American Chemical Society* **2009,** 132, 76-78.

[6] Chen, Y.; Hu, Z.; Gupta, K. M.; Jiang, J., Ionic Liquid/Metal–Organic Framework Composite for CO_2 Capture: A Computational Investigation. *The Journal of Physical Chemistry C* **2011,** 115, 21736-21742.

470543 Strontium-Based Bimetallic Metal Organic Frameworks (MOFs) for CO_2 Separation

Abstract:

Metal organic frameworks (MOFs) are promising candidates for separation and purification applications due to their tunable chemistry, high surface area, and porosity [1]. An interesting class of nanoporous bimetallic compounds known as strontium-based metal-organic frameworks have been synthesized on the organic basis of malonate. Cobalt and copper were employed to form MOFs with the general formula of $[SrM(C_3H_2O_4)_2\,(H_2O)_n]$, (M=Co, n=7 and M=Cu, n=4). These MOFs are designated with the refcodes of FOHQUO and VAMQOP in the Cambridge Structural Database, respectively.

Haldoupis and co-workers [1] examined a new category of MOFs using atomic simulation for gas sorption including FOHQUO. According to their results, FOHQUO with pore limiting diameter of 3 Å has a binary adsorption selectivity above 190 with water-free molecular lattice [1]. Having removed all the guest water molecules between 70°C to 190°C, they report that FOHQUO's framework integrity remain stable up to 350°C [1]. A molecular simulation study on mixed matrix membranes with FOHQUO as the dispersive phase was conducted by Yilmaz and Keskin [2] for CO_2/N_2 separation. According to their predictions, FOHQUO has a CO_2 permeability of about 500000 Barrer and CO_2/CH_4 selectivity of about 5000.

One of the challenges in MOFs nature is their thermal, chemical and physical stability under heat treatment and in presence of acids and moistures. The crystalline lattice of as-synthesized MOFs are typically filled with guest molecules of solvents [3]. Solvent removal can pose obstacles in crystallinity of MOFs and collapse the framework integrity and convert to amorphous materials [4].

In this study, we have synthesized FOHQUO and VAMQOP to investigate their potential for use in CO_2 separation and mixed matrix membrane formation. We examined their thermal and structural stability, surface area, and gas sorption capacity. Interestingly, X-ray diffraction (XRD) analysis proved that FOHQUO framework was not stable above 80°C. However, TGA analysis reflected the dehydration step at 80°C. On the other hand, VAMQOP has a stable structure up to 140°C. XRD analysis revealed that, after an amorphous period above 170°C, phase transformation occurs.

This work is supported by TUBITAK (Turkish Scientific and Technological Research Council) through Project No. 113M776.

References:

1. Haldoupis, E., Nair, S., Sholl, D. S. (2012), Finding MOFs for Highly Selective CO2/N2 Adsorption Using Materials Screening Based on Efficient Assignment of Atomic Point Charges, J Am Chem Soc, 134, 4313-4323.
2. Yilmaz, G., Keskin, S. (2014). Molecular Modeling of MOF and ZIF-Filled MMMs for CO2/N2 Separations, Journal of Membrane Science, 454, 407-417.
3. Qiu, S., Xue, M., Zhu, G. (2014), Metal-Organic Framework Membranes: From Synthesis to Separation Applicatio,. Chem Soc Rev, 43, 6116-6140.
4. Mondloch, J. E., Karagiaridi, O., Farha, O. K., Hup, J. T. (2013,. Activation of Metal–Organic Framework Materials, CrystEngComm, 15, 9258.

474890 Oxygen Selectivity of RPM3-Zn Gate-Opening Metal-Organic Frameworks

Abstract:

Commercial cryogenic distillation process is applied to air separation, based on different normal boiling points of nitrogen (77.4 K) and oxygen (90.2 K), but the process comprises of 30% of capital costs due to significant energy consumption. This study begins to explore the potential for kinetic separations of air via exploiting the gate-opening (GO) process of various metal organic frameworks (MOFs). Here, we report the gas adsorption capacities of nitrogen and oxygen on the GO-MOF RPM3-Zn (a.k.a. $Zn_2(bpdc)_2(bpee)$; bpdc = 4,4â€™-biphenyldicarboxylate; bpee = 1,2-bipyridylethene) at temperatures ranging from 77 K to 273 K up to 20 bar. Given sufficient equilibration time, the nitrogen and oxygen GO pressures are indistinguishable at 77 K, 87 K, 195 K, and 273 K. Yet, somewhat unexpectedly, RPM3-Zn exhibits higher oxygen adsorption capacity than nitrogen at low temperatures. The heats of adsorption in gate opening RPM3-Zn with oxygen and nitrogen are calculated as 12 kJ/mol and 8 kJ/mol, respectively, further evidence for oxygen selectivity. The selectivity is enhanced significantly when kinetic effects are also included. Efforts to deduce the role of various adsorption sites on oxygen selectivity will be discussed.

475638 Redox Active Metal Oxide-Based Thermochemical Processes for Producing Solar Fuels and Storing Thermal Energy (Invited Talk)

Abstract:

Reversing combustion and recycling carbon dioxide and water back to liquid hydrocarbons is an attractive option for storing solar energy and mitigating the growth of atmospheric CO_2 concentrations. For any such process, high solar-to-fuel efficiency and material availability is critical for large scale viability and favorable economics. Redox active metal oxide-based thermochemical approaches for solar-to-fuel have the potential to be highly efficient as they avoid inherent limitations of photosynthesis and sidestep the solar-to-electric conversion necessary to drive electrolytic reactions. This presentation highlights the challenges and progress from multidisciplinary and international efforts that has been progressing down a technical path for systems, novel reactors, and materials design and discovery for making hydrogen, liquid hydrocarbon fuels, and storing thermal energy from concentrated sunlight based on a two-step metal oxide redox cycle motivated to address the dual challenges posed by the strategic and economic importance of petroleum and the increasing concentration of atmospheric carbon dioxide.

454582 Doped Ceria for Solar Thermal Water Splitting: What Works, What Doesn't, and How to Improve It

Abstract:

Solar thermal water splitting (STWS) has the potential to achieve high solar to hydrogen efficiencies because of its use of the entire solar spectrum to split water. Two-step STWS uses a metal oxide oxygen carrier to couple two reduction/oxidation half reactions, one of which generates O_2 and the other which splits water to generate H_2. However, an optimal material which facilitates these reactions has not been identified. Among the candidate materials that are known to be capable of driving STWS, CeO_2 is considered the most promising. However it requires very high reduction temperatures (≥1500°C) to achieve acceptable H_2 production capacities. Many researchers have tried to dope ceria to bring down the reduction temperature, to varying degrees of success. In this work we use *ab initio* calculations based on density functional theory to develop a fundamental understand and explanation for why only Hf, and Zr have "improved" the STWS abilities of ceria. Based on this understanding we propose a novel set of dopant combinations and modifications strategies that, based on DFT results, improve the hydrogen production capacity of ceria at lower temperatures and/or result in higher H_2O conversion extents during the oxidation reaction.

454960 Stabilizing SiC for Solar Thermal Water Splitting Applications

Abstract:

Stabilizing SiC for Solar Thermal Water Splitting Applications

Amanda Hoskins[1], Aidan Coffey[1], Charles B. Musgrave[1,2], Alan W. Weimer[1]

[1]Department of Chemical and Biological Engineering, University of Colorado, Boulder, CO 80303, USA.

[2]Department of Chemistry and Biochemistry, University of Colorado, Boulder, CO 80309, USA.

Solar thermal water splitting (STWS) has been proposed as a means for hydrogen production which has the benefit that both the energy source and the fuel are renewable and readily available. However, considerable technological challenges must first be addressed in regards to hydrogen production, storage and transport in order for hydrogen to become an economically viable fuel on a large enough scale to significantly contribute to future energy demands. A major limiting factor to the viability of this technology is the lack of high temperature materials able to withstand the extreme chemical and physical environment. Specifically, materials that can be used to fabricate efficient heat exchangers for steam/steam heat recuperation will significantly impact the progress of this technology.

Silicon Carbide (SiC) is an ideal material for high-temperature heat exchange due to its stability at much higher temperatures than the current state-of-the-art material (Inconel) which cannot be used above ~1000°C. Using SiC for heat exchangers involves challenges related to the lifetime of SiC in an oxidative environment. Silicon carbide degrades in water-rich environments through the formation and subsequent degradation of a silica (SiO_2) layer at the exposed surface. In a dry environment, the silica has a low permeability to oxygen, however the presence of water, especially at high temperatures, increases the rate of oxidation, causing cracking and volatilization of the silicon from the structure. We propose that the application of nanostructured films grown with atomic layer deposition (ALD) will significantly improve the oxidation resistance of SiC in extreme environments. We have targeted a variety of high temperature coating materials chosen based on thermal properties, structural characteristics, and stability in oxidative environments. We have grown ALD layers of our desired materials on SiC particles using a fluidized bed particle ALD reactor. We have characterized all films using energy dispersive X-ray spectroscopy (EDS) mapping, scanning electron microscopy (SEM), transmission electron microscopy (TEM), and inductively coupled plasma mass spectrometry (ICP). Using thermogravimetric analysis, it has been shown that these coatings improve the oxidation resistance of SiC.

We have applied density functional theory (DFT) to model the performance of these coatings in an oxidative environment. Computational modeling of the chosen coating material properties, as well as oxygen diffusion through these materials can identify promising coatings to further extend the lifetime of SiC. We expect to use this understanding to develop more advanced stabilizing coatings for SiC and the myriad of other applications that require high-temperature oxidation resistant materials. This presentation will focus on both the fundamental DFT calculations and the experimental results for thin ALD films being developed to protect SiC surfaces from water oxidation at temperatures above 1000°C.

460329 Development and Test Operation of a Demonstration Plant for Sulfuric Acid Splitting at the Dlr Concentrating Solar Power Tower Facility

Abstract:

Sulfuric acid splitting is a key step of the hybrid sulfur cycle (HyS) for solar thermochemical hydrogen production. This exothermal reaction can be divided into two steps: firstly, the evaporation of liquid sulfuric acid (H_2SO_4) at about 400 Â°C forming sulfur trioxide (SO_3), and secondly, the decomposition of SO_3 to sulfur dioxide (SO_2) and oxygen (O_2) at 800 â€“ 1000 Â°C. While the first sub-reaction has fast kinetics, the second one is rather slow and requires the introduction of catalysts to achieve sufficient conversion. Since 2004 the concept of a solar receiver-reactor for sulfuric acid splitting has been developed by DLR and operated in its solar furnace during the European projects HYTHEC and HycycleS. In the follow-up European project SOL2HY2, a scale-up of this concept has been designed, developing a solar tower demonstration plant. This demonstrator has a design flow rate of 1Â l/min of sulfuric acid (50Â w%) and consist of four main components arranged in series and connected by Joule heated piping: a 60 kW electrical evaporator for vaporization of the liquid acid, a solar receiver for superheating the SO_3 to about 1000 °C, an adiabatic reactor with a fixed bed of an iron(III) oxide catalyst and a scrubber. The Joule heated evaporator consists of six vertical steel tubes with a siliconized silicon carbide (SiSiC) inner tubes filled with porous SiSiC foam structures for enhanced heat transfer. Liquid acid is injected at the bottom and vaporizes while passing upwards through the pipes. The vapors are collected in a steel manifold and, subsequently, conveyed to the solar receiver with an outer shell also made of steel. Concentrated radiation from the solar field passes through a quartz glass window closing the receiver and heats up a sectioned absorber composed of porous SiSiC foam structures. After superheating in the receiver, the process gas passed through the catalyst bed for an adiabatic reaction forming SO_2. Before neutralization with sodium hydroxide solution in the scrubber, the SO_2 concentration is measured by a customized gas analysis

system via UV/Vis spectroscopy. The pilot plant was constructed and assembled on the research platform of the DLR concentrating solar power tower facility in Juelich, Germany. The layout of the plant was accompanied and supported by thermo-mechanical modelling of the most important components like the evaporator and solar receiver. Initial operation of the demonstrator was performed with air and water as process fluids. During water operation, the solar receiver reached the predicted design parameters achieving a gas outlet temperature of 1000 °C at an absorber front temperature of 1200 °C and a solar power on aperture of 50 kW. In the adiabatic reactor, however, temperatures of only about 400 °C were measured which are too low for SO_3 decomposition. Therefore, the system was modified placing the catalytic fixed bed directly behind the solar absorber in the outlet section of the receiver. In the following test runs, the temperatures of this adiabatic reaction zone were sufficiently high with a minimum temperature in excess of 750 °C below which the catalyst would be deactivated due to sulfate formation. As a result, testing could proceed with sulfuric acid as the feed successfully demonstrating decomposition of SO_3. A detailed analysis of all results of the systematic on-sun test series is given in the present paper.

466891 Modelling of a Continuous Solar Tubular Aerogel Reactor for the Thermochemical Reduction of CeO_2: The Effect of Residence Time, Particle Diameter and Particle Loading

Abstract:

A simulation is developed and implemented to describe a continuously operated and indirectly irradiated solar reactor for the thermochemical reduction of CeO_2 to Ce_2O_3. An aerogel, consisting of N_2 and CeO_2 (particle diameter between 1-20 microns), enters the reactor at 900 K. The aerogel flows upward through a 30 cm long by 2.54 cm diameter tube, made of Tungsten coated with Silicon Carbide at 2500 K. The model aims at calculating the concentration and temperature profiles within the reaction tube at steady state, the conversion of CeO_2 particles and the Ce_2O_3 particles mass flow exiting the reactor.

The radiative transfer model within the reactor is implemented by considering a gray, non-isothermal, absorbing, emitting and scattering media. Scattering is considered to be anisotropic, as described by the Henyey-Greenstein phase function. Optical parameters where calculated from Mie Scattering theory, weighted according the blackbody spectra at 2500 K. The thermochemical reduction of CeO_2 is modeled by considering an endothermic, first order, source term. Temperature dependence follows the Arrhenius equation. Convective heat and mass transport mechanisms disregard any thermo-hydrodynamic effects, and assume instead plug flow. Finally, gas conduction is accounted and CeO_2 particle diffusivity is estimated from the Einstein-Smoluchowsky equation for brownian motion of spherical particles.

Simulations consider axisymmetric fields, i.e. dependent on the axial and radial position within the tube. First, the radiation heat transfer mechanisms where implemented using an implicit hybrid Monte Carlo-Finite Volume scheme, linearizing the fourth order emission term using a Taylor series expansion. On the other hand, the heat and mass convection, diffusion and reaction where implemented using an explicit iterative scheme. The steady state distributions of all combined physical phenomena are obtained by using the false transient-state approximation, while considering an iterative-splitting scheme.

The parameters considered within this study where the average flow velocity, the CeO_2 particle loading and the particle diameter. Simulations are carried out considering particle loadings of 10 and 100 mg/l, particle diameter of 1,5,10 and 20 microns and average flow velocities of 1,5 and 10 mm/s. Obtained results suggest that high particle loadings tend to favour temperature gradients, which hinder the overall reactor performance. Higher gas velocity lowers the conversion, but slightly increases the mass Ce_2O_3 flow rate. Finally, it is interesting to point out, that higher particle diameter, presents low specific extinction coefficients, which allow relatively high radiation heat transfer rates throughout all the reactor, minimizing temperature gradients and yielding overall higher CeO_2 conversion and mass Ce_2O_3 flow rate.

468882 Conceptual Analysis of Process Alternatives for Solar Thermochemical Methanol Production: The Role of Chemical Storage

Abstract:

Conceptual Analysis of Process Alternatives for Solar Thermochemical Methanol Production:

The Role of Chemical Storage

Bruno A. Calfa[1] and Christos T. Maravelias NOTEREF OLE_LINK1 \h * MERGEFORMAT 1
08D0C9EA79F9BACE118C8200AA004BA90B02000000080000000A0000004F004C0045005F004C0049004E004B0031000000

In this talk, we study the production of fuels and chemicals from a renewable source, namely solar energy. In a solar refinery, solar energy is utilized to convert raw materials, such as CO_2 emitted from fossil fuel power plants (and sometimes with H_2O), into value-added products; e.g., methanol and Fischer-Tropsch (F-T) fuels (Herron et al., 2015). While different solar-based technologies have been investigated to carry out the conversion, in this paper, we consider the solar thermochemical splitting of CO_2 and H_2O (see Meier and Steinfeld (2013) and references therein), which has some similarities with Concentrated Solar Power (CSP) technology in the electric power industry.

The viability, based on systems-level analyses, of a solar-driven (thermochemical cycle technology) process to produce methanol and/or F-T fuels from syngas produced from CO_2 and H_2O splitting has recently been evaluated in the literature (Kim et al., 2011; Kim et al., 2012; Kim et al., 2013; Rytter et al., 2016; Falter, Batteiger, and Sizmann, 2016). However, all aforementioned works considered a continuously operated process. However, one fundamental aspect of the production of solar fuels is the inherent *intermittency* of solar energy. Therefore, since the splitting reactions cannot take place when there is not sufficient solar irradiance (i.e., cloudy days or at night), there is a need to consider *chemical storage*.

In this work, we provide a systems-level analysis of three process configurations, a base case and two alternatives, which employ solar-thermal technology for the conversion of CO_2 and H_2O to syngas, which is further upgraded to methanol. We propose and assess the use of chemical storage to cope with the intermittency of solar-based chemical technologies. The base case requires intermediate gas storage, since all streams are treated as continuous from a high-level perspective (typically assumed in the literature).

First we determine the number of compression stages that minimizes the total annualized cost of compression and storage of CO_2/CO and H_2 streams. Material and energy balances followed by an economic analysis indicate that gas compression and storage significantly contribute to the production cost in the base case, which makes it less economically favorable than its alternatives. We perform sensitivity analysis for key process parameters (e.g., chemical conversion of the solar reactors, cost of the solar field and solar reactors, and solar refinery capacity) to investigate their effect on the production cost of methanol.

The two alternative process configurations do not require gas storage, since the syngas production step is operated entirely intermittently. One process alternative uses a reverse water-gas-shift reactor to further convert CO_2 to CO and correct the stoichiometric number (H_2:CO ratio) required for the methanol synthesis step, whereas the other process alternative has a membrane-based CO_2/CO separation that is operated intermittently. Our analyses show that the process alternatives have overall lower total capital investment and operating costs than the base case, whose compression, storage, and amine-based CO_2/CO separation costs represent major components of the production cost.

References

Falter, C.; Batteiger, V.; and Sizmann, A. 2016. Climate Impact and Economic Feasibility of Solar Thermochemical Jet Fuel Production. Environmental Science & Technology. 50(1):470-477.

Herron, J. A.; Kim, J.; Upadhye, A. A.; Huber, G. W.; and Maravelias, C. T. 2015. A General Framework for the Assessment of Solar Fuel Technologies. Energy & Environmental Science. 8(1):126-157.

Kim, J.; Henao, C. A.; Johnson, T. A.; Dedrick, D. E.; Miller, J. E.; Stechel, E. B.; and Maravelias, C. T. 2011. Methanol Production from CO2 Using Solar-Thermal Energy: Process Development and Techno-Economic Analysis. Energy & Environmental Science. 4(9):3122-3132.

Kim, J.; Johnson, T. A.; Miller, J. E.; Stechel, E. B.; and Maravelias, C. T. 2012. Fuel Production from CO_2 Using Solar-Thermal Energy: System Level Analysis. Energy & Environmental Science. 5(9):8417-8429.

Kim, J.; Miller, J. E.; Maravelias, C. T.; and Stechel, E. B. 2013. Comparative Analysis of Environmental Impact of S2P (Sunshine to Petrol) System for Transportation Fuel Production. Applied Energy. 111:1089-1098.

Meier, A., and Steinfeld, A. 2013. Solar Energy. New York, NY: Springer New York. chapter Solar Energy in Thermochemical Processing, 521-552.

Rytter, E.; Sou__kov‡. F; Lundgren, M. K.; Ge, W.; Nannestad, . D.; Venvik, H. J.; and Hillestad, M. 2016. Process Concepts to Produce Syngas for Fischer-Tropsch Fuels by Solar Thermochemical Splitting of Water And/Or CO_2. Fuel Processing Technology. 145:1-8.

---

[1] Department of Chemical and Biological Engineering. University of Wisconsin-Madison. Madison, WI 53706. USA.

471163 Hercynite Reduction and ITM Membrane Oxygen Removal System for Solar Thermal Water Splitting

Abstract:

Solar Thermochemical Water Splitting Processing (STWS) is a promising technology for producing renewable H_2[1]. A metal oxide, such as hercynite ($FeAl_2O_4$), is first reduced with high temperature concentrated sunlight to produce O_2[2]. Then, the reduced material is exposed to steam to re-oxidize it and produce H_2. Low O_2 partial pressure is required to drive the reduction reaction. An intriguing potentially highly efficient process is the recycle of inert sweep gas at near atmospheric pressure using a high temperature O_2 ionic transport membrane (ITM)[3], since the recycled gas is already at the required high temperature. In this work, we present preliminary experimental results for hercynite reduction using a TGA and separately the removal of O_2 from 1% O_2/inert down to ppm levels using a laboratory high temperature ITM. The impact of separation energy requirements is evaluated for the overall solar to H_2 cycle.

1. Muhich, C.L., et al., *A review and perspective of efficient hydrogen generation via solar thermal water splitting.* WIREs Energy Environ 2016, **5**:p. 261â€"287. doi: 10.1002/wene.174

2. Muhich, C.L., et al., *Predicting the solar thermochemical water splitting ability and reaction mechanism of metal oxides: a case study of the hercynite family of water splitting cycles.* Energy & Environmental Science, 2015. **8**(12): p. 3687-3699.

3. Ehrhart, B.D., et al., *System efficiency for two-step metal oxide solar thermochemical hydrogen production â€" Part 3: Various methods for achieving low oxygen partial pressures in the reduction reaction.* Int. J Hydrogen Energy (online) http://dx.doi.org/10.1016/j.ijhydene.2016.07.106.

474467 Monte Carlo method to calculate the lifetime efficiency of a solar reactor for reduction of zinc oxyde

Abstract:

Monte Carlo method to calculate

the lifetime efficiency of a solar reactor for reduction of zinc oxyde

O. Farges, J.J. B´ezian, and M. El Hafi

Universit´e F´ed´erale de Toulouse Midi-Pyr´en´ees, Mines Albi, UMR CNRS 5302, Centre Rapsodee, Campus Jarlard,

f-81013 Albi ct cedex 09, France

Concentrated solar plant are ommonly used to produce electricity through ther- modynamic cycles (Gemasolar Power Plant, Ivanpah Solar Power Plant, ...) and can also be seen as an alternative to fossil fuel-based methods for H2 generation [5]. High temperature solar thermochemical processes make use of concentrated solar energy to realize metal oxydes reduction. These processes can be considered as a solar energy storage : products may be stored and transported (H2 in this case). Among all metal oxydes reduction, the zinc oxyde to zinc (ZnO/Zn) redox reactions are particularly attractive as the first part of a two step water splitting cycle (eqs. (1a) and (1b)).

As presented on fig. 1, a thermochemical reactor *R*, placed at the top of a solar tower receives concentrated solar energy emited from the sun *S* and reflected by heliostats *H*. Thus, the reactor, feeded with zinc oxyde, reaches high temperature, *ie* above 1400 K, to achieve reduction of zinc oxyde and to procude zinc Zn.

$$ZnO \rightarrow Zn + \frac{1}{2}O_2 \quad (1a)$$
$$Zn + H_2O \rightarrow ZnO + H_2 \quad (1b)$$

In order to evaluate the performance of a thermochemical solar plant and to allow an optimally design of both the reactor and the whole facility (heliostat field,

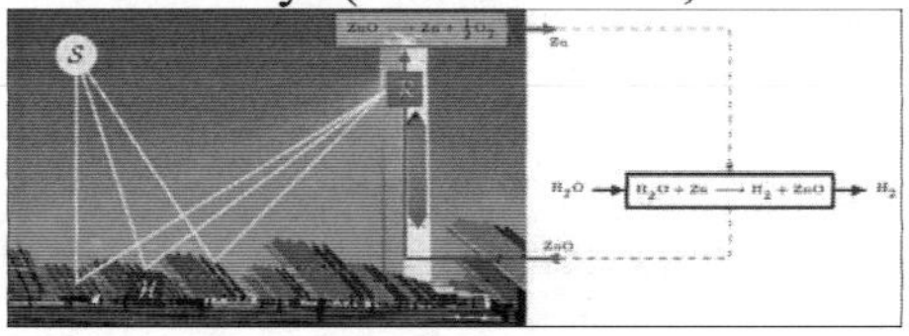

Figure 1: H2 solar production process

tower), numerical modelling of this process is required. We present here a Monte Carlo algorithm that allows the estimation of the solar plant annual productivity which is a non-linear problem. This non-linearity is due to the coupling between photon transport and the metal oxyde reduction reaction. The widespread opinion is currently that "Monte Carlo methods are not generally effective for nonlinear problems mainly because expectations are linear in character"[3]. This difficulty has been overcome by Dauchet who proposed a methodology in his Phd thesis [1].

To evaluate the quantity of interest, *ie* the annual solar-to-chemical conversion rate (ηr), this Monte Carlo algorithm firstly computes the instantaneous power re- ceived at the entrance of the reactor at an instant $\boldsymbol{i}$, $(Pth(\boldsymbol{X}\,|\boldsymbol{i}))$. A Monte Carlo algorithm previously presented in [2] is used as a first step of the proposed algo- rithm to evaluate this quantity. The random variable $Pth(\boldsymbol{X}\,|\boldsymbol{i})$ is the contribution of optical paths $\boldsymbol{X}\,|\boldsymbol{i}$ to the thermal power collected at instant $\boldsymbol{i}$. Then, $(Pth(\boldsymbol{X}\,|\boldsymbol{i}))$, the expectation of this random variable, is the instantaneous thermal power col- lected at moment $\boldsymbol{i}$. A part of the power received by the reactor, the instantaneous useful power $Pu = f\,(Pth(\boldsymbol{X}\,|\boldsymbol{i}))$, represents the efficient energy used to achieve redox reactions with an instanteneaous reaction rate described by a zero-order Arrhenius-type law, strongly non-linear. Integration over time of the instanta- neous conversion rate leads to the lifetime conversion rate $(\eta r\,((Pth(\boldsymbol{X}\,|\boldsymbol{i}))))\boldsymbol{i}$ as presented eq. (2).

$$\langle \eta_r \rangle = \int_{LifeTime} \frac{P_u(t)}{P_{th}(t)} \quad (2)$$

To treat this non-linearity, the Monte Carlo algorithm is based on a Taylor expansion of (ηr) around $Pth0 > Pth\ (\boldsymbol{X})$. Independent and equally distributed optical-path random variables $\boldsymbol{Xj}\ |\boldsymbol{I}$ are introduced and the infinite power series is statistically formulated thanks to the discrete random variable J whose realization

j is the order of Taylor expansion. The Taylor expansion is stopped through a Bernoulli process in order to decide whether the algorithm stops or continues without introducing a statistical bias.

Simulations are performed for the 1 MW reactor presented in [4]. A simplified model of thermal losses is applied and a comparison of the nominal performance between results obtained with the Monte Carlo algorithm and experimental results presented in [4] are presented in table 1. The difference between the measured and calculated results are due to the simplified thermal losses model as well as uncertainties of the measurements.

Table 1: Size and performance of a 1 MW reactor

Parameters	Units	Values
Aperture diameter	mm	600
Window diameter	mm	790
Cavity diameter	mm	1600
Cavity length	mm	2400
Conversion rate η_r [4]	%	54
Conversion rate estimated η_r	%	40.2

We have shown in this work that we are able to calculate the efficiency by a Monte Carlo Method, of a solar chemical reactor involving non-linearities between the heat source and the products of the reaction. A futur work could be the improvement of the heat loss model.

References

[1] J. Dauchet. *Analyse radiative des photobioracteurs*. PhD thesis, 2012.

[2] O. Farges, J. B´ezian, H. Bru, M. E. Hafi, R. Fournier, and C. Spiesser. Life- time integration using monte carlo

methods when optimizing the design of concentrated solar power plants. *Solar Energy*, 113(0):57 – 62, 2015.

[3] M. H. Kalos and P. A. Whitlock. *Monte Carlo Methods*. John Wiley & Sons,

2008.

[4] L. Schunk, W. Lipin´ski, and A. Steinfeld. Heat transfer model of a solar receiver-reactor for the thermal dissociation of zno-experimental validation at

10 kw and scale-up to 1 mw. *Chemical Engineering Journal*, 150(23):502 – 508,

2009.

[5] A. Steinfeld. Solar thermochemical production of hydrogen--a review. *Solar*

Energy, 78(5):603-615, 2005.

461151 Carbon Dioxide Reduction and Carbon Dioxide ReuseÂ in Advanced Biorefineries

Abstract:

An advanced biorefinery is effectively a polygeneration facility with integrated, efficient and flexible conversion of wastes, through a combination of physical, chemical, biochemical and thermochemical processes, into multiple products[1]. The biorefinery concept was analogous to the complex crude oil refineries adopting the process engineering principles applied in their designs, such as feedstock fractionation, multiple value-added productions, process flexibility and integration[1]. Enhancing the economic performance of biorefinery is the key challenge during the commercialisation stage. Energy products such as heat, electricity and biofuel are low value high volume products, while biochemicals and biomaterials (e.g. polymers) are high value low volume products. Biofuel or bioenergy production needs subsidies or economic incentives via supportive regulatory frameworks for feasibility. Hence, the aim of this study is to devise a strategy to realise an inherently economically competitive biorefinery system by added-value co-productions using stillage streams including carbon dioxide in offgases or exhaust gases, which can advocate a circular economy. Therefore, waste streams including those containing carbon dioxide from biofuel or bioenergy system must be reused to produce added-value products for achieving sustainability of biorefineries. Carbon dioxide reuse into biofuels, chemicals and biomaterials can be realised via three main routes: carbon dioxide reduction at cathode in microbial electrosysthesis system (MES)[2]; reusing carbon dioxide from exhaust gas of a thermochemical platform[3] and carbon dioxide capture by growing algae in algae biorefinery systems[1]. The case studies in this paper focus on the production of biopolymer, biochemical, biofuel by CO_2 reduction in MES; and biofuel production by trigeneration and Sabatier reactions. The economic performances using techno-economic modelling and environmental performances using life cycle assessment are conducted to result in integrated and advanced biorefineries that can enhance the three measures of sustainability. Producing biomaterial together with biofuels and bioenergy is the route for competitive biorefinery systems.

---

[1] Sadhukhan, J., Ng, K.S., Hernandez, E.M. 2014. *Biorefineries and Chemical Processes: Design, Integration and Sustainability Analysis.* Wiley.

[2] Sadhukhan, J., Lloyd, J., Scott, K., Premier, G.C., Yu, E., Curtis, T., Head, I. 2015. A Critical Review of Integration Analysis of Microbial Electrosynthesis (MES) Systems with Waste Biorefineries for the Production of Biofuel and Chemical from Reuse of CO_2, *Renewable & Sustainable Energy Reviews,* 56, 116-132.

[3] Ng, K.S., Zhang, N., Sadhukhan, J. 2013. Techno-economic analysis of polygeneration systems with carbon capture and storage and CO2 reuse, *Chemical Engineering Journal*, 219, 96-108.

463161 Hydrogen Storage and Carbon Dioxide Valorization through Reductive Calcination of Mineral Carbonates

Abstract:

Hydrogen storage and carbon dioxide valorization through reductive calcination of mineral carbonates

Georg Baldauf-Sommerbauer, Susanne Lux, Matthaeus Siebenhofer

Institute of Chemical Engineering and Environmental Technology, Graz University of Technology

The calcination of mineral carbonates is a widely used preparation reaction for metal oxides in industry (see Eq. 1). Carbon dioxide that has been fixed in mineral form is released, with the well-known detrimental effect on the global carbon balance and climate. This drawback of calcination can be overcome in reductive calcination. Carbon dioxide can be used as a means of storage for renewably produced hydrogen as methane and/or synthesis gas CO/H_2.

$MCO_3 = MO + CO_2$ (1)

The conversion of carbon monoxide and dioxide into methane is feasible for temperatures below 550 °C from a thermodynamic point of view. Due to kinetic hindrance, however, heterogeneous catalysis is needed to reach high yields. The Gibbs enthalpy for the reverse water gas shift reaction (RWGS: $CO_2 + H_2$ à $CO + H_2O$) decreases roughly linear from approximately +30 kJ mol^{-1} at 0 °C to approximately -6 kJ mol^{-1} at 1000 °C. The RWGS is also kinetically hindered and needs to be catalyzed for conversions near the thermodynamic equilibrium. If the carbon dioxide emitting step, namely calcination, is already performed in reductive hydrogen atmosphere, Eq. 2 can be formally derived for the direct production of carbon monoxide and methane from mineral carbonates. Calculations of

the Gibbs enthalpy revealed that methane and carbon monoxide production by reductive calcination of the industrially relevant carbonates of iron, magnesium, and other metal carbonates is possible.

$(a+b+c)MCO_3 + (b+4c)H_2 = (a+b+c)MO + aCO_2 + bCO + cCH_4 + (b+2c)H_2O$ (2)

Numerous experiments were conducted to elucidate the relationship between product gas composition, mineral carbonate, and process temperature and pressure. Experimental investigations were conducted using a tubular reactor setup with mineral iron (= siderite, 5-10 mm size fraction), magnesium (= magnesite, 5-8 mm size fraction), and magnesium/calcium (= magnesite/dolomite, 5-8 mm size fraction) carbonates as a fixed bed. The inlet gas composition was set to 90-100 vol.% hydrogen and 0-10 vol.% nitrogen at a total flow of 30 L h^{-1} at standard temperature and pressure.

Thermogravimetric experiments showed that reductive calcination proceeds at lower temperatures than calcination in inert nitrogen atmosphere. The minimum temperature needed for reductive calcination at isothermal conditions using the tubular reactor setup increases from siderite (~330 °C), through magnesite (~430 °C) to magnesite/dolomite (~475 °C).

Reductive calcination experiments with magnesite in hydrogen atmosphere (430-530 °C, 1-13 bar total pressure) revealed a maximum methane yield of ~37 % (475 °C, 13 bar). Methane formation is facilitated by increase of pressure at moderate temperatures below 475 °C. The yield of carbon monoxide ranges from 27 to 47 %. It increases from 430 to 505 °C and decreases at higher temperatures. Pressures above 1 bar slightly decrease carbon monoxide yield. When the magnesium carbonate fraction of a mixed magnesite (M) - dolomite (D) (M:D = 1.0:2.3) is calcined reductively (475-535 °C, 1-13 bar total pressure), only traces of methane could be detected in the product gas. Compared to pure magnesite, the carbon monoxide yield is enhanced. Up to 68 % of the total carbon can be released as carbon monoxide. Siderite ($FeCO_3$) differs from magnesite when the methane yield is compared. Experiments in the temperature region 330-380 °C at 1-11 bar pressure did show that the formation of methane is facilitated by low temperatures and elevated pressures. Optimization of the process parameters temperature, pressure, feed flow, size fraction, and bed length revealed that methane yields above 60 % without any additional catalyst are obtainable.

467141 Carbon Dioxide Sequestration Via Zn-Catalyzed Urea Conversion to Dimethyl Carbonate

Abstract:

Dimethyl carbonate (DMC) is emerging as an attractive electrolyte in advanced battery technologies and as a non-toxic solvent in several applications. The production of DMC via a number of pathways is currently under investigation; we report here the reaction of urea with methanol to form DMC over a Zn-based catalyst. Urea is a low-cost commodity, commonly used as nitrogen fertilizer, produced by the reaction of carbon dioxide with ammonia. The formation of DMC from urea thus represents a carbon sequestration route to a commodity chemical.

The primary pathway for urea to DMC involves two reactions through the intermediate methyl carbamate (MC). The first reaction of urea with methanol to MC is rapid and proceeds without catalyst; DMC formation is therefore kinetically limited by the Zn-catalyzed second reaction involving MC and methanol. Prior literature suggests that Zn interacts with the amino groups in urea to form a complex that is the active species in the reaction. In this work, we show that the nature of this complex changes over the course of reaction, and that a substantial fraction of the Zn complex exists as a dissolved homogeneous catalyst. The yield of DMC is further limited in reaction by the formation of byproducts including N-methyl-methyl carbamate (NMMC), formed by the methylation of MC with DMC. A kinetic model of the overall reaction network allows prediction of optimum conditions for DMC formation, and clarifies the role of the Zn catalyst in producing DMC from urea.

462623 The Stability Performance of Synthetic Ca-Based Sorbents Doped By Zirconium-Based Ceramics in Cyclic CO_2 capture Operations

Abstract:

Calcuim (Ca) based sorbents as an option for carbon dioxide (CO_2) capture process at high temperatures have displayed considerable activity loss during extended operation, due to sintering phenomena. Refractory zirconium-based ceramics have shown excellent performance as thermal barrier coatings in many high-tech applications; and, the recent doping of zirconia in Ca-based sorbents has exhibited improvement in the stability in cyclic CO_2 capture operations. In this work, sorbents were doped, at loadings of 10 wt%, with a zirconium-based ceramics, such as alkaline earth metal zirconate ($MZrO_3$, M= Mg, Ca and Ba) and aluminum zirconate ($Al_2Zr_3O_9$), which have the chemical structure of perovskite, and rare earth metal zirconates ($Ln_2Zr_2O_7$ and $LnZrO_4$, Ln = La, Ce and Sm), with the chemical structure of pyrochlore. These materials were prepared using the wet co-precipitation method. The results indicated significant sorbent stability and capacity improvement for mixed metal oxide zirconia-stabilized sorbents over those of pure CaO and natural Cadomin. Among the tested sorbents, the $CaZrO_3$-doped sorbents showed the best performance and remarkable stability due to their high resistance to sintering. The sorbent demonstrated a CO_2 capture capacity of 12 mole CO_2 / kg sorbent, with an activity loss of 19.6% through 30 cycles of carbonation at 675°C and calcination at 850°C. Among the rare earth metal zirconate ceramics, samarium zirconate ($Sm_2Zr_2O_7$) presented the best stability with an activity loss of 23.4 % loss, but at a lower uptake of 7.2 mole CO_2 / kg sorbent at the same conditions. In comparison, Cadomin (a natural calcium carbonate) at the same conditions resulted in a CO_2 capture capacity of 13 mole CO_2 / kg sorbent and an activity loss of 69%. Surface area measurements indicated that addition of the ceramics dopants to the Ca-based sorbents greatly inï¬‚uenced both the surface area and pore volume of the sorbents. X-ray diffraction results showed that the improved resistance of the sorbent is due to the formation of

structures such perovskite-type $CaZrO_3$, which provided a stable framework to the sorbent during the carbonation-calcination process.

464577 Improvement of Sn Electrocatalytic Reaction for CO_2 Conversion into Formate

Abstract:

Carbon capture and utilization is an interesting strategy to contribute to reduce CO_2 emissions and mitigate climate change, since it may involve the conversion of CO_2 into added-value products, such as liquid fuels. Among the different possibilities for CO_2 conversion, its electrochemical reduction appears as an interesting option because it could be used to store the intermittent and unpredictable energy from renewable sources, such solar or wind power, in form of chemical products. One of the most interesting products obtained from CO_2 electroreduction is formate. This product is used as raw material in several industrial processes with growing demand and it has also been proposed as a suitable fuel for fuel-cells and as a renewable energy hydrogen carrier.

One of the electrocatalysts that allow CO_2 conversion into formate on aqueous solutions is Sn [1]. This metal could be used in electrodes with different configurations, such as plate electrodes or in the form of particles of different sizes deposited on porous supports. The aim of this work is to analyze the performance of Sn electrodes with different configuration and to remark the improvements achieved using Sn nanoparticles in gas diffusion electrodes (GDEs) with respect to our previous studies, taking account the faradaic efficiency, formate concentration in the product, rate of formate production and the current density of the process.

Starting with conventional Sn plate electrodes for CO_2 electrochemical reduction, a maximum formate rate of $4.4 \cdot 10^{-4}$ mol m^{-2} s^{-1} was achieved with a 70% faradaic efficiency and formate concentration of 140 ppm at current densities around 12 mA cm^{-2} [2]. This low performance is due to the small electrocatalytic surface of the plate electrode and the mass transport limitations. To increase the specific surface and mass transport on the electrode, Sn particles were deposited on carbon paper to manufacture particulated electrodes. In these electrodes, gaseous CO_2 is directly fed to the cell and flows through the electrode. Sn commercial particles of 150 μm [3] achieved a formate concentration of 1.3 g L^{-1} with 70% faradaic efficiency working at 40 mA cm^{-2}. When the size of particle was reduced to 150 nm [4], it was possible to keep a 70% faradaic efficiency with a current density of 90 mA cm^{-2}, increasing the concentration to 1.5 g L^{-1} with a formate rate of $3.21 \cdot 10^{-3}$ mol m^{-2} s^{-1}.

Recent research efforts are focused on the development of Sn Gas Diffusion Electrodes (Sn-GDEs) which allow working at high current density with high faradaic efficiencies . Sn-GDE electrodes consist of a carbon paper as carbonaceous support, a microporous layer and the catalytic layer formed by the Sn carbon supported nanoparticles. Using GDEs we were able to achieve a 70% faradaic efficiency to formate at a current density of 150 mA cm^{-2}, obtaining a formate concentration of 2.5 g L^{-1}, which outperforms our results obtained with particulated electrodes. When the current density is increased and with a low catholyte flow, concentrations over 16 g L^{-1} were achieved, but at the expense of a 42% faradaic efficiency. However, despite the promising results obtained with Sn-GDEs, more research efforts are still required in terms of stability of the electrodes and optimization of the process.

[1] D. Kopljar, A. Inan, P. Vindayer, N. Wagner, E. Klemm., J. Appl. Electrochem (2014) 44, 1107-1116

[2] Alvarez-Guerra M, Del Castillo A, Irabien A., Chem. Eng. Res. Des. (2014) 92, 692â€“701

[3] A. Del Castillo, M. Alvarez-Guerra, A. Irabien, AIChE J., (2014) 60, 3557-3564.

[4] A. Del Castillo, M. Alvarez-Guerra, J. Solla-GullÃ³n, A. SÃ¡ez, V. Montiel, A. Irabien., Appl. Energy (2015) 157, 165-173

Innovative CO_2 Utilization by Carbonation Curing of Lightweight Concrete Made with Portland Limestone Cement
Paper Number: 451499

Hilal El-Hassan[a,*] and Yixin Shao[b]

[a] *Department of Civil and Environmental Engineering, United Arab Emirates University, Al Ain, United Arab Emirates, P.O. Box 15551*
[b] *Department of Civil Engineering and Applied Mechanics, McGill University, 817 Sherbrooke Street West, Montreal, Quebec, Canada, H3A 2K6*

Abstract
The atmospheric concentration of carbon dioxide gas has immensely increased with escalation in the world's greenhouse gas (GHG) emissions. Through CO_2 recovery, storage and utilization, emission induced global climate changes can be alleviated. The use of Portland limestone cement (PLC) as binder in lightweight concrete can effectively improve net gains in carbon emission reductions by partial replacement of cement with limestone. To further mitigate CO_2 discharge, carbonation curing was examined as a potential curing mechanism of lightweight concrete made with PLC. Slab samples were cast and initially cured at 25°C and 50% relative humidity for durations up to 18 hours. Carbonation was then carried out for 4 hours in a sealed chamber at an absolute pressure of 1 bar. The degree of reactivity was assessed by the CO_2 uptake based on PLC content and excluding preexisting limestone. The resulting uptake of fresh lightweight concretes without initial curing and after 18 hours of initial curing reached 6% and 18% respectively, corresponding to 13% and 39% reaction efficiency. The effect of carbonation curing on the mechanical properties, microstructure and reaction products of concrete was then investigated. A comparison between the optimum carbonated sample and hydrated reference was performed. While both concretes showed similar early-age compressive strength, late strength of carbonated sample was only comparable if water compensation through surface spray was performed. Hydration products were characterized as well crystalline calcium carbonates with the transformation of carbonate polymorphs into more stable calcite. In the process, PLC could replace OPC in carbonation curing of lightweight concrete while maintaining similar carbon sequestration potential.

Keywords: Carbon Capture & Storage, Materials and Sustainability & Environment.
* Corresponding author. Tel: +971(3)713-5156; Fax: +971(3)713-4343
E-mail address: helhassan@uaeu.ac.ae (Hilal El-Hassan)

1. Introduction

Portland limestone cement (PLC) is produced by adding limestone and calcium sulfate to Portland cement clinker in the grinding phase. It is different than ordinary Portland cement (OPC) in that cement replacement by limestone may reach 20%, thus reducing the carbon footprint associated with cement production. Such a replacement may also affect the compressive strength and hydration of concrete. Early research believed that fine limestone particles acted as nucleation sites, thus increasing the rate of hydration of the calcium silicates at early age and possibly improving the distribution of hydrates [1-2]. Ramachandran had reported that the hydration rate of C_3S also increased with an increase in the fineness and amount of $CaCO_3$ [3]. The formation of hydration products as ettringite had also been accelerated in the presence of $CaCO_3$, while C-S-H incorporated significant amount of calcium carbonates into its structure and formed calcium silicocarbonated hydrates [3]. Additionally, $CaCO_3$ will react chemically with aluminate phases to form carboaluminate phases [4]. Finally, because limestone is softer than clinker, when interground, it will achieve a finer particle size, thus refining particle packing and improving particle size distribution [5].

To further reduce the carbon footprint from cement production, CO_2 could be sequestered in PLC concrete through carbonation curing. Other work presented the physical performance and reaction products of carbonated OPC concrete [6, 7]. To investigate the effect of carbonation process parameters on the microstructure and characteristics of the reaction products of PLC concrete, fresh and air cured concretes are carbonated in a static carbonation setup. Three batches were selected based on the previous study: hydration reference without initial air curing (0a), carbonated concrete without water compensation (18a+4c), and carbonated concrete with water compensation (18+4c+sp). The reaction products of the concretes are characterized utilizing X-Ray Diffraction (XRD) and Thermogravimetry/Differential Thermogravimetry (TG/DTG). The potential replacement of OPC with PLC in manufacturing lightweight concrete by carbonation curing could allow the sequestration of CO_2 while also using lesser cement in the production.

2. Experimental Investigations

2.1. PLC concrete sample preparation and curing

Rectangular concrete samples 127 mm long, 76 mm wide, and 38 mm thick were cast for the purpose of this study. The lightweight concrete mix consisted of PLC as binder and saturated-surface dry expanded slag aggregates with 5% water content. Mixture proportions are shown in Table 1. The chemical composition of PLC was determined by infrared carbon analyzer and presented in Table 2. The limestone content was measured as 15.9%, equivalent to a CO_2 content of 7.0%. The raw materials were mixed in a pan mixer and samples were compact formed using a vibrating hammer. PLC concrete was then demolded right after casting. Three curing schemes were investigated: (1) Sealed hydration in a plastic bag served as reference (0a) without water loss; (2) Initial air curing of 18 hours in an environmental chamber of 50% relative humidity (RH) and 25°C followed by 4-hour carbonation and subsequent hydration (18a+4c); (3) Initial air curing of 18 hours in an environmental chamber of 50% relative humidity (RH) and 25°C, followed by 4-hour carbonation, water compensation and subsequent hydration (18a+4c+sp). Water compensation through spray was introduced immediately after carbonation to restore all the curing-induced water loss. Subsequent hydration was carried out in a sealed bag at a relative humidity of 80±5% and room temperature (24±1°C) until 24 hours and 28 days for compressive strength tests following ASTM C140 [8].

The static carbonation setup employed in this work is schematized in Fig. 1. Within a few seconds, the pressure is reduced to 0.7 bars below atmospheric prior to carbonation by means of a low-capacity vacuum. A heater installed at the inlet brings up the 99% pure carbon dioxide gas to room temperature as it flows into the sealed chamber. A pressure regulator is connected to the system to retain the pressure at 1 bar. The whole system is placed on a digital scale to monitor the mass increase due to the carbonation reaction.

Table 1: Mixture proportion of PLC concretes

	Slab (g)	Mass (kg/m^3)	Percent (m.-%)
Portland Limestone Cement	88	241	13
Mixing Water	35	96	5
SSD Expanded Slag	554	1502	82
Concrete Sample	677	1839	100

Table 2: Chemical composition of PLC and OPC

Cement	Constituent (%)								
	CaO	SiO_2	Al_2O_3	Fe_2O_3	MgO	Na_2O	K_2O	SO_3	CO_2
PLC	59.8	21.6	4.5	2.7	1.7	0.8	-	3.2	7.0

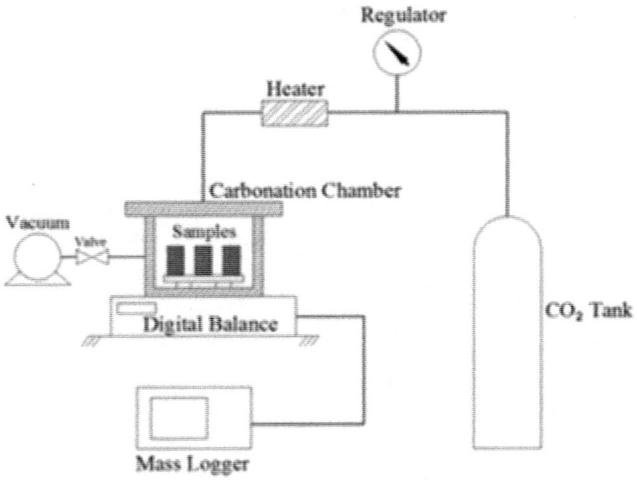

Fig. 1. Schematic of carbonation setup

2.2. Measurement of CO_2 uptake in PLC concrete

Three methods were used to quantify the carbon uptake of carbonated PLC concrete: mass gain, mass curve, and furnace decomposition. The mass gain method compares the sample's mass before and after carbonation in accordance with Equation 1. As the chemical reaction is exothermic, the increase in temperature induced water evaporation. Such water was collected and added to the final mass to ensure a closed system.

CO_2 uptake (%) = (Final mass + Mass of water loss – Initial mass) / (Mass of cement) (1)

The mass curve method was performed by placing the carbonation setup on a digital balance, which was zeroed after vacuuming the chamber. A mass curve was recorded as mass versus time until the end of the process at which CO_2 was released and the residual mass, M, was measured. The system was calibrated by repeating the tests using CO_2-insensitive styrofoam samples of the same volume to obtain a second residual mass, m. The difference between M and m represented the CO_2 uptake by concrete (Equation 2). Data collected by mass gain and mass curve methods are two simultaneous measurements from the same process and therefore should be comparable. They are independent from any CO_2 content existing before carbonation.

CO_2 uptake (%) = $(M - m)$ / (Mass of cement) (2)

An independent furnace decomposition test was employed to serve as a third check. A furnace of maximum temperature of 1100°C was employed to test large concrete samples with mass range of 35-70 g. PLC concrete samples were ignited up to 105°C, 470°C, and 950°C to quantitatively measure the evaporable water, bound water in hydration products, and carbon dioxide in carbonates respectively [9].

2.3. Microstructure evaluation

XRD and TG/DTG were employed to characterize the microstructure of the concrete specimens. A 5-mm thick concrete surface layer were extracted using a hammer and sieved to pass a 125-µm sieve to remove coarse aggregates. No attempts were made to separate the fine aggregates from the cement. Therefore, powder samples were a mix of paste and fine aggregates. XRD and TG analysis was performed on the same concrete powder to determine the reaction products.

At first, the concrete powder was analyzed using XRD. It was performed using a Philips PW1710 Powder Diffractometer (Cu, Kα radiation, X'celerator proportional detector, scan interval 10-100°, 0.02°, and 0.5 seconds per step) to relatively quantify the amount of calcium hydroxide, calcium silicates, and calcium carbonates in carbonated and hydrated samples.

The same concrete powder to that used in XRD analysis was used in a thermogravimetry (TG/DTG) test. TG was utilized to estimate the respective contents of C-S-H and ettringite (105-420°C), calcium hydroxide (420-470°C), and calcium carbonate (470-950°C) [9]. A thermal analyzer (NETZSCH, TG 449 F3 Jupiter) with a resolution of 0.01 mg was employed with a sample mass of 120 mg for each test. The TG/DTG curves were thus obtained by plotting the mass loss between 25°C and 950°C at a heating rate of 10°C/min.

3. Experimental Results and Discussion

3.1. CO_2 uptake of PLC concrete

The CO_2 uptake and content calculated based on the mass gain, mass curve, furnace analysis, and TG are shown in Table 3. The first two methods, which are insensitive to the initial carbon content, were used to determine the carbon uptake (% by mass) of the 4-hour carbonated concrete. The results show a consistency among non-sprayed and sprayed concrete at 18.3%. In order to determine the initial carbon content at the concrete surface, the furnace analysis method was employed. The hydrated reference (0a) resulted in CO_2 content of 6.68% initially present prior carbonation. The equivalent limestone content is 15.20%. However, after 4-hour carbonation curing, the carbon content was measured as 26.14%.

The cement content of concrete powder was determined using past experimental work and equations [6]. All reaction products are expressed in terms of dry cement content by mass. The cement contents (CC) of different concrete batches served as correction factors and were on the range of 0.25. The TG analysis of the hydrated reference resulted in approximately 7% carbon content. The obtained carbonates are to be deducted from those of the carbonated samples to obtain the CO_2 uptake. The average carbon uptake is then calculated based on that of the four methods and is summarized in the last column of Table 3.

Table 3: CO_2 content and uptake (%) in PLC concretes

	CO_2 Uptake		CO_2 Content		CO_2 Uptake
Sample	Mass gain	Mass curve	Furnace	TG 28-day	Average
0a (Ref)	n/a	n/a	6.68	6.62	0.00
18a + 4c	18.34 ± 1.6	18.23	26.14	26.06	18.87
18a + 4c + sp	18.29 ± 1.7	18.09	26.92	26.54	19.14

3.2. Compressive strength of PLC concrete

Compressive strength results for 1-day and 28-day concretes are shown in Fig. 2. The 24-hour duration comprises initial air curing followed by 4-hour carbonation and subsequent hydration to 24 hours. For instance, sample 18a+4c underwent 18 hours of air curing followed by 4 hours of carbonation and 2 hours of subsequent hydration. In comparison to hydrated counterparts, carbonated concrete exhibited 13.4±0.9% higher strength after 1 day. It was obvious that carbonation was an accelerated hydration process. Even though only part of the lost water was restored, within 2 hours of subsequent hydration, water compensation improved the compressive strength by 29.5±2.1% compared to the non-sprayed carbonated concrete.

After 28 days, the carbonated concrete (18a+4c) exhibited lower compressive strength than the hydrated reference owing to the water loss in early air curing. Water spray after carbonation was therefore essential to restore the curing-induced water loss. Hence, the strength increased by 11.7±0.8% after water compensation. The results indicate that carbonation followed by water compensation is beneficial. While early carbonation does not hinder subsequent hydration, the water loss during early air curing and carbonation curing reduces the degree of hydration. With water compensation through surface spray, it is possible to make carbonation strength comparable to hydration counterpart.

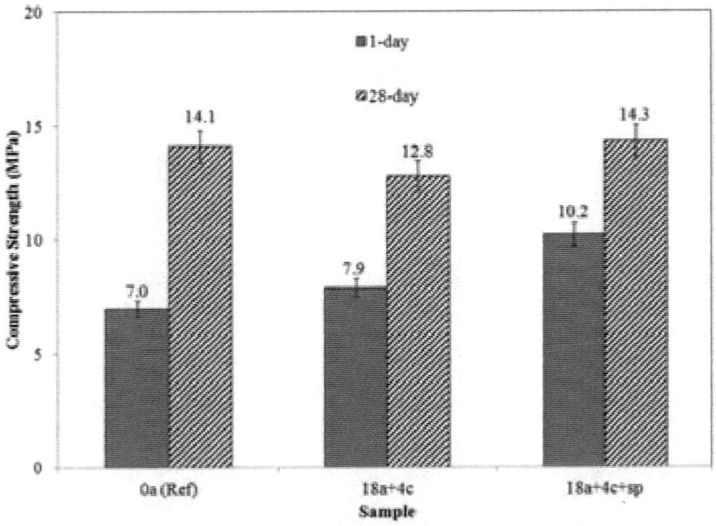

Fig. 2. Compressive strength of PLC concretes

3.2. Microstructure analysis of PLC concrete

Fig. 3 shows the 1-day XRD patterns for carbonated and hydrated PLC concretes. Dominant and strong peaks of calcium carbonate polymorphs including calcite, vaterite, and aragonite are present throughout carbonated samples. Higher carbonate peaks at 22°, 29.4° and 43° (2θ) are detected in the carbonated batches in comparison to the hydrated reference (0a). The coexistence of hydration and carbonation products is noticed in the carbonated samples. Calcium hydroxide is distinctively identified at 18° in hydrated samples. However, it is conclusive that calcium hydroxide is consumed given the absence of this peak in the carbonated PLC concrete. In addition, anhydrous calcium silicates are consumed, evidenced by the decrease in peak intensity at 34.5° and 41.5°, indicating the existence of C_3S. Ettringite has also been consumed verified by the lack of the 52.5° peak after carbonation.

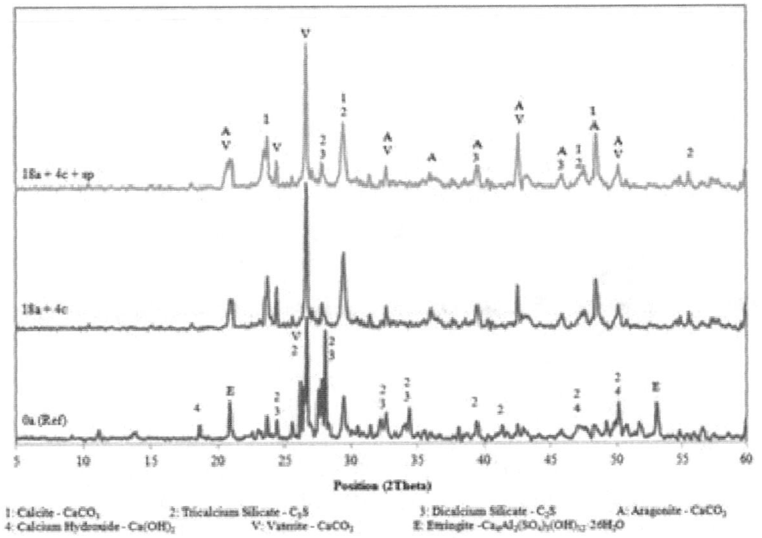

Fig. 3. XRD patterns of PLC concretes after 1 day

XRD patterns of the PLC concretes after 28-day subsequent hydration are shown in Fig. 4. The amount of vaterite and aragonite, shown at 21.5° and 43°, have decreased with reference to the patterns obtained after 1 day. While aragonite is suggested to form when the sample is allowed to dry out as in the case of 18a+4c [10], vaterite forms due to the rapid carbonation reaction and high heat release [10-11]. The hydrated reference does not differ between 1 and 28 days other than higher calcium hydroxide peaks with 27 more days of hydration. The 15% added limestone in the cement was detected as calcite peaks at 26.8° and 29.4° in the hydrated reference.

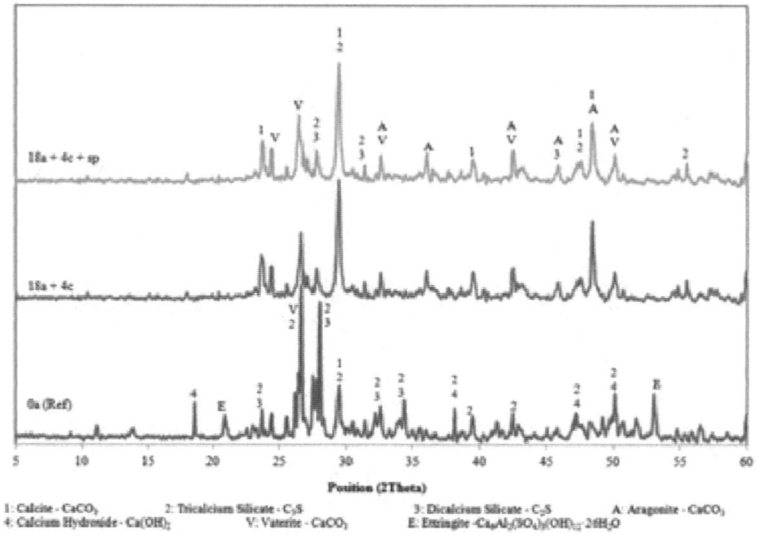

Fig. 4. XRD patterns of PLC concretes after 28 days

According to Ramachandran and Beaudoin (2001), the mass loss along thermogravimetric curves can be classified into 6 categories: Below 105°C, the mass loss is associated with the evaporable water and possibly the poorly formed C-S-H. The water loss between 105°C and 200°C represents the bound water in low temperature C-S-H and ettringite, while the loss in the range of 200°C-420°C is associated to the bound water in well-formed hydration products including C-S-H and C-A-H. Bound water in calcium hydroxide (CH) is decomposed between 420°C and 470°C [14]. While CO_2 in poorly crystalline calcium carbonates designated by aragonite and vaterite are decomposed between 470°C and 680°C, that in well crystalline calcite lies in the range of 680°C-950°C [9,11-12].

Based on TG/DTG analysis, detailed mass loss is presented in Fig. 5. Following the previously mentioned mass loss distribution, it can be noted that the hydrated reference had the lowest amounts of C-S-H and C-A-H and exceeded by the carbonated counterparts. As carbonation is an accelerated hydration process, the anhydrous calcium silicates in PLC were consumed to form C-S-H following 4-hour carbonation [14]. This explains the increase in C-S-H and C-A-H contents after carbonation. Further formation of C-S-H was identified in the sprayed carbonated sample (18a+4c+sp) after two hours of subsequent hydration. CH formation was a direct indication of the degree of hydration of any sample. The hydrated reference formed the highest amount of CH associated with no water loss during 1-day curing. The sprayed carbonated sample (18a+4c+sp) showed a comparable amount and was followed by its drier counterpart (18a+4c). Even with 0.8% loss in the range of 420-470°C, it is possible that CH was not detected by XRD possibly due to the presence of amorphous forms of CH [15]. Evidently, a sufficient water supply promoted the hydration of unreacted calcium silicates in the carbonated PLC concrete. Beyond 470°C, the hydrated reference exhibited some loss due to decarbonation of pre-existing carbonates. The respective CO_2 loss of carbonated concretes represented the CO_2 content, and included the added limestone and produced carbonates. The mass loss between 470-680°C, representing the carbonates produced from carbonation of C-S-H, showed the least thermal stability, suggesting poorly crystalline calcium carbonate designated by aragonite and vaterite [9,12,15]. The carbonates with better crystallinity, such as calcite, required higher temperatures to decompose and were lost above 680°C [10].

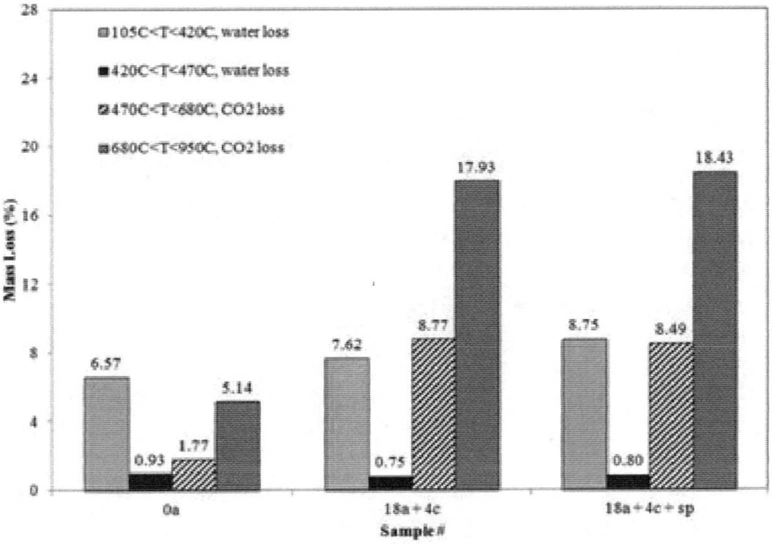

Fig. 5. Mass loss due to dehydration and decarbonation in 1-day PLC concretes

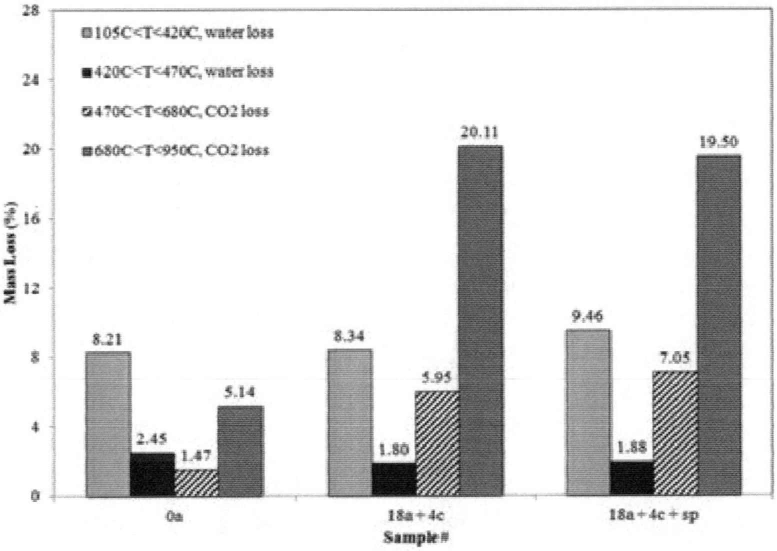

Fig. 6. Mass loss due to dehydration and decarbonation in 28-day PLC concretes

Hydration and carbonation products of 28-day PLC concretes are summarized in Fig. 6. The major change between 1 and 28 days was the hydrated reference concrete, in which increase in mass loss between 105-470ºC was indicative of increase in hydration products. Sealed curing conditions in hydration reference prevented water evaporation, which is an essential factor for subsequent hydration. The increase in compressive strength (Fig. 2) is associated with the increase in C-S-H and C-A-H content. Similar behavior was reported in the two carbonated concretes with higher losses attributed to 28-day subsequent hydration. It was suggestive that carbonation of PLC may slightly hinder the hydration of silicates and aluminates, thus affecting the development of compressive strength.

The percentage of poorly crystalline and well crystalline carbonates were calculated in order to study the stability of carbonates throughout subsequent hydration. Table 6 summarizes the poorly crystalline (470-680ºC) and well crystalline (680-950ºC) proportions of carbonation-induced carbonates in each batch. After 1 day, 33-35% of the total 4-hour reaction carbonates were poorly crystalline, while 65-67% were well crystalline. After 28 days, the percent of well crystalline increased to 72-77%. It was conclusive that unstable carbonates (aragonite and vaterite) transformed into more stable polymorphs (calcite) in PLC concretes.

Table 6: Percentage of carbonation-induced poorly and well crystalline carbonates in PLC concretes

	Poorly Crystalline		Well Crystalline	
Sample	1-day	28-day	1-day	28-day
18a+4c	35.4	23.0	64.6	77.0
18a + 4c + sp	33.6	28.0	66.4	72.0

4. Conclusions

The carbonation curing of PLC concrete was investigated. Its effect on the compressive strength and microstructure of PLC concrete was examined. Based on the findings of this study, the following conclusions may be drawn:

1. In the presence of initial air curing, a CO_2 uptake of about 19% in terms of cement mass was achieved in a lightweight PLC concrete with expanded slag aggregates. It was corresponding to a carbonation degree of 38%.
2. Compressive strength of PLC concretes varied depending on curing method. Carbonated samples exhibited higher strength at early age in comparison to hydrated counterparts. After 28-day subsequent hydration, carbonated concretes could only compare to those cured using hydration when water compensation through surface spray was employed.
3. Carbonation consumed calcium hydroxide, calcium silicates, and ettringite to produce calcium carbonates in its three polymorphs. The XRD peaks showed evidence of this phenomenon. Even though a mass loss between 420-470°C was detected in the carbonated PLC concrete, calcium hydroxide was not identified by XRD possibly due to its amorphous nature.
4. The carbonation-induced calcium carbonates are distinguished between poorly and well crystalline. For 18-hour air cured and 4-hour carbonated PLC concrete, 33-35% of the total carbonation-induced carbonates were poorly crystalline throughout after 1 day, while 65-67% remained well crystalline. After 28 days, 72-77% of carbonates became well crystalline. It was conclusive that phase transformation of carbonates occurred in PLC concrete during subsequent hydration.

5. Acknowledgements

The authors gratefully acknowledge the support by the Natural Science and Engineering Research Council (NSERC) of Canada, Canadian Concrete Masonry Producers Association (CCMPA), and Lafarge Canada.

6. References

[1] Soroka I, Setter N. The Effect of Fillers on Strength of Cement Mortars. Cem. and Concr. Res. 1977; 7(4): 449-456.

[2] Bonavetti V, Donza H, Menéndez G, Cabrera O, Irassar, E.F. Limestone Filler Cement in Low w/c Concrete: A Rational Use of Energy. Cem. and Concr. Res. 2003; 33(6): 865-871.

[3] Ramachandran V, Zhang C. Influence of $CaCO_3$ on hydration and microstructural characteristics of tricalcium silicate. Cemento 1986; 83(3): 129-152.

[4] Matschei T, Lothenbach B, Glasser F.P. The Role of Calcium Carbonate in Cement Hydration. Cem. and Concr. Res. 2007; 37(4): 551-558.

[5] Tsivilis S, Chaniotakis E, Kakali G, and Batis G. An analysis of the properties of Portland limestone cements and concretes. Cem. and Concr. Composites 2002; 24(3): 371-378.

[6] El-Hassan H, Shao Y, Ghouleh Z. Reaction Products in Carbonation Cured Lightweight Concrete. Am. Soc. of Civ. Eng. Mater. J. 2013; 25(6): 799-809.

[7] El-Hassan H, Shao Y, Ghouleh Z. Effect of initial curing on carbonation of lightweight concrete masonry units. Am. Concr. Inst. Mater. J. 2013; 110(4): 441-450.
[8] ASTM C140. Standard Test Methods for Sampling and Testing Concrete Masonry Units and Related Units. ASTM International 2004.
[9] Ramachandran V. S, Beaudoin J. J. Handbook of Analytical Techniques in Concr. Sci. and Technol., William Andre Publishing/Noyes 2001.
[10] Goodbrake C. J. Reaction of beta-dicalcium silicate and tricalcium silicate with carbon dioxide and water. J. of the Am. Ceram. Soc. 1979; pp. 168-171.
[11] Bukowski J. M. Reactivity and strength development of CO_2 activated non-hydraulic calcium silicates. Cem. and Concr. Res. 1979; 9: 57-68.
[12] Goto S. Calcium silicate carbonation products. J. of the Am. Ceram. Soc. 1995; pp. 2867-2872.
[13] Li D, Shi, H, Deng J, Xu Y. Study on the thermal decomposition kinetics of nano-sized calcium carbonates. J. of Zhejiang University 2003; 4(3): 363-368.
[14] Chang C, Chen J. The experimental investigation of concrete carbonation depth. Cem. and Concr. Res. 2006; 36: 1760-1767.
[15] Young J. F, Berger R. L, Breese J. Accelerated Curing of Compacted Calcium Silicate Mortars on Exposure to CO_2. J. of the Am. Ceram. Soc. 1974; 57(9): 394-397.

473956 Surface Behaviors of Electrodeposited Tin Oxide on Selected Substrates and Their Effectiveness in CO_2 Electro-Reduction

Abstract:

Carbon dioxide (CO_2) effects on greener energy remain the topic of discussion in the need to direct new ways to sustainability. Electro-sequestration of CO_2 is being pursued as a promising route leading to many studies in this area. Findings have shown the effectiveness of electrocatalytic conversion of CO_2 to useful chemicals. However, high over-potentials to produce efficient current densities, coupled with a vast range of faradaic efficiencies, and issues on stability and longevity have been the major drawbacks. Post transition metal oxides have shown promise to CO_2 reduction, among which tin oxide (SnO_2) is the most cost effective and least toxic. But the reported range of current density, and faradaic efficiency using a SnO_2 catalyst is sporadic. In this work, the reduction of CO_2 is investigated by focusing on electrode materials, specifically, on surface behaviors of the anode structures. Since SnO_2 has shown great potential for CO_2 reduction, our aim is to investigate this material further. Thus, the objective is to study the cathodic electrodeposition characteristics of some notable substrates- glassy carbon, indium tin oxide, tin, and their alloys- as they interact with SnO_2 catalyst. The result of our study and the effectiveness of these materials to reduce CO_2 will be presented.

473500 Thermochemical Characterization of Sugarcane Bagasse for the Production of Combustibles Via Pyrolysis Using N2 As Carrier Gas

Abstract:

Global warming is a large-scale issue that has led humanity to seek for renewable energy sources. Biomass thermal gasification is one the new technologies to produce clean fuels using biomass as feedstock. The resultant product from this process is a renewable neutral carbon fuel. Temperature parametric results obtained from a sugarcane-bagasse (CH1.581O0.671N0.00306S0.00117) pyrolysis experimental study carried out at pyrolysis temperatures ranging between 400 C and 900 C and at a heating rate of 4.30 C/min ± 7.8%, are discussed. The experiments were developed in a static batch reactor with nitrogen as carrier gas. Sugarcane bagasse samples were thermo-chemically characterized in order to determine thermodynamic properties such as enthalpy of formation, high heating value, empirical formulas of biomass, volatile matter, etc. The chemical formula of the pyrolysis oil obtained was estimated using gas composition, given by a gas analyzer, and atom balance on components. The effect of temperature on the production of liquid and gaseous fuels are presented along with the HHV of the syngas produced. The following results were obtained: chemical empirical formulas for biomass and volatile matter were CH1.581O0.671N0.00306S0.00117 and CH1.807O0.767N0.00349S0.00133 respectively; high heating values (DAF basis) of biomass and volatile matter 18,925.08 kJ/kg and 17,472.77 kJ/kg, respectively; average chemical empirical formula of liquid fuel CH1.942O0.484N0.005S0.002.

464366 The Effect of Torrefaction Pretreatment, Temperature, and Lignin Content of Hybrid Poplar on the Quality of Fast Pyrolysis Bio-Oil

Abstract:

Introduction: Fast pyrolysis of woody biomass has been identified as a potential means for the production of advanced transportation fuels. During fast pyrolysis the three main components of biomass (cellulose, hemicellulose, and lignin) thermochemically breakdown to produce bio-oil. Several studies have isolated these individual components and looked at their bio-oil distributions. However the interactions between these three main components and how variations in feedstock composition affect product distribution are not well understood. The purpose of this work is to assess the quality of bio-oil as lignin content is varied in the hybrid poplar, how temperature of pyrolysis affects these results, and how the addition of a torrefaction pretreatment will affect the composition of bio-oil. The quality of bio-oil is assessed by calculating bio-oil yield relative to dry biomass and relative to char and gas, changes in relative abundance of oxygen in the bio-oil, and energy density of the bio-oil. Eight genetically different poplar samples with varying lignin content were pyrolyzed at 500ºC using a CDS 5200 HP micro-pyrolysis unit, which was directly connected to a ThermoFisher Trace GC coupled with a DSQII MS. Four of these hybrid poplar samples over a range of lignin content were then pyrolyzed at temperatures of 550ºC and 600ºC to determine effects of pyrolysis temperature. Finally, two of those samples, a high and low lignin concentration, were then torrefied at 300ºC for 30 minutes and then pyrolyzed at 550ºC for 20 seconds.

Results: An increase of lignin concentration, by approximately 10% relative to the low lignin content sample, decreased the abundance of phenolic species in bio-oil by 3% and decreased the relative abundance of oxygen by 8% at 500ºC. The bio-oil yield similarly decreased by 5%, due to an increase char yield at high lignin levels. With an increase in pyrolysis temperatures, there was an increase in phenolics produced overall, but with respect to increasing lignin content, it continued the trend of decreasing phenolic species in bio-oil. Again with respect to lignin content at higher temperatures, the bio-oil yield decreased as the char yield increased. The addition of torrefaction as a pretreatment step, decreased the overall yield of bio-oil produced by 10% and in agreement with previous literature, removed low molecular weight highly oxygenated species, thereby decreasing bio-oil yield but improving the quality of bio-oil produced. From these results over a limited range of lignin content, higher temperatures are needed to increase the recovery of phenolic species from pyrolysis of poplar.

465258 Effect of Operating Parameters on Ablative Pyrolysis of Wood

Abstract:

One of the main requirements for the production of bio-oil from fast pyrolysis of wood is the use of small feedstock particle sizes (1-3 mm), in order to minimize temperature gradients within the particles and therefore avoid undesired reactions that take place in the range 200-400°C. The only exception to the rule is ablative pyrolysis. In ablation, heat transfer to the wood is carried out via direct contact with a hot, moving metallic surface. This situation creates a steep temperature gradient within the particle, such that pyrolysis takes place within a thin liquid layer contacting the solid as opposed to the entire wood particle. For this reason, in principle there is no limit to the particle size in ablative pyrolysis, and this feature is of great value, since grinding costs represent 7-9 % of the total process costs in pyrolysis. Though this characteristic has been speculated in the past, there were no experimental data to back up this claim. Recently, our group developed an ablative pyrolysis unit that is able to convert entire wood chips (about 1â€™â€™ size) and even entire wood rods directly into bio-oil in one single step, with yields similar to those in a fluidized bed. In this work, we evaluated the effects of the main operating parameters involved in ablative pyrolysis of whole wood chips: the temperature of the hot plate, the relative speed between the wood chips and the hot plate, the pressure applied by the plate on the chips, and the initial thickness of the assembly of wood chips. Comparisons involve the bio-oil yield, LHV, water content, composition as detected by GC-MS, and functional groups detected by FTIR. In addition, differences in the char produced is evaluated by elemental composition and FTIR.

467137 A Combined Pyrolysis – Gasification Study to Increase Syngas Production Yield from Beer Bagasse Waste

Abstract:

Extensive resources and efforts have been dedicated to assess opportunities for using locally available biomass to produce energy, reduce CO_2emissions, and lower reliance on petroleum. The use of lignocellulosic biomass is promising since it is readily available in vast amounts and there are processes for converting different biomass sources to useful products of particularly interest in liquid transportation fuels. Energy based in biomass is a source of renewable energy from organic and inorganic matter formed in a biological or mechanical process. Although fossil fuels, such as petroleum, oil and coal, have their foundation in ancient biomass, they have been "out" of the carbon cycle for a long time. Their combustion consequently disturbs the carbon dioxide content in the atmosphere.

One promising technology that has gained interest in the past few years is pyrolysis and gasification. Although fast pyrolysis has been known and studied for many years; due the complexity of fast pyrolysis reaction mechanisms there are still a lot more to investigate to increase the understanding on fast pyrolysis of biomass.

While it is scientifically viable to exploit lignocellulosic materials and organic wastes into energy, chemicals and fuels, it is far from an economic process and the cost needs to be lowered. It also must be demonstrated that a commercial scale biomass utilization process is environmentally sound, and there is a need to find an inexpensive and widely available lignocellulosic source of biomass

The core objective of this research is to study the influence of pyrolysis as a pretreatment on spent grain from brewers' waste to improve syngas production from gasification. Biochar produced from the biomass pyrolysis was used in order to obtain compounds with higher carbon content to subsequent undergo a gasification process. In addition, commercial activated carbon was gasified to compare the yield of syngas produced in the gasification of biochar.

Brewers waste was pyrolyzed under a nitrogen environment in different operating conditions; heating rates, reaction times, and pyrolysis temperatures. Activated carbon and biochar gasification was done at 850 °C and 1 atm of pressure for two hours using water as gasifying agent in a fluidized bed reactor.

Pyrolysis at 300 °C produced a greater yield of biochar leading to a higher concentration of syngas produced during gasification of the waste biomass studied compared with the commercial activated carbon products obtained.

Finally, the processes of chemical thermo-decomposition pyrolysis and gasification applied to brewers' waste, allow to obtain products with higher added value, making possible the production of synthesis gas from the gasification of solid compounds obtained from pyrolysis of beer bagasse waste.

These studies lay the groundwork for longer-term efforts aimed at using waste biomass as a sustainable way to add value to solid waste biomass.

468387 Product Distribution and Kinetics of Hydropyrolysis of Agro Residues Via Py-GC/MS and Py-FT-IR

Abstract:

Approximately 140 billion tons/yr of agricultural waste is generated globally, which is equivalent to 50 billions tons of crude oil. (UNEP Report 2011). In India, a major fraction of the agro residues is unutilized and simply burnt in the farm, while the rest are either used as fodder or co-fired with coal in furnaces. These are large volume, yet low value solutions. Generating liquid bio-crude/biofuel from the agro residues via thermochemical techniques is a promising option to reduce the dependency on fossil-based fuels. Although fast pyrolysis of biomass residues is a promising technique that is being deployed to produce bio-oil/bio-crude, there are inherent difficulties associated with the utilization of bio-oil. These include high oxygen content, low heating value, poor storage and thermal stability, which are attributed to the presence of moisture and organics like carbonyl compounds, carboxylic acids, alcohols and phenols. While this stability issue can be partially addressed by catalytic fast pyrolysis that produces a fairly deoxygenated bio-oil, the bio-oil still requires further upgradation via catalytic hydrodeoxygenation process. Hydropyrolysis is a single step process wherein biomass is fast pyrolyzed at c.a. 500 °C in presence of hydrogen gas at moderate pressures (<35 bar). This results in the production of a high quality liquid fuel with very low O/C ratio. It is, therefore, necessary to thoroughly characterize the evolution of pyrolysates from fast hydropyrolysis and compare it with conventional fast pyrolysis process to understand the mechanism of formation of products. In this study, analytical pyrolysis (Py) of rice straw, empty palm fruit bunch (EFB) and pine wood biomass is conducted and the products are characterized using gas chromatograph/mass spectrometer (GC/MS) and Fourier transform infrared spectrometer (FT-IR).

The biomass characteristics such as volatile matter, fixed carbon, ash content, elemental C, H, N, S and O, and heating value were evaluated using standard techniques. It is important o note that the biomasses investigated in this study vary mainly in ash composition, and follow the trend: rice straw (19%) > EFB (7.5%) > pine wood (0.7%). Fast pyrolysis and hydropyrolysis experiments were conducted in a Pyroprobe® 5200 pyrolyzer in trap mode. Approximately 0.5 mg of the agro residues were pyrolyzed at a heating rate of >10,000 °C/s to the preset temperature of 400-700 °C and held for 60 seconds in hydrogen atmosphere. The vapors were characterized using Agilent two-dimensional GC/MS. The effect of hydrogen pressure from 1 to 25 bar on pyrolysate composition was investigated, and compared with pyrolysate composition in presence of He environment at atmospheric pressure. Experiments were also conducted wherein the time evolution of key functional groups of the pyrolysates was analyzed using an online FT-IR spectrometer (Agilent Cary 660). A significant reduction in the formation of oxygenated organics, especially carboxylic acids, furan derivatives and phenolics, and a concomitant increase in the production of linear and aromatic hydrocarbons were observed during fast hydropyrolysis compared to normal fast pyrolysis of Indian rice straw, even at medium pressures of hydrogen. High temperatures slightly improved the yield of linear hydrocarbons, but greatly increased the yield of aromatics during hydropyrolysis. Py-FT-IR study revealed the time evolution of the pyrolysates, and the reaction completion time. The time taken for maximum production of oxygenates varied from 10-15 s at 500 °C, which was similar for all three biomass varieties. More interesting results on (a) the evolution of individual functional group vibrations, (b) product characterization, and (c) char formation, for different biomass varieties will be discussed during the presentation. The implications of the results on the quality of bio-oil and its properties will also be addressed.

468153 The Top-Ref Approach to Improve the Resource Efficiency of Energy Intensive Industrial Processe

Abstract:

TOP-REF project aims at developing and validating indicators, methodologies and non-invasive tools devoted to the improvement of resource efficiency in energy intensive continuous industrial processes in the agro-chemical, chemical and petrochemical sectors. These methodologies and tools will constitute to the following targets:

- To path the way for achieving the SPIRE2030 objective of reducing non-renewable, primary raw material intensity up to 20 % and fossil energy intensity up to 30%, both compared with current levels.
- Reduce production costs up to 15 % compared with the current levels.
- A significant reduction of the environmental impacts (CO2 emissions, water footprint, pollutants, hazardous emissions, etc.).

To achieve these impacts, TOP-REF develops and demonstrates a robust, resource-efficiency-focused and cross-sectorial methodology based on novel exergy-based resource indicators. This methodology is implemented in three industrial demonstration plants: solid fertilizer production, crude distillation and fractionation unit of a refinery and a naphtha steam cracker. The tools developed are validated by demonstration under real conditions in three pilots.

The methodology developed by TOP-REF consists of the following steps:

1. Development of Key Resource Indicators (KRIs) to measure the performance in resource efficiency
2. Modeling and simulation of the three industrial pilots
3. Identification of the critical process parameters (CPPs) influencing the KRIs
4. Optimization of the CPPs to improve the resource efficiency

5. Demonstration and validation under real conditions

The approach to identify the critical process parameters for the production processes investigated is based on detailed process models validated by the industrial partners. These process models are analyzed using global sensitivity analysis to identify the critical process parameters. As the full simulation models are very complex, simplified mathematical surrogate models are derived prior to global sensitivity analysis. A variance-based global sensitivity analysis is performed for all process parameters to identify the most influential operating parameters on the key resource indicators (KRIs) developed within the TOP-REF project. The use of a variance-based sensitivity analysis enables the consideration of interaction effects among the process parameters. Finally, the identified critical process parameters are optimized using the KRIs as objective. The optimized operating points will be validated under real conditions in the three production plants.

Within the contribution we will explain the TOP-REF approach developed, show its application to the three industrial processes and highlight experiences made during the application.

Acknowledgment

The research leading to the results presented in this contribution has received funding from the European Communityâ €˜s 7th Framework Program for research, technological development and demonstration under grant agreement No 604140, TOP-REF.

468313 Novel Data Envelopment Analysis Approach for Sustainability Assessment: Application to Electricity Generation Technologies

Abstract:

Moving towards sustainable development implies implementing actions, projects, programs, plans and policies, which involve the balanced integration of economic, environmental, and social objectives. However, setting sustainability goals requires some knowledge and understanding of the current level of sustainability. To this end, sustainability assessment approaches can be applied for measuring the extent of progress towards sustainability and for defining quantitative targets that ensure a more sustainable development.

In this work we explore the use of data envelopment analysis (DEA) to quantify the level of sustainability attained by a system and identify targets for improvements. DEA is a non-parametric linear programming (LP) technique that measures the efficiency of a set of entities, termed decision-making units (DMUs), each transforming multiple inputs into multiple outputs (Charnes et al., 1978). By solving the DEA LP problem, it is possible to distinguish between 'efficient units' (i.e., efficiency score of 1) and 'inefficient units' (i.e., non-frontier units with an efficiency score strictly lower than 1), for which DEA provides in turn specific guidelines to become efficient by projecting them on the efficient frontier.

In the context of sustainability assessment, DEA presents two major limitations particularly critical when a wide range of economic, environmental and social indicators are considered in the analysis. First, the standard DEA makes no distinction between the units deemed efficient, which makes it difficult to select a final alternative in the absence of a ranking scheme. Secondly, DEA is very sensitive to the set of inputs and outputs, leading to very different results depending on the sustainability criteria considered in the analysis.

To overcome the limitations of the standard DEA as applied to sustainability studies, this work proposes an enhanced DEA methodology tailored to carrying out sustainability assessments. The methodology proposed integrates DEA with the concept of "order of efficiency", originally introduced to rank Pareto optimal solutions from a set of many Pareto points (Das, 1999). In essence, the main idea behind the new methodology consists of repeating the DEA calculations iteratively for each and every possible combination of inputs/outputs, to subsequently aggregate the results into efficiency metrics that assess each sustainability dimension separately and finally determine an overall sustainability level integrating the three scores (obtained in each dimension).

The capabilities of the methodology proposed are illustrated through its application to the assessment of the electricity-generation technologies expected to play a major role in the future electricity mix of United Kingdom (i.e., nuclear, gas, coal with and without carbon capture and storage, wind, solar photovoltaics and biomass with wood and *Miscanthus* pellets). Each technology is defined as a DMU that is characterised by a set of economic, environmental and social sustainability indicators determined using a life cycle assessment approach by Stamford and Azapagic (2014). We consider 18 sustainability indicators: three economic (capital cost, operation and maintenance cost, and fuel cost), nine environmental (freshwater ecotoxicity, marine ecotoxicity, global warming, ozone depletion, acidification, eutrophication, photochemical smog, land occupation, and land ecotoxicity) and six social (direct employment, worker injuries, human toxicity potential, radiation, depletion of elements, and depletion of fossil fuels).

The results generated with our approach show that gas, nuclear and wind technologies are efficient in each dimension of sustainability when all the indicators within each such dimension are considered. Gas is also the most economically and socially efficient (highest economic and social efficiency scores), while nuclear is the best in the environmental dimension. From an overall sustainability point of view, the gas technology presents the highest overall sustainability efficiency (mainly because of its very good performance in the social dimension), followed by nuclear and wind. On the other hand, coal is found economically inefficient; coal, solar PV and both biomass options are environmentally inefficient; and coal and biomass with *Miscanthus* are found socially inefficient. For these technologies deemed inefficient in a particular sustainability dimension, an inefficiency assessment is performed which provides targets that should be attained to reach the efficiency in each specific dimension.

The enhanced DEA proposed can facilitate the transition to a more sustainable future, since it enables ranking alternatives according to the extent to which they adhere to sustainability principles; and also pinpoints in a systematic manner the main sources of inefficiency and sets targets for improvements.

REFERENCES

Charnes, A., Cooper, W.W., Rhodes, E., 1978. Measuring the efficiency of decision making units. Eur. J. Oper. Res. 2, 429–444. doi:10.1016/0377-2217(78)90138-8

Das, I., 1999. A preference ordering among various Pareto optimal alternatives. Struct. Optim. 18, 30–35.

Stamford, L., Azapagic, A., 2014. Life cycle sustainability assessment of UK electricity scenarios to 2070. Energy Sustain. Dev. 23, 194–211. doi:10.1016/j.esd.2014.09.008

471458 Optimal Design of Sustainable Agricultural Water Networks

Abstract:

This work proposes a multi-period mixed integer non-linear programming model for the optimal design of agricultural water networks, which involves the optimal reuse of resources such as water, fertilizers and pesticides. This model is based on a new superstructure that includes attractive options for the use, reuse, recycle, regeneration and storage for the considered resources. The model incorporates simultaneously the maximization of the net annual profit and the minimization of the environmental impact. The profit is formed by the economic incomes due to the sale of crops minus the corresponding capital and operating costs associated to the fresh water, fertilizer, storage, piping, pumping and water treatment. Whereas the environmental impact is determined through the Eco-indicator 99 and the water footprint to evaluate the effect over the global warming, acidification, eutrophication, green water footprint, gray water footprint and blue water footprint. Two case studies from the Mexican state of Sinaloa were solved with the proposed model. The first one considers four crops of corn and it requires five irrigation periods; while in the second one twelve parcels are included where there are four crops of corn, four crops of bean, two crops of wheat and two crops of alfalfa, and there are needed seventeen irrigation periods. The operating, economic and design parameters are representative for the State of Sinaloa in Mexico; and the results are presented in three dimensional Pareto diagrams. From these diagrams is possible to find the optimal relationship between profit, eco-indicator and water footprint.

459908 Efficiency of Bimetallic Nanoparticles for Treatment of Azo Dyes in Water

Abstract:

The development of novel materials for water treatment applications remains a significant need as water resources become increasingly strained and contaminant removal challenges persist. While current commercial water treatment materials and technologies are able to remove a wide range of contaminants, the addition of functionality or performance through material design enables improved contaminant removal and the ability to address particularly recalcitrant contaminants. In particular, the development of nanostructured, reactive materials for water treatment applications holds great promise due to the reactive degradation and improved removal processes that have been demonstrated. Both laboratory-scale and pilot-scale research studies have demonstrated that iron-based nanostructured materials, such as iron nanoparticles, are able to remove a wide variety of water contaminants, including chlorinated organics, pharmaceutical compounds, and heavy metals, through both oxidative and reductive reaction mechanisms. However, iron nanoparticle lifetime remains a challenge particularly in oxidative environments, where the nanoparticles quickly react with water and oxygen to form non-reactive iron oxide compounds. Thus, there remains a need to investigate approaches to extend the lifetime as well as to enhance the performance of iron-based nanoparticles.

In our work, we have focused on the synthesis of bimetallic nanoparticles, which are then used to treat water contaminated by an azo dye (Orange G). These nanoparticles are synthesized in an aqueous solution and result in an alloy morphology. Variation of synthesis parameters can be used to optimize nanoparticle reactivity and understand which parameters are critical to controlling nanoparticle morphology. In our synthesis method, we focus on different combinations of metals, including iron-nickel and iron-cobalt. Thus far, our results have shown that different molar ratios of the metals result in different rates of removal of Orange G from solution, as well as in the overall removal of Orange G from the solution. Within a range of 0.1 to 0.4 mol Ni:mol Fe, lower molar ratios of nickel to iron in the nanoparticle metal combinations have shown higher removal of Orange G. In addition, changing the metals used in synthesis (e.g., from FeNi to FeCo) is a viable option to optimize the reactivity of iron-based nanoparticles. Finally, the types of stabilizers used during aqueous synthesis of the metal nanoparticles may impact the morphology and reactivity of the particles. Currently, stabilizers used in this work are poly(N-vinyl-2-pyrrolidone) (PVP) and amino tris (methylene phosphonic acid) (ATMP). We are beginning to investigate other novel types of stabilizers, such as peptides and peptoids, which have proven to be successful in stabilizing nanoparticles in nanoparticle formation. However, it is unclear how the use of such specialized stabilizers impacts the reactivity of the nanoparticles in water treatment applications.

In this talk, we will present results on several synthesis parameters that have been identified as key to impacting nanoparticle reactivity towards the target water contaminant Orange G, including nanoparticle composition and metal:metal molar ratio in the alloy morphology. Results will be presented for nickel to iron molar ratios in the range of 0 to 0.4 as well as iron to cobalt with a similar range of molar ratios. We will also present results on the characterization of these nanoparticles using techniques such as electron microscopy, x-ray diffraction, and x-ray photoelectron spectroscopy. Finally, performance results will be discussed as a function of nanoparticle concentration and Orange G concentration.

453872 Ultrasensitive Quantification of Pesticide Contamination and Drift Using Silica Particles with Encapsulated DNA

Abstract:

The impact of pesticides and other potentially harmful and environmentally applied chemicals on ecosystems and human health are a constant issue of an ongoing public and scientific debate. Besides undeniable benefits, some pesticides also pose risks to human health and environment. For some agricultural practices, contamination with certain pesticides needs to be reduced or completely avoided. One prominent example is organic farming that does not rely on synthetic pesticides. As a consequence, there is a urgent need to develop tools that are able to universally quantify trace amounts of pesticides enabling the evaluation of contamination, drift and persistence on non-treated areas or in the environment.

Recently, we have developed a bioinspired tracer particles in our lab, silica particles with encapsulated DNA (SPED).[1] SPED consist of short deoxyribonucleic acid (DNA) tags having a length of ~100 base pairs. These DNA tags are incorporated into a chemically inert spherical silica capsule in the nano- to submicron size range (100 â€“ 250 nm) and are thus protected from degradation even under irradiation and harsh radical or heat treatments. The DNA label can be selectively released from the silica using fluoride chemistry and quantified with matching primers/probes using real-time polymerase chain reaction (qPCR), a widely-applied, specific and ultra-sensitive technique for DNA analysis. Under ideal conditions, SPED can be quantified down to a 1 ppt (ng/L) range.[2] SPED have already been successfully applied in tagging oil products, food, activated sludge, or ecological food webs.[3-6] They pose no risk to the environment, and could be readily extracted and quantified from many environmental or biological matrices.

Here, we would like to present our recent findings of using SPED as an ultrasensitive environmental tracer tool for pesticides and show that SPED fulfill the above mentioned requirements of an ideal tracer tool.[7] SPED could be incorporated and robustly recovered from a large range of pesticides. In field experiments on obstacle-free fields and in an apple orchard, pesticide deposits down to 1 nL cm^{-2} could be quantified after spraying a SPED-labeled test liquid containing 5.8 ppm (mg L^{-1}) SPED. Based on the analysis of the SPED distribution, wind and field-related patterns were clearly traceable. Due to their low material costs and enormous number of available different DNA tags, SPED cannot only be employed to identify and elucidate pesticide drift patterns down to ultra-low concentrations, but also allow cost-effective multiplexing experiments. Additionally, SPED could be also used as a unique chemical barcode for pesticides and other environmentally applied chemicals, helping, e.g., to identify the source of a contaminating pesticide. Misuse of pesticides could be therefore better investigated and prosecuted. Overall, the use of SPED improves and simpliï¬es pesticide drift assessments for agricultural or environmental protection purposes, thus facilitating the use of new agricultural techniques and decreasing pesticide-related safety risks.

[1] Paunescu, D.; Puddu, M.; Soellner, J. O. B.; Stoessel, P. R.; Grass, R. N. Reversible DNA encapsulation in silica to produce ROS-resistant and heat-resistant synthetic DNA 'fossils'. *Nature Protocols* **2013**, 8, 2440-2448.
[2] Paunescu, D.; Mora, C. A.; Querci, L.; Heckel, R.; Puddu, M.; Hattendorf, B.; Günther, D.; Grass, R. N. Detecting and number counting of single engineered nanoparticles by digital particle polymerase chain reaction. *ACS Nano* **2015**, 9, 9564-72.
[3] Puddu, M.; Paunescu, D.; Stark, W. J.; Grass, R. N. Magnetically recoverable, thermostable, hydrophobic DNA/silica encapsulates and their application as invisible oil tags. *ACS Nano* **2014**, 8, 2677-2685.
[4] Bloch, M. S.; Paunescu, D.; Stoessel, P. R.; Mora, C. A.; Stark, W. J.; Grass, R. N. Labeling milk along its production chain with DNA encapsulated in silica. *Journal of Agricultural and Food Chemistry* **2014**, 62, 10615-10620.
[5] Grass, R. N.; Schälchli, J.; Paunescu, D.; Soellner, J. O. B.; Kaegi, R.; Stark, W. J. Tracking trace amounts of submicrometer silica particles in wastewaters and activated sludge using silica-encapsulated DNA barcodes. *Environmental Science & Technology Letters* **2014**, 1, 484-489.
[6] Mora, C. A.; Paunescu, D.; Grass, R. N.; Stark, W. J. Silica particles with encapsulated DNA as trophic tracers. *Molecular Ecology Resources* **2015**, 15, 231-241.
[7] Mora, C. A.; Schärer, H.-J.; Oberhänsli, T.; Ludwig, M.; Stettler, R.; Stoessel, P. R.; Grass, R. N.; Stark, W. J. Ultrasensitive Quantification of Pesticide Contamination and Drift Using Silica Particles with Encapsulated DNA. *Environmental Science & Technology Letters* **2016**, 3, 19-23.

454131 Novel $TiO_2/ZnO/Bi_2O_3$ Composite Nanofibers for Air Purification - Optimization, Kinetic Modeling and Mechanisms

Abstract:

Semiconductor heterojunction structures can effectively enhance the separation efficiency of photogenerated electron/hole pairs and the subsequent photocatalytic performance. Novel $TiO_2/ZnO/Bi_2O_3$ composite nanofibers, synthesized by a simple sol-gel assisted electrospinning method, exhibited much higher photocatalytic activity for the oxidation of nitrogen monoxide (NO) under simulated solar irradiation than commercial TiO_2 nanoparticles. These composite nanofibers have increased absorption in both UV and visible range when compared with TiO_2 nanoparticles. The enhanced photocatalytic activity of $TiO_2/ZnO/Bi_2O_3$ is attributed to the difference in the energy band positions of anatase, rutile, zincite and bismuth oxide, resulting in both lower band-gap energy and reduced recombination rate of photogenerated electron/hole pairs. Moreover, the photocatalytic performances are more stable for TZB nanofibers than that of TiO_2 nanoparticles, which are easily deactivated. In addition, a new kinetic model, taken into account of flow retention time and physical-chemical kinetics, is used to shed light on the behavior of the photocatalytic reaction. Faster kinetics (resulting in higher reactor throughput) and higher conversion efficiency of NO may be realized by optimizing the bismuth concentration in the composite nanofibers. The degradation pathway of o-xylene by TZB has also been investigated.

Keywords:

$TiO_2/ZnO/Bi_2O_3$ composite nanofibers, solar-light driven photocatalyst, kinetic model, o-xylene degradation

454961 Filtration of Compositionally Heterogeneous Aerosols Using Micro- and Nano-Fiber Media

Abstract:

< style="padding-left: 17pt; text-align: center;">2016 AIChE Annual Meeting in San Francisco, CA, November 13â €“18, 2016.

Subject Areas (Order of preference): (a) Separations, (b) Particle Technology, (c) Environmental Topics< style="padding-left: 17pt; text-align: center;">Deadline for submissions: May 9, 2016, Word Limit: 1500

< style="padding-left: 17pt; text-align: center;">Filtration of compositionally heterogeneous aerosols using micro- and nanofiber media

Saptarshi Chattopadhyay1, **Junli Hao**1*, Koli Taghizadeh2 and Gregory C Rutledge1# 1Chemical Engineering Department,2Center for Environmental Health Sciences Massachusetts Institute of Technology

*-Presenting author, #Corresponding author. Tel.: +1 617 253 0171; Fax: +1 617 324 3127, E-mail address: rutledge@mit.edu (G. C. Rutledge)

< style="padding-left: 5pt; text-align: left;">Key words:

Particle, Condensation, Mobility Diameter, Maximum Penetrating Particle Size, Composite Filter, toxicity, Kelvin Equation, Vapor Pressure

< style="padding-left: 5pt; text-align: left;">Abstract:

Many naturally occurring aerosols exhibit compositional heterogeneity as a function of particle diameter (*d*p). They are formed by condensation from mixtures of volatile gases with different vapor pressures. Tobacco smoke1, coal gas2 and combustion products3 are examples of aerosol particulates whose chemical compositions vary with particle size. Typically, the portion of chemical compounds with higher boiling point increase with *d*p in such aerosol particulates 4. Because some of the higher boiling point (BP) compounds are toxic to human health5, a method for selectively removing these compounds is desired. Particulate aerosol filters have a wide range of applications targeting human health and environment, among them are respirators, high volume air samplers, engine exhaust filters, and clean room filters. A good filter medium is characterized by low particle penetration (fraction of particles present downstream of the filter medium) and small pressure drop. Theoretically and empirically, particle penetration is found to be a function of *d*p, permitting a filter to be selective with respect to incoming particle sizes. The aerosol size at which the probability of particle penetration through the filter is maximal is called the most penetrating particle size (MPPS)6. This work seeks to capitalize on our previous research on size-selective filter media and filtration design solutions7, for the purpose of reducing the transmission of high BP constituents in particulate aerosols. Our work focuses on producing model aerosols exhibiting compositional heterogeneity with *d*p, and application of selective filter media to reduce the transmission of higher BP constituents of the aerosol particulate phase.

In this study, we first generated novel aerosol surrogates for naturally occurring heterogeneous aerosols. A commercial *Sinclair-La Mer* type8 condensation mono-dispersed aerosol generator (CMAG) was used for this purpose. Condensation of vapor on the surface of salt particles, acting as nuclei, was controlled by varying the vapor pressure and the concentration of nuclei6. Many different hydrocarbons have been studied by Japuntich D.A. (1991)9 to produce either solid particles (made of steric acid or carnauba wax etc) or liquid particles (made of sebacate oil, phthalate oil etc) using CMAG. However, the CMAG has been used so far to condense only single component vapor on a salt nucleus. We generated bi- component oil aerosol particles, whose compositional heterogeneity is a function of *d*p. We generated aerosols comprised of two miscible oils: bis(2-ethylhexyl)sebacate (DEHS) and

diethylhexyl phthalate (DEHP), with BP of 212 °C and 383 °C, respectively. The objective was to investigate if the fraction of higher boiling oil (DEHP) is preferentially enriched within aerosols with larger *d*p. Controlled *d*p distribution can be achieved in the range of ~100 nm to

~500 nm by regulating the temperature at which the vapor is generated (saturator temperature or Tsat) in CMAG. We chose to compare the overall compositions of aerosols generated at different Tsatâ€™s as a way to obtain the composition-size correlation. Sizing of aerosol particles was done using a differential mobility analyzer (DMA) and condensation particle counter (CPC). For composition analysis, aerosols were collected on a glass fiber filter and assayed using Gas Chromatography Mass Spectrometry (GC-MS). The mass of oils in the aerosols was estimated from the GC-MS peak areas using a calibration correction. The mass ratio of DEHS to DEHP was then related to the mean diameter at the corresponding Tsat. Mathematical models based on the Kelvin equation were employed to understand the formation of these heterogeneous aerosols. In the second part of this work, nanofiber-microfiber composite filter media were developed for size-selective separation of aerosol particulates. In a previous study7 we examined performances of commercial microfiber filters (MFc) and cellulose acetate (CA) nanofiber filters as stand-alone uniform filter media, and demonstrated how the MPPS of CA nanofiber filters could be regulated by tuning the fiber properties. In this study, we deposit nanofibers on one face of MFc to obtain improved filter performance of nanofiber media without incurring large pressure drop. The nanofibers were generated by the electrospinning process using a solution of cellulose acetate in acetone and N,N-dimethylacetamide mixture. For measurement of separation selectivity, aerosols downstream of the filter media were collected by a glass fiber filter for

composition analysis in GC-MS.

Our results showed that the ratio of DEHP to DEHS exhibited an almost linear dependence on *d*p at smaller *d*p range, but the increase leveled off as the mean *d*p becomes larger. Nonetheless, these observations accord with the existing theory in general, which predicts an increase in the fraction of higher BP compounds in aerosol particles of larger *d*p. Combined with applications of size-selective filter media, this finding demonstrates the potential of compositionally selective separation for particulate aerosols through air filtration.

< style="padding-left: 5pt; text-align: justify;">References:

1. Jenkins RW, Francis RW, Flachsbart H, Stober W. Chemical Variability of Mainstream Cigarette Smoke as a Function of Aerodynamic Particle-Size. *Journal of Aerosol Science.* 1979;10(4):355-362.

2. Schirmer RE, Springer DL, Phelps DW, Pelroy RA, Mahlum DD. Variation of Composition with Particle-Size in Coal Liquid Aerosols Generated for Inhalation Toxicology Studies. *American Industrial Hygiene Association Journal.* 1985;46(1):28- 33.

3. Liu C, Hsu PC, Lee HW, et al. Transparent air filter for high-efficiency PM2.5 capture.

   Nature Communications. Feb 2015;6.

4. Ishizu Y, Kaneki K, Okada T. A New Method to Determine the Relation between the Particle-Size and Chemical-Composition of Tobacco-Smoke Particles. *Journal of Aerosol Science.* Apr 1987;18(2):123-129.

5. U.S. Surgeon General. *Centers for Disease Control and Prevention (US); National Center for Chronic Disease Prevention and Health Promotion (US); Office on Smoking and Health (US). How Tobacco Smoke Causes Disease: The Biology and Behavioral Basis for Smoking-Attributable Disease: A Report of the Surgeon General. Atlanta (GA):*

   Centers for Disease Control and Prevention (US); 2010. 3, Chemistry and Toxicology of Cigarette Smoke and Biomarkers of Exposure and Harm. Available from: http://www.ncbi.nlm.nih.gov/books/NBK53014/2010.

6. Hinds WC. *Aerosol Technology: Properties, Behavior, and Measurement of Airborne Particles.* Hoboken, NJ.: John Wiley & Sons; 1999.

7. Chattopadhyay S, Hatton TA, Rutledge GC. Aerosol filtration using electrospun cellulose acetate fibers. *Journal of Materials Science.* Jan 2016;51(1):204-217.

8. Sinclair D, Lamer VK. Light Scattering as a Measure of Particle Size in Aerosols - the Production of Monodisperse Aerosols. *Chemical Reviews.* 1949;44(2):245-267.

9. Japuntich DA. *Clogging of Fibrous Filters with Monodisperse Aerosols* [A Doctoral Thesis]. Loughborough Chemical Engineering, Loughborough University of Technology; 1991.

450165 Inhibitory Effect of Condensed and Non-Condensed Phenolic Moieties in Lignin on Enzymatic Hydrolysis

Abstract:

The residual lignin plays an inhibitory role in the enzymatic hydrolysis of cellulosic biomass. In this study, we examined the structure change of the isolated lignins from hot water hydrothermal pretreatment of aspen and their inhibitory effects on the enzymatic hydrolysis of Avicel. The functional groups of the isolated lignins were determined by quantitative ^{13}C, 2D HSQC and ^{31}P NMR. The increase of pretreatment temperature significantly increased the condensed and non-condensed syringyl and guaiacyl OH group contents in the isolated lignins, and decreased the aliphatic OH, p-hydroxybenzoate OH and carboxylic OH group contents. A compelling adverse association (r^2=0.998) was observed between the condensed syringyl and guaiacyl phenolic OH group contents in lignins and their inhibitory effects on enzymatic hydrolysis. Langmuir adsorption isotherms showed that higher pretreatment temperature resulted in higher binding ability between isolated lignins and enzymes, which led to more non-productive binding. It is hypothesized that condensed syringyl and guaiacyl phenolic units are mainly responsible for the inhibitory effect of lignin on enzymatic hydrolysis, in which the condensed aromatic rings enhance the hydrophobic interactions and the phenolic OH group boost the hydrogen bonding. The combination of hydrophobic interactions and hydrogen bonding can further strengthen the non-productive binding.

467057 Measurement and Evaluation of Recalcitrance Changes for Organic Solvent Pretreatment Methods

Abstract:

Biomass recalcitrance is largely responsible for the high cost of chemical and fuel production from lignocellulose. To enhance biodegradation for these applications, it is crucial to determine how a pretreatment method affects the physical characteristics of lignocellulose[1,2]. In this work, we show the efficacy of Direct Yellow 11 as a probe to measure the cellulosic accessibility to cellulase enzymes. Together with crystallinity and delignification measurements, this accessibility measurement technique provides insight into recalcitrance changes that enhance biomass susceptibility towards enzymatic conversion to glucose[2,3]. Here we elucidate how a variety of organic solvents affect the key physical characteristics of biomass. These physical characteristic changes provide an increased understanding of biomass recalcitrance and its effect on cellulase kinetics.

1. Chandra, R., et al. (**2008**). "The characterization of pretreated lignocellulosic substrates prior to enzymatic hydrolysis, part 1: a modified Simons' staining technique." Biotechnology progress 24(5): 1178-1185.
2. Kang, Y. et al. (**2013**). "SO2 catalyzed steam explosion: The effects of different severity on digestibility, accessibility, and crystallinity of lignocellulosic biomass." Biotechnology progress 29.4: 909-916.
3. Hall, M. et al. (**2010**). "Cellulose crystallinity–a key predictor of the enzymatic hydrolysis rate." FEBS journal 277.6: 1571-1582.

473939 One Step Breaking Softwood Recalcitrance By Cellulose Solvent-Based Lignocellulose Fractionation Followed By Enzymatic Saccharification

Abstract:

In our efforts to shift away from conventional petroleum-based feedstock to supply transportation fuels, plant biomass has emerged as an attractive renewable alternative. Softwoods have great potentials as renewable feedstocks for biorefineries. However, softwoods are more recalcitrant compared to hardwood counterparts. Hence, the use of softwoods in biorefineries has been viewed as limited. In the present study, we applied a cellulose solvent-based pretreatment/fractionation process on southern yellow pine and douglas fir at a mild condition: 50 °C and 1 atm for 90 min. Pretreated softwood samples were hydrolyzed fast by cellulase (15 FPUs) and > 85% glucose yield was achieved by 24 h. Cellulose accessibility to cellulase of pretreated samples was probed by the fusion protein and was found to increase by 39 folds. After fractionation, degree of crystallinity of pretreated samples decreased and fibrous structure appeared to be disrupted as observed by solid-state ^{13}C Cross-Polarization Magic Angle Spinning (CP/MAS) nuclear magnetic resonance (NMR) and scanning electron microscopy. No changes in lignin functionalities after fractionation were observed by 2D Heteronuclear Single-Quantum Correlation (HSQC) NMR, giving rise to the potential use of lignin as a co-product in biorefineries.

472908 A Novel Acid Catalyst for Conversion of Ethanol to Butanol

Abstract:

The extensive consumption of fossil fuels contributes to a long-term threat to the ecosystem on which people highly depend. . Accordingly, people are increasingly interested in developing biorenewable fuels, among them is N-butanol. Due to its desired noncorrosivity, immiscibility with water, and relatively high energy density, preparing N-butanol has gained increasing attention. The major common challenge of the recently reported N-butanol production technologies lies in their low N-butanol selectivity. The novel acid catalyst (instead of base catalysts used in literature) developed in this work has achieved the highest selectivity to date. The effects of different factors on N-butanol selectivity and yield are evaluated. In addition, the associated reaction mechanism is understood with various material and process characterization methods.

470370 Developing a Simulation Framework for Optimizing CO2-Capture Processes

Abstract:

DEVELOPING A SIMULATION FRAMEWORK

FOR OPTIMIZING CO_2-CAPTURE PROCESSES

Stuart Higgins and Y. A. Liu

PetroChina-AspenTech Center of Excellence in Process System Engineering,

Department of Chemical Engineering, Virginia Polytechnic Institute and State University,

Blacksburg, VA 24061

We have reported[1] a highly efficient, amine-based CO_2-capture unit design, achieving a thermal regeneration energy of merely 1.67 GJ/(tonne CO_2 captured). While our results look promising, the optimization of our process superstructure required a significant amount of effort, largely due to the number of optimization parameters and our use of computationally expensive rigorous models.

This experience motivated us to develop a more practical simulation framework. We focus on a modular design that allows us to use commercial modeling tools, and develop an overall simulation structure to take advantage of multi-

threading, distribution, custom convergence methods, error recovery, and other features for increasing productivity.

Our new software framework enables faster, more reliable, more productive process optimization, while still using the same rigorous modeling tools validated by respected vendors. Multi-threading allows us to take advantage of the increasing core counts on modern CPUs, while distribution is similarly important as computational devices become more numerous.

In this talk, we will discuss the simulation framework and its application to optimize generalized process designs. We expect these improvements to enable optimizations that would otherwise require far too much time and effort to complete. We are particularly interested in superstructure process designs that generalize base-case process designs with potential improvements, because we can optimize these superstructures to find new process designs.

---

[1] Stuart J. Higgins and Y. A. Liu, "CO_2 Capture Modeling, Energy Savings, and Heat Pump Integration", *Industrial and Engineering Chemistry Research*, 54, 2526-2553 (2015).

461580 The Development and Demonstration of Best-Practice Guidelines for the Start-up Injection of CO_2 into Highly-Depleted Gas Fields

Abstract:

Highly-depleted gas fields represent prime potential targets for large-scale storage of captured CO_2 emitted from industrial sources and fossil-fuel power plants. Given the potentially low reservoir pressure as well as the unique thermodynamic properties of CO_2, especially in the presence of stream impurities, the injection process presents significant safety and operational challenges.

The most effective way of transporting the captured CO_2 for subsequent sequestration using high-pressure pipelines is in dense phase. The CO_2 arriving at the injection well will typically be at pressure greater than 70 bar and at temperature between 4 and 8 °C. Given the substantially lower pressure at the wellhead, the injection of CO_2 will result in its rapid, quasi-adiabatic Joule-Thomson expansion leading to significant temperature drops. This could pose several risks, including hydrate and ice formation around the wellbore and thermal shocking of the wellbore casing steel, leading to its fracture and ultimately escape of CO_2.

Despite its importance, the modelling of the flow dynamics taking place during the start-up injection of CO_2 has been devoted mainly to either steady-state operations[1] or flows where only pure CO_2 is injected[2]. The steady-state flow model is inappropriate as it ignores important phenomena, such as expansion wave propagation and bubble nucleation, occurring during the rapid depressurisation process. In addition, there is little work in the literature regarding the impact of the injection rate and the upstream inlet conditions (e.g. temperature and pressure) on the degree of cooling along the well[3]. Moreover, all previous work has been confined to pure CO_2. In practice the injected CO_2 will have a range of different impurities which will have a profound impact on the CO_2 fluid phase behaviour and hence its injectivity[4].

The aim of our work is twofold. First, we present in detail a homogeneous relaxation flow model for the numerical simulation of the highly transient phenomena taking place in the wellbore. Mass, momentum, and energy conservation equations are considered in the tubing. Wall friction, gravitational force, and heat transfer between the fluid and the surrounding formation are also taken into account. The phase equilibrium and thermodynamic properties for CO_2 are determined using the Peng-Robinson equation of state, which guarantees both accurate and computationally efficient predictions of VLE data for CO_2[5]. Operational envelopes for the injection well are obtained from data already available in the literature[3]. Second, by performing a sensitivity analysis on various injection rates and temperatures, we propose optimal transient-operation design and develop best-practice guidelines for the minimisation of the risk associated with start-up injection of CO_2 into highly-depleted gas fields.

1. Hughes D., *Carbon storage in depleted gas fields: Key challenges*, Energy Procedia, 1, 2009, 3007-3014.

2. Galic H., Cawley S. and Bishop S., *CO_2 injection into depleted gas reservoirs*, SPE 123788, 2009, 1-8.

3. Li X., Xu R., Wei L. and Jiang P., *Modeling of wellbore dynamics of CO_2 injector during transient well shut-in and start-up operations*, International Journal of Greenhouse Gas Control, 42, 2015, 602-614.

4. Brown S., Beck J., Mahgerefteh H. and Fraga E. S., *Global sensitivity analysis of the impact of impurities on CO_2 pipeline failure*, Reliability Engineering & System Safety, 115, 2013, 43-54.

5. Oosterkamp A. and Ramsen J., *State-of-the-art overview of CO_2 pipeline transport with relevance to offshore pipelines*, Technical Report, PolyTec.

464433 Application of a Systematic Methodology for Sustainable Carbon Dioxide Utilization Process Design

Abstract:

Application of a systematic methodology for sustainable carbon dioxide utilization process design

Cristina Calvera, Rebecca Frauzem*, Rafiqul Gani

Department of Chemical and Biochemical Engineering, Technical University of Denmark, Søltofts Plads Building 229, 2800 Kgs. Lyngby, Denmark

* rebfra@kt.dtu.dk

As concerns about the environment are growing, new efforts are needed to achieve more sustainable processes. One such environmental concern is global warming, which is primarily caused by the greenhouse effect or the increase in concentration of greenhouse gases [1]. The most significant greenhouse gases are carbon dioxide, methane and nitrous oxide, of which carbon dioxide is the highest constituent at 82%. Furthermore, the amount of carbon dioxide emissions is growing with time. These trends make it evident that there is a need for methods to reduce these greenhouse gases emissions. While there are two methods of reducing carbon dioxide emissions, carbon capture and storage (CCS) and carbon capture and utilization (CCU), CCU is considered promising as it makes further use of the carbon dioxide as a solvent, raw material, and reagent to produce valuable products [1]. Using such utilization processes, the emissions can be reduced as they are being utilized and profit can be obtained, or the cost of operation for the carbon dioxide treatment can be returned, through this utilization process.

In order to systematically reduce such emissions, carbon capture and utilization is considered rather than carbon capture and storage. To achieve this a methodology is developed to design sustainable carbon dioxide utilization processes. First, the information on the possible utilization alternatives is collected, including the economic potential of the process and the carbon dioxide emissions. The carbon dioxide emissions can be classified as direct and indirect emissions in a chemical process. The net carbon dioxide is determined for the utilization processes as the indirect carbon dioxide emissions minus the carbon dioxide utilized. Processes that presents zero or negative net carbon dioxide emission are desired in order to reduce the carbon dioxide emissions. Using this estimated preliminary evaluation, the top processes, with the most negative carbon dioxide emission are investigated by rigorous detailed simulation to evaluate the net carbon dioxide emissions. Once the base case design is established and evaluated, targeted improvements are made by exploiting opportunities, for example, optimization, heat integration and improved design decisions so that more sustainable and lower net carbon dioxide emission alternatives are obtained. .

This method is applied to various processes where carbon dioxide is used as raw material. First, the process data are collected and compared. The economic feasibility is evaluated. From this, five processes are selected and analyzed in detail: the production of dimethyl carbonate, succinic acid, propylene carbonate, dimethyl ethylene and methanol. Not all the studied processes could be designed for zero or negative net carbon dioxide emission. Propylene carbonate production is found to have a negative net carbon dioxide, where, implementing targeted process improvements minmized the net carbon dioxide emission to -0.389 kg of carbon dioxide per kg of propylene carbonate. On the other hand, for succinic acid production, even after targeted improvements, the net carbon dioxide remained positive, even though compared to the existing industrial processes there is a relative reduction of upto 85%.

What this study shows is that for meaningful net carbon dioxide reduction, the carbon dioxide utilization processes need to be selected very carefully to obtain the best results. However, the opportunity to potentially reduce the net carbon dioxide emissions for the production of some bulk chemicals with carbon dioxide as feedstock exists.

[1] Yuan, Z., Eden, M.R., Gani, R., 2016. Toward the Development and Deployment of Large-Scale Carbon Dioxide Capture and Conversion Processes. Ind. Eng. Chem. Res. 55, 3383–3419. doi:10.1021/acs.iecr.5b03277

466690 Dynamic Modeling and Simulation of the Carbon Capture System Using Blended Amine-Based Solvent

Abstract:

Control of greenhouse gas emission has become an important issue due to the rising concerns for global warming. Among the various potential modes of carbon capture, post-combustion capture (PCC) is an appropriate option for capturing carbon dioxide from existing coal-fired power plants because the plants could be retrofitted easily to accommodate PCC systems at commercial scale. However, extra capital and operating cost required to install and operate the PCC system is the main hurdle for its widespread application. The amine-based absorption process causes significant loss in power production due to the large amount of regeneration energies needed in the desorption stage. Therefore, the blended amine solvents, which are having bigger absorption capacity and fast reactions, have been actively searched and developed. The various efficient operation methods such as flexible operation have been also suggested in line with decreasing the cost of energy loss in the process. Furthermore, there have been some attempts to develop a dynamic model of MEA (Monoethanolamine) absorption process in order to investigate the operational issues related to the dynamics of the process. However, most dynamic models have not been validated due to lack of data.

In this research, the absorption process based on a newly developed blended-amine solvent is simulated in *gPROMS* and the process operability and dynamic issues are examined. Thermodynamics and rigorous process modeling for the absorption process using the new solvent are developed in simulator. Dynamic models are compared with pilot plant data for validation. Furthermore, dynamic behaviors with control loops are observed under some potential operational scenarios to optimize the operational strategy to reduce the regeneration energies. Important dynamics, disturbances are analyzed and control structure is selected and tested.

464107 Continuous-Time Model for Retrofitting of Power Plants and Optimal Carbon-Capture and Storage

Abstract:

Carbon capture and storage (CCS) is one of the options for mitigating the adverse effects of CO_2 emissions and the resulting greenhouse gas and climate change effects. CCS involves a two-step process of directly capturing CO_2 from different sources (such as flue gas emissions of power plants), using one of the carbon capture technologies, and then transporting and storing CO_2 in natural sinks (such as geological reservoirs or exhausted oil/gas wells). The different carbon capture technologies include pre-combustion capture in integrated gasification combined cycle (IGCC), post-combustion capture based on flue gas scrubbing, oxy-fuel combustion, and chemical looping combustion (CLC). Other alternatives for reducing CO_2 emissions include improving energy efficiency of power plants or using low-carbon fuel technologies [1-4].

CCS is essential for continued use of fossil fuel based technologies to mitigate the environmental concerns related to carbon footprint. Several approaches based on pinch analysis and mathematical programming have been developed in the literature for source-sink matching of CO_2 emissions. All the three operations: carbon capture, transportation, and storage consume some energy to perform them, again resulting in additional CO_2 emissions. On one hand we are capturing CO_2 and storing into allocated geological sink, while on the other hand we are allowing additional CO_2 emission into atmosphere in order to ensure feasibility of whole process, leading to compensatory power generation in case of power plants. Enabling of CCS requires retrofitting of existing power plants to allow for capturing of CO_2 in pure form, thus reducing the power output of these plants, which are modeled in the literature as energy penalties.

Tan et al [2] proposed a CO_2 source-sink continuous time mixed integer nonlinear programming (MINLP) model considering physical and temporal effects of planning CCS. The temporal effects include time

matching of start and end times of operating times of sources with sinks. Later they used linearization of bilinear terms and converted the model into MILP. Fixed operating lives and fixed flow rates were considered for CO_2 sources. Lee and Chen [3] presented an improved version of Tan et al [2] model with reduction in number of variables and constraints. Lee et al [4] extended this work using discrete time intervals to include decisions related to different carbon capture technologies for retrofitting of power plants, variable flow rates for CO_2 sources, and finite injection rates for CO_2 sinks.

In this work, we present an improved version of the literature models [2-3] for optimal source-sink matching in carbon capture and storage systems, without requiring bilinear terms and the resulting linearization. Hence, it directly results in an MILP model with reduced problem size in terms of number of variables and constraints. Two case studies from literature will be presented to illustrate the proposed advantages based on these modifications.

Then, we present an MILP model for retrofitting of power plants based on different carbon capture technologies using continuous time representation. The objective is to maximize CO_2 emissions reduction by retrofitting power plants and matching CO_2 sources and sinks, while keeping electricity cost within a specified limit and satisfying all temporal and physical constraints. A case study from literature [4] was solved using the proposed continuous time MILP model, yielding orders of magnitude reduction in problem size, as expected. Additionally, we will present sensitivity analysis by varying the minimum duration of viable connection between sources and sinks to study its effect on the total CO_2 captured.

References

[1] Shenoy, A. U.; Shenoy, U. V. (2012). Targeting and design of energy allocation networks with carbon capture and storage, *Chem. Eng. Sci.*, 68, 312–327.

[2] Tan, R. R.; Aviso, K.B.; Bandyopadhyay, S.; Ng, D. K. S. (2012). Continuous-time optimization model for source-sink matching in carbon capture and storage systems, *Ind. Eng. Chem. Res.*, 51, 10015–10020.

[3] Lee, J.-Y.; Chen, C.-L. (2012). Comments on 'Continuous-time optimization model for source-sink matching in carbon capture and storage systems', *Ind. Eng. Chem. Res.*, 51, 11590–11591.

[4] Lee, J.-Y.; Tan, R. R.; Chen, C.-L. (2014). A unified model for the deployment of carbon capture and storage. *App. Energy*, 121, 140–148.

467083 Heat Integration of Absorption-Based Carbon Capture Processes in the Industrial Sector

Abstract:

The industrial sector is struggling with the difficult switch to renewable feedstocks, since the generated carbon is sometimes inherent in their processes or a part of their product. Therefore, carbon capture and storage (CCS) performed in the industrial sector has gained increasing attention recently, and most research is aimed at absorption-based capture processes. The largest cost for post-combustion CCS when applied at a power plant is the cost of regeneration of the absorbent, but industrial plants often have vast amounts of excess heat that can be utilized in the capture process. The focus of this paper is, therefore, heat integration of absorption-based carbon capture in the industrial sector. Models of two capture processes, using monoethanolamine (MEA) or ammonia as absorbents, were evaluated on the basis of how suitable they are for generic, industrial applications, depending on the availability and quality of the excess heat at the industrial site, and the temperature in the stripper reboiler. Models of the two processes have been constructed in Aspen Plus using a temperature interval of 90 – 120 °C for MEA and 105 – 155 °C for ammonia, and have been presented in previous work.

The study then continues with a heat integration case study of an oil refinery situated on the west coast of Sweden, emitting 1.8 Mt CO_2/yr from the four main chimneys. Over 100 coolers, using air or water as a cooling medium, are present at the refinery, providing cooling in the proximity of 500 MW. Because a system where over 100 new units are installed would be very complex and expensive, the possibility of installing fewer heat exchangers while having an acceptable loss of the available excess heat was investigated. Both the MEA and the ammonia process were evaluated based upon the levels of heat that resulted from this investigation.

The results point at a maximum heat recovery temperature of 108 °C for ammonia, and for MEA there is a boundary solution at the lower boundary of the investigated temperature interval, 90 °C. MEA could capture more CO_2 than ammonia using only excess heat, and techno-economic calculations were then made for the implementation of the MEA process in the oil refinery, including the cost for excess heat collection. The calculations comprise of the cost of the MEA process at both boundary temperatures in the stripper-reboiler, 90 °C and (standard) 120 °C, at two different sizes, and CO_2 concentrations. Compared to estimates where excess heat is not considered, the cost for carbon capture was decreased by up to 40 %.

461901 Study of Mass Transfer and Sorption Phenomena in CO_2/Brine/Rock Systems

Abstract:

Geological sequestration of CO_2 is a promising technique to reduce the emissions of greenhouse gases and to mitigate global warming. Understanding of mass transfer and sorption characteristics of CO_2 in relevant brines, and in brine-saturated reservoir rock samples at realistic reservoir conditions is considered essential for predicting the gas distribution after injection, and for potentially estimating the cumulative amount of trapped gas. In this study, Mount Simon cores from Illinois Basin were characterized via measurements of their porosity and permeability. Their internal surface area and pore size distribution were also measured via probe-gas sorption experiments using two different gases (N_2 at 77 K and CO_2 at 195 K) in order to provide additional information about the microstructure of the cores. A series of high-pressure core flooding experiments were conducted to understand interactions between core, brine and CO_2 in terms of sorption, mass transfer and displacement characteristics. In addition, mathematical models are developed to describe the observed mass transfer and sorption behavior at the core-scale.

472020 Numerical Modeling of CO_2 Sequestration in Deep, Saline, Dolomitic-Limestone Aquifers

Abstract:

Geological sequestration of CO_2 is one of the most promising technologies for large-scale CO_2 mitigation. As per U. S. Department of Energy [1] the United States has an estimated potential of 1.8–20 trillion metric tons of underground CO_2 storage. To assess the suitability of a potential CO_2 storage site, we must understand the thermodynamics [2] and kinetics of the CO_2-brine system under the relevant conditions. Using TOUGHREACT [3], we have estimated the changes in physical and chemical properties of limestone-dolomitic aquifers over a simulated period of 100 years after injecting high pressure (160-200 bars) CO_2 for 10 years under isothermal conditions. The objectives of the numerical model are to 1) estimate the amount of CO_2 sequestered by solubility trapping, mineral trapping, and residual trapping; 2) estimate the dissolution and precipitation of different minerals in the system; 3) estimate the concentration of the major chemical species and pH change in the aquifer brine; and 4) estimate the changes in porosity and permeability of the aquifer. Also, effects of other parameters like temperature, pressure, and salinity on CO_2 sequestration have been studied for long-term feasibility of the process.

Results from the study suggest that solubility trapping accounts for the highest amount of CO_2 sequestered as compared to mineral and residual trapping. Due to formation of carbonic acid, the pH of the brine dropped to below 5. This drop in pH led to dissolution of minerals like calcite and dolomite. Also, there was precipitation of minerals like gypsum. Changes in porosity and permeability due to dissolution and precipitation of minerals have been estimated. Although the study was conducted primarily for calcite, dolomite, and gypsum, it may be applicable to other formations as well. The study gives an overall estimate of the amount of CO_2 sequestered under different conditions (salinity, pressure, temperature, and mineralogy). It also predicts the changes in chemistry and permeability of the system under different

conditions. The study thereby helps to choose the feasibility of sites for CO_2 sequestration more efficiently.

References

1. NACAP (2012) The North American Carbon Storage Atlas, The U.S. Department of Energy (DOE), Natural Resources Canada (NRCan), and the Mexican Ministry of Energy (SENER).
2. Thomas, M. W., Stewart, M., Trotz, M., and Cunningham, J. A. (2012). Geochemical modeling of CO_2 sequestration in deep, saline, dolomitic-limestone aquifers: Critical evaluation of thermodynamic sub-models. Chemical Geology, 306, 29-39.
3. Xu, T., E.L. Sonnenthal, N. Spycher, and K. Pruess, (2006) TOUGHREACT - A simulation program for non-isothermal multiphase reactive geochemical transport in variably saturated geologic media: Applications to geothermal injectivity and CO2 geological sequestration, Computers & Geosciences, v.32, p.145-165.

469496 Effects of Recycling Regenerated Heat Carrier on the Performance of an Auger Pyrolysis Reactor

Abstract:

Auger pyrolysis reactors offer unique advantages over traditional fluidized bed pyrolyzers. Some advantages include the minimal requirement of an inert gas and the flexibility of heat carrier materials with varying thermophysical properties. Previous work used a lab-scale auger pyrolyzer to test the effect of heat transfer properties of various heat carriers such as stainless steel shot, silicon carbide and sand. While high liquid yields were achieved (~65 wt. %), the long-term performance of these heat carriers is not well understood. If auger pyrolyzers are to be commercially viable, the heat carrier must be recycled to minimize the cost of system inputs and solids handling. The objective of this research was to evaluate the effect of heat carrier regeneration and recyclability on the long-term performance of an auger pyrolysis reactor. The effect on product yields and changes to the thermophysical characteristics of the heat carrier materials were of specific focus.

A twin-screw, lab-scale auger reactor was used to pyrolyze up to 1 kg/hr of red oak at 515°C. The heat carriers of interest in this study included fine (250-600 μm) and coarse sand (600-1000 μm). Pyrolysis trials using recycled heat carrier were conducted at up to 5 recycles. The heat carrier was regenerated following each trial to remove any carbon residue. A cold gas quench system was used to collect the bio-oil into two fractions: a heavy phase and an aqueous phase. Product characterization included ultimate and proximate analysis of the bio-oil and char. Gas chromatography of the bio-oil fractions was also conducted. Product yields of bio-oil (~60 wt. %), char (~17 wt. %) and gas (~18wt %) were consistent across the limited number of recycle trials. Heat carrier attrition varied from 4 to 8 wt. %, and the mean particle size decreased following the recycled trials. These results indicate that heat carrier properties may change significantly over a short number of rotations, but their impacts on biomass pyrolysis yields is small. Long-term tests would be necessary to determine heat carrier replacement rates in order to avoid significant impacts on bio-oil yields and quality.

469939 Pyrolysis of Biomass Using Blast Furnace Slag As Catalyst

Abstract:

Catalytic pyrolysis of biomass using various acidic zeolite catalysts have shown promise of producing stable biomass pyrolysis oils that can be easily upgraded to fuels and biobased products. However, zeolite catalysts are relatively expensive and tend to produce low liquid yields, large fractions of gas and aqueous phase and therefore there is a need to develop other catalysts that will improve liquid yields. Pine sawdust was pyrolyzed in a bench scale bubbling fluidized reactor using fine blast furnace slag as the pyrolysis medium and nitrogen as fluidizing gas. The pyrolysis was conducted at 500 C and feed rate of 600-700 g/h. The pyrolysis reaction showed an initial exotherm during the first 20 min and there was significant production of hydrogen. The reaction became endothermic like most conventional pyrolysis reactions and hydrogen production decreased. The organic liquid yields ranged from 28 to 48 wt% depending on the run time and were higher than those reported for acidic zeolite catalysts. The char/coke yields ranged from 14 -18 wt%; the non-condensable gas yields ranged from 9 to 13 wt% and pyrolytic water yields ranged from 8 to 18 wt%. The oxygen content of the oils was 27-30 wt%, and the total acid number (TAN) was surprisingly low (23-26 mg KOH/g oil). The oils were less viscous than the sand pyrolysis oils. Thus the blast furnace slag appears to be a promising catalyst for the catalytic pyrolysis of biomass.

471364 Recycling of Used Railroad Ties Via a Thermochemical Process Using a Semi-Pilot Scale Auger Reactor System

Abstract:

A thermochemical process, an initial thermal treatment followed by an intermediate pyrolysis step, was investigated to recover wood preservatives and produce bio-oil and biochar containing low amount of polycyclic aromatic hydrocarbons (PAHs) from used creosote-treated railroad ties. The material was first treated using a semi-pilot scale auger pyrolysis reactor at 250, 280, and 300 °C for 15min. The yield of the liquid fraction was 28, 35, and 38 wt% while the solid fraction yield was 65, 57 and 54 wt%, respectively. The liquid fraction had two phases: aqueous phase in the upper and organic phase (enriched in creosote) in the bottom. As thermal desorption temperature increased, the aqueous phase exhibited a decrease in water content from 95 to 88% and pH from 3.2 to 2.6 and an increase in total acid number (TAN) from 20 to 63 mgKOH/g. The yield of the creosote phase increased from 2.0 to 4.5 wt% of total creosote content of 7 wt% found in the wood tie. As thermal desorption temperature increased, in addition to creosote, the organic phase also contained larger amount of carbohydrates- and lignin-derived compounds, decreasing the purity of the recovered creosote. The thermally treated samples were subsequently pyrolyzed using the same semi-pilot scale auger pyrolysis system at 500 oC for 72s residence time. The yield of bio-oil produced from the untreated, 250, 280, and 300 °C-treated materials was 61, 42, 35, and 30 % while the yield of biochar was 21, 25, 32, and 33 %, respectively. The characterization of the bio-oil produced from the thermally treated ties showed that as the desorption temperature increased, water content decreased from 36 to 32 %,TAN value decreased from 127 to 113 mg KOH/g, and density increased from1.1 to 1.2 g/mL. PAHs traces found in biochar from the thermally treated wood tie were dramatically lower (less than 0.2 wt%) than that (0.8 wt%) in biochar from the untreated ties. This result demonstrates that a thermal preservative-removal step can recover valuable creosote for re-use as preservatives and subsequently supply a feedstock without significant levels of contaminant hazardous pollutants for pyrolysis to produce bio-oil and biochar.

471572 Effect of Feedstock Formatting on Gasification Performance for Hyprid Poplar, Corn Stover and Low-Quality Materials

Abstract:

This work reports the effects of moisture content and particle size distribution on the gasification performance of hybrid poplar, corn stover and municipal solid wastes (MSW) in a modular plasma-gasification system. The MSW tested includes construction wastes, plastics and refuse derived fuels. To date, there has been very little data in the literature to support the feasibility of MSW gasification. In addition, while gasification has been demonstrated at large scale (over 200 tons per day), it is unknown whether smaller scale (2-40 tons per day) deployment of gasification technologies will be economical. It is at these scales where many of the applications of MSW gasification will reside since there are relatively few areas in the US with large amounts of MSW available. The presented work may serve as a basis for techno-economic analyses and scale up research.

Hybrid poplar was tested at 9 different conditions using a 3^2 matrix with respect to particle size and moisture content. Corn stover was tested at varying particle sizes, and MSW were tested at 4-6 mm particle size and â€˜as isâ€™ moisture. In addition, the different samples were air classified and assigned a measure of buoyancy to delineate between particle size affects and buoyancy. Inorganic and organic mass balances were done for the initial feedstock materials and the resulting slag, when possible, using ICP and CHNO proximate/ultimate analyses. CO_2, CO and H_2 production were measured inline and from collected product gas samples. The presence of other hydrocarbons was tested using GC-MS, but none were detected. Results suggest that hybrid poplar produced syn gas between 30 and 40% H_2, and between 35 and 45% CO, with some dependency on formatting conditions. Less than 2% of the material was recovered as slag, and it generally consisted of silica and alkali and alkaline earth metals. Initial testing of the corn stover and MSW suggest successful gasification, but that formatting plays a more critical role in MSW gasification than in hybrid poplar gasification.

472096 Effects of Catalyst Properties on Biomass Conversion By Catalytic Fast Pyrolysis and Hydropyrolysis

Abstract:

Effects of Catalyst Properties on Biomass Conversion by Catalytic Fast Pyrolysis and Hydropyrolysis

David P. Gamliel and Julia A. Valla

Department of Chemical & Biomolecular Engineering, University of Connecticut

191 Auditorium Road, Unit 3222, Storrs, CT 06269-3222, USA,

Phone: +1-860-486 4602, e-mail: ioulia.valla@.uconn.edu

Biomass is a clean and renewable carbon source, capable of providing sustainable carbon for conversion to fuels and platform chemicals. Catalytic fast pyrolysis (CFP) of biomass is the thermochemical conversion of biomass to liquid bio-oil, gas and char under inert atmosphere at intermediate temperatures and high heating rate. Recently, catalytic pyrolysis of biomass under a hydrogen atmosphere at elevated pressure, termed hydropyrolysis, has been studied as an effective method for production of deoxygenated high-value bio-oil with plentiful aromatic and aliphatic hydrocarbons [1]. The objective of this work is to determine how catalyst properties correlate to CFP and hydropyrolysis product yields, and propose the ideal catalyst for the production of value-added products from biomass resources.

A wide variety of catalysts were prepared and characterized for this study, such that the effects of transition metal choice, metal loading, support type and support acidity could be determined. Each catalyst was completely characterized using a variety of techniques to find the surface area, morphology, pore structure, metal oxidation state and crystal structure. Pd, Ru and Ni were chosen as candidate metals due to their high activity for deoxygenation and hydrogenation [2]. These metals were deposited via dry impregnation on a variety of alumino-silicate supports including alumina, silica and ZSM-5 zeolite. Furthermore, the Si/Al ratio of the ZSM-5 support was varied in order to determine the effects of acidity on the product distribution.

CFP was performed in a pyrolysis gas chromatograph (PyGC) unit, at 600 °C, under inert conditions and at atmospheric pressure. Increasing ZSM-5 catalyst acidity was found to increase yields of mono-aromatic hydrocarbons (MAHs), naphthalenes and CO, and decrease char formation. However, once the zeolite Si/Al ratio was lower than 25, coke formation reactions became dominant. It was found that significant increase of catalyst acid site density resulted in enhanced formation of coke and char precursors, such as poly-aromatic hydrocarbons (PAHs). The results are in agreement with the theory that the primary mechanism of acid catalyzed deoxygenation of pyrolysis vapors is decarbonylation and aromatization [3]. Impregnation of 3% of each transition metal slightly favored permanent gas, primarily in the form of CO, and decreased char yield. CFP with non-zeolitic support alone was found to have high yields to oxygenates and char.

Catalytic hydropyrolysis of biomass was performed in the same unit in pure H¬2 atmosphere, at 600 °C and 450 psig pressure. When hydropyrolysis was performed in the presence of ZSM-5 zeolite, a similar product distribution to CFP was observed. When Ni was incorporated on the catalyst, high amounts of CH4 were produced at the expense of solids, and alkanes were present in the liquid product, as shown in Figure 1. As the zeolite acidity was increased naphthalenes, PAHs and carbon oxide yields increased significantly at the expense of CH4 and liquid alkanes. This demonstrates the competition between acid-catalyzed decarbonylation, decarboxylation and aromatization and metal-catalyzed methanation and hydrogenation. As with CFP, hydropyrolysis with alumina or silica produced few liquids and high char yields. When Ni was incorporated onto silica, alkanes yield totaled about 6 wt.% carbon, and CH4 yield totaled over 50 wt.% carbon. Alternatively, when Ni was incorporated on the alumina support, lower CH4 yields and almost no alkanes were observed, along with higher solid yields, proving that significant Lewis acidity and large pore structure are detrimental for value-added product formation.

Catalyst design is essential should the thermochemical conversion of biomass reach large-scale implementation. Proper catalyst acidity aids in the formation of MAH compounds, while highly acidic catalysts result in high coke production. Incorporation of metals in the hydropyrolysis environment aids in the creation of alkanes and CH4, at the expense of char. In particular, metal-catalyzed methanation and hydrogenation reactions were found to compete with acid-catalyzed deoxygenation when both were present.

Acknowledgement

The study was funded by the National Science Foundation Award CBET-1236738 and the University of Connecticut GK12 program.

References

[1] V.K. Venkatakrishnan, W.N. Delgass, F.H. Ribeiro, R. Agrawal, Green Chem. 17 (2015) 178–183.

[2] T. Prasomsri, M. Shetty, K. Murugappan, Y. Román-Leshkov, Energy Environ. Sci. 7 (2014) 2660.

[3] D.P. Gamliel, S. Du, G.M. Bollas, J.A. Valla, Bioresour. Technol. 191 (2015) 187–196.

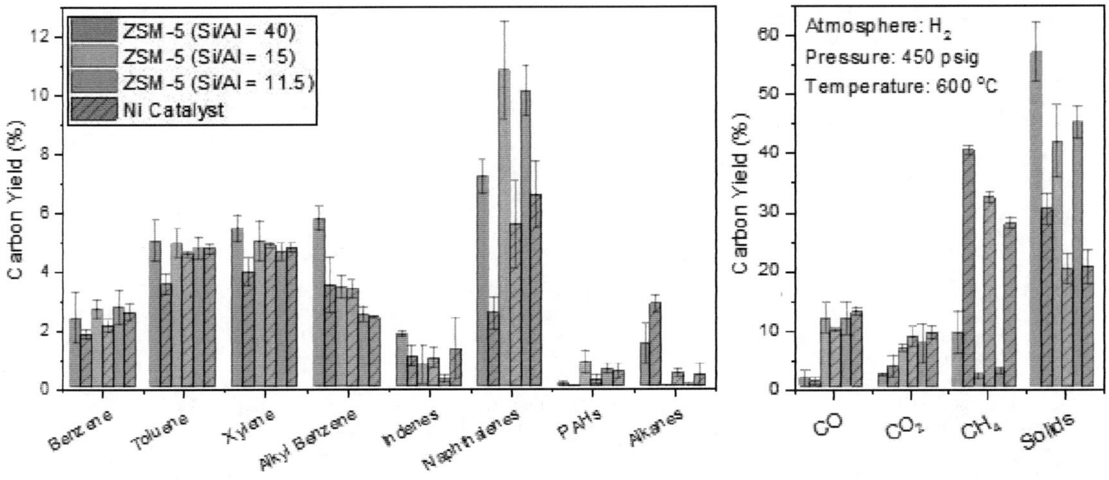

Figure 1. Bio-oil product distribution (left) and permanent gas product yields (right) for hydropyrolysis with ZSM-5 zeolites of various acidity with and without Ni incorporated.

471815 Effects of Feedstock on the Product Yields and Selectivity of Lignocellulosic Biomass Pyrolysis Using HZSM-5 Catalyst

Abstract:

Effects of Feedstock on the Product Yields and Selectivity of

Lignocellulosic Biomass Pyrolysis Using HZSM-5 Catalyst

Brendon Shea, Charles Coe, Justinus Satrio*

Department of Chemical Engineering

Villanova University, Villanova, PA 19085

One promising process of utilizing lignocellulosic biomass for the production of fuels and chemicals is fast pyrolysis. In this approach, the biomass is thermally treated at temperatures around 400-600°C in the absence of oxygen. The products formed include a solid char, a liquid bio-oil, and a mixture of non-condensable gases. The liquid bio-oil is composed of a multitude of organic compounds; many of them are oxygenated, such as ketones, phenols, and carboxylic acids. The high presence of these oxygenated compounds is not desirable since they cause bio-oil to be highly acidic, highly polar, chemically unstable, and low in heating value products. The use of solid acid catalysts, such HZSM-5, an aluminosilicate zeolite, can alter the liquid product distribution in favor of non-oxygenated compounds such as aromatics. These hydrocarbon compounds are desirable in bio-oil, especially when the bio-oil is used as intermediate feedstock for the production of liquid transportation fuels.

The product distribution is heavily affected by the feedstock of biomass that is pyrolyzed. A study on fast pyrolysis of 5 different biomass feedstocks was conducted with and without H-ZSM-5(40) as a solid acid catalyst. Feedstocks tested included pinewood, mixed hardwood, switchgrass, horse-barn bedding, and phragmites (common reeds). Fast pyrolysis experiments were conducted using a micropyrolyzer system which was connected to a GC/MS and a TCD gas analyzer for analyzing chemical composition of the pyrolytic vapors and non-condensable gases respectively. Catalytic pyrolysis experiments were performed in *ex-situ* mode, in which catalytic vapors were passed through a heated catalyst bed before being sent to the vapor and gas analyzers. The *ex-situ* catalytic pyrolysis configuration allowed multiple catalytic pyrolysis runs were done in series on the same catalyst sample to observe the effect of biomass feedstock on the catalyst activity. For completing mass balance, the amount of char formed was also measured in each run.

Experimental results indicated that biomass feedstock has a significant effect on pyrolysis product yields and selectivity as well as on the catalyst activity. Switchgrass and phragmites produced significantly more char than the other biomass feedstock. In non-catalyzed pyrolysis, these two grass-type feedstock produced less CO and more CO_2. The composition of chemical moieties of bio-oil products also was affected by the type of feedstock used for the fast pyrolysis reaction. In catalytic pyrolysis, pinewood and mixed hardwood were found to produce the most CO. while mixed hardwood and phragmites produced the most CO2. There was little correlation in the light gas formation as the catalyst deactivated. The bio-oil composition did vary as the catalyst deactivated; however the light gas formation showed very little correlation with the catalyst deactivation. For all biomass types, as the catalyst deactivated, the amounts of aromatics produced decreased while the amounts of ketones, phenols, and carboxylic acids produced increased. It appeared that pinewood had the greatest rate of change in bio-oil composition as the catalyst deactivated. While catalytic pyrolysis of pinewood showed the largest hydrocarbon yields initially, after several runs the yield difference compared to those from catalytic pyrolysis of different biomass feedstock became negligible.

*Contact: Justinus Satrio, Ph.D. Email: justinus.satrio@villanova.edu

473194 Physiochemical Characterization of Lignocellulosic Biomass Dissolution By Flowthrough Pretreatment

Abstract:

Comprehensive understanding of biomass solubilization chemistry in aqueous pretreatment such as water-only and dilute acid flowthrough pretreatment is of fundamental importance to achieve the goal of valorizing biomass to fermentable sugars and lignin for biofuels production. In this study, poplar wood was flowthrough pretreated by water-only or 0.05% (w/w) sulfuric acid at different temperatures (220-270 °C), flowrate (25 mL/min), and reaction time (8-90 min), resulting in significant disruption of the lignocellulosic biomass. Ion chromatography (IC), Fourier transform infrared (FTIR) spectroscopy, X-ray diffraction (XRD) analysis, and solid state cross polarization/magic angle spinning (CP/MAS) ^{13}C nuclear magnetic resonance (NMR) spectroscopy were applied to characterize the pretreated biomass whole slurries in order to reveal depolymerization as well as solubilization mechanism and identify unique dissolution structural features during these pretreatments.

465220 Comparison of Ethanol Yields from Fermentation of Hydrolysates Produced By CO-Solvent Enhanced Lignocellulosic Fractionation (CELF) and Dilute Acid Pretreatments of Switchgrass

Abstract:

For decades, numerous environmental, economic, and strategic concerns have motivated research into sustainable, domestic sources of fuels to replace those derived from petroleum. Lignocellulosic biomass represents the most abundant renewable resource and provides the only known route to sustainably produce liquid fuels on a large scale and low cost. Due to its complex and recalcitrant structure, raw biomass hinders biological conversion of polysaccharides, thus necessitating pretreatment to deconstruct the plant with high yields through making cellulose and hemicellulose more accessible to enzymes for release of fermentable sugars. Several pretreatment methods, including hydrothermal and dilute acid approaches, have been developed to improve enzyme accessibility, but each is limited in their ability to remove lignin, a key contributor to biomass recalcitrance, and require expensive enzyme loadings to achieve high yields. Recently, we developed a novel pretreatment called Co-solvent Enhanced Lignocellulosic Fractionation (CELF) that applies aqueous tetrahydrofuran (THF) with dilute sulfuric acid to remove a large portion of the lignin, simultaneously achieve high yields of five carbon sugars from hemicellulose during pretreatment, and subsequently hydrolyze most of the glucan at very low enzyme loadings. In this study, we applied CELF to switchgrass to determine optimum pretreatment conditions for maximum sugar release in the combined operations of pretreatment and enzymatic hydrolysis at low solids loadings and compared the results to those for DA pretreatment performed at conditions that realized maximum sugar yields from pretreatment coupled with enzymatic hydrolysis. For each pretreatment, we then established ethanol yields from fermentation of the hydrolysates produced at the optimum conditions, the degree to which fermentation was inhibited by dissolved compounds, and the effectiveness of various conditioning strategies in reducing inhibition.

466651 Understanding the Effects of Ammonia and Organosolv Pretreatments on Physicochemical Characteristics of Lignin

Abstract:

Pretreatment is an important step in biological conversion of biomass. It renders structural changes of biomass to reduce the biomass recalcitrance, thus enhancing its sugar release performance. Lignin is one of major barriers for lignocellulosic biomass utilization; therefore, various pretreatment methods have been studied to reduce the recalcitrance of biomass by removal and/or modification of lignin in the biomass. Among many pretreatment methods, two pretreatment methods using ammonium hydroxide and organic solvents were selected to understand how these methods reduces the recalcitrance and improves sugar release. In this study, the crude lignins isolated from poplar and switchgrass were subject to aqueous ammonia and organosolv pretreatments, and the effects of each pretreatment on lignin structural changes were investigated. Physicochemical characteristics of the lignins during the pretreatments were analyzed by diverse analysis methods. Chemical compositions of soluble fraction and residual solid fraction were analyzed by HPLC and GCMS. Molecular weights of both fractions were investigated using GPC. 2D ^{13}C-^{1}H HSQC NMR and ^{31}P NMR analyses were conducted to reveal the structural and chemical transformation of lignin. Adsorption of enzymes on the pretreated lignin was also investigated.

466953 How Solvent Properties Determine Pretreatment Effects on Key Lignocellulosic Characteristics

Abstract:

Chemical pretreatment efficacy is largely dependent on the solvent employed. Therefore, picking a solvent is one of the most important decisions in the early stages of chemical process design[1]. Traditionally, biomass pretreatment decisions have been more guided selection than rational design. Here we present a study of different organic solvent pretreatments on steam exploded bagasse. We demonstrate how certain chemical moieties affect the delignification capacity of a solvent. In addition to delignification, we describe the effect of organic solvent properties towards enhancing cellulose conversion to glucose. These methods provide insights into the chemistry of organic solvent pretreatments and a foundation for the molecular design of improved organic solvent pretreatment methods.

1. Shuai, L. and J. Luterbacher **(2016).** "Organic Solvent Effects in Biomass Conversion Reactions." *ChemSusChem* 9.2: 133-155.

469494 Comparison of Celf, Cu-AHP, and HS Ionic Liquid Pretreatment of Hardwoods: Hydrolysis Yields, Comprehensive Mass Balances, and Lignin Properties

Abstract:

In this work, three pretreatments under investigation at the DOE Bioenergy Research Centers (BRCs) are subjected to a side-by-side comparison to assess their performance on model bioenergy hardwoods. These will include co-solvent enhanced lignocellulosic fractionation (CELF), high-solid loading pretreatment with ionic liquid (HS-IL), and Cu-catalyzed alkaline hydrogen peroxide pretreatment (Cu-AHP). The pretreatments will be individually optimized for the feedstock and common performance metrics as assessed by a single lab will be used as a basis of comparison. The first component of this will be to assess the susceptibility of the pretreated biomass to hydrolysis by cellulolytic enzymes. This reactivity will be related to both structural and compositional changes to plant cell wall as a consequence of pretreatment. The second component will be to determine comprehensive mass balances on the pretreatments. This will include assessing the solubilization, depolymerizaiton, and conversion of cell wall biopolymers as well as the solvent requirements, their degradation, and their potential for reuse. An important feature of the pretreatments in this study is that all three have shown preliminary evidence of yielding lignins with properties amenable to further valorization. As the final component of this work, we will characterize the yields, structural properties, and reactivity of the soluble and insoluble lignin fractions generated by these pretreatments.

471752 Understanding the Influence of Hemicellulose and Lignin Removal on Deconstruction of Switchgrass By *Clostridium Thermocellum* Consolidated Bioprocessing *Vs.* Conventional Fungal Enzymatic Hydrolysis

Abstract:

Low cost sustainable fuels are urgently needed to support human prosperity. Currently, almost all liquid fuels are derived from petroleum that is finite in quantity and contributes significantly to climate change. According to the US Energy Information Administration, transportation accounts for about 70% of the total US oil consumption and about 95% of which is from fossil fuels consumption. Alternatively, biofuels would provide a sustainable energy source and help reduce our dependence on imported petroleum. Ethanol made from corn starch and cane sugar is presently the largest biotechnology-based product and commands a large market share of the alternative fuels market. However, currently the trend is to produce ethanol from lignocellulosic biomass that, unlike corn starch and cane sugar, does not directly compete with our food infrastructure. Ethanol is a high octane, cleaner burning fuel that, when made from lignocellulosic biomass, offers a low carbon footprint and the potential to be further upgraded to other drop-in fuels, such as butanol, jet fuel, gasoline, and diesel. Ethanol production from lignocellulosic biomass involves numerous important operations including size reduction, pretreatment, enzyme production, enzymatic hydrolysis, fermentation, and product recovery. The production of cellulolytic enzymes, typically from a fungus such as *Trichoderma reesei*, is one of the most expensive steps because of the high concentrations of enzymes needed to achieve economically attractive ethanol yields from recalcitrant lignocellulosic biomass. Consolidated bioprocessing (CBP) is a simplified, integrated bioprocess for directly converting lignocellulosic biomass into ethanol by utilizing thermophilic anaerobes that can produce their own enzyme consortium to hydrolyze biomass polysaccharide chains into fermentable sugars after which the same organism metabolizes the sugars to produce ethanol and other products. *Clostridium*

thermocellum, in particular, is a very promising CBP organism. However, pretreatment may still be required to assist *C. thermocellum* in biomass deconstruction for high yields. Dilute acid and hydrothermal are leading pretreatment technologies that reduce biomass recalcitrance by removing some of the lignin and most of the hemicellulose. Alkali pretreatments, on the other hand, remove some hemicellulose and a lot of the lignin. By comparison, co-solvent enhanced lignocellulosic fractionation (CELF), a novel pretreatment technology recently invented at the University of California, Riverside, which makes use of tetrahydrofuran (THF) as a co-solvent in water, enhances the solubilization and recovery of both hemicellulose sugars and lignin from lignocellulosic biomass beyond that of conventional pretreatment technologies. Because different pretreatment technologies produce solids with different physical and compositional characteristics, it is important to evaluate how these features impact CBP performance. This study, therefore, compared the effects of CELF pretreatment combined with *C. thermocellum* CBP on total sugar release from switchgrass to those of hydrothermal, dilute acid, and alkali pretreatments. The objective was to identify pretreatment conditions that give the highest possible sugar release in pretreatment (Stage 1) and biological conversion (Stage 2) steps combined for each pretreatment technology. The pretreatments were performed over a range of temperatures and times, and the resulting washed solids were then subjected to *C. thermocellum* CBP at 5 g/L glucan loading in 50 mL bottle reactors incubated at 60ËšC with a shaking speed of 180 rpm and fungal enzymes mediated enzymatic hydrolysis at 50ËšC and 150 rpm with the same solids loading to establish overall switchgrass deconstruction and sugar recovery.

457412 Separation of Lignin from Organosolv Spent Liquor-from Phase Behavior to Continuous Processing

Abstract:

Worldwide pulping mills combust several hundred million tons of lignin every year in their recovery boilers to recycle the pulping chemicals and provide process heat [Cazacu et al. 2013]. The most applied Kraft and Sulfite pulping techniques are focused on pulp production whereas lignin is utilized as cheap energy provider.

Lignin is build up from phenylpropane monomers with hydroxyl and methoxy functional groups, making it a valuable renewable resource for the chemical and material industry. Lignocellulose biorefineries are supposed to sustainably provide renewable platform chemicals for the future by fractionating woody biomass with appropriate processes. Organosolv pulping techniques can fractionate lignocellulosic biomass into mainly pulp, lignin and (oligo-) sugars by using an organic solvent and water as pulping liquor at temperatures near 200 °C. The resulting pulp is of reasonable quality [Zhang et al. 2016] and the dissolved lignin can be separated from the spent liquor as precipitate by removing the organic solvent, which is recycled to pulping liquor. The (oligo-) sugar fraction remains dissolved in the aqueous phase. Organosolv lignin is known to be of high quality with low ash, sulfur and carbohydrate contents. The separation processes in a lignocellulose biorefinery utilizing organosolv pulping are essential for the economic viability [Viell et al. 2013].

The Fraunhofer Center for Chemical-Biotechnological Processes CBP in Leuna, Germany, operates a lignocellulose pilot plant consisting mainly of a batch forced circulation reactor for ethanol organosolv pulping, an enzymatic digestion unit for cellulose to sugar hydrolysis, as well as ethanol evaporation and lignin precipitation facilities. Fraunhofer CBP and MPI DCTS initiated a cooperation for investigation and optimization of the lignin separation process especially in terms of energy consumption, fouling prevention and downstream processability in filtration and drying.

In this contribution the path from fundamental phase behavior determination to an advanced continuous lignin precipitation process is presented [Schulze et al. 2016]. Systematic solubility measurements in the ternary lignin-ethanol-water system at different temperatures have been performed in order to elucidate attractive process possibilities. Thereby lignin was found to form a liquid phase under certain conditions as result of a decreased glass transition temperature. The liquid lignin phase potentially causes fouling on the reactor internals during separation processes if the process conditions are not carefully adjusted. On the other hand, softened lignin particles can agglomerate and coalesce to form bigger particles. A compromise between particle enlargement and incrustation formation has to be made in selection of process conditions. Enlarged lignin particles contribute to a better filterability. A (semi-) continuous process for simultaneous lignin precipitation, ethanol recovery and particle enlargement was developed on the knowledge from phase behavior. Experimental results from lab scale continuous processing of spent liquors produced at different pulping conditions indicate the influence of lignin properties (like molecular weight distribution) on the separation process. Semi-continuous precipitation experiments were performed in pilot scale confirming the scalability of the process to the next level. Inline process analytic techniques (Mettler Toledo FBRM®, PVM and ATR-FTIR) were utilized for monitoring and controlling lignin particle size distribution, lignin particle shape and aqueous phase composition. Agglomeration kinetics were derived from the monitored data. Filtration experiments with the produced lignin dispersions provided filter cake resistance data to assess filterability.

References:

Cazacu, G., Capraru, M., Popa, V. (2013). Advances concerning lignin utilization in new materials. In: Thomas, S., Visakh, P.M., Mathew, A.P. (Eds.), Advances in Natural Polymers. Springer, Berlin Heidelberg, pp. 255–312.

Schulze, P., Lorenz, H., Seidel-Morgenstern, A., Leschinsky, M., Unkelbach, G. (2014). Verfahren zur Fällung von Lignin aus Organosolv-Kochlaugen. DE priority patent application filed in 20th October 2014.

Schulze, P., Lorenz, H., Seidel-Morgenstern, A., Leschinsky, M., Unkelbach, G. (2016). Advanced process for precipitation of lignin from ethanol organosolv spent liquors. Bioresource Technology, 199, 128–134.

Viell, J., Harwardt, A., Seiler, J., Marquardt, W., (2013). Is biomass fractionation by Organosolv-like processes economically viable? A conceptual design study. Bioresource Technology, 150, 89–97.

Zhang, K., Pei, Z., Wang, D. (2016). Organic solvent pretreatment of lignocellulosic biomass for biofuels and biochemicals: a review. Bioresource Technology, 199, 21–33.

462650 Cell Lysis and Lipid Recovery from Oleaginous Yeast

Abstract:

Single-cell oils, or lipids derived from oleaginous microbes, have seen recent interest as a potential renewable source of transportation fuels. However, one of the main technical challenges involved in single-cell-oil-derived products is due to the intracellular nature of the oils: the microbial cells must be lysed and the lipids extracted from the resulting complex matrix. We have screened several approaches for cell lysis approaches using the oleaginous yeast *Lipomyces starkeyi*, and have found that a dilute acid pretreatment compares favorably to a number of alternative strategies including high-pressure homogenization, alkaline pretreatment, bead beating, enzymatic hydrolysis or thermolytic approaches. Extraction of the lysed cells with hexane allows for recovery of the lipids in yields above 90% and in high purity. The acid pretreatment is also effective on strains of *Cryptococcus curvatus* and *Rhodosporidium toruloides*, and is highly scalable.

471016 Transport and Reaction Modeling of Biomass Pretreatment and Topochemical Evolution Using Actual 3D Structure of Plant Cell Walls

Abstract:

Plant biomass is a renewable raw material for a wide variety of bio-based products which include pulp and paper, bio-plastics, biofuels, wood plastic composites etc. The process of conversion of plant biomass to other useful products is heavily dependent on our understanding of the physio-chemical structure of plant cell walls. Plant cell walls are composed of a complex network of pore spaces and fibers of cell wall components that has a significant influence on the efficiency of the biomass conversion process, the properties of the bioproducts and their end-use applications. Hence it is important to characterize the 3D plant cell wall structure. 3D characterization techniques such as X-ray computed tomography (XRCT) and Transmission electron microscopy computed tomography (TEM-CT) have been popular. Although, the X-ray CT provides micron scale visualization, it is not sufficient to get a complete understanding of the 3D structure of plant cell walls. Hence, a more powerful technique of TEM-CT is used to probe the nanoscale 3D ultrastructure. The computed tomography technique is based on capturing a series of images by successive tilts around a single axis and reconstructing the 3D image by back projection. The structural information obtained from TEM-CT can be correlated with topochemical information from Raman spectroscopy to fully understand the physiochemical structure of plant cell walls.

The first step in the conversion process is the pretreatment step, which disrupts the recalcitrant structure of lignocellulose and increases the efficiency of the subsequent hydrolysis. One or more of the cell wall components are dissolved during this process. The changes in structural characteristics such as porosity, specific surface area and 3D pore size distribution during the pretreatment process can be determined from the processed TEM-CT images. The structural information from TEM-CT can be correlated with the topochemical distribution from Raman Spectroscopy to develop a transport-reaction model based on a hybrid random walk and reaction. This model involves a stochastic dynamic approach that keeps track of the change in structure and spatial concentration of the cell wall components simultaneously. A fixed number of particles of the reagent used for pretreatment, which is representative of its concentration, diffuse through the pore spaces of the plant cell walls till it encounters a pore-fiber interface. This path can be monitored by a stochastic hybrid random walk approach. At the interface, the reagent used for pretreatment can either react with one or more of the cell wall components or undergo reflection back into the pore space, based on the reaction probability. The, spatial concentration profile of the pretreating agent at different times can be determined by keeping track of the number of particles of the pretreating agent at each position as time proceeds. The spatial concentration profiles of the cell wall components and their evolution during biomass conversion can also be determined from the corresponding rate equations.

Thus, with such a model, a complete understanding of the 3D structure and the physicochemical distribution and its evolution can be understood. This provides a fundamental insight into biomass recalcitrance and how they can be addressed in developing efficient conversion strategies.

472309 Lignin-Derived Products from Biocrude As Building Block Chemicals

Abstract:

The production of chemicals as bioproducts in an integrated biorefinery could enhance the economic viability of biomass-to-liquid fuel technologies and afford a higher return on investment. RTI International is developing an advanced biofuels technology that integrates a catalytic biomass pyrolysis step and a hydroprocessing step to produce infrastructure compatible biofuels. In recent efforts, RTI is exploring and developing processes that could be used to produce bioproducts in addition to liquid fuel from the biocrude intermediate.

The present study explores distillation as a separation method for recovery of lignin-derived chemicals from biocrude. The biocrudes used were relatively thermally stable and distillable; they were produced from catalytic pyrolysis with a non-zeolite, alumina-based catalyst. Gas chromatography-mass spectrometry (GC-MS) analysis indicated that the biocrude contained reasonable amount of lignin-derived compounds. The distillation studies were performed with a laboratory distillation unit (PETRODIST 300 CC) and the biocrude were fractionated into four distillate cuts: Cut 1 (IBP-110 °C), Cut 2 (110-200 °C), Cut 3 (200-300 ° C), and Cut 4 (300-400 °C). The yields, physicochemical properties, and chemical composition of the fractions as well as the overall distillation separation efficiency will be discussed.

461248 Continuous Butanol Extraction Using Supercritical Carbon Dioxide

Abstract:

Butanol has promise as a drop-in biofuel or as a fuel additive that can be blended with gasoline at much higher proportions than ethanol. To date, economical production of butanol has been hampered largely by its cytotoxicity, which becomes limiting at levels as low as several weight percent. In this work, we demonstrate n-butanol extraction from dilute aqueous solutions used to simulate fermentation broths using supercritical carbon dioxide ($scCO_2$). In this talk, we will focus on process engineering of the n-butanol extraction. Our process engineering work is complemented by microbiological and metabolic engineering work that has identified a $scCO_2$-tolerant bacterial strain and genetically modified it for biofuel production. Compared to other butanol recovery approaches, $scCO_2$-extraction has sterilization and potential energy balance advantages; because $scCO_2$ selectively extracts butanol instead of water, a highly concentrated butanol stream can be recovered which requires minimal post-processing purification.

Specifically in this work, we studied n-butanol extraction performance and material/energy balance analysis. For butanol extraction performance, we used batch-wise extraction to study the effects of initial n-butanol concentration, extraction vessel pressure, and $scCO_2$ volumetric flow rate on n-butanol extraction rate. Additionally, we modeled the data using a standard liquid-liquid mass transfer model to determine values for the mass transfer coefficient, $k_l a$. Figure 1 shows representative extraction data compared to the model predictions using fit mass transfer coefficients. In all cases, the mass transfer model adequately described the experimental data. Best-fit values of $k_l a$ did not vary within our estimated limits of uncertainty for variation in extraction pressure (from 10.3 to 13.7 MPa – the range over which the $scCO_2$-tolerant bacterial strain has exhibited growth); therefore, we conclude that operation at lower pressures should be favored to achieve better process economy. Similar analysis was performed to interpret the mass transfer coefficient from correlations developed for gas-liquid and liquid-liquid extraction and compared to interfacial area results obtained from $scCO_2$-droplet size analysis. In addition to n-butanol, we investigated extraction of other alcohol products (e.g., iso-butanol, pentanol, and hexanol) as these are potential metabolic targets for a modified $scCO_2$-tolerant organism. Finally, we studied the extraction of the aldehyde intermediate, butanal, to set limits on its accumulation in the fermentation products.

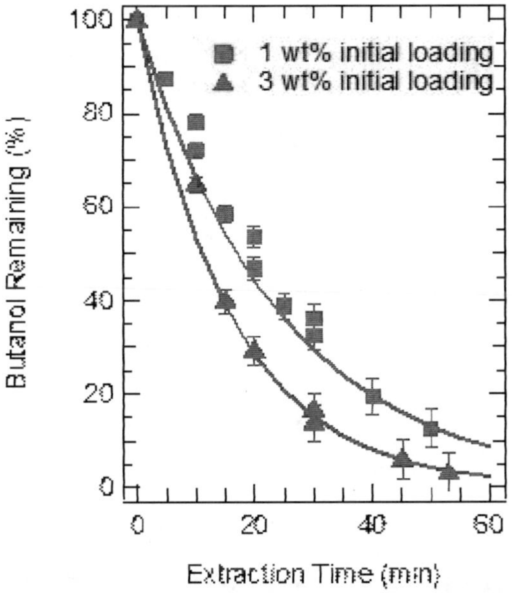

Figure 1. Representative n-butanol extraction data obtained at 10.3 MPa and 40 °C.

In addition to extraction experiments and modeling, we performed preliminary material and energy balance calculations of a conceptual continuous process. Thermodynamic and extraction data were used to model n-butanol removal from the fermenter and fermentation kinetics were modeled based on activity of similar organisms. The Peng-Robinson equation of state was used for modeling thermodynamic properties. To improve energy efficiency, the $scCO_2$ extraction stream was only partially de-pressurized from 10 to 5 MPa, resulting recovery of n-butanol at 79 wt% purity. Energy was recovered during the de-pressurization, which partly offset the energy required for CO_2 compression. Analysis was performed assuming the availability of either 10 or 0.1 MPa CO_2. Overall, we found that n-butanol with 95% purity could be produced at an energy cost of 3.9 MJ kg^{-1} (assuming that a 1 bar CO_2 source is available) or 3 MJ kg^{-1} (assuming a 10 MPa CO_2 source). These figures compare well with existing methods of butanol recovery, which range from 9 to more than 20 MJ kg^{-1}, and the lower heating value of the butanol fuel itself (33.1 MJ kg^{-1}). Further analysis is needed to account for heat loss and heat integration and to optimize process efficiency.

466873 *Development of an Integrated Aquaculture System for Clean Water and Animal Feed Production*

Abstract:

The concept of a "simulated lichen system" was recently developed by our UMN research group in which we can select different desired microalgae and fungal combinations which can be grown on the surface of specific polymers to form a biofilm. This technology can be extended to aquaculture wastewater bioremediation in order to recycle the nutrients as a proteinaceous microbial biomass feed for the reared aquatic animals. This process modification could make aquafarming more efficient as commercial feeds account for up to 50% of total production costs. The "mycoalgae biofilm" utilizes photosynthetic aeration to replace the conventional and energy intensive mechanical aeration for biological processes. By using the mycoalgae biofilm in aquaculture production, farmers will realize lowered costs for feed and reduced costs for maintenance and cleaning of the aquaculture water. Microalgae cultivation has been heavily researched to treat nutrient-rich water; however, the process is limited by problems associated with the settling and separation of small microalgae cells downstream of the treatment site. The proposed methodology of using a bioaugment of filamentous fungi in the lichen biofilm overcomes this limitation. The fungi efficiently retain algae and can also aid in the recovery and recycling of nutrients in the wastewater. The fungi retaining the algae allows for a less energy intensive harvesting process as a singular solid biofilm is produced. The strains complement each other by the exchange of required respiratory gases. The biofilm composition can be tailored based on the influent stream components, feasibility of the strains to grow together and its nutritional value as animal feed.

The effect of different initial levels of ammonia were tested in synthetic aquaculture wastewater to study the removal efficiency of nutrient pollutants by our biofilm. Starting ammonia levels were chosen based on a literature review of effluent discharge from various cultured aquatic species. Ammonia levels were reduced to undetectable limits in the medium to low initial levels of ammonia trials. At high levels of ammonia a set ammonia removal rate was observed, independent of total ammonia present in the system. We suspect the removal rate is dependent on nutrient availability and can be optimized in the future. The mycoalgae biofilm outperformed mono-cultures of fungi and algae at every level of ammonia tested in regards to total nutrients removed from the media. Combined biomass production was also higher in the co-cultures than in mono-cultures at all ranges of ammonia tested. The biomass has added value as feed for the cultured fish, thus making the process more economically advantageous. The biomass generated will be high in polyunsaturated fatty acids, proteins, and carbohydrates thus enabling robust fish growth. The biofilm's nutritional properties, low energy harvesting method, and effective nutrient removal abilities make it a promising system for wastewater remediation.

470075 Carbon Dioxide (CO_2) Capture and Utilization Pathways â€“ Assessing the Synergies Between Biological Use and Geological Sequestration of CO_2

Abstract:

As an important greenhouse gas (GHG), the emission of CO_2 to the atmosphere needs to be reduced in order to mitigate climate change. Designing of proper of CO_2 utilization pathways are imperative to reduce GHG emission. There are several ways to restrict CO_2 emission into the atmosphere. Two such promising technologies include geological sequestration of CO_2 in depleted oil and gas fields, saline aquifers and others, and biological sequestration of CO_2 via microalgae. The primary focus of the present study is to assess the integration potential of CO_2 geological sequestration and CO_2 biological sequestration via growth, harvest and conversion of microalgae to value-added bioproducts. Different ways of microalgae conversion, including lipid extraction, utilization of the spent algal biomass as well as hydrothermal liquefaction of the whole microalgal biomass, are considered for the production of value-added products. The present study aims to quantify the techno-economic tradeoffs and optimize the entire supply-chain of CO_2 utilization to ensure the most economically feasible alternative of CO_2 usage. A multi-scale supply chain optimization framework is considered which takes into account CO_2 capture, compression, pipeline transport and geologic storage as well as different processing pathways of microalgae. The state of Texas in USA is considered as the case study to evaluate the optimal design of CO_2 capture, storage and/or utilization pathways. The present study will help inform decision and policy makers the optimal pathway of CO_2 storage and utilization procedures for a particular location.

Chevron
human energy

Sustainable Development and Footprint Optimization in Shale Development

Janet Peargin, Footprint Program Manager
Hong Jin, Lead HES Engineer

Chevron Energy Technology Company

November 17, 2016

The challenge
footprint of shale/tight rock development

Many wells & moving parts

Footprint changes over time

- Jobs
- Community GDP
- Desired energy

- Land/habitat
- Emissions
- Trucks
- Noise
- Waste
- Resource competition

Chevron's environmental principles

- Include the environment in decision making
- Reduce our environmental footprint
- Operate responsibly
- Steward our sites

Footprint optimization: everybody plays a role

We can't solve problems by using the same kind of thinking we used when we created them.

– Albert Einstein

Shifting your perspective is so much more powerful than being smart.

– Astro Teller, Google "X" 2016 TED Talk

The opportunity: sustainable development

SEE across the asset life to create value and protect people and the environment

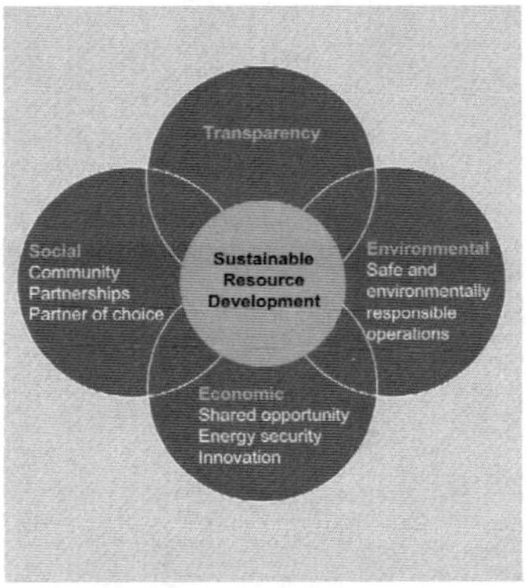

SEE triple bottom line value

SEEk win-win (all-win) ideas
Social, Environmental, Economic

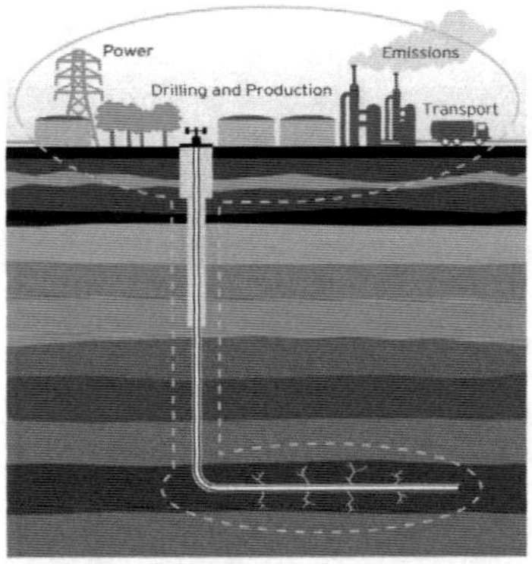

Optimize footprint

Integrated system of materials and energy that changes over time

Design with the end in mind

Decommission, restore, reuse

The win-win (all-win): optimize footprint to enhance value and reduce cost/risk

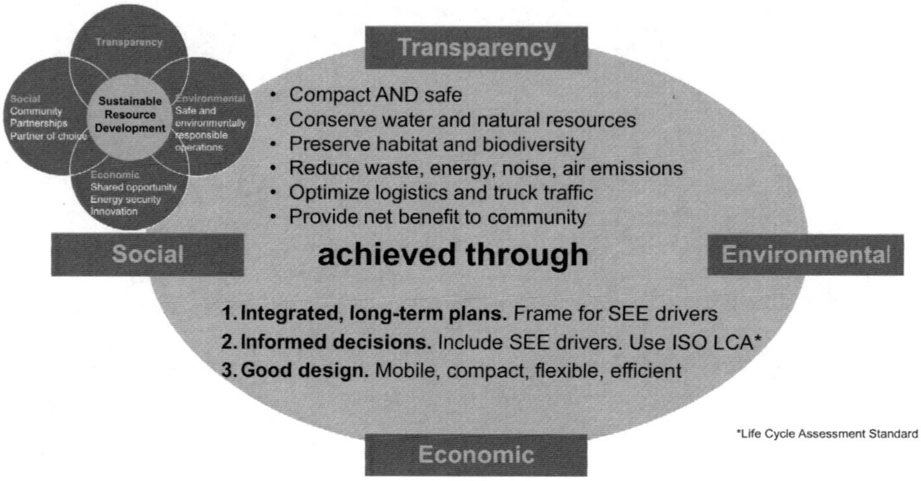

*Life Cycle Assessment Standard 14040-14044

Improve planning: SEE through the lens of sustainable development and SEE win-win (all-win) opportunities

Look at the big picture (asset, region)
Look at the long-term and change over time
Include diverse stakeholder perspectives

Take a systems view. Integrate across functions, activities, hydrocarbon/water/power
Plan and design for social, environmental, and economic (SEE) value drivers

Consider:

- Pace and scale of development
- Sequencing of events
- Degree of centralization
- Logistics and supply chain
- Ability to share infrastructure
- Power/fuel options
- Integration points

Improve decision quality: SEE more broadly and longer-term

Compliance is expected.

A traditional financial analysis is narrow and short-term.

SEE more broadly and longer term to identify risks and opportunities ... and enhance decision quality.

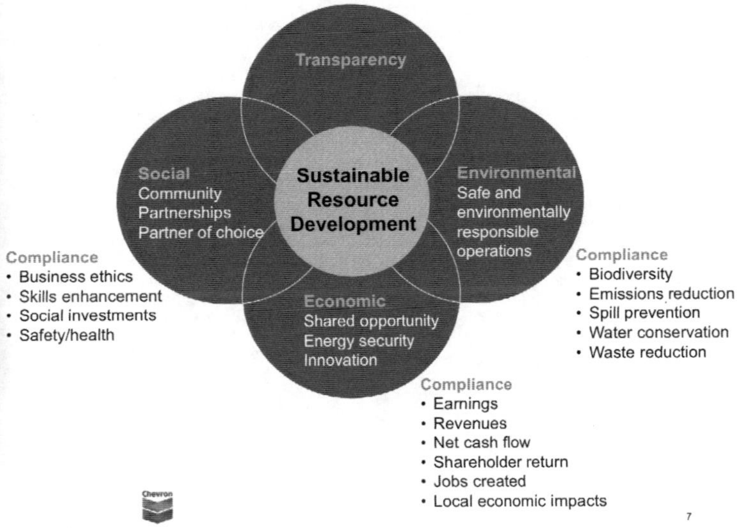

Chevron

7

Improve decision quality: quantify environmental tradeoffs with ISO life cycle assessment (LCA)

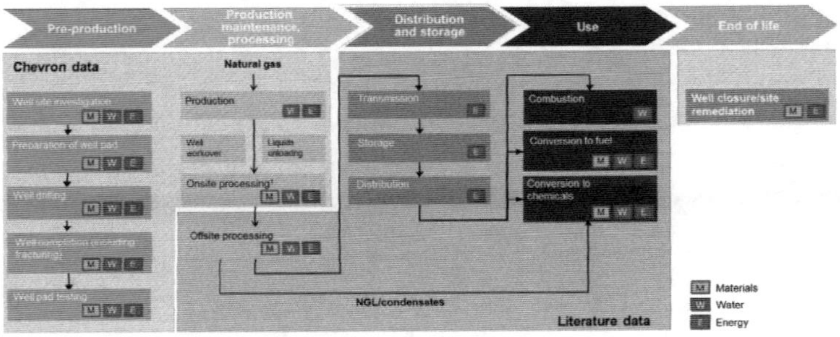

- What is our environmental footprint?
- Where are our big sources?
- What is the "best" fuel for power?
- What if I use this technology vs. that?
- What are the on and offsite impacts?

Chevron

8

Improve design: apply win-win (all-win) principles

Plan/design for sustainable development	• Develop long-term integrated plans • Frame for SEE (social, environmental, economic)
Integrate and innovate	• Look across the asset life at hydrocarbons/water/power & different activities (drilling, operations, etc.)
Question business as usual	• What do stakeholders want? Is there a better way?
Do *more* with *less*	• Reduce – reuse – recycle – be (energy/materials) efficient • Eliminate no-value steps & equipment
Disturb less land	• Fewer pads; share infrastructure/right-of-ways; preserve habitat corridors
Conserve water	• Use less & lower quality; treat to quality needed for use; consider reuse
Reduce waste	• Flares/emissions, discharges, energy use, no-value steps/equipment
Green the process	• Consider non-toxic chemistry, power/fuel efficiencies, fewer steps
Leverage technology	• Look to downstream, offshore, suppliers, competitors, others • Think mobile, modular, compact, flexible, prefab, energy efficient, automated, standard
Leverage supply chain	• How can suppliers help? Can we hire and source locally?
Make informed decisions	• SEE the big picture. Consider short- *and* long-term costs and value. • Use life cycle assessment to quantify environmental aspects of alternatives

Chevron

Win-win example: electricity from field gas

Better emissions/safety/reliability/production/cash flow

Problem

Need cost-effective, low-emissions power source at remote site

Solution

Micro-turbines make electricity from field gas*

Win-Win **Do *more* with *less***

MORE

Safety, flexibility, schedule certainty, reliability (8–10%), production/cash flow (sooner)

LESS

Equipment, permitting, O&M, trucks, emissions, material transfer/spill potential

** Great for ESPs. Rod pumps doable with more systems engineering*

Chevron

Everyone plays a role: ask questions

- What are our stakeholder concerns?
- What perspectives are we missing?
- What are our footprint boundaries?
- What are the priorities and tradeoffs?
- What does good (enough) look like?
- How can we minimize environmental and community impacts?
- How do we do more (product) with less (material, equipment, waste, etc.)?
- How do we design for change over time to operations, region, community?
- What can we learn from other processes?
  - LeanSigma®
  - Environmental, social, health impact assessment
  - Energy Efficiency and Waste Minimization assessments

LeanSigma is a federally registered service mark of TBM Consulting, Inc.

Key takeaways

- We all play a role in creating our company footprint – a system of materials and energy that changes over time
- Think beyond compliance. Include SEE (social, environmental, economic) drivers in planning, designs, and decisions.
- SEE through the lens of sustainable development to SEE opportunities. SEE the difference.
  - Cost savings from less equipment, optimized layout/network
  - Win-win ideas to save $ AND protect people & the environment
  - Decisions that are more robust
- A financial analysis is only part of the risk/uncertainty. We can quantify and include SEE drivers in decision-analysis.
- Sustainable development is good business. SEE it reduce cost and risk, add value, and improve performance.
- Everyone plays a role. You can contribute. Just change your perspective to SEE the win-win (all-win) opportunities to do more (product, energy) with less (energy, materials, waste, emissions, trucks, etc.)

References

- Hardisty, P.E. *Environmental and Economic Sustainability.* CRC Press, New York. ISBN 978-14200594891-56679-643-2, 378 pp. 2010.

- Sivapalan, M.; WorleyParsons. *Sustainable Development of Unconventional Oil & Gas Resources.* Prepared for presentation at *GasTech* 2012 Conference, London, 2012.

- Tanzil, D.; Dragan, M.; Bourque, B.; Seel, K.; Taylor, T.; Golder Associates. *Evaluating Social and Environmental Risk and Opportunities in Site Selection and Routing.* Copyright 2015, Society of Petroleum Engineers. SPE-173488-MS. Prepared for presentation at the SPE E&P Health, Safety, Security, & Environmental Conference – Americas held in Denver, Colorado, USA, 16–18 March 2015.

- Thomas, Charlotte. *Industrial Ecology: Moving Beyond Traditional Engineering Borders.* Society of Women Engineers magazine, spring edition, pp. 24-30. 2010

Chevron

461852 Green Net Value Added As a Sustainability Metric Based on Life Cycle Assessment: An Application to Bounty ® Paper Towel

Abstract:

Sustainability measurement in economics involves evaluation of environmental and economic impact in an integrated manner. In this study, we combine system level economic data with environmental impact from a life cycle assessment (LCA) of a common product. We are exploring a costing approach that captures traditional costs and incorporates externality costs to provide a convenient, easily interpretable sustainability metric. Green Net Value Added (GNVA) is a type of full cost accounting that incorporates total revenue, the cost of materials and services, depreciation, and environmental externalities. It adjusts the traditional economic value added for environmental externality costs. Two, but not all, of the potential environmental impacts calculated by the standard LCIA method (TRACI) could be converted to externality cost values. We compute externality costs disaggregated by upstream sectors in order to accurately reflect the externality costs associated with different industries along the supply chain. We preset full cost, green profit and GNVA to evaluate the relative sustainability of Bounty® paper towels manufactured at two production facilities. We found that the longer running, more established line had a higher GNVA than the newer line. The dominant factors contributing to externality costs are to come from the stationary sources in the supply chain: electricity generation (27-35%), refineries (20-21%), pulp, and paper making (15-23%). Health related externalities from Particulate Matter (PM2.5) and Carbon Dioxide equivalent (CO_2e) emissions appear largely driven by electricity usage and emissions by the facilities, followed by pulp processing and transport. Supply chains for other facility purchases, product distribution, and end-of-life management (landfilling/incineration) contribute the least to the environmental health-related costs.

474653 Sustainability Metrics for Solar Powered Charging Infrastructure for Electric Vehicles

Abstract:

This presentation will include an analysis of sustainability metrics associated with the installation of solar panels in parking lots designed to provide electricity for commercial use and shade for those who park under the solar panels. The installation will also include electric vehicle supply equipment to charge batteries in electric vehicles. The system is connected to the electrical grid so that the electricity generated can be used efficiently and electric vehicle batteries can be charged at all times. Important sustainability metrics are cost of electricity generated, cost of vehicle charging, value of shaded parking, convenience of vehicle charging, land area, water, air quality, greenhouse gas emissions, and materials for the installation and equipment. Social metrics include impact on health status of the community, psychological well being, harmony, values, agency, and environmental justice.

458893 Evaluation of Low-Energy Recovery of N-Methyl-2-Pyrrolidone Solvent from Waste Streams in Specialty Resin Manufacture

Abstract:

N-methyl-2-pyrrolidone (NMP) is a common solvent used as a reaction medium in the manufacture of specialty polymeric resins and their precursors. NMP is used primarily because of its thermal stability, dipolar aprotic properties, and ability to produce a high molecular weight resin precursor. Since NMP is not consumed by the reaction, an aqueous hazardous waste stream is generated. A previous case study at a DuPont specialty chemical plant by the authors, shows that a distillation recovery system that can effectively purify NMP from the waste stream. The distillation recovery process reduces operating cost by 83%, but only reduces the life cycle carbon footprint by 40%. This is due to the high energy demand of the distillation column. Additionally, high NMP purity levels in the resin manufacture process and the complex multicomponent nature of the waste stream, required a more extensive design for the recovery operation.

Alternative distillation designs were explored, but none reduced energy consumption and life cycle carbon emissions as significantly as the economic life cycle impact. More sustainable low energy separations were investigated. Various membrane, adsorption, and extraction separation strategies have the potential to reduce energy demand, but each has its own limitations. An extraction operation was selected as the best candidate for separation, and the proposed process considers alternative uses for the recovered NMP waste in other manufacturing scenarios.

Results with a chloroform extraction operation to produce NMP of suitable quality for alternative use, provides the best option for solvent recovery. The new process can not only yield environmental life cycle improvements, based on various impact factors, but provide economic benefits as well. Yearly projections of waste reductions, life cycle improvements, energy utilization, and cost savings are presented. Life cycle carbon emissions are reduced by over 60% compared to the base case. Since NMP is a prominent solvent for several commercial sectors, methods aimed at reducing waste help promote broader sustainability goals. This study shows how sustainability metrics can be used to guide decision making when evaluating other processing alternatives and uses of the final product. (This work is partially supported by a grant from the EPA Pollution Prevention program).

468272 Thermoelectric Power Technology Choices Based on Water Availability

Abstract:

Thermoelectric Power Technology Choices Based on Water Availability National Energy Technology Laboratory (NETL) has developed a prototype water-energy model to investigate issues surrounding the energy-water nexus as it relates to the effect of power generation plant design choices on water availability and the overall costs of supplying water for thermoelectric power generation.Â The model utilizes the database developed by a team of researchers with the Water Security Program at Sandia National Laboratory (SNL). The database includes water availability, cost, and water use for electric power, agriculture, municipal and industrial for 38 conterminous states representing various Interconnects within the U.S. Data on current and future water use for thermoelectric power are derived from NETL databases on water use by technology and projections from the Energy Information Administrationâ€™s National Energy Modeling System (NEMS). Current and forecasted water availability data from the World Resources Institute are used to develop three water availability scenarios. The prototype water-energy model is used to project the amount of water available for power generation, to project water demand for future power generation based on a NEMS forecast, to identify regions that experience potential water shortfalls, and to estimate the cost of meeting water demand where water shortfalls exist for each water availability scenario. A case study of the Brazos River Basin in Texas is used to illustrate a specific application and to highlight the capabilities of the prototype water-energy model.

462306 Experimentally Driven Choices in the Fabrication of a 45 Kw Solar Simulator

Abstract:

Solar-thermal technology has been successfully used to drive biomass gasification, metals reduction and electrical generation in a renewable manner. The development of these applications relies on high-flux solar simulators: assemblies of high power lamps whose irradiance mimics concentrated sunlight. These platforms provide controlled environments for the evaluation solar-thermal technologies at the laboratory scale. To date more than seven high-flux solar simulators are available worldwide for academic studies.In conjunction with Alpha-Omega Power Technologies (LLC), we evaluated choices in high-flux solar simulator construction. This included the effect of different reflector coatings â€“ uncoated, silvered, and aluminized â€“ on lamp efficiency. Additionally, the importance of automated filament positioning was quantified. These results educated the installation of a new, 45 kW, solar simulator at the University of Colorado Boulder.

Carbon Nanotube (CNT) Based Multifunctional Optoelectronic Rods for Efficient Energy Generation and Transport

Sayeeda T. J. Aishee; Evelyn Gonzalez; Erika Hernandez; Glenn Grissom; H. Justin Moore, M. Jasim Uddin*

Department of Chemistry, College of Science, University of Texas at Rio Grande Valley

1 West University Blvd. Brownsville, Texas 78520

*Corresponding Author: jasim.uddin@utrgv.edu, Phone: 956-882-7772

Abstract

To keep up with the world's increasing demand for energy, a clean and perpetual energy source is vital to sustain this planet. This research activity focused on producing a flexible, inter-aligned, thermally stable, highly conductive, carbon based, and ultra-strong photovoltaic (PV) power wire from carbon nanotubes yarn (CNY) functionalized with quantum dots and N719 dye sensitizer. These innovative three dimensional PV wires, with enhanced photoconversion efficiency, have been developed using CNY as the counter electrode (CE) and TiO_2 grafted CNY as a working electrode (WE). The CEs were twisted around the WEs to collect and transmit the photogenerated electrons from the outer circuit. The optimized photovoltaic cell shows a 3.9% photo-energy conversion efficiency.

Introduction

With increasing global awareness for facilitating the energy demand of the world in which climate is constantly changing and pollution negatively affecting many areas, much research has been focused upon finding alternative sources for energy that are both inexhaustible and clean. Among these alternative sources for energy, photovoltaic cells have drawn much attention in the scientific community due to their potential for highly efficient photovoltaic energy conversion [1]. Photovoltaic power wires could revolutionize the traditional solar cells and supply a significant portion of the world's energy consumption. Much work has been done in order to increase the efficiency of the photovoltaic cells; in fact, the energy conversion efficiency of dye-sensitized solar cells has the potential to reach up to 12.1% utilizing platinized transparent conductive oxide anodes [2]. There is a particular interest in creating PV cells that have low internal resistance and low heat transport. It is important to highlight that PV cells are open to textile application, thus greater cell stability is required. Photovoltaic cells constructed from CNY are seen around the world as promising and extraordinary sources for propelling technological advancement in fields involving electronic materials and others, as their ability to convert light to energy is highly favorable and unique [5]. The research focuses on optimizing the energy conversion and improving the mechano-optoelectronic conditions of photovoltaic cells. Enhancing the efficiency of dye-sensitized carbon nanotube yarn shaped photovoltaic cells through the incorporation of MoS_2 is most definitely an innovative project with the potential to affect many fields such as those of textiles, electronics, and engineering in a greatly beneficial manner.

This paper has focused on developing a PV cell from CNY functionalized with molybdenum disulfide and cadmium sulfide as an addition to cell configurations previously reported [1-3, 4]. In order to optimize the efficiency of PV cells, we must first understand its molecular structure of the photoactive phase. The PV microwire was then coated with TiO_2 which acts as a working electrode (WE). Although most PV cells are assembled with a liquid electrolyte, the CNY was used to test the suitability and electronic potential without using any conventional liquid electrolyte. Incorporating CNY around the WE provides a greater sustainability than most traditional dye-sensitized solar cells (DSSC), and it works as a counter electrode. Indeed, the carbon nanotube yarns demonstrate excellent catalytic properties that provide high electronic performance; when the CNY was twisted around the WE, it became the primary conductor of electrons. This power microwire could lead to the widespread use of three dimensional PV wirses in daily life such as clothing, engineering structures, or electronic products.

Experimental Method

In order to develop photovoltaic cells from dye-sensitized CNY functionalized with molybdenum disulfide and attain larger fill factors with an enhanced efficiency thus leading to increased PV energy conversion, the following procedures were undertaken:

Carbon Nanotube Yarn Preparation

The CNY was prepared by removing the polymer from the CNY by soaking it in acetone, and then it was baked at 350°C in a tube furnace with light air flow to facilitate the removal of ash.

Further, the CNY was treated with nitric acid and sonicated to remove any remaining polymer ash.

Working Electrode Treatment

The WE was submerged in Triton-X and rinsed with ethanol. Five yarns were twisted together utilizing their mechanical properties; this process resulted in an assembly of 5 carbon nanotube yarns for the WE. A TiO_2 nano and macro-sol was applied and calcined at 350^0C.

The CNY was then soaked in a solution of $TiCl_4$ and ethanol. Finally, the WE underwent chemical bath deposition. The WE was functionalized with solutions containing quantum dots (molybdenum disulfide and or cadmium disulfide) and dimethyl sulfoxide, taking care to see that these processes did not ruin the integrity of the TiO_2 nano/macro coats. This was confirmed through observation in the atomic force microscope (AFM).

The WE was also placed in mixed solutions of PCBM (Phenyl-C61-butyric acid methyl ester) and P3HT (Poly(3-hexylthiophene-2,5-diyl)) (Scheme 1) , and it was then dried at RT. The PCBM and P_3HT sensitizer based photovoltaic cells facilitate a high efficiency rate in photon to electron conversion. Carrier conversion efficiency (η_c) and energy conversion efficiency (η) of CNY based photovoltaic devices are improved when semiconducting polymers incorporating Bis(1-[3-(methoxycarbonyl)propyl]-1-phenyl)-[6.6]C62 (mixture of isomers), Bis[60] PCBM (PCBM 62) or its functionalized derivatives are used in conjunction with CdS and CdSe. These processes readied the WE to facilitate photoelectron transfer. The WE was then placed in N719 with *tert*-butanol and acetonitrile solution, which is vital to the photosensitive function of the CNY. Any electrodes were kept in the dark prior to assembly, including the WE, which was ready to be assembled into the PV microwires.

Counter Electrode Preparation

The counter electrode was assembled as both single strand CNY and triple strand CNY; this readied the CE for the Pt nanoparticles deposition by sputtering. The sputtering process was done with care to minimize possible damage to the thin film of the CNY.

Electrolyte Preparation

The electrolyte was then prepared with Poly(vinylidene fluoride-hexafluoropropylene) (PVdF-HFP), I_2, and NaI using a chemical bath deposition process while making sure to obtain a smooth homogenous solution. This prepared the electrolyte to be applied to the cells by pipette.

PV Cell Construction

Finally, the PV cell was constructed with a predefined twisting (1 mm pitch distance) process and using conductive silver paste to hold electrodes connection on the slides and connect the ends of the individual electrodes [4]. Approximately thirty PV cells were developed, with half of those having quantum dot incorporated and half of them without it. These CNY cells were compared through analysis of data to confirm the significance of the use of quantum dot in enhancing the efficiency of PV cells with dye-sensitized CNY.

The materials of construction were characterized using Atomic Force Microscope (AFM), the RAMAN Spectroscopy, Infrared (IR) Spectroscopy, and Ultraviolet-Visible Spectroscopy (UV-Vis). The AFM allowed for high-resolution images of the PV cells to be developed and also allowed for sensitive force measurements to be collected. The RAMAN and IR spectroscope yielded much vital data such as exact composition of material, stress/strain state, symmetry and orientation, quality, and amount of material. The IR spectroscope analyzed the infrared light interacting with the PV cell materials by measuring the absorption, emission, and reflection. The UV-Vis yielded data relating to the wavelength and amount of light absorbed by the PV cells. A potentiostat (VERSASTAT 300) was utilized in order to control the current and voltage of the CNY powerwires.

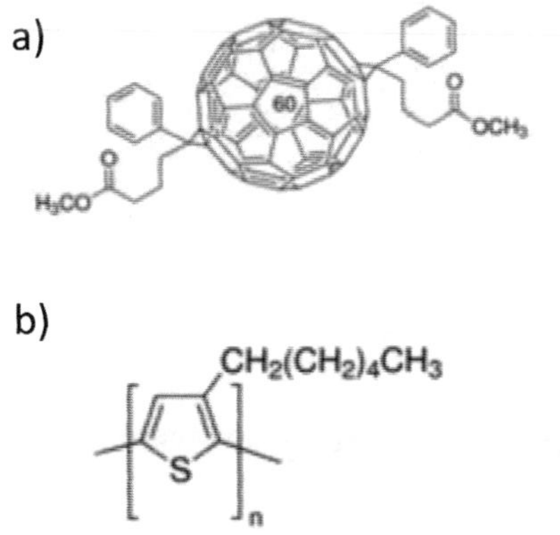

Scheme 1: a) Bis(1-[3-(methoxycarbonyl)propyl]-1-phenyl)-[6.6]C62 (mixture of isomers), Bis[60]PCBM (PCBM 62) and b) Poly(3-hexylthiophene-2,5-diyl) (P3HT).

Results and Discussion

Inventor Computer Aided Design (CAD) Modeling Software was used to better exhibit the construction of the various chemical interlayers of the PV microwires.

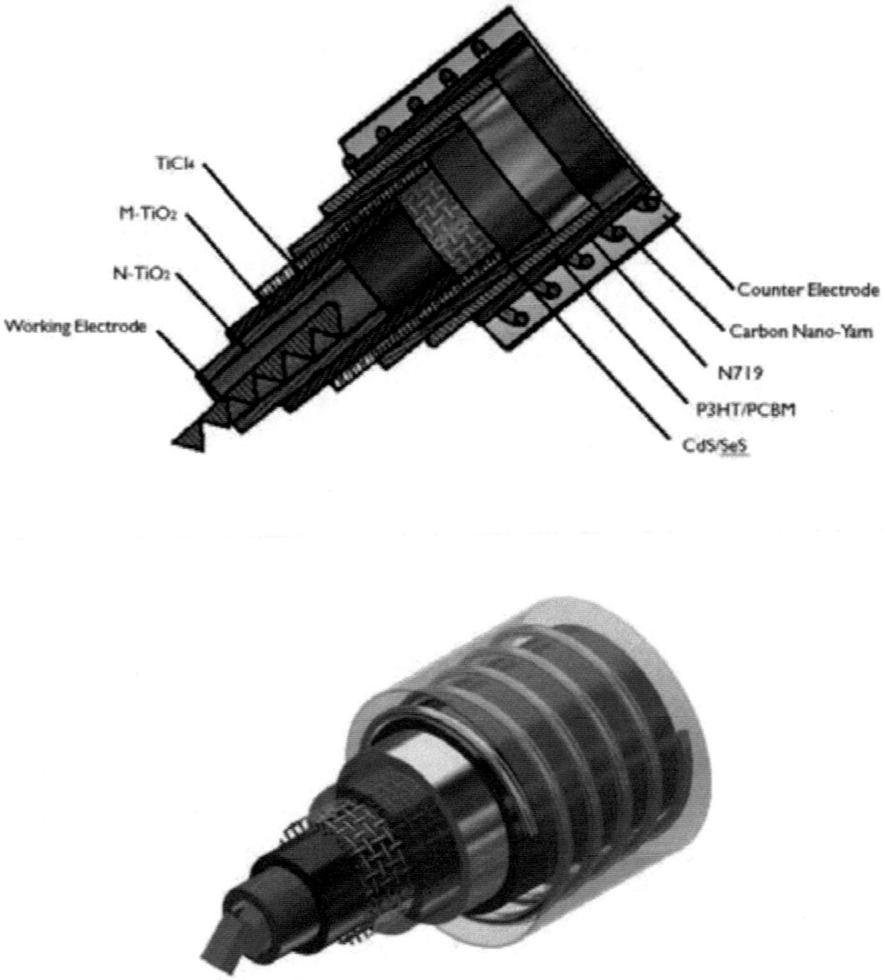

Figure 1a-b: The schematic view of the CNY PV cell.

The PV cells were constructed as shown in Figure 1 and the morphology of the CNY demonstrated with AFM images in Figure 2. The orientation of each individual layer enhanced the energy sensitivity and overall functionality of the cell. As can be seen in figures 2, the CNY clearly exhibits a homogeneous structure without agglomeration. Both images show the underlying carbon nanotube-yarn clearly despite their thick coating and difference in scale. The

CNT yarns are flexible, which enable our WEs and CEs to be twined easily with any of twisting parameters toward the WE and can maintain its structural and chemical properties unchanged after being folded/twisted or tied for multiple times.

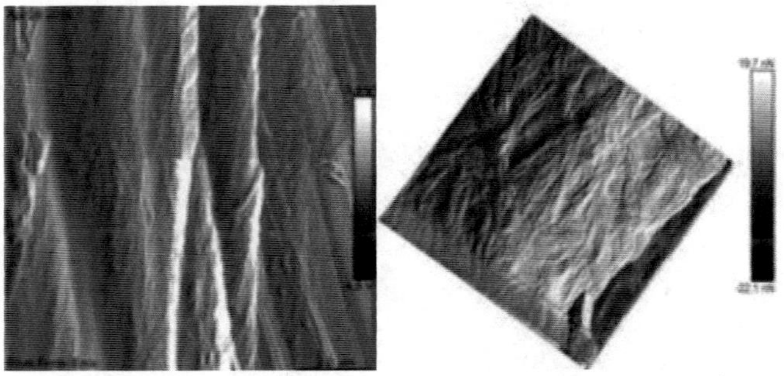

Figures 2 a-b) AFM images of Pristine CNY.

The *I-V* curve exhibits the efficiency of the photovoltaic energy conversion of the cells. This efficiency was calculated by taking the ratio of the maximum electrical power output compared to the solar power input into the PV cell. The fill factor, the measure of quality of the PV cell, was also calculated through the comparison of the maximum power to the theoretical power output at both the short circuit and open circuit voltages. The data obtained from AFM, Raman spectroscopy, and UV-Vis-DRS (Ultra Violet Visible-Diffuse Reflectance spectroscopy) are in-agreement with determining the nanostructured feature of this optoelectronic cells (*the figures are not shown here for the sake of brevity*). The RAMAN spectroscope yielded data that was analyzed in order to determine the structural phase of the WE and CE. The results infer that the titania nanophase are certainly indicative of anatase phase. Finally, data relating to the wavelength and amount of ultraviolet light absorbed by the PV cells was analyzed through the use of the UV-Vis-DRS (figure not shown for the sake of brevity.

The UV–Vis reflectance spectra were carried out at room temperature on a Perkin-Elmer UV–Vis–DRS to detect UV absorption intensities, the location of absorption edges and the quantum size effects of the nanostructured TiO_2 grafted on individual CNTs. The UV–Vis spectra have been obtained in the reflectance mode on the dried samples. This procedure is aimed to avoid systematic errors in evaluating the intensities of the specific UV–Vis absorption bands of the

TiO_2 and CNTs, because, as it is known, different fillings of the intra-crystalline voids by the adsorbed solvent molecules have an important effect on the scattering properties of the supporting phase.

The efficiency of 3D cells is largely dependent upon their fill factor, the measure of quality of the solar cell. We found the fill factor of the cells is ~0.5. Fully functioning photovoltaic cells constructed from dye-sensitized carbon nanotube yarns functionalized with CdS, and MoS_2 (QD) were developed. These PV cells have an enhanced capability for efficient conversion of solar energy, with an increase in efficiency of ~Δ1-2% from that of cells not functionalized with Cds and MoS_2.

Ultimately, the cells constructed from CNY using CdS and MoS_2 had higher fill factors than those without CdS and MoS_2. An increase in fill factor due to the addition of MoS_2 is highly beneficial because it signifies an increase in quality of the PV cells. The energy conversion efficiency of the cells constructed from CNY functionalized with MoS_2 had 3.9% photovoltaic energy conversion efficiency at 0.7 V, as shown in figure 3. Therefore, the PV powerwires were synthesized from CNY using QDs had better efficiency than those without incorporating QDs.

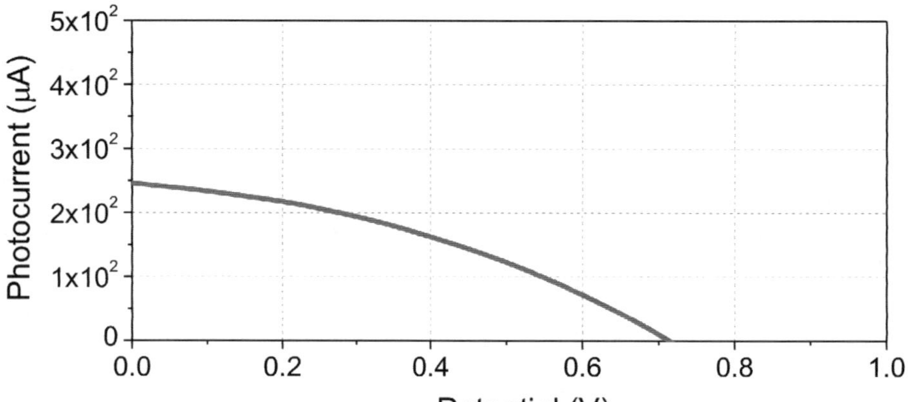

Figure 3: The *I-V* curve of the synthesized cells. The PV cell shows productive usage of power, thus indicating a high efficiency of 3.9% at 0.7 V.

Conclusion

The development of a photovoltaic cell from CNY is advantageous due to the low cost of the materials and the eco-friendly nature of the energy source. The significance of this project exists in the fact that the developed three dimensional PV cells result in much more efficient photovoltaic energy conversion, which makes the future use of photovoltaic cells in everyday life. The application of CdS, MoS_2, and CdSe enhanced the photovoltaic energy conversion productivity of the PV cell. These research outcomes helped maximize the photoenrgy conversion efficiency up to 3.90%. There are many advantages to continuing research on microware structured PV cells, since PV cells are a strong candidate for consideration for future application in a variety of fields such as those involving textiles, electricity, and other forms of engineering.

Acknowledgement:

The authors gratefully acknowledge the research grant (D01_W911SR-14-2-0001-0003) from Department of Defense through MSRDC, Washington, DC 20005.

References

[1] Yan, Jin, Uddin, M. Jasim, Dickens, Tarik J., Daramola, Deborah E., Olawale, David, and Okoli, Okenwa I. 2013. Tailoring the efficiency of 3D wire-shaped photovoltaic cells (WPVCs) by functionalization of solid-liquid interfacial properties. *Phys. Status Solidi* A, 1-7.

[2] Uddin, M.J., Davies B., Dickens, T.J., and Okoli, O.I. 2013. Self-aligned carbon nanotubes yarns (CNY) with efficient optoelectronic interface for microyarn shaped 3D photovoltaic cells. *Solar Energy Materials & Solar Cells* 115, 166-171.

[3] Uddin, M.J., Dickens, T., Yan, J., Chirayath, R., Olawale, D.O., and Okoli, O.I. 2013. Solid state dye-sensitized photovoltaic microwires (DSPMs) with carbon nanotubes yarns as counter electrode: Synthesis and characterization. *Solar Energy Materials & Solar Cells* 108, 65-69.

[4] Uddin, M. Jasim, Grissom, Glenn, Leal, Miguel, Galvez, Veronica, Trad, Tarek, Touhami, Ahmed, Islam, Nazmul, Parsons, Jason, and Moore, H. Justin. 2016. Self-aligned carbon nanotube yarns for multifunctional optoelectronic applications. *Proceedings of the ASME's International Mechanical Engineering Congress and Exposition (IMECE)* 2016-67441, 1-5.

[5] Nonoguchi, Yoshiyuki, Hata, Kenji, and Kawai, Tsuyoshi. 2015. Dispersion of Synthetic MoS_2 Flakes and Their Spontaneous Adsorption on Single-Walled Carbon Nanotubes. *ChemPlusChem* 80, 1158-1163.

464625 Thermal Ageing Characterization of Materials Under Concentrated Solar Power

Abstract:

Concentrating solar power (CSP) technologies have the potential of replacing global natural gas and liquid fuels consumptions by producing solar hydrogen. There are many factors that hinders the competitive low production price of this technology such as degradation of given properties of a solar absorber material over time due to temperature and solar radiation (Brunold *et al.*, 2004). In case of solar absorber materials, the most important factor for aging is degradation of reflectance/absorptance over time and its mechanical integrity (Boubault *et al.*, 2014a). Furthermore, sudden variation of temperature over a short period of time due to variation of solar radiation is another cause of aging (Boubault *et al.*, 2012). It is proposed (Boubault *et al.*, 2014a) that measurement of thermal performance of a material can serve as an ageing criterion (Boubault *et al.*, 2014b), which can be achieved by quantification of its absorption and heat transfer ability. In this work, we will investigate the thermal performance and the optical properties of stainless steel, as a reference material, using a high flux solar simulator and a low power light source. The testing facility consists of a high flux solar simulator which is capable of irradiating Gaussian distributed variable flux (Sarwar *et al.*, 2014). A light collimator was used to homogenize the incident flux. The flux characterization was carried out before and after homogenization using a water cooled Gardon-type circular foil radiometer. The material was exposed to the homogenized flux for given time periods, while we were monitoring continuously the temperature of the material and the heat flux at the back side of the material. The optical properties of the material such as absorption and reflectance were measured and compared before and after exposure to the high concentrated irradiance. Here we present the results of the characterization and the method validation against a total integrated reflectance/transmittance method employing a spectrometer with an integrating sphere.

Acknowledgements

This publication was made possible by a PDRA award [PDRA2-1105-14044] from the Qatar National Research Fund (a member of The Qatar Foundation). The statements made herein are solely the responsibility of the authors.

References

Boubault, A., B. Claudet, O. Faugeroux and G. Olalde, 2014a: Aging of solar absorber materials under highly concentrated solar fluxes. *Solar Energy Materials and Solar Cells*, **123**:211-219.

Boubault, A., B. Claudet, O. Faugeroux and G. Olalde, 2014b: Aging of solar absorber materials under highly concentrated solar fluxes. *Solar Energy Materials and Solar Cells*, **123**:211-219.

Boubault, A., B. Claudet, O. Faugeroux, G. Olalde and J.-J. Serra, 2012: A numerical thermal approach to study the accelerated aging of a solar absorber material. *Solar Energy*, **86**:3153-3167.

Brunold, S., M. Köhl, K. Möller and B. Carlsson, 2004. Chapter 4.4 - Accelerated Indoor Testing. In: M. K. C. J. W. Czanderna (Ed.), *Performance and Durability Assessment* (pp. 175-195). Elsevier: Amsterdam.

Sarwar, J., G. Georgakis, R. LaChance and N. Ozalp, 2014: Description and characterization of an adjustable flux solar simulator for solar thermal, thermochemical and photovoltaic applications. *Solar Energy*, **100**:179-194.

467866 Development and Thermal Test Results of a 10 Kw Porous Media Solar Reactor

Abstract:

Solar volumetric receivers are based on high porosity materials as radiation absorbers, like open cell foams or monolithic honeycombs. These structures allow the volumetric absorption of concentrated solar radiation within them. Solar heat absorbed in the porous media is then transferred to a working fluid passing through it. In this work, a 10 kW volumetric solar reactor has been developed for the reduction of metal oxides. Within the reactor there is a ceramic porous media that supports the metal oxide to be reduced. This structure is directly irradiated with concentrated solar energy in order to reach reaction temperatures above 1300 K.

The developed solar reactor was designed for experimentation with different operational conditions at the IER-UNAM Solar Furnace. The design was accomplished based on detailed ray tracing of the furnace, to obtain the concentrated flux distributions at different planes in the focal zone. The design is based on a cavity with a ceramic conical receiver, and allows for temperature measurements by thermocouples and pyrometers at different spots. Thermal simulations were carried out to determine the reactor body specifications and insulation requirements. The device has been built and preliminary tests have been carried out at the solar. These tests are aimed at evaluating the thermal performance of the system in the absence of chemical reaction. Different porous media that could be used in thermochemical processes have been used in these tests.

468047 Thermal Model of a Solar Porous Media Receiver for a Thermochemical Reactor

Abstract:

Solar receivers containing high porosity materials are a promising option due to higher efficiency compared with tubular receivers. In this system, concentrated solar radiation is absorbed on the porous media in a volumetric manner, then the absorbed heat is transferred to a working fluid when passes through the porous chanels. One critical issue when designing or optimizing volumetric solar absorbers is the knowledge of the temperatures of the solid matrix and fluid. These temperatures are linked with the efficiency and operability of the solar receiver. Hence, this work presents a 3D model of heat transfer and computational fluid dynamics in a porous media receiver subject to concentrated solar flux. This model allow us to analyze the temperature distribution both in the fluid phase and the solid phase, by applying the local thermal non- equilibrium model. The macroscopic model was then used to analyze the impact of geometric properties and operational conditions on the temperature distribution.

474675 Effect of Dust on Straight Tube Receiver of a Parabolic Trough Concentrator

Abstract:

The objective of present investigation is to enhance the performance of parabolic trough collector system. To improve its optical efficiency a dust collector /dirt accumulator has been modelled. This system is being used as parabolic trough based system for power plant application. The experiments were performed to analyze the optical efficiency and optical degradation rate at the roof top of Centre for Energy and Resources Development, Indian Institute of Technology (BHU), Varanasi (25.3176°N,82.9739°E),India. A parametric studies has also been done to estimate the effect of wind flow/air flow, pitch angle, shadow of wall, wind direction ,ambient temperature, solar radiation and collector area for high temperature, high concentration ratio application.

464032 Renewable Transportation Fuel Production from Algal Biocrude Oil Converted Via Hydrothermal Liquefaction

Abstract:

The goal of this project is to understand the properties of biocrude oil converted from Spirulina (SP), which can thrive in wastewater. Hydrothermal liquefaction (HTL) is a promising technology to convert wet biowaste into biocrude oil. Under high temperature and high pressure, HTL technology uses water in biomass as an excellent solvent and catalyst to break down and reform organic materials into biocrude oil products. Our previous study showed that the biocrude oil converted from wet biowaste via HTL contains comparable heating values (HHV) to that of petroleum crude oil. However, due to the lack of up-scaled production and proper separation of HTL biocrode oil, the application of HTL biocrude oil is limited.

In this study, the biocrude oil converted from Spirulina is separated into different fractions by the distillation process. The preliminary distillation test shows that 35-40 wt.% of HTL algal biocrude oil can be distilled at about 175 °C, indicating that the HTL algal biofuel is promising to be used as transportation fuels including kerosene and diesel. In addition, compared to other types of HTL biocrude oil, it is noticed that 25 wt.% and 10 wt.% of the biocrude oil converted from swine manure and food processing waste is respectively distilled at around 175 °C. This also suggests that the HTL algal biocrude contains compounds with lower boiling points.

Detailed physicochemical characterizations, such as acidity, viscosity, boiling point distribution, and chemical composition are examined by the viscometer, thermogravimetric analysis (TGA), CHN Analyzer, Fourier transform infrared spectroscopy (FTIR) and gas chromatography mass spectrometry (GC-MS). The physicochemical properties of the distillates are also compared to those of petroleum transportation fuel, such as jet fuel and diesel, and those of HTL biocrude oil converted from other biomass, such as swine manure and food processing waste.

465013 Biochemicals Production By Genetically Engineered Cyanobacteria Using Photobioreactors

Abstract:

The 3rd generation biorefinery research using microalgae is ongoing in many countries because fuel/chemicals production from carbon dioxide by photogenic microorganisms is essentially sustainable. Cyanobacteria is recognized as a proper platform for chemicals production due to their high growth rate, but it requires genetic modification because the original strain never produce usefulchemicals. In this presentation, I would like to introduce the current status of our research concerning the production of iso-propanol (IPA) and 1,3 propandiol (1.3PDO) by genetically engineered cyanobacteria using photobioreactor system with LED irradiation equipment, which can selectively supply various wavelength of light with low power consumption.

To achieve IPA production in cyanobacteria, we constructed a metabolically engineered Synechococcus elongatus strain TA1741 which produces IPA from acetyl-CoA by four steps. Successful cell growth was observed for blue, and blue and red LED, however, IPA was not detected for all cases. On the other hand, higher intracellular glycerol accumulation was observed, which suggests that the carbon dioxide fixed in a Calvin cycle is converted primarily into Glycogen under light condition. So we thought that, by changing from light to dark condition, accumulated glycogen may be converted to IPA. To verify this idea, Cells grown in a jar fermentor was harvested, re-suspended in new medium and purged by nitrogen gas. Then the tubes were maintained under anaerobic and dark conditions for 5-10days and analyzed. After five or 10 days cultivation, the production of IPA was confirmed for all cases showing that the genes concerning IPA biosynthesis were working in TA1741. Maximum IPA concentration was approx. 23 mg/L under anaerobic and dark conditions, however the concentration is rather low and complicated operation is required for practical IPA production.

Then, we examined another product, 1.3-Propandiol (1.3-PDO) production by genetically engineered cyanobacteria. In cyanobacteria, metabolic flux passing through acetyle-CoA is very narrow under light condition. On the other hand, PDO can be produced from DHAP which is main intermediate product in Calvin cycle. To achieve PDO production in cyanobacteria, five genes were introduced into the host strain, which produces 1.3 PDO from DHAP through glycerol. We call this strain TA2984. Using this strain, batch experiments were performed in the same way. Cell growth was successful and maximum PDO concentration was approximately 0.7 mM/L. It should be noted that the higher production of glycerol was observed. Their production was clearly associated with cell growth. This characteristics allows us to try continuous process. So we are now examining the possibility of continuous process using immobilized cyanobacteria. As a bioreactor for continuous process, an airlift fermenter with an internal draft tube was used. Successful growth was observed using an airlift bioreactor, maximum PDO concentration increased to approximately 1.8mM.

466401 Kinetic Modelling of Microalgae Growth for the Optimization of Starch and Lipid Production

Abstract:

Kinetic Modelling of microalgae growth for the optimization of starch and lipid production

Gonzalo M. Figueroa-Torres [a,b], Jon Pittman[c] and Constantinos Theodoropoulos[a,b]

[a]School of Chemical Engineering and Analytical Science, University of Manchester, M13 9PL, UK

[b]Centre for Process Integration, University of Manchester, M13 9PL, UK

[c]Faculty of Life Sciences, University of Manchester, M13 9PL, UK

Microalgal biomass has been regarded as a sustainable and renewable feedstock for biofuel production due to its ability to naturally accumulate carbohydrates - mainly in the form of starch – and lipids. These carbon-based compounds can potentially be used as raw substrates for biofuel production (Brennan & Owende, 2010). Studies have shown that carbohydrate and lipid content in microalgae cells is positively influenced under conditions of nutrient limitation. However, this increase in starch and lipid content is coupled with a decrease in biomass production (Markou, Angelidaki, & Georgakakis, 2012), an adverse effect on the overall cultivation process. An efficient approach towards improving starch and lipid production by means of a cultivation strategy involves the use of kinetic models capable of predicting the main compositional elements of microalgae growth (Bernard, 2011). However, there have been very limited modelling efforts in the literature regarding the simultaneous production of starch and lipids by microalgal biomass. Thus, the aim of this work is to develop a multi-parameter kinetic model for the optimization of starch and lipid production. The proposed model considers a set of intracellular carbon flows between two main cellular compartments: a pool of active biomass and a storage pool comprising of starch and lipids. Model parameters were fitted against experimental datasets generated from lab-scale cultivation of *Chlamydomonas reinhardtii* CCAP 11/32C in standard Tris-Acetate-Phosphate (TAP) media under different concentration regimes. Fitting was carried out through an in-house developed optimization algorithm. Validation of the model was then carried out against different experimental datasets. The model can be used to predict three carbon-based cellular pools – starch, lipids, and active biomass – as well as nutrient consumption and pH evolution, useful factors for the establishment of optimal cultivation conditions.

Bernard, O. (2011). Hurdles and challenges for modelling and control of microalgae for CO2 mitigation and biofuel production. *Journal of Process Control*, *21*(10), 1378–1389. doi:10.1016/j.jprocont.2011.07.012

Brennan, L., & Owende, P. (2010). Biofuels from microalgae—A review of technologies for production, processing, and extractions of biofuels and co-products. *Renewable and Sustainable Energy Reviews*, *14*(2), 557–577. doi:10.1016/j.rser.2009.10.009

Markou, G., Angelidaki, I., & Georgakakis, D. (2012). Microalgal carbohydrates: an overview of the factors influencing carbohydrates production, and of main bioconversion technologies for production of biofuels. *Applied Microbiology and Biotechnology*, *96*(3), 631–645. doi:10.1007/s00253-012-4398-0

469321 Enhanced Biomass and Lipid Productivities of Outdoor Alkaliphillic Microalgae Cultures through Increased Media Alkalinity

Abstract:

Alkaliphillic microalgae have the ability to thrive under high pH and alkalinity conditions. The phototrophic biomass productivity of alkaliphillic microalgae can be improved by increasing availability of bicarbonate (HCO_3^-) at extreme pH conditions (pH >10.0) through increase in media alkalinity. The high alkalinity would allow the maintenance of sufficient â€œHCO_3^-Â bufferâ€ in solution while high media pH would allow rapid transfer of atmospheric CO_2 to compensate for HCO_3^-Â uptake by photosynthetic mechanism. Thus, growth would not get restricted by inorganic carbon limitations. Further, extreme pH conditions would permit mixotrophic cultivation with sugars present in the medium.

In the present study, phototrophic and mixotrophic cultivation strategies were implemented under high alkalinity and pH conditions for alkaliphillic *Chlorella Sp.* (strain SLA-04) in 1000L outdoor raceway ponds. Under the phototrophic conditions, when the culture concentrations were below 0.6 g/L, biomass productivities of ~20 g/m^2/day were obtained. While some of the HCO_3^- was utilized, ~ 100% of the carbon accumulated in biomass came from atmospheric CO_2. Under mixotrophic conditions, SLA-04 cultures grew without any measurable signs of contamination. In addition, due to availability of additional carbon source as well as the reducing equivalents, biomass productivity was significantly higher than phototrophic conditions (>57 g/m^2/day). Lipid productivities were also higher for cultures grown under mixotrophic conditions (7 g/m^2/day) when compared to phototrophic SLA-04 cultures (2 g/m^2/day). Thus, the use of alkaliphillic microalgae for cultivation, would result in increased carbon capture from atmosphere and thereby reduces the costs accompanied with CO_2 sparging (usually associated with mesophillic

cultures).

469333 Macroemulsion Mixture Assisted Organic Solvent Extraction of Algal Lipids from Unconcentrated Algal Culture

Abstract:

Petroleum fuels are non-renewable resources that is rapidly depleting. To meet the demand for energy while slowing the growth of petroleum fuel consumption, there is a growing interest in algal biofuels as a sustainable alternative form of energy. Algal lipid extraction is considered a major bottleneck to biofuels derived from microalgae due to their extremely small size, tough cell wall, suspended in dilute cultures, and cell density similar to that of the surrounding growth medium (*1*). Conventional methods of algal lipid extraction generally requires long extraction times, dried algal biomass with little water content, and high energy inputs (*2*). Therefore, the development of an effective and energy efficient algal lipid extraction process is critical for a successful implementation of microalgae based biofuel.

This study demonstrates a procedure for lipid extraction from microalgae culture using a mixture of macroemulsion and organic solvents at room temperature. This method does not require harvesting and dehydration of the algal biomass. The effects of extraction variables were investigated. The results showed that the macroemulsion mixture assisted solvent extraction method extracted 83-87% of total recoverable lipids at a mixing time of less than 1 min. No significant differences in terms of distribution of fatty acid profile have been obtained when compared to the Bligh and Dyer method. Evaluation of energy consumption indicates a substantial reduction in extraction costs compared to other methods.

Reference

(*1*) **N. Uduman**, **Y. Qi**, **M. K. Danquah**, and **A. F. A. Hoadley**, "Marine microalgae flocculation and focused beam reflectance measurement," *Chem. Eng. J.*, **162** (3), pp. 935–940, (2010).

(*2*) **R. Halim**, **M. K. Danquah**, and **P. A. Webley**, "Extraction of oil from microalgae for biodiesel

production: A review," *Biotechnol. Adv.*, **30** (3), pp. 709–732, (2012).

469398 An Ecological and Engineering Approach to Optimizing Algal Biofuels

Abstract:

Algae are ubiquitous in natural ecosystems and have been studied extensively due to their versatility for biofuel production. Most studies to date have been conducted on the grounds of synthetic biology and process engineering with few industrial scale projects considering algal community interactions. Such interactions have often indicated the propensity of increasing overall productivity and reducing community invasability, both important characteristics for scalable projects. Over 30,000 species of algae have currently been identified with another 20,000 estimated to exist. Within this context, elucidation of these relationships remains extremely resource and time intensive. This research describes a strategy for rapid, high-throughput screening of algal community combinations using a microfluidic platform to synthesize millions of parallel, nanoliter-scale algal communities for analysis of biomass accumulation. Model communities were first studied in a bench scale flask experiment and then examined using microfluidic droplets. These experiments showed consistent results on both positively and negatively interacting algal bicultures. Specifically, these include better performance of two bicultures, *Ankistrodesmus falcatus* and *Chlorella sorokiniana*, *Chlorella sorokiniana* and *Selenastrum minutum*, and reduced performance of a biculture consisting of *Selenastrum capricornutum* and *Scenedesmus ecornis*. While the holistic success of algal biofuels hinges on an amalgamation of scientific fields, rapid screening of algal communities will prove imperative for discovering community interactions and ideally facilitating a mechanistic analysis of how such interactions arise in natural communities.

471918 Ecosystems As Unit Operations: Designing Integrated Networks of Technological and Ecological Systems

Abstract:

One of the requirements to meet the goal of sustainable design is to ensure that impacts of engineering systems do not exceed the regenerative capacity of ecological systems [1], in order to prevent unintended harm. Industries striving towards sustainable development also aim to become carbon and emissions neutral, with zero water withdrawal. Many companies aim to go even beyond such goals by having "net positive" manufacturing systems. While the conventional approach to meeting this goal has been to increase the use of renewable fuels, optimize the system efficiency to reduce emissions and rely on end-of-pipe pollution control methods, such solutions tend to neglect the ecological overshoot resulting in some perverse outcomes.

In this work, we present an approach to designing coupled techno-ecological systems in order to minimize ecological overshoot within an "industrial-serviceshed", by including ecological systems in a manner analogous to unit operations within the engineering design framework. As a result, simultaneous decisions are made regarding the design of engineering systems to minimize the ecosystem service demand and the design of supporting ecological systems to supply the necessary ecosystems services demanded by industrial sites. Sustainability indices for such systems are quantified using the TES (Techno-Ecological Synergy) framework [1], as a function of the design variables and the objective is to identify optimal coupling between Technological and Ecological systems, to minimize the overshoot.

Traditional engineering unit-operation models capture the demand for ecosystem service, while the supply is quantified by spatially explicit unit-operation models of ecosystems. Similar to process level constraints regarding the plant operating conditions, feedstock availability etc. ecological constraints like land availability, type of ecological systems and other geo-spatial constraints based on remote-sensing and aerial survey are identified and included within the design problem. The objective of TES system design is to

maximize the net ecosystem service supply, quantified by the ecosystem service supply less the demand as opposed to the conventional approach of minimizing only the ecosystem service demand.

The TES design framework is applied to the design of a biofuel production system. Ecological systems included within the design framework include a forest ecosystem captured using the Urban Forest Effects model [2] and a horizontal subsurface flow wetland systems for treating the wastewater, designed using steady-state first order equations of an ideal plug flow reactor [3]. The problem is formulated as a multi-objective nonlinear program (Mo-NLP) and tradeoffs are identified between maximizing the net present value of the system and the net ecosystem service demand for multiple ecosystem services. The dynamic behavior of ecological systems and changes in the design and operation of systems are captured using a multi-period approach. Seasonal variations due to weather fluctuations are also bound to cause a change in ecological processes and thus robust designs are identified at each time-period accounting for these external disturbances. The problem formulation also accounts for the uncertainty associated with the critical design and operating variables.

Preliminary results from the case study demonstrate that systems designed by accounting for ecosystem service supply results in novel designs that have a higher economic value, and higher sustainability index, resulting in "win-win" designs. Thus, coupled TES systems are expected to have economic and environmental advantages over conventional systems design, mainly due to the reduction in external inputs by moving towards a closed loop production system.

References

[1] Bakshi, Bhavik R., Guy Ziv, and Michael D. Lepech. "Techno-ecological synergy: A framework for sustainable engineering." Environmental science & technology 49.3 (2015): 1752-1760.

[2] Nowak, David J., and Daniel E. Crane. "The Urban Forest Effects (UFORE) Model: quantifying urban forest structure and functions." (2000).

[3] Rousseau, Diederik PL, Peter A. Vanrolleghem, and Niels De Pauw. "Model-based design of horizontal subsurface flow constructed treatment wetlands: a review." Water research 38.6 (2004): 1484-1493.

459809 Economic, Environmental and Social Cost Optimization of Biomass Supply Chain for Electricity Generation in WI, USA

Abstract:

There is an increasing interest in identifying the optimal biomass supply chain for the renewable energy system, due to relatively low energy density and high procurement cost of biomass feedstock. The goal of this study is to develop an integrated framework to determine the optimal supply chain network for bioelectricity generation in Wisconsin, USA. First, forest residues collection sites, road and rail network across WI, MN and upper MI were mapped in Arc-GIS. Next, the optimal location of co-generation facility was identified by screening all potential sites by government policy, local community interest, land availability, distance to transmission lines and road/rail network. Finally, biomass supply chain was optimized by minimizing the economic, environmental and social cost of shipping forest residues from forest to the facility. Upstream (cradle-to-gate) GHG and criteria pollutants emissions of feedstock were also quantified for all potential facility sites. The result shows that Superior WI is the best location to build a bioelectricity facility in terms of economic, environmental, and social sustainability. In addition, the quantities, modes and distances of forest residues transportation were determined in order to meet the 0.5 million ton (current capacity) and 1 million ton (future capacity) demand of the bioelectricity facility.

472967 Renewable/Fossil Fuel Supply Selection and Supply Chain Optimization for a Synergized Upstream, Downstream and Power Economy System

Abstract:

The current environment of low oil prices constitutes a substantial headwind for much of the energy value chain, especially for energy systems where the upstream, downstream, and power sectors are synergized with the aid of governmental subsidies. In such cases, the economic viability of the downstream and power sectors depends heavily on the technology costs and the government regulation on oil prices, which are often fixed. In the current decade, energy systems with an abundant supply of oil were able to achieve sizable margins in the downstream sector within the high oil price environment. However, the uncertainty of future oil prices makes it difficult to put forward strategies for the energy and petrochemical sectors. An energy model that takes into account the incorporation of renewable resources mixed with fossil fuels, a dynamic selection of hydrocarbon prices, and a flexible downstream sector that is capable of liquid or gas feedstock inlets can lessen the burden that oil price fluctuations have. In this work, we present a model where static and dynamic optimization techniques are applied to the dynamic identification of prices that optimize supply decisions and minimize fuel and technology costs within a synergized upstream, downstream and power economy system. Emphasis is placed on understanding the role of modeling in the energy sector and understanding energy balances to develop a modern integrated energy planning model.

471652 Synthesis of Renewable Energy Supply Networks Considering Different Frequencies of Fluctuations in Supply and Demand

Abstract:

The demand for and dependency on fossil resources, coupled with growing environmental consciousness and new economic opportunities and developing/utilising clean and renewable energy sources on a larger scale, are among the major challenges and priorities of our society. In 2015 at least 164 countries had renewable energy targets, and most also had renewable energy support policies (Sawin et al., 2015). Among the most promising technologies to help achieve these goals are those using solar irradiance, wind and biomass (Davis and Martín, 2014).

In this study, a biorefinery supply network integrating wind and solar energy (Čuček et al., 2015), involving several biomass and waste sources for the production of biofuels and food (Čuček et al., 2014), is further extended by: i) applying more detailed models for wind and solar energy with more realistic data, and ii) better accounting for intermittent renewable energy sources, especially of wind and solar energy, and electricity demands. The dynamic model corresponds to a multi-period model, and considers different frequencies in supply and demand. In those cases where there is a more pronounced dynamic, the time interval selected is shorter. For supply and demand with longer-term fluctuations, the time interval is longer. Different time periods are applied leading to supply network models with a large number of variables and constraints, and to long computational times. For biomass and biofuels the time periods are longer, while for solar, wind and electricity consumption, the time intervals are shorter.

The proposed model is demonstrated on a case study of a relatively small region in Europe. Several scenarios are considered regarding possible percentage share of renewables in the future energy mix, preferably reaching 100% of renewable energy systems. Furthermore, a sensitivity analysis is performed regarding the availability of energy sources, demand, and prices of biomass, energy technologies and products.

Acknowledgement

The authors acknowledge the financial support of the Slovenian Research Agency (programs P2-0032 and P2-0377), from a bilateral project between Slovenia and the United States entitled "Development of Methodology for the Synthesis of (Bio)chemical Supply Chains," from the Ministry of Science, Education and Sports of the Republic of Croatia and the EC under project CARBEN (NEWFELPRO Grant Agreement No. 39), and from a bilateral project between Slovenia and Croatia entitled SLOCRORES, "Planning of 100% RES communities by use and combination of Total Site Integration and Renewislands methodologies."

References

Čuček, L., Martín, M., Grossmann, I.E., Kravanja, Z., 2014. Multi-period Synthesis of Optimally-Integrated Biomass and Bioenergy Supply Network. Computers and Chemical Engineering 66, 57-70.

Čuček, L., Martín, M., Kravanja, Z., 2015. Integration of Wind and Solar Energy within Continental Biorefinery Supply Network. Proceedings of 2015 AIChE Annual Meeting, November 8 – 13, 2015, Salt Lake City, UT, USA.

Davis, W., Martín, M., 2014. Optimal year-round operation for methane production from CO_2 and water using wind and/or solar energy. Journal of Cleaner Production 80, 252-261.

Sawin, J.L., Sverrisson, F., Rickerson, W., Lins, C., Musolino, E., Petrichenko, K., Rickerson, W., Sawin, J., L., Seyboth, K., Skeen, J., Sovacool, B., Williamson, L.E., 2015. Renewables 2015 - Global status report, Key findings 2015, Paris, France.

469724 A Holistic Look to Carbon Capture and Use in Sustainable and Economically Enhanced Bio-Based Supply Chains

Abstract:

Carbon capture and storage is considered worldwide as an alternative to take into account for cost-effective mitigation of CO_2 emissions. Notwithstanding, recent years have shown significant drawbacks associated with carbon capture and storage (CCS) to secure the CO_2 geological storage from industrial sources or power plants. Geological storage faces the possibility of leakage, liability issues, long-term problems with public acceptance of storage sites on land, and limited cost-effective storage capacity in some key regions (Cuéllar and Azapagic, 2015). Otherwise, sustainability concerns, evidence that a more holistic view is necessary. This includes the whole energy supply chain and a deeper insight on the protagonist role of carbon capture and utilization (CCU), specifically for the chemical industry, and as a part of the share of the environmentally friendly technological options of the future.

CCU provides a CO_2 management solution for a limited amount of CO_2. For instance, the CO_2 generated by a single coal power plant could potentially supply much of the CO_2 demand for CO_2-based methanol plants that supply a growing demand in Europe. This would avoid the use of fossil fuel, like natural gas, which is used nowadays in the conventional synthesis of methanol (Pérez-Fortes et al., 2016).

This work combines a bio-based energy supply chain with CO_2 capture and utilization, for economic optimisation based on a CO_2 emissions constraint. The problem is formulated as a mixed-integer linear program in GAMS. This study presents a complete and flexible supply chain superstructure, as well as the supporting framework developed to answer pending issues on the actual economic and environmental benefits achievable in this complex scenario for both, the biomass-coal plant and the CO_2 utilization plant; i.e. location of intermediate steps, fraction of CO_2 emissions captured per plant, or size of the CO_2 utilization plants, among others. Moreover, eventual incentives and environmental revenues are discussed, to make an economically viable business case. A large-size, real case study is used in this analysis. Two key scenarios are taken into account: (i) No use of biomass, neither capture nor utilization of CO_2 is contemplated; (ii) Biomass, capture and CO_2utilization are all contemplated. Finally, conclusions and guidelines are given.

References

Cuéllar-Franca R, Azapagic A. Carbon capture, storage and utilisation technologies: A critical analysis and comparison of their life cycle environmental impacts. *Journal of CO2 Utilization*. Elsevier. Volume 9, March 2015, Pages 82â€"102.

Available from: doi:10.1016/j.jcou.2014.12.001

Pérez-Fortes M, Laínez-aguirre JM, Puigjaner L. Economic and environmental assessment of CO_2 capture and utilisation technologies in bio-based supply chains. *Computers and Chemical Engineering*. 2016; in press.

.

459900 Optimal Integration of Renewable Based Processes for Fuels and Power Production: Case Study in Spain

Abstract:

Although renewable resources are plentiful, their sources are scattered. Therefore, to meet a certain demand of power and fuels using renewable sources, and mitigate the variability in their availability, process integration at local or greater scale has become an interesting engineering and technical challenge. Weekman (2010) and Yaun and Chen (2012) presented overviews regarding the integration possibilities as a perspective for the future combination of different sources of energy. In order to help with those decisions, process system engineering has the tools to compare sources and allocations in search for the optimal options at local and wider scales. Most of the studies either focus on fuels supply (Cucek et al 2014, Elia et al 2012) or power supply based on the unit commitment problem (Saravanan, et al 2013). However, power is most of the times produced by using a fuel, which is a chemical that can also be synthesized. Therefore, both supply networks are linked and must be addressed simultaneously. Over the last years plant size integration of renewable resources has been considered (Vidal and Martín, 2015, Martín and Davis, 2016, Martín and Grossmann, 2016). However, the integration at a larger scale, beyond the plant level, has not been attempted.

In this work we have developed a framework for the optimal integration of renewable sources of energy to produce fuels and power. A network is developed using surrogate models for various technologies that use solar energy, PV solar, CSP or algae to produce oil, wind technology, biomass based syngas to ethanol, methanol, FT-liquids and thermal energy, hydroelectric power and waste based power plant via biogas production. Hydrogen can be produced if there is a surplus of energy. It can be stored, by producing methanol or methane. Methane can be further used to produce power at need, while methanol is employed for oil transesterification to produced biodiesel. In both cases, the carbon source is CO_2, that has been captured during the gasification based processes or it imported from outside the network. The corresponding processes are modeled using an input- output approach based on the results of detailed optimization studies previously developed by the authors. However, we do not model an entire process from biomass to fuels as a black box, but we break it into sections that produce intermediates that can be used for a different purpose, i.e. syngas can be used to produced ethanol FT-liquids, etc, or because various technologies are available for that; i.e. dry or wet cooling for power plants. The model is formulated as an MILP that allows determining the optimal selection of technologies to meet certain demand. Both cost and environmental objective functions are considered.

We apply the network to evaluate process integration at different scales, from region level, where we also evaluate the effect of uncertainty in renewable resources, up to the level of an entire country (Spain). Solutions under uncertainty are more robust and expensive, about 25% higher requiring the use of further technologies to meet the demand. For Spain, and using 1% of the area, it is possible to substitute 20% of fossil fuels for transportation and 100% of the power demand. The solution suggests the use of Hydropower, and oil production in all regions, while bioethanol biodiesel plants are allocated close to demand points such as large population areas. Up to 44% of fuels and total power can be substituted with the current technology development status. However, we can only reach 90% substitution by using 20% area with the availability and efficiency of current processes. The investment required for this option is more than twice that to reach 20% fossil fuels substitution.

References

Cucek, L., MartÃn, M., Grossmann, I.E., Kravanja, Z (2014) Large-Scale Biorefinery Supply Network â€" Case Study of the European Union. Comp. Aid. Chem. Eng. Vol33 , 319-324.

Elia, J.A., Baliban R.C., Floudas, C.A. (2012) Nationwide energy supply chain analysis for hybrid feedstoch processes with significant CO2 emissions reduction. AIChE J. 58 (7) 2142-2154.

Martín, M., Grossmann, I.E. (2016) Optimal integration of a self sustained algae based facility with solar and/or wind energy. submitted *Energy*.

Martín, M, Davis, W. (2015) Integration of wind, solar and biomass over a year for the constant production of CH_4 from CO_2 and water Comp Chem Eng 84, 314-325,

Saravanan, B., Da, S., Sikri, S., Kpothari, D.P. (2013) A solution to the unit commitment problem-a review. Frointiers in Energy. 7(2) 223-236

Vidal, M., Martín, M.(2015) Optimal coupling of biomass and solar energy for the production of electricity and chemicals.Comp. Chem. Eng 72, 273-283

Weekman, V.W., 2010. Gazing into an energy crystal ball. CEP June 2010; 23-27

Yuan, Z., Chen, B., 2012. Process Synthesis for Addressing the Sustainable Energy Systems and Environmental Issues. AIChE J. 58 (11), 3370-3389

471238 Technoeconomic Analysis of Conversion of Corn Fiber to Ethanol Using Engineered Thermophilic Bacteria

Abstract:

Corn fiber - containing arabinose, xylose, and cellulose in roughly equal amounts and not currently fermented - could in principle be converted to ethanol with little or no loss in the value of residual solids as an animal feed. The challenge is to solubilize and ferment this unusual biomass feedstock in a cost-effective manner. We present data indicating that corn fiber can be nearly completely solubilized by *Clostridium thermocellum* whereas solubilization yields using controls with industry-standard fungal cellulase are on the order of 50%. Data on conversion of arabinose and xylose by engineered strains of *Thermoanaerobacter thermosaccharolyticum* will also be presented.

In the main part of this presentation, we present a technoeconomic analysis of thermophilic conversion of corn fiber. In particular, we analyze the parametric impact of key performance parameters on payback period, including solubilization yield (fractional sugars solubilized), fermentation yield (ethanol produced per sugar fermented), and ethanol titer. Integrated separation of a dilute ethanol stream resulting from thermophilic corn fiber fermentation with a larger flow of more concentrated ethanol produced from corn starch fermentation is analyzed. It is concluded that economic viability can be achieved at lower ethanol titer in a bolt-on configuration as compared to a stand-alone configuration. Results support corn fiber conversion as a low-capital/high return point-of-entry for industrial application of engineered thermophilic microbes.

468353 Assessing Technological, Economic and Environmental Performance of Bio-Based Chemicals

Abstract:

Biomass is a renewable resource which can be converted into various marketable products using several conversion processes. As the research is taking strides in populating techniques to produce chemicals and fuels via multiple biochemical routes, it is critical to appraise the processes for its long term viability as a sustainable source. Sustainable chemical targets must be identified prior to investing time, money and manpower resources to nurture these. This unique contribution focuses on developing a multi-sector, multi-scale framework which can be used to evaluate new technologies based on its economic and environmental potential. The framework facilitates channeling information across multiple scales throughout the workflow of process development. Generated parameters are coupled to an optimization framework which is capable of accessing processes with respect to the contemporary technologies based on fossil derived feedstock. The resulting decision support tool can be used to study the existing market and test viability of bioprocesses while accounting for inter and intra sector interactions in the value chain. The traits of this tool, bolstered with detailed simulation models, make it an appropriate venue to test various renewable process technologies, provide insights towards future production strategies and guide the research to develop bio-based products with commercial potential.

468351 The Effect of Technology Choices on the Biorefinery Value Chain: Facile Analysis Using the Value Chain Assessment Platform (VCAP)

Abstract:

Due to the high production cost of ethanol from lignocellulosic biomass, it is anticipated that lignocellulosic conversion will focus on biorefining to numerous fractions for the chemical production industry. Lignocellulosic biomass consists mainly of three key polymers, lignin, cellulose and hemicellulose which can be further converted to a number of feedstock and end user chemicals. To this end, value chain optimisation analysis may be used to determine the most profitable scenario. Previously, a Value Chain Assessment Platform (VCAP) was presented which is a platform that facilitates the assessment of biorefinery value chains.[1] VCAP combines a user friendly interface and an optimisation algorithm into a web service that is easy to use. This work focuses on the application of VCAP for analysis of technology choices concurrently with value chain analysis for lignocellulosic biomass.

Currently, in most configurations of lignocellulosic conversion, the hemicellulose and cellulose sugar streams are used for conversion to ethanol while the lignin stream is either recycled for energy or discarded. In order to move toward a sustainable, circular bioeconomy, the utilisation of each stream must be considered. However the conversion of lignin to high value products often occurs at cost to the process either through more expensive technologies required or lower yields of products from hemicellulose and cellulose. Hence, in an ethanol refinery, lignin is used as an energy source rather than a feedstock for chemical products. This work will explore the conversion of lignocellulose to ethanol using two different biorefinery configurations. Firstly, a lignocellulosic ethanol plant, with steam pretreatment will be considered, which recycles solids and lignin for heat and electrical energy production.[2] Secondly, an ionic liquid biorefinery will be considered, which is able to produce high value lignin alongside ethanol.[3] VCAP will then be used to evaluate these two options, as well as the options for lignin use as heat or for conversion to value added products, with respect to the holistic assessment of the value chain for a softwood biorefinery. The value chain analysis

considers options with respect to technology location to either capitalise on waste heat from a power station or to minimise minimises transport cost with alternate technology configurations in the context of the technology configuration options. The implications of these results will be discussed in the context of the needs of biorefinery value chain modelling toward a circular bioeconomy configuration.

1. Trokanas N, Bussemaker M, Day K, Drage G, Cecelja F. (2015) 'VCAP: A Software Platform for the Assessment of Biorefinery Value Chains'. Salt Lake City, UA, USA: 2015 *AIChE Annual Meeting*
2. Huang HJ, Ramaswamy S, Al-Dajani W, Tschirner U, Cairncross RA. Effect of biomass species and plant size on cellulosic ethanol: A comparative process and economic analysis. *Biomass and Bioenergy*. 2009;33(2):234-246. doi:10.1016/j.biombioe.2008.05.007.
3. Klein-Marcuschamer D, Simmons BA, Blanch H. Techno-economi analysis of a lignocellulosic ethanol biorefinery with ionic liquid pre-treatment. *Biofuel Bioprod and Bioref.* 2011;5:562-569.

452201 Estimated Economic Impacts of Biofuel Production - a Comparative Analysis of Three Cellulosic Biofuel Pathways

Abstract:

The development of a cellulosic biofuel industry utilizing domestic biomass resources is expected to create opportunities for economic growth resulting from the construction and operation of new biorefineries. We applied an economic input-output model to estimate potential economic impacts, particularly gross job growth, for three different biofuel conversion pathways: 1) cellulosic ethanol via biochemical conversion in Iowa, 2) renewable diesel blendstock via biological conversion in Georgia, and 3) renewable diesel and gasoline blendstock via fast pyrolysis in Mississippi.

Combining direct, indirect (revenue- and supply chain-related), and induced effects, capital investment associated with the construction of a biorefinery processing 2,000 dry metric tons of biomass per day (DMT/day) could yield between 5,960 and 8,470 full-time equivalent (FTE) jobs during the construction period. Fast pyrolysis biorefineries produce the most jobs on a project level thanks to the highest capital requirement among the three pathways. Normalized on the scale of one million dollars of capital investment, the renewable diesel and fast pyrolysis biorefineries are estimated to yield the same number of jobs, i.e., 11.8 jobs. This level is slightly higher than cellulosic ethanol biorefinery, which results in 11.5 jobs per million dollar capital investment.

While operating biorefineries is not labor-intensive, the annual operation of a 2,000 DMT/day biorefinery could support between 720 and 970 jobs when the direct, indirect, and induced effects are considered. The major factor, which results in the variations among the three pathways, is the type of biomass feedstock used for biofuels. Unlike construction jobs, these operation-related jobs are necessary over the entire life of the biorefineries. Our results show that indirect effects stimulated by the operation of biorefineries are the primary contributor to job growth. The agriculture/forest, services, and trade industries are the primary sectors that will benefit from the ongoing operation of biorefineries.

467950 Total Acid Number (TAN) Analysis of Bio-Oil and Tan Balance over Bio-Oil Neutralization

Abstract:

Bio-oil or pyrolysis oil has the potential to replace fossil fuels. However, its low pH or high acidity limits its applications. To overcome this challenging property of bio-oil, we investigated an approach to neutralize bio-oil by adding alkaline solutions or solids. To evaluate the neutralization of bio-oil, measuring the acidity of samples beyond a simple pH measurement is essential. While pH can represent the concentration of hydrogen ions, it cannot identify or quantify acidic components in samples. Thus, chemical analyses, such as high-performance liquid chromatography and gas chromatography-mass spectrometry, become necessary. Yet, there is lack of consistency among different laboratories from the round-robin testing. Moreover, chemical analysis often underestimates the amount of total acids. An analysis of the total acid number (TAN)â€"the amount of potassium hydroxide needed to titrate one gram of an oil sampleâ€"was initially developed for analyzing petroleum products but has also been used for analyzing the acidity of biofuels including biodiesel and bio-oil. The objectives of this study are to investigate the relationship between the total acid number and the concentrations of acidic components in bio-oil samples; to explore whether a variety of chemicals present in bio-oil have any influence on TAN analysis; and to verify whether TAN balance can be achieved over the neutralization of bio-oil, which was performed by adding different alkaline solutions and solids in batch and continuous-flow systems using a static mixer and a centrifugal contactor. The TAN values of standard solutions (e.g., acetic acid, propionic acid, phenol, vanillic acid) were analyzed according to the ASTM D664 standard method with a minor modification of the electrode cleaning procedure. Aqueous bio-oil samples with known amounts of acetic acid added were also analyzed. To examine the TAN balance, the aqueous bio-oil was neutralized by adding alkaline solutions or solids, and the TAN values of neutralized aqueous and organic phases formed were measured. The results showed that the molar concentrations of carboxylic acids (e.g., acetic acid and propionic acid) had the same linear relationship with the TAN values. However, a different linear relationship was found for a phenolic acid possibly due to the presence of functional groups that contribute to the TAN value. Also, bio-oil samples with added acetic acid showed a proportional increase in the TAN value, and demonstrated that the bio-oil components did not interfere with the TAN analysis. The results from the neutralization in batch systems were comparable with those from continuous-flow systems. The results of this study indicate that the TAN of a sample with the carboxylic acids without phenolic acid can be converted to the molar concentration of total acids. A sample with phenolic acid may not be converted to the molar concentration of total acids because the phenolic acid has a higher influence on the TAN values. The TAN balance on the neutralization of bio-oil did not close. This behavior may be due to the exothermic reaction during neutralization. Specifically, heat during the neutralization process may have accelerated the aging process of bio-oil, thus increasing the acidity and the water content of the system.

472158 Development of a Thermoreversible Medium to Culture and Efficiently Harvest Microalgae

Abstract:

With the ability to generate different types of biofuels and a variety of valuable chemicals, microalgae represent a promising alternative to fossil fuels. However, the development of high throughput cultivation and efficient harvesting techonologies is necessary to make microalgal biomass production a more viable option. Herein, a new Tris-Acetate-Phosphate-Pluronic (TAPP) medium with thermoreversible sol-gel transition properties developed for cultivation and efficient harvesting of microalgae will be presented.

This sol-gel transition phenomenon refers to the transformation from a low-viscosity fluid to an elastic solid upon heating. For the TAPP medium developed, this gelation process is completely reversible upon cooling. The thermo-rheological properties of the new pluronic-based medium as well as the resulting pluronic-microalgae matrix after cultivation are systematically characterized. moreover, a framework is proposed to efficiently harvest the microalgal biomass produced through small variations of temperature.

Microalga *Chlamydomonas reinhardtii* is successfully cultivated in the new TAPP medium and leads to equivalent production of microalgal biomass with similar lipid and carbohydrate composition than that obtained from cultivation in the traditional TAP medium. The harvesting process of the microalgal biomass produced is also highly simplified with the TAPP system. In fact, confinement of microalgal cells in the pluronic matrix leads to a cluster distribution that increases the settling velocity by a factor of ten. Through small variations of temperature, microalgae are allowed to settle, the broth subsequently jellified and ninety percent of microalgal biomass are harvested simply through scrapping off the surface.

The implications of this new cultivation and harvesting technology as well as the potential for varieties of large scale

applications will be discussed.

472266 Bioprocessing Strategies for Engineered Coproduction of Biofuels and Medical Biopolymers from Diatom Microalgae

Abstract:

Photosynthetic diatom microalgae have significant capacity for biosynthesis of energy-dense biofuel molecules, as well as unique valued co-products not found in other types of algae, including metal oxide nanomaterials for advanced material applications, and glucosamine biopolymers or monomers for nutraceutical and biomedical applications. Diatoms biomineralize soluble silicon to nanostructured biosilica, and require dissolved silicon (Si) as a required substrate for cell wall biosynthesis and division. Bioprocess engineering strategies have the potential to guide the cellular biosynthesis of three aforementioned product streams, all within the same diatom cell. To exploit diatom silicon metabolism for eliciting the biosynthetic pathways of selected products, a two-stage cultivation process was developed to induce high levels of lipid and chitin production by the centric marine diatom *Cyclotella* within under tightly controlled conditions where light and CO_2 delivery are not limiting. Carbon dioxide delivery and consumption were measured in real time to determine carbon flux into biomass, lipid, and chitin biopolymer products. The two-stage cultivation process synchronized *Cyclotella* diatom cells to silicon-starved state in Stage I. In Stage II, the cells were maintained in a nominal silicon-starved state under controlled nutrient perfusion, which allowed for continued cell division, carbon dioxide assimilation, and product formation. In this cultivation mode, cell biomass, lipid, and chitin production were elicited and linearly maintained at silicon depletion without the need for nitrogen depletion. Intracellular accumulation of lipid and extracellular extrusion of chitin microfibrils were evident. Optimal stage II product yields associated with the biomass were 34 wt% total lipid and 16 wt% chitin, with 60 mol% of assimilated carbon dioxide allocated to these two products. From this information, a process flowsheet and material balance on the diatom-based photosynthetic bio-refinery for production of the nutraceutical glucosamine with co-production of biodiesel and biosilica was developed to illustrate the productivity at scale. This analysis shows that the diatom-based photosynthetic biorefinery has significant potential as a future platform for biofuels and unique, valuable co-products.

454396 Effects of Microalgal Polycultures on Quality of Biomass for Biocrude Oil Production Via Hydrothermal Liquefaction

Abstract:

Rising atmospheric carbon dioxide levels have created a continuously increasing demand for sustainable energy. Algal biomass as a feedstock for sustainable energy production is a compelling option, as algae cultivation does not necessitate arable land and algae have higher energy density than terrestrial plants [1]. Moreover, one biomass-conversion technology called hydrothermal liquefaction (HTL) can convert wet microalgae into an energy-dense biocrude oil without incurring energy penalties associated with drying the biomass [2, 3].

Raceway ponds are a primary method of growing microalgae, although the open nature of these ponds subjects algal cultures to external perturbations like light and temperature variability, predators, diseases, and invasive species, leading to crop instabilities [4, 5]. Multi-species communities, or polycultures, could potentially be more resistant to these environmental factors and thus more stable over time [6, 7]. In a recent study, we used HTL to assess the biocrude productivity and stability of 2-, 4-, and 6-species polycultures of *Ankistrodesmus falcatus, Chorella sorokiniana*, *Pediastrum duplex, Scenedesmus acuminatus, Scenedesmus ecornis,* and *Selenasatrum capricornutum*, compared to the monocultures of these species [8]. This study found that polycultures generally produced biocrude more stably than monocultures but not in higher amounts. However, the quality of the biomass feedstock and the resulting biocrude have yet to be considered, but are also important bases of comparison. Most importantly, these factors affect the extent of catalytic upgrading required before the biocrude can be inserted into existing refining infrastructure [9].

In this presentation, we expand upon earlier work [8] and compare the performance of polycultures and monocultures on several quality-based metrics, including biomass fatty acid content and biocrude yield, atomic ratios, and higher heating value (HHV). Microalgae richer in fatty acid content and biocrude with lower nitrogen and oxygen content and higher carbon and hydrogen content are more desirable. We found

that some polycultures met or exceeded the best monocultures for each metric we examined; moreover, the best monocultures and polycultures were generally different for each metric, suggesting that optimization of polyculture composition based on the desired biocrude specifications is possible. These metrics of quality provide additional bases of comparison for monocultures and polycultures beyond those of biocrude productivity and stability reported in earlier work by [8], allowing for a more comprehensive evaluation of the benefits of biodiversity on biocrude oil production via microalgal HTL.

References

1. Patil, V.; Tran, K. Q.; Giselrod, H. R. *International Journal of Molecular Sciences* **2008**, 9, 1188–1195.
2. Brown, T. M.; Duan, P.; Savage, P. E. *Energy & Fuels* **2010**, 24, 3639–3646.
3. Toor, S. S.; Rosendahl, L.; Rudolf, A. *Energy* **2011**, 36, 2328–2342.
4. Pienkos, P. T.; Darzins, A. Biofuels, *Bioproducts and Biorefining* **2009**, 3, 431–440.
5. Georgianna, D. R.; Mayfield, S. P. *Nature* **2012**, 488, 329–335.
6. Hooper, D. U.; Chapin, F. S.; Ewel, J. J.; Hector, A; Inchausti, P; Lavorel, S; Lawton, J. H.; Lodge, D. M.; Loreau,

   M; Naeem, S; Schmid, B; Setälä, H; Symstad, A. J.; Vandermeer, J; Wardle, D. A. *Ecological Monographs* **2005**,

   75, 3–35.
7. Cardinale, B. J.; Matulich, K. L.; Hooper, D. U.; Byrnes, J. E.; Duffy, E.; Gamfeldt, L.; Balvanera, P.; O'Connor,

   M. I.; Gonzalez, A. *American Journal of Botany* **2011**, 98, 572–592.
8. Narwani, A.; Lashaway, A.; Hietala, D.; Savage, P.; Cardinale, B. J. *Algal Research, Submitted.*
9. Duan, P.; Savage, P. E. *Bioresource technology* **2011**, 102, 1899–906.

455130 Enhanced Autotrophic Growth of *Nannochloris Sp.* with Buffer Chemicals for Sustainable Carbon Recycle

Abstract:

In this study, a potential ability of buffer chemicals for CO_2 gas capture and the resultant solutions were investigated for the autotrophic growth of *Nannochloris sp.* in a modified Shuisheng-4 culture medium. Sodium-based buffer chemicals used in this study well dissolve in water due to their high solubility and the buffer solutions can be converted to sodium bicarbonate solution after absorbing CO_2(g). After absorbing CO_2(g), both total inorganic carbon (TIC=HCO_3^-, CO_3^{2-}, H_2CO_3, CO_2(aq)) and dissolved inorganic carbon (=DIC=HCO_3^-+CO_2(aq)) in the solution significantly increased and the pH of the solution is equilibrated at pH=8.3 suitable for the growth of *Nannochloris sp.* As a result, high DIC concentration dramatically increased the autotrophic growth rate of *Nannochloris sp.* when the Na^+ concentration was less than 0.12 M. In addition, *Nannochloris sp.* was cultivated in the buffer chemicals system using simulated flue gas mixed with CO_2, N_2, SO_2, NO, and HCl gases. As a result, the buffer solutions could suppress the possible inhibition effect of acid gases such as SO_2, NO, and HCl on the growth of *Nannchloris sp.* This study suggests that the sodium-based buffer chemicals can be effectively used for biological CO_2 conversion and sustainable carbon recycle with pure CO_2 or flue gas from a coal-fired power plant.

455395 Effect of Process Parameters and Extractive Solvents on Hydrothermal Liquefaction of Microalgae

Abstract:

Hydrothermal Liquefaction processes proves to be an attractive technology from the view point of energy consumption, renewability and process integration for converting wet biomass into energy dense bio-oil. The yield and quality of bio-crude oil is influenced by process conditions like liquefaction temperature, solid-loading ratio, residence time, type of solvent for extraction. Therefore, we investigated the influence of these parameters on the hydrothermal liquefaction of microalgae *Micractinium sp*. The temperatures and solid-loading ratios were varied from 200 to 350°C and 0-20 % respectively in a stainless steel batch reactor. Experiments were also performed with residence times varied from 0 to 60 minutes (~15 minutes interval). The maximum bio-oil yield was obtained at 300°C and at a solid loading ratio of 10 % for 30 minutes residence time. Experiments were performed with different solvents including hexane, acetone and dichloromethane for the separation of bio-oil from product phase. Energy recovery calculations were performed with the analysis of the products (bio-oil, biochar, water soluble compounds, and gaseous phase) through ultimate and proximate analysis. Products were also analyzed by Gas Chromatography Mass Spectrometry (GC-MS) and calorimetric techniques.

Keywords: Microalgae, Hydrothermal Liquefaction, Renewability, Bio-oils/fuels, GC-MS

455641 Re-Use of Hydrothermal Liquefaction Bio-Char to Grow *Galdieria Sulphuraria*

Abstract:

Hydrothermal liquefaction is one of the promising methods to convert wet algal biomass into biofuels. Maximum biocrude oil was observed at 300 C when *Galdieria Sulpuraria* (GS) was liquefied using water as solvent. But a substantial amount of algal biomass was also converted into bio-char. The biochar produced at 300 C and 30 min of reaction time was analyzed for phosphorous content using inductively coupled plasma optical emission spectrometry (ICP-OES) and for nitrogen content using CHNOS elemental analyzer. It was observed that 1 gram of HTL biochar has 6.59mg of phosphorous and 6.36mg of nitrogen. The phosphorous and nitrogen present in this biochar can be used to cultivate microalgae. In this study, the macronutrients were leached using a 40 ml reactor at varying temperatures (23 C, 39 C and 55 C) and pH (2.5, 4.7 and 7.0) into cyanidium media without phosphates in it. This method was able to extract 54 mg/L of reactive phosphates and 16 mg/L of ammoniacal nitrogen at the optimum conditions. These leached phosphate and ammoniacal nitrogen were tested for the toxicity over 15 days and then used to grow GS in a Tissue Culture Roller drum apparatus with 16 mm borosilicate glass tube photo reactor and confirmed using a Microplate Assay at 40 C and 2-3 % CO in an incubator. A better growth was observed when leached nutrients were used compared to that of regular nutrients in the growth media.

461315 Constructing a Framework for Measuring National Energy Security

Abstract:

Energy policy in many countries, driven by concerns about energy security and environmental damage, is promoting a shift from fossil fuels to a variety of renewable sources. We examine risks to energy security arising from resource discovery and fuel processing, conversion, and use. We analyse the whole of the supply chain for renewable and non-renewable fuels, both current and potential sources, in a national context. The result is a set of indicators summarising the issues which are important for energy security.

Sustainability and energy security are concepts which share common features, some of which are poorly defined or lacking good data. We use the Process Analysis Method for systematically selecting indicators to measure a nationâ €™s energy security which includes environmental, human/social, and economic elements within a single self-consistent framework. This is a triple-bottom-line method of assessment, particularly suitable for complex systems where quantitative numerical models cannot be constructed. To use the indicators in a particular case we require a description of the energy supply and demand system, and we show how this can be constructed to meet the needs of the analysis. Appropriate energy and economic data are needed, together with assessments of issues such as political stability, international relations, and demand management. Although ill-defined, these must be included because of their potential importance. The method underlines the need for assessments and data relating to many issues which are commonly not considered as part of energy security.

Sustainability analysis of the whole value chain of bio-syngas production and its potential as chemical platform for syngas fermentation

Baudine Gevers Deynoot, Delft University of Technology, Delft, the Netherlands
Eduardo Almeida Benalcázar, Delft University of Technology, Delft, the Netherlands
Henk Noorman, DSM Biotechnology Center, Delft, the Netherlands
Patricia Osseweijer, Delft University of Technology, Delft, the Netherlands
John A. Posada Duque[1], Delft University of Technology, Delft, the Netherlands

Introduction

As the world's energy demand is growing and fossil fuels are depleting, concerns are raising on how this demand can be continuously supplied. Solar and wind energy are renewable resources that may contribute to replace fossil energy, However, they cannot provide the carbon needed to produce fuels and chemicals. This fact, coupled to uncertain oil prices and need of lowering the atmospheric CO_2, have led the attention towards biomass and other short-cycled carbon raw materials, such as CO, CO_2 and CH_4 [1]. These feedstocks can yield fuels and chemicals with improved environmental performance (compared to fossil-based products), while preventing the impacts over food security that are argued against first generation feedstocks [2].

Commonly, bio-refineries are conceptualized to produce fuels and chemicals through two types of processing platforms that utilize biomass as feedstock [3,4]: i) the sugar platform, in which biochemical pre-treatment of cellulose yields a fermentable mixture of 5 and 6-carbon carbohydrates; and ii) the syngas platform, which through thermochemical treatment, at 750 – 1000 °C, decomposes biomass into a mixture of gases, commonly referred to as syngas, formed by carbon monoxide, carbon dioxide, hydrogen and small fractions of methane and higher hydrocarbons, hydrogen sulfide and ammonia [5]. The syngas platform is currently under use for heat and electricity generation, and as source of raw materials for the production of fuels via Fischer-Tropsch synthesis (FTS) [6]. Syngas, in recent years, has also been seen as possible feedstock for fermentation [7].

The thermochemical conversion of biomass offers potential advantages over the biochemical route, for instance: i) higher yields, ii) simpler process configuration, iii) more flexibility on feedstock admission and product composition, and iv) higher scale-up possibilities [8]. Moreover, fermentation also shows possible advantages over the catalytic conversion of syngas since microorganisms can: i) work under moderate temperatures and pressures; ii) be more tolerant than catalysts to poisoning by sulphur, tars and chlorine; iii) accept a wider range of gas compositions; and iv) achieve higher conversions and yields [6,7]. Therefore, a hybrid process, combining thermochemical conversion of biomass and syngas fermentation, could drive to reduction on capital and operation costs as well as environmental impacts to the produce fuels and chemicals.

Currently, only few studies have been found to simultaneously address the technical, economic and environmental performance of large-scale production of syngas, as potential feedstock for fermentation [7]. Therefore, in is study we have performed a sustainability

[1] Corresponding author at: van der Maasweg 9 (Room: C0.360), 2629 HZ, Delft, the Netherlands. E-mail address: J.A.PosadaDuque@tudelft.nl (J.A. Posada)

analysis (*i.e.* techno-economic and life cycle assessments) [9] of the syngas production process, from biomass production, its delivery and conversion into a theoretically fermentable gas, in order to evaluate and compare the effects of different processing configurations and operating conditions at multiple stages of the process.

Methodology

The syngas production scale was defined to supply the need of an average second generation bioethanol production plant of 20 million gallons/year, *i.e.* an average of 388 kton of syngas per year, since each type of biomass has different yields to syngas.

Biomass supply

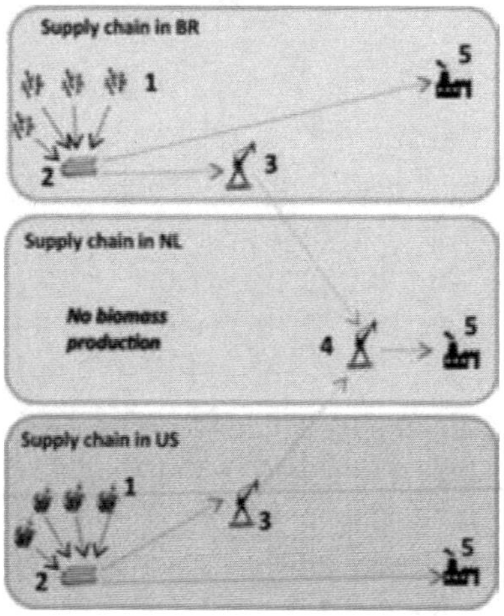

Figure 1. Configuration of supply chains according to the geographical location. Arrows represent transport by truck (black), train (red), and ship (blue). Transfer points: 1. biomass production site, 2. central gathering point, 3. export terminal, 4. import terminal, 5. conversion plant (biomass to syngas and syngas to chemicals)

Different biomass supply configurations were designed considering four different feedstocks: forestry residues (FR) and corn stover (CS), produced in the United States of America (US); and sugarcane bagasse (SB) and eucalyptus wood (EW), produced in Brazil (BR). The Netherlands (NL) has been included only as location for processing, which means biomass is imported from the other two locations and then processed first into syngas and later into one of the bio-based products. The four feedstocks were selected based on availability on the production site, price and composition.

Here, the approach proposed by Hamelinck et al. (2005) [10] was adopted to obtain supply costs, environmental impacts and energy requirements for the delivery of the lignocellulosic biomass. Three main process operations were considered to structure the (inter)national biomass supply, they are: i) biomass production (type of feedstock produced, land establishment, land maintenance and collection/harvesting); ii) conditioning (chipping, baling, pelletization and drying); and iii) transportation (truck, train and ship transport) [10]. The sequence of operations was defined based on generic theoretical chains which contain five transfer points [1] as shown in

Figure **1**: 1) the production site, 2) a central gathering point, 3) a transport terminal for export, 4) a transport terminal for import, and 5) the conversion plant. In the specific case of international transport, pelletizing of biomass is included to increase its density and makes long distance transportation more efficient. The location of the central gathering points was defined to keep minimum costs for transport between the sites with the highest biomass productivities and the next transfer points (considering road infrastructure and cost-efficiency of the transport type). Determination of the ports was mainly based on infrastructure and scale; ports were selected that can take up Panamax ships, assuring large scale biomass transportation [10]. The moisture content of the biomass entering the gasification unit, is considered to be 10 wt.%, which is pointed as the optimum to minimize size and operation costs of the gas-to-liquid plant. When biomass required partial drying before international transport, the costs were accounted for.

Thermochemical conversion

The gasification process here considered was based on the technology described by NREL [11], which is able to produce syngas suitable for fermentation. The process takes place at atmospheric pressure and uses three fluidized bed reactors: an indirectly heated (allothermal) gasification reactor, a combustion reactor (as heat source) and a gas conditioning reactor. The gasification and combustion reactors function with circulating fluidized beds (CFB) while the gas conditioning reactor uses a bubbling fluid bed (BFB). Olivine, which is used as bed material, circulates through the three reactors and is (re)heated in the combustion reactor where char is burned with air.

Table 1. Biomass data: physicochemical characterization, production requirements and production costs

Parameter	***Forestry residues (PW)***	***Agricultural residues***		***Energy crops (EW)***
		SB	***CS***	
Region of production	US	BR	US	BR
Costs in \$/ton$_{wet}$ (@year)	15 (2011)	17 (2012)	24 (2011)	31 (2011)
Moisture content$_{Green\ Biomass}$ (wt.%)*	45	51	24	45
C (wt.%)	49.7	45.2	47.3	46.0
H (wt.%)	7.8	5.5	5.1	5.8
O (wt.%)	41.8	40.7	40.6	44.5
N (wt.%)	0.1	0.1	0.8	0.3
S (wt.%)	0.05	0.05	0.22	0.0

* Green biomass is defined as wet biomass at its harvesting phase

Table 2. Gasification and tar reforming reactions

Gasification	
$C + H_2O \rightarrow CO + H_2$	(1)
$C + O_2 \rightarrow CO_2$	(2)
$2H_2 + O_2 \rightarrow 2H_2O$	(3)
$C + CO_2 \rightarrow CO$	(4)
$C + 2H_2 \rightarrow CH_4$	(5)
$CO_2 + H_2 \rightarrow CO + H_2O$	(6)
$N + 1.5H_2 \rightarrow NH_3$	(7)
$S + H_2 \rightarrow H_2S$	(8)
Reforming	
$CH_4 + H_2O \rightarrow CO + 3H_2$	(9)
$CH_4 + CO_2 \rightarrow 2CO + 2H_2$	(10)
$2CO + O_2 \rightarrow 2CO_2$	(11)
$2NH_3 \rightarrow 3H_2 + N_2$	(12)

The gasification process was simulated in Aspen Plus V8.8 using the physicochemical characterization obtained from literature data (see Table 1) for all considered feedstocks. The developed model is based on a general approach to represent a dual fluidized bed gasifier, known as fast internally circulating fluidized bed [12]. The model considers the following assumptions: ideal gas law applies for the gas phase; operation takes place isothermally and at steady state; distribution of all gases within the emulsion phase is homogenous; drying and pyrolysis are instantaneous; there are not pressure drops or heat losses; chemical equilibrium is assumed; since CH_4 represents all hydrocarbons; all nitrogen and sulfur in biomass are converted into NH_3 and H_2S; and char only consists of carbon and ash.

For the gasification process, steam (at 450 °C) enters at the gasification reactor. This steam has a double function, it acts as a fluidizing medium and as oxidizing agent for the gasification and tar reforming reactions, as shown in Table 2. The mass flow rate of steam is defined by the steam to biomass ratio which was set at 0.75 [12]. The gasification reactions (see Table 2) take place when high temperatures are reached (850 °C). The temperature of the air stream fed to the gasification reactor was set at 450 °C.

In the tar reforming process, the gas coming from the gasifier is sent to the catalytic tar reformer, where methane and higher hydrocarbons are oxidized into CO and H_2, CO into CO_2, and NH_3 into both: N_2 and H_2 (Table 2). Steam addition to the tar reformer was excluded from the model since the water content in the input gas was already sufficient to drive the reactions and avoid additional costs and gas dilution.

Economic evaluation

The economic assessment of the designed processes was made from the perspective of the syngas production costs, considering capital and operational expenditures (CAPEX and OPEX, respectively). CAPEX was constructed from several factors linked to the equipment purchase costs (EPC) [12]. The EPC's of the gasification plant were adapted from and NREL's report on syngas production [11]. The 6/10 scaling factor rule was used for capacity corrections [12]. Prices were updated to 2015 by using the Chemical Engineering Plant Cost Index (CEPCI). A project lifetime was fixed at 15 years. OPEX was assumed to be formed by [12]: facility-dependent costs (depreciation, maintenance, insurance, local taxes and overhead), process-related costs (raw materials and utilities) and labor.

Furthermore, the economic performance for each scenario was adjusted to the specific geographical locations (NL, US and BR) by using specific prices for fuels (used for defining prices of utilities) and for lignocellulosic biomass.

Life Cycle Assessment

The LCA was performed in agreement with the procedures stated in the ISO 14040 and 14044 [14,15] norms. System boundaries were defined from a cradle-to-gate approach. Therefore, the system includes from feedstock's production and supply until syngas production. The assessment was performed from the perspective of two environmental impact categories [16]: the global warming potential (GWP) and the primary energy use (calculated as non-renewable energy use, NREU). The functional units for the inventory analysis and the impacts assessment is the production of 1 kg of syngas and 1 MJ of its low heating value (see Table 3).

Individual impact characterization factors for materials and energy carriers were taken from the Ecoinvent V2.2 database, literature and internet references. The environmental impact data for sugarcane bagasse and corn stover were adopted from literature by applying economic allocation since they are considered as waste materials. Factors for electricity production were defined as site-specific for each country where location was considered.

Results and discussion

Process design

The supply chains were designed to produce a syngas that complies with all technical requirements for its further fermentation into chemicals. Table 3 shows the resulting yields of syngas on biomass for each feedstock as well as the composition of the resulting syngas for each case. The highest syngas yield is obtained from pine wood as a result of its high hydrogen content and low presence of ash and nitrogen compared to the other feedstocks. However, it is worth to notice that the syngas yield does not have significant variability among the four feedstocks; only a variation of 5.3 % is observed.

Table 3. Yields of biomass and final products, and composition of resulting syngas at a gasification temperature of 850 °C

Parameter	PW	SB	CS	EW
Syngas composition (vol. % $_{dry}$)				
H_2	60.9	57.0	55.4	56.5
CO	28.1	28.6	30.0	29.0
CO_2	9.9	13.1	13.0	13.2
N_2	1.0	1.2	1.5	1.2
CH_4	593.2 ppmv	379 ppmv	412 ppmv	386 ppmv
NH_3	12.4 ppmv	12 ppmv	13 ppmv	12 ppmv
H_2S	160.6 ppmv	189 ppmv	822 ppmv	0
H_2/CO	2.2	2.0	1.8	1.9
LHV (kJ/kg)	5101	5451	5378	5446
Yield (kg syngas/kg biomass$_{dry}$)	1.20	1.14	1.18	1.19

Economic evaluation

Figure 2 shows the costs distribution for syngas production, for the four feedstocks, with their respective region of origin and according to the location for biomass gasification. When syngas production takes place in the region of the feedstock origin, the lowest production costs are obtained from CS in the US. However, when syngas production takes place in NL after biomass import, PW led to the lowest syngas production costs of all feedstocks. The costs related to biomass delivery to the gasification plant represent the largest share of the total syngas production costs. For instance, when syngas is produced in NL, the cost of biomass represents around the 50 % of production costs; this makes production in NL is the most expensive option. Utilities are the second largest contributor to the syngas production costs; for example, when syngas is produced in NL, utilities represent between 21 and 24%; while for syngas production in US and BR these contributions range between 19 and 25%.

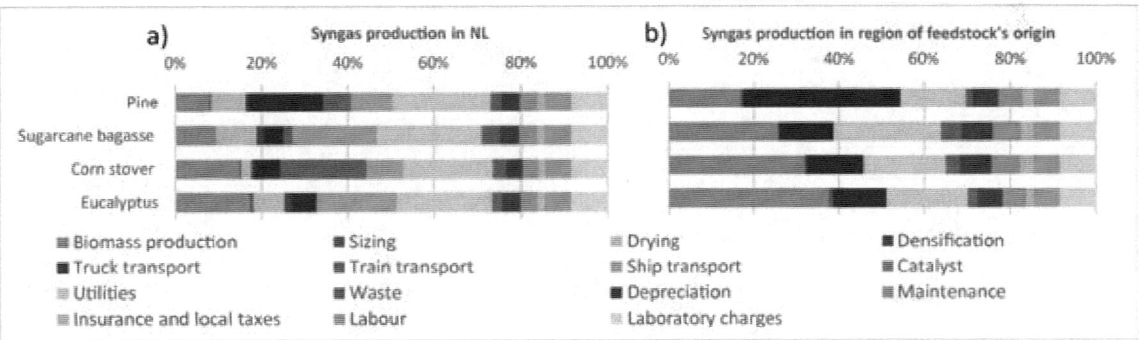

Figure 2. Cost breakdown of syngas production for: a) syngas production in NL, and b) syngas production at the region of the feedstock production

Generally, syngas production costs are found to be higher than the ones obtained by the present study. However, literature data consider syngas as feedstock for FTS, which has larger requirements on gas composition than fermentation. Nevertheless, production costs of bio-syngas from corn stover with a H_2/CO ratio of 2 for FTS, is reported to have a value of \$221.5/ton [17]. This is 1.3 – 85.7 % higher than the values obtained in this study.

Table 4. Total production costs of syngas: contribution analysis by feedstocks and location

Parameter	Costs contribution ($/ton syngas)							
	PW (US)		SB (BR)		SC (US)		EW (BR)	
	(NL)*	(local)**	(NL)	(local)	(NL)	(local)	(NL)	(local)
Production costs ($/ton)	190.7	144.5	198.2	130.0	218.7	119.3	198.0	150.0

* NL case: syngas production takes place in NL

** Local case: syngas production takes place in the region where the feedstock is produced

Life cycle assessment

The obtained environmental impacts (GWP and NREU) for the production of syngas in the three locations (NL, BR and US) from PW, SB, CS and EW are shown in Table 5. Of all feedstocks, eucalyptus returns the lowest impacts during syngas production, while corn stover yields, by far, the highest; additionally BR is the best location while NL produces the highest impacts, mainly due to international transport. Figure 3 shows a contribution analysis for GWP and NREU. Two key factors contributing to the environmental impacts of syngas production are steam for gasification and biomass production; since the latter includes the conditioning of biomass and its delivery, its contribution is larger for the cases in which gasification is carried out in NL.

Again for the environmental impacts, comparing the results of this study with literature data is rather difficult due to the differences in syngas requirements and process configurations. When comparing against averaged results obtained for syngas produced as feedstock for FTS [17], this study returned averaged 44% lower GWP. On the other hand, when impacts are compared against syngas derived from coal in India, this study returned 55 and 73 % lower GWP and NREU.

Table 5. Environmental impacts (GWP and NREU) of syngas production

Impact category	Functional unit	NL				BR		US	
		PW	SB	CS	EW	SB	EW	PW	CS
GWP	kg syngas	0.26	0.26	0.32	0.21	0.23	0.20	0.21	0.29
	MJ syngas	0.051	0.047	0.059	0.038	0.042	0.036	0.041	0.053
NREU	kg syngas	5.87	3.71	5.20	5.99	3.13	3.74	3.46	4.09
	MJ syngas	1.15	0.68	0.97	1.10	0.57	0.69	0.68	0.76

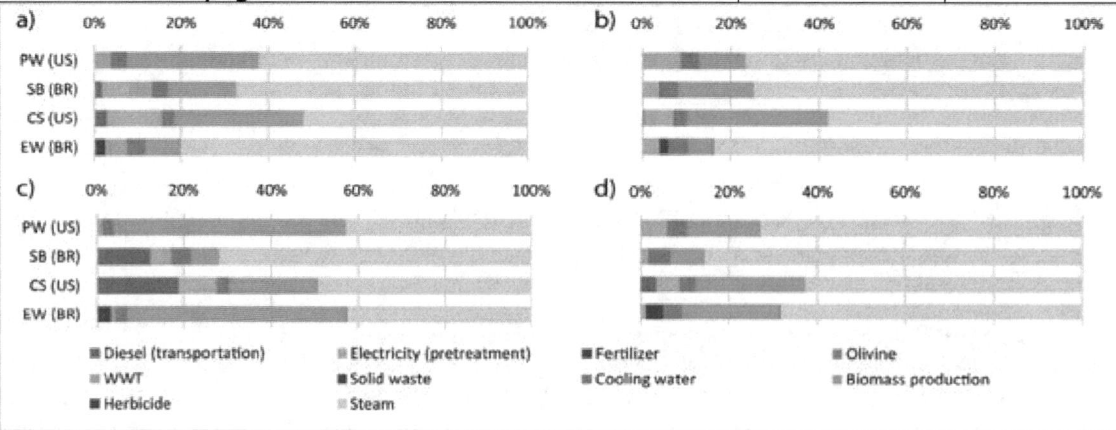

Figure 3. Contribution analysis for environmental impacts of syngas production: a) GWP in NL; b) GWP in the region of feedstocks origin; c) NREU in NL; d) NREU in the region of feedstocks origin

Conclusions

All four feedstocks (pine, sugarcane bagasse, corn stover and eucalyptus) proved to be able to produce syngas suitable for fermentation, with different production cost, environmental impacts and technical performance. Additionally, no specific feedstock-plant location combination performs best on all aspects: economic, environmental and technical performance. However, corn stover when gasification takes place in the US, proved to have the lowest production cost: $119.27/ton. This configuration also delivered syngas with the highest CO and H_2 concentration, plus the lowest nitrogen and ash content; this resulted in the highest syngas yield while keeping low CO_2 concentration. Environmentally, syngas production in Brazil returned the lowest GWP of 0.20 $kgCO_2eq/kg$ when eucalyptus is used as feedstock, and the lowest NREU of 3.13 MJ/kg when sugarcane bagasse is the feedstock. In general, location of gasification plant in the same region where feedstock production takes place is, economically and environmentally, the most optimum configuration, mainly due to the large impacts of biomass transport.

Optimum gasification temperatures were obtained in terms of syngas composition (high CO and H_2, and low CH_4 and CO_2). These temperatures range between 850 and 1000 °C, depending on the feedstock. At these temperatures theoretically, the tar reformer could be excluded from the process configuration. This resulted in reduction of syngas production costs, GWP and NREU by 20.1, 14.0 and 16.0 %, respectively; however, the NH_3 concentration at these conditions is too high for syngas fermentation, according to literature. Finally, heat recovery through steam generation is highly recommended to increase the overall efficiency of the system: maximum 29 and 23 % reductions on syngas production costs and GWP, were respectively observed for pine gasification in the US.

The results of this analysis, among others [18], are used as input data to study the economic and GHG performance of the the syngas fermentation platform for the production of ethanol, 2,3-butanedion and hexanoic acid [19].

References

1. Daystar, Jesse; Gonzalez, Ronalds; Reeb, Carter; Venditti, Richard; Treasure, Trevor; Abt, Robert and Kelley, Steve (2014), "Economics, Environmental Impacts, and Supply Chain Analysis of Cellulosic Biomass for Biofuels in the Southern US: Pine, Eucalyptus, Unmanaged Hardwoods, Forest Residues, Switchgrass, and Sweet Sorghum", BioResources, pp. 393–444.
2. Posada, John A. and Osseweijer, Patricia (2016) "Socio-economic and environmental considerations for sustainable supply and fractionation of lignocellulosic biomass in a biorefinery context", In: Biomass fractionation technologies for a lignocellulosic feedstock based biorefinery, 1st ed. Mussatto SL, p. 674.
3. Cheali, Peam; Posada, John A.; Gernaey, Krist V. and Sin, Gürkan (2015), "Upgrading of lignocellulosic biorefinery to value added chemicals: Sustainability and economics of bioethanol-derivatives", Biomass and Bioenergy, April, pp.282–300.
4. Cheali, Peam; Posada, John A.; Gernaey, Krist V. and Sin, Gürkan (2016), "Economic risk analysis and critical comparison of optimal biorefinery concepts: Economic risk analysis and critical comparison of optimal biorefinery concepts", Biofuels, Bioproducts and Biorefining, May, pp. 435-445.

5. Munasinghe, Pradeep C. and Khanal, Samir K. (2010), "Biomass-derived syngas fermentation into biofuels: Opportunities and challenges", Bioresource Technology, July, pp. 5013–5022.
6. Liew, Fung M.; Köpke, Michael and Dennis, Séan (2013), "Gas Fermentation for Commercial Biofuels Production", In: Liquid, Gaseous and Solid Biofuels - Conversion Techniques, InTech.
7. Daniell, James; Köpke, Michael and Simpson, Séan D. (2012), "Commercial Biomass Syngas Fermentation", Energies, December, pp. 5372–5417.
8. Siedlecki, marcin; De Jong, Wiebren and Verkooijen, Adrian H. (2011), "Fluidized Bed Gasification as a Mature And Reliable Technology for the Production of Bio-Syngas and Applied in the Production of Liquid Transportation Fuels—A Review", Energies, March, pp. 389–434.
9. Palmeros-Parada, Mar; Osseweijer, Patricia; Posada, John A. (2016) Sustainable biorefineries, an analysis of practices for incorporating sustainability in biorefinery design. Industrial Crops and Products http://dx.doi.org/10.1016/j.indcrop.2016.08.052
10. Hamelinck, Carlo; Suurs, Roald S. and Faaij, André P. (2005), "International bioenergy transport costs and energy balance", Biomass Bioenergy, August, pp. 114-134.
11. Worley, M. and Yale, J. (2012), "Biomass Gasification Technology Assessment", NREL.
12. Doherty, Wayne; Reynolds, Anthony and Kennedy, David (2013), "Aspen Plus Simulation of Biomass Gasification in a Steam Blown Dual Fluidised Bed", In: Materials and processes for energy: communicating current research and technological developments, Formatex Research Centre.
13. Peters Max S. and Timmerhaus, Klaus D. (1991), "Plant design and economics for chemical engineers", 4. ed. McGraw-Hill.
14. ISO, (1997), "ISO 14040 Environmental management - Life cycle assessment - Principles and framework".
15. ISO, (2006), "ISO 14044 Environmental management - Life cycle assessment - Requirements and guidelines".
16. Posada, John A; Brentner, Laura B; Ramirez, Andrea; Patel, Martin K. (2016), "Conceptual design of sustainable integrated microalgae biorefineries: parametric analysis of energy use, greenhouse gas emissions and techno-economics", Algal Research, July, pp. 113-131.
17. Moncada, Jonathan; Posada, John A. and Ramírez, Andrea (2015), "Early sustainability assessment for potential configurations of integrated biorefineries. Screening of bio-based derivatives from platform chemicals", Biofuels, Bioproducts and Biorefining, November, pp. 722–748.
18. Posada, John A; Cardona, Carlos A. (2012). Propionic acid production from raw glycerol using commercial and engineered strains. Industrial & Engineering Chemistry Research 51:2354–2361.
19. Almeida-Benalcázar, Eduardo F.; Gevers-Deynoot, Baudine; Noorman, Henk; Osseweijer, Patricia; Posada, John A. (2016). Syngas fermentation for the production of bulk chemicals: early-stage techno-economic and life cycle assessment. 2016 AIChE Annual Meeting. November 13-18, 2016. San Francisco, CA

462372 Design and Operation of Biofuel Supply Chains with Variable Regional Depot and Biorefinery Locations

Abstract:

The transportation, handling and storage of bulky raw biomass is one of the key challenges in biofuel supply chains. To address this challenge, the concept of regional biomass processing depot [1] (referred to in this work as 'depot') has been introduced. At the depot, bulky biomass is pretreated and/or densified into a stable and dense intermediate, which can then be transported economically over long distances. Despite all the extensive research in the area [2,3], the design of biofuel supply chains with depots has received limited attention; depots have not been studied at all or assumed to be installed in predetermined locations. However, the efficiency of the biofuel supply chain (SC) can be improved substantially by optimizing the location of depots and biorefineries.

Accordingly, in this work, we propose a multi-period mixed-integer linear programming (MILP) model after briefly reviewing a mixed-integer non-linear programming (MINLP) model developed by Ng and Maravelias [4]. The new MILP model accounts for biomass selection and allocation, technology selection and capacity planning at depots and biorefineries, inventory and shipment planning, and, importantly, variable depot and biorefinery locations. Unlike other biofuel SC approaches, which precalculate the distances based on the predetermined depot and biorefinery locations, we develop a series of approximations and reformulations that allow us to overcome the bilinearities (shipments*transportation_distance) in the calculation of transportation cost, based on the following assumptions:

1) The biomass from a harvesting site is sent to only one downstream node, either a depot or a biorefinery; and, similarly, the intermediates from a depot are shipped to a single biorefinery. This is a realistic assumption because it is difficult to have multiple pickups from a harvesting site or a depot and also to "dynamically" change where and how much to ship.

2) Biomass and intermediates are stored at the origin (harvesting site and depot, respectively). This assumption has minimal effect because inventory unit costs and material deterioration coefficients are practically independent of location.

3) We approximate the exact timing of the shipment as all biomass from selected harvesting site is eventually shipped to the same depot or biorefinery. Thus, we can replace the variable shipments with the total availability of biomass. A correction factor though is introduced to account for the fact that the actual shipments will be slightly lower than the total availability of biomass due to biomass deterioration during storage.

Based on the above assumptions, we first precalculate the range of correction factors based on the possible inventory profiles. We can choose the average value or a factor that results from a reasonable inventory profile. Second, we disaggregate all variable distance components into "actual" and "dummy" distance variables, which are bounded by variable upper bound constraints using binary arc selection variables. The actual distance variable is equal to zero if the corresponding arc is not selected. Finally, we reformulate the transportation cost, which can be calculated based on the actual distance variables and approximated shipments. The latter are treated as parameters, thereby resulting in a linear transportation cost calculation.

We illustrate the applicability of the proposed model by considering a biofuel SC in South Central Wisconsin that includes Columbia, Dane, Dodge, Grant, Green, Iowa, Jefferson, Lafayette, Richland, Rock, and Sauk counties, with a total of 64 harvesting sites. We consider two types of biomass: corn stover and switchgrass, both of which are abundant biomass resources in Wisconsin. At most three depots can be installed in the studied region. We employ certain criteria for screening the potential biorefinery locations: Candidate biorefinery locations were restricted to cities/towns/villages with populations between 3,000 and 10,000 in 2014 for a maximum of one biorefinery per county [5] and total biomass availability more than 120,000 dry ton per year. Among the locations that met the screening criteria, we preselect two towns, which are located near the center of the studied region, because they are likely to lead to smaller transportation cost. The coordinates of the biorefinery are constrained by the convex hull of the selected town, therefore, the biorefinery can be installed at various points within the selected region, instead of at the centroid of the region.

The objective is to minimize the total annual cost (TAC), which includes biomass acquisition, inventory, operating and capital costs. The total capital investment for the ethanol production level of 340 million liters per year, is $ 592 million, leading to a TAC of $ 251.44 million. In this case study, two depots are selected, while the biorefinery is installed at West Oregon. The SC configuration with two depots has lower TAC as compared to the one without depots. The transportation cost in the configuration with three depots or more is lower than in the configuration with two depots, but the higher capital costs leads to higher TAC.

We also calculate the relative error of our approximation using the exact biomass/intermediate shipments obtained. The calculated relative error is found to be insignificant. Finally, for comparison purposes, we extend the MINLP model [4] to account for variable biorefinery locations and resolve the case study. The proposed model is shown to yield better solutions in a faction of the time required by the MINLP model.

References:

[1] P.L. Eranki, B.D. Bals, B.E. Dale, Advanced regional biomass processing depots: a key to the logistical challenges of the cellulosic biofuel industry, Biofuels, Bioprod. Biorefining. 5 (2011) 621–630.

[2] B. Sharma, R.G. Ingalls, C.L. Jones, A. Khanchi, Biomass supply chain design and analysis: Basis, overview, modeling, challenges, and future, Renew. Sustain. Energy Rev. 24 (2013) 608–627.

[3] D. Yue, F. You, S.W. Snyder, Biomass-to-bioenergy and biofuel supply chain optimization: Overview, key issues and challenges, Comput. Chem. Eng. 66 (2014) 36–56.

[4] R.T.L. Ng, C.T. Maravelias, Design of Cellulosic Ethanol Supply Chains with Regional Depots, Ind. Eng. Chem. Res. 55 (2016) 3420–3432.

[5] W. Alex Marvin, L.D. Schmidt, S. Benjaafar, D.G. Tiffany, P. Daoutidis, Economic Optimization of a Lignocellulosic Biomass-to-Ethanol Supply Chain, Chem. Eng. Sci. 67 (2012) 68–79.

462459 Process Synthesis of Natural Gas to Liquid Transportation Fuels Under Uncertainty: A Robust Optimization Approach

Abstract:

As discussed in a recent perspective [1], natural gas has been explored as an exciting feedstock for the production of liquid transportation fuels [2], olefins [3], and aromatics [4] through advanced process synthesis and global optimization techniques in recent years. The growing abundance of natural gas as a feedstock, coupled with the importance of these products in society, provides substantial motivation for the discovery of optimum refinery topologies and product distributions. Focusing on natural gas to liquid transportation fuels (GTL) specifically, there are many different natural gas conversion routes using synthesis gas as an intermediate and processes such as methanol synthesis or Fischer-Tropsch conversion for final fuel production in the form of gasoline, diesel, and kerosene. Identifying the most cost-effective method for fuel production is imperative.

Process synthesis as a whole, however, has many areas in which uncertainty can appear in model parameters. This uncertainty can be detrimental to the optimal solutions, which may become infeasible or give objective function values worse than expected due to uncertain parameter realizations. To include parameter uncertainty in the model, uncertain constraints are reformulated using robust optimization [5]. The robust counterparts ensure that constraints will feasible for an uncertainty set of parameter realizations; the size of the uncertainty set can be determined using probabilistic bounds, in which considerable advances have recently been made [6-9]. These probabilistic bounds are utilized *a priori* and *a posteriori*to give robust solutions with minimal conservatism and known levels of risk.

Uncertainty has been included in the GTL process synthesis superstructure in order to incorporate price uncertainty from the key feedstocks and products through reformulation of the objective function. Uncertain parameters such as investment costs are also considered and discussed. The non-convex mixed-integer nonlinear optimization problems are solved to global optimality to give optimal solutions at known probabilities of constraint violation [10]. As the uncertainty appears in the objective function alone, probabilistically guaranteed levels of profit can be found with conservative assumptions about the probability distributions for uncertain parameters. An iterative method will be utilized to provide high quality robust solutions at low probabilities of constraint violation using the box, interval + ellipsoidal, and interval + polyhedral uncertainty sets [11]. The guaranteed and expected profit levels, optimal product distributions, and overall investment costs of the robust solutions will be discussed at varying probabilities of constraint violation. These insights will provide key information on the best GTL refinery topologies moving forward under uncertainty.

[1] Floudas, C. A.; Niziolek, A. M.; Onel, O.; Matthews, L. R. Multi-scale systems engineering for energy and the environment: Challenges and opportunities. AIChE Journal 2016, 62 (3), 602-623.

[2] Baliban, R. C.; Elia, J. A.; Floudas, C. A. Novel Natural Gas to Liquids Processes: Process Synthesis and Global Optimization Strategies. AIChE Journal 2013, 59 (2), 505-531.

[3] Onel, O.; Niziolek, A. M; Floudas, C. A Optimal Production of Light Olefins from Natural Gas via the Methanol Intermediate. Industrial & Engineering Chemistry Research 2016, 55 (11), 3043-3063.

[4] Niziolek, A. M.; Onel, O.; Floudas, C. A. Production of benzene, toluene, and xylenes from natural gas via methanol: Process synthesis and global optimization. AIChE Journal 2016, 62(5), 1531-1556.

[5] Li, Z.; Ding, R.; Floudas, C. A. A Comparative Theoretical and Computational Study on Robust Counterpart Optimization: I. Robust Linear Optimization and Robust Mixed Integer Linear Optimization. Industrial & Engineering Chemistry Research 2011, 50, 10567-10603.

[6] Li, Z.; Tang, Q.; Floudas, C. A. A Comparative Theoretical and Computational Study on Robust Counterpart Optimization: II. Probabilistic Guarantees on Constraint Satisfaction. Industrial & Engineering Chemistry Research 2012, 51 (19), 6769-6788.

[7] Guzman, Y. A; Matthews, L. R; Floudas, C. A New a priori and a posteriori probabilistic bounds for robust counterpart optimization: I. Unknown probability distributions. Computers & Chemical Engineering 2016, 84, 568-598.

[8] Guzman, Y. A; Matthews, L. R; Floudas, C. A New a priori and a posteriori probabilistic bounds for robust counterpart optimization: II. A priori bounds for known symmetric and asymmetric probability distributions. 2016, In Preparation.

[9] Guzman, Y. A; Matthews, L. R; Floudas, C. A New a priori and a posteriori probabilistic bounds for robust counterpart optimization: III. A posteriori bounds for known probability distributions. 2016, In Preparation.

[10] Baliban, R. C; Elia, J. A; Misener, R.; Floudas, C. A. Global optimization of a MINLP process synthesis model for thermochemical based conversion of hybrid coal, biomass, and natural gas to liquid fuels. Computers & Chemical Engineering 2012, 42, 64-86.

[11] Li, Z.; Floudas, C. A. A Comparative Theoretical and Computational Study on Robust Counterpart Optimization: III. Improving the Quality of Robust Solutions. Industrial & Engineering Chemistry Research 2014, 53 (33), 13112-13124.

465584 Multi-Scale Operational Planning and Optimal Sizing of Hybrid Renewable Energy System

Abstract:

Decentralized and hybrid renewable energy system has recently received attention for providing energy in a small and remote area with its own grid (referred to as micro-grid). In this study, *operational planning and optimal sizing* of a micro-grid having large wind farm and battery storage device is studied. Operational planning is generally made in two time-scales; unit commitment decisions on which source should be on-line at what time are pre-determined one day ahead; and dispatch decisions on the output of on-line sources are then decided on hourly time basis. Two stage stochastic programming (2SSP) is conventionally formulated for this one-day operation with a number of intra-day uncertainty scenarios. However, the finite horizon setting is a main factor to limit sustainable operation due to end-effects. At the same time, the yearly investment problem for sizing energy generation and storage resources is typically performed to minimize future operating costs as well as capital cost. These argue yearly operation model beyond one-day is required. Therefore in this study, 2SSP model for one-day operation is combined with daily-evolving Markov decision process (MDP). That is, the value of being in a state of current commitment and battery at the end of a day (called value function) is included in the objective function to account for longer term implications of the decisions. In the MDP formulation, inter-day variation of wind is captured, and the value function is approximated with a linear model. The coefficient vector of the linear model is recursively updated with a sampled observation estimated from the one-day operation model. With this value function capturing all future operating costs, optimal sizing of the wind farm and battery devices is determined based on a surrogate function optimization. Meanwhile, for the integration of the decision hierarchy, a wind model consistent across multiple timescales from seasonal to hourly is developed to provide both intra-day scenarios capturing hourly ramping and inter-day transition information. The results of the proposed method are compared to those of the original 2SSP model through a case study and real wind data.

461974 Modelling the Gradual Expansion of Integrated Wind-Hydrogen-Electricity Networks That Decarbonise the Domestic Transport in Great Britain By 2050

Abstract:

The evolution of integrated wind-hydrogen-electricity networks that satisfy the demands of the domestic transport sector is optimised using a detailed, spatially-distributed mixed integer linear programming model that accounts for hourly operational issues, such as intermittency and dynamics of energy storage, simultaneously with long-term strategic decisions.

In our previous recent work [1], the design and operation of the network were determined using representative wind speed and hydrogen demand data for a whole year, to meet the full demand for hydrogen without considering the gradual transition from the current state with no hydrogen demand. However, in reality the full network will be established over many years as the penetration of fuel cell vehicles increases. Therefore, the model presented in Samsatli et al. (2016) [1] was extended to consider staged capacity investments and retirements, such that the demands for hydrogen, based on an assumed trajectory for the penetration of fuel cell vehicles from now to 2050, are met at all times and that the net present cost of doing so is minimised, subject to the available land area for wind turbines.

The potential technologies and infrastructures that may become part of the network include wind turbines; electrolysers, fuel cells, compressors and expanders; pressurised vessels and underground storage for hydrogen storage; hydrogen pipelines and electricity overhead/underground transmission lines; and fuelling stations and distribution pipelines.

The model determines the optimal spatial structure and the gradual expansion of the integrated networks across Great Britain, i.e. the optimal combinations of conversion and storage technologies invested in at every 5-year period, where to locate them, what their capacities should be and how to connect them by means of the transmission infrastructure networks. Simultaneously with this, the model also determines the hourly operation of each technology that forms the integrated network and the interactions between hydrogen and electricity networks.

In this conference, we will discuss the model and a decomposition method that enables this very large optimisation model to be solved. We will also present some case studies to illustrate the applicability of the model to scenarios in Great Britain.

References:

[1] S. Samsatli, I. Staffell, and N.J. Samsatli. Optimal design and operation of integrated wind-hydrogen-electricity networks for decarbonising the domestic transport sector in great britain. International Journal of Hydrogen Energy, 41:447 â€“475, 2016.

465831 A Computer-Aided Scenario Planning of Future Regional and National Energy Systems Based on Feasible Technology Options in Japan

Abstract:

The design of energy systems has become an issue all over the world. A single optimal system cannot be suggested because the availability of infrastructure and resources, and the acceptability of the system should be discussed locally with the involvement of all related stakeholders in the energy systems. In particular, researchers and engineers of technologies related to energy systems should be able to perform the forecasting and roadmapping of future energy systems and indicate quantitative results of scenario analyses. A simulation-based approach is strongly needed for such a purpose.

We developed a nation-wide energy flow model developed for analysing scenarios of future energy systems implementing a variety of feasible technology options [Kikuchi et al. 2014]. Decentralized power source technologies have been also modeled and analyzed with scenarios installing them in various countries [Shimizu et al. 2015; Kikuchi et al. 2016a]. A case study was conducted with a model for industrial symbiosis in a remote island in Japan, where excess energy and resources from cane sugar mills are effectively utilized in local energy systems [Kikuchi et al., 2016b]. These models were modularized and represented as functionals of appropriate technology options, which enables the aggregation and disaggregation of energy systems by defining functionals for individual technologies, packages integrating multi-technologies, and mini-systems. For scenario planning for actual energy systems, such a computer-aided approach should be able to address the issues related with system design and decision by stakeholders. It includes not only technological aspects, but also socio-economic ones such as the acceptability by energy consumers.

In this study, we are tackling with the scenario planning of energy systems in regional and national boundaries of Japan by developing and using computer-aided tools. Based on the latest energy technology roadmaps by the Society of Chemical Engineers, Japan (SCEJ) [Kato et al. 2016], feasible technology options are modeled for simulating micro- and macro energy grids in Japan. As case examples of microgrid simulation, the energy systems in remote islands were examined and scenarios were planned considering locally available resources. The introduction of energy carriers in Japan was also examined on the basis of energy demand/supply statistics. Regarding socio-economic aspects, questionnaire surveys were conducted to the energy consumers in remote islands and metropoles in Japan to scrutinize the consumers' preferences on energy. Based on the case studies, the requirements of a computer-aided approach for scenario planning of energy systems were defined.

The categorization of feasible technologies options listed in the latest energy technology roadmap [Kato et al. 2016] could be considered in the model framework quantifying energy demand/supply. In the regional grid case study, the energy integration in the rural area was examined on the systems centering on two types of power plant, the existing combined heating and power (CHP) in the sugar mill and new CHP for district heating and cooling in the central town area. The utilization of resources available in regional area can be an effective option for mitigating greenhouse gas (GHG) emission, which is also supported by the socio-economic analyses. Regarding the macro level introduction of energy carriers in Japan, their advantages have a strong connection with the site-specific condition on energy demand/supply, for example, the places where the unused or renewable resources are available have a priority in mitigation of GHG emission by introducing energy carrier-related technologies. Through case studies based on the developed models, it was demonstrated that the contribution of technologies to, e.g., the reduction in GHG emissions, should be carefully examined by quantitative analyses of interdependencies of the technology options for actual decision-making on energy systems. A simulation-based approach can contribute to the achievement of " be smart" by considering the benefit and risk of future energy-system options.

Acknowledgements:

Part of this study is financially supported by grants from the project JSPS KAKENHI Grant Numbers 16H06126 (Grant-in-Aid for Young Scientists A), 15H01750 (Scientific Research A) and the Environment Research and Technology Development Fund (1RF–1503). Activities of the Presidential Endowed Chair for "Platinum Society" in the University of Tokyo are supported by the KAITEKI Institute, Inc., Nippon Telegraph and Telephone Corporation, Fujifilm Holdings Corporation, Mitsui Fudosan Co., Ltd, LIXIL Corporation, and Shin-Etsu Chemical Co., Ltd..

References:

Kato, Y., Koyama, M., Fukushima, Y., Nakagaki, T. (Edited), Future Energy Systems Based on Feasible Technologies Beyond 2030, Springer: Tokyo, ISBN13: 9784431559498 (2016)

Kikuchi, Y., Kanematsu,Y., Sato, R., and Nakagaki, T., Distributed cogeneration of power and heat within an energy management strategy for mitigating fossil fuel consumption, Journal of Industrial Ecology, 20(2) 289-303 (2016)

Kikuchi, Y., Kanematsu,Y., Ugo, M., Hamada, Y., and Okubo, T., Industrial symbiosis centered on a regional cogeneration power plant utilizing available local resources: A case study of Tanegashima, Journal of Industrial Ecology, 20(2) 276-288 (2016)

Kikuchi, Y., Kimura, S., Okamoto, Y., and Koyama, M., A scenario analysis of future energy systems based on an energy flow model represented as functionals of technology options, Applied Energy, 132, 586-601 (2014)

Shimizu, T., Kikuchi, Y., Sugiyama, H., Hirao, M., Design method for a local energy cooperative network using distributed energy technologies, Applied Energy, 154 781-793 (2015)

466300 Numeric Optimization of Power Consumption for Simultaneous Saccharification and Fermentation of Cellulose to Ethanol

Abstract:

Using second-generation bioethanol, cellulose, the most abundant raw material in nature, can be converted to ethanol. Despite intensive research efforts at laboratory scale, the secondary effects of scale-up to large systems are still largely uncertain. A threedimensional numerical model of a laboratory scale fermenter is used to predict the effects of particulate mixing and reaction kinetics. This allows for case studies or optimizations of the reactor design to be carried out. The potential design variables are manifold ranging from agitation speed to impeller geometry while objectives are conversion rates and power consumption.

A two step reaction kinetics for simultaneous saccharification and fermentation (SSF) of Avicel (microcrystalline cellulose) particles to ethanol was presented by [van Zyl, 2012]. This separation of the two primary cellulase enzyme-kinetics gives the capability to predict the heterogeneous behavior of the enzyme-substrate interactions. This model improves the understanding of these systems while maintaining sufficient simplicity for implementation alongside a commercial computational fluid dynamics environment.

Viscosity is a mayor factor for the designing of a mixing vessel, as viscosity influences both: the shear rates and fluid dynamics of the system. To get the viscosity of the particle-fluid mixture an apparent dynamic viscosity model is used.

Cross-coupling the reaction models with computational fluid dynamic simulations provide a novel approach to capture the secondary effects of substrate conversion and particle distribution on the performance of the fermentation vessel. The first effect is the reduction in the average particle diameter due to the conversion of the cellulose. The lower drag forces of these smaller particles allow â€“ in a second effect - for a better distribution in the vessel. It is important to maintain a fully suspended particle distribution to maximize the surface contact with the rest of the mixing fluid.

These effects enable a reduced impeller rotation rate at later stages of the conversion decreasing the power consumption.

Difficulties in the modeling arise due to the different time scales in the simulation. The mixing with rotation rates of 150 rpm compared to the fermentation lasting for several hours. An approach to couple these phenomena in one application is presented.

In this contribution we show how to model this analysis efficiently using the commercial CFD software STAR-CCM+by CD-adapco.

References:
van Zyl, Josebus Maree. *Three-dimensional modelling of simultaneous saccharification and fermentation of cellulose to ethanol.* Diss. Stellenbosch University, 2012.

467208 Furfural Production By Continuous Reactive Extraction

Abstract:

1. Introduction

The bio-based economy grows rapidly during the 21st century. So far, oil and natural gas dominate the industrial world and are widely considered as raw materials. During the last few years, the use of biomass as raw material has increased significantly (1).

One important renewable bio-based chemical is furfural. It is widely used as solvent for saturated hydrocarbons because of its chemical structure. Moreover, it has the ability to dissolve aromatics and other unsaturated olefins (2) and is extensively used in oil and gas industry. Finally, furfural has a high potential as a platform chemical.

The objective of this study is to investigate the reaction mechanism and kinetics of xylose conversion towards furfural production by a continuous reactive extraction process. Moreover, the reaction pathway will be further elucidated and the kinetic rates of the intermediate steps are determined. Finally, the kinetic model for xylose dehydration to furfural is used to maximize furfural yield.

2. Experimental section

2.1. Reaction mechanism

The conversion of xylose to furfural is quite complex. Several studies (2), (3) have been performed in order to analyze and interpret the mechanism of the xylose dehydration. The scheme below shows the suggested mechanisms.

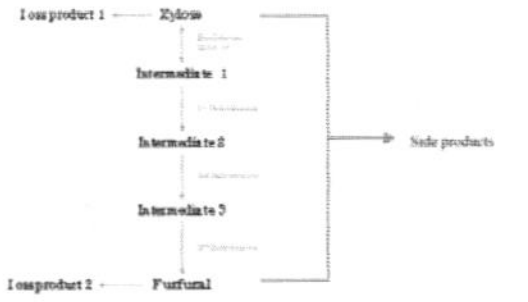

Scheme 1 Reaction mechanisms of xylose dehydration (2), (3)

According to Scheme 2 (blue), xylose is in equilibrium with intermediate I1. I1 undergoes three dehydration reactions to furfural. Byproducts are formed directly from either xylose or furfural (orange). The furfural degradation to byproduct 2 is significant and influences the selectivity. Increasing the temperature and acidity will further catalyze the degradation. An additional reaction is mentioned in (3), where xylose and furfural react and form side products (green).

The formation of byproducts from furfural can be reduced by in situ removal of furfural from the aqueous solution, by evaporation (4) or by extraction. Various studies (3), (4) have been performed to investigate which solvent performs best. Toluene and MIBK seem the most promising candidates with high affinity to furfural. Therefore, in this study, toluene is selected.

2.2. Experimental Setup

The experiments are performed in a biphasic continuous reactor as shown in Scheme 2. The configuration includes 3 pumps for sulfuric acid, xylose solution, and toluene. The maximum total flow is 15 mL/min and the minimum flow is 4 mL/min, with the flow ratio organic/aqueous of 2:1.

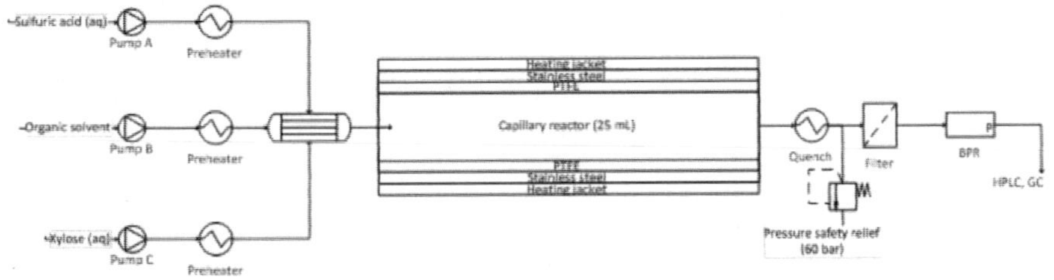

Scheme 2 Experiment lay out

The three streams flow separately in stainless steel tubes of 1.5 m length and 0.5 mm inner diameter through a heat exchanger to acquire the desired temperature. The preheat system is followed by two consecutive T-joints that allow the capillaries to be inserted in each other. Therefore the three streams are injected and mixed at the reactor's entrance (Scheme 3).

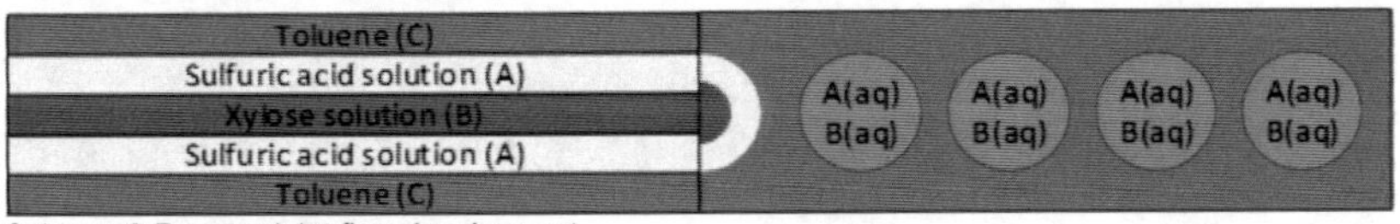

Scheme 3 Reactor inlet flow development

The reactor of 1.5 m in length and an inner diameter of 3.1 mm consists of a PTFE wall and a stainless steel mesh jacket to provide structural integrity. The PTFE wall is hydrophobic and wetted by toluene only, which prevents solid deposition at the reactor wall. The mixture is then quenched followed by a pressure safety relief and a filter with 2 µm pores that protects the back-pressure regulator (BPR) from blocking by solids formed during the reaction. Samples are taken after the BPR at atmospheric pressure and ambient temperature and analyzed with HPLC (aqueous phase) and GC (toluene phase).

The temperature range of the experiment is 100-170°C and the pressure is 30 bar so the presence of any vapor or gas at the working temperature is avoided. The selected catalyst is sulfuric acid (98%, Emsure) at 0.1 M concentration. The xylose inlet concentration is 12 wt %.

3. Results

3.1. Product characterization

In Figures 1 and 2 examples of the water phase HPLC are presented at τ=100s and 500s, respectively. Two signals are presented; (a) from the RI detector and (b) from the UV-vis detector. The compounds known from the calibration are indicated in Figures 1 and 2. In the RI signal, two unknown peaks are present: I1 and I2. In the UV signal more peaks are present. The most significant are termed A, B, C and D. Peaks B_a and B_b have a similar retention time as peak I1, who appears to have a shoulder and probably consists of two components. Component C is close to the retention time of I2. However, its concentration increases while I2 decreases in concentration.

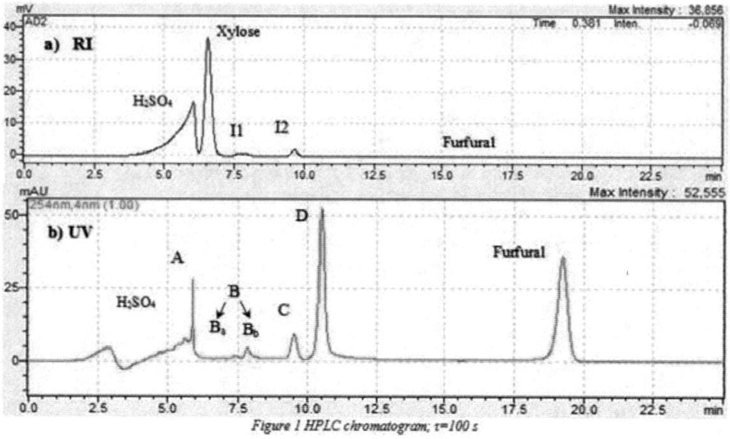

Figure 1 HPLC chromatogram; τ=100 s

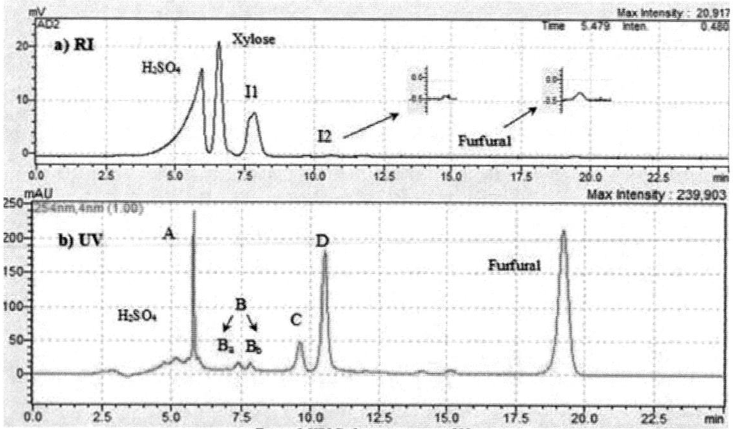

Figure 2 HPLC chromatogram; τ=500 s

Regarding UV chromatograms (Figures 1(b) and 2(b)), peaks corresponding to compounds B_a, B_b increase from 2 and 4 mAu to 10 and 8 mAu respectively with residence time. Moreover, peak of compound C rises from 10 mAu to 50 at the end of the residence time. Finally, peak of D has a raise from 52 to 180 mAu at 500s.

3.2. Reactor performance

In Figure 3 the reactor outlet concentration of furfural and xylose are presented. The highest xylose conversion is 44.52 % and the overall furfural yield is 14.45 %. The selectivity towards furfural appears to be relatively constant at 35% with increasing residence time.

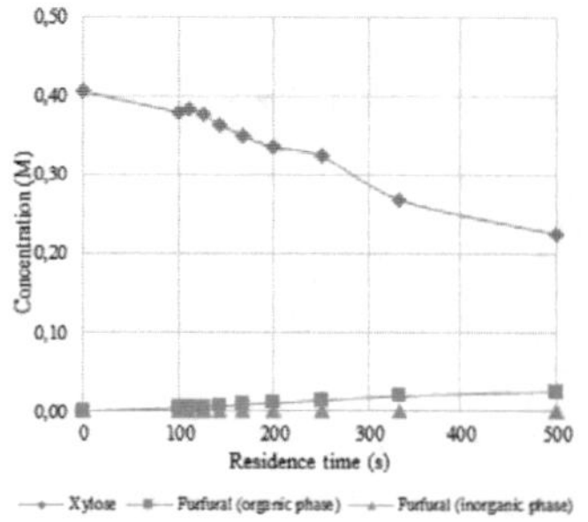

Figure 3 Concentration Profiles

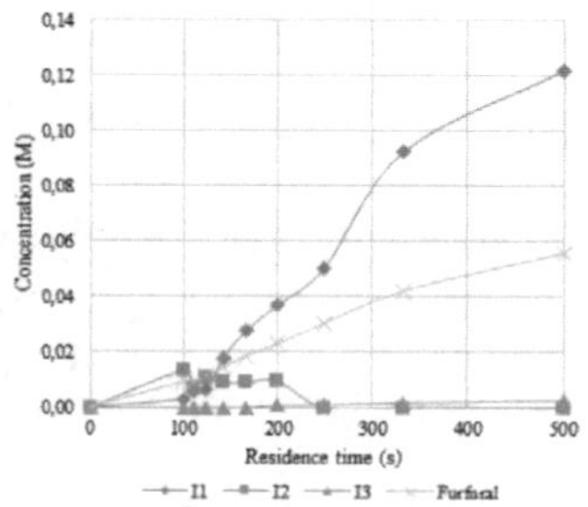

Figure 4 Products Overview (inorganic phase)

In Figure 4, the concentrations of the unknown compounds are shown as a function of residence time, with the concentration of furfural. The intermediates' concentrations are based on the RI chromatograms (Figures 1(a) and 2(a)). However, the intermediates have not been isolated, characterized, and calibrated in the HPLC. Their concentrations are based on an estimate from their RI peak areas and overall mole balance. The RI sensitivity of I1, I2, and I3 are assumed equal and is estimated from the mole balance (Eq. (1)) for the highest conversion. The balance accuracy is more than 99.9%.

$$C_i^{aq} = \left(C^{aq}_{xylose,in} - C^{aq}_{xylose,out} - C^{aq}_{furfural,out} - 2C^{org}_{furfural,out}\right)\frac{A_i}{\sum_i A_i}$$

Byproducts that are not detected by the HPLC are neglected, although a very small amount of solid residue is noticed in the samples with higher residence times.

4. Discussion

It can be observed from Figures 1(a), 2(a) and 4 that component I1 shows a similar trend as furfural at higher concentration. This cannot be explained by the irreversible consecutive reaction scheme 1. It is very likely that I1 corresponds to intermediate 3 in Scheme 1 and is in equilibrium with furfural. Alternatively, all dehydration reactions (Scheme 1) are equilibrium reactions and I1 corresponds to any of the three intermediates. In Figures 1 and 2 the RI does not respond to components C and D and they are presented only in UV. Thus, it can be assumed that their concentration is low. Possibly, C and D correspond to the intermediates

2 and 3 (Scheme 1).

The concentration of component I2 increases before the lowest residence time and decreases to zero after 250s. I2 is probably a precursor to the solid residue formed and reacts with itself and the solid residue. Thus, the declining concentration with residence time can be explained. The concentration of I3 remains low and follows the same trend as furfural, which would indicate that it is a byproduct in equilibrium with I1.

5. Conclusions

By the continuous reactive extraction at 150°C, 44.52 % xylose conversion is achieved and 14.45 % total furfural yield. Compounds in equilibrium with furfural have been detected validating the theory about reaction mechanism presented earlier (Scheme 1). Moreover, additional compounds are presented in the system potentially in equilibrium with furfural intermediates. Finally, solids are formed from furfural degradation in aqueous phase even in short residence times.

6. Outlook

Future work includes additional experiments in the temperature range of 100°C to 170°C and shorter and higher residence times. Moreover, the influence of the catalyst will be investigated. Regarding the extraction part, variations will also be made on the nature and the ratio of the solvent. Finally, a kinetic model will be developed to describe the reaction rates as a function of temperature and concentrations.

7. References

1. B. Kamm, P. R. Gruber, M. Kamm. *Biorefineries-Industrial processes and Products.* s.l. : Wiley, 2006.

2. *The Chemistry and technology of furfural production in modern lignocellulose-feedstock biorefineries.* Marcotullio, Gianluca. s.l. : Delft University of technology, 2011.

3. *Kinetics of furfural production by dehydration of xylose in a biphasic reactor with microwave heating.* R. Weingarten, J. Cho, Wm. C. Conner, Jr. and G.W. Huber. 1423-1429, s.l. : Green Chemistry, 2010, Vol. 12.

4. *Extraction of furfural from aqueous solution uding Alcohols.* J. L. Cabezas, L. A. Barcena, J. Coca, M, Cockrem. 435-437, s.l. : J. Chem. Eng. Data, 1983, Vol. 33.

5. *The chemistry and technology of furfural and its many by-products.* Zeitch, K. J. s.l. : Elsevier, 2000.

469286 An Investigation of the Degradation of Lignin in Sub-Super Critical Water in a Continuous Flow Reactor

Abstract:

Lignin, after cellulose, is the most abundant organic compound on the planet and its conversion has significant potential as a source for the sustainable production of fuels and bulk chemicals. However, Lignin conversion into more meaningful products is a challenge due to the inherent complexity of lignin molecule. Despite many efforts in this field, degradation of lignin has not been entirely successful especially at large scale due to the plugging of catalysts with tars due to unwanted oligomerization reactions. To avoid this phenomenon, continuous flow systems appear to be a better option since they allow for reduced residence times, which may inhibit tar formation and because they are more likely to lead to viable industrial scale processes.

In this research, the degradation of lignin was investigated in a continuous flow reactor in super/sub-critical water at 25, 35 and 45 ml/min and 250, 300 and 320 . Solid, liquid, and gas phase products were collected and analyzed to determine the near optimum reaction conditions and performance results that can be expected from this type of system.

456337 Interaction Between Model Compounds in Lignocellulosic Biomass Under Supercritical Water Conditions

Abstract:

Supercritical water gasification (SCWG) is a conversion technique to transform biomass into energy without necessity of drying process because water in supercritical state (temperature and pressure above 374 °C and 22.1 MPa, respectively) behaves as an organic solvent so that a high conversion can be achieved. Plant biomass or lignocellulosic biomass is a good candidate for a feedstock utilized in this process due to its large availability on Earth. It consists of cellulose, hemicellulose, which are easy to be gasified and lignin, which inhibits gasification. Previously, reaction pathways of cellulose, hemicellulose and lignin were individually proposed by many researchers. However, the study of the reaction mechanisms of the overall biomass is still rare. Therefore, this research is to study the interaction between these compounds in supercritical water aiming for better understanding in a behavior of lignocellulosic biomass in the gasification. To simplify the reaction mechanisms, model compounds for each component are employed instead of the complex structure of those. The interaction between glucose that is a model compound of cellulose, xylose that is a model compound of hemicellulose and guaiacol that is a model compound of lignin is investigated using tubular reactor system composed of high-pressure pumps, a preheater, a reactor made of stainless steel SS316 with inner and outer diameters of 1 mm and 1.59 mm (1/16 inch), respectively, heat exchangers, a back-pressure regulator, and a gas- and liquid-sampling system. Molten salt is used as heating medium for heating pre-heated water and the reactor. Mixture solutions of glucose, xylose and guaiacol were prepared in designated concentrations: 0.3 % wt.: 0.0 % wt.: 0.1% wt., 0.2 % wt.: 0.2 % wt.: 0.0 % wt., 0.0 % wt.: 0.3 % wt.: 0.1 % wt. and 0.15 % wt.: 0.15 % wt.: 0.1 % wt. Reaction temperatures are above 300 °C. A pressure of the system is 25 MPa for all conditions. The effect of each compound in the feedstock, temperature and residence time on product yields as well as interaction networks of glucose, xylose and guaiacol will be presented.

464614 A Novel Single Microsphere-Microreactor Approach Modeling and Experimental for Study of Biomass Fast Pyrolysis

Abstract:

Many technologies are being investigated and developed for chemicals and fuel production from renewable resources. Biomass fast pyrolysis is emerging as a promising approach. Practical question still remain, however, regarding the kinetics of fast pyrolysis; mainly due to the very fast, three-phase reactions that occur. It appears that the pyrolysis heating rate, solid/gas phase residence time at temperature and biomass particle size have an important impact on char, condensable and gas yields and selectivity. In an effort to elucidate reaction kinetics, a novel micro-particle micro-reactor system has been developed allowing reaction and quantification of rates for single manufactured spherical particles of nominally 10 microgram. The very fast rate of reaction for pyrolysis makes all pyrolysis experiments dynamic at some level. This new approach, wherein a single 10 microgram particle is injected into a hot furnace, achieves heating rates as high as 5000 ^{o}C/s. Experimental data is captures in the time-domain wherein time and gas-phase response (using fast flame ionization) are trapped at rates as high as 10,000 Hz. Translation of this time-domain data to temperature-domain can only be done by a model-based approach. This presentation focuses on modeling the reactor dynamics and correlation of time-temperature-reaction rate datasets. Prior studies show that for fast pyrolysis of single microspheres, the amount of hydrocarbon and char production increases linearly as the biomass particle size increases, however the reaction rate decreases. Furthermore, the amount of volatilized hydrocarbon appears to increase linearly as the temperature increases. Finally, the rate of reaction decreases as the size of biomass particle increases and increases as the temperature increases.

467108 Computational Fluid Dynamics Study of Full-Scale Aerobic Bioreactors: Evaluation of Dynamic Oxygen Distribution, Gas-Liquid Mass Transfer, and Reaction

Abstract:

If advanced biofuels are to measurably displace fossil fuels in the near term, they will have to operate at levels of scale, efficiency, and margin unprecedented in the current biotech industry. For aerobically-grown products in particular, scale-up is complex and the practical size, cost, and operability of extremely large reactors is not well understood. Put simply, the problem of how to attain fuel-class production scales comes down to cost-effective delivery of oxygen at high mass transfer rates and low capital and operating costs. To that end, very large reactor vessels (>500 m^3) are proposed in order to achieve favorable economies of scale. Additionally, techno-economic evaluation indicates that bubble-column reactors are more cost-effective than stirred-tank reactors in many low-viscosity cultures.

In order to advance the design of extremely large aerobic bioreactors, we have performed computational fluid dynamics (CFD) simulations of bubble-column reactors. A multiphase Euler-Euler model is used to explicitly account for the spatial distribution of air (i.e., gas bubbles) in the reactor. Expanding on the existing bioreactor CFD literature (typically focused on the hydrodynamics of bubbly flows), our simulations include interphase mass transfer of oxygen and a simple phenomenological reaction representing the uptake and consumption of dissolved oxygen by submerged cells.

The simulations reproduce the expected flow profiles, with net upward flow in the center of column and downward flow near the wall. At high simulated oxygen uptake rates (OUR), oxygen-depleted regions can be observed in the reactor. By increasing the gas flow to enhance mixing and eliminate depleted areas, a maximum oxygen transfer

(OTR) rate is obtained as a function of superficial velocity. These insights regarding minimum superficial velocity and maximum reactor size are incorporated into NRELâ€™s larger techno-economic models to supplement standard reactor design equations.

AUTHOR INDEX

AUTHOR INDEX

AUTHOR INDEX

AUTHOR INDEX

AUTHOR INDEX

AUTHOR INDEX

AUTHOR INDEX

AUTHOR INDEX

AUTHOR INDEX

AUTHOR INDEX

AUTHOR INDEX

AUTHOR INDEX

AUTHOR INDEX